CW00729315

Fjord Systems and Archives

Geological Society books refereeing procedures

The Society makes every effort to ensure that the scientific and production quality of its books matches that of its journals. Since 1997, all book proposals have been refereed by specialist reviewers as well as by the Society's Books Editorial Committee. If the referees identify weaknesses in the proposal, these must be addressed before the proposal is accepted.

Once the book is accepted, the Society Book Editors ensure that the volume editors follow strict guidelines on refereeing and quality control. We insist that individual papers can only be accepted after satisfactory review by two independent referees. The questions on the review forms are similar to those for *Journal of the Geological Society*. The referees' forms and comments must be available to the Society's Book Editors on request.

Although many of the books result from meetings, the editors are expected to commission papers that were not presented at the meeting to ensure that the book provides a balanced coverage of the subject. Being accepted for presentation at the meeting does not guarantee inclusion in the book.

More information about submitting a proposal and producing a book for the Society can be found on its web site: www.geolsoc.org.uk.

It is recommended that reference to all or part of this book should be made in one of the following ways:

Howe, J. A., Austin, W. E. N., Forwick, M. & Paetzel, M. (eds) 2010. *Fjord Systems and Archives*. Geological Society, London, Special Publications, **344**.

Baeten, N. J., Forwick, M., Vogt, C. & Vorren, T. O. 2010. Late Weichselian and Holocene sedimentary environments and glacial activity in Billefjorden, Svalbard. *In*: Howe, J. A., Austin, W. E. N., Forwick, M. & Paetzel, M. (eds) *Fjord Systems and Archives*. Geological Society, London, Special Publications, **344**, 207–223.

GEOLOGICAL SOCIETY SPECIAL PUBLICATION NO. 344

Fjord Systems and Archives

EDITED BY

J. A. HOWE
Scottish Association for Marine Science,
Scottish Marine Institute, UK

W. E. N. AUSTIN
University of St. Andrews, UK

M. FORWICK
University of Tromsø, Norway

and

M. PAETZEL
Sogn & Fjordane University College, Norway

2010
Published by
The Geological Society
London

THE GEOLOGICAL SOCIETY

The Geological Society of London (GSL) was founded in 1807. It is the oldest national geological society in the world and the largest in Europe. It was incorporated under Royal Charter in 1825 and is Registered Charity 210161.

The Society is the UK national learned and professional society for geology with a worldwide Fellowship (FGS) of over 10 000. The Society has the power to confer Chartered status on suitably qualified Fellows, and about 2000 of the Fellowship carry the title (CGeol). Chartered Geologists may also obtain the equivalent European title, European Geologist (EurGeol). One fifth of the Society's fellowship resides outside the UK. To find out more about the Society, log on to www.geolsoc.org.uk.

The Geological Society Publishing House (Bath, UK) produces the Society's international journals and books, and acts as European distributor for selected publications of the American Association of Petroleum Geologists (AAPG), the Indonesian Petroleum Association (IPA), the Geological Society of America (GSA), the Society for Sedimentary Geology (SEPM) and the Geologists' Association (GA). Joint marketing agreements ensure that GSL Fellows may purchase these societies' publications at a discount. The Society's online bookshop (accessible from www.geolsoc.org.uk) offers secure book purchasing with your credit or debit card.

To find out about joining the Society and benefiting from substantial discounts on publications of GSL and other societies worldwide, consult www.geolsoc.org.uk, or contact the Fellowship Department at: The Geological Society, Burlington House, Piccadilly, London W1J 0BG: Tel. +44 (0)20 7434 9944; Fax +44 (0)20 7439 8975; E-mail: enquiries@geolsoc.org.uk.

For information about the Society's meetings, consult *Events* on www.geolsoc.org.uk. To find out more about the Society's Corporate Affiliates Scheme, write to enquiries@geolsoc.org.uk.

Published by The Geological Society from:
The Geological Society Publishing House, Unit 7, Brassmill Enterprise Centre, Brassmill Lane, Bath BA1 3JN, UK

(*Orders*: Tel. +44 (0)1225 445046, Fax +44 (0)1225 442836)
Online bookshop: www.geolsoc.org.uk/bookshop

British Library Cataloguing in Publication Data

A catalogue record for this book is available from the British Library.
ISBN 978-1-86239-312-7

Typeset by Techset Composition Ltd, Salisbury, UK
Printed by MPG Books Ltd, Bodmin, UK

Distributors

North America
For trade and institutional orders:
The Geological Society, c/o AIDC, 82 Winter Sport Lane, Williston, VT 05495, USA
Orders: Tel. +1 800-972-9892
Fax +1 802-864-7626
E-mail: gsl.orders@aidcvt.com

For individual and corporate orders:
AAPG Bookstore, PO Box 979, Tulsa, OK 74101-0979, USA
Orders: Tel. +1 918-584-2555
Fax +1 918-560-2652
E-mail: bookstore@aapg.org
Website: http://bookstore.aapg.org

India
Affiliated East-West Press Private Ltd, Marketing Division, G-1/16 Ansari Road, Darya Ganj, New Delhi 110 002, India
Orders: Tel. +91 11 2327-9113/2326-4180
Fax +91 11 2326-0538
E-mail: affiliat@vsnl.com

Contents

Dedication to Harald Svendsen

Harald Svendsen (1936–2009).

In his 73rd year, Professor Harald Svendsen sadly passed away on November 30th 2009. Harald was a pioneer in investigating the Arctic fjords of Svalbard. He used his extensive knowledge of mainland Norwegian fjords, to investigate the importance of rotational dynamics in broad fjords. He also saw very clearly the importance of multi-disciplinary research to understand the links between the geology, physics and biology of fjords. By his attitude he gained a broad network of collaborators who will all greatly miss him, both as a scientist and as a friend.

V. Tverberg
P. M. Haugan
T. Gammelsrod

Geophysical Institute, University of Bergen, Norway

Fjord systems and archives: an introduction

WILLIAM E. N. AUSTIN[1]*, JOHN A. HOWE[2], MATTHIAS FORWICK[3] & MATTHIAS PAETZEL[4]

[1]*University of St Andrews, UK*

[2]*Scottish Association for Marine Science, Scottish Marine Institute, UK*

[3]*University of Tromsø, Norway*

[4]*Sogn & Fjordane University College, Norway*

**Corresponding author (e-mail: bill.austin@st-andrews.ac.uk)*

The current volume brings together a selection of papers which have variously, but not exclusively, been presented in recent years at one of three international meetings on the theme of *Fjords*. The first of these meetings on '*Fjord Environments: Past, Present and Future*' was held as a workshop following the Challenger Society Conference hosted by The Scottish Association for Marine Science, Scottish Marine Institute, Oban, UK in June 2006. The second meeting was convened as a formal session (CGC-13) entitled '*Fjords: Climate and Environmental Change*' during the 33rd International Geological Congress, Oslo, Norway in August 2008. The third of these meetings, representing the 2nd international workshop on the theme '*Fjord Environments: Past, Present and Future*' was held at the University of Bergen, Norway in May 2009. The aims of these meetings were to bring together physical oceanographers, biogeochemists, biologists and earth scientists who could contribute to an improved understanding of fjord systems, both in terms of modern processes and as palaeoenvironmental archives.

This publication consists of 23 papers and a glossary, dealing with various aspects of fjord systems and their associated archives. Three of the papers focus on the physics of fjord hydrography and circulation; three address aspects of fjord biology; two concern modern sediment processes; six consider fjord sediments and their depositional architecture; and eight highlight fjords as depositional archives and their palaeoenvironmental significance. The current volume therefore represents a significant contribution of both original research and reviews which we hope provides a timely update to the most recent publications in this field, namely *Fjords: Processes and Products* (Syvitski *et al.* 1987) and *Fjord Oceanography* (Freeland *et al.* 1979).

The volume begins with a dedication to **Harald Svendsen**, who sadly passed away on 30th November 2009. Harald made a significant contribution to the investigation of Arctic fjords and it is an honour to dedicate this volume to his memory. An overview by **Howe *et al.***, highlighting some of the key features of the fjord system from modern processes to palaeo-records, provides an accessible introduction to the topic. Aspects of the major physical, chemical and biological processes within fjord systems are highlighted. The volume is subsequently divided into five main themes, comprising (a) physics and physical oceanography; (b) biology and biological indicators; (c) sediment dynamics and processes; (d) sediments and depositional architecture; and (e) depositional archives and palaeoenvironments.

The paper by **Inall & Gillibrand** provides a useful, non-mathematical review of the dominant physical processes of mid-latitude fjords; this is likely to prove useful to non-physical oceanographers who may wish to gain an insight into some of the key fluid dynamical processes. It is followed by a review of Arctic fjords by **Cottier *et al.***, who highlight the role of fjords as 'critical gateways' by which glaciers are influenced by ocean conditions; the latter is a subject of increasing significance at a time of rising global sea surface temperatures. These authors show how fjord salinity plays a critical role in controlling exchange processes between the fjord and coastal ocean. The third paper on the physics of fjords by **Skarðhamar & Svendsen** considers short-term variability from the wide semi-enclosed Van Mijenfjorden on Svalbard. The influence of the Coriolis effect upon circulation and the development of cross-fjord gradients is illustrated.

Three short papers address aspects of biological proxies as palaeoenvironmental indicators within fjords. **Howe *et al.*** review the potential of dinoflagellate cysts collected from sediment traps within Arctic (Svalbard archipelago) fjords as proxies for palaeoceanographic conditions, highlighting the significance of seasonal and inter-annual differences in cyst production. Continuing the theme of dinoflagellate cysts as palaeoceanographic proxies,

From: Howe, J. A., Austin, W. E. N., Forwick, M. & Paetzel, M. (eds) *Fjord Systems and Archives*. Geological Society, London, Special Publications, **344**, 1–3.
DOI: 10.1144/SP344.1 0305-8719/10/$15.00

Harland *et al.* highlight how a major change in the flora from the Gullmar Fjord, Sweden, in the late 1960s to early 1970s appears to coincide with a change in the North Atlantic Oscillation. This suggests that fjord hydrography and dinoflagellate cysts respond to this large-scale atmospheric forcing. Finally, **Austin & Cage** present a preliminary investigation of benthic foraminifera from a maerl (gravels derived from calcareous red seaweed) bed in western Scotland, highlighting the very high species counts in such deposits and their potential in regional studies of foraminiferal distribution and ecology.

Next, two papers consider modern fjord sediment processes. The first of these by **Trusel *et al.*** provides an overview of glaciomarine processes in the inner Kongsfjorden, Svalbard, and highlights sediment dynamics near glacier termini with relatively rare data on sediment flux and yield. These authors speculate on glacier dynamics in a warming world suggesting that, as sediment fluxes increase close to glacier termini, grounding-line deposits may initially act to stabilize glaciers. The second paper by **Loh *et al.*** considers organic carbon budgets in a mid-latitude fjord, providing valuable data on the sources and sinks of organic carbon into the natural sediment trap of the fjord basin.

A further seven papers develop the theme of fjord sediments and their depositional architecture. **Overeem & Syvitski** use numerical models to explore the deglacial history and sediment architecture of fjord systems under differing regimes of deglaciation, highlighting that recently deglaciated fjords provide potentially valuable modern analogues of forced regressive systems. **Dallimore & Jmieff** review the fjords (inlets) of the Canadian west coast, comparing fjord regimes, depositional processes and conditions under which annually laminated sediment sequences form. **Forwick *et al.*** continue the depositional theme by considering the sedimentary environments in two Svalbard fjords, with particular focus on the influence of glaciers and rivers on the sedimentation pattern since the Late Weichselian to present. **MacLachlan *et al.*** continue the theme of Arctic fjord development since the Last Glacial Maximum (LGM) in a study of fjord bathymetric and shallow seismic data from Kongsfjorden and Krossfjorden, Svalbard. The depositional style and cross-cutting relationships of landforms linked to glacial processes are highlighted. This same approach is developed by **Baeten *et al.***, who use swath bathymetry and sediment stratigraphy to reconstruct the glacial history of Billefjorden, Svalbard, focusing in particular on the growth and decay of the tidewater glacier Nordenskiöldbreen during the Late Weichselian and the Holocene. One final paper in this section considers evidence for slope instability within deglaciated mid-latitude fjords. Finally, **Stoker *et al.*** consider the post-glacial evidence for sliding and mass-flow processes within Little Loch Broom, NW Scotland, highlighting the potential significance of paraglacial landscape readjustment and slope instability driven by seismic activity in response to glacio-isostatic unloading.

A further eight papers develop the theme of fjord sediments as depositional archives with palaeoenvironmental significance. **Hass *et al.*** focus on the climate evolution of the South Shetland Islands, West Antarctica, highlighting the role of sediment-laden meltwater plumes on vertical sediment fluxes and providing evidence for contrasting depositional regimes during the past two millennia. **Filipsson & Nordberg** consider a similar time period in a record of benthic foraminiferal carbon isotopes from the Gullmar Fjord, Sweden. They suggest that changes are potentially driven by land-use changes, basin ventilation history and, during the late 20th century, by the oceanic Suess effect. Continuing the theme of the ventilation of fjord basin waters, **Paetzel & Dale** compare sediment records from the periodically anoxic Inner Barsnefjord with the oxic Outer Barsnefjord and Sogndalsfjord, West Norway, outlining the potential sensitivity of the sediment records from these locations to climate variability. **Skirbekk *et al.*** focus on a palaeoceanographic reconstruction spanning the last 12 000 years from the entrance to the warm Arctic Kongsfjorden, Svalbard, tracing the history of Atlantic water masses as they have penetrated northwards over this time period. **McIntyre & Howe** provide a review of the available evidence for the depositional history of Scotland's fjords during the last deglacial transition and Holocene. **Cundill & Austin** provide an example of a Holocene pollen record from a Scottish fjord, highlighting the potential to link terrestrial vegetation and marine-based fossil evidence from a common stratigraphy. Two final papers by **Mokeddem *et al.*** and **Baltzer *et al.*** focus on Loch Sunart, a fjord in west Scotland, where proxy evidence (pollen, foraminifera) provides the basis for palaeoenvironmental reconstructions since the last deglacial transition. These papers highlight the sensitivity of the fjord basins on the maritime margins of NW Europe to long-term, natural climate variability.

We would like to thank all the participants of the *Fjord* meetings held in Oban, Oslo and Bergen for their enthusiasm for this project; we are particularly grateful to R. Powell (Northern Illinois University, USA) and T. Vorren (University of Tromsø, Norway) for their encouragement and support at an early, critical stage in the development of this volume. Staff at the Geological Publishing House, notably A. Hills and T. Anderson, have provided invaluable guidance in the production of this volume and we particularly acknowledge the expert advice of our referees, who gave their time to review the papers published here. Finally, we acknowledge the generous support of the Marine Studies Group of the Geological Society of London, the Scottish Association for Marine Science and the University of Bergen, Norway.

In the immortal words of Slartibartfast, reflecting on the meaning of happiness in Douglas Adams' (1979) *The Hitchhiker's Guide to the Galaxy*, we leave you with a final thought and hope that you enjoy this volume: "*In this replacement Earth we're building they've given me Africa to do and of course I'm doing it with all fjords again because I happen to like them, and I'm old-fashioned enough to think that they give a lovely baroque feel to a continent.*"

References

ADAMS, D. 1979. *The Hitchhiker's Guide to the Galaxy*. Pan Books, London.

FREELAND, H. J., FARMER, D. M. & LEVINGS, C. D. (eds) 1979. *Fjord Oceanography*. NATO Conference Series, Series IV: Marine Sciences, **4**.

SYVITSKI, J. P. M., BURRELL, D. C. & SKEI, J. M. 1987. *Fjords: Processes and Products*. Springer-Verlag, Berlin.

Fjord systems and archives: a review

JOHN A. HOWE[1]*, WILLIAM E. N. AUSTIN[2], MATTHIAS FORWICK[3], MATTHIAS PAETZEL[4], REX HARLAND[5] & ALIX G. CAGE[2,6]

[1]*Scottish Association for Marine Science, Scottish Marine Institute, Oban, Argyll, PA37 1QA, Scotland, UK*

[2]*Department of Geography and Geoscience, St Andrews University, Irvine Building, St Andrews, KY16 9AL, Scotland, UK*

[3]*Department of Geology, University of Tromsø, N-9037 Tromsø, Norway*

[4]*Sogn & Fjordane University College, Postbox 133, 6851 Sogndal, Norway*

[5]*Department of Earth Sciences, Göteborg University, PO Box 460, SE 405 30, Göteborg, Sweden and 50 Long Acre, Bingham, Nottingham, NG13 8AH, UK*

[6]*Department of Environmental and Geographical Sciences, Manchester Metropolitan University, Chester Street, Manchester, M1 5GD, UK*

**Corresponding author (e-mail: john.howe@sams.ac.uk)*

Abstract: Fjords are glacially over-deepened semi-enclosed marine basins, typically with entrance sills separating their deep waters from the adjacent coastal waters which restrict water circulation and thus oxygen renewal. The location of fjords is principally controlled by the occurrence of ice sheets, either modern or ancestral. Fjords are therefore geomorphological features that represent the transition from the terrestrial to the marine environment and, as such, have the potential to preserve evidence of environmental change. Typically, most fjords have been glaciated a number of times and some high-latitude fjords still possess a resident glacier. In most cases, glacial erosion through successive glacial/interglacial cycles has ensured the removal of sediment sequences within the fjord. Hence the stratigraphic record in fjords largely preserves a glacial-deglacial cycle of deposition over the last 18 ka or so. Sheltered water and high sedimentation rates have the potential to make fjords ideal depositional environments for preserving continuous records of climate and environmental change with high temporal resolution. In addition to acting as high-resolution environmental archives, fjords can also be thought of as mini-ocean sedimentary basin laboratories. Fjords remain an understudied and often neglected sedimentary realm. With predictions of warming climates, changing ocean circulation and rising sea levels, this volume is a timely look at these environmentally sensitive coastlines.

Supplementary material: The Glossary is available at: http://www.geolsoc.org.uk/SUP18440.

Definition of fjords

Fjords typically occur in so-called fjord belts along mid to high latitudes in both hemispheres (Fig. 1) and are steep-sided, coastal erosional troughs which have been inundated by the sea. Fjord is the typical Scandinavian term; in Sweden they can be referred to as *fjärd* and in Iceland as *fjordur*. Scottish fjords are referred to as *sea-lochs* or, for larger open bodies of water, *firths*. The derivation of these terms would appear to be the Norse word *fjörðr* meaning a lake-like body of water. The use of fjord is therefore typically Scandinavian, being used to describe a narrow inlet of the sea. This is close to the original Norse definition and is applied to features in Norway, Sweden, Denmark and Finland. The term *sund* (sound) is applied to the fjords of New Zealand and North America. South American fjords, especially Chilean examples, are often referred to as *canals*. In Germany the term *förde* is used slightly differently, referring to glacial troughs carved by ice tongues in smooth, flat-lying coastal areas.

Origin of fjords

Fjords are fundamentally the product of glacial erosion, although contributing factors can also include any geological heterogeneity and structural trends in the underlying bedrock (Gjessing 1956; Holtedahl 1967; Nesje & Whillans 1994). A recent

From: Howe, J. A., Austin, W. E. N., Forwick, M. & Paetzel, M. (eds) *Fjord Systems and Archives.* Geological Society, London, Special Publications, **344**, 5–15.
DOI: 10.1144/SP344.2 0305-8719/10/$15.00

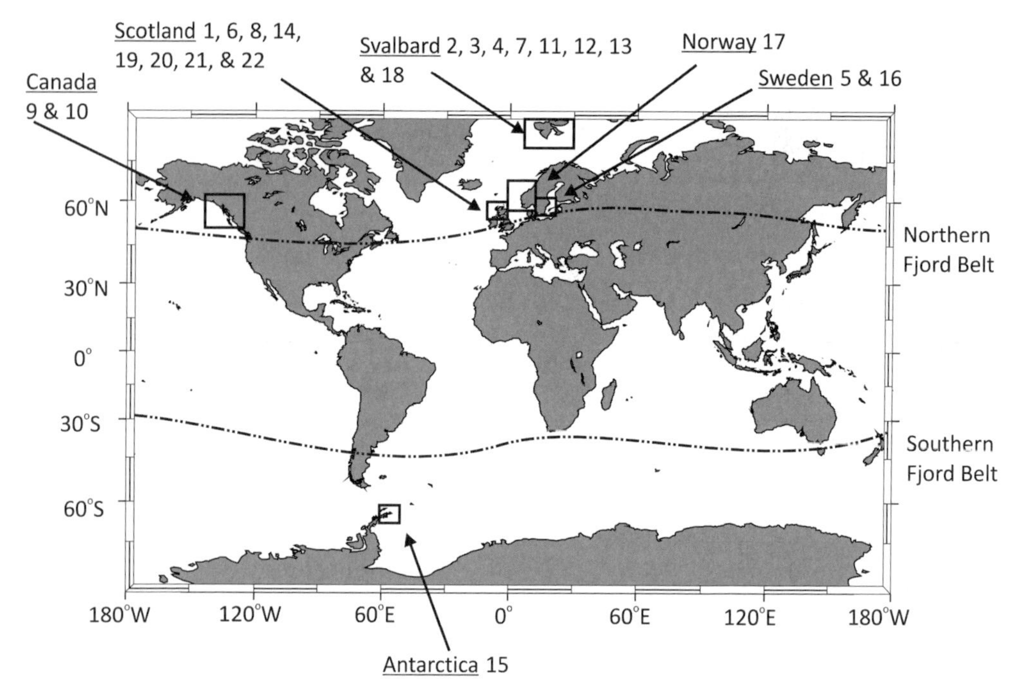

Fig. 1. Location of the fjord studies described in this volume. Also shown are the northern and southern fjord limits adapted from Syvitski *et al.* (1987). Key to numbers: 1, Inall *et al.* (mid-latitude); 2, Cottier *et al.* (Arctic); 3, Skardhamar & Svendsen (van Mijenfjorden, Svalbard); 4, Howe *et al.* (Kongsfjord & Rijpfjord, Svalbard); 5, Harland *et al.* (Gullmar Fjord, Sweden); 6, Austin & Cage (Clyde, UK); 7, Trusel *et al.* (Kongsfjord, Svalbard); 8, Loh *et al.* (Loch Creran, UK); 9, Overeem & Syvitski (All); 10, Dallimore & Jmieff (Western Canada); 11, Forwick *et al.* (Sassenfjorden & Templefjorden, Svalbard); 12, Maclachlan *et al.* (Kongsfjord, Svalbard); 13, Baeten *et al.* (Billefjorden, Svalbard); 14, Stoker *et al.* (Little Loch Broom, UK); 15, Hass *et al.* (Maxwell Bay, Antarctica); 16, Filipsson & Nordberg (Gullmar Fjord, Sweden); 17, Paetzel & Dale (Sognefjord, Norway); 18, Skirbekk *et al.* (Kongsfjord, Svalbard); 19, McIntyre & Howe (Western Scotland, UK); 20, Cundill & Austin (Loch Etive, UK); 21, Mokeddem *et al.* (Loch Sunart, UK); 22, Baltzer *et al.* (Loch Sunart, UK).

study by Kessler *et al.* (2008) found that fjords have their origin in the topographic steering of ice originating from Late Cenozoic ice sheets. This significant study used an ice-sheet model to demonstrate that ice and erosion proportional to discharge is enough to produce a kilometre-deep fjord in one million years with enhanced erosion below thicker and faster ice. The effect of sea-level change is also significant with the down cutting of the ice sheets greater and more extensive during lowstands.

Distribution and classification

The classification of fjords can be based on a variety of parameters including climate regimes, glacier regimes and environmental factors influencing the fjord setting.

Fjords are presently located in *polar*, *subpolar* and *temperate* climate regimes. *Polar* fjords occur for example on east and north Greenland, the Canadian Arctic and large parts of Antarctica. They are almost permanently covered with sea ice or an ice shelf and possess a resident glacier (Fig. 2a; Domack & McClennen 1996; Gilbert 2000). Sediment supply occurs mainly from the glacier bed, as well as from subglacial rivers and icebergs (Powell 2005). *Subpolar* fjords are located in areas where the summer mean air temperatures are above 0 °C (Domack & McClennen 1996; Gilbert 2000), as for example on Svalbard, west Greenland, the Canadian Arctic and on the Antarctic Peninsula. Sea ice is present in winter, but usually breaks up during summers. Glaciers may or may not be present. The main sediment sources include subglacially derived material, subglacial meltwater runoff, icebergs and terrestrial rivers (Powell 2005). Sea ice is generally absent in *temperate* fjords (Domack & McClennen 1996; Gilbert 2000). Some, but not exclusively all, temperate fjords are presently non-glaciated (Fig. 2b). Ice can, however, occur

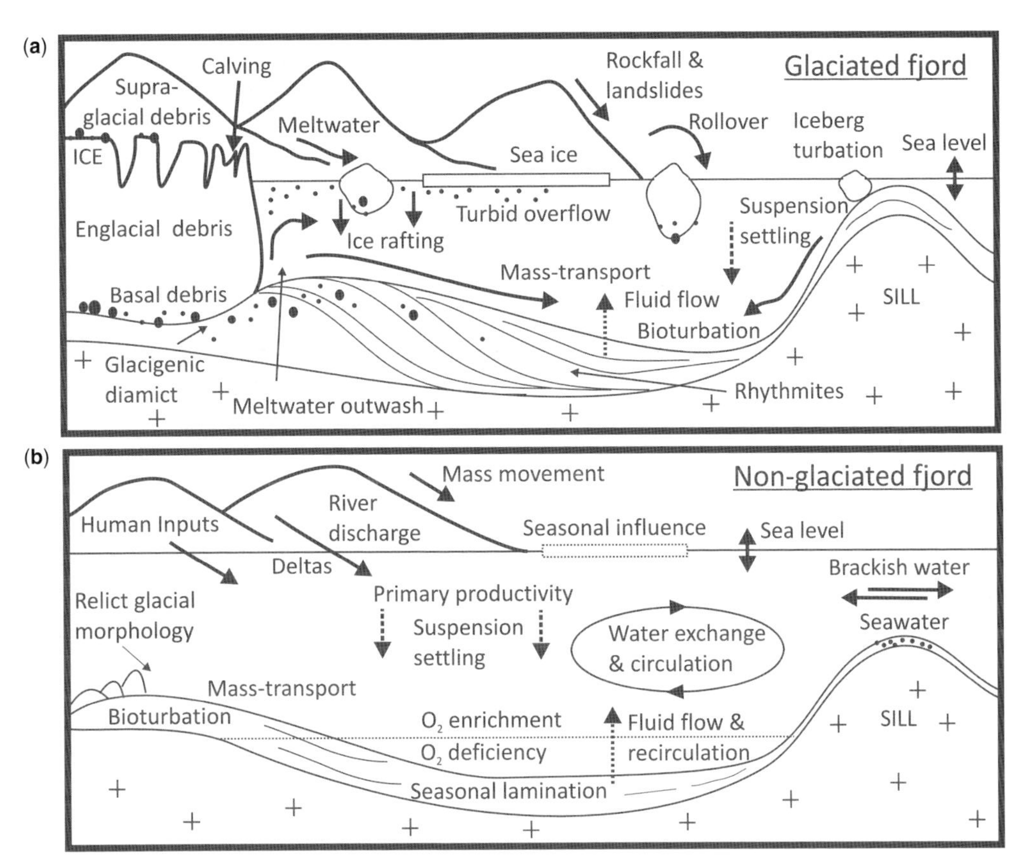

Fig. 2. Principle fjord processes and deposits: (**a**) glaciated fjords (adapted from Hambrey 1994) and (**b**) non-glaciated fjords.

locally, for example off river mouths. In those fjords where tidewater glaciers occur, subglacial meltwater runoff acts as the main sediment source (Powell 2005). Other main sources include terrestrial rivers and mass wasting. *Temperate* fjords are found for example in Iceland, Alaska, Chile, Norway, Scotland and New Zealand.

Based on glacier types (temperate to cold), their basal conditions (grounded or floating), sediment supply and glacier dynamics, Hambrey (1994) describes *Alaskan*, *Svalbard*, *Greenland*, *Antarctic maritime* and *Antarctic arid regimes* in presently glaciated fjords. *Alaskan regimes* are characterized by highly dynamic and grounded temperate glaciers and accordingly high sedimentation rates. In the *Svalbard regime*, dynamic, grounded and slightly cold glaciers terminate in relatively shallow fjords (<200 m). The sedimentation is influenced by large amounts of meltwater during a short summer season. The *Greenland regime* typically includes dynamic and floating outlets from ice sheets, ice caps and highland ice fields that terminate in deeper fjords (>200 m). Dynamic, cold and mainly grounded glaciers that extend to near the mouths of short fjords comprise the *Antarctic maritime regime*. Rock exposure is limited and surface melting restricted. The *Antarctic maritime regime* is, on the other hand, characterized by sluggish and very cold floating glaciers that extend almost to the mouths of fjords. Fast ice off their fronts may survive for several seasons (Hambrey 1994).

Syvitski & Shaw (1995) propose a classification of fjords based on physical regime. They describe *ice-influenced fjords* where sediments are dispersed by ice contact and glaciofluvial processes, as well as by ice rafting. In *river-influenced fjords*, the fjord circulation and sediment transport depend on the hydrological cycle. Fjord depth and the exchange with open-ocean water masses influence the sediment architecture in *wave- and tide-influenced fjords. Fjords dominated by slope failure* contain large amounts and a great variety of mass-transport deposits. The environmental conditions in *anoxic fjords* depend heavily on water exchange and the rate of utilization of organic matter (Syvitski & Shaw 1995).

Importance of fjords

Fjords are fascinating features for study, being viewed either as a proxy miniature ocean basin laboratory or as inshore sheltered deep-water bodies which often possess a unique biogeochemistry, fauna, hydrography and sedimentation (Skei 1983*a*; Syvitski *et al.* 1987; Syvitski & Shaw 1995). They are often the focus of locally enhanced sedimentation, which can provide high-resolution temporal records of late- and postglacial palaeoceanographic change. Historically, fjords have provided a sheltered, fertile place for communities since at least the early Neolithic (9.5 ka), allowing easy access to good farming, fishing and a mode of transport between isolated peoples. The pristine fjords of Scotland, Norway and Chile among others have come under increasing pressure from both economic and tourist demands over time. More recently, with the advent of modern aquaculture techniques, industries such as fish farming have provided these communities with a valuable source of income and social stability. Economically, the sheltered, accessible, locally deep waters of fjords are well suited for providing the favourable environmental conditions for mussel and fish farming. However, in some places the restricted exchange with coastal waters can lead to an accumulation of pollutants beneath fish farms. In Canada, for example, this reduced flushing of fjord waters is used to capture and store pollutants such as mine tailings. Concentrations of people living in towns, villages and (in some cases) cities beside fjords can also lead to localized eutrophication from effluent discharges. Deforestation and intensive farming in the catchment may result in increased runoff and possibly slope instability. These examples illustrate the environmental challenges facing fjords in the future. Inevitably, there is a delicate balance to be sought between preserving what are effectively geologically short-lived, dynamic marine systems and the need to provide employment to remote, fragile communities.

Fjord processes

Morphology

Modern high-resolution multibeam sonars are becoming increasingly common as means of surveying fjords. The navigational accuracy and submetre resolution of these systems permits the detailed mapping of the morphology of the fjord seabed, reflecting both modern and relict depositional and erosional processes (Howe *et al.* 2003). Subglacial features such as glacial scouring, drumlins, mega-scale glacial lineations and channels can be, in some cases, preserved in temperate, presently non-glaciated fjords (Stoker & Bradwell 2005). Sediments in the fjords on the west coast of Scotland are in essence in disequilibrium with the present Holocene transgression, the seabed of these fjords still displaying features preserved from the last glacial re-advance and subsequent retreat, between 15–11.5 ka. During the retreat of the ice, rapid sedimentation produced glaciomarine fans, terminal and recessional moraines. The Early/Mid-Holocene also produced changes in glacio-isostatic sea levels, particularly influencing the fjords of the northern hemisphere where major continental ice sheets were concentrated (Smith *et al.* 2007). In Norway, the rapid disappearance of ice led to crustal unloading and, in some places, the reactivation of faults and an increase in regional seismicity. Sub-aerial and submarine landslides, as well as rockfalls, become more frequent, increasing sedimentation in the fjord (e.g. Blikra & Longva 2000; Forwick & Vorren 2002, 2007; Bøe *et al.* 2003). In temperate fjords, postglacial deposition is dominated by high organic input. Pockmark fields are testament to subsurface fluid flow through dewatering or from gas (Hovland & Judd 1988; Plassen & Vorren 2003; Forwick *et al.* 2009). Any glaciomarine and deglacial sediments become subdued in relief by the deposition of overlying modern riverine and coastal marine sediments.

Hydrography

The hydrographic regime, the processes of tide, Coriolis force, internal waves and jets, are all fundamental to the distribution of sediment within fjords and can profoundly influence the biogeochemical reactions taking place within the sediment–porewaters and at the sediment–water interface. Some fjords are silled systems with many possessing a number of sills restricting the exchange of coastal and basin water. Restricted water exchange subsequently leads to oxygen depletion resulting in periodically or even permanently anoxic fjord bottom water conditions. This effect has already been described for temperate Norwegian fjords at the beginning of the last century (Gaarder 1916; Strøm 1936). Sea-level change will enhance or restrict this water exchange. Exchange provides oxygenated and nutrient-rich coastal waters to the often stagnant deep water in the basins – an essential process for both biological and geochemical processes. ‘Overturning’ or exchange with coastal

ERRATUM

Special Publication **344**

FORWICK, M., VORREN, T.O., HALD, M., KORSUN, S., ROH, Y., VOGT, C. & YOO, K.-C. 2010. Spatial and temporal influence of glaciers and rivers on the sedimentary environment in Sassenfjorden and Tempelfjorden, Spitsbergen. *In*: HOWE, J. A., AUSTIN, W. E. N., FORWICK, M. & PAETZEL, M. (eds) *Fjord Systems and Archives*. Geological Society, London, Special Publications, **344**, 163–193. DOI: 10.1144/SP344.13

There is an error in the Figure 2 which appears on page 166. The indicators for the glacier fronts were swapped in error. The corrected Figure 2 appears below.

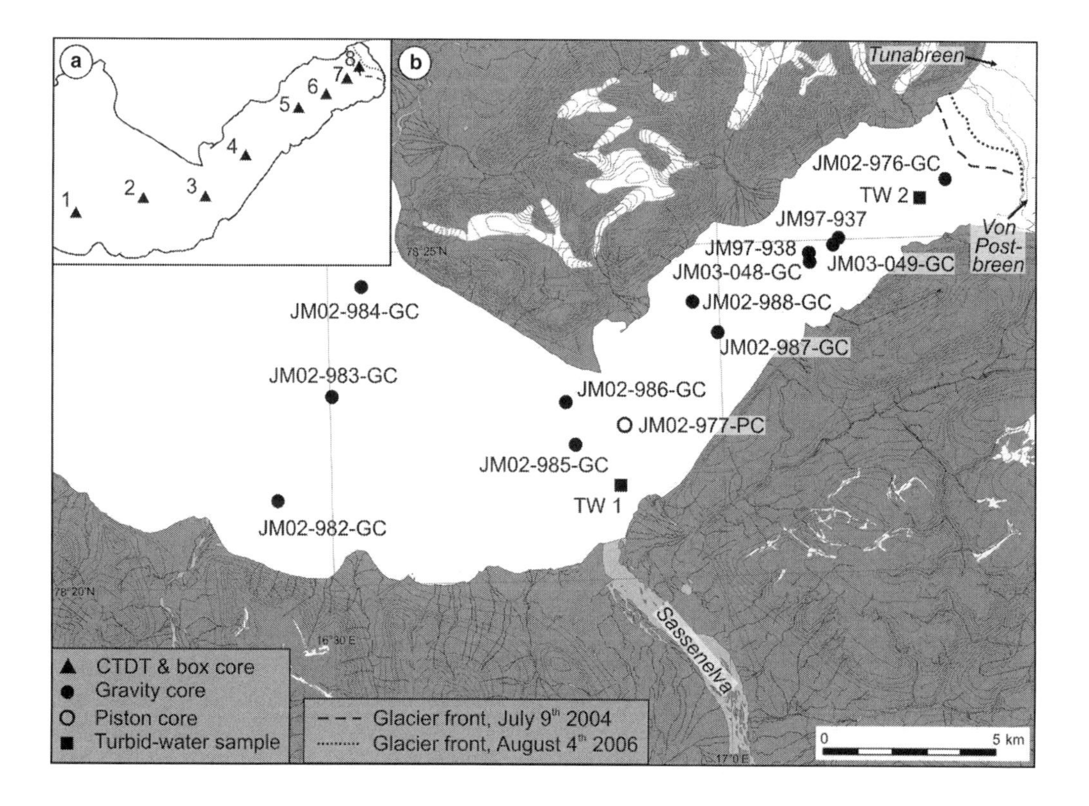

waters can occur at regular intervals or aperiodically, depending on a number of factors such as tide, freshwater input, weather and season. Tides and internal tides can also influence the redistribution of sediments in the fjord; the internal circulation of bottom waters can promote preferential sedimentation such as enhanced sedimentation on the slopes of the fjord, terminating in a tidally created scour. Fjords are essentially estuarine in nature with the flux of freshwater and coastal water creating a salinity trend. This in itself promotes so-called estuarine circulation, the stratified water column containing brackish water at the surface flowing out of the fjord and the more saline coastal waters beneath flowing into the fjord from the mouth.

Sediment supply

Sediment is generally riverine (including runoff) in origin for temperate fjords and glacier-derived for polar fjords (e.g. Powell 2005). The distribution of sediment within the fjord is dependent on its bathymetry, depth and hydrographic regime. Depositional features such as deltas and river-derived fans can be produced on the slopes at the discharge points of rivers and streams. In some more exposed fjords, waves can play a significant role in redistributing sediments. On delta-fronts and fans, channels often aid in the transport of sediment to the basins. Sediment waves can also occur, indicative of current reworking and bedload transport which is the consequence of river outflow into the fjord. Slope instability is also common as sediment creep, slides, debris flows and turbidity flows, resulting from high sediment supply on steep slopes (e.g. Syvitski *et al.* 1987). Postglacial sediment reworking occurs most commonly as downslope mass-wasting. This can result in fjord stratigraphies that contain up to 50% reworked sediments (Holtedahl 1975). In Norwegian fjords this can result from crustal unloading as the ice retreated, reactivating basement faults leading to an increase in seismic activity. Recent human activity, notably land use, can also play a significant role in influencing fjord sedimentation. Deforestation of the fjord hinterland can increase runoff and hence the volume of sediment entering the fjord. Typical modern sedimentation rates can be of the order millimetres to centimetres per year, depending on the availability of organic and mineral matter. High-productivity, anoxic temperate Norwegian fjords revealed limited amounts of mineral matter in the Nordåsvannet resulting in sedimentation rates as low as 0.4 mm a^{-1} (Paetzel *et al.* 1994); in contrast, enhanced mineral matter supply gained sedimentation rates of 2 cm a^{-1} and more in the Barsnesfjord (Paetzel & Schrader 1992). In polar fjords adjacent to a glacier this value can be as high as 100 cm a^{-1} (Shimmield 1993; Svendsen *et al.* 2002). Sedimentation rates of up to 13 m a^{-1} have been recorded in Alaskan fjords (Powell & Molnia 1989; Cowan & Powell 1991).

Biogeochemistry

High sedimentation rates combined with a high organic input from surrounding terrestrial plant material can lead to the production of gas-rich sediment. This is mostly methane gas from the decay of organic material within the sediment (Hovland & Judd 1988). The surface expression of these processes are pockmarks, typically some metres deep and up to more than a hundred metres wide (Forwick *et al.* 2009). Escaping gas and water produce low density sediment which migrates to the surface as a result of the weight of the surrounding sediment overburden (Fader 1997). The distribution of pockmarks in subpolar fjords is generally related to tectonic lineaments, the lithological composition and lateral outcrop of bedrock, the orientation of glacial lineations and exceptionally rapid deposition of debris lobes related to glacier surges (Forwick *et al.* 2009). Fjords are also important carbon sinks, the carbon being trapped as gas (methane) or directly as organic material. Carbon enters the system through riverine discharge and runoff or directly from primary production in the surface waters, subsequently settling in the basins. The frequently encountered anoxic bottom water conditions combined with the high organic content sediments in fjord basins can also lead to high levels of hydrogen sulphide in the sediment porewaters. This can result in metal sulphide formation (e.g. framboidal pyrite or zinc sulphide; Skei 1983*b*; Overnell *et al.* 1995).

Biological

Foraminifera are single-celled marine protozoa (Sen Gupta 1999) which live either in the photic zone of the water column (planktonic foraminifera) or in or on the sediment (benthic foraminifera). Foraminiferal assemblages in fjords are dominated by benthic species. Since planktonic foraminifera do not typically inhabit shelf sea regions, probably due to the low salinities (Arnold & Parker 1999), planktonic juvenile tests found in shelf sea sediments have usually been transported into the environment (Murray 1991). Species diversity and abundance of benthic foraminifera can be moderate to high in fjords (e.g. >500 specimens per cm^3 and Shannon-Weaver values >1.5; Cage 2005; Murray 2003), especially in basins with a stable depositional environment, fine sediment and high bottom water salinities (e.g. Alve & Nagy 1990; Klitgaard-Kristensen & Buhl-Mortensen 1999). Foraminiferal assemblages from these basins have high proportions of hyaline taxa with a calcium carbonate test ($CaCo_3$) which are typically well-preserved.

In fjord basins which have a restricted exchange with coastal waters (and hence a lower salinity), foraminiferal assemblages tend to be dominated by taxa with an agglutinated wall structure (see Murray 2003 for a review of foraminifera in marginal marine environments) and typically have a lower species diversity and foraminiferal abundance. Hyaline specimens present in these environments are often etched and have thin shells (Cage 2005), likely due to low pH or low calcium carbonate availability. Either process creates a stressful environment for calcareous taxa, which must spend 'considerable energy recalcifying their tests' (Boltovskoy & Wright 1976). The low abundance and diversity of porcellaneous taxa in fjord assemblages is also typical of fjordic or low salinity environments (e.g. Alve & Nagy 1990; Austin & Sejrup 1994; Murray *et al.* 2003; Cage 2005). Murray (1973) reports that miliolina taxa are unable to maintain pseudopodial reticulum activity in hyposaline waters and, unlike hyaline taxa which efficiently use available $CaCO_3$ during test formation, porcellaneous taxa are thought to have a less organized method which requires readily available $CaCO_3$ (Boltovskoy & Wright 1976).

Historically, studies have used the biological proxy pathway of benthic foraminifera (i.e. abundance of individuals or assemblage composition) to gain a qualitative handle on environmental conditions. Increasingly, however, the chemical proxy pathway of foraminifera such as stable oxygen and carbon isotopes are being used to investigate environmental parameters and obtain quantitative reconstructions of past conditions in fjords.

The majority of studies utilizing benthic foraminifera from fjord sediments focus on Scandinavian fjords which tend to be either deep with deep sills or anoxic (e.g. Höglund 1947; Thiede *et al.* 1981; Qvale *et al.* 1984; Alve & Nagy 1990; Klitgaard-Kristensen & Buhl-Mortensen 1999; Gustafsson & Nordberg 2000, 2001; Mikalsen *et al.* 2001; Murray *et al.* 2003; Filipsson & Nordberg 2004; Husum & Hald 2004*a*, *b*; Klitgaard-Kristensen *et al.* 2004; Nordberg *et al.* 2009). However, there has recently been an increase in environmental and Quaternary studies using benthic foraminifera in fjord sediment archives from locations such as Alaska (e.g. Ullrich *et al.* 2009), the Arctic (Lassen *et al.* 2004; Lloyd *et al.* 2007; Majewski & Zajaczkowski 2007; Hald & Korsun 2008; Nørgaard-Pedersen & Mikkelsen 2009), Canada (e.g. Riveiros & Patterson 2009), Chile (Hromic *et al.* 2006) and Scotland (Murray *et al.* 2003; Cage 2005; Nørgaard-Pedersen *et al.* 2006; Cage & Austin 2008; Cage & Austin 2010).

The ubiquity, abundance, species diversity and preservation of benthic foraminifera in fjord sediments make them the most widely studied marine microfossils. They have played, and will continue to play, an important role in environmental and palaeoenvironmental investigations in fjord environments.

Primary productivity in the water column to benthic epifauna and infauna on and in the seafloor is an important contributor to the sedimentation of fjords. In particular, the association between fjord sedimentation and dinoflagellate research dates back to at least the late 19th century when Aurivillius (1898) documented plankton, including dinoflagellates (important primary producers), from the Skagerrak and several fjords along the west coast of Sweden. The association between the fjords and the cysts of dinoflagellate may not have been made until 1945, however, when Braarud (1945) described cyst formation from Oslofjord. Nordli (1951) later described the cysts of *Goniaulax polyedra* from the same source. Erdtman (1954) published an important account of living dinoflagellate cysts from Gullmarfjord, Sweden and in 1976 Dale provided a seminal paper on dinoflagellate cysts from Trondheimsfjord, Norway linking them to known plankton records. This relationship is especially important to the study of palynology in fjords, especially sill fjords, providing an environment of deposition that is most like a closed system than any other found in the shelf seas or open ocean. Lack of much water movement or overturn within the deep fjord basins often leads to the development of anaerobic or hypoxic bottom conditions and minimal bioturbation: a situation ideal for the preservation of all organic material including dinoflagellate cyst types, whether normally susceptible to oxidation or not.

The sediment diatom record has been used to identify signals from recent environmental change: in west Norway, the successively changing diatom composition suggested the deposition of up to six seasonal laminae per year in the Barsnesfjord (Paetzel & Schrader 1992); the shifting dominance of diatom species was related to changes in sewage supply in the Nordåsvannet (Paetzel & Schrader 1995); and decadal climatic variation resulted in the sedimentation of alternating layers of planktonic and recycled benthic diatoms in the Store Lungegårdsvannet (Paetzel & Schrader 2003).

Pioneering studies into the benthic macrofaunal diversity of fjords revealed two habitats typically colonized by invertebrates: soft muds and muddy sands (Gage 1972). The diversity of these habitats is principally controlled by the exchange of waters from the sea; the greater the exchange of water, the higher the benthic diversity. Limited exchange increases the residence time of freshwater in the fjord, inhibiting the survival of larval stages.

Fjord sedimentation and depositional geometry

Glaciated fjords: role of ice and glaciomarine environments

Polar fjords are inherently dominated by glacial sedimentation (see Fig. 2a). Two types of glacier can influence polar fjords: (a) tidewater glacier and marine ice sheets where the terminus is grounded below sea level and the calving line coincides with the grounding line, and (b) floating ice shelves or tongues. Both of these are influenced by adjacent glaciers, glacial meltwaters and sea-level change. The glaciomarine fjord environment also includes pack ice, sea ice and seasonal fast ice (Elverhøi *et al.* 1983; Svendsen *et al.* 2002).

The glaciomarine environment is influenced by: (1) the glacier margin being either grounded or floating (this will affect the transport of debris); (2) the amount of meltwater (leading to sediment transport via suspension): copious (temperate setting), moderate (subpolar) or minimal (arid polar); (3) sea-bottom relief and (4) nature of the oceanic currents and wind regime.

Glaciomarine sediment facies can be very broadly divided into laminites and diamicts. These facies can be proximal or distal to the ice. Laminites are predominantly mudstones and include either varve-type formed in fresh or brackish water (also termed *rhythmites*) and deposits without the rhythmic structure deposited formed offshore. The laminites consist of alternating subcentimetre scale sands, silts and clays. Deposited within these sequences can be ice-rafted debris and some sedimentary structures (e.g. cross bedding) depending on the depositional environment. The laminites originate from meltwater plumes carrying suspended sediment out from one or several glacier termini, but also from mass wasting in glacier-proximal environments (e.g. Powell & Molnia 1989; Cowan *et al.* 1998, 1999; Forwick & Vorren 2009). Depending on their sand/mud ratios, laminae couplets deposited in glaciomarine environments can also be termed as cyclopels (low sand/mud ratio) or cyclopsams (high sand/mud ratio; Mackiewicz *et al.* 1984).

Non-glaciated fjords: role of rivers

In modern temperate mid-latitude fjords, sediment supply is controlled by river discharge (see Fig. 2b). Riverine sediment transport and progressive deposition of the transported material is the dominant sedimentation process in silled fjords where the exchange with coastal waters is restricted. River catchment can be vast; for a modest Scottish fjord such as Loch Etive, a large catchment of 1350 km^2 provides an annual average discharge of *c.* 3×10^9 m^3 of freshwater to the fjord (Nørgaard-Pedersen *et al.* 2006). This quantity of freshwater into the fjord has the result of producing a highly stratified water column and deltas adjacent to the river mouths. Relling & Nordseth (1979) examined the processes associated with river sedimentation into a fjord. These include density currents, flocculation and turbulence that can all inhibit the deposition of fine sediments; they produce a sustained plume that can carry sediment far into the fjord within the halocline. Coarser, sand-sized sediment is transported as bedload and deposited more rapidly primarily as a delta.

Fjord archives

High-resolution climate records

Fjords often act as large natural traps preserving sediment deposited over considerable lengths of time without disruption. Sedimentation rates are usually high enough to enable ultrahigh-resolution sampling sometimes close to a daily frequency, creating ideal natural laboratories for high spatial and temporal resolution archives. Modern palaeoceanographic studies using fjord sediments have focused on resolving the timing and impact of rapid climate change events, particularly within the Holocene (Mayewski *et al.* 2004). Sediment deposition within the fjord can produce archives which reflect changing marine, terrestrial and atmospheric environments. Gillibrand *et al.* (2005) modelled the hydrographic influence of the North Atlantic Oscillation (NAO) on Scottish west coast fjords. This atmospheric effect, occurring every $<15-90$ a, is the result of the pathway of low pressure air in the NE Atlantic, altering the strength of westerly storms and precipitation. During increased storminess, the sediment record preserves the effects of increased runoff in the sediment texture, oxygen and carbon isotope values. Such high-resolution Holocene climate records are not typical for marine sediment records, particularly in coastal sediments where the sediments are commonly coarser-grained and can contain hiatuses and erosive horizons.

Fjord sediment records can contain Holocene climate records that closely match terrestrial archives such as tree-rings and lake varves (Cage & Austin 2010), with the advantage of having been deposited continuously through time. In Arctic fjords, the Holocene sediment can preserve a pre-historical seasonal extent of glacial and sea ice.

Foraminiferal records have been demonstrated to reflect warm Atlantic water incursions into north Norwegian fjords (Hald *et al.* 2003; Husum & Hald 2004*a*, *b*). New proxies are now being used to estimate temperature, salinity and ice from the Holocene. The use of lipid biomarkers is becoming increasingly common, these being particularly useful in fjords where the high organic matter input provides a ready source of necessary lipids and carbon. In polar fjords this technique is also being used. The lipid present in the algae in sea ice has the potential to provide estimates of polar ice and temperature throughout the Holocene (Bendle *et al.* 2009). Given the uncertainty in estimating the rate and impact of present global warming, Holocene fjord records can be a valuable predictive tool through the high-resolution insights they can provide on previous rapid climatic change.

Anthropogenic impacts

Fjords are naturally places of high and continuous sediment deposition. Economically, the sheltered water and easy access of the fjord environment makes them particularly attractive to industry. A recent study by Richerol *et al.* (2007) illustrates the typical anthropogenic influences and pressures now faced by fjords. In this work from Labrador on the east coast of Canada, Arctic fjords were revealed to contain elevated levels of PCBs (PolyChloroBiphenyl) as well as mining effluent and waste from shipping. The local Inuit people depend on the fjords for hunting and fishing. This work highlights the increasing impacts of human activities in Arctic fjords, especially with respect to the impacts of climate change, industrialization and contamination of the food chain.

Other anthropogenic pollutants have been detected in Norwegian fjords. Trace metals were used to identify a variety of environmental impacts including the reconstruction of sedimentation processes from sewage deposition (Nordåsvannet – Paetzel & Schrader 1995); the sedimentation of elevated trace metal levels from natural bedrock erosion (Barsnesfjord – Paetzel & Schrader 1992); the deposition of trace metals due to the impact of climate variability on the fjord waters (Store Lungegårdsvannet – Paetzel & Schrader 2003); and artificially restricted water circulation due to building activity, resulting in enhanced sedimentation of pollutants originating from dental amalgam and anti-fouling boat paint (Vågen fjord, Bergen harbour – Paetzel *et al.* 2003). Smittenberg *et al.* (2004) used three groups of terrestrial, general marine and marine productivity biomarkers to identify changes in land use, water circulation and eutrophication, respectively, over the last 400 years in the Kyllaren fjord, west Norway. Tanhua & Olsson (2004) measured bioaccumulation and modelled the hydrography of Framvaren in southern Norway. In this anoxic fjord, CFC's (chlorofluorocarbons) were found to be bioaccumulating at much higher levels compared to the open ocean. This is a result of the high flux of organic matter combined with the restricted exchange with coastal waters, typical of fjord environments. Direct industrial discharge obviously contributes to pollution in fjords. One of the most intensively-studied fjords, Saguenay Fjord leading into the St Lawrence estuary in Canada, has well-established heavy industry along its shores. Gagnon *et al.* (1997) revealed high levels of mercury relating to persistent chemical plant discharges. Although the chemical plant is now closed, the levels of mercury remain high. In an extensive study of SW Scottish fjords, Shimmield (1993) revealed the legacy of low-level radionuclide discharges from Sellafield, locally from military activity in the Firth of Clyde and from the accident at Chernobyl in 1986. Paetzel & Schrader (1991) were the first to use the Chernobyl ^{137}Cs sediment record for dating purposes when analysing historical environmental change in the Barsnesfjord, west Norway. In the same study, they identified sediment signals of historical slide events, confirming the caesium dating basis. Also detectable in the geochemical record is the onset of lead in petrol, using the geochemical 'fingerprint' of stable lead isotope ratios ($^{206}Pb/^{207}Pb$). This fingerprint, unique to the origin of the petrol, can be used as a marker chronologically and also a tracer for water mass pathways into the fjord. Aside from the record of anthropogenic contaminants in fjords, physical change in the catchment can be reflected in recent sedimentation. Scottish fjord catchments were heavily deforested during the industrial revolution to supply wood and charcoal but also due to the demand for more land for farming. The result has been increased runoff reflecting an increase in the sedimentation rate and a corresponding increase in grain size (Nørgaard-Pedersen *et al.* 2006). Cage & Austin (2010) interpret a post-AD 1900 shift in benthic foraminiferal $\delta^{13}C$ of *c.* $-0.6‰$ as evidence of the Oceanic $\delta^{13}C$ Suess Effect, that is the change in $\delta^{13}C$ of dissolved inorganic carbon driven by the release of isotopically light carbon (^{12}C) into the atmosphere from fossil fuel combustion since the industrial revolution.

Fjords in the future?

Fjords are, by their very nature, dynamic and geologically ephemeral systems. Their morphology and sediment fill will most probably become modified, if not removed, over time by a subsequent glaciation. Sea-level change will also influence the rate and supply of sediment. Fjords presently face increasing pressure from both anthropogenic and environmental change. This is particularly acute for high-latitude fjords where activities such as shipping, mining and hydrocarbon exploration utilize fjords which are already facing sea-level rise, warm water incursions, loss of seasonal ice and glaciers in retreat. Fjords as coastal geomorphological marine features essentially face perpetual change. However, in the short-term, these unique pristine habitats and environments need careful monitoring and management to avoid long-term damage.

References

Alve, E. & Nagy, J. 1990. Main features of foraminiferal distribution reflecting estuarine hydrography in Oslo Fjord. *Marine Micropaleontology*, **16**, 181–206.

Arnold, A. J. & Parker, W. C. 1999. Biogeography of planktonic foraminifera. *In*: Gupta, B. K. S. (ed.) *Modern Foraminifera*. Kluwer Academic Publishers, London, 103–122.

Aurivillius, C. W. S. 1898. Vergleichende tiergeographische Untersuchungen über die Plankton-Fauna des Skageraks in den Jahren 1893–1897. *Kungliga Svenska Vetenskapsakademien*, **30**, 1–427.

Austin, W. E. N. & Sejrup, H. P. 1994. Recent shallow water benthic foraminifera from western Norway: Ecology and palaeoecological significance. *In*: Sejrup, H. P. & Knudsen, K. L. (eds) *Late Cenozoic Benthic Foraminifera: Taxonomy, Ecology and Stratigraphy*. Cushman Foundation for Foraminifera Research, Special Publication, **32**, 103–125.

Bendle, J. A., Rosell-Melé, A., Cox, N. J. & Shennan, I. 2009. Alkenones, alkenoates and organic matter in coastal environments of N.W. Scotland: assessment of potential application for sea-level reconstruction. *Geochemistry, Geophysics and Geosystems (G-cubed)*, **10**, Q12003, doi: 10.1029/2009GC002603.

Blikra, L. H. & Longva, O. 2000. Gravitational-slope failures in Troms: indications of palaeoseismic activity? *In*: Dehls, J. & Olesen, O. (eds) *Neotectonics in Norway: Annual Technical Report*. NGU Rapp. 2000.01, 31–40.

Bøe, R., Rise, L., Blikra, L. H., Longva, O. & Eide, A. 2003. Holocene mass-movement processes in Trondheimsfjorden, Central Norway. *Norwegian Journal of Geology*, **83**, 3–22.

Boltovskoy, E. & Wright, R. 1976. *Recent Foraminifera*. Junk, W., The Hague.

Braard, T. 1945. Morphological observations on marine dinoflagellate cultures (*Porella perforate*, *Goniaulax tamarensis*, *Protoceratium reticulatum*). *Avhandlinger utgitt av det Norske Videnskap-Akademi i Oslo, 1. Mat.-Naturv. Klasse 1944*, **11**, 1–18.

Cage, A. G. 2005. *The modern and Late Holocene Marine environments of Loch Sunart, N.W. Scotland*. PhD thesis. School of Geography & Geosciences, University of St Andrews.

Cage, A. G. & Austin, W. E. N. 2008. Seasonal dynamics of coastal water masses in a Scottish fjord and their potential influence on benthic foraminiferal shell geochemistry. *In*: Austin, W. E. N. & James, R. H. (eds) *Biogeochemical Controls on Palaeoceanographic Environmental Proxies*. Geological Society, London, Special Publications, **303**, 155–172.

Cage, A. G. & Austin, W. E. N. 2010. Marine climate variability during the last millennium: the Loch Sunart record, Scotland, UK. *Quaternary Science Reviews*, doi: 10.1016/j.quascirev.2010.01.014.

Cowan, E. A., Cai, J., Powell, R. D., Seramur, K. C. & Spurgeon, V. L. 1998. Modern tidal rhythmites deposited in a deep-water estuary. *Geo-Marine Letters*, **18**, 40–48.

Cowan, E. A. & Powell, R. D. 1991. Ice-proximal sediment accumulation rates in a temperate glacial fjord, southeastern Alaska. *In*: Anderson, J. B. & Ashley, G. M. (eds) *Glacial Marine Sedimentation; Paleoclimatic Significance*. Geological Society of America, Special Paper, **261**, 61–73.

Cowan, E. A., Seramur, K. C., Cai, J. & Powell, R. D. 1999. Cyclic sedimentation produced by fluctuations in meltwater discharge, tides and marine productivity in an Alaskan fjord. *Sedimentology*, **46**, 1109–1126.

Dale, B. 1976. Cyst formation, sedimentation, and preservation: factors affecting dinoflagellate assemblages in Recent sediments from Trondheimsfjord, Norway. *Review of Palaeobotany and Palynology*, **22**, 39–60.

Domack, E. W. & McClennen, C. E. 1996. Accumulation of glacial marine sediments in fjords of the Antarctic Peninsula and their use as Late Holocene Paleoenvironmental indicators. *Foundations for ecological research west of the Antarctic Peninsula – Antarctic Research Series*, **70**, 135–154.

Elverhøi, A., Lønne, Ø. & Serland, R. 1983. Glaciomarine sedimentation in a modern fjord environment, Spitsbergen. *Polar Research*, **1**, 127–149.

Erdtman, G. 1954. On pollen grains and dinoflagellate cysts in the Firth of Gullmarn, SW Sweden. *Botaniska Notiser*, **2**, 103–111.

Fader, G. B. J. 1997. The effects of shallow gas on seismic reflection profiles. *In*: Davies, T. A., Bell, T. et al. (eds) *Glaciated Continental Margins: An Atlas of Acoustic Images*. Chapman and Hall, London.

Filipsson, H. L. & Nordberg, K. 2004. A 200-year environmental record of a low-oxygen fjord, Sweden, elucidated by benthic foraminifera, sediment characteristics and hydrographic data. *Journal of Foraminiferal Research*, **34**, 277–293.

Forwick, M. & Vorren, T. O. 2002. Deglaciation History and Postglacial Sedimentation in Balsfjord (North Norway). *Polar Research*, **21**, 259–266.

Forwick, M. & Vorren, T. O. 2007. Holocene mass-transport activity in and climate outer Isfjorden, Spitsbergen: marine and subsurface evidence. *The Holocene*, **17**, 707–716.

Forwick, M. & Vorren, T. O. 2009. Late Weichselian and Holocene sedimentary environments and ice rafting in Isfjorden, Spitsbergen. *Palaeogeography, Palaeoclimatology, Palaeoecology*, **280**, 258–274.

Forwick, M., Baeten, N. J. & Vorren, T. O. 2009. Pockmarks in Spitsbergen fjords. *Norwegian Journal of Geology*, **89**, 65–77.

Gaarder, T. 1916. De vestlandske fjorders hydropgrafi, 1. Surstoff i fjordene. *Bergens Museumets Årbøker 1915–1916, Naturvidenskapelig Rekke*, **2**, 1–200.

GAGE, J. 1972. Community structure of the benthos in Scottish sea-lochs. I. Introduction and species diversity. *Marine Biology*, **14**, 281–297.

GAGNON, C., PELLETIER, E. & MUCCI, A. 1997. Behaviour of anthropogenic mercury in coastal marine sediments. *Marine Chemistry*, **59**, 159–176.

GILBERT, R. 2000. Environmental assessment from the sedimentary record of high-latitude fiords. *Geomorphology*, **32**, 295–314.

GILLIBRAND, P. A., CAGE, A. G. & AUSTIN, W. E. N. 2005. A preliminary investigation of the basin water response to climate change in a Scottish fjord: evaluating the influence of NAO. *Continental Shelf Research*, **25**, 571–587.

GJESSING, J. 1956. Om iserosjon, fjorddal – dalendedannelse. *Norsk Geografisk Tidsskrift*, **15**, 243–269.

GUSTAFSSON, M. & NORDBERG, K. 2000. Living (Stained) benthic foraminifera and their response to the seasonal hydrographic cycle, periodic hypoxia and to primary production in Havstens Fjord on the Swedish west coast. *Estuarine Coastal and Shelf Science*, **51**, 743–761.

GUSTAFSSON, M. & NORDBERG, K. 2001. Living (stained) benthic foraminiferal response to primary production and hydrography in the deepest part of the Gullmar Fjord, Swedish West Coast, with comparisons to Hoglund's 1927 material. *Journal of Foraminiferal Research*, **31**, 2–11.

HAMBREY, M. 1994. *Glacial Environments*. UCL Press, London.

HALD, M. & KORSUN, S. 2008. The 8200 cal. Yr BP event reflected in the Arctic fjord, Van Mijenfjorden, Svalbard. *Holocene*, **18**, 981–990.

HALD, M., HUSUM, K., VORREN, T. O., GRØSFELD, K., JENSEN, H. B. & SHAPAROVA, A. 2003. Holocene climate in the subarctic fjord Malangen, northern Norway: a multi-proxy study. *Boreas*, **32**, 543–559.

HÖGLUND, H. 1947. Foraminifera in the Gullmar fjord and the Skagerrak. *Zoologiska bidrag från Uppsala*, **26**, 328.

HOLTEDAHL, H. 1967. Notes on the formation of fjords and fjord-valleys. *Geografiska Annaler Series A, Physical Geography*, **49**, 188–203.

HOLTEDAHL, H. 1975. The geology of the Hardangerfjord, West Norway. *Norges Geologiske Undersøkelse*, **323**, 1–87.

HOVLAND, M. & JUDD, A. G. 1988. *Seabed Pockmarks and Seepages. Impact on Geology, Biology and the Marine Environment*. Graham & Trotman Ltd., London.

HOWE, J. A., MORETON, S. G., MORRI, C. & MORRIS, P. 2003. Multibeam bathymetry and the deposition environments of Kongsfjorden and Krossfjorden, western Spitsbergen, Svalbard. *Polar Research*, **22**, 301–316.

HROMIC, T., ISHMAN, S. & SILVA, N. 2006. Benthic foraminiferal distributions in Chilean fjords: 47 degrees S to 54 degrees S. *Marine Micropaleontology*, **59**, 115–134.

HUSUM, K. & HALD, M. 2004*a*. A continuous marine record 8000–1600 cal. Yr BP from the Malangenfjord, north Norway: foraminiferal and isotopic evidence. *The Holocene*, **14**, 877–887.

HUSUM, K. & HALD, M. 2004*b*. Modern foraminiferal distribution in the subarctic Malangenfjord and adjoining shelf, northern Norway. *Journal of Foraminiferal Research*, **34**, 34–48.

KESSLER, M. A., ANDERSON, R. & BRINER, J. P. 2008. Fjord insertion into continental margins by topographic steering of ice. *Nature Geoscience*, **1**, 365–368.

KLITGAARD-KRISTENSEN, D. & BUHL-MORTENSEN, L. 1999. Benthic foraminifera along an offshore-fjord gradient: a comparison with amphipods and molluscs. *Journal of Natural History*, **33**, 317–350.

KLITGAARD-KRISTENSEN, D. K., SEJRUP, H. P., HAFLIDASON, H., BERSTAD, I. M. & MIKALSEN, G. 2004. Eight-hundred-year temperature variability from the Norwegian continental margin and the North Atlantic thermohaline circulation. *Paleoceanography*, **19**, art. no.-PA2007.

LASSEN, S. J., KUIJPERS, A., KUNZENDORF, H., HOFFMANN-WIECK, G., MIKKELSEN, N. & KONRADI, P. 2004. Late-Holocene Atlantic bottom-water variability in Igaliku Fjord, South Greenland, reconstructed from foraminifera faunas. *Holocene*, **14**, 165–171.

LLOYD, J. M., KUIJPERS, A., LONG, A., MOROS, M. & PARK, L. A. 2007. Foraminiferal reconstruction of mid- to late-Holocene ocean circulation and climate variability in Disko Bugt, West Greenland. *Holocene*, **17**, 1079–1091.

MACKIEWICZ, N. E., POWELL, R. D., CARLSON, P. R. & MOLNIA, B. F. 1984. Interlaminated ice-proximal glaciomarine sediments in Muir Inlet, Alaska. *Marine Geology*, **57**, 113–147.

MAJEWSKI, W. & ZAJACZKOWSKI, M. 2007. Benthic foraminifera in Adventfjorden, Svalbard: Last 50 years of hydrographic changes. *Journal of Foraminifera Research*, **37**, 107–124.

MAYEWSKI, P. A., ROHLING, E. E. *ET AL.* 2004. Holocene climate variability. *Quaternary Research*, **62**, 243–255.

MIKALSEN, G., SEJRUP, H. P. & AARSETH, I. 2001. Late-Holocene changes in ocean circulation and climate: foraminiferal and isotopic evidence from Sulafjord, western Norway. *Holocene*, **11**, 437–446.

MURRAY, J. W. 1973. *Distribution and Ecology of Benthic Foraminiferids*. Heinemann Educational Books, London.

MURRAY, J. W. 1991. *Ecology and Palaeoecology of Benthic Foraminifera*. Longman, Harlow, 397.

MURRAY, J. W. 2003. Foraminiferal assemblage formation in depositional sinks on the continental shelf west of Scotland. *Journal of Foraminiferal Research*, **33**, 101–121.

MURRAY, J. W., ALVE, E. & CUNDY, A. 2003. The origin of modern agglutinated foraminiferal assemblages: evidence from a stratified fjord. *Estuarine, Coastal and Shelf Science*, **58**, 677–697.

NESJE, A. & WHILLANS, I. M. 1994. Erosion of the Sognefjord. *Geomorphology*, **9**, 33–45.

NORDBERG, K., FILIPSSON, H. L., LINNE, P. & GUSTAFSSON, M. 2009. Stable oxygen and carbon isotope information on the establishment of a new, opportunistic foraminiferal fauna in a Swedish Skagerrak fjord basin, in 1979/1980. *Marine Micropaleontology*, **73**, 117–128.

NØRGAARD-PEDERSEN, N. & MIKKELSEN, N. 2009. 8000 year marine record of climate variability and fjord dynamics from Southern Greenland. *Marine Geology*, **264**, 177–189.

NØRGAARD-PEDERSEN, N., AUSTIN, W. E. N., HOWE, J. A. & SHIMMIELD, T. 2006. A Holocene sea level record in Loch Etive, Western Scotland: hydrographic changes inferred from sediment properties and benthic foraminifera assemblages. *Marine Geology*, **228**, 55–71.

NORDLI, E. 1951. Resting spores in Goniaulax polyedra Stein. *Nytt Magasin Naturvidenskapene*, **88**, 207–212.

OVERNELL, J., HARVEY, M. S. & PARKES, R. P. 1995. A biogeochemical comparison of sea loch sediments. Manganese and iron contents, sulphate reduction and oxygen uptake rates. *Oceanologica Acta*, **19**, 41–55.

PAETZEL, M. & SCHRADER, H. 1991. Heavy metal (Zn, Cu, Pb) accumulation in the Barsnesfjord, Western Norway. *Norsk Geologisk Tidsskrift*, **71**, 65–73.

PAETZEL, M. & SCHRADER, H. 1992. Recent environmental changes recorded in the anoxic Barsnesfjord sediment: Western Norway. *Marine Geology*, **105**, 23–36.

PAETZEL, M. & SCHRADER, H. 1995. Sewage history in the anoxic sediments of the fjord Nordåsvannet, Western Norway: (2) the sources of the sedimented organic matter fraction. *Norsk Geologisk Tidsskrift*, **75**, 146–155.

PAETZEL, M. & SCHRADER, H. 2003. Natural vs human induced facies change in recent, shallow water sediments of the Store Lungegårdsvannet (Western Norway). *Environmental Geology*, **43**, 484–492.

PAETZEL, M., SCHRADER, H. & CROUDACE, I. 1994. Sewage history in the anoxic sediments of the fjord Nordåsvannet, Western Norway: (I) Dating and trace metal accumulation. *The Holocene*, **4**, 290–298.

PAETZEL, M., NES, G., LEIFSEN, L. Ø. & SCHRADER, H. 2003. Sediment pollution in the Vågen, Bergen Harbour, Norway. *Environmental Geology*, **43**, 476–483.

PLASSEN, L. & VORREN, T. O. 2003. Fluid flow features in fjord-fill deposits, Ullsfjorden, North Norway. *Norwegian Journal of Geology*, **83**, 37–42.

POWELL, R. D. 2005. Subaquatic landsystems: fjord. *In*: EVANS, D. J. A. (ed.) *Glacial Landsystems*. Hodder Arnold, London, 313–347.

POWELL, R. D. & MOLNIA, B. F. 1989. Glacimarine sedimentary processes, facies and morphology of the south-southest Alaska shelf and fjords. *Marine Geology*, **85**, 359–390.

QVALE, G., MARKUSSEN, B. & THIEDE, J. 1984. Benthic foraminifers in fjords; response to water masses. *Norsk Geologisk Tidsskrift*, **64**, 235–249.

RELLING, O. & NORDSETH, K. 1979. Sedimentation of a river suspension into a fjord basin. Gaupnefjord in western Norway. *Norsk geogr. Tidsskr.*, **33**, 187–203.

RICHEROL, T., PIENITZ, R. & ROCHON, A. 2007. Paleoceanographic studies of the impacts of natural and anthropogenic perturbations in three Labrador fjord ecosystems (Nachvak, Saglek and Anaktalak): Preliminary results. *ArcticNet 2007 Annual Scientific Meeting*. December 11–14, 2007, Collingwood, Ontario.

RIVEIROS, N. V. & PATTERSON, R. T. 2009. Late Holocene paleoceanographic evidence of the influence of the Aleutian Low and North Pacific High on circulation in the Seymour-Belize Inlet Complex, British Columbia, Canada. *Quaternary Science Reviews*, **28**, 2833–2850.

SEN GUPTA, B. K. 1999. Systematics of modern Foraminifera. *In*: SEN GUPTA, B. K. (ed.) *Modern Foraminifera*. Kluwer Academic Publishers, Dordrecht, 7–36.

SHIMMIELD, T. 1993. *A study of radionuclides, lead and lead isotope ratios in Scottish sea loch sediments*. PhD thesis, University of Edinburgh.

SKEI, J. M. 1983*a*. Why sedimentologists are interested in fjords. *In*: SYVITSKI, J. P. M. & SKEI, J. M. (eds) Sedimentology of Fjords. *Sedimentary Geology*, **36**, 75–80.

SKEI, J. M. 1983*b*. Geochemical and sedimentological considerations of a permanently anoxic fjord – Framvaren, south Norway. *In*: SYVITSKI, J. P. M. & SKEI, J. M. (eds) Sedimentology of Fjords. *Sedimentary Geology*, **36**, 131–145.

SMITH, D. E., CULLINGFORD, R. A., MIGHALL, T. M., JORDAN, J. T. & FRETWELL, P. T. 2007. Holocene relative sea level changes in a glacio-isostatic area: new data from south-west Scotland, United Kingdom. *Marine Geology*, **242**, 5–26.

SMITTENBERG, R. H., PANCOST, R. D., HOPMANS, E. C., PAETZEL, M. & SINNINGHE DAMSTÉ, J. S. 2004. A 400-year record of environmental change in an euxinic fjord as revealed by the sedimentary biomarker record. *Palaeogeography, Palaeoclimatology, Palaeoecology*, **202**, 331–351.

STOKER, M. S. & BRADWELL, T. 2005. The Minch palaeo-ice stream, NW sector of the British-Irish Ice Sheet. *Journal of the Geological Society*, **162**, 425–428.

STRØM, K. 1936. Land-locked waters. *Det Norske Videnskapsakademiets Srifter, Oslo, Mat.-Nat- Klasse*, **7**, 1–85.

SVENDSEN, H., BESZCZYNSKA-MOLLER, A. ET AL. 2002. The Physical environment of Kongsfjorden-Krossfjorden, an Arctic fjord system in Svalbard. *Polar Research*, **21**, 13–166.

SYVITSKI, J. P. M. & SHAW, J. 1995. Sedimentology and geomorphology of fjords. *In*: PERILLO, G. M. E. (ed.) *Geomorphology and Sedimentology of Estuaries. Developments in Sedimentology*, **53**, 113–178.

SYVITSKI, J. P. M., BURRELL, D. C. & SKEI, J. M. 1987. *Fjords – Processes and Products*. Springer, New York.

TANHUA, T. & OLSSON, K. A. 2004. Removal and bioaccumulation of anthropogenic, halogenated transient tracers in an anoxic fjord. *Marine Chemistry*, **94**, 27–41.

THIEDE, J., QVALE, G., SKARBØ, O. & STRAND, J. E. 1981. Benthonic foraminiferal distributions in a southern Norwegian fjord system: a re-evaluation of Oslo Fjord data. *Special Publications of the International Association of Sedimentologists*, **5**, 469–495.

ULLRICH, A. D., COWAN, E. A., ZELLERS, S. D., JAEGER, J. M. & POWELL, R. D. 2009. Intra-annual Variability in Benthic Foraminiferal abundance in sediments of disenchantment Bay, an Alaskan Glacial Fjord. *Arctic Antarctic and Alpine Research*, **41**, 257–271.

The physics of mid-latitude fjords: a review

M. E. INALL[1]* & P. A. GILLIBRAND[2]

[1]*Scottish Association for Marine Science, Dunstaffnage Marine Laboratory, Oban, Scotland, UK*

[2]*National Institute of Water and Atmospheric Research, Christchurch, New Zealand*

**Corresponding author (e-mail: mark.inall@sams.ac.uk)*

Abstract: A rich and wide variety of fluid dynamic processes occur in fjords. Although a fjord may at one level be simply defined a glacially formed coastal inlet, this simple definition belies a huge range of geomorphological manifestations and environmental forcing conditions. It is the interplay between geomorphology and environmental forcing which defines the relative importance of differing physical fluid processes within a given fjord. In this chapter we present a non-mathematical review of the dominant physical processes which are found to occur in fjordic systems, how their relative importance may depend on geomorphology and forcing, and how, in turn, the dominant physical processes effect circulation and sediment distribution. Our aim is to provide the non-physical oceanographer with an insight into the rich and varied fluid dynamical processes presented to us by the fascinating 'mini-ocean' geo-type generically referred to as a fjord.

Why should the flow of water within a fjord be of interest to the geologist? After all, fjordic geomorphology has changed little over the last few thousand years. One reason is that attention has recently focused on fjords as archives of Holocene climate change. We must therefore understand the sedimentary depositional environment, that is, the flow of water within the fjord, before deciphering the climate signals in a sedimentary record. In reviewing our knowledge of the physical processes in fjords, this chapter puts some emphasis on those processes relevant to the redistribution of sediments (Table 1).

The word fjord derives from Old Norse and is commonly used to describe a long, narrow, deep inlet of coastal water with steep sides and one or more submarine sills. Alternative names are sea loch, inlet, arm, sound and voe. Creation of fjords is associated with glacial carving; they are therefore a common feature of high-latitude mountainous coastlines. Farmer & Freeland (1983) (FF83) give a long (72 pages) and comprehensive review of the physical oceanography of fjords. More recently, Inall (2005) has reviewed the literature on turbulent mixing within fjords. Much other research on fjords has been published since FF83 and, although we do not go into the same level of detail, we try to distil much of the fjordic oceanography research published before and after FF83.

The reader is also referred to the largest published collection of fjordic oceanography papers (Freeland *et al.* 1980) and all the fjordic papers of Anders Stigebrandt, who has perhaps contributed the most to our understanding of the physics of fjords (Stigebrandt 1976, 1977, 1979, 1980, 1981, 1999*a*, *b*, 2001; Stigebrandt & Aure 1989; Stigebrandt & Molvaer 1996). By coincidence, the total volume of water in all the world's fjords is similar to that of freshwater lakes (Syvitski *et al.* 1987). Many similar processes occur and studies in lakes have informed our understanding of fjordic processes; a relatively recent compendium is available (Imberger 1998).

It is because such a wide variety of the geo-type 'fjord' exists along the mid to high-latitude coastlines of the world that the concept of a 'typical' fjord is fraught with problems; there are always exceptions to any particular generalization and there may not be a fjord which is 'typical' in every aspect. For simplicity we use the word fjord to designate a 'typical fjord' in this chapter, while recognizing that no two fjords are truly alike. This chapter does have a Scottish flavour, partly due to the location of the workshop from which this book emerged and partly due to the bias in the authors' own research, but we take care to emphasize that the physical processes discussed apply to mid-latitude fjords globally.

Topographic constraints: a restricted environment

From the point of view of water movement, topographic (or geomorphological) constraints define the behaviour of a fjord given the particular external forcing; tides, wind, freshwater input and the vertical temperature and salinity structure of the adjacent seas. Topographic constraints often include both a relatively shallow sill and a relative narrowing of the coastline at the entrance to a fjord (Fig. 1). Both these constraints affect the flow in different and various ways. Both reduce the vertical

From: HOWE, J. A., AUSTIN, W. E. N., FORWICK, M. & PAETZEL, M. (eds) *Fjord Systems and Archives*. Geological Society, London, Special Publications, **344**, 17–33.
DOI: 10.1144/SP344.3 0305-8719/10/$15.00

Table 1. *Qualitative summary of the effects of physical processes on sediment deposition and redistribution*

Process	Typical maximum current speeds ($m\ s^{-1}$)	Location of current maximum	Sedimentary effects: deposition, distribution, redistribution
Riverine in-flows	Highly variable: 0.1–4	Surface. Often deflected to the right/left (in the direction of flow) in the northern/southern hemisphere.	Main terrestrial sedimentary input. Finer particles transported greater distances from river mouth before settling. Spate flows may resuspend and redistribute sediments accumulated near the river mouth during normal river flows. Slumping may also occur at other times due to sediment over-accumulation near the river mouth.
Barotropic tides I: sills	Strong depth and regional dependence: 0.1–4	Over the shallowest portion of the sill.	Main source of marine-origin particulates. Sills are generally well scoured of all sediment. Currents decay away from sill with increasing water depth. In 'jet-type' basins, strong surface currents may persist over deep water many hundreds of metres from the sill.
Barotropic tides II: away from sills	Generally weak: 0.05–0.2	Currents generally decay to zero at head of fjord, although constrictions may cause local maxima and current rectification.	Redistributive effects lessen towards the head. Residual eddies associated with headlands may locally trap some sedimenting material.
Estuarine circulation	0.01–0.2	Uniform along axis, although cross-axis distribution may be affected by Earth rotation.	An extension of riverine inflow at the surface. Subsurface arm draws marine-origin particulates into the fjord interior, although flows are generally weak. Surface outflow may exceed tidal inflow, such that the surface layer flow does not reverse with the tidal cycle.
Intermediate water exchange	0.01–0.2	Below surface layer and above sill depth. Exchange flow which can reverse (i.e. upper and lower layer flows always oppose, and can reverse).	Marine origin euphotic zone materials import increased when upper intermediate flows are inwards; resuspended material may be exported from fjord. Reverse flows export fjord-origin euphotic materials and import deeper marine load.
Deep water renewal	0.1–0.4	Near bed, in deeper waters close to sills.	Agent of resuspension. In time will redistribute material along the bottom of a deep basin. May displace deoxygenated waters upwards, resulting in periodic or intermittent biodegredation and/or fish kills.
Wind driven currents	A few percent of wind speed	Surface layer, decaying with depth.	Winds are generally aligned along axis and therefore either enhance or oppose the estuarine circulation. A systematic local correlation between up-basin wind and rainfall may result in a local wind-enhanced retention of riverine sediment towards the head of the fjord. The opposite is also a possibility. Upon the relaxation of winds, internal seiching may have an effect akin to internal tides.
Internal tides	0.1–0.3	Near bed and near surface. Maximum shear at mid depths.	The most dramatic effect may be on the sloping boundaries below sill depth. Shoaling depths cause internal tide to break and create local current maxima, resuspending material and allowing it to intrude into the basin interior as intermediate nepheloid layers.

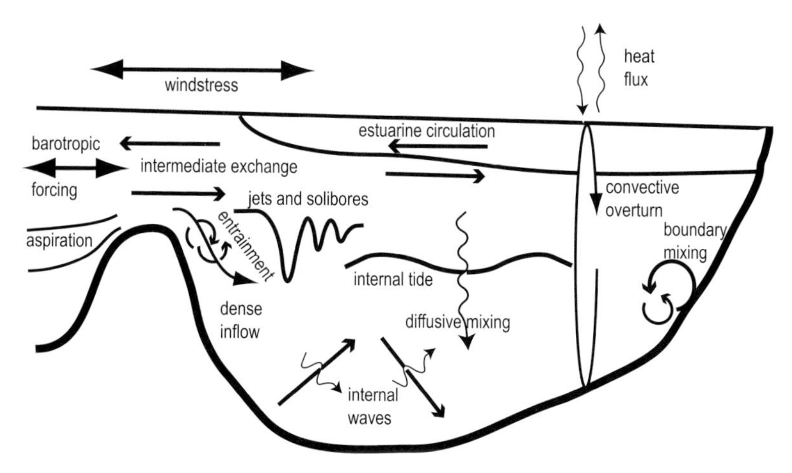

Fig. 1. A cartoon of the physical processes in fjords.

cross-sectional area leading to a local acceleration of the flow. This increases the friction of the flow against the seabed (termed 'skin friction') roughly as the square of the flow speed; the overall volume of water flowing through the entrance therefore diminishes as the vertical cross-sectional area decreases. This can be followed to its logical end-point where there is zero volume flow through a closed entrance. The lateral constriction may additionally impede the flow through the shedding of energetic eddies where the coastline widens from the narrows, a phenomenon called barotropic form drag which is discussed in more detail later in this chapter. The presence of density-stratified water and a sill may induce another form of frictional drag: (a) if the mean flow is faster than some critical value determined by the degree of stratification, and/or (b) if the flow is tidally oscillating. These two phenomena are generically referred to as baroclinic form drag (see later). In a given fjord, the relative proportions by which these frictional processes absorb energy from the tidal flow will vary considerably. There are some straightforward formulae to estimate their relative contributions, however (Deyoung & Pond 1989; Stigebrandt 1999*b*; Inall *et al.* 2004).

Physical inputs and processes

Stratification: setting and changing

The water column structure of fjords typically differs from the adjacent coastal seas in one important aspect: the structure and temporal variation of density stratification. Firstly, coastal and shelf seas generally exhibit a seasonal cycle of solar-driven thermal stratification, with the exception of Regions of Fresh Water Influence or ROFIs (Simpson 1997). Within a fjord, however, the flow constrictions referred to above, combined with a high freshwater input, result in a pattern of vertical density stratification often dominated by salinity throughout the year and often with no clear seasonal pattern. Secondly, the isolated deep basin waters of fjords may see a complete change in its temperature and salinity characteristics on quasi-periodic time-scales ranging from weeks (Allen & Simpson 1998*a*) to years (Edwards & Edelsten 1977).

Freshwater input. Almost by the definition of their glacial origins, fjords have a high freshwater supply. Often the major freshwater input is riverine and at the head of the fjord, but large riverine inputs may also occur at any point along their coastline as determined by the surrounding terrain. Freshwater catchment areas are easily defined by reference to cartographic maps, and rainfall can be simply scaled to give a total freshwater supply based on measured rainfall (a figure of around 70% is often used). Less straightforward is an estimate of the terrestrial freshwater residence time, although catchment models can give predictions based on knowledge of the soil type, vegetation and underlying geology. Direct rainfall on the water surface has a negligible effect in most fjords, since catchment area typically exceeds surface area by at least an order of magnitude. Of relevance to sediment redistribution is the occurrence of turbidity currents in the deep basins, which have been linked to high freshwater inputs (Bornhold *et al.* 1994) via a mechanism whereby sediment accumulated in the shallow upper reaches is displaced during flood events.

Surface heat fluxes. There is nothing peculiar in fjords with regard to surface heat fluxes. As in all marine environments, the net heat flux across the air/water interface is composed of shortwave solar radiation, net longwave radiation and sensible and latent heat fluxes. Shortwave radiation is determined by latitude, season, time-of-day and cloud cover; as in all coastal seas the seasonal cycle is clearly seen in the thermal stratification of the upper *c*. 50 m of the water column. Net longwave radiation accounts for the largest heat loss and, combined with large sensible heat losses in late autumn (when air/sea temperature differences tend to be greatest), there exists the possibility in some fjords of complete breakdown of water column stratification driven by large surface heat losses (e.g. the Clyde Sea, west Scotland; Rippeth & Simpson 1996). Latent heat flux is generally the smallest term in the heat flux budget.

Renewal and baroclinic exchange. In principle, deep water renewal (see the later section on this subject) can lead to stratification changes in the fjord without either of the two buoyancy sources discussed above. Deep water renewal is the process whereby water external to the fjord, which is denser than water within the fjord and below sill depth, is able to cross the sill (e.g. by tidal advection) and displace upwards some or all of the isolated water within the fjord. Baroclinic exchange may also modify the stratification of water within the fjord but above sill depth. Externally driven changes to the stratification outside the fjord (e.g. by coastal upwelling) can modify the density-driven exchange between the fjord and adjacent seas. Deep water mixing driven by internal wave breaking slowly modifies the deep water stratification. All these processes are covered in greater detail in the section on baroclinic exchanges.

Barotropic exchanges

Use of the words 'barotropic' and 'baroclinic' deserve brief clarification. Barotropic is used to describe flows which are driven by variations in sea surface height between two locations (relative to the geoid). A sea surface height difference gives rise to a pressure gradient force on the water column which does not vary with depth. This does not necessarily imply that barotropic *flow* is invariant with depth, since barotropic flows (and all flows) will be preferentially impeded at the bed through friction. Nor will barotropic flow be horizontally uniform, for example, due to the action of eddy shedding around headlands (barotropic form drag). The word baroclinic is used to describe flows which are driven by differences in density *stratification* between two locations. Stratification differences give rise to pressure gradient forces which vary with depth (and may even change sign) and therefore, generally speaking, flows which vary with depth even in the absence of any frictional effects. Both barotropic and baroclinic pressure forces and flows may be quasi-steady in time or periodically oscillating.

Tides: choking. A solid Earth rotating underneath two gravitationally created bulges of water is the simplest representation of the so-called 'equilibrium' tide. Consideration of the interaction between the equilibrium tide and ocean basin scale bathymetry gives rise to the 'dynamic' response. Broadly speaking the dynamic response can be pictured as very long (*c*. 1000 km) wavelength surface waves propagating anticlockwise around a point called an amphidrome (clockwise in the southern hemisphere). Most major amphidromes are sited far into the ocean basin interior and possess the property of zero tidal elevation amplitude. The dynamic response is so named because the position of the amphidromes, and the resultant pattern of ocean tides, is governed by friction between the tidal currents and the Earth's solid surface (and the Earth's rotation). Shelf sea tides are a continuation of the oceanic dynamic response to the equilibrium tide. As ocean tides interact with the shallow coastal and shelf seas, currents increase, the speed of the tide wave decreases (with the square root of the water depth), friction increases (with the square of the current speed) and the resulting pattern of amphidromes and tidal phase lines becomes correspondingly more complex.

From the perspective of a fjord, the tides can best be viewed as a periodic oscillation of the height of the water surface at the mouth. There will generally be some phase difference between the height of the tide and the speed of the tidal currents. For most systems the minimum tidal flow (slack water) will occur close to high and low water, with maximum flows close to mid-tide. The greater the frictional resistance of the fjordic system to barotropic tidal flow, the later slack water becomes in relation to the time of high and low waters. In addition to this effect, friction causes the amplitude of the tide to be smaller within the fjord than outside. A third phenomenon of a slightly higher mean sea level within the fjord than outside combines with the two phenomenon described above to create tidal choking (Stigebrandt 1980).

Storm surges: atmospheric pressure, wind stress, rainfall. Tides describe a periodic and deterministic variation in the sea level at the mouth of a fjord. Other phenomena may also cause sea-level variations which are a-periodic and non-deterministic. Low pressure atmospheric weather systems (cyclones) cause a rise a sea level beneath their centre of *c*. 10 mm per millibar drop in pressure (called the inverse barometer effect). Winds associated with weather systems tend to induce surface currents in the ocean, directed to the right (left) of the wind in the northern (southern) hemisphere. Low pressure weather systems tend to travel in an easterly direction, and so a low pressure system approaching a coastal boundary will tend to have an elevated sea surface height associated with it. This is particularly true if the trajectory of the weather system gives rise to winds parallel to the coastline. This bulge of water associated with the cyclone is termed a storm surge. Once it has developed it is then able to propagate, like a single wave of elevation, along the coastline with the coast to the right (left) in the northern (southern) hemisphere. Storm surges can be dramatic at a particular location if, by chance, they coincide with high water. Storm surges are a common feature of the NE European shelf seas, propagating around the coast of the UK and into and around the North Sea. Storm surges within a fjord may commonly be accompanied by heavy rainfall, further exacerbating the increased sea level. Sediment-laden waters from rivers in spate could be retained within a fjord by a combination of a high tide, prolonged by the effects of the storm surge.

Mean sea-level change. Since mean sea level determines the extent to which a given fjord is restricted in its exchange with the shelf waters, changes over time of mean sea level of the order of metres are likely to have a profound effect on the fjordic hydrodynamics. By 'mean sea level' we mean sea level averaged over periods longer than tidal and storm surge periods. As glaciers retreat to reveal a fjord, the ice load on the continental land mass decreases, the continent undergoes an 'isostatic rebound' and relative mean sea level falls. Isostatic rebound can be significant; mean sea level around the northern UK is some 10 m lower now than at the Last Glacial Maximum. Sea level is rising globally due to thermal expansion of the warming oceans and volume increase from melting land ice. This is termed eustatic sea-level change and is currently (1993–2003) estimated to be +3.1 mm a^{-1} (IPCC 2007). Future predictions for eustatic sea-level change have a median value of about 480 mm over the next 100 years. However, under the scenario of total loss of the Greenland ice cap, eustatic sea-level rise would be *c.* 7 m.

On centennial to millennial timescales, we might therefore expect variations in mean sea level that are a significant fraction of the sill depth of a fjord. A major impact of a significant change in sill depth is on the salinity gradient between the fjord and the outer sea, that is, on the freshwater residence time and the mean time between deep water renewal events. We might therefore anticipate significant changes in fjordic salinity over centennial to millennial timescales. Little or no work has been published on this effect, although the potential hydrodynamics effects in fjords on sedimentary records of past climate are beginning to receive some attention (Austin & Inall 2002; Gillibrand *et al.* 2005). Estimates on the magnitude of this effect, based on a simplified numerical model of a fjord, suggest that the deep basin waters of Loch Etive, for example, may have been more saline by several psu during a 10 m higher mean sea level 8000 ka ago.

Baroclinic exchanges

To recap, the word baroclinic is used to describe flows which are driven by a difference in density stratification between two locations. Stratification differences give rise to pressure gradient forces which vary with depth (and may even change sign) and therefore, generally speaking, flows which vary with depth even in the absence of any frictional effects. The strong influence of freshwater in fjords gives rise to the potential for strong internal density gradients within fjords and between fjords and the adjacent shelf waters.

Estuarine circulation. In the mean (averaged over a tidal cycle), fresher water will flow out of a fjord on the surface and more saline water will enter the fjord beneath this out-flowing layer. Bottom boundary induced mixing and current shear between the fresher outflow and more saline inflow will generate mixing between the two layers. The outflow will not be entirely fresh but include some salt mixed up from below. To conserve the volume of water in a fjord (averaged over a tidal cycle), saline water must be drawn into the fjord to replace that which is mixed upwards. This averaged circulation pattern is termed estuarine circulation. Estuaries (fjords without sills) can be broadly categorized by the ratio between the volume of freshwater input (R) and the volume of external seawater brought in to the fjord (V). Fjords generally fall into the category of highly stratified estuary, where the ratio $R{:}V$ is greater than or equal to one. In highly stratified estuaries, the strength of estuarine circulation is generally weaker than in slightly stratified estuaries (where $R{:}V$ is small). In other words, the higher the value of $R{:}V$, the stronger the stratification, the weaker the mixing and the weaker the consequent estuarine circulation. For a given fjord, $R{:}V$ can readily be calculated from a knowledge of the mean salinity of the basin, that of the external water and of river inflow to the fjord (Stigebrandt 1981).

Intermediate water exchange. If the sill depth is greater than the base of the upper layer of the estuarine circulation, then an additional type of baroclinic exchange flow may occur. This is termed intermediate exchange, and is a result of the differing stratification between the intermediate waters of the fjord and the water outside. A simple definition of intermediate water is that lying beneath the upper brackish layer and above the sill depth. As already implied, this layer may be absent in a shallow-silled fjord or many tens of metres thick in deeper silled fjords. In fjords with weak tidal forcing, such as most of those along the southern Norwegian and Swedish coasts and all those in the Baltic Sea, surface layer and intermediate layer exchanges are dominated by baroclinic currents caused by vertical movement in the pycnocline outside the fjord (e.g. Aure *et al.* 1997; Arneborg 2004). In this way, surface and intermediate exchange (flushing) rates in these types of systems are determined by the strength, duration and periodicity of wind-driven coastal upwelling/downwelling.

Deep water renewal: aspiration and blocking. Fjord water beneath sill depth is often referred to as the deep water or isolated water; the former is used here. Since water is generally stratified on both sides of the sill and there is no direct connection on a geopotential surface between deep water on either side of a sill, the deep water within the fjord can in some sense be regarded as isolated. However, as a stratified barotropic flow accelerates the water over the sill there is a tendency to bring more dense water from below sill depth, rather than simply advect the waters shallower than sill depth over the sill. This is because the pressure gradient driving the flow is independent of depth. The process of barotropic flows drawing water from beneath the sill depth up and across the sill is termed aspiration and is a common feature of all fjords. If the aspirated water is not mixed completely as it travels over the sill, as is often the case when the barotropic flow is not too strong and the sill is deep and/or short enough, then it will return to its equivalent density level beneath sill depth on the other side of the sill. If the density of the aspirated water is greater than the resident deep water's maximum density, then the intruding water will displace the resident deep water upward and a deep water renewal event is said to have occurred (Edwards & Edelsten 1977; Gade & Edwards 1979; Allen & Simpson 1998*a*).

When and why might the density of the inflowing aspirated water exceed that of the deep fjord water? Generally speaking, there is a horizontal salinity (and density) gradient from coast to ocean. Large tidal excursions, at an equinoxal spring tide for example, therefore bring water of greater density towards a fjord sill which is perhaps sufficiently dense to displace the deep water within the fjord. In addition to this, the deep water of a fjordic basin is continually being made fresher by the gradual downward mixing of freshwater from above (Gade 1973). This latter process essentially preconditions the deep water to quasi-periodic renewal. The rate of downward buoyancy flux sets the timescale for renewal in a given fjord, and is discussed in greater detail in a later section. Two further points should be noted. Firstly, partial renewals can occur when the aspirated water is not sufficiently dense to displace the most dense water in the fjord, but is still dense enough to displace some portion of the water beneath the intermediate water. Secondly, full deep water renewals may take place over a number of tidal cycles on a making tide (moving from neaps towards springs). Fast bottom currents of short duration are often found during full renewal events.

Biochemical consequences. One consequence of deep water isolation and stagnation is that of dissolved oxygen depletion and hypernutrification of the deep basin waters (Aure & Stigebrandt 1989; Gillibrand *et al.* 1996). Both these factors may affect the sedimentary environment, particularly if sediment interface waters become anoxic.

Basin waters in fjords can contain relatively high levels of particulate organic matter (POM), especially when a fjord supports aquaculture development. Sources of POM in the marine environment include phytoplankton growth and decay, bacteria, seaweeds and sea grasses and natural and anthropogenic inputs from rivers and point discharges.

In situ production of POM depends on the availability of nutrients and the residence time of surface (euphotic) waters and can therefore be highly variable. Surface water residence time is physically determined, largely through a balance between river inflow, tidal exchange and wind-modified estuarine exchange. Nutrient supply is determined by riverine and coastal end-member concentrations and the magnitude of physical delivery of these waters into the fjord. As a result, the details of nutrient supply are likely to be highly system specific. For example, in Loch Linnhe, west Scotland, silicate supply is largely derived from riverine input while the dominant nitrate and phosphate sources are marine (Watts *et al.* 1998). The spring bloom in this system is often triggered by the uplifting of intermediate marine-origin waters into the euphotic zone by a deep water renewal event. Vertical mixing processes may also exert a control on production through vertical upward flux of recycled and/or marine-origin dissolved nutrients into the euphotic zone throughout the productive season. The physics of exchange and mixing therefore intimately affects the supply and distribution of nutrient availability and production of POM.

The breakdown of POM creates a Biochemical Oxygen Demand (BOD), which can lower oxygen levels in the sediment and water column. Lowered oxygen levels occur naturally on a range of spatial scales in the marine environment, including in some deep fjord basins. Mobile fauna such as fish can sense oxygen concentration gradients and avoid these, but less mobile or sedentary species display a range of tolerances to hypoxia and, in some cases, change their behaviour to compensate. Anoxia (i.e. the absence of dissolved oxygen) in bottom waters excludes all metazoan life. Anoxic waters typically have high concentrations of toxic sulphide species and redox sensitive metals such as manganese. Episodic overturning of such waters can result in widespread mortality including fish kills, thus exporting the ecological consequences of anoxia (Rosenberg *et al.* 2001).

The likelihood of hypoxia in the basin water of a fjord is determined by the ratio of two timescales: (1) the period between successive deep water renewal events (which provide oxygen-rich water to the basin); and (2) the time taken for all the oxygen in the basin water following a renewal event to be consumed. Clearly, if the latter timescale is shorter than the former, the fjord basin will be prone to hypoxia. If the two timescales are similar, the basin may be subject to periodic oxygen depletion or occasional hypoxia, which may also have deleterious effects. Only in the case where the residence time is much shorter than the hypoxia time can adverse impacts be effectively ruled out.

The two timescales are dependent on four key factors: (1) the rate of mixing of the deep water of the fjord basin, which reduces the water density and primes the basin water for renewal (Gade 1973) and also provides a source of oxygen to the deep water; (2) the absolute reduction in basin density needed for deep water renewal to occur; (3) the rate of oxygen consumption in the basin water, which is related to the supply of organic matter (Aure & Stigebrandt 1989; Stigebrandt *et al.* 1996); and (4) the volume of the fjord basin and the initial oxygen concentration, which determines the mass of oxygen to be consumed. These factors are all highly basin-dependent and ultimately controlled by the processes described in this paper.

To give an example, for a fjord with a mean renewal interval of 8 weeks (Loch Ailort, west Scotland), dissolved oxygen saturation levels as low as 1% have been reported following deep water isolated for 56 days duration (Gillibrand *et al.* 1996).

Wind-driven circulation. Due to the presence of a shallow freshwater layer, wind-induced surface velocities can be significant in fjords. There are also many secondary effects of wind-driven surface currents; those which affect circulation will be discussed here, while the effect of wind on vertical mixing is dealt with later. Winds in fjords tend to be strongly steered by the surrounding terrain and therefore predominantly along-axis. Sustained up-fjord winds tend to pile the freshwater towards the head of the basin, inducing an opposite flow (outflow) in the intermediate layer beneath. The converse is true for down-fjord winds. In this case, estuarine circulation may be enhanced by the increased vertical mixing at the base of the wind-driven surface layer. In other cases this wind-enhanced circulation contributes to deep water renewal (Gillibrand *et al.* 1995). Frequently, this wind-driven exchange will dominate over the density-driven baroclinic exchange (Svendsen & Thompson 1978). Up-fjord winds tend to produce a downward slope in the pycnocline towards the head of the fjord, while down-fjord winds tend to induce less of a pycnocline response since there is a lesser boundary (the sill) at the seaward end. When the wind relaxes, any wind-induced pycnocline slope will tend to flatten out and a baroclinic sieche often results (Arneborg & Liljebladh 2001). If the fjord basin length matches a half-integral number of internal seiche wavelengths, then the seiche may oscillate for many periods after the cessation of the wind (Cushman-Roisin *et al.* 1989).

Effects of Earth's rotation. All baroclinic exchanges are potentially modified by the Earth's rotation. A length scale called the baroclinic Rossby radius is defined as $R = c/f$, where f is the Coriolis parameter which increases sinusoidally with latitude and c is the fastest internal wave phase speed which increases with the square root of stratification. If the fjord width is similar to or greater than the Rossby radius, then the effects of the Earth's rotation begin to modify the baroclinic flows in the fjord. Typically R is of the order of a few kilometres, which is larger than a typical fjord width and so the Earth's rotation is generally not important. There are documented exceptions where fjords are particularly wide in, for example, Scotland (Janzen *et al.* 2005) and Norway (Cushman-Roisin *et al.* 1994) or more commonly at high latitudes where R is smaller due to the larger value of f, for example, Kongsfjorden (Svendsen *et al.* 2002; Cottier *et al.* 2005). R will of course vary with time within a given fjord as the stratification changes, thus some fjords may feel the effects of the Earth's rotation at some times and not at others. Rotational effects manifest themselves by deflecting baroclinic flows (estuarine circulation and intermediate exchange flows) to their right (left) in the northern (southern) hemisphere. Since the baroclinc flows are largely along the axis of the fjord, this deflection of the flows proceeds until flow entering (leaving) the fjord becomes concentrated on the right (left) of the fjord (looking towards the head) (Janzen *et al.* 2005). This deflection may affect the transport pathways of material, biotic or abiotic, suspended in the water column (Gillibrand & Willis 2007). Internal tides are, by definition, baroclinic; the effect of Earth's rotation is to modify the internal tide such that the wave amplitude becomes a maximum on the right-hand boundary, decaying exponentially into the interior. This modified wave then propagates around the basin anticlockwise (clockwise) from the sill, with the boundary to its right (left) in the northern (southern) hemisphere. This is directly analogous to the description already given of the external tide propagating around an ocean basin.

Mixing processes and vertical exchanges

Turbulent mixing takes place at the smallest (millimetre) scales, and yet fundamentally shapes the largest scale flows. This statement is as true for fjords as it is for global ocean circulation. To discuss mixing we must first define what this word means in the present context; to do so we have to introduce the concepts of the kinetic and potential energy of a body of water. Kinetic energy may simply be expressed as being proportional to the square of the velocity of the flow. Potential energy may be simply related to the stratification of the water column. A stratified water column has less potential energy than a well-mixed water column (of the same depth and mean density), since to raise up dense water by mixing requires work to be done. (Vertical) mixing may then be defined as the irreversible processes of homogenizing a layer of previously stratified water; in this process heavy water is mixed upwards and light water mixed downwards, thus increasing the potential energy of the water column. Mixing in this manner takes place at very small scales (cascading from metres to millimetres), and the energy required to increase the water column potential energy comes from the kinetic energy of turbulence. Turbulent kinetic energy generally derives from the instability of mean flows (including flows associated with internal waves) or convective instability through surface cooling. One final definition is useful here: the efficiency of mixing is defined as the portion of the turbulent kinetic energy which is ultimately converted to increased water column potential energy. Its value is generally between 3% and 20%, the remainder dissipating to heat.

Wind-induced mixing. In fjords with weaker tidal forcing, internal mixing may be dominated by wind-induced effects. For example, studies in Gullmar fjord (Swedish west coast) found that wind forcing, including contributions from an external seiche, was responsible for about 60% of the basin water mixing compared to 40% for tidal forcing (Arneborg & Liljebladh 2001). Along-axis wind energy can directly excite external and internal seiching through the set-up of water either at the head of the fjord or near the sill. Additionally, wind energy can be fed into internal seiches by the direct action of wind on coastal systems, causing oscillations in the coastal stratification which may propagate into relatively deep-silled fjords.

Internal tide: interior and boundary mixing. If barotropic flow is slower than some critical value (see the section on near sill mixing below), the consequent process of aspiration essentially imparts a vertical velocity on the flow near the sill. Furthermore, if the barotropic flow causing aspiration is oscillating (e.g. tidal), then the pycnocline on either side of the sill will be forced up and down every tidal cycle. Provided the fjord is not located polewards of 74.9°, then the internal wave created by the periodic vertical oscillation of the pycnocline is free to propagate away from the sill in both directions at a speed roughly proportional to the square root of the stratification. This is called an internal tide; phase speeds typically range from 0.2 to 1 m s^{-1}. The vertical structure typically has oscillating flows in both upper and lower layers that are 180° out of phase, with current amplitudes typically 0.01–0.3 m s^{-1}. Energy contained within the internal tide comes from the barotropic tide (the transformation is called baroclinic wave drag) and must ultimately be converted either to raising the potential energy within the basin via mixing (i.e. reducing stratification) or to heat (Allen & Simpson 2002). It should be noted here that the energy available cannot raise the temperature by a measureable amount, due to the high specific heat capacity of water. Internal tides therefore present an important source of energy which can result in strong vertically sheared currents in the interior or at the sloping seabed boundaries all around the fjord, not just in the vicinity of the sill. Polewards of 74.9°, the Coriolis acceleration is strong enough to prevent these waves propagating freely.

Although energy is fed into the internal wave field almost exclusively at the semi-diurnal and/or diurnal frequencies, the energy cascades to higher frequencies through a variety of processes. Higher frequency waves have short wavelengths, are more likely to break and are also able to propagate at steeper angles in the vertical. Breaking internal waves can therefore cause mixing throughout the interior of a fjord basin, below sill depth and many kilometres distant horizontally. Just how much of the internal tide energy is dissipated in this way is uncertain, although it is likely to be considerably less than half (Inall *et al.* 2004). The efficiency at which internal wave breaking converts kinetic energy to changes in water column potential energy is thought to be around 20% (Osborn 1980). However, many fjordic studies suggest that the overall ratio between the energy in the internal tide and the increase in water column potential energy may be closer to 5–7% (Stigebrandt 1976; Arneborg *et al.* 2004).

What then is the fate of the remaining internal tide energy which is not dissipated in the interior? By analogy to surface waves breaking on a sloping beach, internal waves are likely to break as they encounter the sloping sides of a fjordic basin. Recent studies have shown, however, that this is an inefficient process in terms of mixing (Inall 2009). The partition of internal tide energy dissipation between interior and boundary mixing may not therefore reflect their relative contribution to mixing of the interior basin.

It is the rate of vertical mixing below the sill depth, to which both of the above contribute, which has important consequences for the frequency of deep water renewal. The greater the rate of downward mixing of lighter (fresher) surface waters, the more frequently deep water renewal takes place.

Near-sill mixing. In so-called 'jet-type' fjords (Stigebrandt & Aure 1989) where the barotropic flow speed over the sill is equal to or greater than the phase speed of the internal tide, the rate of conversion from barotropic to internal tidal energy is much reduced and asymmetric between ebb and flood (Stashchuk *et al.* 2007). During the flood tide on the landward side of the sill, intense localized vertical mixing has been predicted and observed (Inall *et al.* 2005) by the formation of an internal hydraulic jump. Elevated near-sill mixing may be further enhanced through interaction with internal tide energy which has been reflected (Allen & Simpson 1998*b*) from the head of the basin, and then back towards the sill (Arneborg & Liljebladh 2009).

Overall, the spatial pattern of strong internal wave-induced currents and shear will vary significantly from fjord to fjord. Patterns of sediment resuspension induced by these currents will similarly vary from system to system. It is only with detailed observation, coupled with 3D modelling, that a detailed map of these processes might emerge for a given system.

Convective overturning. Convection is an instability arising from the introduction of dense water over lighter water or light water beneath denser water. Both situations can in theory arise in fjords, with the former being considerably more common. The latter, a process termed tidal straining, can lead to bottom-driven convection in weakly stratified estuaries (Rippeth *et al.* 2001), although the process has not been documented in fjords and is less likely to occur due to the generally strong stratification.

Both surface cooling and the introduction of brine during sea ice formation can cause surface convection; the latter is confined to high-latitude fjords and is discussed elsewhere in this volume. Due to the non-linear nature of the equation of state for seawater (an equation which describes density as a function of temperature, salinity and pressure), a given temperature change has a decreasing effect on density as temperature decreases. Thus, temperature-driven surface convection is increasingly unlikely in colder (higher latitude) fjords. Coupled to this, the strong salinity stratification in fjords makes it rare for convective overturning to penetrate into the deep water of a fjord. Forced convection, a combination of surface heat loss and wind forcing, can penetrate to full depth in some fjords (e.g. the Clyde Sea, west Scotland; Rippeth & Simpson 1996) particularly in late autumn when warmer water is trapped beneath colder fresher water. Forced convection can then temporarily raise the temperature of the surface layer, driving a greater heat loss at a water temperature more conducive to larger density change.

Numerical models of fjordic circulation and exchange

Computer models of varying type and complexity have been developed over the past four decades to simulate aspects of the water circulation and exchange, both within fjords and between fjords and the adjacent coastal ocean. To provide insights into the transport and deposition of suspended sediments in fjords, a model must resolve the key features of the circulation and dynamics of the environment that have been described above. The exchange between fjord and ocean affects the type and concentration of suspended sediment in the system; local circulation fields coupled with the locations of sediment supply determine the pattern of depositional footprints and near-bed current speeds control the resuspension and redistribution of settled sediments. A model must provide information on some, or all, of these dynamic features if it is to explain the distribution and characteristics of the fjord sedimentary regime. Of particular interest in the analysis of records of sediment characteristics and other proxy variables preserved in long cores are: changes in near-bed current speed; current direction; water temperature; and water salinity over long timescales (i.e. decades to centuries).

In order to perform lengthy simulations to reconstruct long-term variability, models must be coded efficiently. Compromises over spatial resolution may be necessary.

Two-layer and other analytical models

Early models of fjord dynamics sought mathematical solutions to simplified equations of mass and momentum conservation in order to ascertain the dominant mechanisms controlling fjord circulation and hydrography. Rattray (1967) and Winter (1973), for example, utilized similarity solutions to gain insights into the vertical structures of salinity and velocity in fjord basins and their response to the gravitational and wind-driven circulation. The dominant dynamical balance was found to be between the horizontal pressure gradient and the vertical gradient of turbulent stress (Rattray 1967). Subsequently, a series of analytical studies, typically approximating fjords as two-layer systems, developed better understanding of the response of the surface layer salinity and thickness to meteorological forcing (Rattray 1967; Winter 1973; Welander 1974; Long 1975; Pearson & Winter 1978; Klinck *et al.* 1981; Stigebrandt 1981). The underlying paradigm here was that narrow entrance regions with shallow sills exert a hydraulic control on the fjord–ocean exchange and consequently on the hydrographic conditions and circulation in the fjord interior. This hypothesis was extended by Klinck *et al.* (1981) who demonstrated, also through development of a two-layer model, that the geostrophic coastal response to along-shore wind forcing provided an alternate control mechanism on transport into and out of the fjord.

All the above models neglected tidal forcing and the solutions described steady-state conditions. Other simplifications to the mathematical equations were also required in order to achieve analytical solutions. Although providing much insight into fjord dynamics, these models do not offer an opportunity to investigate the time-dependent response of fjord basins to fluctuating forcing conditions.

Two-dimensional models

Increasing computer power in the 1970s heralded the development of models capable of simulating the time-dependent circulation in estuaries and fjords in greater detail. Since fjords are typically narrow relative to their length, lateral variations in density and circulation can often be justifiably ignored. Two-dimensional (2D) width-averaged models, resolving the water column vertically and longitudinally, were used to explore the details of first the tidal circulation (Dunbar & Burling 1987; Falconer & Cox 1988; Stacey *et al.* 1991, 1995) and subsequently the dynamics of deep water renewal (Lavelle *et al.* 1991; Gillibrand *et al.* 1995; Stacey *et al.* 2002) and horizontal exchange (Vlasenko *et al.* 2002). These models solve the equations of motion and mass conservation on finite-difference grids, with boundary forcing of tides, wind stress and river discharge. The models typically provide realistic, rather than idealistic, simulations of circulation and hydrography. Deep water renewal was generally shown to be driven by a combination of factors including tidal range, wind stress, external density fluctuations, sill dimensions, freshwater discharge and pre-conditioning of the basin water. These 2D models provided detailed information on the time-varying nature of tidal and residual currents along the central axis of the water body (see Fig. 2). For example, changes in deep inflowing current speeds inside entrance sills during deep water renewal events could be resolved, offering the potential to distinguish high-energy resuspension events from intervening periods of quiescent conditions in basins. The models of Stacey and Vlasenko also revealed the basin-wide influence of internal waves generated by the tidal oscillations of stratified water over sill topography. The interactions between stratified flow and sill topography have been studied using very high spatial resolution 2D models (Cummins *et al.* 2003; Xing & Davies 2006*a*, *b*, 2007; Stashchuk *et al.* 2007) which revealed the non-hydrostatic nature of the response to sub- and super-critical flow over sills.

With present-day computing power, combined with efficient numerical algorithms, a typical year-long simulation may take a few minutes. Two-dimensional models therefore offer the potential to characterize the water properties and sedimentation environment over relatively long (e.g. decadal) timescales (Gillibrand *et al.* 2005). What 2D 'slice' models are unable to provide is spatial discrimination of depositional and erosional areas, and simulation of the transport of sediment between them. For that, the domain of interest must be resolved in all three spatial dimensions.

Three-dimensional models

The horizontal transport and distribution of sediments can only be mapped through three-dimensional (3D) time-dependent model simulations. Although computationally expensive to run, such models deliver unprecedented synoptic detail of hydrodynamic environments. In recent years, 3D circulation models have informed studies of flow regimes in Norwegian (Utnes & Brors 1993; Asplin *et al.* 1999; Eliassen *et al.* 2001), Canadian (Forman *et al.* 2009), New Zealand (Bowman *et al.* 1999) and Scottish (Gillibrand & Amundrud 2007) fjords. In the Scottish study, a 3D

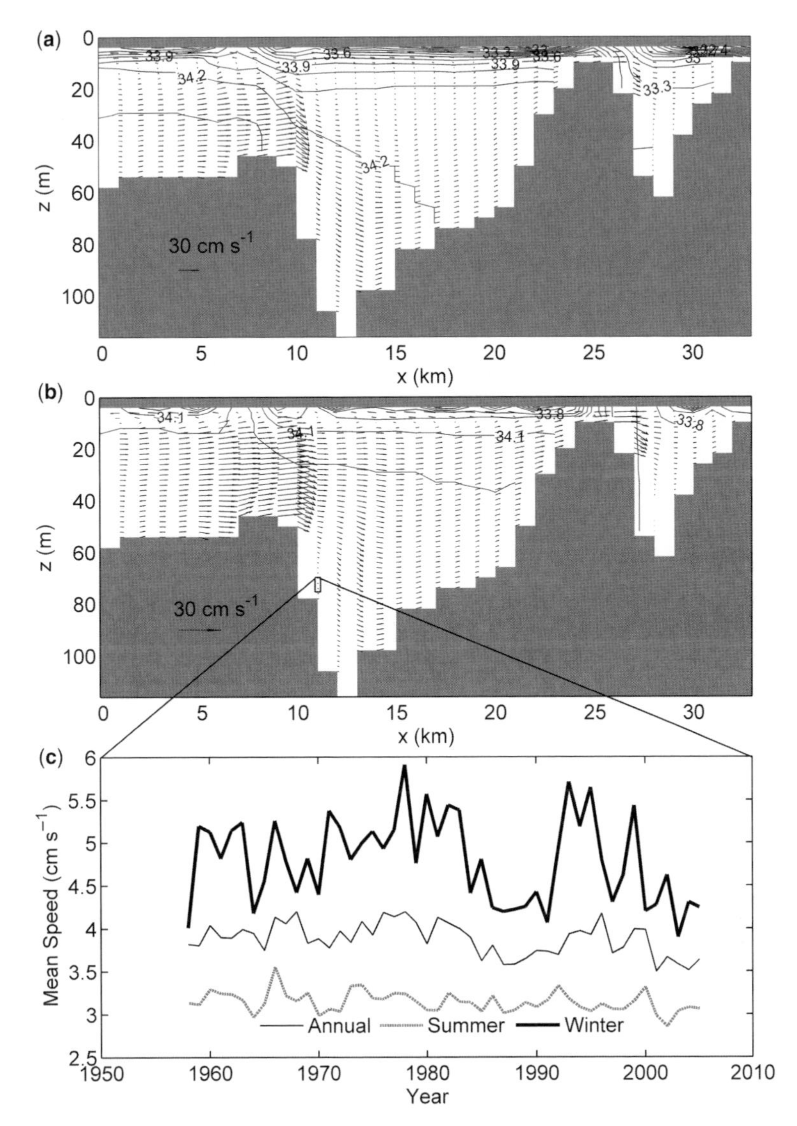

Fig. 2. Typical output from a 2D fjord circulation model: snapshots of modelled velocity and salinity fields during flood tide in Loch Sunart, western Scotland: (**a**) with deep water renewal occurring in the main basin, and (**b**) without deep water renewal. The horizontal axes are distance from the mouth (x, km); the vertical axes are depth (z, m). In (**c**), time series of near-bed basin current speed from a 48-year long simulation of Loch Sunart are shown (see Gillibrand *et al.* 2005). The model output of hourly current speed has been averaged into annual-, winter- and summer-mean values.

model study of the circulation in Loch Torridon revealed the influence of the Coriolis acceleration on the near-surface flows (Fig. 3), which modified the transport pathways of larval sea lice in the region (Murray & Gillibrand 2006; Amundrud & Murray 2009).

The spatial resolution of the circulation delivered by 3D hydrodynamic models permits coupled modelling of transport of pelagic organisms (Brooks & Stucchi 2006; Murray & Gillibrand 2006; Amundrud & Murray 2009) and, by analogy, simulation of suspended sediment transport is clearly possible. Water properties such as temperature, salinity and density are also simulated at the same spatial resolution (Fig. 4). However, this spatial and temporal detail comes at a cost: a year-long 3D simulation of a fjord might take several hours, making multi-centennial simulations impractical except on the largest computers. The difficulty is exacerbated by the quantity of data needed to force

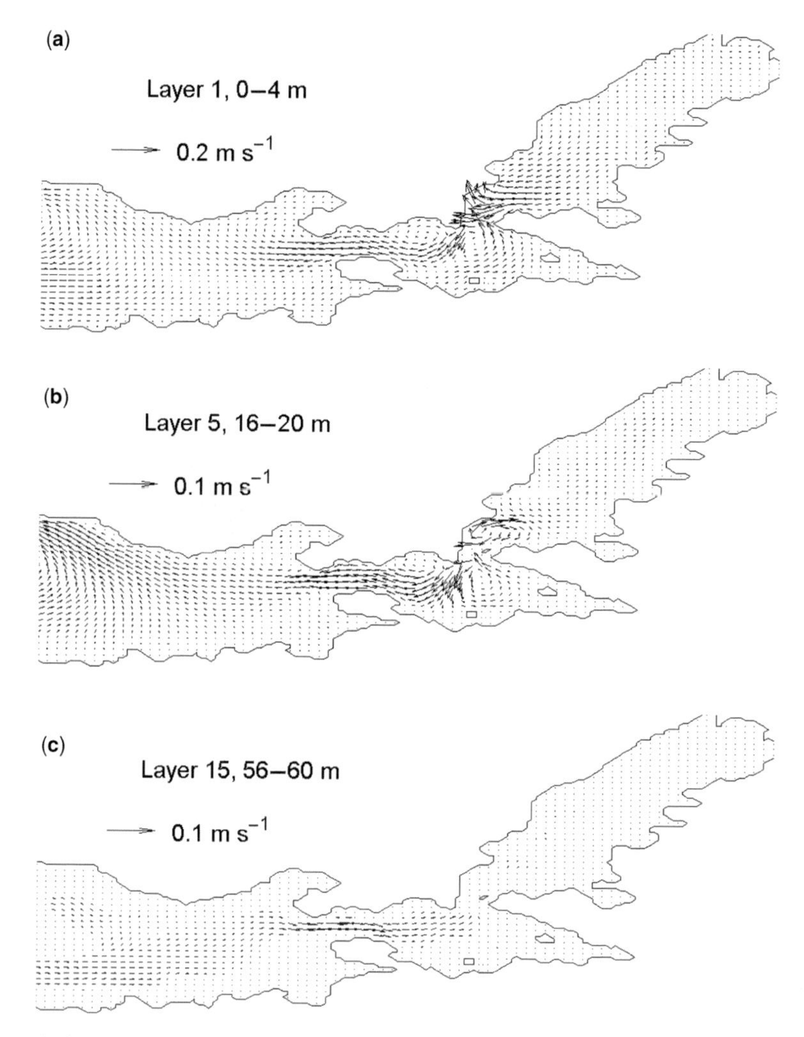

Fig. 3. Mean velocity vectors from a 3D model simulation of circulation in Loch Torridon, NW Scotland, from model layers at depths of (**a**) 0–4 m, (**b**) 16–20 m, (**c**) 56–60 m. The vectors are averaged over a 3-month long simulation over April–July 2001 (see Gillibrand & Amundrud 2007). To assist clarity, only 25% of the model vectors are plotted. Note the steering in the surface layer currents due to topography and geostrophy.

the simulations; 3D models typically require spatially and temporally varying wind, river flow and oceanic temperature and salinity data to produce accurate simulations, data that are simply not available over timescales longer than a few decades.

Box models

Relatively simple box models of fjord exchange offer a potential solution to the problem of reconstructing past records of basin properties over long (i.e. greater than decadal) timescales. In recent years, fjord box models have provided insights into the seasonal cycle of stratification and mixing in the Clyde Sea (Simpson & Rippeth 1993; Rippeth & Simpson 1996), the circulation in the Strait of Georgia (Li *et al.* 1999) and Puget Sound (Babson *et al.* 2006), and physical and biological cycles in Loch Creran (Tett *et al.* 2010). All these models take a similar philosophical approach: a simplified spatial representation of the fjord system combined with empirical or analytical parameterizations of physical processes. The models thus forego spatially variable predictions of water properties in return for fast and efficient simulations and clearer insight into the dynamical balance of the fjord system.

The fjord basin, or series of interlinked basins, is typically represented as a number of discrete homogeneous boxes or layers (e.g. Fig. 5). This approach

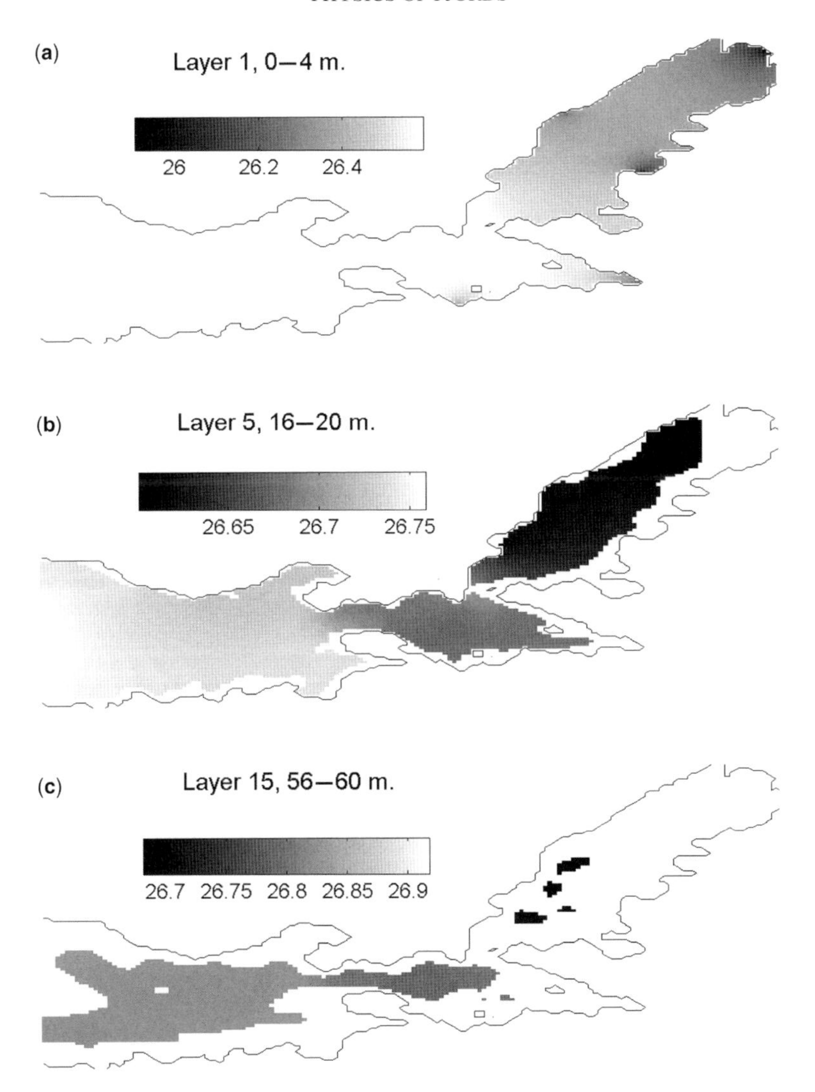

Fig. 4. Mean density (σ_t) from a 3D model simulation of circulation in Loch Torridon, NW Scotland, from model layers at depths of (**a**) 0–4 m, (**b**) 16–20 m and (**c**) 56–60 m. The density values are averaged over a 3-month long simulation during April–July 2001 (see Gillibrand & Amundrud 2007).

is generally justified on the basis that both longitudinal and lateral variability in fjord waters is usually small compared to the vertical gradients. Exchange of water between the layers is represented as the sum of a number of advective and diffusive processes, which are in turn parameterized using empirically and/or analytically derived relationships. The model is stepped forward in time and the exchange of water between layers, and between the fjord and the adjacent coastal ocean, leads to the evolution of layer properties over time in response to the applied forcing (Fig. 5).

The limited spatial resolution and simplified representation of key processes permits long simulations to be performed on desktop computers; these simulations may capture, for example, the timing and strength of deep water renewal events and therefore changes in mean current speeds and the associated scouring of sediment layers on the inner slopes of sills. Water properties such as bottom temperature and salinity can potentially be reconstructed over decadal, or even longer, timescales. Clearly, careful calibration of the model against available data is necessary before performing such long simulations, and the feasibility of the approach depends critically on the availability and quality of the forcing data. Sources of forcing data include direct observation, re-analysis datasets

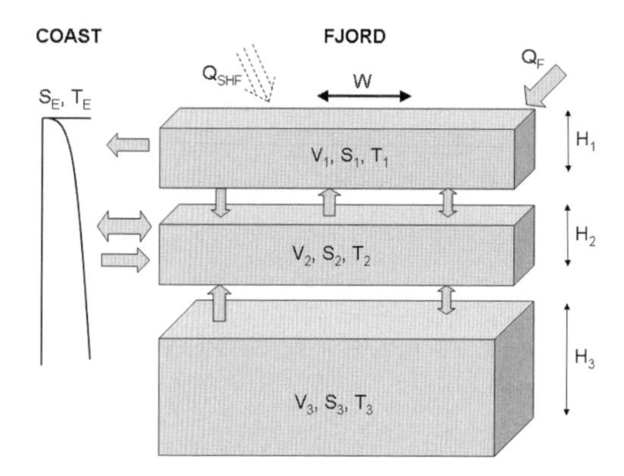

Fig. 5. Schematic of the interactions between parameters in a new box model of fjord exchange. The fjord is represented by three layers, $i = 1, 2, 3$, of mixed fluid, with the thickness, volume, salinity and temperature of each layer denoted by H_i, V_i, S_i, T_i, respectively. The model is forced by external profiles of temperature and salinity T_E and S_E, wind stress W, surface heat flux Q_{SHF} and river discharge Q_F. Mass and water properties are exchanged between layers as a result of advective and diffusive processes (represented by uni- and bi-directional arrows respectively), which are parameterizations of the processes identified in Figure 1 using empirically and/or analytically based solutions.

or coupled global climate models (Piani *et al.* 2005). Sediment cores, provided they are collected from appropriate depositional locations identified through 3D modelling of circulation patterns, may provide an archive of local current speed which can be linked to environmental forcing conditions through long-term modelling using such simple box models. Such an approach is currently being tested in Loch Sunart, western Scotland, through comparisons between sediment core proxies and model simulations of the fjord hydrography and deep water renewal.

Summary

There has been a considerable amount of research into the physics of fjords since FF83; 75% of the literature cited here has been published since this last review in 1983. Geographically, effort has continued to focus on Scandinavian and NW American fjords, with a growing body of literature on the Scottish fjords. As foretold in FF83, much progress in understanding has come about through the combined application of more sophisticated modelling techniques and advanced measurement techniques, particularly acoustic doppler velocimetry and shear microstructure profiling. Dye tracers have recently yielded new insights into the mixing regimes near the boundaries, but long-lived CFC-based tracers have yet to be used to extend observational based mixing estimates beyond a few days. The current generation of fjordic oceanographers look forward to using unmanned underwater vehicles (UUVs) to better describe spatial patterns of circulation and mixing as well as non-hydrostatic and adaptive grid models to focus on spatial hot-spots of turbulent mixing and entrainment flows. One thing, however, remains unchanged. To quote FF83: '...a fjord coastline should attract oceanographers not only because of the practical problems associated with man's interaction with his environment, but also because it provides an excellent laboratory for studying natural flows'.

References

Allen, G. L. & Simpson, J. H. 1998*a*. Deep water inflows to Upper Loch Linnhe. *Estuarine Coastal and Shelf Science*, **47**, 487–498.

Allen, G. L. & Simpson, J. H. 1998*b*. Reflection of the internal tide in Upper Loch Linnhe, a Scottish fjord. *Estuarine Coastal and Shelf Science*, **46**, 683–701.

Allen, G. L. & Simpson, J. H. 2002. The response of a strongly stratified fjord to energetic tidal forcing. *Estuarine Coastal and Shelf Science*, **55**, 629–644.

Amundrud, T. L. & Murray, A. G. 2009. Modelling sea lice dispersion under varying environmental forcing in a Scottish sea loch. *Journal of Fish Diseases*, **32**, 27–44.

Arneborg, L. 2004. Turnover times for the water above sill level in Gullmar Fjord. *Continental Shelf Research*, **24**, 443–460.

Arneborg, L. & Liljebladh, B. 2001. The internal seiches in Gullmar Fjord. Part II: contribution to basin water mixing. *Journal of Physical Oceanography*, **31**, 2567–2574.

Arneborg, L. & Liljebladh, B. 2009. Overturning and dissipation caused by baroclinic tidal flow near the sill of a fjord basin. *Journal of Physical Oceanography*, **39**, 2156–2174.

Arneborg, L., Erlandsson, C. P., Liljebladh, B. & Stigebrandt, A. 2004. The rate of inflow and mixing during deep-water renewal in a sill fjord. *Limnology and Oceanography*, **49**, 768–777.

Asplin, L., Salvanes, A. G. V. & Kristoffersen, J. B. 1999. Nonlocal wind-driven fjord-coast advection and its potential effect on plankton and fish recruitment. *Fisheries Oceanography*, **8**, 255–263.

Aure, J. & Stigebrandt, A. 1989. On the influence of topographic features upon the oxygen consumption rate in sill basins of fjords. *Estuarine Coastal and Shelf Science*, **28**, 59–69.

Aure, J., Molvaer, J. & Stigebrandt, A. 1997. Observations of inshore water exchange forced by a fluctuating offshore density field. *Marine Pollution Bulletin*, **33**, 112–119.

Austin, W. E. N. & Inall, M. E. 2002. Deep-water renewal in a Scottish fjord: temperature, salinity and oxygen isotopes. *Polar Research*, **21**, 251–257.

Babson, A. L., Kawase, A. & MacCready, P. 2006. Seasonal and interannual variability in the circulation of Puget Sound, Washington: a box model study. *Atmosphere-Ocean*, **44**, 29–45.

Bornhold, B. D., Ren, P. & Prior, D. B. 1994. High-frequency turbidity currents in British-Columbia Fjords. *Geo-Marine Letters*, **14**, 238–243.

Bowman, M. J., Dietrich, D. E. & Mladenov, I. 1999. Predictions of circulation and mixing in Doubtful Sound, arising from variations in runoff and discharge from the Manapouri power station, Coastal Ocean Prediction. *Coastal and Estuarine Studies*, **56**, 59–76.

Brooks, K. M. & Stucchi, D. J. 2006. The effects of water temperature, salinity and currents on the survival and distribution of the infective copepodid stage of the salmon louse (Lepeophtheirus salmonis) originating on Atlantic salmon farms in the Broughton Archipelago of British Columbia, Canada (Brooks 2005) – a response to the rebuttal of Krkošek *et al.* (2005). *Reviews in Fisheries Science*, **14**, 13–23.

Cottier, F., Tverberg, V., Inall, M., Svendsen, H., Nilsen, F. & Griffiths, C. 2005. Water mass modification in an Arctic fjord through cross-shelf exchange: the seasonal hydrography of Kongsfjorden, Svalbard. *Journal of Geophysical Research*, **110**, C12005, doi: 10.1029/2004JC002757.

Cummins, P. F., Vagle, S., Armi, L. & Farmer, D. M. 2003. Stratified flow over topography: upstream influence and generation of nonlinear internal waves. *Proceedings of the Royal Society of London Series A – Mathematical Physical and Engineering Sciences*, **459**, 1467–1487.

Cushman-Roisin, B., Tverberg, V. & Pavia, E. G. 1989. Resonance of internal waves in fjords – a finite-difference model. *Journal of Marine Research*, **47**, 547–567.

Cushman-Roisin, B., Asplin, L. & Svendsen, H. 1994. Upwelling in broad fjords. *Continental Shelf Research*, **14**, 1701–1721.

Deyoung, B. & Pond, S. 1989. Partition of energy-loss from the barotropic tide in fjords. *Journal of Physical Oceanography*, **19**, 246–252.

Dunbar, D. S. & Burling, R. W. 1987. A numerical-model of stratified circulation in Indian Arm, British-Columbia. *Journal of Geophysical Research – Oceans*, **92**, 13 075–13 105.

Edwards, A. & Edelsten, D. 1977. Deep water renewal of Loch Etive: a three basin Scottish fjord. *Estuarine and Coastal Marine Science*, **5**, 575–595.

Eliassen, I. K., Heggelund, Y. & Haakstad, M. 2001. A numerical study of the circulation in Saltfjorden, Saltstraumen and Skjerstadfjorden. *Continental Shelf Research*, **21**, 1669–1689.

Falconer, R. A. & Cox, I. C. S. 1988. Deterministic modelling of vertical circulation in shallow lakes and reservoirs. *Research and Management*, **2**, 583–596.

Farmer, D. & Freeland, H. 1983. The physical oceanography of fjords. *Journal of Physical Oceanography*, **12**, 147–220.

Forman, M. G. G., Czajko, D. J. & Guo, M. 2009. A finite volume model simulation for the Broughton Archipelago, Canada. *Ocean Modelling*, **30**, 29–47.

Freeland, H. J., Farmer, D. M. & Levings, C. D. (eds) 1980. *Fjord Oceanography*. NATO Conference Series, Series IV: Marine Sciences. Plenum Press, New York.

Gade, H. 1973. Deep water exchanges in a sill fjord, a stochastic process. *Journal of Physical Oceanography*, **3**, 213–219.

Gade, H. G. & Edwards, A. 1979. Deep water renewal in fjords. *In*: Freeland, H. J., Farmer, D. M. & Levings, C. D. (eds) *Fjord Oceanography*. NATO Conference Series. Plenum Press, NewYork, NY, 453–489.

Gillibrand, P. A. & Amundrud, T. L. 2007. A numerical study of the tidal circulation and buoyancy effects in a Scottish fjord: Loch Torridon. *Journal of Geophysical Research – Oceans*, **112**, C0503, doi: 10.1029/2006JC003806.

Gillibrand, P. A., Turrell, W. R. & Elliott, A. J. 1995. Deep-water renewal in the Upper Basin of Loch Sunart, a Scottish Fjord. *Journal of Physical Oceanography*, **25**, 1488–1503.

Gillibrand, P. A., Turrell, W. R., Moore, D. C. & Adams, R. D. 1996. Bottom water stagnation and oxygen depletion in a Scottish sea loch. *Estuarine Coastal and Shelf Science*, **43**, 217–235.

Gillibrand, P., Cage, A. & Austin, W. E. N. 2005. A preliminary investigation of bottom water response to climate forcing in a Scottish fjord: evaluating the influence of the NAO. *Continental Shelf Research*, **25**, 571–587.

Gillibrand, P. A. & Willis, K. J. 2007. Dispersal of sea louse larvae from salmon farms: modelling the influence of environmental conditions and larval behaviour. *Aquatic Biology*, **1**, 63–75.

Imberger, J. 1998. Physical processes in lakes and oceans. *In*: Bowman, M. J. & Mooers, C. N. K. (eds) *Coastal and Estuarine Studies*. American Geophysical Union, Washington, DC, 668.

Inall, M. E. 2005. Turbulence measurements in fjords. *In*: Baumert, H. Z., Simpson, J. H. & Sunderman, J. (eds) *Marine Turbulence: Theories, Observations and Models*. Cambridge University Press, Cambridge, 340–345.

Inall, M. E. 2009. Internal wave induced dispersion and mixing on a sloping boundary. *Geophysical Research Letters*, **36**, L05604, doi: 10.1029/2008GL036849.

Inall, M. E., Cottier, F. R., Griffiths, C. & Rippeth, T. P. 2004. Sill dynamics and energy transformation in a jet fjord. *Ocean Dynamics*, **54**, 307–314.

Inall, M. E., Rippeth, T. P., Griffiths, C. R. & Wiles, P. 2005. Evolution and distribution of TKE production and dissipation within stratified flow over topography. *Geophysical Research Letters*, **32**, L08607, doi: 10.1029/2004GL022289.

IPCC 2007. *Climate Change 2007: Synthesis Report*. Working Groups I, II and III Fourth Assessment Report, Intergovernmental Panel on Climate Change, IPCC, Geneva, Switzerland.

Janzen, C., Simpson, J. H., Inall, M. E. & Cottier, F. R. 2005. Cross-sill exchange and frontal flows over the sill of a broad fjord. *Continental Shelf Research*, **25**, 1805–1824.

Klinck, J. M., O'Brien, J. J. & Svendsen, H. 1981. A simple model of fjord and coastal circulation interaction. *Journal of Physical Oceanography*, **11**, 1612–1626.

LAVELLE, J. W., COKELET, E. D. & CANNON, G. A. 1991. A model study of density intrusions into and circulation within a deep, silled estuary – Puget Sound. *Journal of Geophysical Research – Oceans*, **96**, 16 779–16 800.

LI, M., GARGETT, A. E. & DENMAN, K. 1999. Seasonal and interannual variability of estuarine circulation in a box model of the Strait of Georgia and Juan de Fuca Strait. *Atmosphere-Ocean*, **37**, 1–19.

LONG, R. R. 1975. Circulations and density distributions in a deep, strongly stratified, two-layer estuary. *Journal of Fluid Mechanics*, **71**, 529–540.

MURRAY, A. G. & GILLIBRAND, P. A. 2006. Modelling dispersal of larval salmon lice in Loch Torridon, Scotland. *Marine Pollution Bulletin*, **53**, 128–135.

OSBORN, T. R. 1980. Estimates of the local rate of vertical diffusion from dissipation measurements. *Journal of Physical Oceanography*, **10**, 83–89.

PEARSON, C. E. & WINTER, D. F. 1978. Two-layer analysis of steady circulation in stratified fjords. *In*: NIHOUL, J. C. J. (ed.) *Hydrodynamics of Estuaries and Fjords*. Elsevier, Amsterdam, 495–514.

PIANI, C., FRAME, D. J., STAINFORTH, D. A. & ALLEN, M. R. 2005. Constraints on climate change from a multi-thousand member ensemble of simulations. *Geophysical Research Letters*, **32**, L23825, doi: 10.1029/2005GL024452.

RATTRAY, M. 1967. Some aspects of the dynamics of circulation in fjords. *In*: LAUFF, G. H. (ed.) *Estuaries*. American Association of Advanced Science, Washington, 52–63.

RIPPETH, T. P. & SIMPSON, J. H. 1996. The frequency and duration of episodes of complete vertical mixing in the Clyde Sea. *Continental Shelf Research*, **16**, 933–947.

RIPPETH, T. P., FISHER, N. R. & SIMPSON, J. H. 2001. The cycle of turbulent dissipation in the presence of tidal straining. *Journal of Physical Oceanography*, **31**, 2458–2471.

ROSENBERG, R., NILSSON, H. C. & DIAZ, R. J. 2001. Response of benthic fauna and changing sediment redox profiles over a hypoxic gradient. *Estuarine Coastal and Shelf Science*, **53**, 343–350.

SIMPSON, J. H. 1997. Physical processes in the ROFI regime. *Journal of Marine Systems*, **12**, 3–15.

SIMPSON, J. H. & RIPPETH, T. P. 1993. The Clyde Sea – a model of the seasonal cycle of stratification and mixing. *Estuarine Coastal and Shelf Science*, **37**, 129–144.

STACEY, M. W., POND, S. & CAMERON, A. N. 1991. The comparison of a numerical-model to observations of velocity and scalar data in burrard inlet and Indian Arm, British-Columbia. *Journal of Geophysical Research – Oceans*, **96**, 10 777–10 790.

STACEY, M. W., POND, S. & NOWAK, Z. P. 1995. A numerical model of the circulation in Knight Inlet, British Columbia, Canada. *Journal of Physical Oceanography*, **25**, 1038–1062.

STACEY, M. W., PIETERS, R. & POND, S. 2002. The simulation of deep water exchange in a fjord: Indian Arm, British Columbia, Canada. *Journal of Physical Oceanography*, **32**, 2753–2765.

STASHCHUK, N., INALL, M. & VLASENKO, V. 2007. Analysis of supercritical stratified tidal flow in a Scottish Fjord. *Journal of Physical Oceanography*, **37**, 1793–1810.

STIGEBRANDT, A. 1976. Vertical diffusion driven by internal waves in a sill fjord. *Journal of Physical Oceanography*, **6**, 486–495.

STIGEBRANDT, A. 1977. On the effect of barotropic current fluctuations on the two-layer transport capacity of a constriction. *Journal of Physical Oceanography*, **7**, 118–122.

STIGEBRANDT, A. 1979. Observational evidence for vertical diffusion driven by internal waves of tidal origin in the Oslofjord. *Journal of Physical Oceanography*, **9**, 435–441.

STIGEBRANDT, A. 1980. Some aspects of tidal interaction with fjord constrictions. *Estuarine Coastal and Shelf Science*, **11**, 151–166.

STIGEBRANDT, A. 1981. A mechanism governing the estuarine circulation in deep, strongly stratified fjords. *Estuarine Coastal and Shelf Science*, **13**, 197–211.

STIGEBRANDT, A. 1999*a*. Baroclinic wave drag and barotropic to baroclinic energy transfer at sills as evidenced by tidal retardation, seiche damping and diapycnal mixing in fjords. *Dynamics of Internal Gravity Waves*, **11**, 73–82.

STIGEBRANDT, A. 1999*b*. Resistance to barotropic tidal flow in straits by baroclinic wave drag. *Journal of Physical Oceanography*, **29**, 191–197.

STIGEBRANDT, A. 2001. Fiord circulation. *In*: STEELE, J., TUREKIAN, K. & THORPE, S. A. (eds) *Encyclopedia of Ocean Sciences*. Academic Press, Oxford, 897–902.

STIGEBRANDT, A. & AURE, J. 1989. Vertical mixing in basin waters of fjords. *Journal of Physical Oceanography*, **19**, 917–926.

STIGEBRANDT, A. & MOLVAER, J. 1996. Evidence for hydraulically controlled outflow of brackish water from Holandsfjord, Norway. *Journal of Physical Oceanography*, **26**, 257–266.

STIGEBRANDT, A., AURE, J. & MOLVAER, J. 1996. Oxygen budget methods to determine the vertical flux of particulate organic matter with application to the coastal waters off western Scandinavia. *Deep-Sea Research Part II – Topical Studies in Oceanography*, **43**, 7–21.

SVENDSEN, H. & THOMPSON, R. 1978. Wind-driven circulation in a fjord. *Journal of Physical Oceanography*, **8**, 703–712.

SVENDSEN, H., BESZCZYNSKA-MØLLER, A. ET AL. 2002. The physical environment of Kongsfjorden-Krossfjorden, an Arctic fjord system in Svalbard. *Polar Research*, **21**, 133–166.

SYVITSKI, J. P. M., BURRELL, D. C. & SKEI, J. M. 1987. *Fjords: Processes and Products*. Springer-Verlag, New York.

TETT, P., PORTILLA, E., GILLIBRAND, P. A. & INALL, M. 2010. Carrying and assimilative capacities: the ACExR-LESV model for sea-loch aquaculture. *Aquaculture*, in press.

UTNES, T. & BRORS, B. 1993. Numerical modelling of 3-D circulation in restricted waters. *Applied Mathematical Modelling*, **17**, 522–535.

VLASENKO, V., STASHCHUK, N. & HUTTER, K. 2002. Water exchange in fjords induced by tidally generated internal lee waves. *Dynamics of Atmospheres and Oceans*, **35**, 63–89.

WATTS, L. J., RIPPETH, T. P. & EDWARDS, A. 1998. The roles of hydrographic and biogeochemical processes in the distribution of dissolved inorganic nutrients in a Scottish sea-loch: consequences for the spring phytoplankton bloom. *Estuarine Coastal and Shelf Science*, **46**, 39–50.

WELANDER, P. 1974. 2-Layer exchange in an estuary basin with special reference to Baltic Sea. *Journal of Physical Oceanography*, **4**, 542–556.

WINTER, D. F. 1973. A similarity solution for steady state gravitational circulation in fjords. *Estuarine Coastal and Shelf Science*, **1**, 387–400.

XING, J. X. & DAVIES, A. M. 2006*a*. Influence of stratification and topography upon internal wave spectra in the region of sills. *Geophysical Research Letters*, **33**, L23606, doi: 10.1029/2006GL028092.

XING, J. X. & DAVIES, A. M. 2006*b*. Processes influencing tidal mixing in the region of sills. *Geophysical Research Letters*, **33**, L04603, doi: 10.1029/2005GL025226.

XING, J. X. & DAVIES, A. M. 2007. On the importance of non-hydrostatic processes in determining tidally induced mixing in sill regions. *Continental Shelf Research*, **27**, 2162–2185.

Arctic fjords: a review of the oceanographic environment and dominant physical processes

F. R. COTTIER[1]*, F. NILSEN[2], R. SKOGSETH[2], V. TVERBERG[3], J. SKARÐHAMAR[3] & H. SVENDSEN[4†]

[1]*Scottish Association for Marine Science, Scottish Marine Institute, Dunbeg, Oban, PA37 1QA, UK*

[2]*The University Centre in Svalbard, Postbox 156, 9171 Longyearbyen, Norway*

[3]*Institute of Marine Research, N-9294 Tromsø, Norway*

[4]*Formerly at: Geophysical Institute, University of Bergen, N-5007 Bergen, Norway*

[†]*Deceased*

**Corresponding author (e-mail: fcott@sams.ac.uk)*

Abstract: Fjords have long been recognized for their value as sites of sediment deposition, recording past climatic conditions. Recently, Arctic fjords have been recognized as the critical gateway through which oceanic waters can impact on the stability of glaciers. Arctic fjords are also used as idealized locations to study ice-influenced physical, biological and geochemical processes. In all cases a clear understanding of the physical oceanographic environment is required to interpret and predict related impacts and linkages. In this review we consider the characteristic elements of Arctic fjords and the important dynamical processes. We show how the intense seasonality of these regions is reflected in the varying stratification of the fjords. In particular, we show that sea ice has a central role in terms of the fjord salinity which ultimately influences the exchange with oceanic waters. When the fjord is ice free, wind forcing from the intense down-fjord katabatic winds gives rise to rapidly changing cross-fjord gradients, upwelling and strong surface circulations. The stratification and dimensions of Arctic fjords mean that they are often classed as 'broad' fjords where rotational effects are important in their circulation. We refer to the link between the physical oceanographic conditions and the related depositional records throughout.

Research on Arctic fjords has been dominated on the one hand by marine biology and ecological studies and on the other by sedimentary processes and records. While physical oceanography clearly plays a central role in each of these research areas, it was not until relatively recently that there has been sustained activity where oceanography is the principle focus. This contrasts with mid-latitude fjordic research where there is an extensive literature pertaining to the physical oceanographic environment, in the context of the fjord itself but also in how fjordic processes can inform our wider knowledge of ocean processes. This surge of interest and understanding in the oceanography of Arctic fjords allows us to interpret the other related aspects of fjord research with greater awareness of the variability in the physical environment.

Biologically, fjords provide a pseudo-mesocosm environment in which to investigate a variety of parameters including: life histories and strategies, trophic linkages, response to abiotic change and the coupling and fluxes between the surface and the benthos. Advection is known to be a major influence on the biological system both through nutrient supply, bloom dynamics and species composition (Basedow *et al.* 2004; Willis *et al.* 2006, 2008). The utility of fjords as sedimentary archives of past marine and glacial conditions is clear. Present efforts to reconstruct local, regional and global climatic histories are exploiting the fjordic archives and this is particularly so in the Arctic where the apparent and predicted rapidity of change is a prime motivation. Appropriate interpretation of such sediment records requires an understanding of the present-day processes and mechanisms operating in the system. The contemporary oceanographic state and processes are essential to establish the palaeoclimatic and palaeoceanographic conditions from sediment records (Azetsu-Scott & Syvitski 1999; Rasmussen *et al.* 2006; Majewski *et al.* 2009). We

From: Howe, J. A., Austin, W. E. N., Forwick, M. & Paetzel, M. (eds) *Fjord Systems and Archives*. Geological Society, London, Special Publications, **344**, 35–50.
DOI: 10.1144/SP344.4 0305-8719/10/$15.00

will make reference to such palaeo-climatic studies from fjords in relation to modern-day processes.

The relevance of studying the physical oceanographic processes pertinent to ice-covered fjords cannot be overstated. Fjords are commonly regarded as the link between the ocean and the land through cross-shelf exchange and the circulation dynamics of the fjord. The oceanic and terrestrial realms set the fjord boundary conditions, and the fjord is free to respond to or exert change on these. In the Arctic, the inshore boundary is usually dominated by glaciers and substantial seasonal freshwater input. The offshore boundary will generally have a relatively warm oceanic component. Currently there is enormous interest and activity in investigating the role of ocean heat on the stability of the glacial boundary (Holland *et al.* 2008; Nick *et al.* 2009; Rignot *et al.* 2010; Straneo *et al.* 2010). Propagation of warm oceanic waters into fjords, with the potential to increase the melt rates of glaciers, has been identified as a likely mechanism leading to the acceleration, thinning and retreat of glaciers (Hanna 2009). The Greenland ice sheet is a critical source of freshwater to the north Atlantic and it is via exchanges through the fjords that this freshwater is transported into the oceanic system. Similar topographically controlled exchange and circulation mechanisms are probably also relevant to melt processes under Antarctic ice sheets (Walker *et al.* 2007). Finally, the modification of the density of ocean water in the Arctic through salt release during sea-ice formation is a critical process in terms of atmosphere–ice–ocean interactions. While this densification process occurs throughout the Arctic, fjords provide a means of investigating the underlying physics in a relatively contained natural laboratory.

The purpose of this review is to discuss how the physical oceanography of fjords is modified at high latitudes compared to mid-latitude fjords, for which a number of good reviews exist (Farmer & Freeland 1983; Stigebrandt 2001). In particular, we see this review as a companion paper to the review of mid-latitude fjords by Inall & Gillibrand (2010). To define the scope, we will focus exclusively on those fjords that have a seasonal sea-ice cover. Given that southern hemisphere fjord systems do not extend further south than about 54°S, this necessarily limits the focus to northern high-latitude fjords. We will identify the principle characteristics of Arctic fjords and highlight those processes that are of particular relevance to glaciological and sedimentological studies. This will provide a background to the oceanographic mechanisms that may act on the system to shape the geological processes and their interpretation.

We start with a brief description of the geographic distribution of fjords in the Arctic. Next, the common physical oceanographic aspects of the fjords will be described. Finally, we shall highlight some of the key physical processes and properties that are pertinent to geophysical or palaeoceanographic research.

Distribution of Arctic fjords

The Arctic islands with fjordic coastlines are shown in Figure 1a. Given the glaciological action required to carve out a fjord, it is not surprising that almost the entire coastline of Greenland is dominated by fjords. The scale of the fjords can range from just a few kilometres to hundreds of kilometres in length. On the east coast, the rather complex system of interconnected fjords comprising Scoresby Sund is the largest fjord system in the world (Ó Cofaigh *et al.* 2001). The fjords of Greenland tend to be deep (>1000 m); Scoresby Sund is the world's deepest fjord at over 1500 m (Dowdeswell & Hambrey 2002). Most of the other large Arctic islands or archipelagos also have fjords. In the Canadian archipelago, Baffin is the largest of the islands. Its east coast contains over 200 fjords yet the west coast has few. Other fjordic islands in the archipelago are Ellesmere Island and Axel Heiberg (Dowdeswell & Hambrey 2002).

In the Eurasian Arctic the main fjordic region is the Svalbard archipelago, shown in Figure 1b, with fjords along the west and north coasts of Spitsbergen and on the island of Nordaustlandet. The longest of these at over 100 km is Wijdefjorden in north Spitsbergen. The west-facing Spitsbergen fjords are rather unusual in the sense that they are adjacent to the West Spitsbergen Current and therefore experience a much greater influence of warm saline Atlantic Water than any other Arctic fjords (Saloranta & Svendsen 2001; Cottier *et al.* 2007). In the Russian Arctic, parts of Severnaya Zemlya and Franz Josef Land have fjords and Novaya Zemlya is heavily indented by fjords.

For simple reasons of availability of literature, this review will be based upon work conducted in the fjords of Svalbard and Greenland. However, despite the inevitable variability that exists in the physical oceanography of these systems, there is also a good deal of commonality in the processes and conditions. It is these common factors that we will focus on.

Physical oceanography overview

Stratification and water masses

In terms of the physical oceanographic conditions alone, the most systematically studied Arctic fjords are those found in Svalbard. The majority of the observations have been focused on (from south to

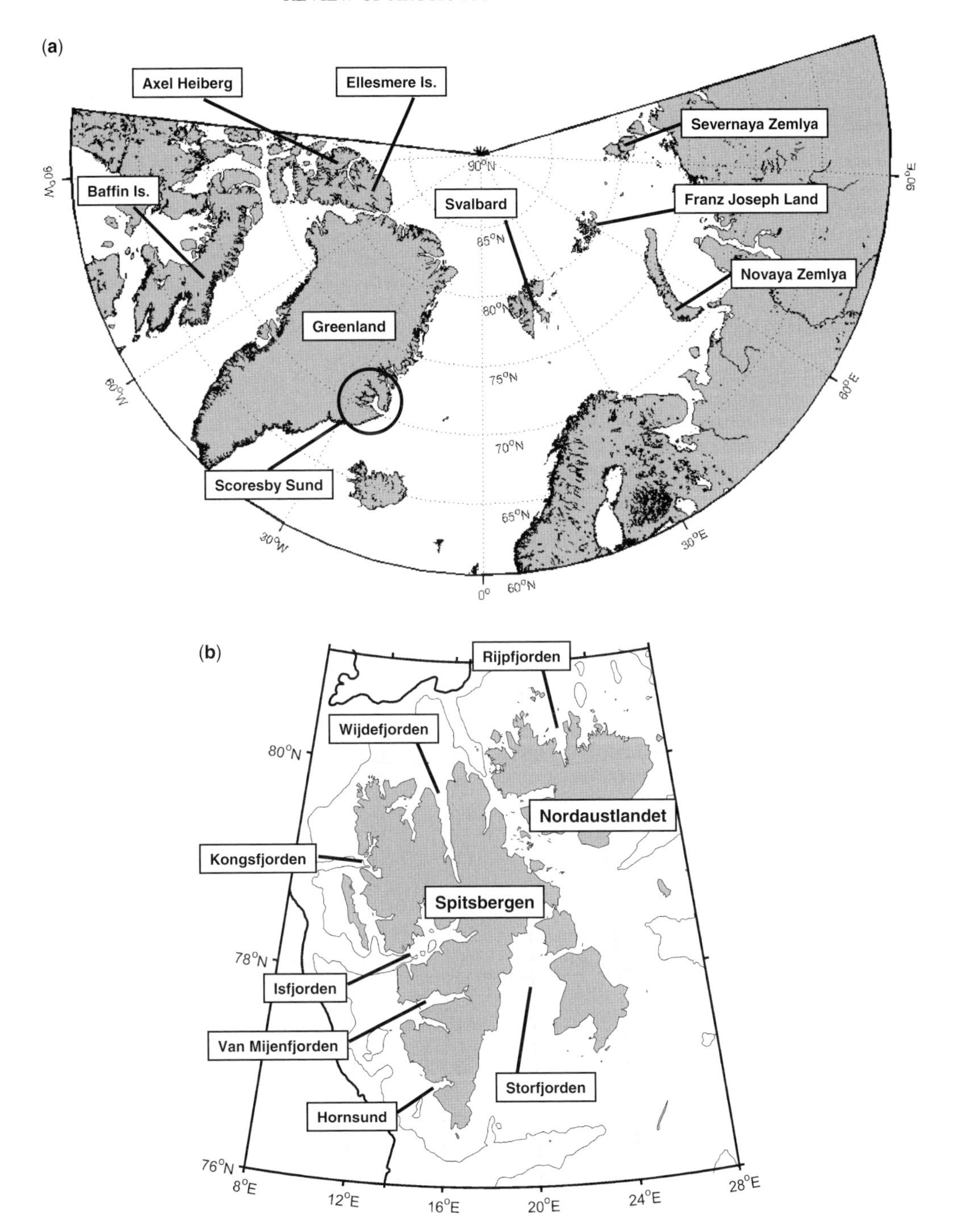

Fig. 1. Maps of (**a**) locations of fjordic regions in the Arctic and (**b**) locations of specific fjords in Svalbard.

north): Storfjorden (Skogseth *et al.* 2004, 2005*a*, 2008), Isfjorden (Nilsen *et al.* 2008) and Kongsfjorden (Svendsen *et al.* 2002; Cottier *et al.* 2005, 2007). Other fjords that have been observed and described include Rijpfjorden (Ambrose *et al.* 2006; Howe *et al.* 2010), Van Mijenfjorden (Skarðhamar & Svendsen 2010) and Hornsund. In the fjords of Greenland most of the published observations are summer time only and have mainly been in support of glaciological, sedimentary or biogeochemical observations. Consequently, they do not go very much beyond simple vertical profiles of temperature and salinity. Recent publications have described the physical oceanography of fjords on the east and west coasts of Greenland, particularly regarding the exchange of warm water (Straneo *et al.* 2010; Rignot *et al.* 2010). Certainly, the majority of the published observations that have captured long-term change using moored systems are from the fjords of Svalbard.

The classic arrangement of water masses found within a fjord with a sill is typically that of three layers (Farmer & Freeland 1983). This would comprise a fresh surface layer, an intermediate layer at the sill depth and a deep layer below. Typical profiles of temperature and salinity showing a classic three-layer structure are presented in Figure 2. These are profiles from late summer in Storfjorden which has a sill at *c.* 120 m. In Figure 2, a well-mixed fresh surface layer extends down to about 40 m separated from the intermediate layer, comprising advected Atlantic Water, by a steep halocline. The deepest layer is a saline water mass at near-freezing point which sits below the sill depth. Even in those fjords in Svalbard without an entrance sill, three layers of water masses are clearly identifiable (Cottier *et al.* 2005; Nilsen *et al.* 2008). Such a three-layer arrangement has also been described for deep-silled fjords in Greenland: Kangerlugssuaq (Azetsu-Scott & Syvitski 1999) and Scoresby Sund (Ó Cofaigh *et al.* 2001). An exception is Sermilik Fjord where the deep water is warm and saline, of North Atlantic origin (Straneo *et al.* 2010).

The upper layer is characteristically fresh (see Fig. 2) and comprises melt from the glacier, basal melting or other terrestrial runoff, for example snowmelt and rivers. A significant contribution to the freshwater budget can arise through melting from the base of icebergs (Azetsu-Scott & Syvitski 1999). In terms of volumes, the meltwater can actually constitute a rather small percentage of the overall fjord water, perhaps $<1\%$ for some Spitsbergen fjords. While sea ice melt also makes a rather minor contribution to the freshwater budget (Azetsu-Scott & Tan 1997; MacLachlan *et al.* 2007; Nilsen *et al.* 2008), some studies have reported that the surface water can itself be capped by a very thin fresh lens, thought to be from ice melt at the time of summer thaw (Rysgaard *et al.* 1999). This lens would be rapidly mixed into the surface layer once sea ice has left the fjord. Surface waters exhibit a very wide range of salinities and temperatures primarily due to variability in discharges and surface warming. The high sediment load and summer insolation can result in rather warm water temperatures $>5\,^{\circ}\mathrm{C}$ in the surface layer (Rysgaard *et al.* 1999; Cottier *et al.* 2005; Nilsen *et al.* 2008). The dominance of meltwater and its role in the formation of a well-defined surface layer generally decreases towards the fjord mouth. This along-fjord gradient is seen in salinity profiles (Cottier *et al.* 2005) and oxygen isotope composition (Azetsu-Scott & Tan 1997) and is reflected in the sediment facies (Ó Cofaigh *et al.* 2001).

The intermediate layers tend to be those of advected water masses, external to the fjord. These are often heavily modified from their original water mass characteristics through mixing with adjacent

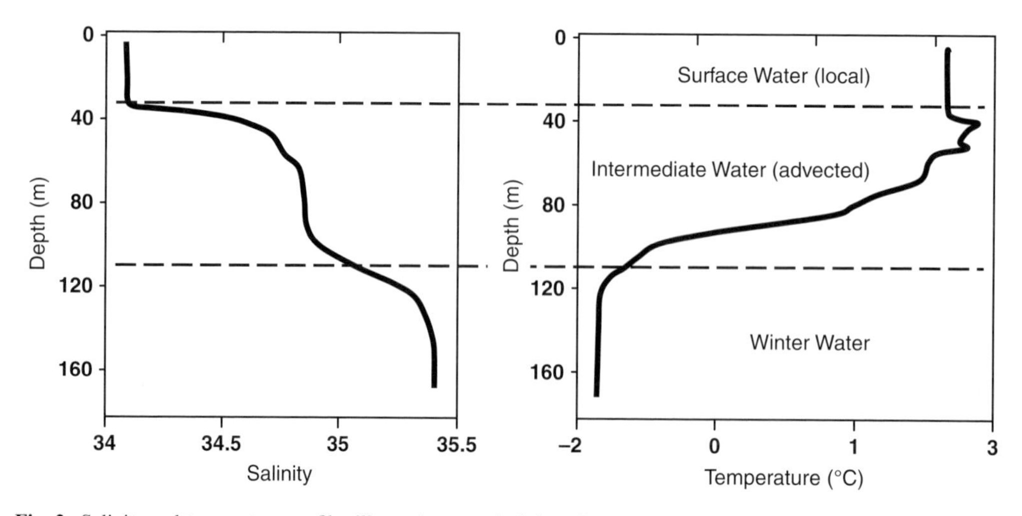

Fig. 2. Salinity and temperature profiles illustrating a typical three-layer structure of water masses within an Arctic fjord. The profiles are adapted from data obtained in Storfjorden, Svalbard (Skogseth *et al.* 2005*a*, fig. 4c, d).

waters (e.g. see figure 7 in Skogseth *et al.* 2005*a*). In the case of east-facing Greenland fjords, the intermediate water is generally termed Polar Water and derives from the adjacent East Greenland Current (Azetsu-Scott & Tan 1997; Azetsu-Scott & Syvitski 1999; Ó Cofaigh *et al.* 2001; Straneo *et al.* 2010). In the Svalbard fjords, it is most likely to be water which has derived from Atlantic Water carried north in the Norwegian Atlantic Current and the West Spitsbergen Current (Cottier *et al.* 2005; Skogseth *et al.* 2005*a*; Nilsen *et al.* 2008). The Atlantic Water has probably mixed to some extent with shelf waters before entering the fjord and is therefore rather cooler and fresher than the source water. Around Greenland, Atlantic Water is carried close to the continental shelf in the Irminger Current. In West Greenland, this has been reported to flow into the silled Jakobshavn fjord (Holland *et al.* 2008). The intermediate layers can often be relatively warm (see Fig. 2) and such subsurface warm layers have been implicated in the release of iceberg debris through bottom melt, thus influencing the formation of ice-rafted debris layers within the fjord sediments (Azetsu-Scott & Syvitski 1999).

The deepest water masses are necessarily the densest and therefore generally have the highest salinity. In the Svalbard fjords, the source of deep water is either from sea-ice formation and brine release (Skogseth *et al.* 2005*a*; Nilsen *et al.* 2008), or it is the result of saline Atlantic Water undergoing intense cooling during winter but without further salinity increase through sea-ice formation (Nilsen *et al.* 2008). In the deep Greenlandic fjords the situation is rather different. Greater freshwater flux means that the surface waters are generally fresher than the smaller and shallower Svalbard fjords. Consequently, winter sea-ice formation only affects the upper layers. This is akin to the cold halocline of the Arctic Ocean (Rudels *et al.* 1996; Kikuchi *et al.* 2004), where the deep water has its origin in Atlantic Water that has either recirculated in the Nordic Seas or which has evolved into Arctic Intermediate water (Azetsu-Scott & Tan 1997; Ó Cofaigh *et al.* 2001).

Critically, these layers can themselves comprise numerous water masses from different sources, such that the composition of the three layers varies spatially within the fjord and temporally (Skogseth *et al.* 2005*a*). Even for two adjacent fjords, the differing meltwater fluxes and exposure to oceanic water sources due to fjord bathymetry leads to strongly contrasting thermohaline structures (Gilbert *et al.* 1998). Given the variety of source waters, attempting to disentangle the structure and history of fjordic hydrography using temperature and salinity alone is very challenging. Oxygen isotope techniques are now widely used as a label for specific components such as glacial melt, Atlantic Water sources and brine released during sea-ice formation (Azetsu-Scott & Tan 1997; Yamamoto *et al.* 2002; MacLachlan *et al.* 2007).

Nutrients have also been used successfully to indicate the sources of the various water masses. For example, combining salinity measurements with those of silicate in Kangerludssuaq (Azetsu-Scot & Syvitski 1999) identified the low-salinity, high-silicate glacial meltwater as distinct from the high-salinity low-silicate Polar Water from the East Greenland Current and the high-salinity high-silicate North Atlantic Water. Polar Water originates outside the fjord within the euphotic zone, where it becomes depleted in silicate before being carried below glacial surface water. Surface water at the head of the fjord was shown to have salinity and nutrient characteristics similar to that of Atlantic Water, indicating significant upwelling at the glacier front drawing the deeper Atlantic Water to the surface.

While it is possible to identify a three-layer arrangement of water masses, most dynamical interpretations of fjords invoke a two-layer approach. This is a consequence of the well-developed surface layer (often of brackish water) that is found in most fjords. At low temperatures, seawater density is controlled primarily by salinity and therefore the surface layer can be regarded as being decoupled from the lower layer(s) by a strong pycnocline. The significance of this two-layered approach in terms of the fjord dynamics will be discussed later in this paper.

Stratification within Arctic fjords will vary interannually and depend on the extent to which the water was modified the previous winter through cooling and sea-ice formation, the seasonality of the freshwater discharge and the variability of any inflowing water masses. In general, however, the classic layered arrangement, depicted by the profiles in Figure 2, will only be fully developed during the summer. Wind mixing and intense winter cooling will tend to break down the stratification. Further, brine release through sea-ice formation will initiate convective overturning that will continue to erode the stratification, giving rise to rather homogeneous water masses during winter.

Seasonality

There are few complete observations of the seasonal cycle of temperature, salinity and sea-ice cover in Arctic fjords. Most observations were made during summer months only, but the use of moorings has enabled us to record the changes and infer the processes giving rise to those changes (Cottier *et al.* 2005, 2007; Skogseth *et al.* 2005*a*; Howe *et al.* 2010). Here we discuss seasonal changes in

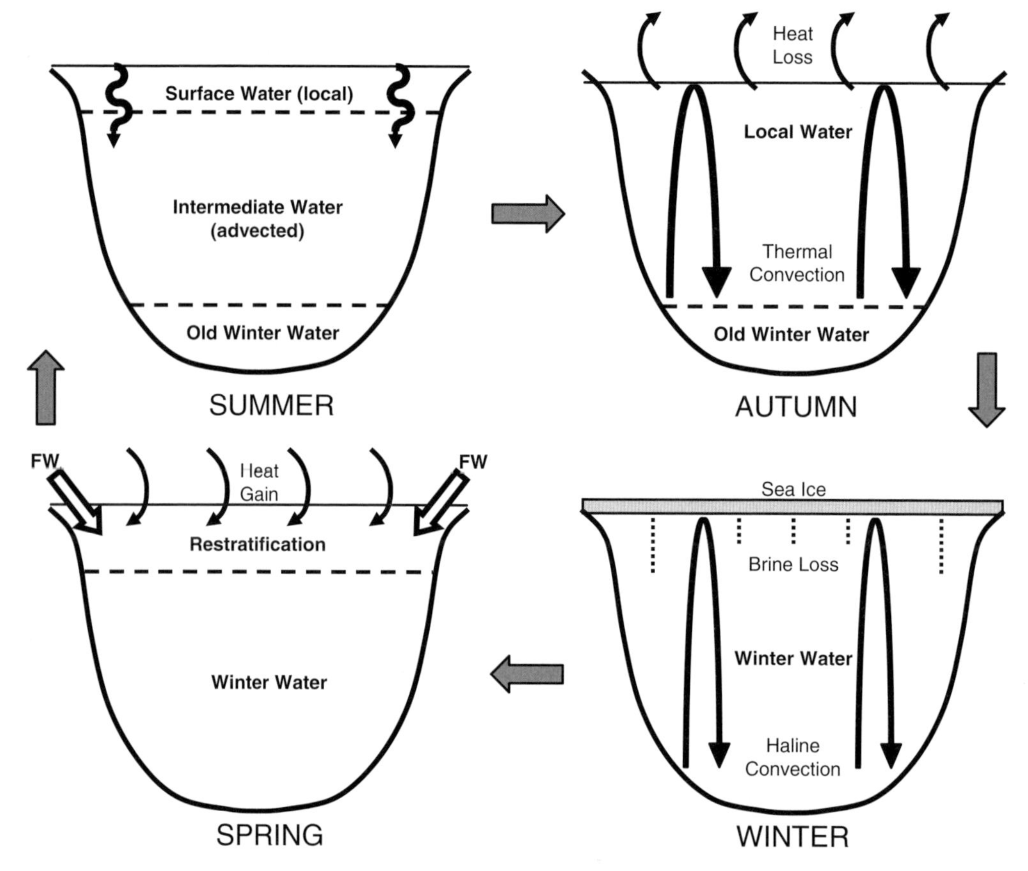

Fig. 3. Illustration of the seasonal cycle of stratification in a shallow (<300 m) Arctic fjord, typical of Svalbard. Significant processes include wind mixing, surface heat fluxes, freshwater (FW) input and brine release during sea ice growth.

hydrography typical of fjords in Svalbard. We will refer to the schematic representation of stratification and processes given in Figure 3. Beginning with a fully stratified system in summer, many fjords will be ice free at this time so that wind mixing will dominate. In Storfjorden for example, intermediate waters in the stratified system generally become freshened and warmed from summer to autumn through mixing with surface waters (Skogseth *et al.* 2005*a*) while the surface waters will become rather more saline and cooler.

Through autumn, the decrease in air temperature and stronger winds leads to rapid extraction of sensible heat from the fjord, decrease in water temperatures and deepening of the surface mixed layer (Cottier *et al.* 2007). The mixing and cooling of the fjord gives rise to what has been termed Local Water (Svendsen *et al.* 2002; Nilsen *et al.* 2008) such that cold, fresher water is overlying more saline waters. Due to the initially freshwater at the surface, intense cooling may not be sufficient to increase the density of the water to a point where it will convect to the bottom of the fjord and permit mixing with the more saline bottom waters. Once the water is at its freezing point, typically in November or December (Gerland & Renner 2007; Nilsen *et al.* 2008) although earlier in Greenlandic fjords (Rysgaard *et al.* 1999), continued surface cooling will lead to the formation of sea ice and the associated brine release (Haarpaintner *et al.* 2001). The effect is to increase the salinity and density in the surface such that the water column becomes unstable to haline convection. A small additional contribution of salt occurs during the subsequent desalination of the sea-ice cover in the fjord. This secondary release of brine has been observed to take place toward the end of the winter in response to increased oceanic heat fluxes (Widell *et al.* 2006). The saline water produced in the fjord has been termed Winter Cooled Water (Svendsen *et al.* 2002; Nilsen *et al.* 2008) yet it is often appropriate to adopt the more general term of Winter Water (first proposed by Nansen; 1915) to represent both the Local Water and Winter Cooled Water. The production of sea ice in winter will govern the salinity of the water column by the end of the winter. Depending on the salt flux and the depth of the fjord, the convection will extend over the entire water column thus renewing and re-oxygenating the bottom water (Nilsen *et al.* 2008). The effect is a rather homogeneous water mass, with only very weak stratification.

By the end of the winter, fast ice comprising one season of growth will typically be 1.0 m thick. The timing of break up of sea ice is highly variable and dependent on wind and swell conditions and the ice concentration outside the fjord. In some Arctic fjords, the presence of warm water masses means that sea ice forms intermittently or is very thin and will break up easily (Cottier *et al.* 2007; Gerland & Renner 2007). In other locations there is a physical barrier to sea ice export, extending the sea-ice season (Skarðhamar & Svendsen 2010). In northern Svalbard, the oceanic pack ice can prevent sea ice being exported and thus extend the ice season (Wallace *et al.* 2010). However, sea ice will typically begin to melt or break up between May and July (Rysgaard *et al.* 1999; Mikkelsen *et al.* 2008). After the ice break-up, there is a gradual increase in temperature through the surface water layer through warming and mixing and a reduction in salinity due to the addition of freshwater as the melt period starts. The strong pycnocline that dominates summer profiles in the fjord is therefore re-established.

Dynamical processes

Fjords at all latitudes experience similar forms of external forcing that cause the water masses to become modified, mixed or exchanged (Inall & Gillibrand 2010). The forcing also determines the extent of vertical and horizontal circulation of water masses within the fjord. Here we describe the most significant and commonly occurring dynamical processes occurring in, though not exclusive to, Arctic fjords. In the previous section we described the broad features of thermohaline stratification, which is constantly evolving through vertical mixing processes within the fjord and lateral exchanges with adjacent coastal waters. We therefore begin this section with a discussion of vertical mixing, particularly the role played through sea-ice formation. The resulting degree of stratification determines the extent to which cross-fjord flows becomes important, that is, there is a rotational aspect to the fjord circulation rather than simple axial flow. Wind is a dominant environmental factor in any fjordic setting, particularly in forcing the surface layers. Glaciers at the head of Arctic fjords means that winds are usually katabatic in nature, being more persistent, stronger and colder than winds in mid-latitude fjords (Beine *et al.* 2001; Svendsen *et al.* 2002; Argentini *et al.* 2003). We therefore discuss the role of wind on the fjord dynamics as this is a unifying factor. Polynyas are not common to all Arctic fjords as they require a mobile sea-ice cover and, in many cases, sea ice in fjords is fast ice. However, the dynamics of polynyas and the resultant ice–ocean interactions are more easily studied and modelled in fjords, thus giving insight into a dynamical process that occurs around the Arctic coastal margin. Finally, the exchange of waters between the fjord and coast is fundamental to their function, particularly in terms of the changes in heat content of the fjord through influx of warmer coastal waters.

Vertical mixing

Much of the seasonal modification of water masses arises through vertical mixing processes and the advection of new water masses into the fjord. Wind stress is a dominant mechanism for mixing in many Arctic fjords, and gives rise to much of the seasonal changes. A simplified, low-resolution wind field within a 1D mixing model can be used successfully to describe the overall seasonal mixing (Cottier *et al.* 2007). However, a high-resolution wind field is required to simulate the wind-induced mixing accurately by resolving the topographic influence of the land on the local wind strength and variability (Skogseth *et al.* 2007).

In contrast to mid-latitude fjords, the presence of a seasonal sea-ice cover means that wind mixing is suppressed in winter and other mixing mechanisms become relatively more important. Exchange of waters by tides and estuarine circulation, for example, can sustain the mixing rates through the action of shear between layers and momentum fluxes at the fjord boundaries. Measurements from Van Mijenfjorden show tidally driven turbulent dissipation and diffusivities under an ice cover that are comparable to values previously found at other ice-free coastal sites with similar topography and forcing (Fer & Widell 2007). Release of salt from sea ice changes the buoyancy of the surface layers and induces convective overturning which will fully mix the fjord. The opposite process also occurs, with the injection of freshwater at depth due to melting of icebergs or glaciers which will generate localized upwelling and vertical mixing.

Like mid-latitude fjords, tides provide some of the energy necessary for mixing both through barotropic flow and baroclinic flows that are sustained due to the stratification within the fjord (Inall & Gillibrand 2010). The occurrence of internal waves can aid the mixing between the upper and lower layers away from the boundaries (Stigebrandt 1976; Perkin & Lewis 1978; Stigebrandt & Aure 1989; Fer 2006). Brine released under an ice cover leads to continual variation in the density structure of the water column. Once there is a match between the frequency of a tidal component and the natural resonant frequency of the internal seiche motion, the vertical mixing rates can become greatly enhanced (Lewis & Perkin 1982).

Rotation

The dynamic effect of the Earth's rotation on fjords depends on the width of the fjord and the vertical stratification (Cushman-Roisin *et al.* 1994). The rotational characteristics of all fjords will therefore vary by season, but potentially more so in the case of Arctic fjords. A gauge of the importance of rotational effects is to compare the width of the fjord to a parameter termed the *internal radius of deformation*, sometimes called the internal Rossby radius. The internal Rossby radius is given as the ratio of the speed of an internal wave (supported by the stratification) to the Coriolis parameter. It is therefore a measure of the relative importance of stratification v. rotation. When the internal Rossby radius is smaller than the width of the fjord then cross-fjord variations in the flow, induced by rotational dynamics, can be expected. In such cases the fjord is often termed 'broad'.

In a two-layer model the definition of the internal Rossby radius (r_i) is given by:

$$r_i = \frac{c_i}{f} \qquad (1)$$

where

$$c_i^2 = \frac{g' H_1 H_2}{H} \qquad (2)$$

is the speed of the internal wave and

$$g' = g \frac{\rho_2 - \rho_1}{\rho_2} \qquad (3)$$

is the reduced gravity with ρ_1, ρ_2 and H_1, H_2 being the densities and depths of the upper and lower layer, respectively, H is the total water depth and f is the Coriolis parameter.

We can gain an appreciation of this parameter in Arctic fjords by calculating the internal Rossby radius under summer conditions for a fjord of water depth 200 m. Using data from Svalbard fjords (Cottier *et al.* 2005; Nilsen *et al.* 2008) we assume a surface melt layer ($T = 4$–7 °C, $S = 30$–32) of thickness 20 m overlying a more saline lower layer ($T = 0$–4 °C, $S = 34$–35), giving an internal Rossby radius ranging from 3.5–6 km. Even in the deeper Greenlandic fjords, where water depths can be in excess of 1000 m, surface and lower layer values from Scoresby Sund (Ó Cofaigh *et al.* 2001) yield a range for r_i of 4–6 km.

The effect of rotation on fjord dynamics is commonly reported for high-latitude fjords (Cushman-Roisin *et al.* 1994; Ingvaldsen *et al.* 2001; Cottier *et al.* 2005; Skogseth *et al.* 2005*b*; Skarðhamar & Svendsen 2010), possibly more so than mid-latitude fjords. Examining the parameters that determine r_i, we note that this contrast between high- and mid-latitude regions is perhaps unexpected. The overriding balance is that between the thickness of the surface layer and the degree of stratification, with the surface layer thickness being the dominant factor under conditions of high stratification. The high stratification that is typical during summer conditions in Arctic fjords actually acts to increase r_i, and the depth of the pycnocline, at 20 m or less, in Arctic fjords is similar to that of mid-latitude fjords. Whilst r_i does decrease as latitude increases, the difference in Coriolis parameter between 60°N and 90°N is only 15%. The main reason for the greater importance of the rotation in the Arctic is simply that the fjord widths, certainly in their outer parts, are typically 10 km or more, which is 2–3 times greater than r_i. Therefore they cannot be regarded as straightforward two-dimensional, estuarine system which narrower, mid-latitude fjords can often be reduced to. Arctic fjords will generally respond as 'broad' fjords.

A consequence of rotation is that the distribution of freshwater is not uniform across the fjord. Surface meltwater entering the fjord at the head is deflected such that the freshest and coolest waters are found to the right in the direction of the outflow (Azetsu-Scott & Tan 1997; Ingvaldsen *et al.* 2001; Svendsen *et al.* 2002; Skogseth *et al.* 2005*a*). To appreciate this, the distribution of salinity needs to be mapped across the surface of the fjord. Such a study was carried out in the inner part of Kongsfjorden where the outflow was colder and fresher with a cross-fjord temperature difference of surface temperature of 2 °C and a salinity difference of 1 (Aliani *et al.* 2004). However, even at 20 m depth, these gradients were much reduced highlighting that the flow was confined to a well-delineated surface layer. This modified estuarine flow can be further intensified by wind effects as described below.

Similarly, inflowing water masses will tend to hold to the right-hand shore. This can be seen clearly in cross-fjord sections from some of the Spitsbergen fjords. Isfjorden (Nilsen *et al.* 2008), Kongsfjorden (Svendsen *et al.* 2002) and Storfjorden (Skogseth *et al.* 2005*a*) all show warmer inflows on the right-hand side of the direction of flow. The full rotational effect can be seen clearly in profiles of water currents. In the outer basin of Kongsfjorden, acoustic Doppler current profile data show a clear development of cyclonic circulation in the upper half of the basin with current velocities strongest in the surface (Basedow *et al.* 2004). Evidence for the dominant circulatory environment of high-latitude fjords can be detected in the sedimentary patterns. For example, in Kongsfjorden Howe *et al.* (2003) report morphologies of sediment deposits, consistent with along-slope cyclonic circulation of the bottom currents in the outer basin which transports the sediments. Additionally, Gilbert *et al.* (1998) find a greater deposition of sediment occurring on the side of the fjord coincident with the outflow of meltwater.

The rotational effects also have an influence on the outflow of dense bottom waters from fjords. In a narrow silled fjord, the outflow would be uniform across the sill and would experience hydraulic control with a critical overflow thickness limiting the overflow transport (Farmer & Armi 1986). In broad fjords the rotational effects are the controlling factor for a dense overflow over a sill (Skogseth *et al.* 2005*a*). The overflow is not uniformly distributed over the sill, rather in a narrower stream at one side of the sill, but it is not limited by the hydraulic conditions.

Wind forcing

One of the most important modifications to simple estuarine dynamics in any fjord is the effect of wind. In Arctic fjords, the impact of strong and persistent katabatic winds is arguably greater. Within a two-layer fjord, the upper layer circulation is consistently shown to be dominated by the wind (Ingvaldsen *et al.* 2001; Svendsen *et al.* 2002; Skogseth *et al.* 2007; Skarðhamar & Svendsen 2010). The reason for this is twofold. First, in light of the discussion in the previous section, many Arctic fjords are classified as 'broad' such that rotation is important. Second, due to the topography of the fjords, the wind tends to be strongly steered with the direction of the strongest winds being essentially bi-modal, down-fjord and up-fjord (see Nilsen *et al.* 2008, fig. 4; Svendsen *et al.* 2002, fig. 4). The primary effects of the wind are to modify the outflow of surface water, either by intensifying it or retarding it depending on the wind direction (Skogseth *et al.* 2007), and to induce upwelling effects.

The simplest case to consider is a strong up-fjord wind in which the surface flow is reversed and brackish water begins to be stacked up at the head of the fjord. This sets up a down-fjord pressure gradient which can develop such that it overcomes the wind and re-establishes the down-fjord flow. Once the wind ceases, there is a relaxation time where the stack-up water flows out with an intensified surface flow (Ingvaldsen *et al.* 2001; Svendsen *et al.* 2002).

The effect of down-fjord winds is regarded as twofold (although closely coupled): the adjustment of the cross-fjord gradients of salinity and temperature and the enhancement of the surface flow due to the resulting cross-fjord pressure gradient. A strong down-fjord wind sets up an Ekman transport of the surface layer to the right of the wind direction (Fig. 4). This will change the surface salinity and temperature distributions and the vertical displacement of isopycnals as water tends to upwell to the left of the wind and downwell to the right. In the case of strong and persistent wind, the cumulative wind stress will lead to outcropping of isopycnals (Fig. 4b). It has been demonstrated that the response of the fresh surface layer to wind can be near instant (Skarðhamar & Svendsen 2010), leading to short-term variability in the stratification of the fjord. The effect is to create substantial cross-fjord gradients.

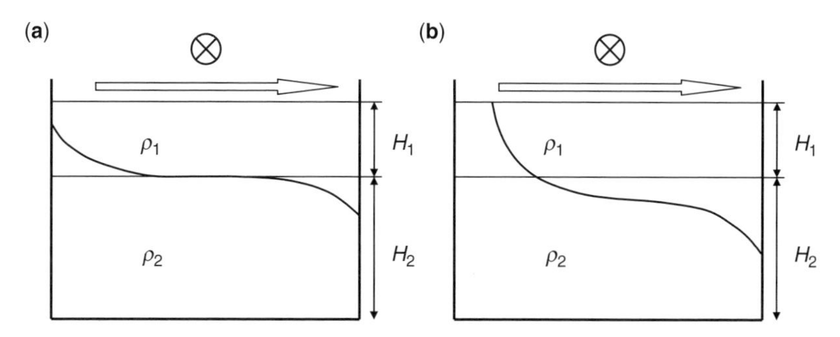

Fig. 4. Schematic illustration of the initial two-layer density structure and the resulting density structure following from down-fjord winds (adapted from Cushman-Roisin *et al.* 1994). The wind is shown as the crossed circle blowing into the page and the large arrow indicates the Ekman transport of surface water: (**a**) the effect on the isopycnals of up- and down-welling; and (**b**) continued down-fjord winds can lead to out-cropping of isopycnals.

The resulting displacement of surface water and rapid establishment of cross-fjord thermohaline gradients can cause very rapid changes in salinity at a given depth. For example, Cottier *et al.* (2005) showed that in Kongsfjorden variations in salinity in the summer at 40 m depth would change by up to 1 salinity unit in 24 hours. The only consistent interpretation of this was a response to down-fjord winds, Ekman transport and a deepening of the halocline. A correlation was established with the wind forcing with the hydrographic response lagging the wind forcing by 3 days. Other studies of the role of wind forcing have noted significant variations in salinity at depth at timescales of the order hours (Skarðhamar & Svendsen 2010). The difference between these observations is most likely due to the width of the fjord and the thickness of the surface layer.

Down-fjord winds will, through the wind stress, intensify the outflow of brackish surface freshwater to the right-hand side of the fjord. Additionally, Ekman transport of surface water to the right of the wind and the cross-fjord pressure gradient generated will enhance the flow pattern of the brackish water (Ingvaldsen *et al.* 2001; Svendsen *et al.* 2002). Establishing the pressure gradient also means that the enhanced surface circulation can persist for many hours after the wind has ceased (Ingvaldsen *et al.* 2001) and will in turn be compensated by an inflow in the lower layers (Nilsen *et al.* 2008). This provides a mechanism for wind-mediated exchange with the shelf water.

In mid-latitude fjords and in lakes there is a substantial literature on wind-induced seiches (e.g. Arneborg & Liljebladh 2001; Lemmin *et al.* 2005). In contrast there have been no observations of wind-induced seiches for Arctic fjords although the conditions for the generation evidently exist. Other internal oscillations that may be induced by wind (or tidal) forcing include Kelvin waves that can propagate around the fjord with the coastline to the right of the direction of travel. Such oscillations have been observed in Kongsfjorden based on current metre data (Daae 2008) and are acknowledge as a mechanism for the advection of water masses into such open fjords (Svendsen *et al.* 2002).

Polynyas

A polynya is 'any non-linear shaped opening enclosed in (sea) ice' (WMO 1970). They generally form in specific regions of the Arctic, but the variability in their size and the duration of open water can show large interannual variation due to their sensitivity to the thermodynamic and dynamic forcing by the ocean and atmosphere (Morales Maqueda *et al.* 2004). Polynyas have two important roles within the Arctic: (1) oceanographically, they modify the underlying water properties through ice formation and brine release leading to the production of dense water and ultimately deepwater formation (Martin & Cavalieri 1989; Haarpaintner *et al.* 2001; Shcherbina *et al.* 2003); and (2) ecologically, they act as oases and enable birds and mammals to overwinter in high latitudes and encourage enhanced primary production in spring (Smith & Barber 2007). In terms of dense water production rates, the most important polynya sites are on the Arctic shelves. However, polynyas are also active within Arctic fjords where they have been studied in terms of their role in water modification; they also provide a natural laboratory for studying the process physics through observations and models.

Polynyas are typically divided into two classes: *sensible-heat polynyas* which are thermally driven through convection and *latent-heat polynyas* which are mechanically driven through divergence and export of sea ice. In sensible-heat polynyas, heat is transferred from the ocean to the atmosphere through upwelling and vertical mixing. Locations where warm water comes to the surface in winter can therefore be regions of low ice concentration: for example, Whaler's Bay Polynya to the north of Spitsbergen forms close to the ice edge over the shelf break of the Eurasian Basin.

In fjords, latent-heat polynyas are the most prevalent form. Given the strong and directional nature of winds within fjords, with the predominance of down-fjord winds, sea-ice divergence can readily occur with open water being maintained at the head of the fjord. Divergence can also be maintained by down-fjord currents linked to the circulation and exchange processes. The open water formed through divergence loses heat and freezes. The new ice that has formed is then carried away by the forcing mechanism, thus maintaining an area of open water. Consequently, the temperature of the water within a latent-heat polynya stays at the surface freezing point. The persistent and cold winds blowing down-fjord can lead to high rates of ice formation as newly formed ice is advected away. Consequently, there is a high rate of brine release into the underlying water column. This means that the water column in a wind-forced latent-heat polynya, as found in fjords, experiences very efficient water mass transformation. A record of the polynya activity in Storfjorden through the Holocene can be inferred from the oxygen isotope composition of selected foraminifera species (Rasmussen *et al.* 2006). The inference to the bottom water temperature implies that the formation of cold, saline BSW has been active in Storfjorden for the past 5000 years.

Given the prevalence of polynyas around the coastal margin of the Arctic, the role of polynyas in Arctic oceanography through brine release and their contribution to the intermediate and deep water cannot be underestimated. Having a means of studying the underlying physics of a polynya in a rather enclosed setting has obvious advantages in terms of access, observations, salt budgets and modelling. Fjords have therefore been found to be ideal locations for detailed polynya studies for understanding and documenting the ice production and brine formation. Such a location is Storfjorden where there is a recurrent wind-driven polynya which forms brine-enriched waters, commonly referred to in the literature as brine-enriched shelf waters (BSW) (Skogseth *et al.* 2005*a*, 2008). Storfjorden has been used as a site for investigating the production and discharge (Fer *et al.* 2003, 2004; Skogseth *et al.* 2008) of BSW, but also as a location to test theories of polynya dynamics (Haarpaintner *et al.* 2001; Skogseth *et al.* 2004).

The polynya is located in the NE of Storfjorden over relatively shallow water which is less than 100 m deep. The polynya is driven by rather persistent NE winds, although the ice production rate in Storfjorden is highly variable giving rise to a varying release of brine (Skogseth *et al.* 2005*a*). The salinity of the BSW is primarily governed by the initial salinity of the surface water and the total ice production. As brine is formed it travels across the shallow polynya region, descending the slope and into the deeper fjord basin. Once formed in the polynya, the BSW accumulates in the deep basin until it reaches the depth of the sill where it overflows. The density difference between the BSW and the ambient water outside the fjord causes the overflow to spill down the continental slope into the deep Norwegian Sea and travel northwards along the eastern slope of Fram Strait towards the Arctic Ocean (Quadfasel *et al.* 1988; Schauer & Fahrbach 1999; Fer *et al.* 2003). Although relatively small, it is estimated that this polynya alone supplies between 5–20% of all the newly formed dense water which enters the Arctic Ocean (Skogseth *et al.* 2005*b*).

Within the fjord the fresher surface waters lie above a cold isothermal layer of high salinity BSW formed from shelf convection. Skogseth *et al.* (2005*a*) draw an analogue of this system with the mode of formation and structure of the cold halocline of the Arctic Ocean (Steele & Boyd 1998). A similar description has been given based on observations in Cambridge Bay in the Canadian Archipelago (Perkin & Lewis 1978). Here, the dense water produced by brine release over shallows descends down the boundary slope into the deeper part of the bay. While the Arctic halocline shows weak vertical mixing between these layers, there is the potential in a fjord setting for the dense water to mix with the upper layer through breaking of internal waves at the boundary (Perkin & Lewis 1978).

Coastal polynya models have been used successfully in fjord environments (Haarpaintner *et al.* 2001; Skogseth *et al.* 2004; Nilsen *et al.* 2008). They are able to quantify the interannual variation in ice production and associated salt flux. They are therefore capable of rationalizing the variation in stratification within the fjord on the basis of water mass modification during the previous winter. Measurements of bottom water temperature and salinity in Billefjorden (a silled fjord with a latent heat polynya) from successive winters have shown that the variation in the bottom water can be linked directly to the wind-driven polynya activity of the fjord (Nilsen *et al.* 2008). The model itself comprises a channel which is partially open at the seaward end due to ice concentration at the mouth. Ice forms in three different ice-class areas: open water, thin ice and fast ice. The actual polynya itself is defined by the area of open water and thin ice and the area of each is determined by the wind forcing such that there is a balance between ice production in the open water and ice removal.

In fjords where there is no sill at the entrance, the polynya characteristics can switch between being a latent-heat and a sensible-heat polynya through interplay between the available oceanic heat in a coastal system and the efficiency in the heat loss and ice production through air-ice-ocean processes. An example of this is Isfjorden which is open to the West Spitsbergen Shelf and therefore can experience inflow of warm Atlantic Water from the West Spitsbergen Current. In winter the upper layer of Isfjorden, influenced by the available Atlantic Water on the shelf, has to be cooled down to freezing point before ice growth can start. Hence, the heat loss from the ocean to the atmosphere has to be large enough in order for the freezing point to be reached and for Isfjorden to function as a wind-driven latent-heat polynya. When production of frazil ice during winter is high, there will be a correspondingly high salt release to the surface water and the resulting formation of Winter Water. This has the potential to be denser than the Atlantic Water residing on the shelf. An unrestricted outflow of Winter Water as a bottom current onto the shelf is compensated by an intrusion of warm Atlantic Water to intermediate and upper layers of the fjord (Nilsen *et al.* 2008). The increase in sensible heat in the fjord switches the system into a sensible-heat polynya, with the intruded warm water keeping the fjord ice free. The climatologically normal state is for the locally produced Winter Water to be less dense than the shelf water during a freezing season. Consequently, Atlantic Water from the shelf will intrude along the bottom of the fjord and Isfjorden system will remain as a latent-heat polynya.

Fjord-shelf exchanges

Fjords act as the link between the terrestrial domain and the oceanic domain. There is therefore the potential for continuous exchange between the fjord and the coastal waters on the adjacent shelf. Oceanic water masses are brought onto the shelf such that they are then available for exchange at the fjord mouth. An essential process is to bring the oceanic water masses, usually transported as a boundary current flowing along the shelf slope, onto the shelf. The interaction between the boundary current and the shelves is a key element in this exchange process; the stronger Coriolis effect at high latitude means that they tend to follow the contours of the slope. Eddy activity at the front between the boundary current and the shelf water is widely considered to be an effective means of cross-shelf transport (Nilsen *et al.* 2006; Tverberg & Nøst 2009), as is upwelling of bottom confined boundary currents (Carmack & Chapman 2003; Dmitrenko *et al.* 2006; Cottier *et al.* 2007).

Inflow of intermediate-depth waters into the fjord is often in response to wind forcing where there is an enhanced estuarine-type flow pattern with down-fjord winds. This will then induce a compensating inflow from the shelf (Lewis & Perkin 1982; Nilsen *et al.* 2008; Straneo *et al.* 2010). Clearly this type of wind-forced exchange will only be active when any sea-ice cover has broken up. An example of this effect was recorded in the entrance to Kongsfjorden, west Spitsbergen (Cottier *et al.* 2007). In winter, surface water on the shelf was forced to stack up against the coast during a period of predominantly southerly winds. This generated a pressure gradient force in the offshore direction. When the southerly winds decreased, there was a relaxation of the system with the surface water moving offshore, thus generating a compensating onshore pulse of Atlantic Water in the lower layer into the fjord entrance.

In shallow-silled fjords, exchange with the shelf waters is impeded by a topographic barrier. Where the sill is deep or non-existent, there is typically no physical barrier to the exchange of intermediate and surface waters. However, the currents will tend to be geostrophic due to lateral density gradients across the fjord. Where a coastal current flows along the shelf and crosses the entrance to a deep-silled fjord, for example in west Spitsbergen, the current is usually identified by sloping isopycnals in the cross-shelf density sections. Such sloping isopycnals indicate that the alongshore coastal current is geostrophic. The resulting density gradients between the shelf and the fjord waters then acts to restrict exchange, thus determining the transport of water into the fjord; this process has been termed geostrophic control (Klinck *et al.* 1981). The degree of control is determined by the stratification of the coastal waters and the along-shelf wind conditions. The steeper the isopycnals at the fjord mouth, the stronger the coastal current and the greater the restriction on exchange. Such density gradients at the mouth of the fjord can cause the fjord waters to become isolated from the shelf waters (Svendsen *et al.* 2002; Cottier *et al.* 2005; Nilsen *et al.* 2008). Any process that perturbs or counters the geostrophic coastal current will tend to reduce the restriction on exchange, ultimately leading to free communication between the shelf and fjord.

An example of geostrophic control is found in Isfjorden (Nilsen *et al.* 2008) which has no distinct shallow sill at its mouth and is therefore linked directly to the shelf and slope area along West Spitsbergen. Here, Atlantic Water can be guided from the West Spitsbergen Current towards the mouth of the fjord, yet the exchange with the fjord is often restricted due to the alongshore coastal current. Due to seabed topography, Atlantic Water is turned south as it approaches the fjord mouth such that it opposes the coastal current, relaxing the geostrophic control. Nilsen *et al.* (2008) further concluded that the critical parameter determining the depth at which Atlantic water entered the fjord was the difference in density between the Winter Water produced in the fjord through seasonal ice formation and the Atlantic Water entering from the shelf. During years of high sea-ice production and greatest brine release, the Winter Water is denser than the Atlantic water causing it to enter in the intermediate layer. With low ice production, the situation is reversed and the Atlantic Water forms a bottom layer in the fjord.

Another example where geostrophic control limits exchange is Kongsfjorden in NW Spitsbergen (Cottier *et al.* 2005), which (like Isfjorden) has no sill at its entrance. The coastal current across the mouth of the fjord and the associated sloping isopycnals limit the extent to which Atlantic Water that is transported onto the shelf can exchange with the fjord waters. Conditions where the geostrophic control on exchange is relaxed can be initiated through wind forcing. Such a situation occurred during January and February 2007 (Cottier *et al.* 2007). A persistent northerly wind blowing counter to the coastal current and the resulting southward flow eased the geostrophic control at the entrance to the mouth of the fjord. Consequently, Atlantic Water that was transported onto the West Spitsbergen Shelf through upwelling was able to penetrate Kongsfjorden. The same process acted along the whole of West Spitsbergen, flooding the adjacent shelf and fjords with Atlantic Water. During winter 2007, the fjords underwent a transition from having a sea-ice cover and being dominated by cold water masses (termed by Cottier *et al.* 2005 an 'Arctic-type' fjord) to one with no sea ice and dominated by warm water masses (termed an 'Atlantic type' fjord) (Cottier *et al.* 2007; Nilsen *et al.* 2008). Succeeding winters therefore have to transform the Atlantic-influenced water in the fjord back to the normal 'Arctic type' in order to rebuild the density front between the shelf and fjord waters, required for geostrophic control.

Concluding remarks

It is clear that the response of Arctic fjords to meteorological and oceanic forcings differ from mid-latitude fjords in a number of ways, and that it is important to appreciate these differences when interpreting the response and functioning of these fjords. Of particular significance is the intense seasonality of Arctic locations leading to the formation of sea ice and associated brine release; the fjords are therefore capable of enhancing salinity above that found in adjacent oceanic waters. The strong down-fjord winds can enhance this effect of increasing salinity through the formation of latent-heat polynyas. Brine release can give rise to full-depth haline convection causing complete vertical mixing, renewal of bottom waters and vertically homogeneous conditions.

At the other end of the salinity scale, glacial discharge, basal melting and snowmelt during summer give rise to substantial seasonal pulses of freshwater into the surface layers. To some extent, this freshwater can be mixed downwards through wind mixing. However, Arctic fjords are typically characterized by steep vertical salinity gradients; a strong pycnocline with fresh surface layers overlies more saline intermediate waters with layers of dense brine-enriched water at the bottom. The delineation between layers by a strong pycnocline makes the surface layer of high-latitude fjords strongly responsive to wind forcing and the resultant development of strong cross-fjord thermohaline and pressure gradients.

Sources of oceanic water into the fjords tend to be relatively warm and saline, having originated as Atlantic Water although usually cooled and freshened through mixing with shelf waters or basin-scale modification. The biggest controls on the inflow of oceanic waters are wind conditions on the adjacent shelf and the prevailing stratification within the fjord. Hydrographically, entry of oceanic waters into Arctic fjords constitutes a huge increase in the heat content of the system and is therefore a critical aspect of the role of the fjord in communication between ocean and cryosphere. Biologically, entry of oceanic waters can lead to a substantial shift in species composition, elements of which can be recorded in the sedimentary record of the fjord. Consequently, the continued research into the relationship between contemporary hydrography, fjord processes and sedimentary archives will have great value in our understanding of environmental change.

Prof. Harald Svendsen is a co-author of this paper, but sadly died before it was published. We feel privileged to share co-authorship of this paper with Harald who was a mentor, field companion and friend to us all. I think we can all picture Harald deploying a CTD under the midnight sun against a backdrop of the Spitsbergen coast. Without Harald's enthusiasm and leadership, research into the physical oceanography of Arctic fjords would not be as advanced or have the status it enjoys now.

The authors would like to thank all colleagues and crews that assisted with the fieldwork necessary for us to have gained such an intimate knowledge of Arctic fjords over the years. The contribution by FC was supported through the European Regional Development Fund, under the Addressing Research Capacity (in the 26 Highlands and Islands) project. It forms a contribution to the NERC-funded project NE/E001440/1 and NERC-funded Oceans 2025 Theme 10. It is also an output to the Research Council of Norway-funded project MariClim 165112/S30 and an output of the Scottish Alliance for Geoscience, Environment and Society.

References

Aliani, S., Bartholini, G. *et al.* 2004. Multidisciplinary investigations in the marine environment of the inner Kongsfiord, Svalbard islands (September 2000 and 2001). *Chemistry and Ecology*, **20**, 19–28.

Ambrose, W. G., Carrol, M. L., Greenacre, M., Thorrold, S. R. & McMahon, K. W. 2006. Variation in *Serripes groenlandicus* (Bivalvia) growth in a Norwegian high-Arctic fjord: evidence for local- and large-scale climatic forcing. *Global Change Biology*, **12**, 1595–1607, doi: 1510.1111/j.1365-2486.2006.01181.x.

Argentini, S., Viola, A. P., Mastrantonio, G., Maurizi, A., Georgiadis, T. & Nardino, M. 2003. Characteristics of the boundary layer at Ny-Alesund in the Arctic during the ARTIST field experiment. *Annals of Geophysics*, **46**, 185–196.

Arneborg, L. & Liljebladh, B. 2001. The internal seiches in Gullmar Fjord. Part I: dynamics. *Journal of Physical Oceanography*, **31**, 2549–2566.

Azetsu-Scott, K. & Tan, F. C. 1997. Oxygen isotope studies from Iceland to an East Greenland Fjord: behaviour of glacial meltwater plume. *Marine Chemistry*, **56**, 239–251.

Azetsu-Scott, K. & Syvitski, J. 1999. Influence of melting icebergs on distribution, characteristics and transport of marine particles in an East Greenland fjord. *Journal of Geophysical Research – Oceans*, **104**, 5321–5328.

Basedow, S. L., Eiane, K., Tverberg, V. & Spindler, M. 2004. Advection of zooplankton in an Arctic fjord (Kongsfjord, Svalbard). *Estuarine Coastal Shelf Science*, **60**, 113–124.

Beine, H. J., Argentini, S., Maurizi, A., Mastrantonio, G. & Viola, A. 2001. The local wind field at Ny-Alesund and the Zeppelin mountain at Svalbard. *Meteorology and Atmospheric Physics*, **78**, 107–113.

Carmack, E. C. & Chapman, D. C. 2003. Wind-driven shelf/basin exchange on an Arctic shelf: the joint roles of ice cover extent and shelf-break bathymetry. *Geophysics Research Letters*, **30**, 1778, doi: 1710.1029/2003GL017526.

Cottier, F. R., Tverberg, V., Inall, M. E., Svendsen, H., Nilsen, F. & Griffiths, C. 2005. Water mass modification in an Arctic fjord through cross-shelf exchange: the seasonal hydrography of Kongsfjorden, Svalbard. *Journal of Geophysical Research – Oceans*, **110**.

Cottier, F. R., Nilsen, F., Inall, M. E., Gerland, S., Tverberg, V. & Svendsen, H. 2007. Wintertime warming of an Arctic shelf in response to large-scale atmospheric circulation. *Geophysical Research Letters*, **34**, L10607.

Cushman-Roisin, B., Asplin, L. & Svendsen, H. 1994. Upwelling in broad fjords. *Continental Shelf Research*, **14**, 1701–1721.

Daae, R. L. 2008. *Long Periodic Vorticity Waves in Kongsfjorden*. University of Bergen, Bergen.

Dmitrenko, I. A., Polyakov, I. V. *et al.* 2006. Seasonal variability of Atlantic water on the continental slope of the Laptev Sea during 2002–2004. *Earth and Planetary Science Letters*, **244**, 735–743.

Dowdeswell, J. & Hambrey, M. 2002. *Islands of the Arctic*. Cambridge University Press, Cambridge, 280.

Farmer, D. & Freeland, H. 1983. The physical oceanography of fjords. *Journal of Physical Oceanography*, **12**, 147–220.

Farmer, D. M. & Armi, L. 1986. Maximum two-layer exchange over a sill and through a combination of a sill and contraction with barotropic flow. *Journal of Fluid Mechanics*, **164**, 53–76.

Fer, I. 2006. Scaling turbulent dissipation in an Arctic fjord. *Deep-Sea Research II*, **53**, 77–95.

Fer, I. & Widell, K. 2007. Early spring turbulent mixing in an ice-covered Arctic fjord during transition to melting. *Continental Shelf Research*, **27**, 1980–1999.

Fer, I., Skogseth, R., Haugan, P. M. & Jaccard, P. 2003. Observations of the Storfjorden overflow. *Deep-Sea Research I*, **50**, 1283–1303.

Fer, I., Skogseth, R. & Haugan, P. M. 2004. Mixing of the Storfjorden overflow (Svalbard Archipelago) inferred from density overturns. *Journal of Geophysical Research – Oceans*, **109**, C01005.

Gerland, S. & Renner, A. 2007. Sea ice mass balance monitoring in an Arctic fjord. *Annals of Glaciology*, **46**, 435–442.

Gilbert, R., Nielson, N., Desloges, J. & Rasch, M. 1998. Contrasting glacimarine sedimentary environments of two arctic fjords on Disko, West Greenland. *Marine Geology*, **147**, 63–83.

Haarpaintner, J., Gascard, J.-C. & Haugan, P. M. 2001. Ice production and brine formation in Storfjorden, Svalbard. *Journal of Geophysical Research – Oceans*, **106**, 14 001–14 013.

Hanna, E., Cappelen, J., Fettweis, X., Huybrecht, P., Luckman, A. & Ribergaard, M. H. 2009. Hydrologic response of the Greenland ice sheet: the role of oceanographic warming. *Hydrological Processes*, **23**, 7–30.

Holland, D., Thomas, R., De Young, B., Ribergaard, M. & Lyberth, B. 2008. Acceleration of Jakobshavn Isbræ triggered by warm subsurface ocean waters. *Nature Geosciences*, **1**, 659–664.

Howe, J. A., Moreton, S. G., Morri, C. & Morris, P. 2003. Multibeam bathymetry and the depositional environment of Kongsfjorden and Krossfjorden, western Spitsbergen, Svalbard. *Polar Research*, **22**, 301–316.

Howe, J. A., Harland, R. *et al.* 2010. Dinoflagellate cysts as proxies for palaeoceanographic conditions in Arctic fjords. *In*: Howe, J. A., Austin, W. E. N., Forwick, M. & Paetzel, M. (eds) *Fjord Systems and Archives*. Geological Society, London, Special Publications, **344**, 61–74.

Inall, M. E. & Gillibrand, P. 2010. The physics of mid-latitude fjords: a review. *In*: Howe, J. A., Austin, W. E. N., Forwick, M. & Paetzel, M. (eds) *Fjord Systems and Archives*. Geological Society, London, Special Publications, **344**, 17–33.

Ingvaldsen, R., Reitan, M. B., Svendsen, H. & Asplin, L. 2001. The upper layer circulation in the Kongsfjorden and Krossfjorden – a complex fjord system on the west coast of Spitsbergen. *Memoirs of National Institute of Polar Research, Special Issue*, **54**, 393–407.

Kikuchi, T., Hatakeyama, K. & Morison, J. H. 2004. Distribution of convective lower Halocline Water in the eastern Arctic Ocean. *Journal of Geophysical Research – Oceans*, **109**, C12030.

Klinck, J. M., O'Brien, J. J. & Svendsen, H. 1981. A simple model of fjord and coastal circulation interaction. *Journal of Physical Oceanography*, **11**, 1612–1626.

Lemmin, U., Mortimer, C. H. & Bauerle, E. 2005. Internal seiche dynamics in Lake Geneva. *Limnology and Oceanography*, **50**, 207–216.

Lewis, E. L. & Perkin, R. G. 1982. Seasonal mixing processes in an Arctic fjord system. *Journal of Physical Oceanography*, **12**, 74–83.

MacLachlan, S. E., Cottier, F. R., Austin, W. E. N. & Howe, J. A. 2007. The salinity: $\delta^{18}O$ water relationship in Kongsfjorden, western Spitsbergen. *Polar Research*, **26**, 160–167.

Majewski, W., Szczucinski, W. & Zajaczkowski, M. 2009. Interactions of Arctic and Atlantic water-masses and associated environmental changes during the last millennium, Hornsund (SW Svalbard). *Boreas*, **38**, 529–544.

Martin, S. & Cavalieri, D. J. 1989. Contributions of the Siberian shelf polynyas to the Arctic Ocean intermediate and deep-water. *Journal of Geophysical Research – Oceans*, **94**, 12 725–12 738.

Mikkelsen, D. M., Rysgaard, S. & Glud, R. N. 2008. Microalgal composition and primary production in Arctic sea ice: a seasonal study from Kobbefjord (Kangerluarsunnguaq), West Greenland. *Marine Ecology Progress Series*, **368**, 65–74.

Morales Maqueda, M. A., Willmott, A. J. & Biggs, N. R. T. 2004. Polynya dynamics: a review of observations and modelling. *Reviews of Geophysics*, **42**, RG1004.

Nansen, F. 1915. *Spitsbergen waters, oceanographic observations during the cruise of the Veslemoy to Spitsbergen in 1912*. Ventenskapsselskapets Skrifter I. Mat. Naturv. Klasse (2).

Nick, F. M., Vieli, A., Howat, I. M. & Joughin, I. 2009. Large-scale changes in Greenland outlet glacier dynamics triggered at the terminus. *Nature Geosciences*, **2**, 110–114.

Nilsen, F., Gjevik, B. & Schauer, U. 2006. Cooling of the West Spitsbergen Current: isopycnal diffusion by topographic vorticity waves. *Journal of Geophysical Research – Oceans*, **111**, doi: 10.1029/2005JC002991.

Nilsen, F., Cottier, F., Skogseth, R. & Mattsson, S. 2008. Fjord–shelf exchanges controlled by ice and brine production: the interannual variation of Atlantic Water in Isfjorden, Svalbard. *Continental Shelf Research*, **28**, 1838–1853.

Ó Cofaigh, C., Dowdeswell, J. & Grobe, H. 2001. Holocene glacimarine sedimentation, inner Scoresby Sund, East Greenland: the influence of fast-flowing ice-sheet outlet glaciers. *Marine Geology*, **175**, 103–129.

Perkin, R. G. & Lewis, E. L. 1978. Mixing in an Arctic fjord. *Journal of Physical Oceanography*, **8**, 873–880.

Quadfasel, D., Rudels, B. & Kurz, K. 1988. Outflow of dense water from a Svalbard fjord into the Fram Strait. *Deep-Sea Research*, **35**, 1143–1150.

Rasmussen, T. L., Thomsen, E., Ślubowska, M. A., Jessen, S., Solheim, A. & Koç, N. 2006. Paleoceanographic evolution of the SW Svalbard margin (76°N) since 20,000 ^{14}C yr BP. *Quaternary Research*, **67**, 100–114.

Rignot, E., Koppes, M. & Velicogna, I. 2010. Rapid submarine melting of the calving faces of West Greenland glaciers. *Nature Geoscience*, **3**, 187–191.

Rudels, B., Anderson, L. G. & Jones, E. P. 1996. Formation and evolution of the surface mixed layer and halocline of the Arctic Ocean. *Journal of Geophysical Research – Oceans*, **101**, 8807–8821.

Rysgaard, S., Nielsen, T. G. & Hansen, B. 1999. Seasonal variation in nutrients, pelagic primary production and grazing in a high-Arctic coastal marine ecosystem, Young Sound, Northeast Greenland. *Marine Ecology Progress Series*, **179**, 13–25.

Saloranta, T. M. & Svendsen, H. 2001. Across the Arctic front west of Spitsbergen: high-resolution CTD sections from 1998–2000. *Polar Research*, **20**, 177–184.

Schauer, U. & Fahrbach, E. 1999. A dense bottom water plume in the western Barents Sea: downstream modification and interannual variability. *Deep-Sea Research I*, **46**, 2095–2108.

Shcherbina, A. Y., Talley, L. D. & Rudnick, D. L. 2003. Direct observations of North Pacific ventilation: brine rejection in the Okhotsk Sea. *Science*, **302**, 1952–1955.

Skarðhamar, J. & Svendsen, H. 2010. Short-term hydrographic variability in a stratified Arctic fjord. *In*: Howe, J. A., Austin, W. E. N., Forwick, M. & Paetzel, M. (eds) *Fjord Systems and Archives*. Geological Society, London, Special Publications, **344**, 51–60.

Skogseth, R., Haugan, P. M. & Haarpaintner, J. 2004. Ice and brine production in Storfjorden from four winters of satellite and in situ observations and modeling. *Journal of Geophysical Research – Oceans*, **109**, C10008.

Skogseth, R., Fer, I. & Haugan, P. M. 2005*a*. Watermass transformations in Storfjorden. *Continental Shelf Research*, **25**, 667–695.

Skogseth, R., Haugan, P. M. & Jakobsson, M. 2005*b*. Dense-water production and overflow from an arctic coastal polynya in Storfjorden. *In*: Drange, H. et al. (eds) *The Nordic Seas: An Integrated Perspective – Oceanography, Climatology, Biogeochemistry, and Modeling*. American Geophysical Union, Washington, DC, 73–88.

Skogseth, R., Sandvik, A. D. & Asplin, L. 2007. Wind and tidal forcing on the meso-scale circulation in Storfjorden, Svalbard. *Continental Shelf Research*, **27**, 208–227.

Skogseth, R., Smedsrud, L. H., Nilsen, F. & Fer, I. 2008. Observations of hydrography and downflow of brine-enriched shelf water in the Storfjorden polynya, Svalbard. *Journal of Geophysical Research – Oceans*, **113**, doi: 10.1029/2007jc004452.

Smith, W. O. Jr & Barber, D. G. 2007. *Polynyas: Windows to the World*. Elsevier, Amsterdam.

Steele, M. & Boyd, T. J. 1998. Retreat of the cold halocline layer in the Arctic Ocean. *Journal of Geophysical Research – Oceans*, **103**, 10 419–10 435.

Stigebrandt, A. 1976. Vertical diffusion driven by internal waves in a sill fjord. *Journal of Physical Oceanography*, **6**, 486–495.

Stigebrandt, A. 2001. Fiord circulation. *In*: Steele, J., Turekian, K. & Thorpe, S. A. (eds) *Encyclopedia of Ocean Sciences*. Academic Press, 897–902.

Stigebrandt, A. & Aure, J. 1989. Vertical mixing in basin waters of fjords. *Journal of Physical Oceanography*, **19**, 917–926.

Straneo, F., Hamilton, G. S. et al. 2010. Rapid circulation of warm subtropical waters in a major glacial fjord in East Greenland. *Nature Geoscience*, **3**, 182–186.

Svendsen, H., Beszczynska-Møller, A. et al. 2002. The physical environment of Kongsfjorden-Krossfjorden, an Arctic fjord system in Svalbard. *Polar Research*, **21**, 133–166.

Tverberg, V. & Nøst, O. A. 2009. Eddy overturning across a shelf edge front: Kongsfjorden, west Spitsbergen. *Journal of Geophysical Research – Oceans*, **114**, C04024.

Walker, D., Brandon, M. A., Jenkins, A., Allen, J., Dowdeswell, J. & Evans, J. 2007. Oceanic heat transport onto the Amundsen Sea shelf through a submarine glacial trough. *Geophysical Research Letters*, **34**, L02602.

Wallace, M., Cottier, F., Brierley, A. S., Berge, J., Tarling, G. A. & Griffiths, C. 2010. Impact of sea ice on zooplankton vertical migration in the Arctic. *Limnology and Oceanography*, **55**, 831–845.

Widell, K., Fer, I. & Haugan, P. M. 2006. Salt release from warming sea ice. *Geophysical Research Letters*, **33**, doi: 10.1029/2006GL026262.

Willis, K. J., Cottier, F. R., Kwasniewski, S., Wold, A. & Falk-Petersen, S. 2006. The influence of advection on zooplankton community composition in an Arctic fjord (Kongsfjorden, Svalbard). *Journal of Marine Systems*, **61**, 39–54.

WILLIS, K. J., COTTIER, F. & KWASNIEWSKI, S. 2008. Impact of warm water advection on the winter zooplankton community in an Arctic fjord. *Polar Biology*, **31**, 475–481.

WMO 1970. *WMO Sea Ice Nomenclature*. World Meteorological Organisation, Geneva.

YAMAMOTO, M., WATANABE, S., TSUNOGAI, S. & WAKATSUCHI, M. 2002. Effects of sea ice formation and diapycnal mixing on the Okhotsk Sea intermediate water clarified with oxygen isotopes. *Deep-Sea Research I*, **49**, 1165–1174.

Short-term hydrographic variability in a stratified Arctic fjord

J. SKARÐHAMAR[1]* & H. SVENDSEN[2†]

[1]*Institute of Marine Research, N-9294 Tromsø, Norway*

[2]*Formerly at: Geophysical Institute, University of Bergen, N-5007 Bergen, Norway*

[†]*Deceased*

**Corresponding author (e-mail: jofrid.skardhamar@imr.no)*

Abstract: Fjords in the Arctic often have a more complex circulation pattern than the classical two-dimensional estuarine circulation. This is due to the effects of the Earth's rotation on stratified waters in wide fjords. Observations from a semi-enclosed fjord basin, Van Mijenfjorden on Spitsbergen, show that the hydrography and circulation vary considerably on short timescales (hours) in the summer season. The depth and distribution of the low salinity upper water layer respond quickly to changes in the wind field. The Coriolis effect has an essential impact on the circulation, inducing eddy-like flow patterns and strong cross-fjord gradients. Within the upper layer, the lowest salinity values and highest temperatures were found on the northern side of the fjord in calm wind periods. When the wind was strong from the west, the cross-fjord gradients were reversed. Internal wave activity contributes to large vertical displacement of water below the upper layer. Knowledge of such strongly variable hydrographic conditions in fjords are important for sampling strategy and interpretation of data, for instance of primary production and sedimentation processes, and for the understanding of fjords as depositional systems.

Sill fjords normally have distinct vertical stratification with an approximate three-layer structure; an upper layer with low salinity above a more saline intermediate layer, and high salinity basin water below the sill depth. Large seasonal variations in air temperature and freshwater discharge lead to significant variations in both the stratification and circulation pattern. In Arctic fjords, the described stratification structure appears only in the summer season (see Svendsen *et al.* 2002); in winter, the water masses are overturned due to cooling. In general the circulation in fjords is forced by a combination of external forces such as wind, freshwater discharge and tides. The motions are modified by topography and friction and, in wide stratified fjords, rotational dynamics (Coriolis effect) may have an important impact on the fjord dynamics. The effect of the Coriolis force depends on the stratification; the impact of the Earth's rotation will therefore vary both seasonally and locally within a fjord. For details of physical processes in fjords, see for example Farmer & Freeland (1983), Svendsen (1986), Cottier *et al.* (2010) and Inall & Gillibrand (2010).

The physical oceanography of most fjords on Spitsbergen is poorly investigated. Marine biological and geological investigations have been carried out in Van Mijenfjorden and several other fjords in the area, but contemporary studies of the physics of the fjords have often been limited to a few CTD stations which is insufficient to give information about complex circulation and exchange patterns. Three exceptions are the fjords Kongsfjorden, Isfjorden and Hornsund, which have been subjects to several oceanographic research projects during the last decade (e.g. Ingvaldsen *et al.* 2001; Svendsen *et al.* 2002; Cottier *et al.* 2005; Nilsen *et al.* 2008; Tverberg & Nøst 2009). Van Mijenfjorden differs from the other Spitsbergen fjords by its mouth being nearly closed by an island, restricting the water exchange between the fjord basin and coastal water masses. This fjord is therefore a good 'laboratory' fjord for process studies; see for example Widell *et al.* (2006) and Fer & Widell (2007) who studied turbulence due to ice freezing. The present work focuses on the effects of wind and the short-term variability of the fjord circulation and hydrography. The investigation was based on field data from a summer season, that is, without ice cover. The semi-enclosed nature of Van Mijenfjorden makes wind effects pronounced and easily distinguishable relative to fjord–ocean exchange processes. However, the wind effects described here are also relevant to other Arctic and sub-Arctic fjords.

Materials and methods

Study area

Van Mijenfjorden is a 50 km long sill fjord at the west coast of Spitsbergen (Fig. 1). The mean width of the fjord is 10 km, and the surface area covers 515 km^2 (Schei *et al.* 1979). The island Akseløya lies across the mouth of the fjord, leaving

From: HOWE, J. A., AUSTIN, W. E. N., FORWICK, M. & PAETZEL, M. (eds) *Fjord Systems and Archives*. Geological Society, London, Special Publications, **344**, 51–60.
DOI: 10.1144/SP344.5 0305-8719/10/$15.00

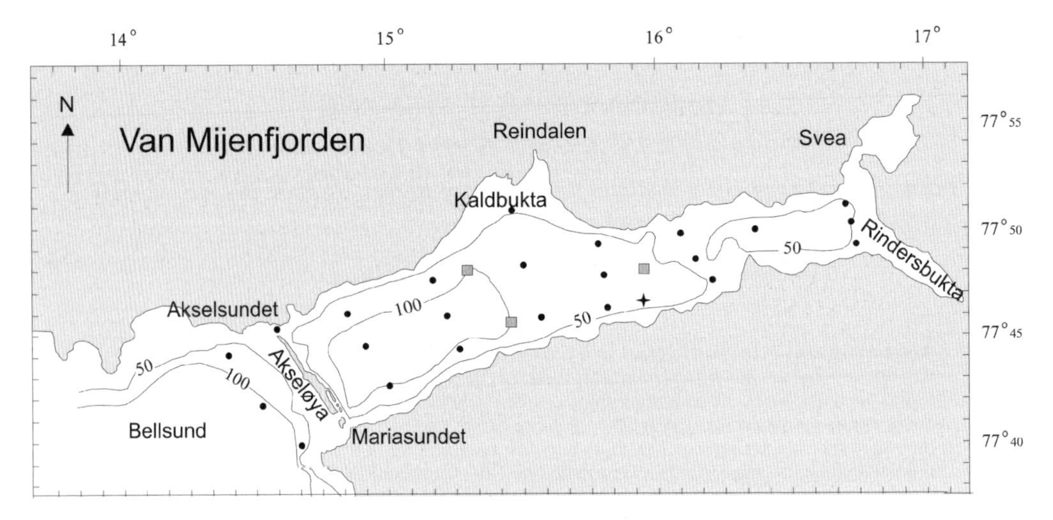

Fig. 1. Map of Van Mijenfjorden with depth contours (m) and positions of CTD stations (dots), time series station (star) and current meter moorings (squares).

two narrow sounds where the water exchange between the fjord and coast takes place. The sound Akselsundet, on the northern side of Akseløya, is 1 km wide with a sill depth of 34 m (Fer & Widell 2007). The sound on the southern side, Mariasundet, is intersected by a small islet, leaving a 600 m wide and 2 m deep passage to the north of the islet and a 500 m wide and 12 m deep passage to the south. The majority of the water exchange takes place through Akselsundet, where the tidal currents are strong with current speed up to 3 m s^{-1} (Norwegian Hydrographic Service 1990). Tidal choking (Stigebrandt 1980) can create tidal jets through the sounds during flood and is an important driving force for the mean circulation in the fjord (Bergh 2004).

Van Mijenfjorden consists of two basins. The outer basin is 115 m deep and is separated from the 74 m deep inner basin by a 45 m deep sill, which is the remains of the moraine after the major surge of the glacier Paulabreen 600–250 years ago (Rowan *et al.* 1982; Hald *et al.* 2001). Van Mijenfjorden is surrounded by tall mountains (800–1200 m) and broad valleys, including one of the largest ice-free valleys on Svalbard: Reindalen. Two glaciers calve in the fjord: Fridtjovbreen in Fridtjovhavna near the fjord mouth and Akselsundet and Paulabreen in Rindersbukta at the head of the fjord.

Observations of wind at Svea in the inner part of the fjord show that the prevailing winds are from NE, that is, katabatic down-fjord wind except in the summer season when up-fjord winds occur nearly as frequently as down-fjord winds (Hanssen-Bauer *et al.* 1990). The climate of west Spitsbergen is relatively dry. The total precipitation varies between 180 and 440 mm a year (Hanssen-Bauer *et al.* 1990), at a minimum in April–June and maxima in August, February and March.

The fjords of west Spitsbergen are normally covered by ice from December to May/June. Van Mijenfjorden is ice-covered for a longer period than the other fjords, because of the protecting effects of Akseløya. Some years the fjord freezes up as early as September, and the fjord is normally not navigable by boat until the beginning of June (Norwegian Hydrographic Service 1990). Little drift ice enters the fjord, so the fjord ice mainly consists of locally frozen fjord water and ice from the glaciers. The sounds Akselsundet and Mariasundet are ice free all winter due to the strong tidal currents.

Field data

The field data were collected during a cruise with RV *Håkon Mosby* in the period from 28 July to 3 August 1996. Repeated hydrographic mapping of Van Mijenfjorden was performed with a SeaBird CTD covering a dense station net of 23 stations (Fig. 1) three times: 28 July, 31 July and 2 August. One of the CTD stations was repeated every hour 22 times from 31 July to 1 August. The CTD was calibrated at the Institute of Marine Research, Bergen, in accordance with the ICES procedure prior to the cruise. The data were averaged over depth intervals of 2 m.

Current measurements were obtained using Aanderaa current meters RCM4 and RCM7 deployed on three moorings; see Figure 1 and Table 1 for positions and depths. The moorings were deployed for a period of 6 days. The measuring interval was 10 min, and the measurements were averaged over a 40 hour Butterworth low-pass filter to

Table 1. *Positions and measuring depth of the current meter moorings in Van Mijenfjorden 28 July–3 August 1996*

Mooring	Position	Measuring depths
Innermost	77°48.00 N, 15°56.4 E	2, 30 and 60 m
Northern	77°48.46 N, 15°15.02 E	2, 10, 30 and 70 m
Southern	77°46.11 N, 15°25.38 E	2, 10, 30 and 70 m

remove the tidal effects (see e.g. Emery & Thomson 1997).

An automatic weather station was installed on the innermost mooring, measuring wind speed and wind direction. In addition, wind speed, wind direction and air temperature were recorded from the weather station on board the ship at every CTD station. Meteorological data were also available from the Norwegian Meteorological Institute's weather station at Svea.

Results

Hydrography

The vertical salinity and temperature distribution revealed a stably stratified fjord with strong vertical salinity and temperature gradients in the upper 10–15 m (Fig. 2). Below this, the salinity increased more slowly with depth from 32 just below the pycnocline to 34 at 80 m in the outer basin and 50 m in the inner basin (Fig. 2a). The temperature decreased from 1 °C just below the pycnocline to -1 °C at 80–90 m in the outer basin and at 50 m in the inner basin (Fig. 2b). The vertical thickness and horizontal distribution of the low salinity layer (salinity <31) varied considerably between the three surveys and between the inner and outer basin. The low salinity layer had a more even distribution along the longitudinal axis of the fjord during the first and last surveys than during the second survey. During the second survey the pycnocline depth was shallower in the outer basin (Fig. 3, lower panels) and deeper in the inner basin compared to the other two surveys (Fig. 3, upper panels). The horizontal across-fjord gradients of salinity and temperature also varied during the field campaign. In general, higher temperatures and lower salinities were measured on the northern side of the fjord compared to the southern side on the first and third survey, with horizontal salinity gradients between 0.4 and 0.7 km^{-1} at 4 m depth in the main basin. On the second survey, the gradients were reversed with the lowest salinities and highest temperatures along the southern side of the fjord (Fig. 4) and horizontal salinity gradients between -0.3 and -0.5 km^{-1}.

The data from the time series station show a large vertical excursion of water properties below the pycnocline. Within four hours, the isoline for salinity 33 ascends from 35 m depth to 15 m depth, before abruptly sinking back down again (Fig. 5).

Currents

The current meter measurements showed strongest current speed near the surface (measured at 2 m depth) and weakest current speeds at depth 60–70 m at all three moorings (Fig. 6). Variable current directions were recorded at all depths at the inner mooring (Fig. 6a). On the northern side of the fjord, the currents were directed out of the fjord at all depths almost throughout the period (Fig. 6b). On the southern side of the fjord the current direction varied between into and out of the fjord at 2 m depth and also 70 m depth, while the currents were constantly directed inwards at depths 10 m and 30 m (Fig. 6c).

Meteorological observations

The air temperature varied between 3.5 °C and 6 °C during the cruise. The wind was blowing with a westerly component (towards the head of the fjord) during the whole cruise period (Fig. 6a), and the wind speed varied between 0.3 m s^{-1} (calm) and 12 m s^{-1}. Weak wind was recorded the first days of the cruise, with wind speed less than 5 m s^{-1} the first day (28 July) followed by one day of calm wind (29 July). The wind speed increased to 6–10 m s^{-1} in the afternoon 30 July, lasting for two days before dropping to less than 5 m s^{-1} 1–2 August. The mean wind speeds and directions for the three CTD-surveys are given in Table 2.

Discussion

The observed horizontal salinity and temperature gradients and their variability within the fjord can be explained by three main factors: the positions of the largest river mouths, the Coriolis effect and wind. Two large valleys have outlets to the northern side of the fjord, and supply large volumes of freshwater on that side of the fjord. In addition, freshwater from Fridthjovbreen is discharged on the northern side, near the fjord mouth. The valleys on the southern side of the fjord are smaller and probably have less freshwater

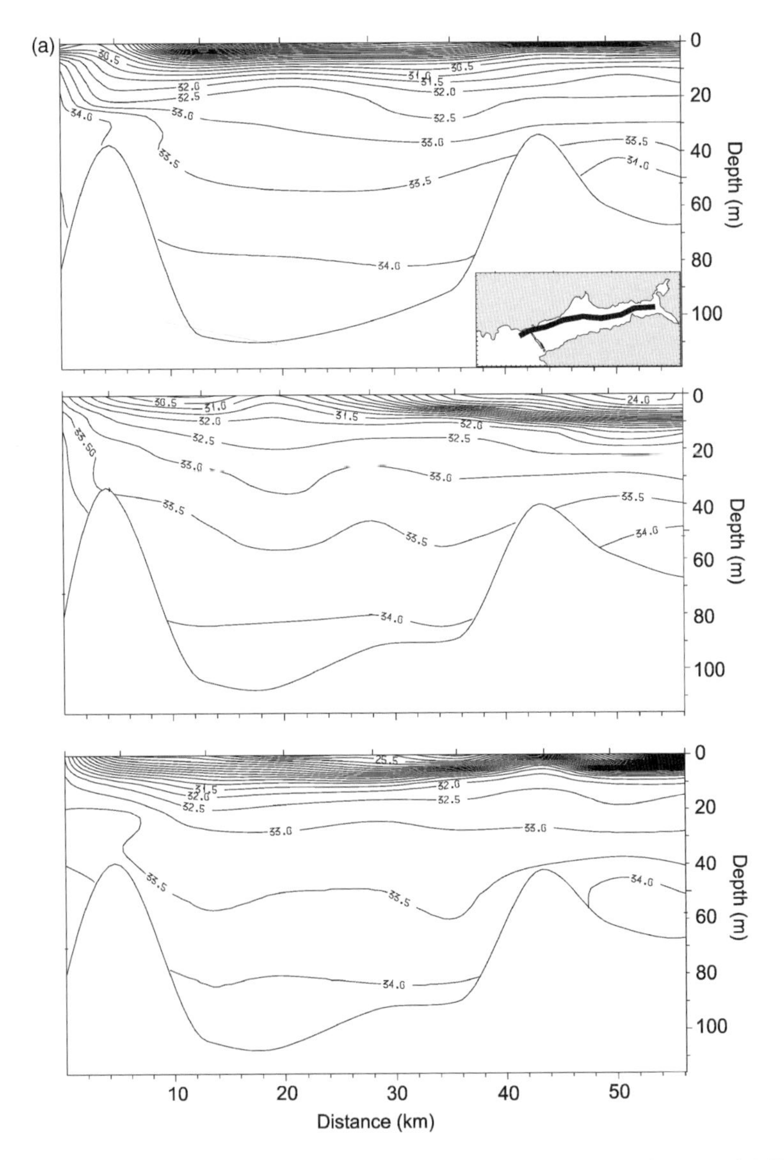

Fig. 2. (**a**) Vertical along-fjord sections of salinity distributions from the first (upper panels), second (middle panels) and third (lower panels) surveys. Seen from south (west is to the left in the figures).

discharge. This alone could explain the lower salinity on the northern side of the fjord during calm wind conditions.

The dynamic effect of the Earth's rotation on fjords depends on the width of the fjord and the vertical stratification (see e.g. Cushman-Roisin *et al.* 1994). If we regard Van Mijenfjorden as a two-layer system with a 7 m thick upper layer of water density 1020 kg m^{-3} and 80 m thick deep layer of density 1026 kg m^{-3} (based on the present field measurements from the outer basin), the baroclinic Rossby radius is 4.3 km. This radius is less than half of the fjord width in the outer basin, and consequently the motions here are strongly affected by the Earth's rotation. Similar considerations for the inner basin, with a 10 m thick upper layer of water density 1019 kg m^{-3} and a 40 m thick deep layer of density 1026 kg m^{-3} (measurements from the inner basin) give a Rossby radius of 5.2 km, which corresponds to the fjord width in this area. The Coriolis force therefore also affects the circulation in the inner basin, but to a lesser degree than the outer part of the fjord. The Earth's rotation deflects the outward-flowing surface current containing

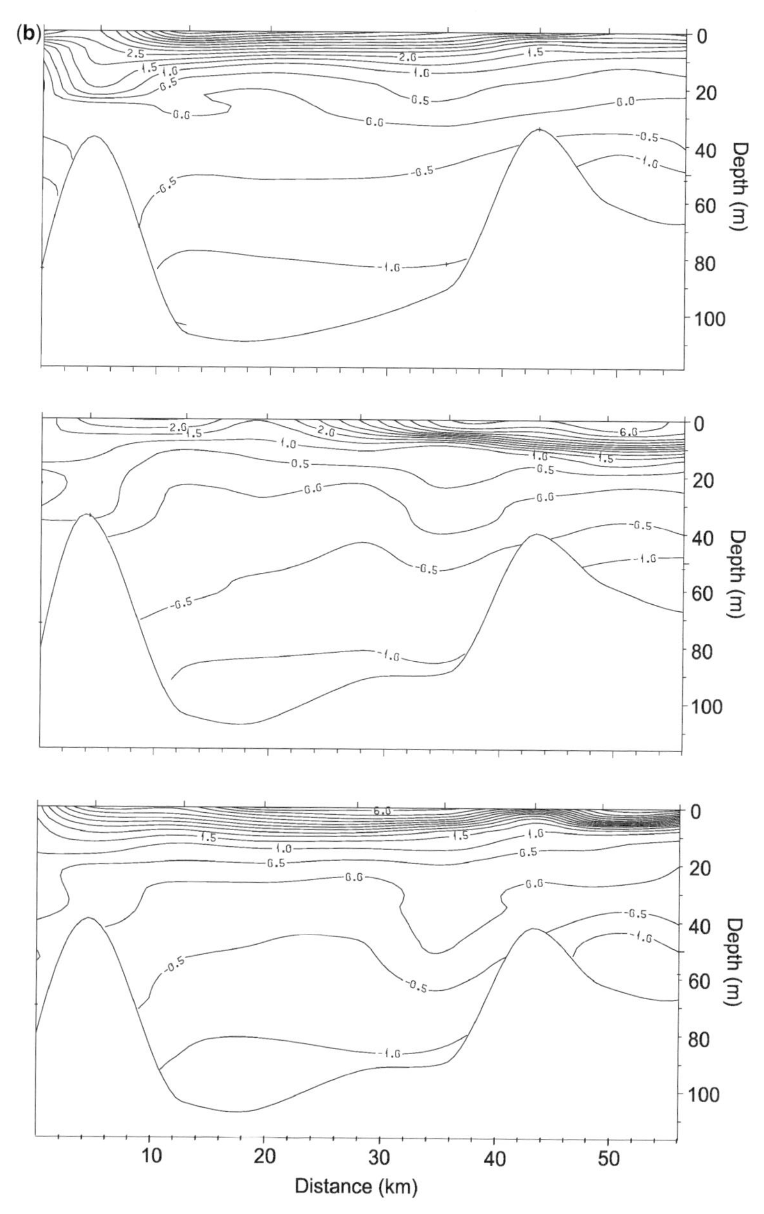

Fig. 2. (**b**) Vertical along-fjord sections of temperature distributions from the first (upper panels), second (middle panels) and third (lower panels) surveys. Seen from south (west is to the left in the figures).

low-salinity water to the right, thus following the northern shore towards the mouth. Consequently, the river water discharged on the northern side of the fjord will not spread evenly over the fjord's surface, but be guided along the northern coast towards the mouth of the fjord. This effect contributes to the maintenance of a salinity gradient across the fjord, with the lowest salinities on the northern side of the fjord as observed.

Wind-driven Ekman transport will also affect the surface salinity and temperature distribution. Easterly winds amplify the estuarine circulation with the out-flowing low-salinity water along the northern side of the fjord, while westerly wind may counteract this circulation. The fjord was surveyed with CTD measurements three times during the cruise. During the first and third survey the wind was weak with mean wind speed 1 m s^{-1}

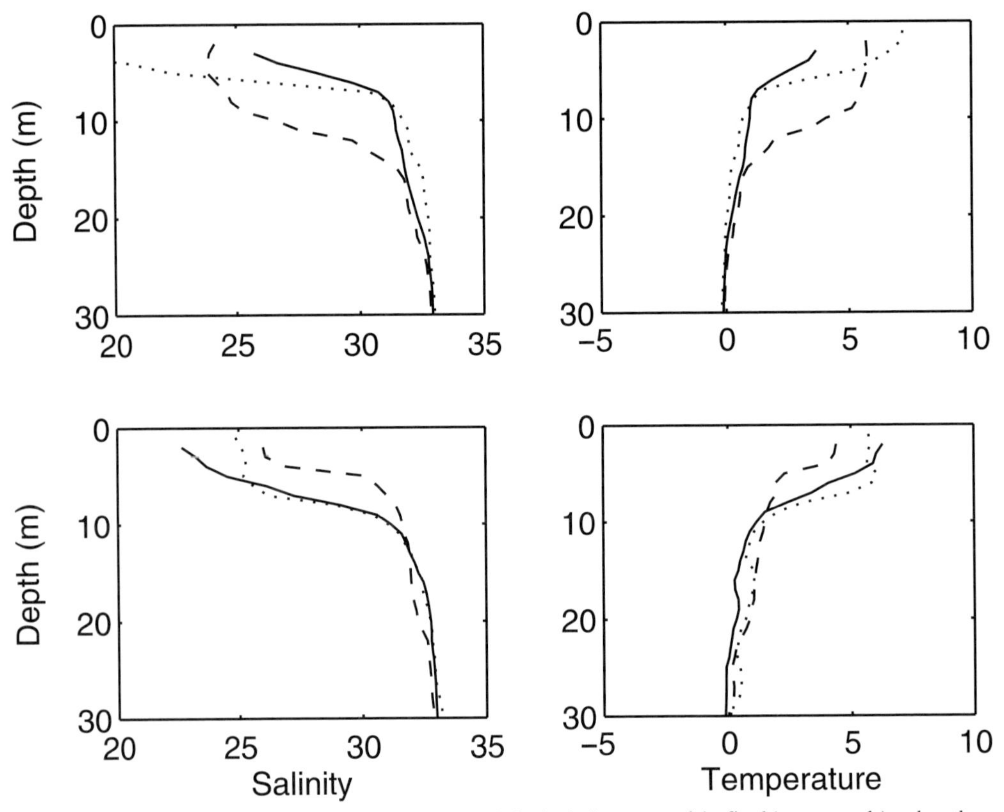

Fig. 3. Vertical profiles of salinity (left) and temperature (right) in the inner part of the fjord (upper panels) and northern side of the outer part of the fjord (lower panels) from the first (solid line), second (broken line) and third (dotted line) surveys.

(Table 2), while stronger westerly wind (7 m s^{-1}) was blowing during the second survey. The different wind conditions were clearly reflected in the horizontal temperature and salinity gradients across the fjord. The surface salinity and temperature gradients across the fjord were reversed on the day of the strongest westerly (up-fjord) winds (31 July) compared to the other two surveys. The stratification was reduced, and the horizontal gradients were evident at larger depths when the wind was strong, indicating deeper vertical mixing. The up-fjord wind forces the 'warm' and low-salinity surface water to the south; upwelling of colder and more saline water reaches the surface along the northern side of the fjord. This fjord response is in accordance with the theory of estuarine circulation in broad fjords of Cushman-Roisin *et al.* (1994).

The current metre data and wind measurements from the innermost mooring show a relationship between westerly winds and current direction towards the east (up-fjord) at all depths. The near-surface current turned 180° when the wind ceased, driven by a down-fjord pressure gradient established during the period of strong up-fjord wind pushing water to the fjord head. The near-surface current direction at the southernmost mooring varied similarly with the wind direction at depth 2 m. The other current meters did not reveal such clear relationships with changing wind. The circulation in the fjord is therefore a complex result of combined effects of estuarine circulation, rotational dynamics and wind effects.

Our time series are too short to detect long period internal waves such as Kelvin waves. However, we have observed large vertical displacement of water masses within the deeper layer, indicating internal wave activity in the fjord. The Ekman transport and piling up of water against the southern shore during up-fjord winds and against the northern shore during down-fjord winds causes, as described above, disturbances of the upper layer thickness. Theoretically, the distortion of the interface may travel as a Kelvin wave with the shore on the right-hand side, looking in the direction of its propagation (Asplin 1995; Svendsen 1995). The mixed water in the narrow entrances to the fjord would prevent

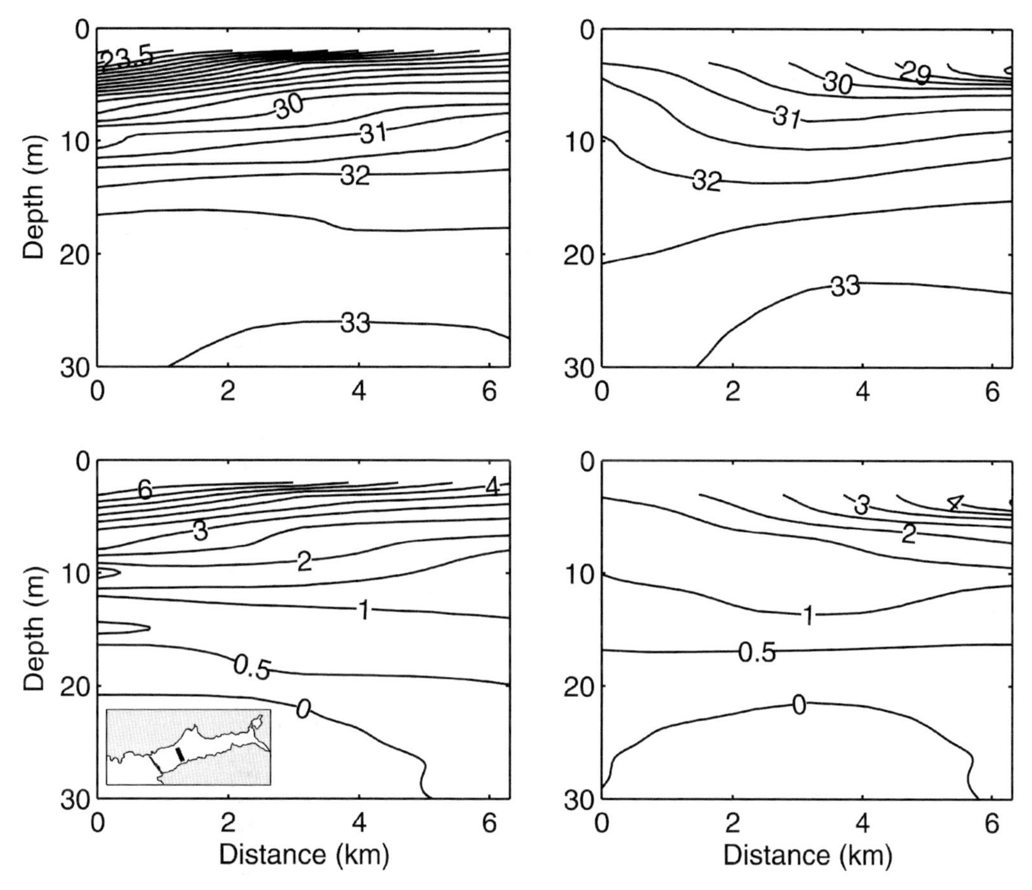

Fig. 4. Distribution of salinity (upper panels) and temperature (lower panels) along a cross-section of the fjord, as measured during the first (left panels) and second (right panels) surveys. Seen from west (north is to the left in the figures).

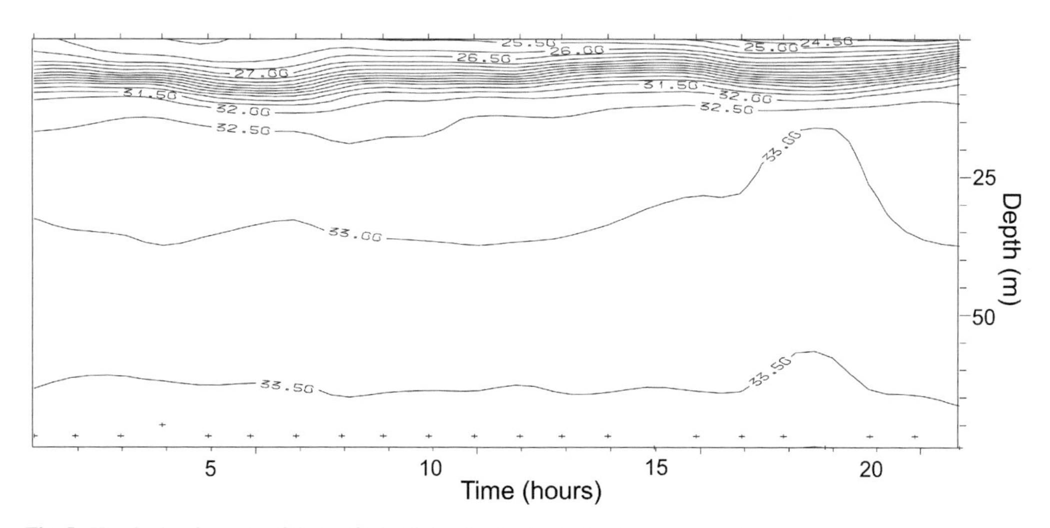

Fig. 5. Hourly development of the vertical salinity distribution at the time-series station.

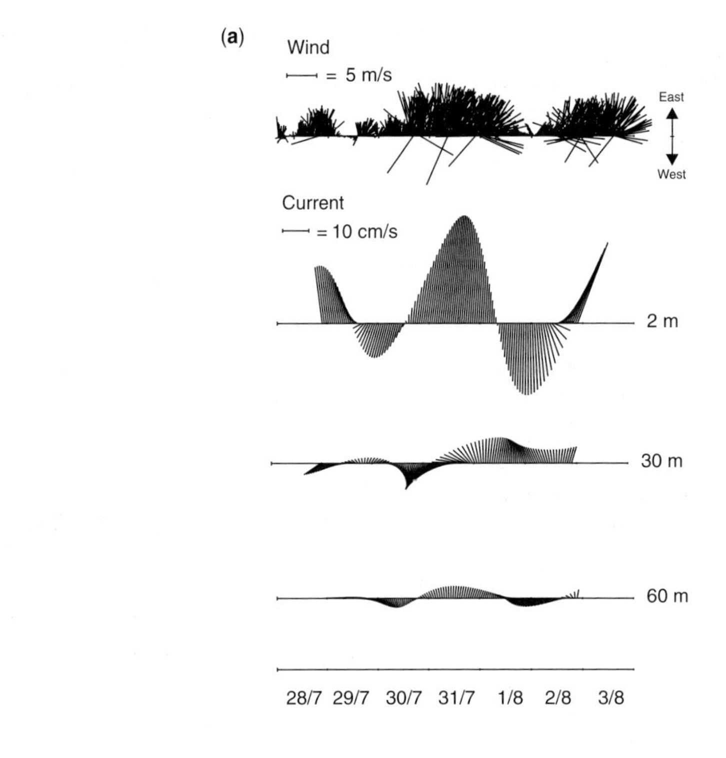

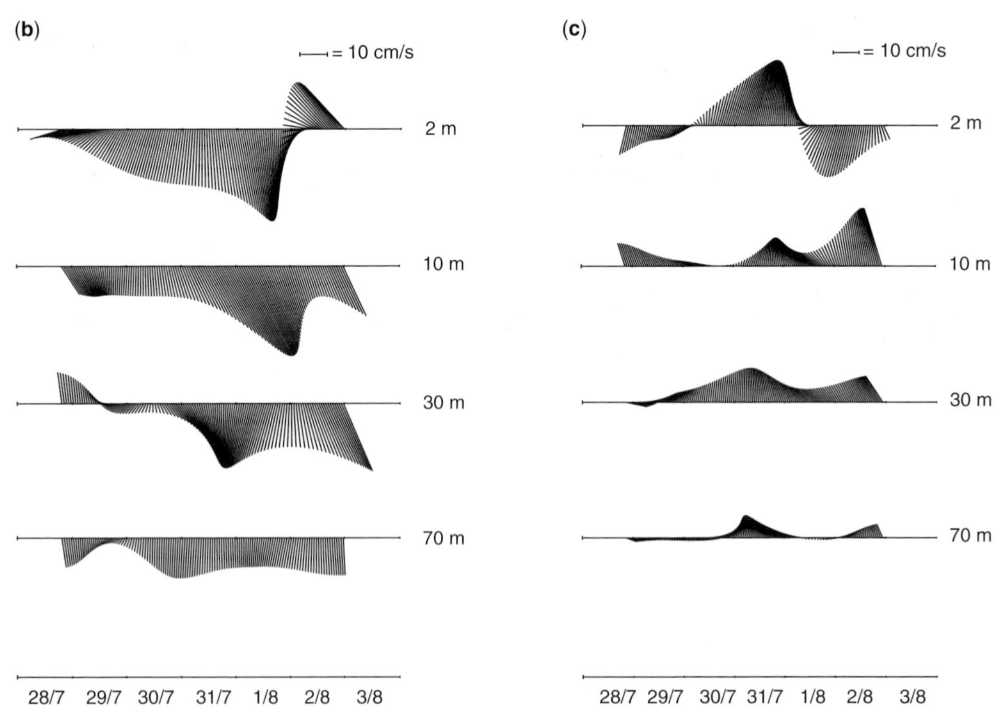

Fig. 6. Measurements of (**a**) wind and currents at the innermost mooring, and current measurements from the moorings on the (**b**) northern and (**c**) southern side of the outer basin. Note that the vertical figure axis is directed east–west, and that all vectors represent the direction towards which the currents and winds are moving.

Table 2. *12-hour mean wind speed and direction for the three periods of CTD surveys*

Survey	Time	Wind speed ($m\ s^{-1}$)	Wind direction
1	27/7 22:00–28/7 10:00	0.93	264
2	31/7 03:00–15:00	6.81	294
3	1/8 17:00–2/8 05:00	1.10	293

Kelvin waves from leaving the fjord; instead, such waves can be guided along the shore around the whole basin. The combined effect of surface elevation and interface displacement could lead to quasi-geostrophically balanced steady-state currents circulating the basin (Asplin 1995). Given the varying wind pattern, and since the currents will persist for some time after a wind event, the flow field at any given time may be related to a superposition of several Kelvin waves circulating the fjord. In order to discuss the influence of the tide on the flow field in the fjord, it is appropriate to estimate the internal Froude number (F_i) for the topographic constriction Akselsundet. F_i is defined as u/c where u is the velocity of the upper layer and c is the phase speed of a long internal wave on the interface between the layers. The phase speed c is defined

$$c = \sqrt{g'h}$$

where g' is the reduced gravity and h is the depth of the upper layer. Using the same two-layer structure as described above, the phase speed of the wave is estimated to be 0.6 $m\ s^{-1}$. Current velocity of the order 2 $m\ s^{-1}$ in the sound (as measured by Bergh 2004) and the phase speed calculated above yields $F_i \gg 1$, that is, supercritical conditions. Supercritical conditions imply that kinetic energy exceeds the potential energy of the field, thus inhibiting the development of wave-like behaviour on the interface. However, the speed of the tidal current varies considerably during the tidal period. Internal flows that are sub-critical may therefore readily occur during a tidal period. This then makes the conditions favourable for internal tides to be generated and appear as 'pulses' travelling in the same direction as the Kelvin waves. Inall *et al.* (2005) found that about 1/3 of the barotropic tidal energy in a sill fjord with supercritical conditions was transformed to internal wave energy. Model simulations by Støylen & Weber (2010) showed that internal Kelvin waves, generated by barotropic tidal pumping in the sounds, can propagate cyclonically around the basin of Van Mijenfjorden. They argue that the associated drift can contribute significantly to the horizontal circulation in the fjord.

The shallow and narrow sounds at the mouth of the fjord and the sill between the two fjord basins prevent free water mass exchange between Bellsund and the fjord and between the two fjord basins. As a consequence, the local freshwater discharge to the fjord strongly affects the salinity distribution within the fjord area in the melting season, when the salinity in the upper layer is markedly lower than that measured in Bellsund. The high salinity and low temperature of the deep water is probably caused by sinking of dense surface water (vertical convection) formed by cooling and brine release during ice freezing in winter. An alternative source of deep water renewal is intrusion of coastal water from Bellsund. However, since the temperature of the deep water was lower than that measured in Bellsund in summer, a possible renewal of the deep water caused by intrusion must have taken place during the preceding winter. The surface water temperatures in the fjord were higher than those measured in Bellsund during our surveys. In addition to direct solar heating of the surface layer, the high temperature can be explained by supply of 'warm' river water that has been warmed up in the shallow river beds and tidal flats on its way to the fjord. Weslawski *et al.* (1991) measured temperatures up to 14 °C in shallow waters over dark sediments in Vestervågen, Bellsund.

Summary and conclusion

Van Mijenfjorden is characterized by short-time variations in current pattern and the horizontal and vertical distribution of temperature and salinity. Wind and the Earth's rotation (Coriolis effect) are the dominating factors determining the pattern and strength of the circulation in the fjord in summer, when a low salinity upper layer is present in the fjord. The major part of the fjord is dominated by a prevailing eddy-like flow pattern, which was reversed several times during our cruise period and related to varying wind strength and direction. Up-fjord westerly wind forces the 'warm' and low-salinity surface water to the south, and upwelling of colder and more saline water reaches the surface along the northern side of the fjord. In periods of calm wind, and probably also of down-fjord wind, the lowest salinity water is found along the northern side of the fjord. The alternating circulation, the corresponding changing of cross- and along-fjord gradients in salinity and temperature and the excitation of internal waves in the fjord ensure that it is subject to high-frequency variations of the hydrographic conditions in time and space. In order to interpret data, it is important that researchers from all disciplines sampling in fjords are aware of such strongly variable conditions. In wide fjords the expected main transport pathway of sediment-rich surface water of terrestrial origin is along the right-hand side of the fjord (i.e. following the northern shore for Van Mijenfjorden), due to the Coriolis effect. It is therefore reasonable to assume the highest sedimentation on that side of the fjord. We have shown here that strong up-fjord winds can disturb this pattern by deflecting the brackish water plume towards the opposite shore, thus reversing the cross-fjord hydrographic gradients. Sedimentation is therefore expected to take place on both sides of wide fjords. Cross-fjord differences in sedimentation rates are likely and will depend upon the wind conditions in the fjord.

References

Asplin, L. 1995. Examination of local circulation in a wide, stratified fjord including exchange of water with the adjacent ocean, due to constant upfjord wind. *In*: Skjoldal, H. R., Hopkins, C., Erikstad, K. E. & Leinaas, H. P. (eds) *Ecology of Fjords and Coastal Waters*. Elsevier Science B.V., Amsterdam, 177–184.

Bergh, J. 2004. *Measured and modelled tidally driven mean circulation under ice cover in Van Mijenfjorden*. MSc thesis, Göteborg University.

Cottier, F., Tverberg, V., Inall, M., Svendsen, H., Nilsen, F. & Griffiths, C. 2005. Water mass modification in an Arctic fjord through cross-shelf exchange: the seasonal hydrography of Kongsfjorden, Svalbard. *Journal of Geophysical Research*, **110**, doi: 10.1029/2004JC002757.

Cottier, F., Nilsen, F., Skogseth, R., Tverberg, V., Skarðhamar, J. & Svendsen, H. 2010. Arctic fjords: a review of the oceanographic environment and dominant physical processes. *In*: Howe, J. A., Austin, W. E. N., Forwick, M. & Paetzel, M. (eds) *Fjord Systems and Archives*. Geological Society, London, Special Publications, **344**, 35–49.

Cushman-Roisin, B., Asplin, L. & Svendsen, H. 1994. Upwelling in broad fjords. *Continental Shelf Research*, **14**, 1701–1721.

Emery, W. J. & Thomson, R. E. 1997. *Data Analysis Methods in Physical Oceanography*. 2nd and revised edn. Elsevier Science B.V., Amsterdam.

Farmer, D. M. & Freeland, H. J. 1983. The physical oceanography of fjords. *Progress in Oceanography*, **12**, 147–220.

Fer, I. & Widell, K. 2007. Early spring turbulent mixing in an ice-covered Arctic fjord during transition to melting. *Continental Shelf Research*, **27**, 1980–1999.

Hald, M., Dahlgren, T., Olsen, T. E. & Lebesbye, E. 2001. Late Holocene palaeoceanography in Van Mijenfjorden, Svalbard. *Polar Research*, **20**, 23–35.

Hanssen-Bauer, I., Solås, M. K. & Steffensen, E. L. 1990. *The climate of Spitsbergen*. Norwegian Meteorological Institute, Oslo, Report **39/90**.

Inall, M. E. & Gillibrand, P. A. 2010. The physics of mid-latitude fjords: a review. *In*: Howe, J. A., Austin, W. E. N., Forwick, M. & Paetzel, M. (eds) *Fjord Systems and Archives*. Geological Society, London, Special Publication, **344**,17–33.

Inall, M. E., Rippeth, T. P., Griffiths, C. R. & Wiles, P. 2005. Evolution and distribution of TKE production and dissipation within a stratified flow over topography. *Geophysical Research Letters*, **32**, L08607, doi: 10.1029/2004GL022289.

Ingvaldsen, R., Reitan, M. B., Svendsen, H. & Asplin, L. 2001. The upper layer circulation in Kongsfjorden and Krossfjorden – a complex fjord system on the west coast of Spitsbergen. *Polar Research Special Issue*, **54**, 393–407.

Nilsen, F., Cottier, F., Skogseth, R. & Mattsson, S. 2008. Fjord-shelf exchanges controlled by ice and brine production: the interannual variation of Atlantic Water in Isfjorden, Svalbard. *Continental Shelf Research*, **28**, 1838–1853.

Norwegian Hydrographic Service 1990. *Arctic Pilot, Sailing Directions Travellers' Guide, Svalbard and Jan Mayen*. **7**, 2nd edn. The Norwegian Hydrographic Service and Norwegian Polar Institute, Stavanger, Norway.

Rowan, D. E., Péwé, T. L., Péwé, R. H. & Stuckenrath, R. 1982. Holocene glacial geology of the Svea lowland, Spitsbergen, Svalbard. *Geografiska Annaler*, **64**, 35–51.

Schei, B., Eilertsen, H. C., Falk-Petersen, S., Gulliksen, B. & Taasen, J. P. 1979. Marinbiologiske undersøkelser i Van Mijenfjorden (Vest-Spitsbergen) etter en oljelekkasje ved Seagruva 1978. *TROMURA, Naturvitenskap nr. 2*. Universitetet i Tromsø, Institutt for Museumsvirksomhet, Tromsø, Norway.

Stigebrandt, A. 1980. Some aspects of tidal interaction with fjord constrictions. *Estuarine Coastal and Marine Science*, **11**, 151–166.

Støylen, E. & Weber, J. E. 2010. Mass transport induced by internal Kelvin waves beneath shore fast ice. *Journal of Geophysical Research*, **115**, C03022, doi:10.1029/2009JC005298.

Svendsen, H. 1986. Mixing and exchange processes in estuaries, fjords and shelf waters. *In*: Skreslet, S. (ed.) *The Role of Freshwater Outflow in Coastal Marine Ecosystems*. NATO ASI Series, Springer, Berlin, G7, 13–45.

Svendsen, H. 1995. Physical oceanography of coupled fjord–coast systems in northern Norway with special focus on frontal dynamics and tides. *In*: Skjoldal, H. R., Hopkins, C., Erikstad, K. E. & Leinaas, H. P. (eds) *Ecology of Fjords and Coastal Waters*. Elsevier Science B.V., Amsterdam, The Netherlands, 149–164.

Svendsen, H., Beszczynska-Møller, A. *et al.* 2002. The physical environment of Kongsfjorden-Krossfjorden, an Arctic fjord system in Svalbard. *Polar Research*, **21**, 133–166.

Tverberg, V. & Nøst, O. A. 2009. Eddy overturning across a shelf edge front: Kongsfjorden, west Spitsbergen. *Journal of Geophysical Research*, **114**, doi: 10.1029/2008JC005106.

Weslawski, J. M., Koszteyn, J., Kwasniewski, S., Swerper, S. & Ryg, M. 1991. Summer hydrology and zooplankton in two Svalbard fjords. *Polish Polar Research*, **12**, 445–460.

Widell, K., Fer, I. & Haugan, P. M. 2006. Salt release from warming sea ice. *Geophysical Research Letters*, **33**, doi: 10.1029/2006GL026262.

Dinoflagellate cysts as proxies for palaeoceanographic conditions in Arctic fjords

JOHN A. HOWE[1]*, REX HARLAND[2,3], FINLO R. COTTIER[1], TIM BRAND[1], KATE J. WILLIS[1,4], JØRGEN R. BERGE[5], KARI GRØSFJELD[6] & ANITA ERIKSSON[1]

[1]*Scottish Association for Marine Science, Scottish Marine Institute, Dunbeg, Oban, PA37 1QA, UK*

[2]*Department of Earth Sciences, Göteborg University, PO Box 460, SE 405 30, Göteborg, Sweden*

[3]*50 Long Acre, Bingham, Nottingham, NG13 8AH, UK*

[4]*National Institute for Water and Atmospheric Research, PO Box 8602, Riccarton, Christchurch, New Zealand*

[5]*The University Centre in Svalbard, Postbox 156, 9171 Longyearbyen, Norway*

[6]*Geological Survey of Norway, Postbox 3006 – Lade, 7491 Trondheim, Norway*

**Corresponding author (e-mail: john.howe@sams.ac.uk)*

Abstract: The potential of using dinoflagellate cysts as proxies for palaeoceanographic conditions and as monitors of the dynamic marine environment of climatically sensitive Arctic fjords was investigated with sediment traps. Dinoflagellate cysts were analysed from three separate deployments in two high Arctic fjords in the Svalbard archipelago. Two deployments in Kongsfjorden on the west coast of Svalbard occurred during 2002 and 2006–2007 and a deployment in Rijpfjorden on the NE coast occurred during 2006–2007. The cyst production displayed peaks of abundance in the spring and late summer with distinct differences in cyst occurrence in different fjords and in different years. The recorded and identified cyst species were consistent both with the hydrography of the fjords and with changes in cyst composition that are comparable to the seasonal shifts in water mass characteristics. The presence of the heterotrophic species *Protoperidinium conicum* in Kongsfjorden during 2002 is of note and may reflect the availability of a particular food source possibly associated with the strong influx of Atlantic Water. Cysts recovered from Kongsfjorden during 2006–2007 were dominated by *Islandinium minutum*, an indicator of cold, polar to subpolar conditions. The temperature and salinity characteristics of the ambient hydrography in this period indicated less influence by Atlantic Water than in 2002, and the cyst production was consistent with regional cyst distribution patterns. In Rijpfjorden, cyst assemblages were dominated by *Pentapharsodinium dalei*, consistent with the fjord being dominated by full Arctic conditions during the mooring deployment and the possible occurrence of stratified water with high productivity during the spring phytoplankton bloom.

Kongsfjorden and Rijpfjorden are glacial fjords on the NW coast of Spitsbergen and the north coast of Nordaustlandet, respectively, both within the Svalbard archipelago (Fig. 1). Although both fjords are located within the Arctic Circle, previous work has demonstrated (Svendsen *et al.* 2002; Cottier *et al.* 2005) that Kongsfjorden can become strongly influenced by warm Atlantic Water (AW). Since January 2006 the shelf and fjords of western Svalbard have seen a much more continuous influence from AW than previously recorded (Cottier *et al.* 2007). These hydrographic changes are fundamental to the seasonal cycle of phytoplankton production and zooplankton community structure in the fjord. In contrast, Rijpfjorden can be considered a true Arctic system with minimal influence from AW (Berge *et al.* 2009).

The purpose of this study was to investigate the seasonality of dinoflagellate cyst production over a complete growing season and to attempt to link their occurrence to the annual and interannual hydrography of the two fjords. Developing the tools to record extensive changes of water mass within the fjords, using dinoflagellate cysts as possible indicators, offers the potential for the recognition of similar changes over the latest Holocene timescale; dinoflagellate cysts are readily fossilized and thus enter the temporal record. Indeed, dinoflagellate cysts have proved to be useful as proxies to chart changing water mass distribution in Nordic

From: HOWE, J. A., AUSTIN, W. E. N., FORWICK, M. & PAETZEL, M. (eds) *Fjord Systems and Archives*. Geological Society, London, Special Publications, **344**, 61–74.
DOI: 10.1144/SP344.6 0305-8719/10/$15.00

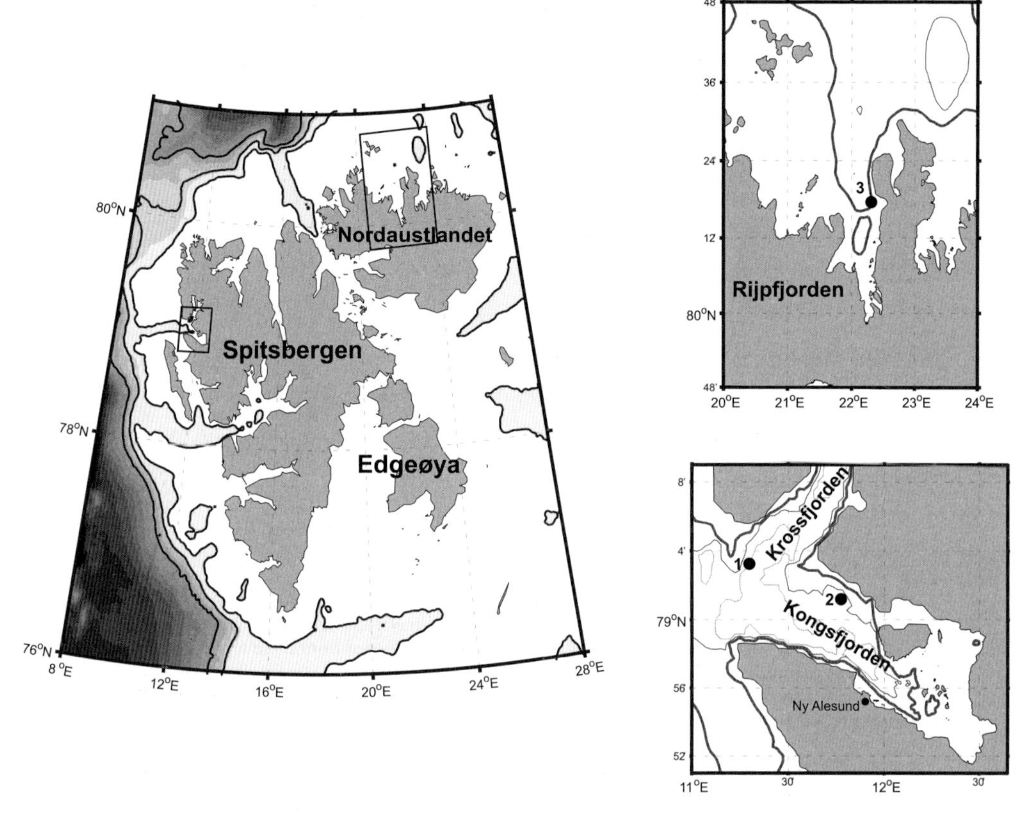

Fig. 1. The location of Kongsfjorden and Rijpfjorden, Svalbard. Insets show the position of the moorings with the sediment traps (black dots) in the outer basin of Kongsfjorden, 2002 (1) and 2006–2007 (2) and in Rijpfjord 2006–2007 (3). Contour interval in metres, 200, 500 and 1000 m for Svalbard; 100, 200 and 300 m for the inset maps.

Seas throughout the Quaternary (see Dale & Dale 1992; Matthiessen *et al.* 2005).

The position of these Arctic fjords, close to the Fram Strait gateway, is important for the potential to recognize of the strength, configuration and extent of the influence of AW across the region as climate changes during the Holocene. In this paper we present the results of a detailed investigation into the seasonal dinoflagellate cyst records from sediment traps in the Kongsfjorden–Krossfjorden system and relate these results to the degree to which AW is present in the system. By way of comparison, a similar suite of sediment trap samples were also investigated from Rijpfjorden. The detection and determination of major AW intrusions into the fjords is potentially of high significance as an indicator of the present day strength and influence of AW and this information may also be used for palaeo-reconstructions.

The use of dinoflagellate cysts as proxies for palaeoceanographic conditions in Arctic sediments is particularly relevant given the scarcity of calcareous fossils of planktonic foraminifers at high latitude. The appearance of carbonate tests on the western Svalbard margin provides an indication of the northern advection of warm AW accompanying high productivity episodes during glacial episodes (Hald *et al.* 2004). The use of dinoflagellate cysts as water mass tracers in the Arctic, using transfer function techniques, was first established by De Vernal *et al.* (2001) who examined dinoflagellate assemblages from 677 sites in the northern North Atlantic with respect to sea-ice cover, salinity and temperature, establishing a major database of modern conditions. Matthiessen *et al.* (2001) examined the temporal record over the last 150 ka of dinoflagellate cysts on the Arctic Eurasian margin, finding a decrease in AW from west to east especially during glacial times. The ultra high-resolution records of fjords can provide dinoflagellate records of rapid climate change over various timescales. Harland *et al.* (2004*a*, *b*) and Filipsson *et al.* (2005) described a 150 year record of seasonal change in a Swedish fjord using dinoflagellate cysts, finding a possible link between the changing dinoflagellate assemblages and the North Atlantic Oscillation as well as local anthropogenic effects including major engineering work. Previously, Dale & Dale (1992) had indicated the potential of dinoflagellate cysts to trace water masses in Nordic Seas; Dale *et al.* (1999) successfully used dinoflagellates as indicators of anthropogenic-induced eutrophication in Oslofjord, Norway.

Dinoflagellates and their cysts

Dinoflagellates are a component of the marine phytoplankton and, together with diatoms and coccolithophores, they make up the most important primary producers in the world's oceans (Marret & Zonneveld 2003). They are eukaryotic, mostly unicellular protists that characteristically possess two flagella that propel them through the surface waters; this mobility ensures their optimum position in the water column for photosynthesis or the capture of prey. As part of their sexual life cycle and environmental stress, dinoflagellates often form a cyst and about 13–16% of species produce a cyst capable of fossilization (Head 1996).

The ecology of dinoflagellates and their cysts is complex due to their life cycles and diverse nutritional strategies. Sediment trap studies on the production of cysts within surface waters, and their recruitment to bottom sediments, offers a way to understand both: (1) the seasonal and interannual variability of cyst production and (2) how the sediment cyst record can be linked to dinoflagellate cyst fluxes and potentially to dinoflagellate primary production.

The use of diverse nutritional strategies means that autotrophic dinoflagellates are light limited but also require nitrogen, phosphorus and sulphur together with some trace elements such as iron (Taylor & Pollingher 1987) for their wellbeing. The heterotrophic and mixotrophic dinoflagellates are largely controlled by the availability of food, which may itself be controlled by various biotic and abiotic controls. Recent research suggests that nutritional strategies are even more complex than originally thought with those dinoflagellates possessing chlorophyll also often employing a mixotrophic strategy (Jeong 2008). Mixotrophy implies the ability of the species to derive both energy from sunlight (via photosynthesis) and cell material from an organic material from the water column.

Plankton primary production in the Arctic shows pronounced seasonality (Hop *et al.* 2002). The active growing season is limited to the biological summer of some 180–200 days close to the southernmost ice cover, and to 70–100 days close to the multiyear ice (Sakshaug 2003) providing a limited window of opportunity for their study. The spring phytoplankton 'bloom' is triggered by increased solar insolation, freshwater runoff, melting sea ice and high concentrations of winter accumulated nutrients (Harland & Pudsey 1999). It is, of course, time transgressive and may begin in April in the low Arctic and as late as September in the northern Arctic (Falk-Petersen *et al.* 2008). Finally, it is limited by nutrient depletion, grazing and stabilization. Within a typical phytoplankton bloom succession the dinoflagellate 'blooms' usually follow after the diatom 'blooms' (Smayda 1980).

Dinoflagellate encystment is related both to periods of favourable growth for the motile dinoflagellates and may also be partly triggered by nutrient depletion (Matthiessen *et al.* 2005). Peak cyst production is a distinct seasonal signal that occurs in association with or slightly delayed from the peak of the vegetative motile cells (Kremp & Heiskanen 1999). The cyst yields are extremely variable from 1–50% of the motile cells (Kremp & Heiskanen 1999) and, therefore, it may be difficult or impossible to infer phytoplankton concentrations and hence primary production from cyst abundances (Montresor *et al.* 1998). Nonetheless, the short biological summer in the Arctic coupled with the important peak cyst production and its relationship with primary production can provide a dinoflagellate cyst signal. This is intimately linked to various biotic and abiotic controls, themselves being directly associated with the hydrography of the surrounding water masses. It is this relationship that provides the potential to look at the seasonal cyst production as a tool to understand changing water mass dynamics in Arctic fjord environments.

Regional setting

Svalbard has very distinct oceanographic realms (Grøsfjeld *et al.* 2009). While it sits at the edge of the Arctic Ocean, the west coast is strongly influenced by the northward-flowing AW which transports warm and saline waters in the West Spitsbergen Current (WSC). This keeps the waters west of the shelf break relatively ice free, even in winter. Until recently, winter sea ice developed regularly on the West Spitsbergen Shelf and in adjacent coastal waters (Gerland & Renner 2007; Nilsen *et al.* 2008). However, in recent years the shelf and fjord waters have experienced increased exchange with water of Atlantic origin resulting in mainly ice-free waters on the shelf, even in winter (Cottier *et al.* 2007). In contrast, the north and east coasts of the archipelago are strongly influenced by water and ice from the Arctic basin and develop extensive sea-ice cover in winter.

Kongsfjorden is one of a double-fjord system on the NW coast of Spitsbergen, the other being Krossfjorden (Svendsen *et al.* 2002) (Fig. 1). Unlike classic glaciated fjords, exchange of water between the shelf and Kongsfjorden is not restricted by a shallow sill. Rather, there is a deep (>250 m) trench that transits the shelf to the shelf break, providing a topographic guide for the exchange of waters between the WSC and the fjord (Svendsen *et al.* 2002). Typically, the density difference between the waters in the fjord and waters on the shelf, established through winter sea-ice formation (Nilsen *et al.* 2008), is sufficient to limit water exchange (Cottier *et al.* 2005). However, during the summer, modification and mixing of waters in the fjord can reduce the density contrast, allowing AW to flood the coastal and fjordic waters causing a rapid shift in the marine climate from Arctic to Atlantic (Fig. 2a, b). Since 2005 this rapid switch in hydrography has not been so apparent with a rather more continuous occupation by AW or Transformed Atlantic Water (TAW) (Cottier *et al.* 2005).

In contrast to Kongsfjorden, Rijpfjorden occurs on the north coast of Nordaustlandet, indicating that it is an Arctic-dominated fjord in terms of its oceanography (Fig. 1). It is *c.* 30 km long and opens out to a width of 10 km. The fjord remains ice-covered for up to eight months of the year (Berge *et al.* 2009). During the melting period, there is limited glacial discharge into the fjord from the surrounding glaciers. The mouth is a further 60 km from the shelf break compared to Kongsfjorden (45 km), and there is no deep connecting trench. Therefore, while AW does circulate along the northern shelf slope of Svalbard (Ivanov *et al.* 2008) there is less Atlantic influence in Rijpfjorden. Previous hydrographic surveys on this northern shelf have found no evidence of AW (Sundfjord pers. comm. 2007). The fjord comprises at least two basins separated by sills, but more detailed knowledge of its bathymetry is very sparse. Water masses are generally close to freezing point for much of the year and salinification of the waters is observed during the formation of the ice cover through brine release.

The zooplankton communities of the two fjords differ in terms of the relative contributions of boreal and Arctic species. The zooplankton fauna in Kongsfjorden is relatively rich and comprises boreal and Arctic species, which respond to variations in the distribution and dynamics of the West Spitsbergen water masses (Kwasniewski *et al.* 2003). Boreal species of Atlantic origin include *C. finmarchicus*, *Pareuchaeta norvegica*, *Oithona atlantica*, *Themisto abyssorum*, *Thysanoessa* spp., *Heterorhabdus norvegicus* and

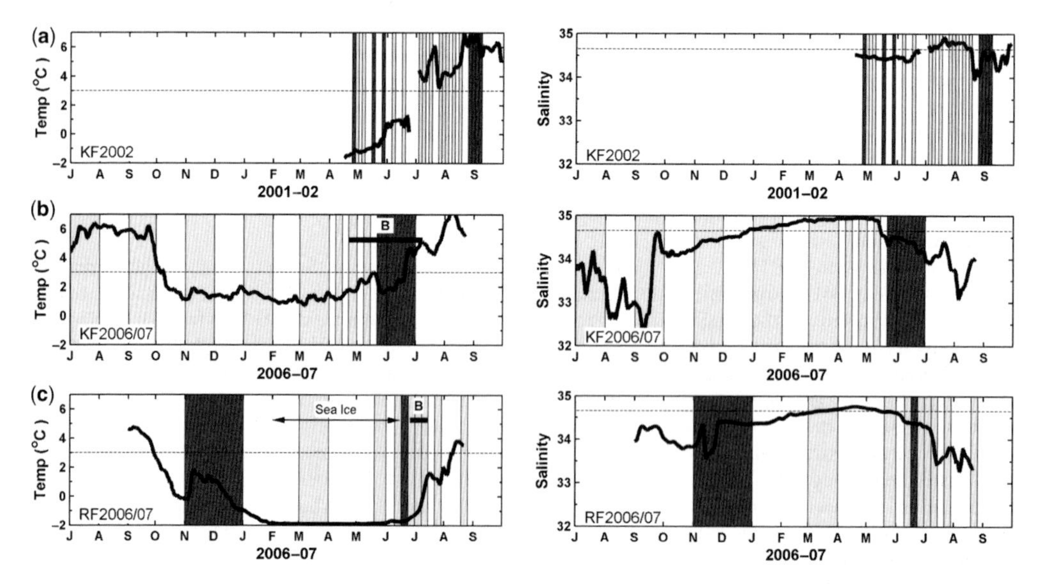

Fig. 2. Temperature and salinity plots for the duration of the moorings in (**a**) Kongsfjorden 2001–2002, (**b**) Kongsfjorden 2006–2007 and (**c**) Rijpfjorden 2006–2007. The shaded sections (dark and light grey) denote periods during which all those sediment trap bottles analysed for this study were open; dark shading is used to highlight those bottles discussed within the text with reference to dinoflagellate blooms. The phytoplankton bloom period (derived from fluorometer records) is denoted by the thick horizontal line labelled B and the period of sea-ice cover is indicated by the horizontal arrows.

Tomopteris helgolandica. Species of Arctic origin in the fjord include *Calanus glacialis*, *Calanus hyperboreus*, *Metridia longa*, *Themisto libellula*, *Clione limacina* and *Limacina helicina*. The zooplankton fauna in Rijpfjorden is predominantly Arctic in origin, with species such as *C. glacialis*, *C. hyperboreus*, *M. longa*, *T. libellula* and *Gammarus wilkitzkii* dominating the community (Falk-Petersen *et al.* 2008).

Methods

Sediment trap moorings

All the data are derived from subsurface moored instruments deployed in the two fjords. The outer Kongsfjorden mooring was deployed between April–September 2002 in the mouth of the Kongsfjorden-Krossfjorden system at position 79°3.25N, 11°18.0E in 215 m water depth (position 1 in Fig. 1). A sequential Paarflux 78H-21 sediment trap was deployed at 65 m depth and sampling encompassed 18 April–23 June and 4 July–8 September in two separate deployments to enable maintenance. The inner Kongsfjorden mooring was deployed from June 2006 to August 2007 on the northern side of Kongsfjorden at position 79°01.2N, 11°46.4E in 210 m water depth (position 2 in Fig. 1). The Paarflux trap was deployed at 115 m with sampling period 1 July 2006–1 June 2007.

The Rijpfjorden mooring was deployed from September 2006 to August 2007 within the fjord at position 80°17.6N, 22°18.8E in 215 m water depth (position 3 in Fig. 1). Samples were collected using a sequential sediment trap deployed at a depth of 110 m with sampling period 2 September 2006–20 August 2007.

The comment on the state of sea ice in the fjords is based on observations during the period of trap deployment.

Each mooring had a Seabird microcat temperature and salinity (TS) sensor at *c.* 25 m depth and 10 m from the bottom. Further temperature sensors were distributed along the length of the mooring. From these, the modification and exchange of water masses could be detected within the fjord systems. The mean temperature and salinity of the water during the sampling period of each sediment trap bottle was calculated from the upper TS sensor. Additionally, fluorescence in the water column was measured in 2006–2007 using Seapoint fluorometers to give a measurement of chlorophyll in the water and hence the timing and duration of the phytoplankton bloom period.

The sediment traps were pre-dosed with filtered seawater, *Analar* grade sodium chloride to provide a density contrast with the ambient water and 2% buffered formalin to preserve the biological samples. Sediment trap bottle opening/closing varied between deployments and between fjords and is summarized in Table 1. Upon return to the laboratory, the trap samples were first prepared by removing zooplankton swimmers (Willis *et al.* 2006). Each sample was then split into ten equal aliquots using a McLane sediment trap sample splitter. All subsequent analyses were carried out on single subsamples from selected periods throughout the deployments.

The sediment trap subsamples were washed and sieved before drying and weighing, so that palynological processing was performed on a known sample weight. The cyst analysis was conducted on the retained 10–212 micron fraction. An aliquot of the sample was prepared using standard processing techniques but without the provision of oxidizing reagents to avoid the loss of fragile and susceptible cysts (see Wood *et al.* 1996). Ultrasound treatment was used to disperse the amorphous organic material but all stages of the processing technique were carried out so that a minimum disturbance of the contained palynomorphs was ensured; cell contents were observed within many of the dinoflagellate cysts. Indeed the presence of an occasional thecate cell is also evidence of the gentle preparatory techniques employed.

The use of a known aliquot weight in sample preparation enabled the calculation of the number of cysts per gram of sediment and hence the daily flux into the sediment trap. Cyst counts are derived by multiplying cyst count by the aliquot divided by the time the trap was open. This provides cysts per gram per day.

Results

Hydrography

The main hydrographic features are shown in Figure 2 with data from each fjord presented for the same 15-month period. The collection period for those trap samples used in this study are indicated and the period of phytoplankton bloom recorded by fluorometery. The typical temperature and salinity (TS) characterization of AW is depicted by dashed lines at $T = 3$ °C and $S = 34.65$. Cottier *et al.* (2005) describe the hydrography of Kongsfjorden in 2002 (Fig. 2a), which is characterized by cold and rather saline conditions in spring. This is followed by a very rapid inflow of AW in July to give warm water temperatures peaking in September with pure AW conditions present in the fjord (the slightly reduced salinity values are due to freshwater pulses; see Cottier *et al.* 2005 for details).

Kongsfjorden in 2006–2007 shows the seasonal cycle of warm summer conditions and cooling

Table 1. *Summary of sediment trap samples and duration*

Sample number	Locality/sample duration	Ref./start date	Sample weight per slide (mg)
1	Kongsfjord (south) 3.5 days	S3 (1) 25/4/02	3.1
2	3.5 days	S4 28/4/02	1.17
3	3.5 days	S6 5/5/02	4.99
4	3.5 days	S9 16/5/02	1.38
5	3.5 days	S12 26/5/02	0.25
6	3.5 days	S15 6/6/02	4.56
7	3.5 days	S18 16/6/02	0.94
8	3.5 days	S1 (2) 4/7/02	2.01
9	3.5 days	S2 7/7/02	4.49
10	3.5 days	S4 14/7/02	1.48
11	3.5 days	S7 25/7/02	9.55
12	3.5 days	S8 28/7/02	6.66
13	3.5 days	S9 1/8/02	6.29
14	3.5 days	S11 8/8/02	1.72
15	3.5 days	S13 15/8/02	4.32
16	3.5 days	S16 25/8/02	5.13
17	3.5 days	S17 29/8/02	7.28
18	3.5 days	S18 1/9/02	5.92
19	3.5 days	S19 5/9/02	4.16
20	Kongsfjord (north) 7 days	S2 1/7/06	121.9
21	1 month	S4 1/9/06	116.5
22	1 month	S6 1/11/06	44.1
23	1 month	S8 1/1/07	575.3
24	1 month	S10 1/3/07	107.8
25	7 days	S12 8/4/07	83.5
26	9 days	S14 22/4/07	75.5
27	7 days	S16 8/5/07	27.0
28	9 days	S18 22/5/07	6.5
29	1 month	S19 1/6/07	12.6
30	Rijpfjord 2 months	S3 1/11/06	240.7
31	1 month	S5 1/3/07	28.4
32	12 days	S8 20/5/07	4.8
33	7 days	S10 10/6/07	1.4
34	7 days	S11 17/6/07	1.7
35	7 days	S12 24/6/07	2.9
36	7 days	S13 1/7/07	3.7
37	7 days	S14 8/7/07	43.5
38	7 days	S16 22/7/07	6.6
39	7 days	S20 19/8/07	15.7

through autumn, leading to an increase in salinity as the water column is mixed. Winter temperatures did not drop below zero and there was a rapid rise in water temperature in July (Fig. 2b). The phytoplankton bloom lasted from mid-April to July and the period of maximum cyst production occurred within the phytoplankton bloom period. During the majority of the period of peak cyst production, temperature and salinity were both below the values for pure AW; this indicates that the cyst production occurred in waters of mixed Arctic and Atlantic Water. The temperature and salinity also decrease just prior to this period indicating a rather cooler, fresher water mass during the dinoflagellate bloom.

In Rijpfjorden, there was relatively warm water in the fjord in summer 2006; this rapidly cooled during autumn leading to sea-ice formation in winter (Fig. 2c). Sea-ice growth caused a steady increase in salinity through brine release. The short bloom period in early July occurred immediately after the sea ice was observed to have retreated out of the fjord and there was a rapid warming in the surface waters.

Figure 3 depicts the mean TS characteristics of the water overlying the trap for each sediment trap bottle analysed in this study. Two water masses are indicated – AW and Winter Cooled Water (WCW) – which represent the Atlantic and Arctic extremes. The change in hydrographic conditions

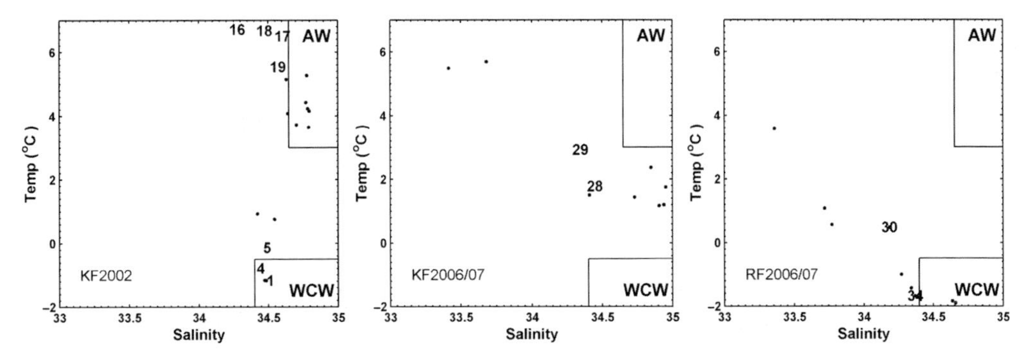

Fig. 3. Plots of temperature v. salinity averaged over the period during which each sediment trap bottle was open for each mooring deployment in Kongsfjorden and Rijpfjorden. Numbers indicate those sample bottles discussed in the text and having high numbers of dinoflagellate cysts; other sample bottles are simply marked with a point. Sample number dates and duration can be found in Table 1. The boxes indicate water masses of Atlantic Water (AW) and Winter Cooled Water (WCW) as defined by Cottier *et al.* (2005).

throughout the sampling season is clearly depicted by the spread in TS properties. In Kongsfjorden in 2002 there is a strong distinction between those samples obtained under Arctic hydrographic conditions and those obtained during dominant Atlantic conditions (Fig. 3). Samples from Kongsfjorden in 2006 are characterized by very mixed hydrographic conditions that span the pure Arctic or Atlantic conditions (Fig. 3). In contrast to 2002, the hydrography was not dominated by pure Atlantic Water during the periods of sample collection. In Rijpfjorden, there was an even greater shift in hydrography away from AW influence (Fig. 3) for each sample period.

Dinoflagellate cysts

Figure 4 shows the main cyst taxa recovered from the sediment trap deployments in the three fjords.

Kongsfjorden 2002

All the samples except three (numbers 7, 8 and 10) yielded cysts, often with cell contents. Many of the samples contained only a few cysts providing low cyst counts. Where the cyst counts were >10 000 the assemblage was considered of value. The Kongsfjorden dataset can be divided into two distinct groups (Fig. 5a).

The first group, occurring between April and May (samples 1, 4 and 5) is characterized by the presence of *Islandinium minutum* (Reid in Harland *et al.* 1980) Head *et al.* 2001 (27.3–54.6%) with occasional round brown *Protoperidinium* spp. and some indeterminate cysts of probable dinoflagellate affinity. This assemblage is typically Arctic in character (Matthiessen *et al.* 2005). *Islandinium minutum* is an established indicator of cold polar to subpolar waters of high northern and southern latitudes (Head *et al.* 2001; Marret & Zonneveld 2003). High percentages (>35%) often occur in regions where there is seasonal ice cover and sea surface temperatures (SSTs) rarely exceed 7 °C (Marret & Zonneveld 2003). As is the case here, this species is often accompanied by round brown *Protoperidinium* cysts. These cysts are produced by heterotrophic dinoflagellates and their distribution is controlled largely by the availability of food. This group is found within the Phase I zooplankton assemblage of Willis *et al.* (2006) and is associated with local WCW/Arctic Water. It is characteristic of Arctic surface water as described by Grøsfjeld *et al.* (2009).

The second group occurs during August–September (samples 16–19) and is characterized by *Protoperidinium conicum* (Gran 1900) Balech 1974 (44–93.6%) with occasional *Pentapharsodinium dalei* Indelicato & Loeblich III 1986, round brown *Protoperidinium* spp., *Islandinium minutum* and *Spiniferites frigidus* Harland & Reid 1980. The presence of the *Protoperidinium conicum* and *Spiniferites elongatus/frigidus* suggests rather more temperate waters than those indicated by the first group. Marret & Zonneveld (2003) characterized them as cold to tropical and cold to temperate, respectively, over a broad range of physical parameters. The species are well-known components of the dinoflagellate cyst assemblages from fjords along the west coasts of Sweden and Norway (Grøsfjeld & Harland 2001; Harland *et al.* 2006).

Kongsfjorden 2006–2007

The 2006–2007 Kongsfjorden samples (Fig. 5b) display a high export production of dinoflagellate

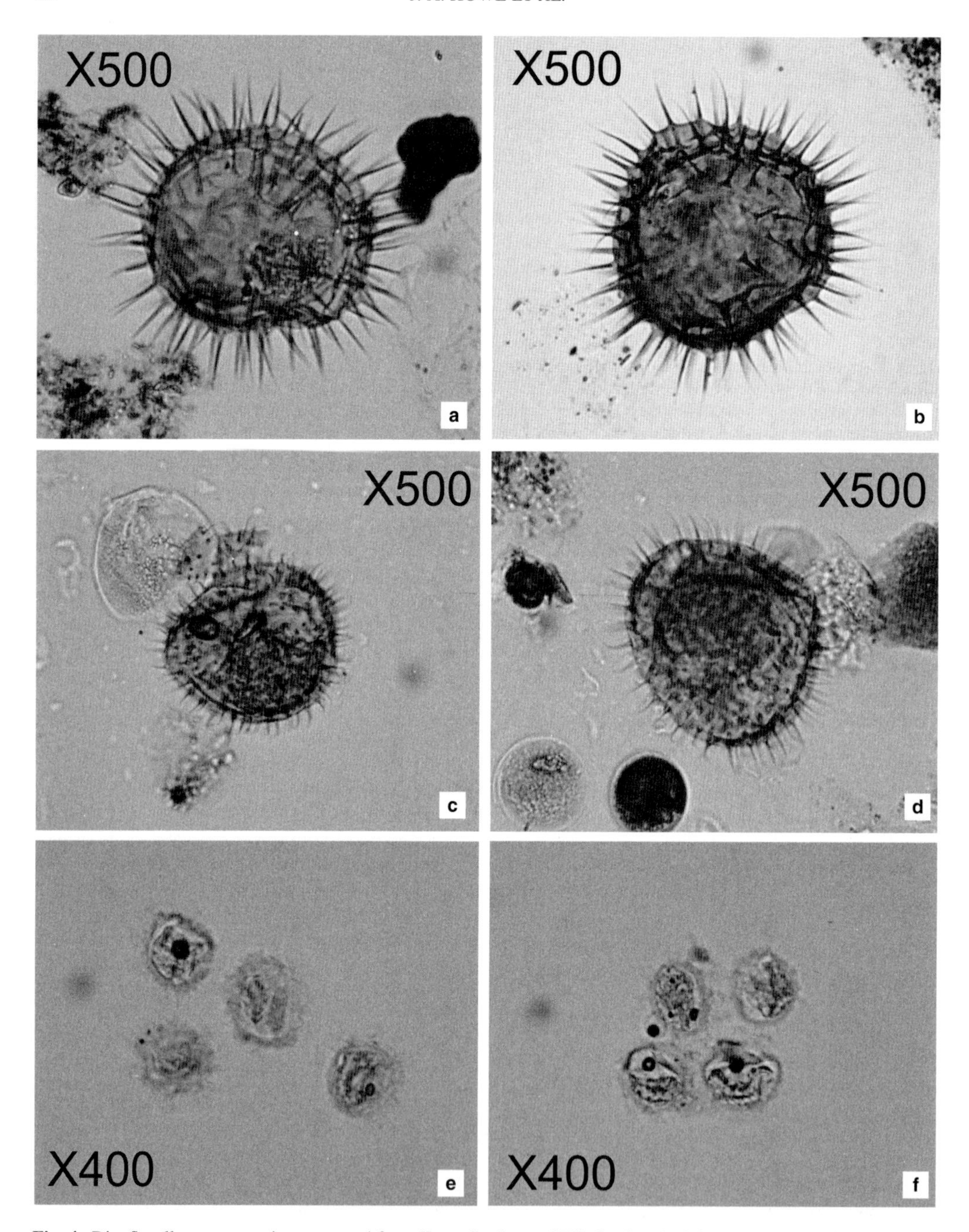

Fig. 4. Dinoflagellate cyst species recovered from Kongsfjorden and Rijpfjorden. (a & b) *Protoperidinium conicum* (Gran 1900) Balech 1974 from Kongsfjorden 2002. (**a**) Sediment trap sample taken on 25/08/2002, sample 07RH16a, English Finder P36/3 and (**b**) sample 07RH16a, English Finder N33/0, magnification ×*c*. 500. (c & d) *Islandinium minutum* (Harland & Reid in Harland *et al.* 1980) Head *et al.* 2001, from Kongsfjorden 2006–2007. (**c**) Sediment trap sample taken on 22/05/2007, sample 07RH38, English Finder U43/0 and (**d**) sample 07RH38, English Finder U31/2, magnification ×*c*. 500. (e & f) *Pentapharsodinium dalei* Indelicato & Loeblich III 1986 from Rijpfjorden 2006–2007. (**e**) Sediment trap sample taken on 17/06/2007, sample 07RH24, English Finder U36/0 and (**f**) sample 07RH24, English Finder U33/0, magnification ×*c*. 400.

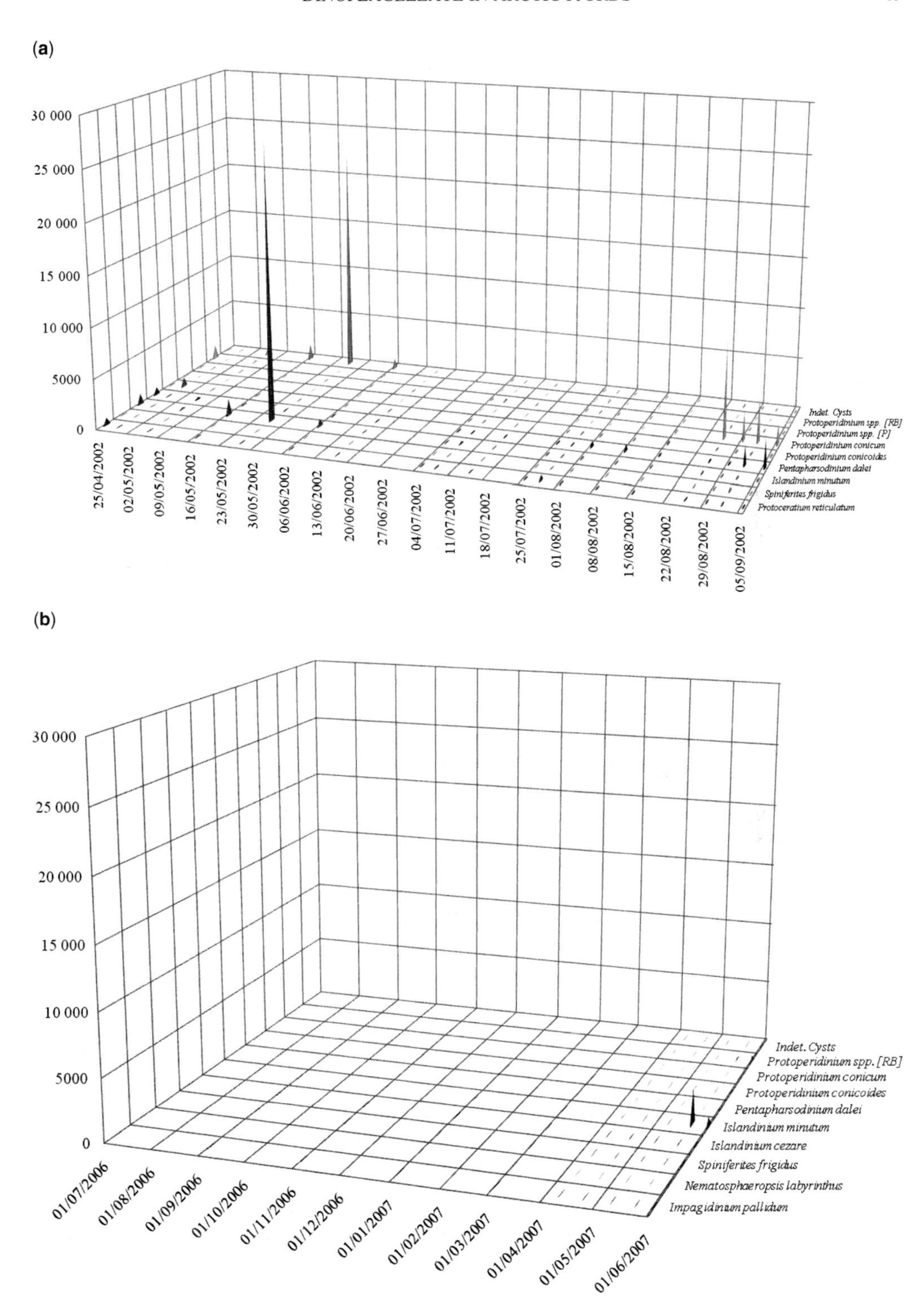

Fig. 5. Cyst flux data for the seasonal cycle of dinoflagellate cyst export production in (**a**) Kongsfjorden 2002, (**b**) Kongsfjorden 2006–2007 and

(c)

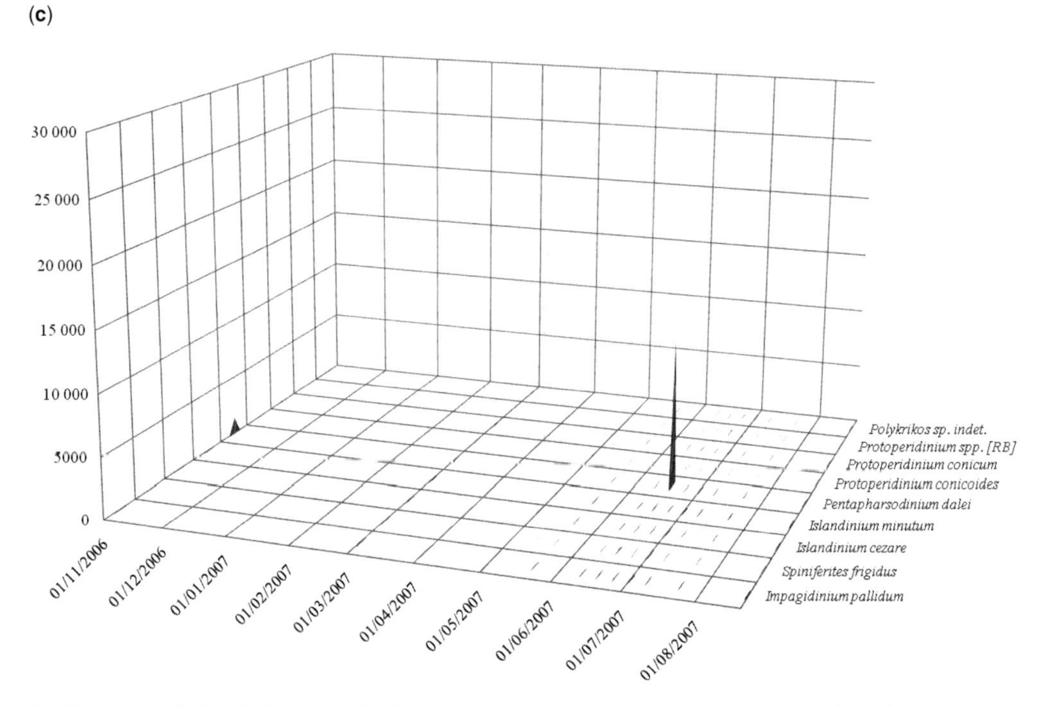

Fig. 5. (*Continued*) (**c**) Rijpfjorden 2006–2007. Data on the vertical axis are expressed as cysts/gram/day.

cysts, again overwhelmingly dominated by *Islandinium minutum* (94.9%), with minor amounts of round brown *Protoperidinium* cysts (5.1%) in the sample from late May (sample 28) and 63% and 36%, respectively, in the samples from early June (sample 29). These two samples document a peak cyst production for the summer biological season in Kongsfjorden, largely characterized by *Islandinium minutum.* This compares well with bottom sediment dinoflagellate cyst assemblages for the fjord (Grøsfjeld *et al.* 2009) and is consistent with normal cyst production in typically Arctic settings.

Rijpfjorden 2006–2007

In contrast to Kongsfjorden, characterized by *Islandinium minutum*, the export production of dinoflagellate cysts in Rijpfjorden (Fig. 5c) was dominated by cysts of *Pentapharsodinium dalei* (92.3%) with minor amounts of round brown *Protoperidinium* cysts (3.6%) and *Spiniferites elongatus/frigidus* (2.9%) in November 2006 (sample 30) and *Pentapharsodinium dalei* (98.2%) with minor amounts of *Spiniferites frigidus* (1.6%) in June 2007 (sample 34). These samples provide evidence of late autumn and summer cyst production in Rijpfjorden for these two years. It should be pointed out, however, that the sample deployment for sample 30 began in November 2006, lasted for two months and most probably was contaminated with allochthonous cysts from autumn equinoxal storms and should perhaps be viewed with caution.

Discussion

This study compliments a similar study from the Svalbard region using seabed surface sediment samples (Grøsfjeld *et al.* 2009). These authors report finding dinocyst assemblages associated with AW, Arctic Water and stratified water with a high productivity. They analysed seabed samples from Kongsfjorden and Krossfjorden and found assemblages dominated by *Islandinium minutum*, interpreted as associated with Arctic Water. Although most seabed samples from the shelf and fjord regions were dominated by *Islandinium minutum*, one of the samples from the inner part of Kongsfjorden collected in 2002 contained cysts of *Pentapharsodinium dalei.* This taxon, which occasionally occurred in significant-to-high percentages in fjords, was interpreted to be more associated with stratified waters and high productivity in the spring. This assumption is supported by Solignac *et al.* (2009). In this sediment trap study we find the export production of cysts in Kongsfjorden to be characterized by *Islandinium minutum.* A mixed assemblage is present in 2002, however, significantly containing *Protoperidinium conicum* which is interpreted as reflecting the inflow of AW. The dominance of *Islandinium minutum* in the 2006–2007 samples is interpreted as being associated with water masses with a reduced Atlantic influence (Fig. 3b) in contrast to the strongly dominant AW conditions in 2002 (Fig. 3a).

Comparing the dinoflagellate cysts recovered from the surface sediment (Grøsfjeld *et al.* 2009), the sediment trap (this study) and the zooplankton communities (Willis *et al.* 2006) indicates the complexity of surface water community response to changing water masses. The study by Willis *et al.* (2006) reports three distinct phases of zooplankton community in Kongsfjorden during 2002, where each phase is associated with different water masses. The first phase is dominated by local WCW, a second phase is associated with an influx of Arctic water from the coastal waters into the fjord and a final phase, following extensive intrusion of warm AW from the WSC, comprises species of both boreal and Arctic origin (Hop *et al.* 2006). Such a close coupled relationship is neither reflected in the dinoflagellate trap samples in this study (although the presence of round, brown *Protoperidinium* is possibly associated with WCW) nor in the surface sediment samples of Grøsfjeld *et al.* (2009). The reason for this dissimilarity is that the seasonal export production of dinocysts occurs over a very short time period and is therefore not reflected in the changing zooplankton communities. Here it is important to note that it is difficult to draw too many conclusions from comparisons between other datasets due to the differences in sampling times. No continuous year-round data have been studied; the period of cyst production (e.g. *Protoperidinium conicum* production) may have been missed. It is possible that the surface sediments do not contain a complete record of cyst production, or may even integrate multi-year records. A useful avenue for future study would be a detailed comparison of year-round sediment trap and surface sediment samples.

Protoperidinium conicum was present in the trap samples from Kongsfjorden in 2002, but not recovered from any surface sediment samples collected in 2001 (Grøsfjeld *et al.* 2009). This is unusual and raises questions as to whether such cyst export production is a regular occurrence in the fjord during times of major AW advection or whether other cyst species can also be involved in other years? It may also be linked to the idea that the dominance of the hydrography by AW in 2002 had no recent precedence.

The Rijpfjorden samples from July 2007 provide evidence for a late spring dinoflagellate production in the high Arctic. However, the occurrence of cyst production in the November samples is questionable and may be the result of a certain amount of resuspension of sediment as discussed above, perhaps through convective mixing of the water column. The dominance of *Pentapharsodinium dalei* is characteristic of spring production in temperate seas (Harland *et al.* 2004*a*), and not the late summer as observed here. However, given the time transgressive nature of the Arctic biological summer, it is perhaps not surprising that this species occurs later in the year at this latitude. In fact, this species seems to characterize Rijpfjorden and is a well-known polar to subpolar species (Dale 1996; Marret & Zonneveld 2003), although it is also known from temperate latitudes. While Rijpfjorden shows no direct measured evidence of AW advection into the fjord during the time of peak production, limited evidence suggests that Rijpfjorden is influenced to some extent by AW. There are only two years' of seasonal hydrographic measurements of full oceanographic data: August 2006–August 2008 (Fig. 2c). However, this data can only provide a broad overview of the local hydrography and shows periods during the summer when the temperature in Rijpfjorden rises by 4 °C in 25–30 days (Fig. 2c). This rate of warming cannot be achieved by simple surface heat fluxes but rather implies advection of warm water. Consequently, although the summer temperature and salinity characteristics of Rijpfjorden do not correspond to any published values of AW, we can be reasonably certain that AW (albeit in a highly modified form through mixing with shelf waters) was present in Rijpfjorden during the summer of 2007.

The relationship between the export production of dinoflagellate cysts and their final preservation is complex. There are issues such as the preferential preservation of certain cysts (Zonneveld *et al.* 2001), scavenging by benthic organisms (Persson & Rosenberg 2003) and the mixing of allochthonous and autochthonous assemblages either due to bioturbation, resuspension or long distance transport to consider if accurate palaeoceanographic reconstruction is attempted. Fjord sediments, with high sedimentation rates and, in some cases high productivity, can produce high-resolution dinocyst archives. Harland *et al.* (2004*a*, *b*) reported on the dinocyst record from Koljö Fjord, Sweden, which closely responds to seasonal changes in the fjord. The primary controlling factor there is the availability of nutrients in the surface waters of the fjord in spring and summer, followed by the establishment of a stable, well-stratified water column to enhance the production of cysts during the seasonal growing cycle. The situation in the Arctic fjords is similar although much more dynamic with potentially rapidly changing conditions, notably the transition of different water masses. This may either enhance or inhibit dinocyst production.

The role of any lateral transport of cysts into the Arctic fjords is unknown, although we noted a quantity of resuspended sediment in the Rijpfjorden trap in November 2007, possibly as a result of storms or convection. Matthiessen *et al.* (2005) discuss the wider role of lateral transport of particulates in bottom nepheloid layers in the Nordic and wider Polar Seas. Most studies have been in deep water and compare sediment trap samples with seafloor sediment samples. Studies from the Antarctic (Harland & Pudsey 1999), Nordic (Dale 1992) and Barents Seas (Honjo *et al.* 1988) reported mixed results suggesting that there is a variable influence on the cyst communities, reflecting local rather than regional conditions. Here we report cysts that show little evidence of transport (i.e. degraded and damaged cysts) yet were recovered from resuspended sediment. These fresh cysts associated with resuspended sediment may reflect localized sediment input during storms.

The relationship between the dinocyst assemblage composition and the fjord hydrography is indicated in Figures 2 and 3. Here the division of samples between those dominated by species characteristic of Arctic water masses are found in the WCW portion of the temperature-salinity diagram (Fig. 3), while those dominated by Atlantic species are found in the AW domain. Cyst production in Kongsfjorden in 2006–2007 occurs after a decrease in both temperature and salinity (Fig. 2b). Due to the circulation pattern of the fjord, water from the inner part of the fjord exits along the northern shore where the mooring was located. Water masses in the inner fjord tend to retain their cold Arctic characteristics (MacLachlan *et al.* 2007) and associated biota. Cyst production in 2006–2007 is therefore likely to reflect the less dominant role of AW compared to 2002 and the more Arctic water masses from the inner fjord that typify an Arctic system. Finally, the temperature–salinity characteristics of Rijpfjorden show that at the time of the phytoplankton bloom, the water column was dominated by Arctic water masses. The relationship between peak cyst production and timing of the phytoplankton bloom in Kongsfjorden and Rijpfjorden is shown in Figures 2b and c. In Kongsfjorden in 2006–2007, peak cyst production occurred in the later part of the phytoplankton bloom in June. Peak production occurred in Rijpfjorden in 2006–2007 before the main phytoplankton bloom period, just after sea-ice cover in the fjord disintegrated allowing for a rapid increase in light intensity in the water column.

Conclusions

Dinoflagellate cysts were examined from three sediment traps from oceanographic moorings in Kongsfjorden, western Svalbard, during 2002 and 2006–2007 and Rijpfjorden, northern Svalbard during 2006–2007. Kongsfjorden located on the western coast of Svalbard experiences seasonal influences on its water mass composition, changing from cold Arctic waters to warmer AW. Rijpfjorden, located on the northern Svalbard coast, has no documented major influence of AW and is presumed to be dominated by Arctic water for most of the year. The cyst assemblages in the sediment traps in Kongsfjorden in 2002 were consistent with spring and late summer phytoplankton blooms typical of the seasonally variable shift of glacial/Arctic type waters to waters of Atlantic origin. However, the cyst production of the heterotrophic species *Protoperidinium conicum* is unusual and may reflect the influx of warmer AW to the fjord or the availability of a particular food source. In contrast with the Atlantic-dominated season of 2002, the hydrography of Kongsfjorden in 2006–2007 showed a high degree of mixing between Arctic Waters and Atlantic Waters. The cysts recovered in this period were characterized by *Islandinium minutum*, consistent with normal cyst production in an Arctic fjord and considered to be an indicator of colder polar conditions.

In Rijpfjorden during 2006–2007, the sediment trap cyst assemblages were dominated by the Arctic *Pentapharsodinium dalei*, consistent with the fjord being dominated by full Arctic conditions with stratification and high productivity.

This study demonstrates the use of dinoflagellate cysts for documenting changing water masses in Arctic fjords over a seasonal cycle and therefore their potential for wider use in high-resolution fjord sediment archives. Of particular note is the role of *Protoperidinium conicum* to differentiate between cyst production during strongly Atlantic dominating conditions (2002) and periods when AW does not dominate the hydrography (2006–2007).

The authors would like to thank the officers and crew of the RRS *James Clark Ross*, RV *Lance* and RV *Jan Mayen* for productive and enjoyable cruises to Svalbard since 2001. The authors would like to thank D. Bodman of MB Stratigraphy, Sheffield, for the preparation of the dinoflagellate cyst samples. We gratefully acknowledge the useful comments of two anonymous reviewers. This paper is a contribution to the Scottish Association for Marine Science, Natural Environment Research Council funded Oceans 2025 programme.

References

Balech, E. 1974. *El genero "Protoperidinium" Bergh, 1881 ("Peridinum" Ehrenberg, 1831, partim).* Revista del Museo Argentino de Ciencias Naturales "Bernardino Rivadiva" Hidrobiologia, **4**, 1–79.

BERGE, J., COTTIER, F. *ET AL.* 2009. Diel vertical migration of Arctic zooplankton during the polar night. *Biology Letters*, **5**, 69–72.

COTTIER, F. R., TVERBERG, V., INALL, M. E., SVENDSEN, H., NILSEN, F. & GRIFFITHS, C. 2005. Water mass modifications in an Arctic fjord through cross-shelf exchange: the seasonal hydrography of Kongsfjord, Svalbard. *Journal of Geophysical Research (Oceans)*, **110**, C12005, doi: 10.1029/2004JC002757.

COTTIER, F. R., NILSEN, F., INALL, M. E., GERLAND, S., TVERBERG, V. & SVENDSEN, H. 2007. Wintertime warming of an Arctic shelf in response to large-scale atmospheric circulation. *Geophysical Research Letters*, **34**, L10607, doi: 10.1029/2007GL029948.

DALE, A. L. & DALE, B. 1992. Dinoflagellate contributions to the sediment flux of the Nordic Seas. *In*: HONJO, S. (ed.) *Dinoflagellate Contributions to the Deep Sea*. Ocean Biocenosis Series 5, Woods Hole (Wood Hole Oceanographic Institution), 45–75.

DALE, B. 1992. Dinoflagellate contributions to the open ocean sediment flux. *In*: HONJO, S. (ed.) *Dinoflagellate Contributions to the Deep Sea*. Ocean Biocenosis Series 5, Woods Hole (Wood Hole Oceanographic Institution), 1–31.

DALE, B. 1996. Dinoflagellate cyst ecology: modelling and geological applications. *In*: JANSONIUS, J. & MCGREGOR, D. C. (eds) *Palynology: Principles and Applications*. American Association of Stratigraphic Palynologists Foundation, Dallas, **3**, 1249–1275.

DALE, B., THORSEN, T. A. & FJELLSA, A. 1999. Dinoflagellate cysts as indicators of cultural eutrophication in the Oslofjord, Norway. *Estuarine, Coastal and Shelf Science*, **48**, 371–382.

DE VERNAL, A., HENRY, M. *ET AL.* 2001. Dinoflagellate cyst assemblages as tracers of sea-surface conditions in the northern North Atlantic, Arctic and sub-Arctic Seas: the new '$n = 677$' data base and its application for quantitative palaeoceanographic reconstruction. *Journal of Quaternary Science*, **16**, 681–698.

FALK-PETERSEN, S., WOLD, A. *ET AL.* 2008. Calanus hyperboreus feeding on an Arctic bloom at 82°N induced by the northernmost location of the ice edges since 1751. *Deep-Sea Research II*, doi: 10.1016/j.dsr2.2008.05.010.

FILIPSSON, H. L., BJØRK, G., HARLAND, R., MCQUOID, M. R. & NORDBERG, K. 2005. A major change in the phytoplankton of a Swedish sill fjord – a consequence of engineering work? *Estuarine, Coastal and Shelf Science*, **63**, 551–560.

GRAN, H. H. 1990. *Hydrographic-biological studies of the North Atlantic Ocean and the coast of Nordland*. Norwegian Fishery and Marine Investigations, Report Vol. 1, No. 5.

GERLAND, S. A. & RENNER, H. H. 2007. Sea-ice mass balance monitoring in an Arctic fjord. *Annals of Glaciology*, **46**, 435–442.

GRØSFJELD, K. & HARLAND, R. 2001. Distribution of modern dinoflagellate cysts from inshore areas along the coast of southern Norway. *Journal of Quaternary Science*, **16**, 651–659.

GRØSFJELD, K., HARLAND, R. & HOWE, J. A. 2009. Dinoflagellate cyst assemblages inshore and offshore Svalbard reflecting their modern hydrography and climate. *Norwegian Journal of Geology*, **89**, 121–134.

HALD, M., EBBESEN, H. *ET AL.* 2004. Holocene palaeoceanography and glacial history of the West Spitsbergen area, Euro-Arctic margin. *Quaternary Science Reviews*, **23**, 2075–2088.

HARLAND, R. & PUDSEY, C. J. 1999. Dinoflagellate cysts from sediment traps deployed in the Bellinghausen, Weddell and Scotia Seas, Antarctica. *Marine Micropalaentology*, **37**, 77–99.

HARLAND, R. & REID, P. C. 1980. Systematics Part 1. *In*: HARLAND, R., REID, P. C., DOBELL, P. & NORRIS, G. (eds) *Recent and Sub-Recent Dinoflagellate Cysts from the Beaufort Sea*. Canadian Arctic, Grana, **19**, 211–225.

HARLAND, R., NORDBERG, K. & FILIPSON, H. L. 2004*a*. A high-resolution dinoflagellate cyst record from latest Holocene sediments in Koljo Fjord, Sweden. *Review of Palaeobotany and Palynology*, **128**, 119–141.

HARLAND, R., NORDBERG, K. & FILIPSON, H. L. 2004*b*. The seasonal occurrence of dinoflagellate cysts in surface sediments from Koljo Fjord, west coast of Sweden – a note. *Review of Palaeobotany and Palynology*, **128**, 107–117.

HARLAND, R., NORDBERG, K. & FILIPSSON, H. L. 2006. Dinoflagellate cysts and hydrographical change in Gullmar Fjord, west coast of Sweden. *Science of the Total Environment*, **355**, 204–231.

HEAD, M. J. 1996. Modern dinoflagellate cysts and their biological affinities. *In*: JANSONIUS, J. & MCGREGOR, D. C. (eds) *Palynology: Principles and Applications*. American Association of Stratigraphic Palynologists Foundation, Dallas, **3**, 1197–1248.

HEAD, M. J., HARLAND, R. & MATTHIESSEN, J. 2001. Cold marine indicators of the late Quaternary: the new dinoflagellate cyst genus *Islandinium* and related morphotypes. *Journal of Quaternary Science*, **16**, 621–636.

HONJO, S., MANGANINI, S. J. & WEFER, G. 1988. Annual particle flux and a winter outburst of sedimentation in the northern Norwegian Sea. *Deep-Sea Research*, **35**, 1223–1234.

HOP, H., FALK-PETERSEN, S., SVEIAN, H., KWASNIEWSKI, S., PAVLOV, V., PAVLOVA, O. & SØREIDE, J. E. 2006. Physical and biological characteristics of the pelagic system across Fram Strait to Kongsfjorden. *Progress in Oceanography*, **71**, 182–231.

HOP, H., PEARSON, T. *ET AL.* 2002. The marine ecosystem of Kongsfjorden, Svalbard. *Polar Research*, **21**, 167–208.

INDELICTO, S. R. & LOEBLICH, A. R. III 1986. A revision of the marine peridinioid (Pyrrhophyta) utilizing hypothecal-angular plate relationships as a taxonomic guideline. *Japanese Journal of Phycology (Sôrui)*, **34**, 153–162.

IVANOV, I. V., POLYAKOV, V. V. *ET AL.* 2008. Seasonal variability in AW off Spitsbergen. *Deep-Sea Research I*, **56**, 1–14.

JEONG, H. J. 2008. The ecological roles of heterotrophic and mixotrophic dinoflagellates in marine food webs. Abstracts. *Eighth International Conference on Modern and Fossil Dinoflagellates*, Montreal, Canada, **23**.

KREMP, A. & HEISKANEN, A.-S. 1999. Sexuality and cyst formation of the spring-bloom dinoflagellate *Scrippsiella hangoei* in the coastal northern Baltic Sea. *Marine Biology*, **134**, 771–777.

Kwasniewski, S., Hop, H., Falk-Petersen, S. & Pedersen, G. 2003. Distribution of Calanus species in Kongsfjorden, a glacial fjord in Svalbard. *Journal of Plankton Research*, **25**, 1–20.

MacLachlan, S. E., Cottier, F., Austin, W. E. N. & Howe, J. A. 2007. The salinity: $\delta^{18}O$ water relationship in Kongsfjorden, western Spitsbergen. *Polar Research*, **26**, 160–167.

Marret, F. & Zonneveld, K. A. F. 2003. Atlas of modern organic-walled dinoflagellate cyst distribution. *Review of Palaeobotany and Palynology*, **125**, 1–200.

Matthiessen, J., de Vernal, A., Head, M., Okolodkov, Y., Zonneveld, K. & Harland, R. 2005. Modern organic-walled dinoflagellate cysts in Arctic marine environments and their (paleo-) environmental significance. *Paläontologische Zeitschrift*, **79**, 3–51.

Matthiessen, J., Knies, J., Nowaczyk, N. R. & Stein, R. 2001. Late Quaternary dinoflagellate cyst Stratigraphy at the Eurasian continental margin, Arctic Ocean: indications for AW inflow in the past 150,000 years. *Global and Planetary Change*, **31**, 65–86.

Montresor, M., Zingone, A. & Sarno, D. 1998. Dinoflagellate cyst production at a coastal Mediterranean site. *Journal of Plankton Research*, **20**, 2291–2312.

Nilsen, F., Cottier, F. R., Skogseth, R. & Mattsson, S. 2008. Fjord-shelf exchanges controlled by ice and brine production: the interannual variation of AW in Isfjorden, Svalbard. *Continental Shelf Research*, **28**, 1838–1853.

Persson, A. & Rosenberg, R. 2003. Impact of grazing and bioturbation of marine benthic deposit feeders on dinoflagellate cysts. *Harmful Algae*, **2**, 43–50.

Sakshaug, E. 2003. Primary and secondary production in the arctic seas. *In*: Stein, R. & MacDonald, R. W. (eds) *The Organic Carbon Cycle in the Arctic Ocean*. Springer, Berlin, 57–81.

Smayda, T. J. 1980. Phytoplankton species succession. *In*: Morris, I. (ed.) *The Physiological Ecology of Phytoplankton*. University of California Press, Berkeley, 493–570.

Solignac, S., Grøsfjeld, K., Giraudeau, J. & de Vernal, A. 2009. A Distribution of modern dinocyst assemblages in the western barents sea. *Norwegian Journal of Geology*, **88**, 109–119.

Svendsen, H., Beszczynska-Møller, A. *et al.* 2002. The physical environment of Kongsfjorden-Krossfjorden, an Arctic fjord system in Svalbard. *Polar Research*, **21**, 133–166.

Taylor, F. J. R. & Pollingher, U. 1987. Ecology of dinoflagellates. *In*: Taylor, F. J. R. (ed.) *The Biology of Dinoflagellates*. Blackwell Scientific Publications, Oxford, 399–529.

Willis, K., Cottier, F., Kwasniewski, S., Wold, A. & Falk-Petersen, S. 2006. The influence of advection on zooplankton community composition in an Arctic fjord (Kongsfjord, Svalbard). *Journal of Marine Systems*, **61**, 39–54.

Wood, G. D., Gabriel, A. M. & Lawson, J. C. 1996. Palynological techniques-processing and microscopy. *In*: Jansonius, J. & McGregor, D. C. (eds) *Palynology: Principles and Applications*. American Association of Stratigraphic Palynologists Foundation, Salt Lake City, **1**, 29–50.

Zonneveld, K. A. F., Versteegh, G. J. M. & de Lange, G. J. 2001. Palaeoproductivity and post-depositional aerobic organic matter decay reflected by dinoflagellate cyst assemblages of the Eastern Mediterranean S1 sapropel. *Marine Geology*, **172**, 181–195.

A major change in the dinoflagellate cyst flora of Gullmar Fjord, Sweden, at around 1969/1970 and its possible explanation

REX HARLAND[1,2], KJELL NORDBERG[1] & HELENA L. FILIPSSON[3]

[1]*Earth Sciences Centre, Göteborg University, PO Box 460, S-405 30, Göteborg, Sweden*

[2]*50 Long Acre, Bingham, Nottingham NG13 8AH, UK*

[3]*GeoBiosphere Science Centre, Quaternary Sciences, Lund University, Sölvegatan 12, S-223 62, Lund, Sweden*

**Corresponding author (e-mail:rex.harland@ntlworld.com)*

Abstract: An ultra high-resolution study of the latest Holocene dinoflagellate cysts from Gullmar Fjord, on the west coast of Sweden, provides evidence for the recognition of at least two major dinoflagellate communities within the fjord over the last 85 years. These communities may result from changes within the North Atlantic Oscillation (NAO) and hydrography of the fjord between the approximate years 1915 and 1999 and/or from the local pollution history. The dinoflagellate cyst populations were compared in detail with hydrographical parameters available from this fjord with its long historical instrumental records. The dinoflagellate cysts fail to demonstrate a convincing ongoing eutrophication for the fjord, although the reduction of *Lingulodinium polyedrum* partly coincides with the curtailment of activity at a sulphite pulp mill at Munkedal and canning activity at Lysekil, together with a cessation in the influx of untreated sewage from water closets. The significant change in the assemblage composition at about the late 1960s/early 1970s coincides with a change in the NAO from a negative phase to its present-day positive phase. The unravelling of local environmental effects from those associated with regional fluctuations is complex and needs to be approached with caution.

Gullmar Fjord is a sill fjord on the west coast of Sweden, north of Göteborg. The fjord (Fig. 1) is some 30 km long and 2 km wide and has a maximum depth of 119 m. It is orientated NE/SW and opens into the Skagerrak across a sill at 42 m depth. The fjord is subjected to strong stratification with pycnoclines occurring between 15 and 20 m, separating upper low salinity waters (24–27 psu) from more saline waters (32–33 psu). Between 50 and 60 m, these waters are separated from the highly saline waters (34–35 psu) in the deeper parts of the fjord (Lindahl *et al.* 1998; Arneborg 2004). This stratification arises from the presence of low salinity Baltic water at the surface in the Skagerrak (Gustafsson 1999; Björk & Nordberg 2003) and higher salinity North Sea and North Atlantic water at depth. The deepest parts of the fjord, below 80 m, are occasionally subjected to hypoxic conditions during the autumn and winter providing the perfect environment to preserve the stratigraphic record without significant disturbance.

Materials and methods

Sampling and processing

The core site G113, situated at latitude 58°18′95N and longitude 11°32′36E (Fig. 1), was sampled in June 1999 by the RV *Skagerak* using a Gemini Corer. The corer provided a virtually undisturbed sediment/water interface and cores some 80 mm in diameter and 610 mm in length. After collection, the cores were X-rayed with an Andrex BV (155 140 Kv/5 mA) portable machine. In the laboratory, the cores were sliced and analysed for their organic carbon content using a Carlo Erba NA 1500 instrument. Radiometric dating using the ^{210}Pb methodology and the constant rate of supply (CRS) model (Appleby & Oldfield 1978) was carried out on two parallel cores at the Department of Radiation Physics, University of Lund, Sweden (see Nordberg *et al.* 2001 and Filipsson & Nordberg 2004 for further details and the age–depth profile). The sediment accumulation rate was calculated at

From: Howe, J. A., Austin, W. E. N., Forwick, M. & Paetzel, M. (eds) *Fjord Systems and Archives.* Geological Society, London, Special Publications, **344**, 75–82.
DOI: 10.1144/SP344.7 0305-8719/10/$15.00

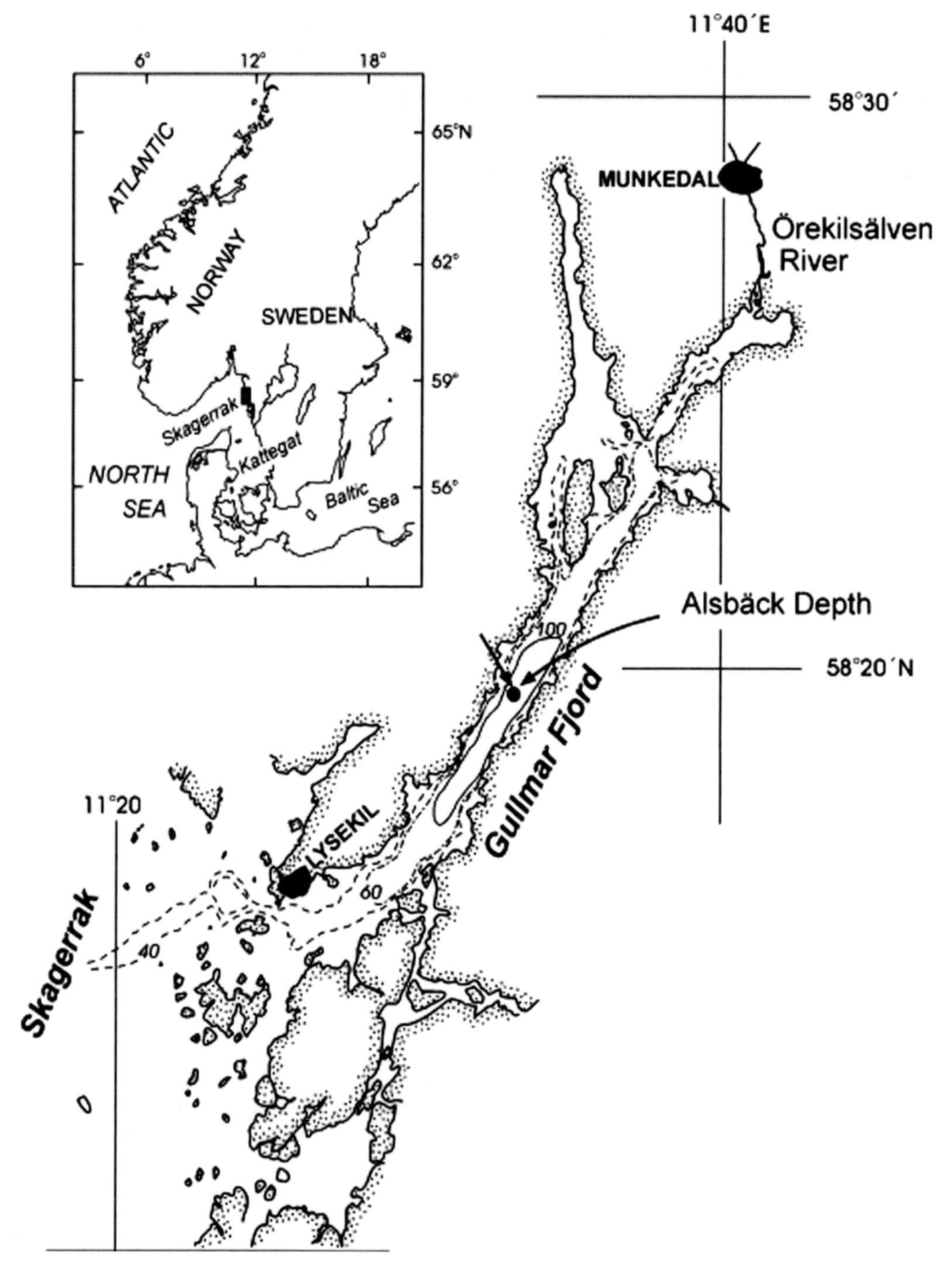

Fig. 1. Location map of Gullmar Fjord showing the core site and the source of local pollution from Lysekil and Munkedal.

0.7 cm a^{-1}. Additional data on the stable isotopes and calculated bottom water temperatures are now available (Filipsson & Nordberg 2010).

Sixty samples were taken from the recovered core GA113-2Ab at a sample interval of 1 cm. The sediment is homogenous, organic-rich clay

(Nordberg *et al.* 2000) and the samples were processed using the normal palynological techniques (Wood *et al.* 1996). The samples were not oxidized during processing, preventing the differential loss of protoperidiniacean cysts (Dale 1976; Zonneveld *et al.* 1997), but were subjected to prolonged washing and filtering together with the use of ultrasound to free the contained dinoflagellate cysts from the inherent amorphous organic material (AOM). The resulting residues were stained with Safranin, mounted in Elvacite and bonded to slides using Petrapoxy 154 resin which has a refractive index of 1.54. The samples were treated quantitatively to calculate the numbers of cysts per gram of sediment (see Harland 1989; Harland *et al.* 2006). TILIA/TILIAGRAPH software was used to construct a selected dinoflagellate cyst spectrum; the age–depth model, the spectrum and all the data were published in Harland *et al.* (2006) and are not repeated here.

Data treatment

The stratigraphically constrained incremental sum of squares cluster analysis using CONISS (Grimm 1987) was employed to differentiate the various cyst assemblages. In addition, the heterotrophic ratio was calculated for each sample, that is, the numbers of cysts from heterotrophic dinoflagellates divided by those identified as derived from autotrophic dinoflagellates and expressed as a log value (see fig. 3 of Harland *et al.* 2006). High values have been assumed to indicate the probable enhancement of nutrients, as in areas of upwelling (see Powell *et al.* 1990). Since there is little research that links this ratio to any established palaeoecological model, it is however advisable to treat its interpretation with caution for the moment.

In order to compare the historical instrumental records of hydrography and atmospheric data with the dinoflagellate cysts the data have been treated differently, especially with respect to averaging over time. All the surface water data were, however, depth averaged between 0–10 m. The yearly hydrographical data is provided in figure 3 of Harland *et al.* (2006), but the larger part of the datasets were divided into two periods: a spring period (March, April and May) and a late summer–autumn (July, August and September) (figs 4 & 5 of Harland *et al.* 2006). Details of all the data handling are provided in full in Harland *et al.* (2006) including upwelling events, air temperature, precipitation, the North Atlantic Oscillation (NAO) and primary productivity and are not repeated here.

Results

Dinoflagellate cysts were recovered from all samples together with bisaccate and angiosperm pollen, copepod egg cases, foraminiferal linings and structured and amorphous organic material. Dinoflagellate cysts generally number about 4000 cysts per gram but rise to over 7000 cysts per gram of sediment in places; this is close to the figure recorded by Persson *et al.* (2000) of 11 000 from Alsbäck, Gullmar Fjord. The cyst species recovered are well known and all occur in the Zone 1 assemblage of Dale's (1985) study of the Skagerrak and the present-day cyst flora from the coastal waters of southern and western Norway (Grøsfjeld & Harland 2001).

The Gullmar Fjord dinoflagellate cyst record, as seen in the dinoflagellate cyst spectrum published by Harland *et al.* (2006), can be divided into the following two major units. Unit I at 0.01–0.21 m (core depth) was deposited during the years 1969–1999 and is characterized by a diversity of 29 cyst species and over 3000 cysts per gram of sediment. In particular, it contains high numbers of *Protoceratium reticulatum*, *Spiniferites bentorii*, *Pentapharsodinium dalei*, *Protoperidinium avellana*, *Protoperidinium conicoides*, *Protoperidinium conicum*, *Protoperidinium pentagonum*, *Protoperidinium* spp. [spiny], *Islandinium* cf. *cesare* and *Polykrikos schwartzii*. Full citations for all the dinoflagellate cysts recovered in the study are provided in Harland *et al.* (2006). The heterotrophic ratio is within positive values throughout.

Unit II at 0.22–0.60 m (core depth) was deposited during the years 1915–1969 and is characterized by a diversity of 31 species and over 4000 cysts per gram of sediment. In particular, it contains high proportions of *Bitectatodinium tepikiense*, *Lingulodinium polyedrum*, *Spiniferites elongatus*, *Spiniferites mirabilis* and peridinioid *Protoperidinium* species. The heterotrophic ratio is mostly within negative values. Unit II itself can be further divided (see Harland *et al.* 2006) but the resulting subdivisions, although of interest, are not as significant in our opinion as that occurring in the late 1960s/early 1970s and are not discussed further in this contribution.

Comparisons with other long-term datasets

The dinoflagellate cyst record was compared to a number of long-term historical datasets (see *Materials and methods*). Comparison of these data to the dinoflagellate cyst record reveals little correlation despite the major cyst change at around 1969/1970 and the accompanying increase in the heterotrophic ratio (Harland *et al.* 2006).

Although there is little differentiation within the temperature data there is some indication that summer sea-surface temperatures (SSTs) were somewhat higher between 1930 and 1970 and particularly between 1931 and 1950; this is also seen in the averaged summer air temperatures recorded locally (see Harland *et al.* 2006). This appears to correlate with the raised numbers of some cysts between the same years (see Fig. 2b). The air temperature data also reveals higher summer temperatures from about 1924 until about 1958. In addition, the histogram of upwelling events over the temporal record (see Harland *et al.* 2006 and Fig. 2c) also shows an increase over these same years, consistent with the influence of the negative NAO index (Fig. 2d) and Ekman pumping along the west coast of Sweden.

Discussion

The dinoflagellate cyst species recovered are consistent with those from modern sediments along the west coast of Sweden (Erdtman 1954; Persson *et al.* 2000). Dinoflagellate cyst distribution patterns are known to be largely governed by several factors given the assumption that there has been no *post mortem* alteration or other taphonomic interference, notably temperature, position within an onshore/offshore gradient and the availability of nutrients (Marret & Zonneveld 2003). Salinity is also another well-known determinant in cyst distribution patterns (Dale 1996).

Since regional climate has not changed markedly over the last 110 years, as can be demonstrated from the temporal instrumental records, much of the dinoflagellate cyst record must be attributable to changes in the availability of nutrients, trace elements and other abiotic and biotic factors on seasonal and annual timescales. However, evidence

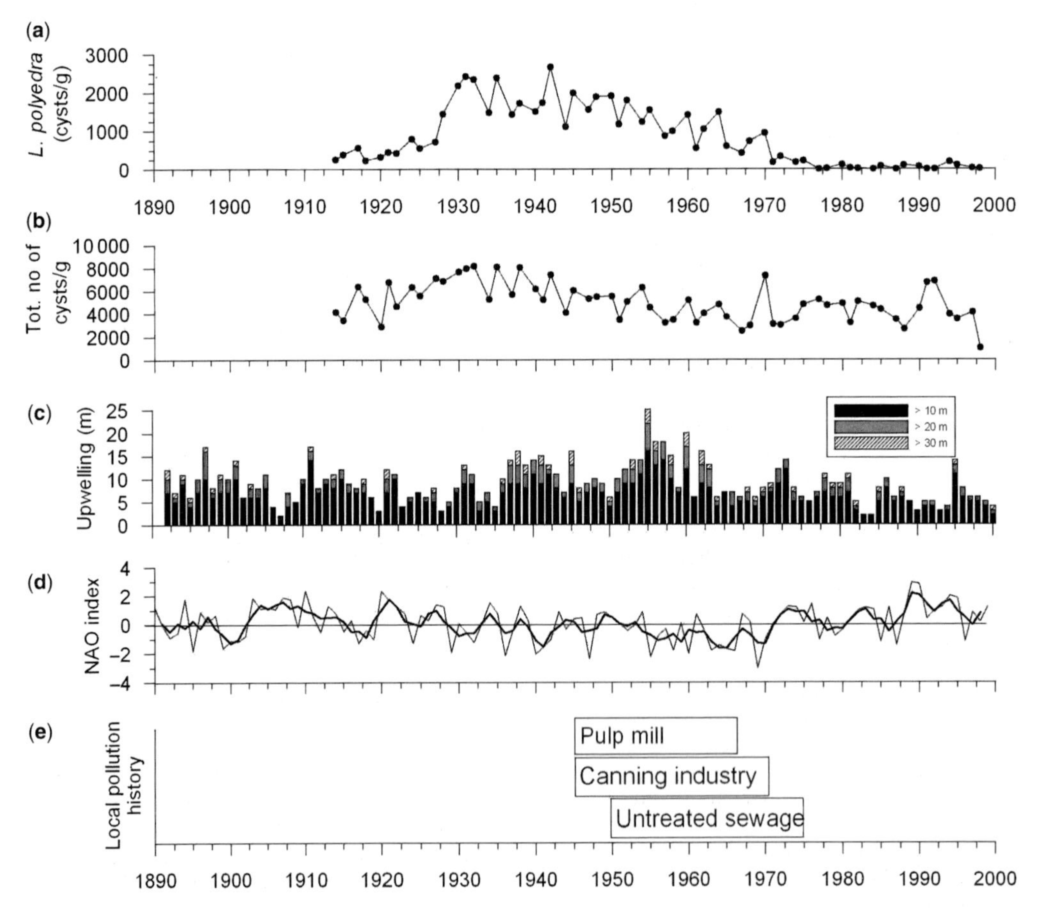

Fig. 2. Composite illustration showing (**a**) *Lingulodinium polyedrum* per gram of sediment against time; (**b**) the total number of cysts per gram of sediment; (**c**) the local upwelling record (from Björk & Nordberg 2003); (**d**) the NAO index, averaged over January to March and the three year running mean (thicker line); and (**e**) the local pollution history. Discharges from the Munkedal pulp mill were greatest between 1945 and 1966, from the local canning industry between 1945 and 1970 and from untreated sewage from the town of Lysekil and surrounds between 1950 and 1975. After the mid-1970s there are no significant local pollution sources in the fjord (Lindahl *et al.* 1998).

is growing that recent climate warming has induced major reorganizations of the North Atlantic pelagic ecosystem from phytoplankton to fish from the 1920s, 1930s and into the late 1980s (Beaugrand *et al.* 2002; Beaugrand 2004; Drinkwater 2006). These changes will have undoubtedly impacted the North Sea and its contiguous waters with the plankton (including the phytoplankton, the dinoflagellates and their cyst production) being a part of these regime shifts.

The availability of nutrients within the water column to sustain the dinoflagellate populations is derived from: the coastal hinterland as runoff; from tidal mixing of sediments in the shallow near-shore zone especially during the winter months and around the vernal and autumnal equinoxes; and from upwelling and advection in the offshore areas. Along the Swedish west coast the availability of nutrients derives from the surface water circulation pattern and the changing water mass types. Tidal energy is low and the effect of runoff is negligible in the more open aspect of Gullmar Fjord (Björk & Nordberg 2003). In this respect the exchange of water with the open sea, the Kattegat and Skagerrak, is important to marine life in the coastal zone (Björk & Nordberg 2003). This water exchange is largely driven by geostrophic winds that have the effect of forcing upwelling and controlling the availability of nutrients within the system. The upwelling history, tuned from a unique long-series hydrographic dataset, suggests: a declining trend between the late 1890s and 1920; an increasing frequency to 1940; and a further decline between the 1950s and 1990 (Björk & Nordberg 2003). These upwelling trends are mirrored in the NAO and, in particular, by the *Lingulodinium polyedrum* curve within the dinoflagellate cyst spectrum as published in Harland *et al.* 2006 and as illustrated in Figure 2. This changing frequency in the availability of nutrients through fluctuations in summer upwelling events may be sufficient to allow a dinoflagellate such as *Lingulodinium polyedrum*, that thrives in stable and stratified water, to take ecological advantage of increased nutrients from depth to increase its population size and, on encystment, produce increased cyst numbers to enter the fossil record (Marret & Zonneveld 2003; Harland *et al.* 2004). Indeed, in their matrix of dinoflagellate life forms Smayda & Reynolds (2001, 2003) classify *Lingulodinium polyedrum* as a Type V (Upwelling Relaxation Taxa) that employ survival strategies to prosper in these particular circumstances.

The dinoflagellate cyst record documented from Core GA113-2Ab taken in the Gullmar Fjord appears to be responding to change occurring in the Skagerrak and possibly further afield in the North Sea. The coastal current running up the west coast of Sweden from the Kattegat is highly variable and responds rapidly to changes in the geostrophic wind patterns (Björk & Nordberg 2003). Consequential upwelling resulting from the wind patterns will control the amount of nutrients coming to the surface through advection and Ekman pumping in the offshore area and hence will have a marked effect on the dinoflagellate populations and on the cyst record.

The dinoflagellate cyst record has a distinctive signature, which points to some environmental change within the marine realm at around 1969/1970. Examination of the various hydrographical parameters, using long-sequence high-resolution instrumental records, has enabled us to match this change to only two parameters. Interestingly, it was not possible to identify any correspondence with temperature, salinity or indeed to nutrient content despite a possible gently rising trend in the nutrients. The only reasonable correspondence was between the dinoflagellate cyst record and the frequency of upwelling in coastal waters, brought about as a consequence of the action of the NAO (Fig. 2c). As the NAO changed from a negative phase to a positive phase around 1972/1973, the frequency of upwelling decreased. This partly corresponds with a statistically significant rise in local summer temperatures (see Harland *et al.* 2006).

The results are somewhat complicated by the history of pollution in the fjord from a sulphite pulp mill and untreated sewage outfalls from Munkedal village (Rosenberg 1976) and untreated sewage from water closets in the town of Lysekil together with effluent from a local canning factory (see Fig. 2e). The pulp mill closed in 1966 and, beyond the mid 1970s, no significant local pollution sources were present within the fjord (Lindahl *et al.* 1998). Interestingly, symptoms of large-scale eutrophication only began to be reported from the 1970s and 1980s after both the pulp mill and the canning factory had ceased production. The pattern of dinoflagellate cyst distribution within the cyst spectrum and the major shift in the dinoflagellate cyst assemblages appears to bear little resemblance to the pollution history, unlike the cyst records from Oslofjord (Dale *et al.* 1999).

We consequently recognize the difficulty in assigning the change in dinoflagellate cyst assemblages at around 1969/1970 to either a higher frequency of upwelling events (because of a change in the NAO regime) or to local pollution difficulties (see Fig. 2d, e). For the moment, we err on the side of the increased frequency of upwelling events as Gullmar Fjord reflects hydrodynamic changes that are happening outside in the Skagerrak (rather than the more local effects within the fjord). Indeed, newly acquired data on a 2500 year record from the fjord (work in progress) describes fluctuating numbers of the species *Lingulodinium polyedrum* at times long before the possibilities of any cultural influences within the fjord. It is perhaps premature to link the occurrence of this dinoflagellate species to problems with eutrophication alone.

In addition to the major change in dinoflagellate cyst assemblages at the end of the 1960s/early 1970s, the published spectrum (Harland *et al.* 2006) shows several other features. One of these is the raised numbers of *Bitectatodinium tepikiense* from the base of the cored sequence until the mid 1950s. Dale (1985) recognized this species in his Skagerrak cores and Marret & Zonneveld (2003) describe it as a fully marine species in cold/temperate environments with winter SST between −2 and 17 °C and summer SST between −2 and 21 °C. Its presence in numbers within Unit II may well signify some cooler water conditions within Gullmar Fjord prior to the mid 1950s. Other changes in the cyst populations are apparent but are not the subject of this contribution.

The interpretation of changing dinoflagellate cyst assemblages is fraught with difficulties and the recognition of environmental proxies using a limited number of cyst species should be approached with caution. Nevertheless, the continuing accumulation of ultra-high-resolution dinoflagellate cyst records will undoubtedly assist in building a picture of the changing hydrography of the west coast of Sweden through the latest Holocene and into the historical timeframe of climate change and the impending consequences to society.

Conclusions

Our ultra-high-resolution study shows a distinctive change in the dinoflagellate cyst populations during around the late 1960s/early 1970s, which is clearly indicative of a major change in the environment. The only change noted in our study of the long-term instrumental records is the changeover from a largely negative to positive NAO at about the same time and a decrease in the frequency of upwelling events in the Skagerrak. The response of the dinoflagellate populations to this change is not easily explained as it is likely to be complex and dynamic. It probably involves differences in the availability of nutrients to the dinoflagellate populations, as the region is variously affected either by north and northeasterly winds (−ve NAO index) or west or southwesterly winds (+ve NAO index). It may also involve major regime changes within the phytoplankton of the North Sea and northeastern Atlantic Ocean. The cyst record shows little correlation with the putative ongoing eutrophication in the fjord.

This paper was presented at the 1st International Workshop on 'Fjord environments: Past, Present and Future' held at the Scottish Association of Marine Science, Oban in May 2006. The authors would like to thank the crew of the RV *Arne Tiselius* and RV *Skagerak* for their assistance in collecting the cores, S. Ellin and A. B. Ibrahim for processing the samples in the Palynology Research Facility, University of Sheffield and two anonymous reviewers for their constructive comments. The study was financed by the Swedish Natural Science Research Council (NFR grant no. VR 621-2007-4369, K. Nordberg), the Futura Foundation, the Oscar & Lili Lamm Foundation, the Carl Trygger Foundation, the Wåhlsröm Foundation and Göteborg University Marine Research Centre (GMF).

Addendum

Since the publication of Harland *et al.* (2006) and the submission of this manuscript, a paper by Dale (2009) reviews the use of dinoflagellate cysts as signals for eutrophication. Several criticisms, and a reinterpretation of our data, were included that question our explanation of the high numbers of *Lingulodinium polyedrum* in Gullmar Fjord as a response to increased nutrients from depth during a negative phase of the NAO, and not simply as an eutrophication signal.

In Harland *et al.* (2006) we commented on the striking difference between the dinoflagellate cyst record from Gullmar Fjord and that from Koljö Fjord (Harland *et al.* 2004) despite their close proximity (some 8 km apart) and the similarity of their temporal archive. The Koljö Fjord sediments reveal increasing numbers of *Lingulodinium polyedrum* from 1938 onwards following engineering work to improve navigation within the fjord, with maximum numbers occurring in 1980. This is in marked contrast to the Gullmar Fjord record described by Harland *et al.* (2006) and in this paper. In addition, a dinoflagellate cyst record from the adjacent Havstens Fjord details this increase in *Lingulodinium polyedrum* numbers (Filipsson *et al.* 2005) with two notable acmes at *c.* 1940 and *c.* 1980. These facts alone are reason enough to recognize that Gullmar Fjord is responding to environmental change over the last 90 years in a manner different to other fjords along the west coast of Sweden, and that invoking eutrophication as the main driver maybe too simplistic. Indeed, comparing Gullmar Fjord with Oslofjord may not be useful as Oslofjord is a more restricted basin with little surface water circulation whereas Gullmar Fjord has an open connection to the Skagerrak and water residence times of only 20–38 days (Arneborg 2004). Gullmar Fjord clearly mirrors the environmental history of the Skagerrak where no increasing trends in nutrient development have been recorded over the last 30 years (Andersson 1996; Jonsson *et al.* 2003). Also, the Oslofjord area is urbanized to a much greater extent with a population of *c.* 1.7 million in contrast to that of *c.* 20 000 around Gullmar Fjord.

Finally, Godhe & McQuoid (2003) counter the view of *Lingulodinium polyedrum* as an eutrophication signal and show (using multivariate analysis) that summer temperature is the environmental parameter which most clearly impacts the present distribution of *Lingulodinium polyedrum* along the west coast of Sweden, with higher temperatures promoting its abundance. We believe the situation more complex than a simple response to eutrophication and that caution is required in interpreting these data. The Gullmar Fjord appears to be an example where *Lingulodinium polyedrum* reacts to climate variations; however, it is indisputable that *L. polyedrum* also reacts positively to nutrient load as a result of anthropogenic activity, as Dale (2009) has argued. We welcome the ongoing discussion.

References

ANDERSSON, L. 1996. Trends in nutrient and oxygen concentrations in the Skagerrak-Kattegat. *Journal of Sea Research*, **35**, 63–71.

APPLEBY, P. G. & OLDFIELD, F. 1978. The calculation of Lead –210 dates using a constant rate of supply of unsupported 210Pb to the sediment. *Catena*, **5**, 1–8.

ARNEBORG, L. 2004. Turnover times for waters above sill level in Gullmar Fjord. *Continental Shelf Research*, **24**, 443–460.

BEAUGRAND, G. 2004. The North Sea regime shift: evidence, causes, mechanisms and consequences. *Progress in Oceanography*, **60**, 245–262.

BEAUGRAND, G., REID, P. C., IBANEZ, F., LINDLEY, J. A. & EDWARDS, M. 2002. Reorganisation of North Atlantic marine copepod biodiversity and climate. *Science*, **296**, 1692–1694.

BJÖRK, G. & NORDBERG, K. 2003. Upwelling along the Swedish west coast during the 20th century. *Continental Shelf Research*, **23**, 1143–1159.

DALE, B. 1976. Cyst formation, sedimentation, and preservation: factors affecting dinoflagellate assemblages in recent sediments from Trondheimsfjord, Norway. *Review of Palaeobotany and Palynology*, **22**, 39–60.

DALE, B. 1985. Dinoflagellate cyst analysis of Upper Quaternary sediments in core GIK 15530–4 from the Skagerrak. *Norsk Geologisk Tidsskrift*, **65**, 97–102.

DALE, B. 1996. Dinoflagellate cyst ecology: Modeling and geological applications. *In*: JANSONIUS, J. & MCGREGOR, D. C. (eds) *Palynology: Principles and Applications*. American Association of Stratigraphic Palynologists Foundation, Salt Lake City, **3**, 1249–1275.

DALE, B. 2009. Eutrophication signals in the sedimentary record of dinoflagellate cysts in coastal waters. *Journal of Sea Research*, **61**, 103–113.

DALE, B., THORSEN, T. A. & FJELLSA, A. 1999. Dinoflagellate cysts as indicators of cultural eutrophication in the Oslofjord, Norway. *Estuarine, Coastal and Shelf Science*, **48**, 371–382.

DRINKWATER, K. F. 2006. The regime shift of the 1920s and 1930s in the North Atlantic. *Progress in Oceanography*, **68**, 134–151.

ERDTMAN, G. 1954. On pollen grains and dinoflagellate cysts in the Firth of Gullmarn, SW. *Sweden. Botaniska Notiser*, **2**, 103–111.

FILIPSSON, H. L. & NORDBERG, K. 2004. Climate variations, an overlooked factor influencing the recent marine environment. An example from Gullmar Fjord, Sweden. *Estuaries*, **27**, 867–888.

FILIPSSON, H. L. & NORDBERG, K. 2010. Variations in organic carbon flux and stagnation periods during the last 2400 years in a Skagerrak fjord basin, inferred from benthic foraminiferal $\delta^{13}C$. *In*: HOWE, J. A., AUSTIN, W. E. N., FORWICK, M. & PAETZEL, M. (eds) *Fjord Systems and Archives*. Geological Society, London, Special Publications, **344**, 261–270.

FILIPSSON, H. L., BJÖRK, G., HARLAND, R., MCQUOID, M. & NORDBERG, G. 2005. A major change in phytoplankton of a swedish sill fjord – a consequence of engineering work? *Estuarine, Coastal and Shelf Science*, **63**, 551–560.

GODHE, A. & MCQUOID, M. R. 2003. Influence of benthic and pelagic environmental factors on the distribution of dinoflagellate cysts in surface sediments along the swedish west coast. *Aquatic Microbial Ecology*, **32**, 185–201.

GRIMM, E. C. 1987. CONISS: a FORTRAN 77 program for stratigraphically constrained cluster analysis by the method of incremental sum of squares. *Computers & Geosciences*, **13**, 13–35.

GRØSFJELD, K. & HARLAND, R. 2001. Distribution of modern dinoflagellate cysts from inshore areas along the coast of southern Norway. *Journal of Quaternary Science*, **16**, 651–659.

GUSTAFSSON, M. 1999. High frequency variability of the surface layers in the Skagerrak during SKAGEX. *Continental Shelf Research*, **19**, 1021–1047.

HARLAND, R. 1989. A dinoflagellate cyst record for the last 0.7 Ma from the Rockall Plateau, northeast Atlantic Ocean. *Journal of the Geological Society*, **146**, 954–951.

HARLAND, R., NORDBERG, K. & FILIPSSON, H. L. 2004. A latest Holocene high-resolution dinoflagellate cyst record from Koljö Fjord, Sweden and its implications for palaeoenvironmental analysis. *Review of Palaeobotany and Palynology*, **128**, 119–141.

HARLAND, R., NORDBERG, K. & FILIPSSON, H. L. 2006. Dinoflagellate cysts and hydrographical change in Gullmar Fjord, west coast of Sweden. *Science of the Total Environment*, **355**, 204–231.

JONSSON, H., HAFSTRÖM, M., LÖNNROTH, M., JOHANNESSON, K., ÖSTERBLOM, H. & VEEM, K. 2003. *Havet – tid för en strategi. Slutbetänkande av Havsmiljökommissionen*. SOU (Statens offentliga utredningar), Stockholm, **73**, 1–273.

LINDAHL, O., BELGRANO, A., DAVIDSSON, L. & HERNROTH, B. 1998. Primary production, climatic oscillations, and physico-chemical processes: the Gullmar Fjord time series data set (1985–1996). *ICES Journal of Marine Science*, **55**, 723–729.

MARRET, F. & ZONNEVELD, K. A. F. 2003. Atlas of modern organic-walled dinoflagellate cyst distribution. *Review of Palaeobotany and Palynology*, **125**, 1–200.

NORDBERG, K., GUSTAFSSON, M. & KRANTZ, A.-L. 2000. Decreasing oxygen concentrations in the Gullmar

Fjord, Sweden, as confirmed by benthic foraminifera, and the possible association with NAO. *Journal of Marine Systems*, **23**, 303–316.

Nordberg, K., Filipsson, H. L., Gustafsson, M., Harland, R. & Roos, P. 2001. Climate, hydrographic variations and marine benthic hypoxia in Koljö Fjord, Sweden. *Journal of Sea Research*, **46**, 187–2000.

Persson, A., Godhe, A. & Karlson, B. 2000. Dinoflagellate cysts in recent sediments from the west coast of Sweden. *Botanica Marina*, **43**, 69–79.

Powell, A. J., Dodge, J. D. & Lewis, J. 1990. Late Neogene to Pleistocene palynological facies of the Peruvian continental margin upwelling, Leg 112. *Proceedings of the Ocean Drilling Program*, Scientific Results, **112**, 297–321.

Rosenburg, R. 1976. Benthic faunal dynamics during succession following pollution abatement in a swedish estuary. *Oikos*, **27**, 414–427.

Smayda, T. J. & Reynolds, C. S. 2001. Community assembly in marine phytoplankton: application of recent models to harmful dinoflagellate blooms. *Journal of Plankton Research*, **23**, 447–461.

Smayda, T. J. & Reynolds, C. S. 2003. Strategies of marine dinoflagellate survival and some rules of assembly. *Journal of Sea Research*, **49**, 95–106.

Wood, G. D., Gabriel, A. M. & Lawson, J. C. 1996. Palynological techniques – processing and microscopy. *In*: Jansonius, J. & McGregor, D. C. (eds) *Palynology: Principles and Applications*. American Association of Stratigraphic Palynologists Foundation, Salt Lake City, **1**, 29–50.

Zonneveld, K. A. F., Versteegh, G. J. M. & De Lange, G. J. 1997. Preservation of organic-walled dinoflagellate cysts in different oxygen regimes: a 10,000 year natural experiment. *Marine Micropaleontology*, **29**, 393–405.

High benthic foraminiferal species counts in a Clyde Sea maerl bed, western Scotland

WILLIAM E. N. AUSTIN* & ALIX G. CAGE

School of Geography and Geosciences, University of St Andrews, St Andrews, Fife, KY16 9AL, Scotland, UK

**Corresponding author (e-mail: bill.austin@st-andrews.ac.uk)*

Abstract: A preliminary investigation of the occurrence of hard-shelled benthic foraminifera in a sample of maerl (gravels derived from calcareous red seaweed) collected from a water depth of 10 m by SCUBA diving in the Clyde Sea, Scotland reveals an exceptionally well-preserved and diverse assemblage. A total of 90 different species were identified, comprising 12 agglutinated, 15 porcelaneous and 63 hyaline types; this represents the highest recorded species count described from the west of Scotland. The assemblage probably represents an integrated record of material derived directly from the maerl (mostly epifaunal types) and material derived from the reworking of adjacent sediments (mostly infaunal types). Maerl deposits represent an important near-shore habitat which appear to have been largely overlooked in regional studies of benthic foraminiferal distribution/ecology.

Maerl beds are calcareous gravels formed by the accumulation of slow-growing unattached non-geniculate Corallinaceae (Rhodophyta), commonly known as calcareous red seaweed. They are globally distributed (e.g. Foster *et al.* 1997) and usually occur at water depths of less than 40 m, particularly where there is some tidal flow. They are therefore of particular importance as a near-shore habitat, typically supporting a high benthic biodiversity and biomass (e.g. Bosence 1979). European maerl beds are of recognized conservation importance (Anon 1992) despite continued threats to these habitats specifically from direct exploitation, eutrophication and towed demersal fishing gear (see Hall-Spencer & Atkinson 1999).

A recent review of foraminifera from the west coast of Scotland is provided by Murray (2003) in his study of the benthic foraminifera on the Hebridean shelf. He reports the most noteworthy distributional/ecological studies from the region as (i) Heron-Allen & Earland (1916): various inner-shelf localities; (ii) Edwards (1982): North Minch Channel; (iii) Murray (1985) and Murray & Whittaker (2001): open shelf and deeps; and (iv) Hannah & Rogerson (1997): Clyde Sea area. The diversity of benthic foraminifera reported in the above studies is typical of NW European shelf-seas. For example, Murray (2003) records a total of 77 species of dead foraminifera from an investigation of 9 sample stations. None of the above records provide an account of benthic foraminifera in maerl deposits from the west of Scotland. However, within a palaeontological context, maerl deposits are known to preserve foraminiferal assemblages from other European settings (e.g. Morhange *et al.* 2003). Additional references which deal with maerl and foraminifera include Blanc-Vernet (1969), Boillot (1964), Freiwald & Henrich (1994) and Rosset-Moulinier (1972); all suggesting relatively high diversity.

The aim of this study is to provide the first systematic listing of hard-shelled modern benthic foraminifera from Scottish maerl beds and to evaluate whether or not this globally distributed near-shore habitat supports an unusually high biodiversity of benthic foraminifera.

Regional setting

The study site (Fig. 1) is a maerl bed, 10 m below chart datum (CD) in Stravanan Bay off the Isle of Bute, Clyde Sea area, Scotland (55°45.323′N 005°04.265′W; site 2 of Hall-Spencer & Atkinson 1999). In the broad sense of the term *fjord*, the Clyde Sea shares many of the physical oceanography characteristics of fjords (e.g. Rippeth & Simpson 1996). Hall-Spencer & Atkinson (1999) report that the sediments at this site are coarse (typically 40–70% gravel in the upper 4 cm), with subsurface sediments composed of *c.* 50% gravel, *c.* 31% coarse sand, *c.* 14% fine sand and *c.* 4% mud. Bottom water salinity at the site did not vary significantly (31–33), bottom currents of up to 11 cm s^{-1} were recorded and temperatures ranged from 6 °C (February 1997) to 14.1 °C (September 1997; Hall-Spencer & Atkinson 1999). Such large

From: Howe, J. A., Austin, W. E. N., Forwick, M. & Paetzel, M. (eds) *Fjord Systems and Archives*. Geological Society, London, Special Publications, **344**, 83–88.
DOI: 10.1144/SP344.8 0305-8719/10/$15.00

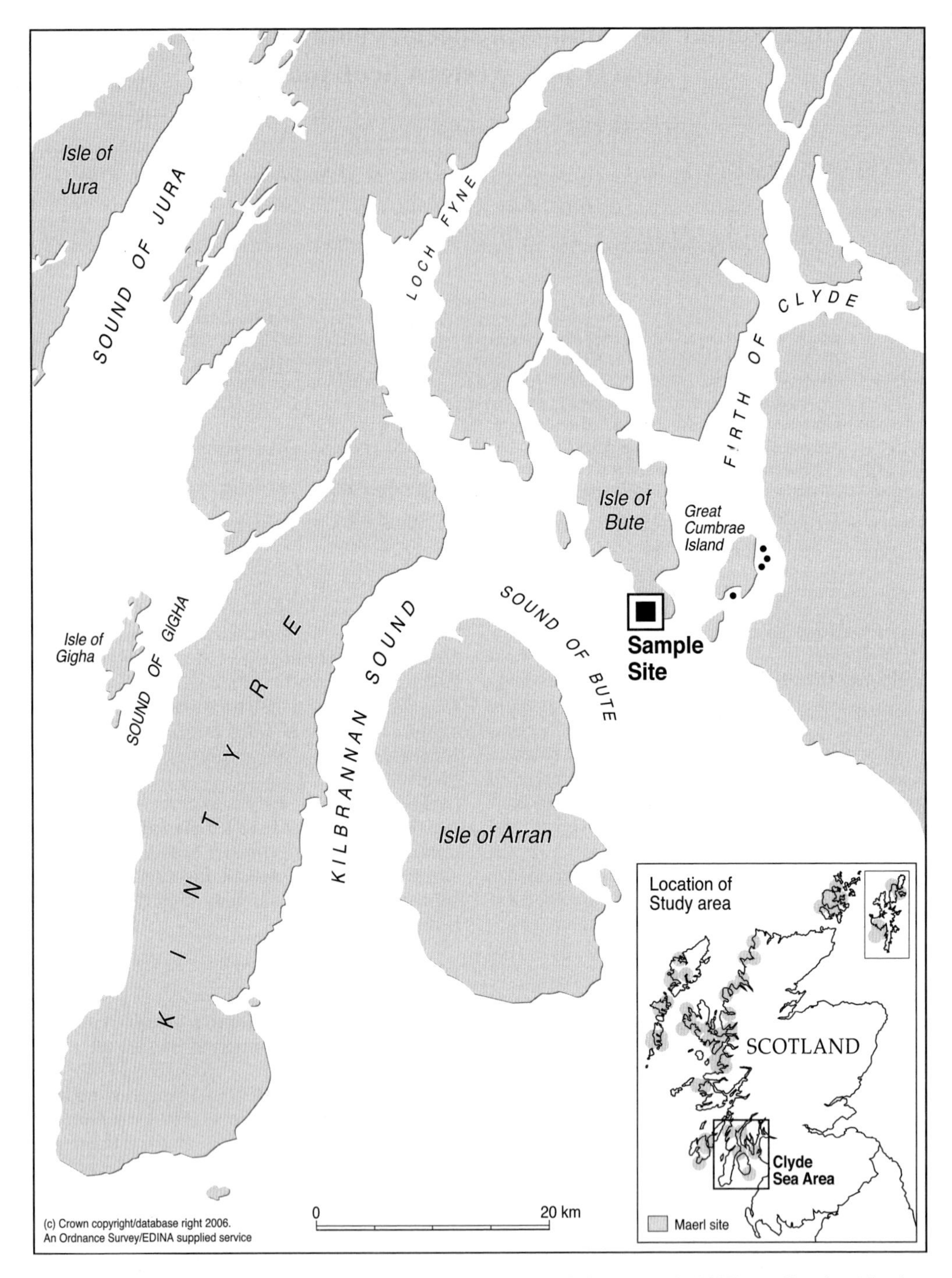

Fig. 1. Location Map, showing the study site (square symbol) and Hannah & Rogerson's (1997) sampling sites (circular symbol). The inset map shows (shaded areas) the approximate location of recorded maerl beds in Scotland (data from Scottish Natural Heritage; Haskoning Ltd. 2006).

intra-annual temperature differences are typical of the effects of the seasonal heating cycle experienced throughout much of the NW European shelf seas (Austin *et al.* 2006).

Material and methods

The sample was collected as a short sediment-water interface core by SCUBA diving in July 2001. The sample was wet sieved at 63 μm, dried at 40 °C overnight and then separated using a heavy-liquid floatation method before foraminifera were picked and sorted. A quantitative assemblage count of randomly encountered hard-shelled benthic foraminifera was undertaken (Table 1), together with a qualitative inspection of the entire assemblage from both the 'floated' and coarse fractions in order to detect additional, rare types (Table 2). Unfortunately, the sample available to us was neither preserved nor treated with the protoplasmic stain Rose Bengal at the time of collection. For this reason, live and dead specimens are not distinguished in this study. Taxonomic assignments are largely based upon Höglund (1947), Haynes (1973) and Murray (2003).

Results, discussion and conclusion

Based on a faunal assemblage count of 402 specimens, a total of 45 hard-shelled benthic foraminifera were identified from the maerl sample (Table 1). The assemblage is dominated by the following species: *Asterigerinata mamilla* (16.4%), *Milionella subrotunda* forma *hauerinoides* (10.7%), *Spiroplectammina wrightii* (10.4%), *Fissurina elliptica* (8.2%), *Quinqueloculina seminulum* (7.0%) and *Elphidium gerthi* (5.2%). Careful inspection of the full 'floated' sample yielded a total of 90 hard-shelled benthic foraminifera, comprising 12 agglutinated, 15 porcelaneous and 63 hyaline types. The relative species abundances of the full assemblages have been estimated (Table 2). In addition, the coarse fraction (including algal fragments) was inspected and revealed a tendency to contain a greater proportion of robust agglutinated and porcelaneous tests; this probably reflects a known methodological bias in the heavy-liquid separation technique which favours the recovery of intact, multi-chambered forms. Preservation of the fauna was generally very good, with the exception of a few of the attached epibenthic species such as *Acervulina inhaerens* and *Cibicides lobatulus*. The latter are likely to have been detached and reworked into an integrated assemblage, combining 'high-energy' types associated with mobile coarse sediments (e.g. heavily built *Elphidium macellum*) with those more likely to have been associated with finer grained, possibly infaunal habitats (e.g. *Stainforthia fusiformis*). This heterogeneity is more than likely a function of the megarippled nature of seabed topography, in which relatively coarse and fine sediments are juxtaposed. Therefore, the total assemblage (i.e. live and dead) described here most likely represents an integration of material

Table 1. *Faunal count list of benthic foraminifera from a maerl bed, 10 m below chart datum (CD) in Stravanan Bay off the Isle of Bute, Clyde Sea area, Scotland (55°45.323′N 005°04.265′W; site 2 of Hall-Spencer & Atkinson 1999)*

Species	Count
Asterigerinata mamilla (Williamson)	66
Milionella subrotunda (Montagu) forma *hauerinoides* Rhumbler	43
Spiroplectammina wrightii (Silvestri)	42
Fissurina elliptica (Cushman)	33
Quinqueloculina seminulum (Linné)	28
Elphidium gerthi van Voorthuysen	21
Planorbulina distoma Terquem	13
Cibicides lobatulus (Walker & Jacob)	12
Pyrgo depressa (d'Orbigny)	11
Rosalina praegeri (Heron-Allen & Earland)	11
Elphidium macellum (Fichtel & Moll)	10
Oolina williamsoni (Alcock)	9
Rosalina cf. *bradyi* (Cushman)	9
Elphidium margaritaceum (Cushman)	8
Spiroloculina rotunda d'Orbigny	8
Quinqueloculina bicornis (Walker & Jacob)	7
Astrononion gallowayi Loeblich & Tappan	6
Cibicides pseudoungerianus (Cushman)	6
Elphidium crispum (Linné)	6
Fissurina marginata (Walker & Boys)	6
Textularia truncata (Höglund)	5
Ammonia batavus (Hofker)	4
Spiroloculina depressa d'Orbigny	4
Acervulina inhaerens Schultze	3
Bolivina variablis (Williamson)	3
Nonion pauperatum (Balkwill & Wright)	3
Quinqueloculina lata Terquem	3
Fissurina lucida (Williamson)	2
Massilina secans (d'Orbigny)	2
Oolina borealis Loeblich & Tappan	2
Pyrgo willamsoni (Silvestri)	2
Buccella frigida (Cushman)	1
Bulimina marginata d'Orbigny	1
Cribrostomoides jeffreysi (Williamson)	1
Eggerelloides scabrum (Williamson)	1
Elphidium albiumbilicatum (Weiss)	1
Fissurina danica (Madsen)	1
Fissurina trigono-marginata (Parker & Jones)	1
Nonion depressulus (Walker & Jacob)	1
Oolina hexagona (Williamson)	1
Oolina melo d'Orbigny	1
Rosalina anomala Terquem	1
Rosalina williamsoni (Chapman & Parr)	1
Triloculina trihedra Loeblich & Tappan	1
Trochammina adaptera Rhumbler	1
Total Specimen Count	402
Total Species Count	45

Table 2. *Faunal list of benthic foraminifera from a maerl bed, 10 m below chart datum (CD) in Stravanan Bay off the Isle of Bute, Clyde Sea area, Scotland (55°45.323′N 005°04.265′W; site 2 of Hall-Spencer & Atkinson 1999). Abundant: ≥2.5%; Common: 0.5–2.5%; Rare: ≤0.5%*

Acervulina inhaerens Schultze	Common
Ammodiscus sp.	Rare
Ammonia batavus (Hofker)	Common
Ammoscalaria runiana (Heron-Allen & Earland)	Rare
Anomalina cf. *globulosa* Chapman & Parr	Rare
Asterigerinata mamilla (Williamson)	Abundant
Astrononion gallowayi Loeblich & Tappan	Common
Astrononion tumidum Cushman & Edwards	Rare
Bolivina pseudoplicata Heron-Allen & Earland	Rare
Bolivina pseudopunctata Höglund	Rare
Bolivina cf. *skagerrakensis* Qvale & Nigam	Rare
Bolivina variablis (Williamson)	Common
Buccella frigida (Cushman)	Rare
Bulimina elongata d'Orbigny	Rare
Bulimina marginata d'Orbigny	Rare
Buliminella elegantissima (d'Orbigny)	Rare
Cassidulina obtusa Williamson	Rare
Cibicides lobatulus (Walker & Jacob)	Abundant
Cibicides pseudoungerianus (Cushman)	Common
Cornuspira involvens (Reuss)	Rare
Cribrostomoides jeffreysi (Williamson)	Rare
Dentalina pauperata d'Orbigny	Rare
Dentalina cf. *flintii* (Cushman)	Rare
Dentalina guttifera d'Orbigny	Rare
Discorbis chasteri (Heron-Allen & Earland)	Rare
Discorbis wrightii (Brady)	Rare
Eggerelloides scabrum (Williamson)	Rare
Elphidium albiumbilcatum (Weiss)	Rare
Elphidium crispum (Linné)	Common
Elphidium gerthi van Voorthuysen	Abundant
Elphidium macellum (Fichtel & Moll)	Abundant
Elphidium margaritaceum (Cushman)	Common
Elphidium cf. *williamsoni* Haynes	Rare
Fissurina danica (Madsen)	Rare
Fissurina elliptica (Cushman)	Abundant
Fissurina lucida (Williamson)	Common
Fissurina marginata (Walker & Boys)	Rare
Fissurina cf. *quadrata* (Williamson)	Rare
Fissurina serrata (Schlumberger)	Rare
Fissurina trigono-marginata (Parker & Jones)	Rare
Globulina cf. *aequalis* (Reuss)	Rare
Guttulina harrisi Haynes	Rare
Lagena laevis (Montagu)	Rare
Lagena perlucida (Montagu)	Rare
Lamarckina haliotidea (Heron-Allen & Earland)	Rare
Lenticulina spp.	Rare
Massilina secans (d'Orbigny)	Common
Milionella subrotunda (Montagu)	Common
Milionella subrotunda (Montagu) forma *hauerinoides* Rhumbler	Abundant
Nonion depressulus (Walker & Jacob)	Rare
Nonion pauperatum (Balkwill & Wright)	Common
Nonionella auricula Heron-Allen & Earland	Rare
Oolina borealis Loeblich & Tappan	Common
Oolina hexagona (Williamson)	Common
Oolina lineata (Williamson)	Rare
Oolina melo d'Orbigny	Rare
Oolina squamosa (Montagu)	Rare

(*Continued*)

Table 2. *Continued*

Oolina willamsoni (Alcock)	Common
Patellina corrugata Williamson	Rare
Planorbulina distoma Terquem	Abundant
Polymorphindae sp.	Rare
Polymorphinidae *formae fistulosae*	Rare
Pseudopolymorphina suboblonga Cushman &Ozawa	Rare
Pyrgo depressa (d'Orbigny)	Abundant
Pyrgo willamsoni (Silvestri)	Common
Quinqueloculina aspera d'Orbigny	Rare
Quinqueloculina bicornis (Walker & Jacob)	Common
Quinqueloculina cliarensis Heron-Allen & Earland	Rare
Quinqueloculina cf. *duthiersi* (Schlumberger)	Rare
Quinqueloculina lata Terquem	Common
Quinqueloculina seminulum (Linné)	Abundant
Rosalina anomala Terquem	Rare
Rosalina cf. *bradyi* (Cushman)	Common
Rosalina neapolitana (Hofker)	Rare
Rosalina praegeri (Heron-Allen & Earland)	Abundant
Rosalina williamsoni (Chapman & Parr)	Rare
Spiroloculina depressa d'Orbigny	Common
Spiroloculina rotunda d'Orbigny	Common
Spirophthalmidium sp.	Rare
Spiroplectammina wrightii (Silvestri)	Abundant
Spiroplectammina earlandi (Parker)	Rare
Stainforthia fusiformis (Williamson)	Rare
Textularia truncata (Höglund)	Common
Trifarina angulosa (Williamson)	Rare
Triloculina trigonula (Lamarck)	Rare
Triloculina trihedra Loeblich & Tappan	Rare
Trochammina adaptera Rhumbler	Rare
Trochammina helgolandica Rhumbler	Rare
Trochammina intermedia Rhumbler	Rare
Trochammina ochrea (Williamson)	Rare

derived directly from the maerl in addition to material derived from the surrounding sediments.

The assemblage described in this study is very similar, in terms of the species present and their relative abundances, to the composition of shelf sea sediments to the west of Scotland and Clyde Sea area (e.g. Hannah & Rogerson 1997; Murray 2003). However, it contains a higher total species count than any single site previously described from the west of Scotland. This is consistent with the high biodiversity generally observed in maerl deposits and the interpretation of the assemblage as an integration of *in situ* and locally reworked material.

In conclusion, our preliminary investigation suggests that Scottish maerl deposits appear to have unusually high benthic foraminiferal species counts and, as such, may act as useful repositories recording (or at least integrating) the regional faunal composition, highlighting species richness and providing an approximation of species diversity. Further work is now required in order to establish the full significance of maerl deposits in the context of benthic foraminiferal biodiversity; this work will involve careful spatial sampling by SCUBA and the use of protoplasmic stains, such as Rose Bengal, to establish live distributions. In a global context, maerl deposits represent an important near-shore habitat which appear to have been largely overlooked in regional benthic foraminiferal distributional/ecological studies.

We thank J. Hall-Spencer for providing us with the sample material for this study and for helpful discussion. N. Nørgaard-Pedersen contributed to the original taxonomic exercise and picked some of the foraminiferal specimens. J. Murray, P. Edwards and J. Gregory provided constructive criticism on an early draft of this manuscript; we are particularly grateful to J. Murray for drawing our attention to some of the early French literature that record foraminifera in maerl deposits. We thank K. McIntyre and M. Sayer for their constructive reviews. This work is a contribution to the EU project *Millennium*.

References

ANON, 1992. *Council directive 92/43/EEC of 21 May 1992 on the conservation of natural habitats and of wild fauna and flora*. Official Journal L206.

AUSTIN, W. E. N., CAGE, A. G. & SCOURSE, J. D. 2006. Mid-latitude shelf seas: a NW European perspective on the seasonal dynamics of temperature, salinity and oxygen isotopes. *The Holocene*, **16**, 937–947.

BLANC-VERNET, L. 1969. Contribution á l'étude des foraminifers de Méditerranée. *Recueil des Travaux de la Station Marine d'Endoume*, **64-48**, 1–279.

BOILLOT, G. 1964. Géologie de la Manche occidentale. *Annals de l'Institut Océanographique*, **42**, 1–219.

BOSENCE, D. W. J. 1979. Live and dead faunas from coralline algal gravels, Co. Galway. *Palaeontology*, **22**, 449–478.

EDWARDS, P. G. 1982. Ecology and distribution of selected foraminiferal species in the North Minch Channel, northwest Scotland. *In*: BANNER, F. T. & LORD, A. R. (eds) *Aspects of Micropalaeontology*. Allen and Unwin, London, 111–141.

FOSTER, M. S., RIOSMENA-RODRIGUEZ, R., STELLER, D. L. & WOELKERLING, W. J. 1997. Living rhodolith beds in the Gulf of California and their implications for paleoenvironmental interpretation. *Geological Society of America*, **318**, 127–139.

FREIWALD, A. & HENRICH, R. 1994. Reefal coralline algal buildups within the Arctic Circle: morphology and sedimentary dynamics under extreme environmental seasonality. *Sedimentology*, **41**, 963–984.

HALL-SPENCER, J. M. & ATKINSON, R. J. A. 1999. *Upogebia deltaura* (Crustacea: Thalassinidea) in Clyde Sea maerl beds, Scotland. *Journal of the Marine Biological Association of the United Kingdom*, **79**, 871–880.

HANNAH, F. & ROGERSON, A. 1997. The temporal and spatial distribution of foraminiferans in marine benthic sediments of the Clyde Sea area, Scotland. *Estuarine, Coastal and Shelf Science*, **44**, 377–383.

HASKONING UK LTD 2006. *Investigation into the impact of marine fish farm deposition on maerl beds*. Scottish Natural Heritage, Report **213** (ROAME No. AHLA10020348).

HAYNES, J. R. 1973. Cardigan Bay recent foraminifera. *Bulletin of the British Museum (Natural History) Zoology*, Supplement 4, 1–245.

HERON-ALLEN, E. & EARLAND, A. 1916. The foraminifera of the West of Scotland. *Transactions of the Linnean Society of London, Series 2 (Zoology)*, **11**, 197–300.

HÖGLUND, H. 1947. Foraminifera in the Gullmar fjord and the Skagerrak. *Zoologiska bidrag från Uppsala*, **26**, 328.

MORHANGE, C., BLANC, F. ET AL. 2003. Stratigraphy of late-Holocene deposits of the ancient harbour of Marseilles, southern France. *The Holocene*, **13**, 323–334.

MURRAY, J. W. 1985. Recent foraminifera from the North Sea (Forties and Ekofisk areas) and the continental shelf west of Scotland. *Journal of Micropalaeontology*, **4**, 117–125.

MURRAY, J. W. 2003. An illustrated guide to the benthic foraminifera of the Hebridean shelf, west of Scotland, with notes on their mode of life. *Palaeontologica Electronica*, **5**, 31, 1.4MB; http://palaeo-electronica.org/paleo/2002_2/guide/issue2_02.htm.

MURRAY, J. W. & WHITTAKER, J. E. 2001. A new species of microforaminifera (*Gavelinopsis caledonia*) from the continental shelf west of Scotland. *Journal of Micropalaeontology*, **20**, 179–182.

RIPPETH, T. P. & SIMPSON, J. H. 1996. The frequency and duration of episodes of complete vertical mixing in the Clyde Sea. *Continental Shelf Research*, **16**, 933–947.

ROSSET-MOULINIER, M. 1972. Etude des foraminifères des côtes nord et ouest de Bretagne. *Travaux du Laboratoire de Géologie, Ecole Normale Supérieur, Paris*, **6**, 1–225.

Modern glacimarine processes and potential future behaviour of Kronebreen and Kongsvegen polythermal tidewater glaciers, Kongsfjorden, Svalbard

LUKE D. TRUSEL[1,2]*, R. D. POWELL[1], R. M. CUMPSTON[1] & J. BRIGHAM-GRETTE[3]

[1]*Analytical Center for Climate and Environmental Change, Department of Geology and Environmental Geosciences, Northern Illinois University, 310 Davis Hall, Normal Rd., DeKalb, Illinois 60115, USA*

[2]*Present address: Graduate School of Geography, Clark University, 950 Main St., Worcester, Massachusetts 01610, USA*

[3]*Department of Geosciences, University of Massachusetts – Amherst, 611 N. Pleasant St., Morrill Science Center, Amherst, Massachusetts 01003, USA*

**Corresponding author (e-mail: ltrusel@clarku.edu)*

Abstract: Glacimarine dynamics and associated sedimentary processes are closely tied to glacial regime and reflect dominant climatic conditions. Quantitative measurements for subpolar glaciers, such as sediment yield, are limited especially near glacial termini where most sediment accumulates. Here we characterize the modern glacimarine environment, quantify sediment flux and yield, document landform genesis and hypothesize potential future behaviour of Kronebreen and Kongsvegen glaciers in inner Kongsfjorden, Svalbard. A minimum of 6.74×10^3 g m^{-2} d^{-1} (at least 300 mm a^{-1}) of glacimarine sediment is building a grounding-line fan via submarine stream discharge from Kronebreen. Average daily sediment flux to the ice-contact basin is recorded to be 2.6×10^3 g m^{-2} d^{-1} or an average annual flux of 1.56×10^5 g m^{-2} a^{-1}. We measure an average annual ice-contact sediment yield of 1.20×10^4 tonnes km^{-2} a^{-1} associated with the rapid genesis of grounding-line landforms. With forecasted warming we expect meltwater volumes and sediment flux to increase. Grounding-line deposits may aggrade above water, tending to stabilize the terminus at least initially if the sediment is sufficient to counteract total terminus ablation. This would hold until either the glaciers next surge or climatic warming ablates the glaciers through surface melting.

The glaciers of western Spitsbergen, Svalbard are highly susceptible to climate fluctuations because of their intermediate position relative to the warm West Spitsbergen Current that transports Atlantic Water from the south and the colder Polar Water that flows from the north (Hald *et al.* 2004). Changes in freshwater flux from western Svalbard glaciers have many implications including influences on fjord sea ice, local biotas, fjord and river hydrology and, in extreme climate warming scenarios, a disturbance in deepwater production on the Svalbard shelf (Hagen *et al.* 2003). Sediment flux to marine grounding lines also has significant influences on glacial stability and potential future retreat or advance rates (Alley 1991; Powell 1991; Powell & Alley 1997; Fischer & Powell 1998; Alley *et al.* 2007). Because of amplified climatic warming at northern high latitudes (e.g. IPCC 2007), the glaciers of Spitsbergen, Svalbard are prime subjects for understanding not only glacial dynamics but also the effects of contemporary climate change.

Modern baseline measurements are fundamental and must be established to quantify the potential effects of climatic variability. Glacimarine processes such as sediment flux and yield, terminus fluctuations, calving rates, freshwater flux and mass-balance are useful for comparison among differing glacial and climatic regimes as well as for overall characterization of the state of the glacimarine system. Additionally, understanding the relationship of complex modern processes to those found in the glacimarine sediment record can enhance interpretation of past glacimarine processes and changes in biological productivity (Powell 1984).

The primary objective of this study is to characterize the modern ice-contact and ice-proximal

From: Howe, J. A., Austin, W. E. N., Forwick, M. & Paetzel, M. (eds) *Fjord Systems and Archives*. Geological Society, London, Special Publications, **344**, 89–102.
DOI: 10.1144/SP344.9 0305-8719/10/$15.00

glacimarine environment of the Kronebreen-Kongsvegen glacier complex (Fig. 1). We employ sediment flux records as well as fjord hydrographic and bathymetric profiling. This glacimarine system has been previously studied (e.g. Elverhøi *et al.* 1980, 1983; Svendsen *et al.* 2002) although extremely ice-proximal (within 1 km) measurements of sediment flux and water column properties are very limited. This understudied proximal zone is consequently the most dynamic, owing to the interplay of fresh meltwater discharge mixing with saline fjord water, rapid sediment flux and deposition, sediment slides and slumps, active calving and debris dumping, among other processes. Further, the confluent glacier complex has recently undergone pronounced changes including the emergence of Kongsvegen from tidewater. Post-surge quiescent phase conditions (cf. Meier & Post 1969) combined with significant meltwater and sediment fluxes have contributed to these conditions. Bathymetry near the termini reveals landforms that serve as records of ice-contact sediment flux. These measurements are the first of their kind in Kongsfjorden and represent the full potential of these subpolar glaciers in landscape denudation. We evaluate the modern system in relation to historical measurements and ultimately assess the ability of these glaciers to potentially influence their own future behaviour.

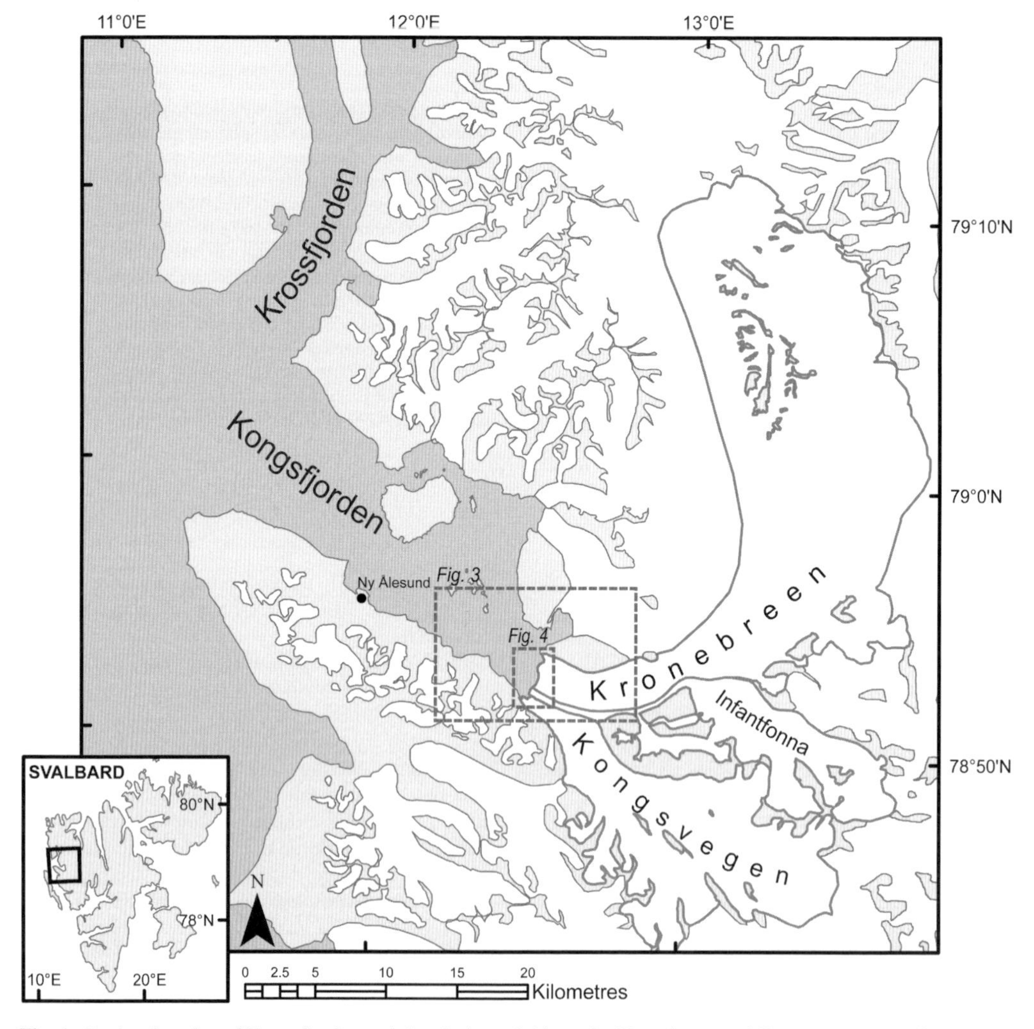

Fig. 1. Regional setting of Kongsfjorden and the glaciers of this study, Kronebreen and Kongsvegen as well as Infantfonna. Location of the large map is shown with respect to the islands of Svalbard in the inset map. Locations of additional figures indicated.

Study area

Physical setting

Kongsfjorden is a 22 km long and 4–12 km wide glacially eroded fjord in western Spitsbergen, Svalbard (Fig. 1). Fjord bathymetry is complex and variable, ranging from <60 m at its head to 400 m in several areas, separating the fjord into multiple distinct basins. Marine processes in Kongsfjorden are both spatially and temporally dynamic, characterized by a high degree of seasonality as is typical of arctic fjords (e.g. Syvitski 1989). Lacking a well-defined sill at its mouth (Howe *et al.* 2003), the fjord is highly interconnected with neighbouring water masses on the West Spitsbergen Shelf, including Atlantic Water (Svendsen *et al.* 2002), with influx particularly pronounced during midsummer (Cottier *et al.* 2005). Weather in the region is commonly unstable since Svalbard lies on the border-zone of cold Polar Basin circulation and warm maritime air masses from the south (Ingólfsson 2004). These conditions, combined with a seasonally abundant glacial meltwater flux, dictate both the physical (Svendsen *et al.* 2002; Gerland *et al.* 2004; MacLachlan *et al.* 2007) and biological (Hop *et al.* 2002) oceanography and make the Kongsfjorden area particularly susceptible to climate change (Lamb 1977).

Numerous subpolar, polythermal and surge-type tidewater glaciers calve into the fjord (e.g. Hagen *et al.* 1993) including the glaciers of this study, Kronebreen (445 km^2) and Kongsvegen (165 km^2) at the southeastern fjord head (Fig. 1). These glaciers join 5 km from their termini, forming a confluent glacier complex with Infantfonna (85 km^2) but with Kronebreen as the dominant tidewater cliff (Fig. 1). A recent widening of Kronebreen and the very slow, post-surge quiescent velocity of Kongsvegen (2.6 m a^{-1}) (Melvold & Hagen 1998) combined with its significant meltwater and sediment fluxes contributed to Kongsvegen emerging from tidewater between 1990 and 2005. Unlike Kongsvegen, Kronebreen has a high velocity of about 2 m d^{-1} (Hagen *et al.* 2003) or about 750 m a^{-1} (Melvold & Hagen 1998), providing evidence for sliding and melting at its bed to produce significant meltwater and sediment fluxes to the fjord (Svendsen *et al.* 2002). Bedrock strata underlying the complex include Devonian sandstone, conglomerate and localized marble, Carboniferous-Permian red sandstone, shale, coal and Proterozoic phyllite, carbonates and schists (Hjelle 1993). The length of the melt season for the Kronebreen-Kongsvegen complex is not directly known; however, the melt season for nearby terrestrial glaciers is documented at 60 days (Hodson *et al.* 1998).

Methodology

Field methods

All samples and data were collected from the inner basin of Kongsfjorden between July 18 and 31, 2005. Sample locations were sited using a hand-held GPS unit. Samples were collected from a 6 m aluminium skiff and 5 m inflatable boat.

Several devices were used to profile the marine water column in order to define its structure and potential paths of sediment transportation. Salinity and temperature were recorded and water density calculated by using a Sea-Bird SBE 19 CTD Profiler. The relative amount of suspended particulate matter within the water column was determined using a SeaTech transmissometer and an optical backscatter (OBS) device. A remotely triggered water sampler (2 L van Dorn bottle) was used to collect water samples from various fjord depths to quantify the suspended particulate matter (SPM) concentrations. Although the OBS and water samples measure total SPM, including both inorganic (siliciclastic) and organic (biologic) components, for simplicity we refer to total SPM as the 'suspended sediment' load because during summers the organic component is relatively very small (Svendsen *et al.* 2002).

Sediment traps were used as the primary measure of sediment flux and consequent rates of sedimentation from the subglacial stream discharges. Each trap was 250 mm × 48 mm and modelled after Cowan (1988), a trap design determined sufficient for quantitative sediment flux studies in Alaskan fjords. Each mooring was anchored to the bottom of the fjord and held relatively vertical using a rigid submersible float at about 10 m below low tide level to avoid iceberg interference. Each mooring had traps set at three depths: one near the surface (*c.* 15 m), a second in the middle of the water column (*c.* 25 m) and the third near the bottom (*c.* 50 m). At each depth an array of four individual cylindrical traps was set to assess error in trap measurements at each level in the water column. The bottom trap was located *c.* 5 m above the fjord floor, presumably high enough to avoid turbidity current activity, but low enough to accurately measure sediment being deposited on the fjord floor at each location.

Fjord bathymetry was determined using a Knudsen 320BP echosounder and sub-bottom profiler deployed along transects run across the fjord and following bearings approximately perpendicular to the ice cliff and the across-fjord lines. Points along transects were recorded using GPS-located coordinates.

Laboratory methods

Initial laboratory analyses were performed at the Arctic Marine Laboratory in Ny Ålesund, and were completed at the home institutions in the US.

Individual samples from sediment traps at each mooring were dried and the volume of collected sediment calculated. The dry sediment from these traps was weighed and used to calculate sedimentation rates per unit time. A Coulter LS-200 laser particle size analyser was used for particle size measurements.

Annual sedimentation rates were linearly extrapolated from trap measurements using the 60 day melt season of Hodson *et al.* (1998), derived from nearby terrestrial glaciers. This time span mediates a potentially pronounced midsummer meltwater flux (e.g. Svendsen *et al.* 2002) and the likelihood of longer melt seasons associated with the much thicker and more dynamic glaciers of this study. Sixty days are also significantly fewer than the average annual positive days between 2001 and 2008 (average 126 days per annum, most of which occur in summer) (Norwegian Meteorological Institute data from Ny Ålesund climate database website: http://eklima.met.no). Further evidence of a prolonged melt season comes from direct observations of meltwater effluent at the Kronebreen terminus during colder annual periods in autumn, winter and early spring, although this flux is thought to be rather infrequent and of low volume (Sund pers. comm. 2008). Therefore, although the Kronebreen-Kongsvegen ablation season is likely longer, for the purposes of conservative estimation a period of 60 days is used to calculate annual sediment flux.

Total volume of water in water samples was measured and the sediment within each sample was vacuum filtered using 63 μm Millipore filters. The resulting dry sediment was weighed and used to calculate suspended sediment concentrations in kg m^{-3}. These concentrations were also used to calibrate and convert OBS values into continuous kg m^{-3} measurements throughout the water column (Fig. 2). Two separate OBS calibration curves were generated based on different backscatter characteristics for sand particles as opposed to clay and silt. An upwelling-proximal calibration was therefore used in the vicinity of the Kronebreen point source with an r^2 of 0.9841; a Kongsvegen deltaic runoff calibration with an r^2 of 0.9844 was used for locations not in direct influence of the glacial upwelling (Fig. 2).

Recorded bathymetric points were interpolated using the Kriging method to create a continuous bathymetric map near the glacial termini. Recent terminus fluctuations were characterized using several datasets including satellite imagery and mapped terminus positions of other studies.

Results

Glacial retreat

The combined Kronebreen-Kongsvegen terminus has fluctuated over time and is most recently characterized by post-surge quiescent phase retreat. Retreat has been variable along the margin (Fig. 3), making the calculation of a retreat rate in kilometres per annum impractical. Available data, however, indicate overall terminus retreat rate has generally decreased following the recent maximum extension during the surge of 1948. The terminus has become almost stable since 1990, including a slight advance between 1999 and 2001.

This retreat has exposed a new sediment basin in the innermost fjord. Combined with recent quasi-stability, this has allowed for the generation of large submarine and subaerial morphological features that act as records of recent glacial sedimentary activity in the proximal basin.

Sedimentary landforms and denudation rates

A previously-undocumented ice-marginal, Gilbert-type delta has emerged in front of Kongsvegen, separating the glacier from tidewater (Figs 1, 3 & 4). The delta likely originated as a grounding-line fan at the former Kongsvegen tidewater terminus and, over time, built upon the southern fjord wall to eventually aggrade above sea level (e.g. Powell 1990). Photographs indicate the initial subaerial exposure of these deposits occurred before July 1996 (cf. Woodward *et al.* 2003, Fig. 1). From bathymetric soundings and satellite imagery we are able to accurately record the delta dimensions. The plain is measured at 0.33 km^2 and the delta-front slope extents 500 m into the fjord at a slope of 8°, terminating 70 m below sea level. Assuming the fjord wall on which the delta formed is at the same angle as the delta slope and a parallelogram delta-foreset configuration, we are able to geometrically estimate the volume. Using an average delta plain width of 350 m and the above dimensions, the cross-sectional area is calculated at 2.45×10^4 m^2. By multiplying by the average delta plain length of 800 m, total deltaic volume is estimated to be 1.96×10^7 m^3. Initial formation likely began immediately after Kongsvegen retreated over the western edge of the modern delta. Based on terminus positions mapped in Lefauconnier *et al.* (1994), initial uncovering of this area occurred in late summer 1983. Assuming growth over 22 years (1983–2005), its building rate is calculated to be 8.91×10^5 m^3 a^{-1}. Given that the total Kongsvegen drainage basin area is 210 km^2 with an average bedrock density of 2500 kg m^{-3} (primarily sedimentary rocks), this estimate provides a sediment yield of 1.06×10^4 tonnes km^{-2} a^{-1} (Table 1). This yield is equivalent to 4.24 mm a^{-1} bedrock erosion if 2500 tonnes km^{-2} is equal to 1 mm bedrock denudation (Table 1), following a similar methodology to Elverhøi *et al.* (1998).

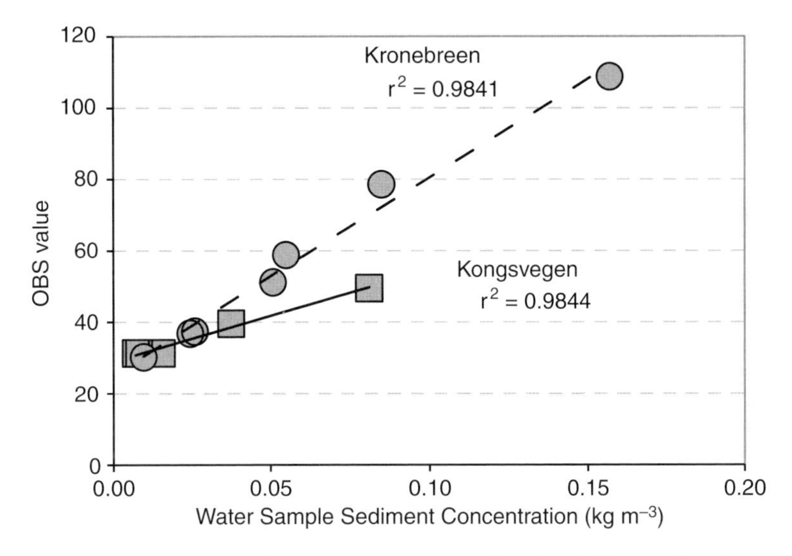

Fig. 2. Optical backscatter v. suspended-sediment concentrations and linear best-fit lines for Kronebreen (dashed line) and Kongsvegen (solid line). By relating measured sediment concentrations to OBS readings at specific depths and locations, we are able to infer sediment concentration from OBS.

Satellite, aerial and time-lapse imagery (Sund 2007) document the dynamic switching of submarine stream discharges at the Kronebreen-Kongsvegen margin. Available imagery from 1987, 1990, 1999 and 2000 reveal turbid upwelling adjacent to a recorded bathymetric high (B in Fig. 4). This submarine landform is interpreted as a grounding-line fan formed by those subglacial discharges (GLF in Fig. 4). In July 2005 the dominant meltwater upwelling was located about 500 m north along the terminus relative to its historical position (A in Fig. 4). High meltwater overflow velocities and frequent calving near the 2005 discharge prevented continuous acquisition of bathymetric data in this region. However, the inferred grounding-line fan appears to extend into the modern calving embayment, creating a continuous deposit with a minimum volume of 6.6×10^6 m^3. Assuming this

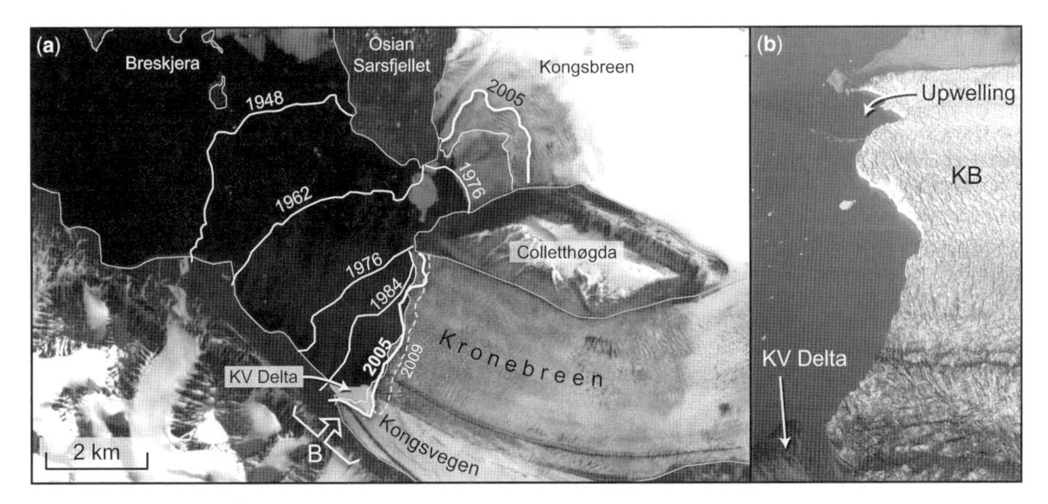

Fig. 3. (**a**) Retreat of the glacial complex after its recent maximum extent during the surge of 1948. Mapping of terminus positions indicates that overall retreat rate has recently considerably slowed. (**b**) Oblique aerial photo taken in July 2005 showing the Kongsvegen delta in the foreground and the prominent submarine meltwater upwelling from Kronebreen in the distance. Arrow in (a) indicates the perspective view shown in (b). Figure extent shown in Figure 1.

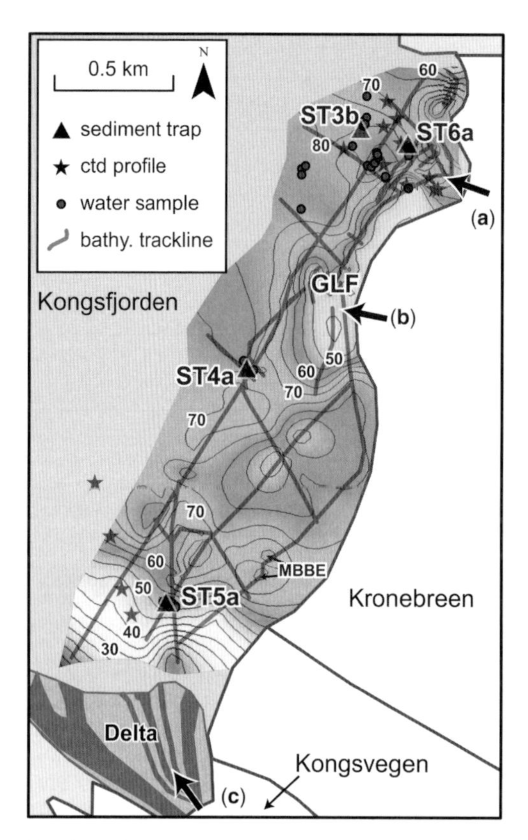

Fig. 4. Bathymetric and sample location map. Water depths are in metres with a 5 m contour interval. Bold arrows indicate modern and historical discharge point sources: (**a**) modern Kronebreen submarine stream discharge and upwelling; (**b**) historical discharge and upwelling location observed in available imagery from 1987, 1990, 1999 and 2000; and (**c**) modern Kongsvegen discharge source (GLF, grounding-line fan; MBBE, morainal bank bulls eyes).

deposit began accumulating *c.* 1984 when the terminus retreated to this location, a growth rate is calculated to be $3.3 \times 10^5\ m^3\ a^{-1}$. This rate equates to a sediment yield of 1.40×10^3 tonnes $km^{-2}\ a^{-1}$, equivalent to $0.56\ mm\ a^{-1}$ of bedrock denudation (Table 1).

A semi-continuous bathymetric high at 60–70 m depth trends nearly north–south, parallel to the glacier termini, extending from the delta slope to the grounding-line fan. This feature may represent glaciotectonically thrusted sediments during winter, when calving is minimized and glacial flow results in net terminus advance (e.g. Elverhøi *et al.* 1980). Conversely, the ridge may indicate former subglacial discharge depocentres. Directly east of this high is a deeper basin extending to the ice face. Morainal bank 'bulls eyes' are especially obvious, associated with supraglacial dumping and englacial meltout of the medial moraine formed where Kronebreen and Kongsvegen join (MBBE in Fig. 4).

Inner Kongsfjorden water column

The primary sediment source and driver of water column stratification is a meltwater upwelling from Kronebreen. As meltwater discharges from the base of the glacier, it entrains high concentrations of sediment and rapidly forms a turbulent jet (cf. Powell 1991). Because of relative density contrasts, the jet rises vertically and forms a buoyant and brackish surface overflow plume (SO water mass in Fig. 5). At the surface the brackish plume spreads laterally, confined mostly above 12 m depth as defined in hydrographic profiles (Fig. 5a). Water samples confirm that suspended sediment settles out of suspension through the entire water column in the glacier marginal zone. Suspended sediment concentrations peak between 0–4 m depth and, on average, decrease approximately logarithmically towards the fjord floor. The highest measured suspended sediment concentration in the overflow was $0.157\ kg\ m^{-3}$; however, calibrated OBS measurements were as high as $0.266\ kg\ m^{-3}$. The greatest variability in suspended concentrations occurs in the turbid overflow, with the highest dispersion occurring at the surface (Table 2; Fig. 5a).

The SO is on average colder, less saline and more sediment rich than ambient fjord water. Near the surface the SO is generally 0.5–1.0 °C and has salinities as low as 22‰ with a mean salinity of 30.96‰ in the upper 4 m. Most hydrographic profiles show a sharp thermocline between 10–15 m depth with temperature increasing to 1.0–2.5 °C. This warmer water below the SO also coincides with the pycnocline and low sediment concentrations, suggesting that water near the glacier between 10–25 m depth is the displaced, solar-radiated and more saline fjord surface water (SW) mass below the overflow (Svendsen *et al.* 2002; MacLachlan *et al.* 2007). Below 15 m depth salinity increases while sediment concentration decreases, both to relatively constant values. In all profiles below at least 28 m depth, temperature slowly decreases to a constant value of around 0 °C and salinity increases to a constant of about 34‰. Svendsen *et al.* (2002) defined a similar water mass as the local water (LW) mass, suggesting it is caused by convection as the warmer SW flows along the glacier face, cools and sinks.

A secondary meltwater source from the glacial complex is a submarine meltwater upwelling originating from Kongsvegen. Subglacial meltwater exits Kongsvegen as a turbulent fountain *c.* 0.5 m in height above the delta plain. This water flows across the recently formed ice-contact delta (Figs 3 & 4c)

Table 1. *Sediment yields and erosion rates calculated from grounding-line landforms deposited over the last c. 20 years*

Glaciers	Total glacier basin area (km^2)	Sediment yield (tonnes km^{-2} a^{-1})	Effective erosion rate (mm a^{-1})
Kronebreen, Infantfonna and tributaries	590	1.40×10^3	0.56
Kongsvegen, Sidevegen, Bontfjellbreen and tributaries	210	1.06×10^4	4.24
Total	800	1.20×10^4	4.80

Note: Glacier basin area represents the entire drainage areas for the main glaciers and their tributaries within a single divide. The Kongsvegen basin area incorporates the area of small, sidewall glaciers to the south that have potentially contributed to delta growth.

before entering the fjord, where it forms a buoyant and brackish overflow (SO water mass) (Fig. 5b).

The SO from Kongsvegen is confined to the upper 6 m and is stratified with respect to temperature, salinity and suspended sediment concentration. Salinity is the dominant stratification, with temperature and sediment concentrations displaying greater variance. Sediment concentrations are greatest at the surface, as is the variance in total suspended sediment. The highest measured suspended sediment concentration was 0.081 kg m^{-3} (Table 3); however, calibrated OBS measurements were as high as 0.392 kg m^{-3} (Fig. 5b). Two secondary peaks in sediment concentration were measured with water samples and hydrographic profiles around 12–15 m depth. If these were an interflow it is expected that the higher density sediment layers would be associated with warmer or fresher water than the surrounding water to maintain appropriate density. We therefore infer these higher suspended sediment concentrations within the fjord water column to represent distinct layers of particles settling from the overflow (cf. Cowan & Powell 1991). The warmer and more saline displaced fjord SW is recorded on most profiles below the SO layer and transitions into LW at depths below 28 m.

At 4–5 km from the glaciers, the water column is dominantly stratified with respect to salinity and suspended sediment concentration; however, the thermocline is more poorly defined than in the innermost basin. The influence of solar radiation is obvious with surface temperatures in excess of 5 °C. Surface salinities are as low as 25‰, likely reflecting a combination of glacial and paraglacial meltwater inputs and iceberg melting. Suspended sediment is much lower than in the proximal basin and peaks around 0.02 kg m^{-3}.

Sediment flux to innermost Kongsfjorden

Sedimentation rates vary greatly both with distance from the Kronebreen upwelling and vertically through the water column. Two traps were proximal to the Kronebreen upwelling: ST6a at *c.* 240 m distance and ST3b at *c.* 470 m distance (Fig. 4). The highest vertical accumulation rate, or actual amount of sediment captured per deployment period, was measured in the upper moorings of ST6a and ST3b (Table 4). As is predictable, the lowest accumulation rates occurred in the bottom traps of these moorings. In order to quantify the volume of sediment being deposited on the fjord floor, minimum annual sedimentation rates were calculated using only data from the bottom traps of each mooring. Using the conservative annual melt season of 60 days previously described, linear extrapolation produces a minimum annual (dry) sedimentation rate for sediment traps ST6a and ST3b, respectively, of 4.05×10^5 g m^{-2} a^{-1} (305.29 mm a^{-1}) and 4.88×10^4 g m^{-2} a^{-1} (63.48 mm a^{-1}) (Table 4).

The highest measured sedimentation rates (14.3 mm d^{-1}) occurred in ST6a at a depth of 10 m. At 5 m above the seafloor the daily vertical accumulation rate was about a third of the upper traps at 5.09 mm d^{-1}, still a significant quantity of sediment. The high flux of sediment suggests the grounding-line fan (Fig. 4 GLF) is aggrading at a minimum of 300 mm a^{-1} and likely an even greater rate closer to the grounding line. At this rate, if all variables such as annual sediment flux, upwelling position, relative terminus stability and fan slope stability remain constant, the grounding-line fan will emerge above water in *c.* 18 years.

Two suspended sediment traps recorded sediment flux in proximity to the delta point source in addition to a distal signature of the Kronebreen plume. Sediment trap ST5a was located about 200 m from the low tide level on the delta plain, and ST4a was located equidistance from both the delta and the Kronebreen upwelling at *c.* 1200 m from each (Fig. 4). Sediment accumulation at both moorings was greatest in the bottom traps indicating a pronounced contribution from sediment gravity

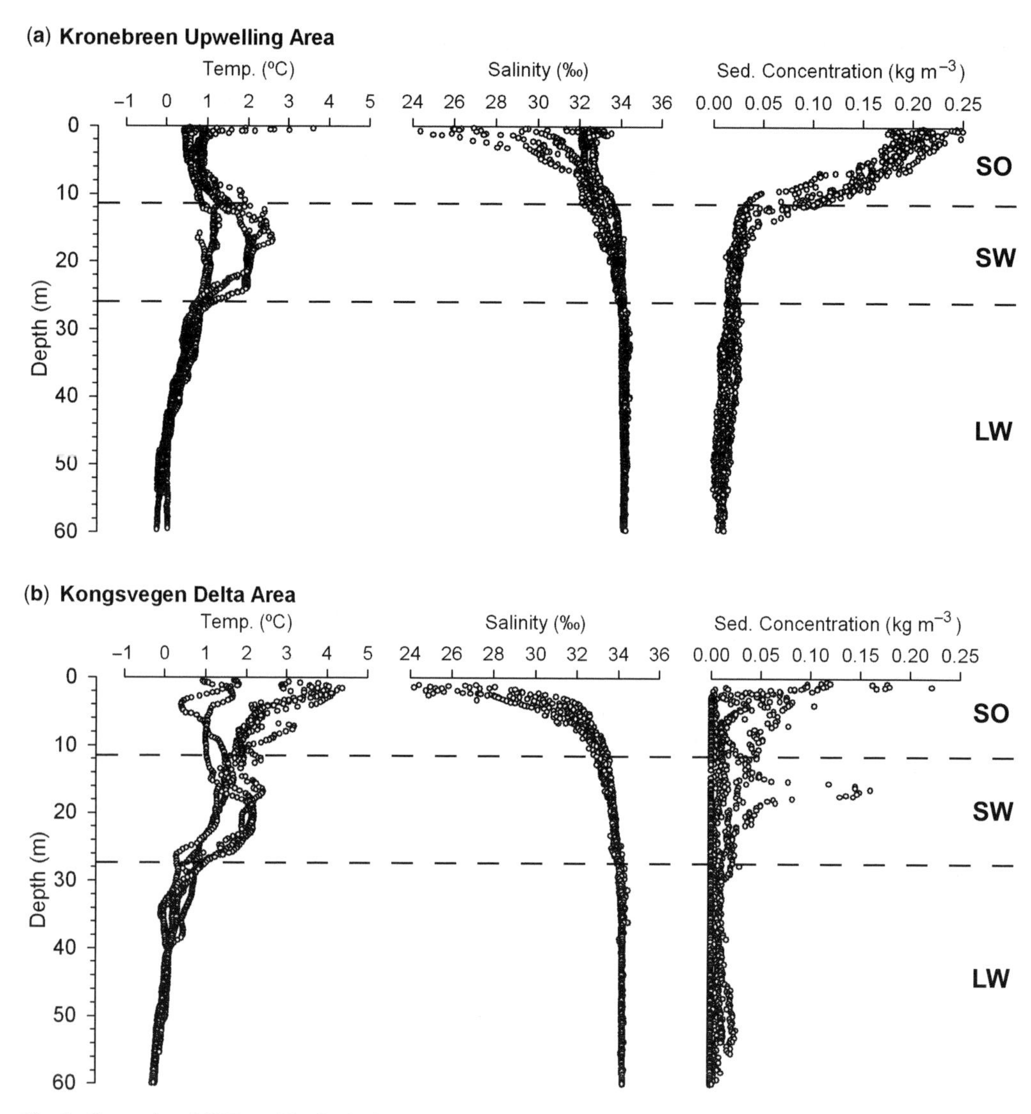

Fig. 5. Composite of CTD profiles for both primary sediment sources with water masses indicated (SO, surface overflow; SW, surface water; LW, local water; nomenclature adapted from Svendsen *et al.* 2002).

flows off the delta front relative to SO contributions (Table 4). ST5a in particular had 72% more sediment in the lower traps compared to the middle traps. Therefore, when calculating a minimum annual sedimentation rates for ST5a, the middle trap data are used. Using the methods outline above, the minimum annual sedimentation rates for sediment traps ST5a and ST4a, respectively, were 7.54×10^4 g m^{-2} a^{-1} (69.38 mm a^{-1}) and 9.50×10^4 g m^{-2} a^{-1} (88.33 mm a^{-1}) (Table 4).

Suspended sediment traps act not only as a gauge for sedimentation rates, but also indicate where sediment is being transported within the water column and the carrying capacity of the surface overflow. As expected, suspended particle size decreases rapidly with distance from the source. At mooring ST6a (240 m distance) mean particle size for the upper sediment traps was 172.9 μm (fine-grained sand) and contained sediment as large as medium-grained sand. At ST3b (470 m distance), mean particle size in the upper traps was only 80 μm (very fine-grained sand). Furthermore, all three trap levels at mooring ST6a captured fine-grained sand which composed 56% of the top traps, whereas the upper traps of ST3b contained only 23% fine-grained sand. This trend reveals

Table 2. *Measured water sample suspended sediment concentrations for the Kronebreen glacial upwelling area*

Collection depth (m)	Number of samples	Average concentration (kg m^{-3})	Minimum concentration (kg m^{-3})	Maximum concentration (kg m^{-3})	Standard deviation (kg m^{-3})
0	4	0.102	0.016	0.140	0.058
2	4	0.079	0.031	0.104	0.034
4	1	0.157	0.157	0.157	–
6	1	0.085	0.085	0.085	–
8	1	0.055	0.055	0.055	–
10	1	0.051	0.051	0.051	–
15	2	0.037	0.024	0.049	0.017
20	5	0.030	0.021	0.049	0.011
30	1	0.026	0.026	0.026	–
35	5	0.017	0.007	0.027	0.007
45	1	0.010	0.010	0.010	–
55	3	0.010	0.008	0.012	0.003

that although sediment as large as medium-grained sand is actively transported up to the fjord surface via the subglacial jet and plume, it quickly falls out of suspension.

ST3b had nearly identical mean particle sizes in the middle and upper traps of 80 μm (very fine-grained sand); however, the lower traps captured a mean particle size of 34.8 μm (silt). This trend shows that the smallest particle size fractions, silt and clay, are held mostly in suspension at 470 m distance and are carried farther out into the fjord. The presence of primarily very fine-grained sand but also coarser particles in the top traps, and their significant lack in the lowest traps, shows that these coarser particles are travelling in suspension beyond 470 m which is greater than the 200–300 m estimate of Elverhøi *et al.* (1980).

For ST5a and ST4a, the traps within direct influence of the deltaic runoff, temporal changes in particle transport were also evident. The trap nearest to the delta, ST5a (200 m distance), captured a mean particle size of silt in the upper and middle traps and very fine-grained sand in the lower traps. The coarser sediment in the lower traps is likely a reflection of sediment gravity flows off the delta front. At the more distal ST4a (1200 m distance), the traps all captured at least 5% very fine sand or larger particles. Because of its distance from both sediment point sources, iceberg rafting is believed to contribute the larger particle fractions. However, suspension settling was the dominant process captured at ST4a, with the upper and middle traps recording mean particle sizes of silt and the lower traps capturing mostly clay.

High-resolution particle size analyses conducted on an upper trap in ST6a reveals a cyclic trend in sediment size (Fig. 6). Mean particle size as measured at discrete intervals in the sediment trap fluctuates according to time of day. Deposition time at different points in the trap sediment column

Table 3. *Measured water sample suspended sediment concentrations for the Kongsvegen delta area*

Collection depth (m)	Number of samples	Average concentration (kg m^{-3})	Minimum concentration (kg m^{-3})	Maximum concentration (kg m^{-3})	Standard deviation (kg m^{-3})
2	3	0.054	0.014	0.081	0.035
4	1	0.008	0.008	0.008	–
8	1	0.007	0.007	0.007	–
15	1	0.037	0.037	0.037	–
20	2	0.019	0.006	0.036	0.018
25	1	0.012	0.012	0.012	–
30	1	0.015	0.015	0.015	–
35	3	0.019	0.013	0.031	0.011
40	1	0.016	0.016	0.016	–
55	1	0.010	0.010	0.010	–

Table 4. *Sedimentation rates measured from inner Kongsfjorden, July 2005*

Mooring number trap location	Sampling period (dates) (total hours) (total days)	Average sediment collected (mm)	Average VAR (mm d^{-1})	Average minimum VAR (mm a^{-1})	Average minimum sed. rate (g m^{-2} d^{-1})	Average minimum sed. rate (g m^{-2} a^{-1})
ST6a top	27/7–30/7	43.8	14.29	305.29	6745	4.05×10^5
ST6a mid	73.6 hours	23.6	7.70			
ST6a bot	3.07 days	15.6	5.09			
ST3b top	23/7–27/7	11.3	2.90	63.48	814	4.88×10^4
ST3b mid	93.65 hours	4.88	1.26			
ST3b bot	3.90 days	4.10	1.06			
ST4a top	23/7–28/7	6.25	1.29	88.33	1583	9.50×10^4
ST4a mid	116.23 hours	5.25	1.08			
ST4a bot	4.84 days	7.13	1.47			
ST5a top	24/7–28/7	3.75	0.94	69.38	1257	7.54×10^4
ST5a mid	96.03 hours	4.63	1.16			
ST5a bot	4.00 days	16.17	4.04			

Note: Averages were obtained using four trap measurements at each depth on a particular mooring. Minimum VAR and sedimentation rates were calculated using data from bottom traps, except for ST5A where the middle trap data were used. Sedimentation rates were calculated using dried sediment weights. Annual numbers were determined assuming a melt season of 60 days.
Abbreviation: VAR, vertical accumulation rate.

is inferred by assuming a linear sedimentation rate between the known times of deployment and recovery. Mean particle size in the trap shows a pattern of coarser particles transported and deposited during mid-to-late afternoon hours and smaller particles accumulating during early morning and night hours. On average, the change in mean particle size for these different peaks is 12.3%. Significantly, this cyclic particle size pattern does not appear to be related to fluctuations in one single size fraction, which would negate a linear sedimentation rate assumption. Rather, the cyclicity represents changes in fractions of sand, silt and clay combined. Furthermore, the fluctuation does not appear related to tidal cycles as has been found in temperate Alaskan fjords with macrotidal fluctuations (Mackiewicz *et al.* 1984; Cowan & Powell 1990).

Although this is an inferred polythermal glacier and the subglacial conduit supplying the suspended sediment is under thick ice, the most logical forcing of the particle size variability based on event timing at this frequency is diurnal temperature

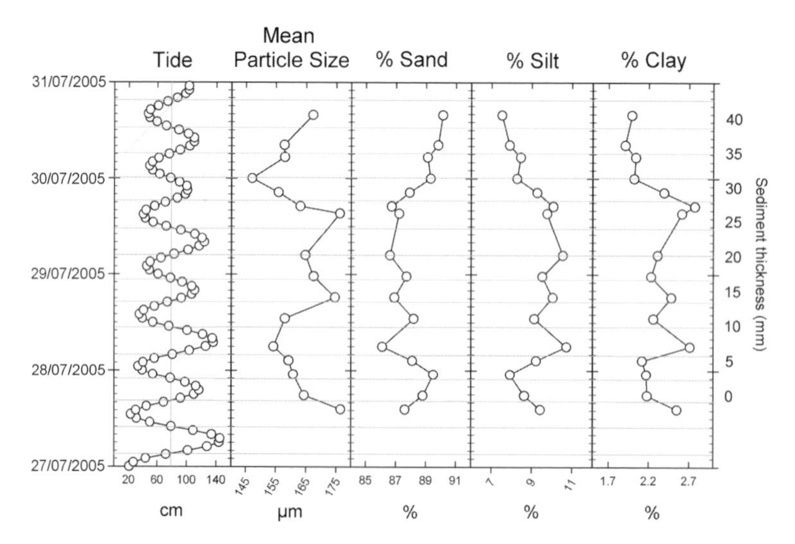

Fig. 6. Cyclic particle size fluctuations as recorded in a top trap from mooring ST6a inferred to represent diurnal discharge variations.

fluctuations. These fluctuations likely drive variations in discharge strength and turbulence with associated changes in travel distance of different particle sizes, resulting in the mean particle size cyclicity associated with time of day.

Discussion

Water column

The Kronebreen meltwater plume and Kongsvegen delta runoff both form buoyant overflows; however, there are several distinctions between the two sources. The SO water mass at the Kronebreen upwelling area has a higher average salinity and more variance than at the delta area. This signature is a result of direct deltaic freshwater flow from Kongsvegen to the surface of the fjord; at the Kronebreen source, however, meltwater turbulently mixes with saline fjord water as it rises to the surface (cf. Syvitski 1989; Powell 1990). The Kronebreen efflux has previously been postulated to mix in crevasses extending back from the grounding line prior to exiting the glacier (Elverhøi *et al.* 1980); however, the current stream when studied was discharging from a shear face at the grounding line. Another distinction between the deltaic and upwelling sources is that water characterized above 12 m water depth in the upwelling area shows more uniform temporal temperature patterns and is 2 °C colder than water from the delta. Also significantly, the Kronebreen upwelling area generally had higher average sediment concentrations and less variance in that parameter over time in the upper 12 m of water compared to those recorded in the delta area. These characteristics show the upwelling from Kronebreen had a more uniform higher velocity and locally concentrated discharge of sediment and meltwater throughout the sampling period relative to the deltaic runoff.

The deltaic overflow plume, with higher variability in temperature and sediment concentration, shows more susceptibility to surface processes as it flows over the delta and into the fjord. Another major distinction in the SO profiles is thickness of the overflow; from the upwelling it is *c.* 12 m thick, whereas the overflow at the delta is *c.* 7 m thick. This difference in thickness indicates a larger volume of water flowing out from the Kronebreen subglacial meltwater upwelling compared to the relatively smaller volume flowing from the marginal Kongsvegen fountain and over the delta.

Spatial and temporal variability also exists in the deeper (12–28 m depth) SW mass. At both sites, two distinct temperature trends exist in the SW water mass: one of lower temperatures *c.* 1.2 °C and a second at *c.* 2.4 °C (see Fig. 5a, b). We speculate this difference reflects surface conditions: solar radiation, meteorological conditions, dominant wind direction and dominant current direction that alter the spatial extent of SW before it is displaced by the SO. Sediment concentration profiles through the SW mass in the Kronebreen area again show a very uniform spatiotemporal distribution, whereas those in the Kongsvegen delta area show variations that reflect predominant surface conditions.

The deepest water mass, LW (28 m depth to the fjord floor), shows uniformity at both locations (Fig. 5a, b). The profiles reflect a stable interval of the water column not affected by surface conditions. It is through this interval that suspended sediments settle calmly to the fjord floor.

Suspended sediment concentrations from calibrated OBS data in July 2005 peaked at 0.39 kg m^{-3}; however, most overflow samples were around 0.20–0.25 kg m^{-3} which are lower than found in other studies (e.g. 0.34 kg m^{-3} in Svendsen *et al.* 2002 and 0.50 kg m^{-3} in Elverhøi *et al.* 1980). However, our measured concentrations were higher than another recent study (maximum 0.1 kg m^{-3} near Kronebreen in Aliani *et al.* 2004). Our actual sediment concentrations were probably much higher, however, as shown by our sediment trap records. The relatively low concentrations in our measured water samples used to calibrate the OBS data may skew our data to the lower end of the suspended sediment spectrum. This would result if our samples were not collected in the highest concentrations of suspended sediments in the plume. Alternatively, although speculative, the lower suspended sediment concentrations but higher sediment fluxes may indicate higher total meltwater with an average sediment load relative to the other studies. This hydrologic scenario would result in dilution and less sediment per unit volume of water, but higher sedimentation rates because of increased discharge.

Sediment flux spatial variation

The interplay and dynamics of the two meltwater sources is indicated by varied suspended sediment loads recorded in sediment traps and hydrographic profiles. Traps proximal to the Kronebreen upwelling plume (ST6a, ST3b) captured most sediment in their upper traps, whereas traps positioned near the deltaic runoff (ST4a, ST5a) predominantly captured sediment in the lowest traps. High suspended sediment load in the Kronebreen overflow resulted in the disproportionate sediment captured in upper traps. Although the Kronebreen SO water is more turbulently mixed relative to the deltaic SO, the plume and overflow velocities are sufficient to maintain high sediment concentrations thus resulting in a smaller relative volume of sediment immediately released from suspension. Rather, this material is transported more distally beyond the lower traps in the immediate upwelling area to a point where overflow velocity sufficiently decreases and causes particle release to the lower water column.

In contrast, traps positioned to primarily capture deltaic runoff (ST4a, ST5a) received a smaller volume of suspended sediment from the SO in their upper traps. Also in these traps, sediment gravity flows appear more frequent, contributing significantly to the relatively larger volume of sediment captured in bottom traps. For example, ST5a anchored to the delta front with bottom traps at 5 m above the floor contained numerous sand laminae from sediment gravity flows. ST4a, in contrast, located farther from the delta and on relatively flat seafloor, contained only silt and clay in its lower traps. Here, we interpret increased sediment volume in the lower traps to predominantly reflect suspension settling. At this location we also infer a small sediment component contributed by distal, low-viscosity gravity flows that compositionally indicate they are sourced from the Kronebreen grounding-line fan. Such slides, slumps and sediment gravity flows are common at Alaskan tidewater termini because of slope over-steepening by high sedimentation rates and disruption of bottom sediment from glacier pushing and iceberg calving (Powell 1981). However, higher overall sedimentation rates at ST4a indicate it is influenced by both sources (Table 4).

Local and global sediment flux perspectives

Daily sediment flux recorded as dry weights and accumulation thicknesses in this study are higher than those from previous studies at nearby glaciers. Our highest sediment flux of 1.86×10^4 g m^{-2} d^{-1} (14.3 mm d^{-1}) was measured at a depth of 10 m at a distance 240 m from Kronebreen. This value is two orders of magnitude higher than the highest fluxes of other studies in inner Kongsfjorden (800 g m^{-2} d^{-1} measured at 15 m depth, 250 m from Kongsbreen in Svendsen *et al.* 2002 and 933 g m^{-2} d^{-1} measured at 15 m depth, 300 m from Kongsbreen in Zajaczkowski 2002). Our minimum fluxes were both from bottom traps: 6.74×10^3 g m^{-2} d^{-1} at 240 m and 813 g m^{-2} d^{-1} at 470 m distance from Kronebreen. The latter was the lowest measured sediment flux, a value comparable to the maximum near-surface values recorded in other studies.

When we extrapolate our measured sedimentation rates to a conservative 60-day melt season, our results are similar to those of some previous studies but are higher than others. Elverhøi *et al.* (1980) measured rates in cores to be 100 mm a^{-1} wet mud or 30 mm a^{-1} dry sediment, and 50–100 mm a^{-1} within 10 km from the glacier (Elverhøi *et al.* 1983). These are similar values to those we estimate, where minimum sediment accumulation rates are 60–90 mm a^{-1} at <0.5 km from point source depocentres. Our annual sedimentation rates at the glacier front are a minimum of 4.05×10^5 g m^{-2} a^{-1} with an inner basin average of 1.56×10^5 g m^{-2} a^{-1}. These annual rates are an order of magnitude higher than those of other studies in inner Kongsfjorden: 1.80×10^4 g m^{-2} a^{-1} measured near Ossian Sarsfjellet using $^{210}Pb_{ex}$ dating of sediment cores (Aliani *et al.* 2004), 2.0×10^4 g m^{-2} a^{-1} measured for the Kongsbreen 'glacier front' using $^{210}Pb_{ex}$ dating of sediment cores (Svendsen *et al.* 2002) and 5.6×10^4 g m^{-2} a^{-1} measured in sediment traps at 15 m depth, 300 m from Kongsbreen (from daily rates in Zajaczkowski 2002 multiplied by a 60-day melt season). Higher rates of this study may indicate a higher relative meltwater and sediment flux of the Kronebreen–Kongsvegen glacier complex compared to the other nearby glaciers measured in other studies. These results are expected given that Kongsbreen drains only about 42% of the total area of Kronebreen. Sediment flux has been shown to generally scale with glacier basin size in Alaska (Powell 1991) and worldwide (Hallet *et al.* 1996), although others argue the basin area–erosion relationship is too simple (Elverhøi *et al.* 1998). It is also possible that measurements in other studies were taken at greater distance and are therefore expected to be significantly lower (e.g. Cai 1994; Cowan *et al.* 1999).

Our measured sediment fluxes from Kronebreen and Kongsvegen fall within the range of those for other subpolar polythermal glaciers when placed within the spectrum of glacial regimes. For example, the maximum sediment flux from temperate glaciers in southeastern Alaska is at least one order of magnitude higher than for Kronebreen. For McBride Glacier in Alaska, Cowan & Powell (1991) measured maximum and average sediment fluxes of 2.4×10^5 g m^{-2} d^{-1} and 5.3×10^4 g m^{-2} d^{-1}, respectively, into ice-proximal McBride Inlet. Both are at least one order of magnitude higher than the ice-proximal rates for Kongsfjorden (Table 4). At the opposite end of the spectrum are polar regime glaciers. A recent sediment flux measured on the Antarctic Peninsula was recorded at $1.7–3.5 \times 10^3$ g m^{-2} a^{-1} (Gilbert *et al.* 2003), one order of magnitude less than the subpolar glaciers of this study (Table 4).

Sediment yields calculated from volumes of ice-contact depositional landforms are 1.40×10^3 tonnes km^{-2} a^{-1} for Kronebreen and 1.06×10^4 tonnes km^{-2} a^{-1} for Kongsvegen (1.20×10^4 tonnes km^{-2} a^{-1}, or 4.80 mm a^{-1} of effective bedrock erosion combined) (Table 1). The majority of this contribution comes from very rapid building of the Kongsvegen ice-contact delta. This yield is about a factor of five greater than the 30-year average estimated by Elverhøi *et al.* (1998), but is still at least an order of magnitude less than temperate Alaskan glaciers (Hallet *et al.* 1996; Elverhøi *et al.* 1998). The Kongsvegen yield represents an ice-contact measurement and, since sediment accumulation rates decrease according to power and exponential decay functions with distance (Cai 1994; Cowan *et al.* 1999), these data are considered compatible with other measurements obtained farther from the glacier termini. Indeed, our own sediment trap data collected farther from the termini produce sediment accumulation rates similar to the longer term averages of nearby studies. Due to their close glacier proximity, these yields are therefore very rare and represent the true erosive ability of these subpolar glaciers.

Conclusions

Retreat of the Kronebreen–Kongsvegen glacier complex has significantly slowed in recent decades due to post-surge quiescent-phase conditions. Hydrographic profiles record the mixing of fjord and glacial waters including the transport and rainout of suspended sediments, closer than previously documented. Sediment trap data show that Kronebreen is now the major suspended sediment source as opposed to Kongsvegen, a relationship that has reversed within the last 25 years. Glacifluvial deltaic runoff from Kongsvegen presently contributes only 20–25% of the total volume of sediment in the inner fjord basin. The submarine discharge from Kronebreen contributes the remainder. Measured daily sediment fluxes are higher than those presented by previous researchers. However, when extrapolated for a full melt season, minimum annual vertical accumulation rates are similar to some previous estimates but higher than others. We expect this difference reflects the larger relative glacierized drainage area of the Kronebreen–Kongsvegen complex compared to other nearby glaciers. Most of the inner fjord basin away from point sources receives at least 60–90 mm a^{-1} of sediment from suspension settling.

Bathymetric profiling reveals a grounding-line fan adjacent to the major submarine meltwater discharge of Kronebreen. Near-bottom sediment traps indicate it is actively building at more than 300 mm a^{-1}. If terminus and subglacial conduit stability ensues, at this rate of aggradation the fan will emerge from the fjord in *c.* 18 years. An ice-contact delta in the southeastern fjord corner has built at a rate of 8.91×10^5 m^3 a^{-1}, providing an important and rare estimate of fully ice-contact glacial debris flux. Combined, these landforms equate to 4.80 mm a^{-1} of effective bedrock erosion for the glacial complex. This is almost an ice-contact value and therefore represents the full capacity of these subpolar polythermal glaciers in landscape denudation.

With continued climatic warming we expect the ablation season of Kronebreen and Kongsvegen to grow. Meltwater and therefore sediment flux may also increase. Under warming conditions we would expect greater sedimentation rates and faster infilling of the inner basin of Kongsfjorden. If the terminus remains relatively stable, it is expected that an ice-contact delta will form in front of Kronebreen as it has for Kongsvegen. We expect these landforms will enhance glacier margin stability and the ability of the terminus to re-advance in future if ice mass-balance permits. This may occur during initial stages of climate warming if precipitation increases, causing mass-balance to become less negative.

This work was completed as part of the Svalbard Research Experience for Undergraduates program, funded by the US National Science Foundation (award 0649006). Logistical support by Kings Bay AS, the Norwegian Polar Institute and UNIS was fundamental to this research. Comments by two anonymous reviewers were of significant benefit to the final product.

References

ALIANI, S., BARTHOLINI, G. ET AL. 2004. Multidisciplinary investigations in the marine environment of the inner Kongsfiord, Svalbard islands (September 2000 and 2001). *Chemistry and Ecology*, **20**, S19–S28.

ALLEY, R. B. 1991. Sedimentary processes may cause fluctuations of tidewater glaciers. *Annals of Glaciology*, **15**, 119–124.

ALLEY, R. B., ANANDAKRISHNAN, S., DUPONT, T. K., PARIZEK, B. R. & POLLARD, D. 2007. Effect of sedimentation on ice-sheet grounding-line stability. *Science*, **315**, 1838–1841.

CAI, J. 1994. *Sediment yields, lithofacies architecture, and mudrock characteristics in glacimarine environments.* PhD thesis, Northern Illinois University, DeKalb, Illinois, USA.

COTTIER, F., TVERBERG, V., INALL, M., SVENDSEN, H., NILSEN, F. & GRIFFITHS, C. 2005. Water mass modification in an Arctic fjord through cross-shelf exchange: the seasonal hydrography of Kongsfjorden, Svalbard. *Journal of Geophysical Research*, **110**, C122005, doi: 10.1029/2004JC002757.

COWAN, E. A. 1988. *Sediment transport and deposition in a temperate glacial fjord, Glacier Bay, Alaska.* PhD Thesis, Northern Illinois University, Illinois, USA.

COWAN, E. A. & POWELL, R. D. 1990. Suspended sediment transport and deposition of cyclically interlaminated sediment in a temperate glacial fjord, Alaska, U.S.A. *In*: DOWDESWELL, J. A. & SCOURSE, J. D. (eds) *Glacimarine Environments: Processes and Sediments.* Geological Society, London, Special Publications, **53**, 75–89.

COWAN, E. A. & POWELL, R. D. 1991. Ice-proximal sediment accumulation rates in a temperate glacial fjord, southeastern Alaska. *In*: Anonymous (ed.) *Glacial Marine Sedimentation: Paleoclimatic Significance.*

Geological Society of America, Special Paper, **261**, 61–73.

Cowan, E. A., Seramur, K. C., Cai, J. & Powell, R. D. 1999. Cyclic sedimentation produced by fluctuations in meltwater discharge, tides and marine productivity in an Alaskan fjord. *Sedimentology*, **46**, 1109–1126.

Elverhøi, A., Liestøl, O. & Nagy, J. 1980. Glacial erosion, sedimentation and microfauna in the inner part of Kongsfjorden, Spitsbergen. *Norsk Polarinstitutt Skrifter*, **172**, 28–61.

Elverhøi, A., Lønne, O. & Seland, R. 1983. Glaciomarine sedimentation in a modern fjord environment, Spitsbergen. *Polar Research*, **1**, 127–149.

Elverhøi, A., Hooke, R. LeB. & Solheim, A. 1998. Late Cenozoic erosion and sediment yield from the Svalbard-Barents Sea region: implications for understanding erosion of glacierized basins. *Quaternary Science Reviews*, **17**, 209–241.

Fischer, M. P. & Powell, R. D. 1998. A simple model for the influence of push-morainal banks on the calving and stability of glacial tidewater termini. *Journal of Glaciology*, **44**, 31–41.

Gerland, S., Haas, C., Nicolaus, M. & Winther, J.-G. 2004. Seasonal development of structure and optical surface properties of fast ice in Kongsfjorden, Svalbard. *Reports on Polar and Marine Research*, **129**, 26–34.

Gilbert, R., Chong, A., Dunbar, R. B. & Domack, E. W. 2003. Sediment trap records of glacimarine sedimentation at Müller Ice Shelf, Lallemand Fjord, Antarctic Peninsula. *Arctic, Antarctic, and Alpine Research*, **35**, 24–33.

Hagen, J. O., Liestøl, O., Roland, E. & Jørgensen, T. 1993. Glacier atlas of Svalbard and Jan Mayen. *Norsk Polarinstitutt Meddeleser*, **129**, 141.

Hagen, J. O., Melvold, K., Pinglot, F. & Dowdeswell, J. A. 2003. On the net mass balance of the glaciers and ice caps in Svalbard, Norwegian Arctic. *Arctic, Antarctic, and Alpine Research*, **35**, 264–270.

Hald, M., Ebbesen, H. *et al.* 2004. Holocene paleoceanography and glacial history of the West Spitsbergen area, Euro-Arctic margin. *Quaternary Science Reviews*, **23**, 2075–2088.

Hallet, B., Hunter, L. & Bogen, J. 1996. Rates of erosion and sediment evacuation by glaciers: a review of field data and their implications. *Global and Planetary Change*, **12**, 213–235.

Hjelle, A. 1993. *Geology of Svalbard*, Norsk Polarinstitutt Oslo.

Hodson, A., Gurnell, A., Tranter, M., Bogen, J., Hagen, J. O. & Clark, M. 1998. Suspended sediment yield and transfer processes in a small high-Arctic glacier basin, Svalbard. *Hydrological Processes*, **12**, 73–86.

Hop, H., Pearson, T. *et al.* 2002. The marine ecosystem for Kongsfjorden, Svalbard. *Polar Research*, **21**, 167–208.

Howe, J. A., Moreton, S. G., Morri, C. & Morris, P. 2003. Multibeam bathymetry and depositional environments of Kongsfjorden and Krossfjorden, western Spitsbergen, Svalbard. *Polar Research*, **22**, 301–316.

Ingólfsson, O. 2004. Outline of the geography and geology of Svalbard. *UNIS Course Packet*. UNIS.

IPCC 2007. Summary for Policymakers. *In*: Solomon, S., Qin, D., Manning, M., Chen, Z., Marquis, M., Averyt, K. B., Tignor, M. & Miller, H. L. (eds) *Climate Change 2007: The Physical Science Basis. Contribution of Working Group I to the Fourth Assessment Report of the Intergovernmental Panel on Climate Change*. Cambridge University Press, Cambridge.

Lamb, H. H. 1977. *Climate: Present, Past and Future*, Methuen, London.

Lefauconnier, B., Hagen, J. O. & Rudant, J. P. 1994. Flow speed and calving rate of Kongsbreen glacier, Svalbard, using SPOT images. *Polar Research*, **13**, 59–65.

Mackiewicz, N. E., Powell, R. D., Carlson, P. R. & Molnia, B. F. 1984. Interlaminated ice-proximal glacimarine sediments in Muir Inlet, Alaska. *Marine Geology*, **57**, 113–147.

MacLachlan, S. E., Cottier, F. R., Austin, W. E. N. & Howe, J. A. 2007. The salinity: del^{18}O water relationship in Kongsfjorde, western Spitsbergen. *Polar Research*, **26**, 160–167.

Meier, M. F. & Post, A. 1969. What are glacier surges? *Canadian Journal of Earth Science*, **6**, 807–817.

Melvold, K. & Hagen, J. O. 1998. Evolution of a surge-type glacier in its quiescent phase; Kongsvegen, Spitsbergen, 1964–95. *Journal of Glaciology*, **44**, 394–404.

Powell, R. D. 1981. A model for sedimentation by tidewater glaciers. *Annals of Glaciology*, **2**, 129–134.

Powell, R. D. 1984. Glacimarine processes and inductive lithofacies modelling of ice shelf and tidewater glacier sediments based on Quaternary examples. *Marine Geology*, **57**, 1–52.

Powell, R. D. 1990. Glacimarine processes at grounding-line fans and their growth to ice-contact deltas. *In*: Dowdeswell, J. A. & Scourse, J. D. (eds) *Glacimarine Environments: Processes and Sediments*. Geological Society, London, Special Publications, **53**, 53–73.

Powell, R. D. 1991. Grounding-line systems as second-order controls on fluctuations of tidewater termini of temperate glaciers. *Glacial Marine Sedimentation; Paleoclimatic Significance*. Geological Society of America, Special Paper, **261**, 75–94.

Powell, R. D. & Alley, R. B. 1997. Grounding-line systems; processes, glaciological inferences and the stratigraphic record. *In*: Barker, P. F. & Cooper, A. C. (eds) *Geology and Seismic Stratigraphy of the Antarctic Margin, 2, Antarctic Research Series*. AGU, Washington, DC, **71**, 169–187.

Sund, M. 2007. Calving front of Kronebreen, Svalbard in July 2007. http://www.geo.uio.no/glaciodyn/Publications/Kronebreen.wmv.

Svendsen, H., Beszczynska-Møller, A. *et al.* 2002. The physical environment of Kongsfjorden-Krossfjorden, an Arctic fjord system in Svalbard. *Polar Research*, **21**, 133–166.

Syvitski, J. P. M. 1989. On the deposition of sediment within glacier-influenced fjords: oceanographic controls. *Marine Geology*, **85**, 301–329.

Woodward, J., Murray, T., Clark, R. A. & Stuart, G. W. 2003. Glacier surge mechanisms inferred from ground-penetrating radar; Kongsvegen, Svalbard. *Journal of Glaciology*, **49**, 473–480.

Zajaczkowski, M. 2002. On the use of sediment traps in sedimentation measurements in glaciated fjords. *Polish Polar Research*, **23**, 161–174.

Sediment fluxes and carbon budgets in Loch Creran, western Scotland

PEI SUN LOH[1,4]*, ALISON D. REEVES[2], AXEL E. J. MILLER[3], S. MARTYN HARVEY[3] & JULIAN OVERNELL[3]

[1]*UHI Millennium Institute, Scottish Association for Marine Science, Scottish Marine Institute, Oban PA37 1QA, Scotland, UK*

[2]*Environmental Systems Research Group, Geography, School of Social and Environmental Sciences, University of Dundee, Dundee DD1 4HN, Scotland, UK*

[3]*Scottish Association for Marine Science, Scottish Marine Institute, Oban PA37 1QA, Scotland, UK*

[4]*Present address: Institute of Marine Geology and Chemistry, National Sun Yat-Sen University, Kaohsiung 80424, Taiwan, R.O.C.*

**Corresponding author (e-mail: psloh@staff.nsysu.edu.tw)*

Abstract: Sea lochs are regions where riverine and marine organic carbon (OC) undergoes decomposition, deposition or transportation to shelf slopes and the deep sea. According to the OC budget presented here, discharge from River Creran (1.44×10^6 kg a^{-1}) and phytoplankton material (0.89×10^6 kg a^{-1}) make up a significant input of OC to Loch Creran while 0.67×10^6 kg a^{-1} OC is from other sources. A total of 1.28×10^6 kg a^{-1} OC is deposited in the loch and 1.14×10^6 kg a^{-1} OC is oxidized in the water column. Discharge to the Lynn of Lorn consists of 0.58×10^6 kg a^{-1} OC. Hence Loch Creran is a sink for OC where 42.7% of the total OC input is buried and 38% and 1.7% decomposed in the water column and subsurface sediments, respectively. River Creran contributes 63% labile and 37% refractory organic matter to the loch. More than 95% of each of the total OC, lignin and organic matter deposited onto the surface sediments is buried in the subsurface sediments. Seventy-five percent of the total organic matter decomposed in the water column is labile. Output to Lynn of Lorn consists of 54.6% refractory organic matter.

Numerous studies have estimated the global riverine export of organic carbon (OC) of which a few examples are given here: 334.5×10^{12} g a^{-1} (Degens *et al.* 1991), 360×10^{12} g a^{-1} (Aitkenhead & McDowell 2000), 380×10^{12} g a^{-1} (Ludwig *et al.* 1996), 430×10^{12} g a^{-1} (Schlunz & Schneider 2000) and 500×10^{12} g a^{-1} (Spitzy & Ittekkot 1991). Of the total OC flux to the oceans, Spitzy & Ittekkot (1991) estimated that half is in particulate form and half in dissolved form. Ludwig *et al.* (1996) stated that of the 380×10^{12} g a^{-1} OC input to the oceans, 210×10^{12} g a^{-1} is dissolved OC and 170×10^{12} g a^{-1} is particulate OC. According to Berner (1982), a total of 126×10^{12} g a^{-1} OC is deposited in deltaic shelf sediments. Of this, 104×10^{12} g a^{-1} is terrestrial and 22×10^{12} g a^{-1} is marine OC. Schlunz & Schneider (2000), on the other hand, found almost equal amounts of 43.4×10^{12} g a^{-1} terrestrial OC and 55.2×10^{12} g a^{-1} marine OC are buried on the shelf. In summary, based on these estimates, the global riverine export of OC ranges from 334.5 to 500×10^{12} g a^{-1} and 10–20% of this OC is deposited on the continental shelves.

Rivers discharge dissolved and particulate material directly to the coastal zone (Wollast 1991). Most of the riverine or terrestrial OC has undergone degradation in the soil (Hurst & Burges 1967; Christman & Oglesby 1971; Martin & Haider 1980; Zeikus *et al.* 1982). Hence most terrestrial OC detected in the sea is old and refractory. For example, the four major rivers (Amazon, Hudson, York and Parker) which discharge into the western North Atlantic Ocean have radiocarbon ages of particulate OC ranging from 316 to 4763 years BP and dissolved OC ages from modern to 1384 years BP (Raymond & Bauer 2000). Because it is highly degraded, riverine OC should undergo minimal decomposition and be preserved in the sea (Hedges *et al.* 1997). In reality, however, riverine OC comprises only a small fraction of the total OC detected in the sea. Only 1–10% and 0.7–2.4% of the total

From: HOWE, J. A., AUSTIN, W. E. N., FORWICK, M. & PAETZEL, M. (eds) *Fjord Systems and Archives.* Geological Society, London, Special Publications, **344**, 103–124.
DOI: 10.1144/SP344.10 0305-8719/10/$15.00

dissolved OC in the Pacific and Atlantic Oceans, respectively, is terrestrially derived (Meyers-Schulte & Hedges 1986; Opsahl & Benner 1997). Many authors have offered insights into the fate of riverine OC in coastal areas and the open ocean.

There are two types of environmental conditions at ocean margins. Some environments favour accumulation of OC due to high sedimentation rate and fine particle size. According to Hedges & Keil (1995), 45% of OC accumulates in the coastal margin due to high sedimentation rates and adsorption of the more labile organic matter onto mineral surfaces. The Norwegian channel and Skagerrak/Kattegat bay sediments serve as sinks for OC due to the presence of fine-grained sediments which protect organic matter from the process of degradation (De Haas & van Weering 1997). OC burial efficiency of up to 62% in the northeastern Skagerrak has also been reported (Stahl *et al.* 2004). Due to their prominence as sites of continental detritus deposition, ocean margins are important in preserving the global carbon balance (Blanton 1991; Mantoura *et al.* 1991; Chester 2000).

The second type of environment favours mineralization of the OC due to repeated cycles of erosion and re-deposition of sedimentary material, plus biological and chemical processes in the water column (De Haas *et al.* 2002). They state that most of the OC in the North Sea, Middle Atlantic Bight, Bering, Chukchi and Amazon shelves is mineralized in the water column. In environments where most of the OC in the shelves is decomposed, continental slopes, canyons and deep-sea fans become the main sinks for OC.

The coastal zone is characterized by high biological productivity as a result of riverine input of nutrients, upwelling and coupling between benthic and pelagic systems (Wollast 1991). Higher rates of organic matter decomposition occur on the shelf, and higher rates of organic matter accumulation occur on the slope of the Gulf of Lions (Accornero *et al.* 2003). The majority of OC in the Celtic Sea, California, Washington and Canadian shelves is buried in the continental slope (De Haas *et al.* 2002). In other environments, sedimentary OC flows into the deep sea. Approximately 53% of the OC from the Sepik River flows to the deep Bismark Sea (Burns *et al.* 2008). Glud *et al.* (2009) showed that a significant input of organic matter in Sagami Bay, Japan, is derived from downslope transport.

Sea lochs are locally important sites for the degradation and mineralization of biomass and the long-term preservation of terrestrial OC (Parkes & Buckingham 1986). Few investigations of biogeochemical cycling in sea loch systems have been conducted (Loh *et al.* 2002, 2008*a–c*; Nuwer & Keil 2005; Walsh *et al.* 2008). Fewer OC budgets have been determined in a sea loch system (Overnell & Young 1995). To our knowledge, the only OC budget study in Loch Creran was previously carried out by Cronin & Tyler (1980). The difference between their study and this is that they collected water samples from the surface and deep water in the upper and main basins whereas, in this study, 10 cm sediment cores were collected across a transect of Loch Creran. In this study, the surface (0–1 cm) and subsurface (9–10 cm layer) sediments were investigated to provide information on the fate of organic matter after burial. Comparison of OC, lignin and organic matter abundances between the surface and subsurface sediments would indicate whether there is burial and subsequent decomposition of organic matter. Budgets for the relatively labile and refractory fractions of organic matter were also quantified. In addition, the input and output of terrestrial OC were quantified by using lignin-derived phenols as biomarkers.

Lignin-derived phenols, oxygen uptake rates, bulk elemental analysis (OC) and organic matter compositions were previously reported for Loch Creran in studies of the fate (Loh *et al.* 2008*c*), biodegradability (Loh *et al.* 2002, 2008*a*) and transport (Loh *et al.* 2008*b*) of terrestrial organic matter in the loch system. In this study, these were used to construct OC, lignin and organic matter (the relatively labile and refractory fractions) budgets for Loch Creran. Sedimentation rate ($g\,cm^{-2}\,a^{-1}$) was obtained from the rate of particles deposited into a sediment trap (Overnell & Young 1995; Pejrup *et al.* 1996). The OC, lignin and organic matter budgets were estimated from the sedimentary fluxes, River Creran flow rate and particle load. The object of this work is to synthesize a budget based upon estimates of inputs and outputs of OC, lignin and organic matter to and from Loch Creran and the amount of OC, lignin and organic matter decomposed and deposited within the loch. This will provide a better insight into the role of sea lochs in regional carbon cycling.

Materials and methods

Study areas

Loch Creran is situated on the west coast of Scotland. This is an ideal location for the study of carbon budgets because it is a small loch system which serves as a direct link between land and sea. It is an enclosed system providing a direct route for a budget study as there is only one major riverine input from River Creran directly to the loch and one output to the Lynn of Lorn. Loch Creran is 12.8 km long and linked to offshore areas by shallow rocky sills. It has a main basin which covers an area of 11.49 km^2. The sill at Creagan encloses a small upper basin with a surface area of 2.04 km^2. Because this is a shallow loch, tidal flushing is sufficient to ensure ventilation of bottom water (Gage 1972). Loch Creran has a relatively small catchment area of 164 km^2 (Edwards & Sharples 1986).

The basin nearest the mouth of Loch Creran is 27 m at the deepest and is separated from the Lynn of Lorn by a sill of *c.* 4 m deep (Fig. 1). A sill of 6 m deep separates the first and second basins. The second basin (where LC3 and LC5 are located), down to 49 m, is the deepest of the four basins and is separated from the third basin by a 10 m sill. This third basin, where LC2 is located, is 27 m at the deepest. The fourth or upper basin, 37 m at the deepest, is the smallest basin and is separated from the previous basin with a 1 m sill at Creagan narrows (Edwards & Sharples 1986). The loch is divided into the upper and main basins throughout this study.

The timetable for sediment sampling and trap deployment was given in Loh *et al.* (2008*c*, table 1). Sampling locations are shown in Figure 1: LC0, LC1, LC2, LC3, LC5 (transect along the loch) and one location in the Lynn of Lorn, LC6. LC0 and LC1 are located in the upper basin of Loch Creran; LC2, LC3 and LC5 are located in the main basin. LC1 was sampled monthly and other locations were sampled in successive order. The frequency and timetable of the sampling and the sampling positions have been reported in Loh *et al.* (2008*a*, *c*).

Sediment and water sampling

Sediment cores with an undisturbed sediment–water interface were collected using a Craib corer (Craib 1965) lined with an acrylic core tube (24 cm long by 5.9 cm internal diameter). In the laboratory, the sediment cores were first subjected to the oxygen uptake rate measurement. After this, the overlying water in the cores was siphoned off and the sediments were sliced at 1 cm intervals from 0 to 10 cm. These sediment slices were kept in plastic containers, frozen overnight and freeze-dried. The dried sediments were ground to fineness using a mortar and pestle for subsequent analyses such as lignin, bulk elemental and loss on ignition.

A water sample was collected from the mouth of River Creran (56°34.27N, 5°13.52W) on 5th December 2001. Water was filtered through ‘Whatman’ 25 mm diameter glass microfibre filters (GF-F). Before filtration, the filters were ashed (500 °C overnight) to eliminate any residual organic matter, cooled and weighed. Water was then filtered under vacuum for a few hours. The

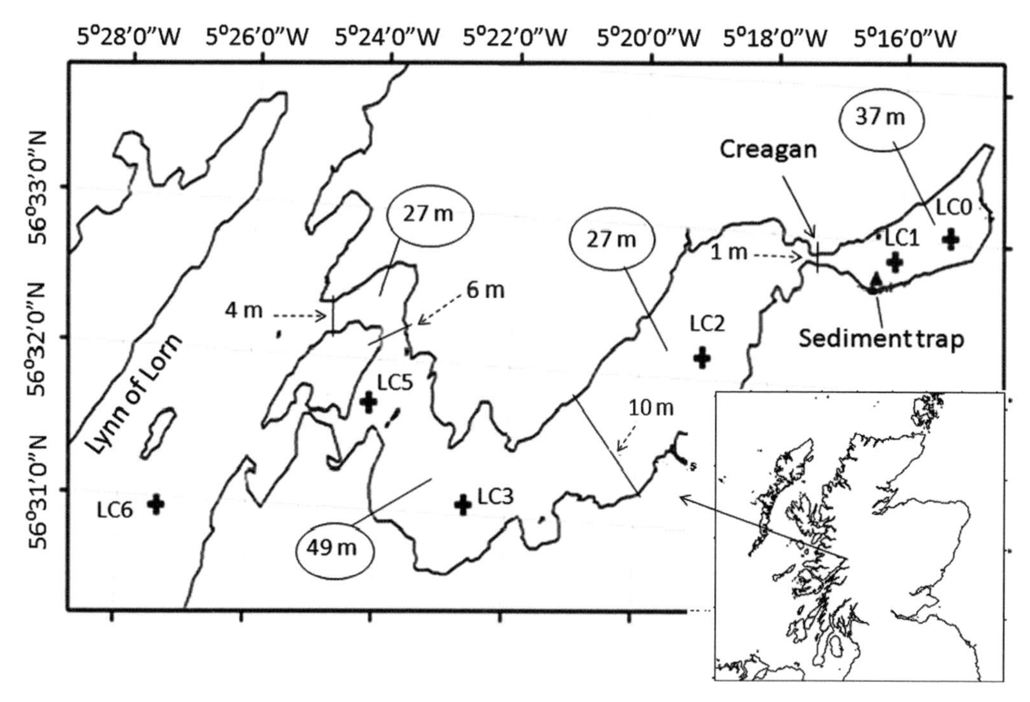

Fig. 1. Map showing the sampling locations in Loch Creran. Inset is map of Scotland showing the location of Loch Creran. The sills are denoted by solid lines and the depths of the sills are indicated by arrows with dashed lines. For example, the sill at Creagan is 1 m deep. The circles show the deepest depth in each basin. For example, the deepest depth at the upper basin of the loch is 37 m.

filters were then heated in an oven at 60 °C for about 2 hours until totally dry and reweighed to obtain the dry weight of the particulate material. The filtrates were kept frozen until analysis. The filtrates were subjected to the CuO oxidation process to determine lignin concentration in the particulate fraction.

Sediment trap

A sediment trap was deployed at 10 m water depth near site LC1 and serviced monthly at the same time as the sediment sampling (Loh *et al.* 2008*c*). Sedimentation rates were determined from December 2001 to October 2002 inclusive. The trap consists of four tubes (1.1 m long × 11 cm internal diameter) with removable clear plastic collecting tubes (Leftley & MacDougall 1991). The sediments in the collecting tubes were not fixed with a conservation agent. In the laboratory, the collecting tubes were allowed to stand until the sediment had settled. The overlying water was siphoned off. Sediment slurry was quantitatively transferred into centrifuge tubes, centrifuged and the supernatant discarded. The sediment fractions were frozen overnight and freeze-dried the following day. The dried sediments were ground to fineness using a mortar and pestle for subsequent lignin, bulk elemental and loss on ignition analyses.

Chemical analyses

The oxygen uptake rate measurement, lignin, bulk elemental and loss on ignition analyses have been explained in details previously (Loh *et al.* 2008*a*–*c*). In this study, the OC and total lignin (Λ, mg/100 mg OC) results were used to construct the respective budgets; O_2 uptake rate was used as an indirect measurement of phytoplankton production and loss on ignition results to construct budgets for the relatively labile and refractory fractions of organic matter.

Briefly, oxygen uptake rates measurement followed the method of Parkes & Buckingham (1986) and Overnell *et al.* (1995*a*, *b*). Oxygen uptake rates were determined for the sediment–water interface layer of the sediment cores collected as mentioned above. The rate was calculated from the difference in oxygen concentration of the overlying water between the start and end of the incubation, based on the method of Skoog *et al.* (2004) and Hansen (1999). Complete details on oxygen uptake rate measurements used in this study are given in Loh *et al.* (2008*a*). During oxygen uptake rate measurements, sediment cores were incubated for 24 hours (Loh *et al.* 2008*a*). During shorter incubation times (for 3–6 hours), we noted inconsistent and negative values; more experiments are needed to determine the reasons. It seems the system becomes stabilized after 20 hours. Since the oxygen concentration never fell below 100 μM in our experiments (according to Hall *et al.* 1989, the flux measurements below this concentration are meaningless as fluxes become non-linear), the validity of using a duration of more than 20 hours was confirmed.

The CuO oxidation method for lignin analysis was based on methods used by Hedges & Ertel (1982), Wilson *et al.* (1985), Readman *et al.* (1986) and Goni & Hedges (1992). Upon CuO oxidation, the product was analysed immediately by a PERKIN-ELMER 8410 gas chromatograph with flame ionization detector (GC-FID) fitted with a 100% dimethylpolysiloxane (ZB-1, Phonomenex, Zebron) column (30 m × 0.25 mm internal diameter). Peak identity was confirmed by a gas chromatograph with mass spectrometer (GC-MS; TRACE MS Thermo Quest, Finnigan). Total lignin is the sum of vanillin, acetovanillone, vanillic acid, syringaldehyde, acetosyringone, syringic acid, p-coumaric acid and ferulic acid, expressed in mg/g or mg/100 mg OC (Λ). Complete details on lignin measurements used in this study are given in Loh *et al.* (2008*a*–*c*).

For the loss on ignition, the percentage weight losses after combustion at 250 °C and 500 °C represent the % labile and % refractory organic matter (Kristensen & Andersen 1987; Sutherland 1998). For total OC determination, sediments were pretreated with sulphurous acid, wrapped in tin capsule and analysed using a LECO CHN-900 analyser.

Calculations

The superscripts in this section are used to show how a parameter is calculated and where this parameter is applied in other calculations.

Sedimentation rate

Sedimentation rate was calculated as the dry mass of particles deposited into the sediment trap:

Sedimentation rate[a] ($g\ m^{-2}\ d^{-1}$)

= total dry weight of sediment collected from the collecting tubes, g/area of the collecting tubes, m^2/total days of trap deployment

= dry weight, $g/0.038\ m^2/d$

Sedimentary fluxes of OC, lignin and organic matter

The surface sediment fluxes of OC, lignin and organic matter were calculated by multiplying the respective parameters (at the surface sediments) by the sedimentation rate. For example:

The sedimentary flux[b] of OC at LC0
= sedimentation rate[a] × mean %OC at LC0

To calculate the subsurface fluxes, the respective parameter at the subsurface sediment (9–10 cm layer) was multiplied with the sedimentation rate.

Accumulation rate

Sediment accumulation rate (w_s, cm a^{-1}) was calculated from the formula:

$$w_s = r/[(1 - \phi)p_s]$$

where r is sedimentation rate (g cm^{-2} a^{-1}), $\phi = 0.7$ is sediment porosity and $p_s = 2.65$ g cm^{-3} is density.

Note that the accumulation mentioned here and in the Results section is not the same as the 'accumulation' mentioned elsewhere in the text. The accumulation rate here and in the Results has the unit 'cm a^{-1}' and was calculated based on the above formula (see Appendix for more detailed calculations). Elsewhere in the OC, lignin and organic matter budgets, the accumulation was calculated as the mean fluxes of the parameters in the 9–10 cm sediment layer multiplied by the total depositional area (see 'Budgets determination' and Appendix).

Budgets determination

Budgets were determined following methods outlined by Cronin & Tyler (1980), Johannessen *et al.* (2003) and Burns *et al.* (2008). The average river flow rate and total OC concentration were used to calculate the OC input to the loch. Cronin & Tyler (1980) calculated the rate of passage of total OC to the sea and found that the net flow of water from the loch is equal to the river flow. Hence the River Creran flow rate is used to calculate the contribution from Loch Creran to Lynn of Lorn. Output is the product of the river flow rate and the mean total OC of the main basin surface sediments. Oxidative loss of OC in the sediment was calculated from the difference in OC fluxes between the surface and buried sediments (Johannessen *et al.* 2003). Total OC deposited at the sediment surface was calculated by multiplying the sediment flux at the surface sediment by the total depositional area. Total OC accumulated in the subsurface sediment was calculated by multiplying the sediment flux at the subsurface sediment by the total depositional area (Burns *et al.* 2008). A brief outline is given here. Details of calculations are given in the Appendix.

River Creran flow rate[c]
= mean of the flow rates reported by Cronin & Tyler (1980), Edwards & Sharples (1986) and Booth (1987).

Particulate concentration[d]
= mass of particulate matter/volume of water

Mass of particulate discharged into Loch Creran[e]
= River Creran flow rate[c] × particulate concentration[d]

Oxygen uptake rate represents the rate at which oxygen is consumed during benthic respiration or organic matter decomposition. We assume that the respiration quotient equals 1.0, that is, the number of molecules of CO_2 liberated is the same as the number of molecules of oxygen used. We also assume that the amount of carbon liberated is the same amount of carbon produced during phytoplankton photosynthesis.

Hence phytoplankton production
= mean oxygen uptake rate in Loch Creran × total depositional area of the loch

Total OC discharged from River Creran into Loch Creran
= mass of particulate discharged into Loch Creran[e] × mean %OC of the trap samples

Total OC deposited (sedimented) onto the surface sediments of Loch Creran[f]
= mean OC fluxes[b] in surface sediments across the loch (from LC0 to LC5) × total depositional area of the loch

Total OC accumulated (remain buried) in the 9–10 cm subsurface sediment[g]
= mean OC fluxes[b] in subsurface sediments across the loch (from LC0 to LC5) × total depositional area of the loch

Note that OC deposition is the amount of OC sedimented onto the surface sediments. OC accumulation is the amount of OC remains buried in the subsurface sediments.

Total OC undergoing degradation in the subsurface sediment
= Total OC deposited onto the surface sediments[f]
− Total OC accumulated in the 9–10 cm subsurface sediment[g]

The difference in OC content between the trap and surface sediment is assumed to be lost through decomposition process.

Total OC decomposed in the water column
= Difference between OC flux in the sediment trap and LC1 surface sediment
× total depositional area of the loch

Total OC discharged from Loch Creran into Lynn of Lorn
= mass of particulate discharged into Loch Creran[e] × mean %OC of the main basin (sediments i.e. LC2, LC3, LC5)

Results

OC, lignin, organic matter and oxygen uptake rates

Complete results of OC, lignin, organic matter and oxygen uptake rates are presented in Loh *et al.* (2008*a–c*). The mean values of these parameters for the surface (0–1 cm) and subsurface (9 to 10 cm) sediments are given in Table 1 (see Loh *et al.* 2008*c* for more details). The figures for the surface sediments are mean values over a period of measurements; the figures for the subsurface sediments (except for LC1) are all based on triplicate measurements of a single sample (Loh 2005).

Sedimentation rates, fluxes and accumulation rates

Sedimentation rates. The monthly total sedimentation rates obtained from the sediment trap samples from December 2001 to October 2002 are as follows: 6.97, 4.10, 8.13, 9.73, 21.38, 6.55, 15.91, 3.67, 31.47, 10.44 and 3.90 g m^{-2} d^{-1}. Hence these rates ranged from 3.67 to 21.38 g m^{-2} d^{-1}, averaging to 11.11 ± 8.66 g m^{-2} d^{-1} (see Appendix for complete results and calculations).

Multiplying the mean percentage OC of 6% for the trap samples by 11.11 g m^{-2} d^{-1}, a mean OC sedimentation rate of 0.67 g m^{-2} d^{-1} (244.55 g m^{-2} a^{-1} OC) is obtained for the upper basin of Loch Creran.

Cronin & Tyler (1980) gave an OC sedimentation rate of 50 g m^{-2} a^{-1} (0.14 g m^{-2} d^{-1}) for the main basin of Loch Creran. Dividing this value, 0.14 g m^{-2} d^{-1}, by the mean %OC of 3.1 for LC2 sediments (Table 1), a total OC sedimentation rate of 4.42 g m^{-2} d^{-1} is obtained. The OC sedimentation rate is converted to total sedimentation rate as a total sedimentation rate is needed to calculate the sedimentary fluxes of OC, lignin and organic matter in other locations. Although a mean sedimentation rate of 11.11 g m^{-2} d^{-1} has been obtained in this study for the upper basin of the loch, a rate of 4.42 g m^{-2} d^{-1} was used for the main basin. This is further discussed in ‘The assumptions’.

Sedimentary fluxes of OC, lignin and organic matter. The sediment flux is the amount of suspended sediment which settles through the water column and is deposited on the sea bed (Lund-Hansen *et al.* 1997). The sedimentary fluxes of OC, lignin and organic matter were calculated by multiplying the respective parameters by the total sedimentation rate. In order to calculate the OC, lignin and organic matter fluxes in LC0 and LC1 (in the upper basin), the total sedimentation rate 11.11 g m^{-2} d^{-1} was used. To determine the fluxes in LC2, LC3, LC5 and LC6, the total sedimentation rate 4.42 g m^{-2} d^{-1} was used.

Results for the OC, lignin and organic matter fluxes in surface and subsurface sediments across the loch are presented in Table 2. All fluxes show a general trend of decreasing order from the head to the mouth of the loch. OC fluxes to the sediment surface, for instance, ranged from 0.05 g m^{-2} d^{-1} at LC6 to 0.53 g m^{-2} d^{-1} at LC0. Lignin fluxes to the surface sediments decreased from 3.67×10^{-3} g m^{-2} d^{-1} at LC0 to 0.22×10^{-3} g m^{-2} d^{-1} at LC6. One interesting observation, however, is that *c.* 50% and 41% of lignin was oxidized in the subsurface sediments at LC5 and LC6 whereas, in the upper loch, none of the lignin was oxidized. (Note that the difference in organic matter content between the surface and subsurface sediments represents the amount of material oxidized or decomposed.) More labile organic matter was oxidized in the subsurface sediments at LC0, and more total OC oxidized in LC0 and LC3 subsurface sediments.

Accumulation rate. The total sedimentation rates of 11.11 g m^{-2} d^{-1} and 4.42 g m^{-2} d^{-1} were used to estimate accumulation rates (cm a^{-1}) for the upper and main basins of Loch Creran. These rates were estimated to be 0.50 cm a^{-1} and 0.20 cm a^{-1}, respectively. These accumulation rates indicate that the 10 cm subsurface sediment layers have been buried between 10 and 50 years. These rates are comparable, for example, to the rates estimated

Table 1. *Lignin, oxygen uptake rate, OC and organic matter contents in Loch Creran*

Locations		Lignin		Oxygen uptake rates (mmol m^{-2} d^{-1})		OC oxidation rate (g m^{-2} d^{-1})	OC (%)	Labile OM (%)	Refractory OM (%)
		mg/g	mg/100 mg OC (Λ)	Range	Mean				
Sediment trap		0.31	0.50	–	–	–	6.0	14.8	8.7
LC0	A	0.33	0.69	15.9–18.7	17.1	0.21	4.8	9.8	7.3
	B	0.33					3.8	8.8	4.8
LC1	A	0.22	0.54	11.8–27.8	20.8	0.25	4.0	9.1	6.8
	B	0.24					3.6	9.5	6.6
LC2	A	0.18	0.56	9.2–9.6	9.4	0.11	3.1	6.8	6.1
	B	0.17					2.9	–	–
LC3	A	0.07	0.37	9.3–14.8	12.5	0.15	1.8	3.5	4.9
	B	0.08					1.3	3.8	4.6
LC5	A	0.10	0.44	12.3–18.7	15.1	0.18	2.3	1.3	3.2
	B	0.05					–	1.7	1.7
LC6	A	0.05	0.44	6.6–14.2	9.4	0.11	1.1	3.6	6.0
	B	0.03					–	–	–

Note: The results in this table have been presented elsewhere (Loh *et al.* 2008*c*). The OC, lignin and labile and refractory OM (organic matter) results in this table multiplied by the sedimentation rates 11.11 and 4.42 g m^{-2} d^{-1} will give the fluxes in Table 2. All the results for surface sediments are mean values over a period of time. For the subsurface sediments, apart from the values for LC1, all other figures represent results from triplicate measurements of only one sample (Loh 2005). Note that 'A' represents surface (0–1 cm) sediments and 'B' represents subsurface (9–10 cm) sediments.

using the ^{210}Pb dating method for the Strait of Georgia (0.28–2.94 cm a^{-1}; Johannessen *et al.* 2003) and the Pacific NW margin (0.15–0.60 cm a^{-1}; Wheatcroft & Sommerfield 2005).

River Creran flow rate

River Creran flow rates were given by Cronin & Tyler (1980), Edwards & Sharples (1986) and Booth (1987) as 0.52, 0.78 and 0.43 × 10^6 m^3 d^{-1}. This gives a mean River Creran flow rate of 0.58 × 10^6 m^3 d^{-1}.

A water sample was collected from the mouth of River Creran on 5th December 2001 and 251.07 mg of particulate matter was obtained from the filtration of 2207 ml water. This gives a particulate concentration of 113.76 g m^{-3}. The lignin content obtained from this particulate fraction was 0.75 mg g^{-1}.

The mass of particulate matter discharged to Loch Creran was calculated by multiplying 113.76 g m^{-3} by 0.58 × 10^6 m^3 d^{-1} × 365 ÷ 1000. This equals to 24.08 × 10^6 kg a^{-1}. This value is used to calculate the OC, lignin and organic matter input and output from Loch Creran (Appendix).

OC, lignin and organic matter budgets

OC budget. The OC and other budgets are estimated from the OC, lignin and organic matter contents at the surface and subsurface sediments and sediment trap samples. Estimates from the present study showed that River Creran and phytoplankton production contributed 1.44 × 10^6 kg a^{-1} and 0.89 × 10^6 kg a^{-1} OC, respectively, to Loch Creran. A total of 1.28 × 10^6 kg a^{-1} OC was deposited at the sediment surface, 1.14 × 10^6 kg a^{-1} was oxidized in the water column and 0.58 × 10^6 kg a^{-1} was discharged to the Lynn of Lorn. Overall, 42.7% of the total OC input was deposited in the loch, 38.0% was oxidized in the water and 19.3% was exported to the Lynn of Lorn. Hence according to this study, 80.7% of the OC was sequestered within the loch through oxidative and depositional processes (Fig. 2).

Lignin budget. River Creran discharged 18.06 × 10^3 kg lignin to the loch annually. Approximately 7.6 × 10^3 kg a^{-1} of the total lignin input from the river was deposited on surface sediments, 4.9 × 10^3 kg a^{-1} was decomposed in the water column and 2.9 × 10^3 kg a^{-1} lignin material was discharged to the Lynn of Lorn. These results show that 41.8% lignin material was deposited onto the surface sediments, 27.3% was decomposed in the water column and 16.0% was exported to Lynn of Lorn; the fate of the remaining 14.8% is not known (Fig. 3).

Lignin materials represent 15–35% of plant tissues (Kratzl 1965) and effectively all the terrestrial organic matter input is of plant origin. There are 18.06 × 10^3 kg a^{-1} of lignin and 1.44 × 10^6 kg a^{-1} OC input to the loch. This

Table 2. *Lignin, organic matter and OC flux parameters at stations in Loch Creran*

Location	Fluxes	Lignin	Labile OM	Refrac OM	Total OM	OC
Trap		3.44×10^{-3}	1.64	0.97	2.61	0.67
LC0	Surface*	3.67×10^{-3}	1.09	0.81	1.90	0.53
	Burial*	3.67×10^{-3}	0.98	0.53	1.51	0.42
	Oxidation rate*	0×10^{-3}	0.11	0.28	0.39	0.11
	% buried	100	89.91	65.43	79.47	79.25
	% oxidized	0	10.09	34.57	20.53	20.75
LC1	Surface*	2.44×10^{-3}	1.01	0.76	1.77	0.44
	Burial*	2.67×10^{-3}	1.04	0.73	1.77	0.40
	Oxidation rate*	-0.23×10^{-3}	−0.03	0.03	0	0.04
	% buried	109.43	102.97	96.05	–	90.91
	% oxidized	−9.43	−2.97	3.95	–	9.09
LC2	Surface*	0.80×10^{-3}	0.31	0.27	0.58	0.14
	Burial*	0.75×10^{-3}	–	–	–	0.13
	Oxidation rate*	0.05×10^{-3}		–	–	0.01
	% buried	93.75	–	–	–	92.86
	% oxidized	6.25	–	–	–	7.14
LC3	Surface*	0.31×10^{-3}	0.16	0.22	0.38	0.08
	Burial*	0.35×10^{-3}	0.17	0.21	0.38	0.06
	Oxidation rate*	-0.04×10^{-3}	−0.01	0.01	0	0.02
	% buried	112.90	106.25	95.45	100	75
	% oxidized	−12.90	−6.25	4.55	0	25
LC5	Surface*	0.44×10^{-3}	0.06	0.14	0.20	0.10
	Burial*	0.22×10^{-3}	0.08	0.08	0.15	–
	Oxidation rate*	0.22×10^{-3}	−0.02	0.06	0.05	–
	% buried	50	133.33	57.14	75	–
	% oxidized	50	−33.33	42.86	25	–
LC6	Surface*	0.22×10^{-3}	0.16	0.27	0.43	0.05
	Burial*	0.13×10^{-3}	–	–	–	–
	Oxidation rate*	0.09×10^{-3}	–	–	–	–
	% buried	59.09	–	–	–	–
	% oxidized	40.91	–	–	–	–

*Units: g $m^{-2} d^{-1}$.

means that terrestrial organic matter contributes 3.6–8.4% of the total OC input to Loch Creran. Similarly, terrestrial organic matter constitutes 1.2–2.9% of the OC oxidized in the water column, 1.7–4.0% of the OC deposited in the sediments and 1.4–3.3% of the total OC transported to the Lynn of Lorn.

Organic matter budget. River Creran discharged 3.56×10^6 kg a^{-1} labile organic matter to Loch Creran. There was a total input of 3.11×10^6 kg a^{-1} labile organic matter to the loch, the fate of which is unknown. Outputs consist of 3.11×10^6 kg a^{-1} oxidized in the water column, 2.62×10^6 kg a^{-1} deposited to the sediment surface and 0.94×10^6 kg a^{-1} exported to the Lynn of Lorn. The input of refractory organic matter to the loch is 2.09×10^6 kg a^{-1}. There was a total input of 2.25×10^6 kg a^{-1} refractory organic matter, the fate of which is unknown. Estimated outputs consist of 2.17×10^6 kg a^{-1} refractory organic matter deposited in the loch, 1.13×10^6 kg a^{-1} exported to the Lynn of Lorn and 1.04×10^6 kg a^{-1} respired in the water column (Fig. 4).

Of the total organic matter input to Loch Creran, 63.0% was comprised of the relatively labile and 37.0% of the refractory fraction. These budgets show some distinctive trends. Most of the labile organic matter undergoes decomposition. As a result of this, more of the refractory fraction will be discharged to the Lynn of Lorn. Of the total organic matter oxidized in the water column, 75.0% was labile. The refractory fraction contributed 54.6% of the total organic matter discharged to the Lynn of Lorn.

Discussion

The assumptions

Sedimentation rates. The OC sedimentation rate obtained in this study (0.67 g $m^{-2} d^{-1}$) is higher than values obtained in previous studies of Scottish sea lochs such as Loch Etive (0.025 g OC $m^{-2} d^{-1}$;

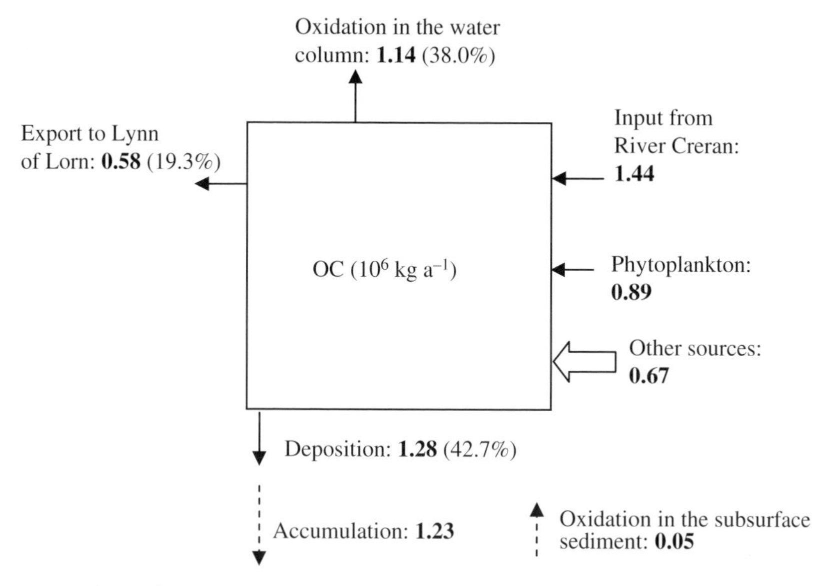

Fig. 2. OC budget ($\times 10^6$ kg a^{-1}). Input is represented by arrow pointing into the box; output is represented by arrow pointing out from the box. Dashed lines represent processes which take place under the surface sediment and are not included in balancing the budget. The imbalance of the budget is represented by a block arrow; in this case there is 0.67×10^6 kg a^{-1} OC input of which the source is unknown. The percentages of OC oxidized in the water column, deposited in the sediments and transported into Lynn of Lorn are calculated based on the total OC input into the loch: 3.00×10^6 kg a^{-1}. These are shown in brackets.

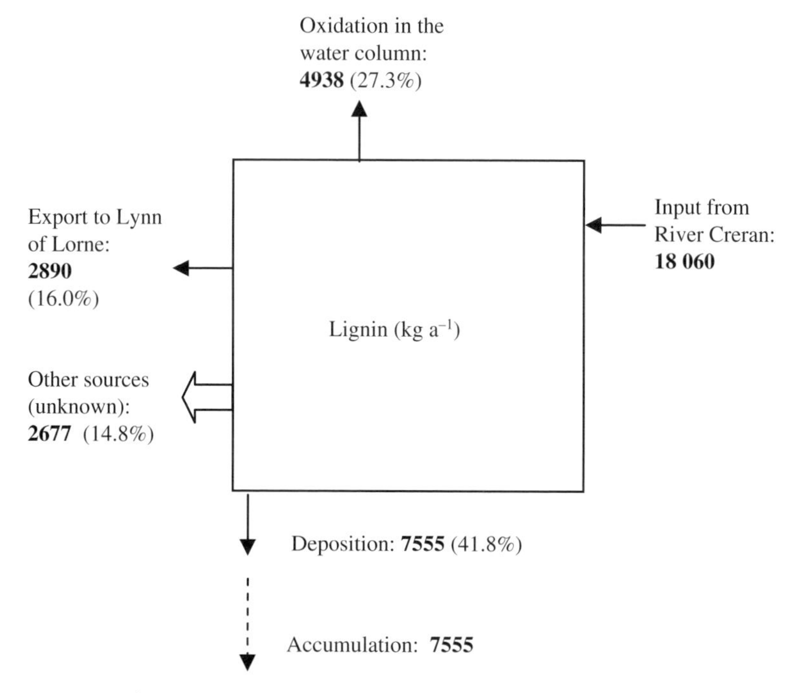

Fig. 3. Lignin budget (kg a^{-1}). The dashed line represents accumulation of lignin material in the subsurface sediments. Block arrow represents a balance of lignin (output) of which the fate is unknown. The percentages are calculated based on the total lignin input into the loch, 18 060 kg a^{-1}.

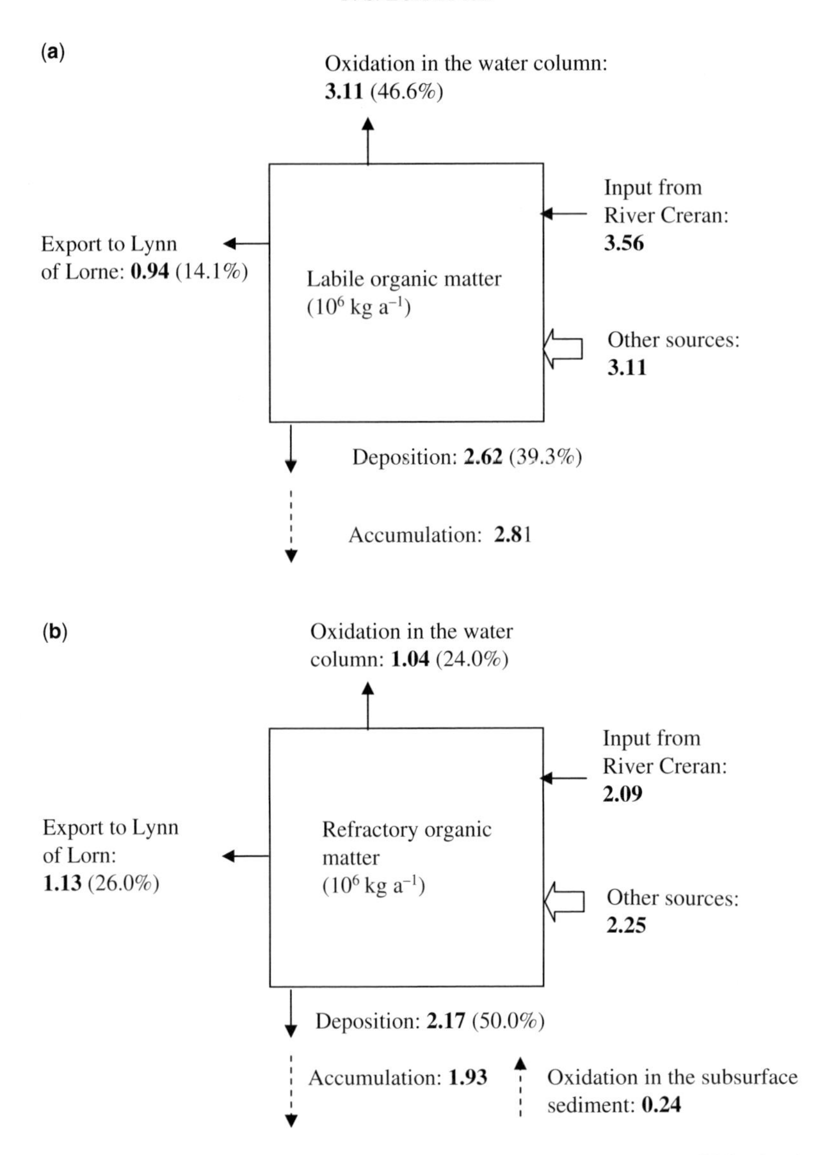

Fig. 4. Labile and refractory organic matter budgets. The dashed lines represent processes which take place in subsurface sediments; these are not included in balancing the budget. Block arrow represents a balance of organic matter of which the source is unknown. (**a**) Labile organic matter ($\times 10^6$ kg a^{-1}). Percentages are calculated based on the total input of 6.67×10^6 kg a^{-1}. (**b**) Refractory organic matter ($\times 10^6$ kg a^{-1}). For the refractory organic matter budget, the percentages are calculated based on the total input of 4.34×10^6 kg a^{-1}.

Ansell 1974), mid Loch Creran (0.14 g OC m^{-2} d^{-1}; Cronin & Tyler 1980) and Loch Linnhe (0.44 g OC m^{-2} d^{-1}; Overnell & Young 1995), as well as the Norwegian fjord Fanafjorden (0.26 g OC m^{-2} d^{-1}; Wassman 1984).

The OC sedimentation rate of 0.14 g m^{-2} d^{-1} (or 50 g m^{-2} a^{-1}) given by Cronin & Tyler (1980) for the main basin of Loch Creran is lower than that presented here (0.67 g m^{-2} d^{-1}). Their estimate is for the main basin whereas, in this study, the rate was determined in the upper basin which receives a higher input of materials from the river. There is also a general decrease in sedimentation rate with distance seaward as observed for several locations such as 1.6 g OC m^{-2} d^{-1} in the Bay of Biscay (Vangriesheim & Khripounoff 1990), 0.128 g OC m^{-2} d^{-1} in the coastal zone of New England, US (Rowe *et al.* 1986) and 0.0128 g OC m^{-2} d^{-1} in the NW Atlantic Ocean (Rowe & Gardner 1979). All of these authors measured the sedimentation rate directly from sediment traps, with the exception of Cronin and Tyler who derived their estimate by modelling.

There is, however, a 23-year gap between Cronin & Tyler (1980) and this study. Besides inter-annual variability, other reasons for this discrepancy could be due to differences in sampling procedures and frequency. Cronin & Tyler (1980) collected water samples from the upper and main basins fortnightly from January to September 1978. In this study, we collected 0–10 cm sediment cores across a transect of the loch from 2001 to 2002.

The total sedimentation rates (11.11 g m^{-2} d^{-1} and 4.42 g m^{-2} d^{-1}) were used to determine (i) the sedimentary fluxes of OC, lignin and organic matter and (ii) sediment accumulation rates (in terms of cm a^{-1}) for the upper and main basins of the loch. The validity of their use is based on the assumption that sedimentation rates normally decrease further from the source of riverine input; the upper basin of Loch Creran can therefore be expected to have a higher sedimentation rate than the main basin. In support of this assumption, the total sedimentation rate of 4.42 g m^{-2} d^{-1}, which was converted from the value 0.14 g OC m^{-2} d^{-1} (Cronin & Tyler 1980), is close to the lower limit of the total sedimentation rates obtained in this study for the upper basin, which ranged from 3.67 to 21.38 g m^{-2} d^{-1}. If the total sedimentation rate obtained in this study (11.11 g m^{-2} d^{-1}) was used to calculate the OC, lignin and organic matter fluxes throughout the loch, this would result in an increase in the respective fluxes at LC2, LC3, LC5 and LC6. This would eventually result in an increase of *c.* 1.4% of the amount of OC, lignin and organic matter (kg a^{-1}) deposited and accumulated in the loch, which is negligible. Moreover, the present estimate of the amount of OC deposited within the loch is already higher than that of Cronin & Tyler (1980).

River Creran flow rate. The River Creran flow rate used in this study (0.58×10^6 m^3 d^{-1}) is a mean value based on the river flow rates reported by Cronin & Tyler (1980), Edwards & Sharples (1986) and Booth (1987) of 0.52, 0.78 and 0.43×10^6 m^3 d^{-1}. The validity of using this value in the budget calculation is that there have been no major topographic changes in the catchment area of the loch. In support of this assumption, comparison between the OC budgets of Cronin & Tyler (1980) and this study shows that the total amount of OC input (and also total OC output) into Loch Creran is the same, 3.00×10^6 kg a^{-1} (Fig. 2).

Particulate concentration. The particulate concentration 113.76 g m^{-3} used in the budget calculation was determined from only one water sample as this was the only attempt which was successful. The representativeness of this value is supported by the fact that the OC budget estimated from this value has the same total OC input (also the total OC output; Fig. 2) into the loch as estimated by Cronin & Tyler (1980) of 3.00×10^6 kg a^{-1}. Also, multiplying the particulate concentration 113.76 g m^{-3} by the mean %OC at LC1 (4.0) and LC2 (3.1), the OC concentrations become 4.55 and 3.53 g m^{-3}. Note that these concentrations are not used in the budget calculations, but serve as a comparison to findings by Cronin & Tyler (1980). These sedimentary OC concentrations are slightly higher than the total OC concentrations in the deep water of the upper (2.04 g m^{-3}) and main basins (1.74 g m^{-3}) of the loch, as determined by Cronin & Tyler (1980).

Balancing the budgets

Cronin & Tyler (1980) estimated that rivers discharged 1.1×10^6 kg a^{-1} of terrestrial OC into the loch. Algae accounted for 1.9×10^6 kg a^{-1} of OC input. The output consists of 1.2×10^6 kg a^{-1} OC discharged to the Lynn of Lorn, 1.1×10^6 kg a^{-1} OC oxidized and 0.7×10^6 kg a^{-1} OC sedimented within the loch. Hence, 40% of the OC was transported to the Lynn of Lorn, 36.7% decomposed in the water column and 23.3% deposited in the sediments. In their study, Cronin & Tyler (1980) considered input from rivers from both upper and main basins, phytoplankton, *in situ* macro-algae and outputs from the then-active seaweed processing plant.

In this study, only the OC from River Creran and phytoplankton production were estimated. For the lignin and labile and refractory organic matter budgets, only the contribution from River Creran was treated quantitatively as an input. Hence the unknown input (Figs 2 & 4) probably represents the contribution from other sources such as macro-algae, other rivers and precipitation.

The most likely source of anthropogenic OC input is from fish farming activities; the station nearest to the fish farm in the loch was LC5. The slightly higher mean %OC at LC5 (2.3) compared to that at LC3 (1.8) could have been due to the contribution from the fish farm. However, total lignin content (Λ, mg/100 mg OC) was also higher at LC5 (0.44) than it was at LC3 (0.37). As there is no direct terrestrial input into LC5, the most likely cause for the increase of sedimentary lignin and OC contents here is due to its location in a bend of the loch, which may facilitate the accumulation of organic matter.

In the lignin budget estimation, the amount of lignin supplied by River Creran is calculated using the lignin concentration of suspended particles in the collected river water of 0.75 mg g^{-1}. This lignin concentration is higher than those in the trap and surface sediments. For the OC, labile and refractory organic matter budgets, the trap sample values were used to estimate their input to the loch as these were not determined in the river water sample (only an attempt to elucidate lignin-derived phenols was made). This may have lead to an underestimation of their contribution by the River Creran.

The role of sea lochs in regional carbon cycling

Site of organic carbon deposition and accumulation/burial. The major pathway which the OC, lignin and refractory organic matter are sequestered within the loch is through deposition onto surface sediments. Of the total OC, lignin and organic matter input into the loch, 42.7% OC, 41.8% total lignin, 39.3% labile and 50.0% refractory organic matter, respectively, were deposited onto the surface sediments. Most of the organic matter deposited onto the surface sediments becomes buried in the subsurface sediments. Of the total OC, lignin and organic matter which are deposited on the surface sediments, *c.* 96% of OC, 100% lignin, 100% labile and 89% refractory organic matter remain buried in the subsurface sediments (Figs 2–4). The fate of this organic matter is shown in Table 2. In the subsurface sediments of the upper basin, some of the labile fraction (10%) and OC (21%) underwent decomposition (the reasons are given in 'Site of organic matter decomposition'). In the subsurface sediments further down the loch, lignin and the more refractory organic matter underwent decomposition (the reasons are given in 'The importance of bioturbation'). Some studies have also found high accumulation rates in the coastal zone. For example, *c.* 79% of the total particulate organic carbon (POC) input was buried in sediments in the Fraser River delta and basin (Johannessen *et al.* 2003), 50% of the total POC was accumulated in the Gulf of Lions and 50% was mineralized (Accornero *et al.* 2003) and 62% OC was buried in the northeastern Skagerrak (Stahl *et al.* 2004). Approximately 41% of the total OC and 42% of the total lignin were buried in Loch Creran.

Areas in the vicinity of a river mouth where there is an increase of nutrients and higher particulate loads favour the accumulation of organic matter (Wollast 1991). This could be the case for the upper basin of Loch Creran. The meeting of the less dense out-flowing freshwater from River Creran with the denser incoming water from the Lynn of Lorn can cause retention of particles within the vicinity of the loch. The mixing of fresh and saline waters also induces flocculation of dissolved (Postma 1980) and fine particulate (Billen *et al.* 1991) material at the freshwater–seawater interface. The formation of larger and denser particles will facilitate their deposition at the sediment surface. Bacteria and phytoplankton also add to the sedimenting particle assemblages. The shallow depth of the loch also means there is a greater probability of the sinking particles reaching the sediments with much of their organic matter intact (Jickells *et al.* 1991). Although the sediment particle size was not determined in this study, based on visual observation the loch sediments are mostly fine-grained clays. Keil *et al.* (1994) has shown that sorption of organic matter to the mineral surfaces of fine-grained sediments can cause some of the labile organic matter to be unavailable for oxidation by micro-organisms. Factors which facilitate OC accumulation include high sedimentation rates and adsorption of organic matter onto mineral surfaces (Hedges & Keil 1995) of fine-grained sediments, thus protecting the organic matter from further degradation (De Haas & van Weering 1997).

Site of organic matter decomposition. The second most important pathway which OC and lignin are sequestered within the loch is through decomposition in the water column: 38% (1.14×10^6 kg a^{-1}) of the total OC input and 27.3% (4938 kg a^{-1}) of the total lignin input were decomposed in the water column. This is the most important pathway in which the labile organic matter was sequestered within the loch, accounting for 46.6% of the total labile organic matter input to the loch.

Wind (Jickells *et al.* 1991), tidal effects and bottom currents can cause resuspension of bottom sediment and maintain this material in the water column. Increased wind and precipitation during winter will increase perturbation of the water column. The shallowness of Loch Creran (with a maximum depth of 49 m in the main basin) and frequent water renewal lead to complete vertical mixing. The narrows at Creagan also create turbulence. All of these factors elevate and facilitate oxygenation and aerobic organic matter degradation in the water column.

The increased ratio of vanillic acid to vanillin, (Ad/Al)v, indicates more highly degraded lignin material (Goni *et al.* 2000). The higher (Ad/Al)v ratio in the sediment trap sample (2.69) compared to surface sediments at LC0 (1.07) and LC1 (0.83) (Loh *et al.* 2008*c*) provides evidence that particulate lignin material is being degraded aerobically in the water column. However, this could have been caused by the fact that the particles in the trap were not fixed with any conservation agent and so there is continued organic matter degradation within the collecting tubes. Even although the organic matter in the water column undergoes degradation while undergoing sedimentation which would lead to the surface sediments having high (Ad/Al)v values (but this was not so), the depth might be too shallow and settling time too short for this to occur. An elevated ratio of the refractory to total organic matter (Rp value) also indicates a more advanced degradation stage of organic matter (Loh *et al.* 2008*a*, *c*). However, it seems that the trap materials have lower Rp value of 0.37 than the LC0 and LC1 surface sediments of 0.43. Therefore, the higher (Ad/Al)v values in the trap could indicate resuspension of previously sedimenting terrestrial material.

Oxygen uptake rates in the upper basin of Loch Creran ranged from 11.8 to 27.8 mmol m^{-2} d^{-1}. Rates from LC2 to LC6 ranged from 6.6 to 18.7 mmol m^{-2} d^{-1}. The higher oxygen uptake rates in the upper basin indicate more rapid organic matter degradation there, most probably due to relatively fresh terrestrially derived organic matter fuelling biogeochemical cycling (Loh *et al.* 2008*c*). Following Stahl *et al.* (2004), the OC oxidation rates were calculated from the mean oxygen uptake rates (Table 1), however, a respiration quotient was not used here. The OC oxidation rates in the upper basin of the loch are lower than the OC fluxes in Table 2 (calculated using the total sedimentation rates); the OC oxidation rates in LC3, 5 and 6 are higher than the OC fluxes, however. This could most probably be attributed to the lower sedimentation rate used to estimate fluxes in the main basin.

A similar phenomenon of oxygen uptake rates in a sea loch environment is given by Overnell *et al.* (1995*b*) who measured oxygen uptake rates across transects of four lochs (Linnhe, Etive, Fyne and Goil). It was found that the rates decreased from the head to the mouth of the lochs. In Loch Linnhe, the highest oxygen uptake rate of 19.0 mmol m^{-2} d^{-1} was found at the head of the loch; the lowest rate of 7.2 mmol m^{-2} d^{-1} was found furthest down the loch. Loch Goil has the highest rate at the head of 23.6 mmol m^{-2} d^{-1} and the rates decreased further down the loch. In Loch Etive, the highest rate at the location furthest down the loch was assumed to be due to the contribution of terrestrial organic matter from the River Awe. In Loch Linnhe, apart from the highest rate at the middle of the loch, the rates decreased overall further down the loch (see table 2 in Overnell *et al.*, 1995*b*).

In shallow depths, a significant fraction of the OC produced by primary productivity rapidly reaches the sediments to yield organic-rich sediments where intensive diagenetic processes occur (Wollast 1991). This, along with the higher rate of organic matter decomposition in the upper basin of Loch Creran, could lead to rapid oxygen depletion below the sediment surface; this in turn elevates the sulphate reduction process (Jørgensen 1982; Holmer 1999). Although the sulphate reduction rates were not measured in this study, sediment cores from the upper basin showed typical darkening of below-surface sediments caused by iron monosulphide formation, indicative of sulphate reducing activities.

In highly depositional environments such as continental margins, oxygen availability is not the primary control of organic matter preservation (Cowie & Hedges 1992). The primary controls are sedimentation rate (Canfield 1994) and quality of organic matter (Kristensen & Blackburn 1987; Henrichs 1992; Canfield 1994; Kristensen *et al.* 1995; Aller & Aller 1998; Kristensen & Holmer 2001). In such environments most of the OC decomposes *via* anaerobic pathways such as sulphate reduction (Westrich & Berner 1984). The 10% labile organic matter and 21% OC oxidized in the subsurface sediments at LC0 (Table 2) were most probably degraded via the anaerobic pathway. It is most likely that due to the absence of oxygen, lignin and refractory organic matter in the subsurface sediments in the upper basin of the loch did not undergo further degradation. The lignin and refractory organic matter in the subsurface sediments at LC5 and LC6 underwent some decomposition (Table 2), most probably due to the role of burrowing organisms exposing the sediments to oxygen ('The importance of bioturbation').

Transportation to Lynn of Lorn. Being a relatively small loch (12.8 km long) and due to the frequency of tidal flushing (3-day cycle) (Edwards & Sharples 1986), OC should be transported rapidly to the Lynn of Lorn. Cronin & Tyler (1980) showed that 40% of the total OC input was transported to Lynn of Lorn, 36.7% decomposed in the water column and 23.3% deposited onto the surface sediment. This study shows that 42.7% OC was deposited, 38.0% decomposed in the water column and 19.3% transported to Lynn of Lorn.

The bulk of sediment and OC supply to Loch Creran most probably occurs during the few days of high river flow during periods of increased rainfall and snowmelt. In Loch Creran this is manifested as a wide range of OC and lignin concentrations, especially at LC5 and LC6 (Loh 2005). Although the flow rates were not measured, occurrences of high and low river water volumes have been observed. When large-scale sediment accumulation occurs there may be a threshold flushing event which will remove a large quantity of sediment (MacKenzie *et al.* 1991). Such an event in Loch Creran would result in some materials being transported further away in the Lynn of Lorn beyond LC6.

The importance of bioturbation

An observed difference between the upper and main basins of Loch Creran is that in the upper basin lignin material in the subsurface sediments did not undergo degradation. Further down the loch at LC5 and LC6, almost half the amount of lignin material in the subsurface sediments underwent degradation (Table 2). At LC0, 100% lignin was buried in the subsurface sediments. Calculations lead to a figure greater than 100% lignin accumulated or buried in the subsurface sediments at LC1 and LC3; this is assumed to be 100% accumulation. The reason for the high calculated values is that sediment fluxes at surface sediments were mean values based on many analyses whereas the values at subsurface sediments were obtained from analyses of a single sample (Table 1). This may have resulted in discrepancies. On average, there was an equal amount of lignin deposition and accumulation (Fig. 3), indicating that overall lignin material tends to be retained within the loch.

Lignin and refractory organic matter were degraded at LC5 and LC6 subsurface sediments. There were 50% and 41% of lignin material oxidized in the subsurface sediments at these sites. This is substantiated by a 42.86% refractory organic matter oxidized at LC5 subsurface sediment. The most probable reason for the increase in decomposition of older and more refractory organic matter is re-exposure to oxygen (Kristensen 2000). Although bioturbation was not measured in this study it is the most likely mechanism for the increase in decomposition.

Increased re-exposure to oxygen by bioturbation can increase the decomposition of buried materials by a factor of 3.6 (Hulthe *et al.* 1998; Kristensen 2000). Even brief and periodic re-exposure to oxygen has been known to result in more complete and, at times, more rapid organic matter decomposition than when similar material is maintained under constant conditions (Aller 1994). The 33.33% increase of labile organic matter in the subsurface sediments at LC5 (Table 2) can most probably be attributed to the activities of burrowing organisms transferring sedimenting particles into deeper sediments (Pearson 2001).

In their study on bioturbation and OC quality, Nickell *et al.* (2003) found maximum macrofaunal biomass and abundances at a station under fish farm cages. However, the animals were very small and did not mix deeply into the sediments. Mixing intensity increased with distance from the farm as a result of increasing oxygen concentration due to a decrease in nutrients and OC content. At the farthest location, mixing intensity was similar to beneath the fish farm as some megafauna had moved material from surface sediments to below the mixed layer. In this study, the location nearest to a fish farm is LC5. Situated *c.* 200 m from the cages, there is likely no great increase in the OC and nutrients and oxygen uptake rates at LC5 due to the fish farm. Burrowing activities are evident here as indicated by the decomposition of lignin and refractory organic matter in the subsurface sediments. The mean oxygen uptake rates at LC5 (15.1 mmol m^{-2} d^{-1}) as well as all other locations in this study are similar to those locations not located directly under the cages (Nickell *et al.* 2003).

Loch Creran and its global significance

Global riverine export of OC ranges from 334.5 to 500 $\times$ 10^{12} g a^{-1} (Spitzy & Ittekkot 1991; Degens *et al.* 1991; Ludwig *et al.* 1996; Aitkenhead & McDowell 2000; Schlunz & Schneider 2000; Bianchi *et al.* 2007*b*). River Creran discharges 1.44 $\times$ 10^6 kg a^{-1} OC into Loch Creran; this represents 0.0003–0.0004% of the global river export of OC.

Louchouarn *et al.* (1999) estimated that, of the total global riverine OC discharged of 200 $\times$ 10^{12} g a^{-1}, 6 $\times$ 10^{12} g a^{-1} (or 3%) is lignin material. This riverine OC input is lower than the above estimates and the lignin input was the sum of six lignin phenols, excluding the cinnamic and ferulic acids. Of the total annual flux of particulate OC of 9.3 $\times$ 10^8 kg a^{-1} to the Gulf of Mexico, total lignin flux (calculated as the sum of eight lignin phenols) represents 0.013% or 1.2 $\times$ 10^5 kg a^{-1} of the total OC (Bianchi *et al.* 2007*b*). Of the total OC flux of 0.13 $\times$ 10^9 kg from Atchafalaya River to the shelf, lignin flux for the Atchafalaya sand was estimated to be 12.4 $\times$ 10^5 kg or 0.95% of the total OC (Bianchi *et al.* 2007*a*). River Creran contributes 18 060 kg a^{-1} lignin to Loch Creran; this is 1.25% of the total OC input to the loch. Overall the percentage of lignin to total OC discharged from River Creran is lower than the percentage lignin to total OC estimated by Louchouarn *et al.* (1999), but higher than the percentages of lignin in particulate OC (Bianchi *et al.* 2007*b*) and sand OC (Bianchi *et al.* 2007*a*).

In Loch Creran, 42.7% of the total OC is deposited within the loch, 38.0% is decomposed in the water column and 19.3% exported to Lynn of Lorn; 41.8% of the total lignin input is deposited, 27.3% is decomposed in the water column and 16.0% is transported to Lynn of Lorn. Both budgets indicate that most of the OC and lignin are lost within the loch through deposition and subsequent accumulation in the sediments and by decomposition in the water column. A similar phenomenon is also observed in most of the continental shelves worldwide. Burial efficiency of

62% and 43% has been observed in the northeastern and south Skagerrak, respectively (Stahl *et al.* 2004). In the Gulf of Lions, 31% and 69% of the particulate OC are accumulated and oxidized in the shelf; 50% is accumulated and 50% oxidized in the slope (Accornero *et al.* 2003). Most of the ocean margins elsewhere serve as a sink of OC due to the processes of accumulation (Hedges & Keil 1995; De Haas & van Weering 1997) and degradation in the water column (De Haas *et al.* 2002; Burns *et al.* 2008).

Conclusions

Bulk OC, lignin-derived phenols and organic matter contents decreased from the head of Loch Creran, seawards, due to organic matter decomposition and deposition. Oxygen uptake rates decreased towards the mouth, indicating higher rates of organic matter decomposition in the upper loch and decreasing downstream. Fresh terrestrial OC is important in fuelling biogeochemical cycling in the upper basin of Loch Creran. The more labile OC, with a fraction of lignin and other refractory organic matter, undergo aerobic degradation in the water column. Only more labile OC is susceptible to anaerobic degradation in the subsurface sediments. Hence this basin is a sink for lignin materials and relatively refractory terrestrial organic matter.

Further downstream, towards the main basin, organic matter becomes increasingly refractory. As a result of the lower rate of organic matter degradation in the water column, oxygenated water is retained in deeper waters and at the sediment–water interface. Under these conditions bioturbation is likely to increase, enhancing the decomposition of lignin and other refractory organic matter in subsurface sediments.

In the context of the regional carbon cycle, Loch Creran is a sink for OC due to organic matter decomposition in the water column and subsurface sediments. OC input to Loch Creran includes 1.44×10^6 kg a^{-1} discharged from River Creran and 0.89×10^6 kg a^{-1} from marine phytoplankton production. OC deposition is 1.28×10^6 kg a^{-1} of which 1.23×10^6 kg a^{-1} OC remains preserved; the remainder undergoes decomposition. Altogether, 39.5% of OC was oxidized in the water column and sediments and 41% of the total OC remained buried in the loch. Loch Creran is also a sink for relatively refractory terrestrial organic matter, as 41% of the total lignin discharged from River Creran is buried within the loch. Terrestrial organic matter, therefore, represents 1.7–4.0% of the total OC buried in the loch.

We thank D. MacAlpine and S. Douglas of the RV *Seol Mara*, A. Wilson for assistance with water sampling, K. Black and H. Orr for advice on GC analysis, T. Brand for advice on CN analysis, S. Gontarek for the map and everyone at the Dunstaffnage Marine Laboratory for useful advice. We are very grateful to the reviewers and editor for their intensive comments which have greatly improved this manuscript. P. S. Loh gratefully acknowledges Professor C.-T. A. Chen and Taiwan National Science Council, as this paper was completed during her time as a postdoctoral fellow at the National Sun Yat-Sen University.

Appendix

Sedimentation rates, fluxes and accumulation rate

(i) Sedimentation rate

Sedimentation rate (g m^{-2} day^{-1})

= total dry weight of sediment collected

from the collecting tubes,

g/area of the collecting tubes,

m^2/total days of trap deployment

This value of mean sedimentation rate, 11.11 g/m^2/d, is then used to calculate the sedimentary fluxes of OC, lignin and organic matter for the locations at the upper basin of Loch Creran, namely LC0, LC1 and sediment trap samples.

For other locations (LC2, LC3, LC5 and LC6), the sedimentation rate 4.42 g/m^2/d (which was calculated based on the OC sedimentation rate by Cronin & Tyler 1980) was used instead. The reasons are given in the text.

(ii) Sedimentary fluxes of OC, lignin and organic matter

A few examples of the calculation are given here:

OC flux at LC0 surface sediment

= sedimentation rate

× mean %OC for LC0 surface sediment

$= 11.11\ \mathrm{g/m^2/d} \times 4.8/100$

$= 0.53\ \mathrm{g/m^2/d}$

OC flux at the LC0 subsurface sediment

= sedimentation rate

× mean %OC for LC0 subsurface

sediment $= 11.11\ \mathrm{g/m^2/d} \times 3.8/100$

$= 0.42\ \mathrm{g/m^2/d}$

(iii) Accumulation rates

In order to estimate the accumulation rate (cm a^{-1}) for the upper basin of Loch Creran, this formula '$w_s = r/[(1-\phi)p_s]$' is used where ϕ = sediment porosity = 0.7; p_s = density = 2.65 g cm^{-3}; and r = sedimentation rate = 11.11 g/m^2/d = 0.41 g/cm^2/a. Where a = annum.

Table A1. *Sedimentation rates*

Date	Weight of dry particle collected in the collecting tubes (g)	Days	Sedimentation rate ($g/m^2/d$)
12.12.2001	12.97	49	6.97
8.1.2002	4.05	26	4.10
6.2.2002	8.96	29	8.13
7.3.2002	10.72	29	9.73
4.4.2002	22.75	28	21.38
7.5.2002	8.21	33	6.55
4.6.2002	16.92	28	15.90
2.7.2002	3.91	28	3.67
1.8.2002	35.88	30	31.47
2.9.2002	12.69	32	10.44
30.9.2002	4.15	28	3.90

Note: The area of the collecting tubes = 0.038 m^2, Mean sedimentation rate = 11.11 $g/m^2/d$.

Substituting these into the equation:

$$w_s = 0.41\,g/cm^2/a/[(1-0.7)2.65\,g/cm^3] = 0.52\,cm/a$$

In order to estimate the accumulation rate ($cm\ a^{-1}$) for the main basin of Loch Creran, this formula '$w_s = r/[(1-\phi)p_s]$' is used where ϕ = sediment porosity = 0.7; p_s = density = 2.65 g cm^{-3}; and r = sedimentation rate = 4.42 $g/m^2/d$ = 0.16 $g/cm^2/a$.

Substituting these into the equation:

$$w_s = 0.16\,g/cm^2/a/[(1-0.7)2.65\,g/cm^3] = 0.20\,cm/a$$

River Creran flow rate and particulate concentration

River Creran flow rates were reported by Cronin & Tyler (1980), Edwards & Sharples (1986) and Booth (1987) as 0.52, 0.78 and 0.43 × $10^6\ m^3\ d^{-1}$. This gives a mean flow rate of 0.58 × $10^6\ m^3\ d^{-1}$.

Water sample was collected from the mouth of River Creran on 5/12/2001. 251.07 mg of particulate matter was obtained from the filtration of 2207 ml water. This gives a particulate concentration of 113.76 g m^{-3}.

Mass of particulate discharged into Loch Creran
= River Creran flow rate × particulate concentration
= $0.58 \times 10^6\ m^3/d \times 113.76\ g/m^3$
= $65.98 \times 10^6\ g/d$
= $24.08 \times 10^6\ kg/a$

Calculations for the OC budget

Mean River Creran flow rate = 0.58 × $10^6\ m^3/d^{-1}$
Particulate concentration of River Creran water = 113.76 g/m^3

Mass of particulate discharged into Loch Creran = 24.08 × 10^6 kg/a

During oxygen uptake rate measurements, sediment cores were incubated for 24 hours (Loh *et al.* 2008*b*). During shorter incubation time (for 3 to 6 hours) we noted inconsistent and negative values; more experiments are needed to determine the reasons. It seems the system becomes stabilized after 20 hours. The validity of using the duration more than 20 hours lies in that in our experiments, the oxygen concentration never fell below 100 μM, as according to Hall *et al.* (1989) the flux measurements below this concentration are meaningless as fluxes become non-linear.

Oxygen uptake rate represents the rate of which oxygen is consumed during benthic respiration or organic matter decomposition. We assume that the respiration quotient equals to 1.0, that is, the number of molecules of CO_2 liberated is the same as the number of molecules of oxygen used. We also assume that the amount of carbon liberated is the same amount of carbon produced during phytoplankton photosynthesis.

Phytoplankton production
= mean oxygen uptake rate at LC1 × depositional area of the loch
= $14.98\ mmole/m^2/d \times 13.53\ km^2$
= $14.98 \times 10^{-3} \times 12 \times 10^{-3} \times 365\ kg/m^2/yr \times 13.53 \times (10^3)^2\ m^2$
= $0.89 \times 10^6\ kg/a$

Total OC discharged from River Creran into Loch Creran
= mass of particulate discharged into Loch Creran × mean % OC of the trap samples
= $24.08 \times 10^6\ kg/a \times 6.0/100$
= $1.44 \times 10^6\ kg/a$

Total OC deposited (sedimented) onto the surface sediments of Loch Creran

= mean OC fluxes in surface sediments across the loch (from LC0 to LC5 : see Table 2) × total depositional area of the loch

$= 0.26\,g/m^2/d \times 13.53\,km^2$

$= 0.26 \times 10^{-3} \times 365\,kg/m^2/a \times 13.53 \times (10^3)^2\,m^2$

$= 1.28 \times 10^6\,kg/a$

Total OC accumulated (remain buried) in the 9 to 10 cm subsurface sediment

= mean OC fluxes in subsurface sediments across the loch (from LC0 to LC5) × total depositional area of the loch

$= 0.25\,g/m^2/d \times 13.53\,km^2$

$= 0.25 \times 10^{-3} \times 365\,kg/m^2/a \times 13.53 \times (10^3)^2\,m^2$

$= 1.23 \times 10^6\,kg/a$

Total OC undergoes degradation in the subsurface sediment

= Total OC deposited onto the surface sediments − Total OC accumulated in the 9 to 10 cm subsurface sediment

$= 1.28 \times 10^6\,kg/a - 1.23 \times 10^6\,kg/a$

$= 0.05 \times 10^6\,kg/a$

Total OC decomposed in the water column

= Difference between OC flux in the sediment trap and LC1 surface sediment# × total depositional area of the loch

$= (0.67 - 0.44)\,g/m^2/d \times 13.53\,km^2$

$= 0.23 \times 10^{-3} \times 365\,kg/m^2/a \times 13.53 \times (10^3)^2\,m^2$

$= 1.14 \times 10^6\,kg/a$

Note that only LC1 surface sediment was compared to the trap samples (to determine the amount of OC decomposed in the water column) because LC1 is situated nearest to the trap. It is assumed that the loss of OC during sedimentation from the water to the surface sediment is due to decomposition.

Total OC discharged from Loch Creran into Lynn of Lorn

= mass of particulate discharged into Loch Creran × mean % OC of the main basin sediments (i.e., LC2, LC3, LC5)

$= 24.08 \times 10^6 kg/a \times 2.4/100$

$= 0.58 \times 10^6\,kg/a$

Cronin & Tyler (1980) calculated the rate of passage of TOC to the sea and found that the net flow of water from the loch is equal to the river flow. Hence the 'mass of particulate discharged into Loch Creran' in the above equation is the same as value used to determine the OC input from River Creran into the loch.

Calculations for the lignin budget

Mean River Creran flow rate $= 0.58 \times 10^6\,m^3/d^{-1}$
Particulate concentration of River Creran water $= 113.76\,g/m^3$
Mass of particulate discharged into Loch Creran $= 24.08 \times 10^6\,kg/a$

Total lignin discharged from River Creran into Loch Creran

= mass of particulate discharged into Loch Creran × mean lignin content of the river water sample

$= 24.08 \times 10^6\,kg/a \times 0.75\,mg/g$

$= 18\,060\,kg/a$

Total lignin deposited (sedimented) onto the surface sediments of Loch Creran

= mean lignin fluxes in surface sediments across the loch (from LC0 to LC5: see Table 2) × total depositional area of the loch

$= 1.53 \times 10^{-3}\,g/m^2/d \times 13.53\,km^2$

$= 1.53 \times 10^{-6} \times 365\,kg/m^2/a \times 13.53 \times (10^3)^2\,m^2$

$= 7555\,kg/a$

Total lignin accumulated (remain buried) in the 9 to 10 cm subsurface sediment

= mean lignin fluxes in subsurface sediments across the loch (from LC0 to LC5) × total depositional area of the loch

$= 1.53 \times 10^{-3}\,g/m^2/d \times 13.53\,km^2$

$= 1.53 \times 10^{-6} \times 365\,kg/m^2/a \times 13.53 \times (10^3)^2\,m^2$

$= 7555\,kg/a$

Total lignin undergoes degradation in the subsurface sediment

= Total lignin deposited onto the surface sediments − Total lignin accumulated in the 9 to 10 cm subsurface sediment

$= 7555 - 7555\,kg/a$

$= 0\,kg/a$

Total lignin decomposed in the water column

= Difference between lignin flux in the sediment trap and LC1 surface sediment × total depositional area of the loch

$= (3.44 - 2.44) \times 10^{-3}\,g/m^2/d \times 13.53\,km^2$

$= 1.00 \times 10^{-6} \times 365\,kg/m^2/a \times 13.53 \times (10^3)^2\,m^2$

$= 4938\,kg/a$

Total lignin discharged from Loch Creran into Lynn of Lorn
= mass of particulate discharged into Loch Creran
× mean lignin content in the main basin sediments (i.e., LC2, LC3, LC5)
$= 24.08 \times 10^6$ kg/a × 0.12 mg/g
= 2890 kg/a

Calculations for the labile organic matter budget

Mean River Creran flow rate = 0.58×10^6 m^3/d^{-1}

Particulate concentration of River Creran water = 113.76 g/m^3

Mass of particulate discharged into Loch Creran = 24.08×10^6 kg/a

Total labile organic matter discharged from River Creran into Loch Creran
= mass of particulate discharged into Loch Creran
× mean % labile organic matter of the trap samples
$= 24.08 \times 10^6$ kg/a × 14.8/100
$= 3.56 \times 10^6$ kg/a

Total labile organic matter deposited (sedimented) onto the surface sediments of Loch Creran
= mean fluxes of labile organic matter in surface sediments across the loch (from LC0 to LC5: see Table2)
× total depositional area of theloch
= 0.53 g/m^2/d × 13.53 km^2
$= 0.53 \times 10^{-3} \times 365$ kg/m^2/a $\times 13.53 \times (10^3)^2$ m^2
$= 2.62 \times 10^6$ kg/a

Total labile organic matter accumulated (remain buried) in the 9 to 10 cm subsurface sediment
= mean fluxes of labile organic matter in subsurface sediments across the loch (from LC0 to LC5)
× total depositional area of the loch
= 0.57 g/m^2/d × 13.53 km^2
$= 0.57 \times 10^{-3} \times 365$ kg/m^2/a $\times 13.53 \times (10^3)^2$ m^2
$= 2.81 \times 10^6$ kg/a

Total labile organic matter undergoes degradation in the subsurface sediment
= Total labile organic matter deposited onto the surface sediments
− Total labile organic matter accumulated in the 9 to 10 cm subsurface sediment
$= 2.62 \times 10^6$ kg/a $- 2.81 \times 10^6$ kg/a
$= -0.19 \times 10^6$ kg/a**

**Note: From these calculations, it seems that the amount of labile fraction of organic matter which was accumulated in the subsurface sediments is higher than the amount sedimented or deposited onto the surface sediment. Table 2 shows that apart from the subsurface sediment at LC0 which underwent decomposition, other locations such as LC1, LC3 and LC5 show more than 100% accumulation of labile organic matter.

The above calculations are based on the mean values of labile organic matter in the surface and only single measurements for the subsurface sediments (except for LC1). It is assumed that the above difference (-0.19×10^6 kg/a) means that overall, the sediments in Loch Creran favour accumulation of the labile organic matter. The higher 33.33% increase in accumulation of the labile organic matter in LC5 subsurface sediment could be due to activities of organisms bringing new materials further down the sediments. See discussion in the text for more details.

Total labile organic matter decomposed in the water column
= Difference between labile organic matter flux in the sediment trap and LC1 surface sediment
× total depositional area of the loch
= (1.64 − 1.01) g/m^2/d × 13.53 km^2
$= 0.63 \times 10^{-3} \times 365$ kg/m^2/a $\times 13.53 \times (10^3)^2$ m^2
$= 3.11 \times 10^6$ kg/a

Total labile organic matter discharged from Loch Creran into Lynn of Lorn
= mass of particulate discharged into Loch Creran
× mean % labile organic matter of the main basin sediments (i.e., LC2, LC3, LC5)
$= 24.08 \times 10^6$ kg/a × 3.9/100
$= 0.94 \times 10^6$ kg/a

Calculations for the refractory organic matter budget

Mean River Creran flow rate = 0.58×10^6 m^3/d^{-1}
Particulate concentration of River Creran water = 113.76 g/m^3
Mass of particulate discharged into Loch Creran = 24.08×10^6 kg/a

Total refractory organic matter discharged from River Creran into Loch Creran
= mass of particulate discharged into Loch Creran
× mean % refractory organic matter of the trap samples
$= 24.08 \times 10^6$ kg/a × 8.7/100
$= 2.09 \times 10^6$ kg/a

Total refractory organic matter deposited (sedimented) onto the surface sediments of Loch Creran
= mean fluxes of refractory organic matter in surface sediments across the loch (from LC0 to LC5: see Table 2) × total depositional area of the loch

$$= 0.44\,\mathrm{g/m^2/d} \times 13.53\,\mathrm{km^2}$$

$$= 0.44 \times 10^{-3} \times 365\,\mathrm{kg/m^2/a} \times 13.53 \times (10^3)^2\,\mathrm{m^2}$$
$$= 2.17 \times 10^6\,\mathrm{kg/a}$$

Total refractory organic matter accumulated (remain buried) in the 9 to 10 cm subsurface sediment
= mean fluxes of refractory organic matter in subsurface sediments across the loch (from LC0 to LC5) × total depositional area of the loch

$$= 0.39\,\mathrm{g/m^2/d} \times 13.53\,\mathrm{km^2}$$

$$= 0.39 \times 10^{-3} \times 365\,\mathrm{kg/m^2/a} \times 13.53 \times (10^3)^2\,\mathrm{m^2}$$
$$= 1.93 \times 10^6\,\mathrm{kg/a}$$

Total refractory organic matter which undergoes degradation in the subsurface sediment
= Total refractory organic matter deposited onto the surface sediments − Total refractory organic matter accumulated in the 9 to 10 cm subsurface sediment

$$= 2.17 \times 10^6\,\mathrm{kg/a} - 1.93 \times 10^6\,\mathrm{kg/a}$$
$$= 0.24 \times 10^6\,\mathrm{kg/a}$$

Total refractory organic matter decomposed in the water column
= Difference between refractory organic matter flux in the sediment trap and LC1 surface sediment × total depositional area of the loch

$$= (0.97 - 0.76)\,\mathrm{g/m^2/d} \times 13.53\,\mathrm{km^2}$$

$$= 0.21 \times 10^{-3} \times 365\,\mathrm{kg/m^2/a} \times 13.53 \times (10^3)^2\,\mathrm{m^2}$$
$$= 1.04 \times 10^6\,\mathrm{kg/a}$$

Total refractory organic matter discharged from Loch Creran into Lynn of Lorn
= mass of particulate discharged into Loch Creran × mean % refractory organic matter of the main basin sediments (i.e., LC2, LC3, LC5)

$$= 24.08 \times 10^6\,\mathrm{kg/a} \times 4.7/100$$
$$= 1.13 \times 10^6\,\mathrm{kg/a}$$

References

Accornero, A., Picon, P., de Bovee, F., Charriere, B. & Buscail, R. 2003. Organic carbon budget at the sediment-water interface on the Gulf of Lions continental margin. *Continental Shelf Research*, **23**, 79–92.

Aitkenhead, J. A. & McDowell, W. H. 2000. Soil C:N ratio as a predictor of annual riverine DOC flux at local and global scales. *Global Biogeochemical Cycles*, **14**, 127–138.

Aller, R. C. 1994. Bioturbation and remineralization of sedimentary organic matter: effects of redox oscillation. *Chemical Geology*, **114**, 331–345.

Aller, R. C. & Aller, J. Y. 1998. The effect of biogenic irrigation intensity and solute exchange on diagenetic reaction rates in marine sediments. *Journal of Marine Research*, **56**, 905–936.

Ansell, A. D. 1974. Sedimentation of organic detritus in Lochs Etive and Creran, Argyll, Scotland. *Marine Biology*, **27**, 263–273.

Berner, R. A. 1982. Burial of organic carbon and pyrite sulfur in the modern ocean: its geochemical and environmental significance. *American Journal of Science*, **282**, 451–473.

Bianchi, T. S., Galler, J. I. & Allison, M. A. 2007*a*. Hydrodynamic sorting and transport of terrestrially derived organic carbon in sediments of the Mississippi and Atchafalaya Rivers. *Estuarine, Coastal and Shelf Science*, **73**, 211–222.

Bianchi, T. S., Wysocki, L. A., Stewart, M., Filley, T. R. & McKee, B. A. 2007*b*. Temporal variability in terrestrial-derived sources of particulate organic carbon in the lower Mississippi River and its upper tributaries. *Geochimica et Cosmochimica Acta*, **71**, 4425–4437.

Billen, G., Lancelot, C. & Meybeck, M. 1991. N, P, and Si retention along the aquatic continuum from land to ocean. *In*: Mantoura, R. F. C., Martin, J.-M. & Wollast, R. (eds) *Dahlem Workshop Reports. Ocean Margin Processes in Global Change. Physical, Chemical, and Earth Sciences Research Report 9*. John Wiley & Sons, Chichester, 19–44.

Blanton, J. O. 1991. Circulation processes along oceanic margins in relation to material fluxes. *In*: Mantoura, R. F. C., Martin, J.-M. & Wollast, R. (eds) *Dahlem Workshop Reports. Ocean Margin Processes in Global Change. Physical, Chemical, and Earth Sciences Research Report 9*. John Wiley & Sons, Chichester, 145–163.

Booth, D. A. 1987. Some consequences of a flood tide front in Loch Creran. *Estuarine, Coastal and Shelf Science*, **24**, 363–375.

Burns, K. A., Brunskill, G., Brinkman, D. & Zagorskis, I. 2008. Organic carbon and nutrient fluxes to the coastal zone from the Sepik River outflow. *Continental Shelf Research*, **28**, 283–301.

Canfield, D. E. 1994. Factors influencing organic carbon preservation in marine sediments. *Chemical Geology*, **114**, 315–329.

Chester, R. 2000. *Marine Geochemistry*. 2nd edn. Blackwell Science, Malden, 506.

Christman, R. F. & Oglesby, R. T. 1971. Microbiological Degradation and the Formation of Humus.

In: SARKANEN, K. V. & LUDWIG, C. H. (eds) *Lignins Occurrence, Formation, Structure and Reactions*. Wiley-Interscience, New York, 769–795.

COWIE, G. L. & HEDGES, J. I. 1992. The role of anoxia in organic matter preservation in coastal sediments: relative stabilities of the major biochemicals under oxic and anoxic depositional conditions. *Organic Geochemistry*, **19**, 229–234.

CRAIB, J. S. 1965. A sampler for taking short undisturbed marine cores. *Journal du Coneid, Conseid permanent international de exploration de la Mer*, **30**, 34–39.

CRONIN, J. R. & TYLER, I. D. 1980. Organic carbon in a Scottish sea loch. *In*: ALBAIGES, J. (ed.) *Analytical Techniques in Environmental Chemistry*. Pergamon Press, Oxford, 419–426.

DE HAAS, H. & VAN WEERING, T. C. E. 1997. Recent sediment accumulation, organic carbon burial and transport in the northeastern North Sea. *Marine Geology*, **136**, 173–187.

DE HAAS, H., VAN WEERING, T. C. E. & DE STIGTER, H. 2002. Organic carbon in shelf seas: sinks or sources, processes and products. *Continental Shelf Research*, **22**, 691–717.

DEGENS, E. T., KEMPE, S. & RICHEY, J. E. 1991. Summary: biogeochemistry of major world rivers. *In*: DEGENS, E. T., KEMPE, S. & RICHEY, J. E. (eds) *Biogeochemistry of Major World River*. Wiley, Chichester, 323–348.

EDWARDS, A. & SHARPLES, F. 1986. *Scottish Sea Lochs: A Catalogue*. Scottish Marine Biological Association/ Nature Conservancy Council, Oban.

GAGE, J. 1972. A preliminary survey of the benthic macrofauna and sediments in Lochs Etive and Creran, sealochs along the west coast of Scotland. *Journal of the Marine Biological Association of the UK*, **52**, 237–276.

GLUD, R. N., STAHL, H., BERG, P., WENZHÖFER, F., OGURI, K. & KITAZATO, H. 2009. In situ microscale variation in distribution and consumption of O_2: a case study from a deep ocean margin sediment (Sagami Bay, Japan). *Limnology and Oceanography*, **54**, 1–12.

GONI, M. A. & HEDGES, J. I. 1992. Lignin dimers: structures, distribution, and potential geochemical applications. *Geochimica et Cosmochimica Acta*, **56**, 4025–4043.

GONI, M. A., YUNKER, M. B., MACDONALD, R. W. & EGLINTON, T. I. 2000. Distribution and sources of organic biomarkers in arctic sediments from the MacKenzie River and Beaufort Shelf. *Marine Chemistry*, **71**, 23–51.

HALL, P. O. J., ANDERSON, L. G., VAN DER LOEFF, M. M. R., SUNDBY, B. & WESTERLUND, S. F. G. 1989. Oxygen uptake kinetics in the benthic boundary layer. *Limnology and Oceanography*, **34**, 734–746.

HANSEN, H. P. 1999. Determination of oxygen. *In*: GRASSHOFF, K., KREMLING, K. & EHRHARDT, M. (eds) *Methods of Seawater Analysis*. 3rd edn. Wiley-VCH, Weinheim, 75–89.

HEDGES, J. I. & ERTEL, J. R. 1982. Characterization of lignin by gas capillary chromatography of cupric oxide oxidation products. *Analytical Chemistry*, **54**, 174–178.

HEDGES, J. I. & KEIL, R. G. 1995. Sedimentary organic matter preservation: an assessment and speculative synthesis. *Marine Chemistry*, **49**, 81–115.

HEDGES, J. I., KEIL, R. G. & BENNER, R. 1997. What happens to terrestrial organic matter in the ocean? *Organic Geochemistry*, **24**, 195–212.

HENRICHS, S. M. 1992. Early diagenesis of organic matter in marine sediments: progress and perplexity. *Marine Chemistry*, **39**, 119–149.

HOLMER, M. 1999. The effect of oxygen depletion on anaerobic organic matter degradation in marine sediments. *Estuarine, Coastal and Shelf Science*, **48**, 383–390.

HULTHE, G., HULTH, S. & HALL, P. O. J. 1998. Effect of oxygen on degradation rate of refractory and labile organic matter in continental margin sediments. *Geochimica et Cosmochimica Acta*, **62**, 1319–1328.

HURST, H. M. & BURGES, N. A. 1967. Lignin and humic acids. *In*: MCLAREN, A. D. & PETERSON, G. H. (eds) *Soil Biochemistry*. Marcel Dekker, London, 260–317.

JICKELLS, T. D., BLACKBURN, T. H. ET AL. 1991. Group report: what determines the fate of materials within ocean margins? *In*: MANTOURA, R. F. C., MARTIN, J.-M. & WOLLAST, R. (eds) *Dahlem Workshop Reports. Ocean Margin Processes in Global Change. Physical, Chemical, and Earth Sciences Research Report 9*. John Wiley & Sons, Chichester, 211–234.

JOHANNESSEN, S. C., MACDONALD, R. W. & PATON, D. W. 2003. A sediment and organic carbon budget for the greater Strait of Georgia. *Estuarine, Coastal and Shelf Science*, **56**, 845–860.

JØRGENSEN, B. B. 1982. Mineralization of organic matter in the sea bed – the role of sulphate reduction. *Nature*, **296**, 643–645.

KEIL, R. G., TSAMAKIS, E., FUH, B., GIDDINGS, J. C. & HEDGES, J. I. 1994. Mineralogical and textural controls on the organic composition of coastal marine sediments: hydrodynamics separation using SPLITT-fractionation. *Geochimica et Cosmochimica Acta*, **58**, 879–893.

KRATZL, K. 1965. Lignin – its biochemistry and structure. *In*: CǑTÊ, W. A. (ed.) *Cellular Ultrastructure of Woody Plants. Proceedings of the Advanced Science Seminar Pinebrook Conference Center Upper Saranac Lake, New York September 1964*. Syracuse University Press, 157–180.

KRISTENSEN, E. 2000. Organic matter diagenesis at the oxic/anoxic interface in coastal marine sediments, with emphasis on the role of burrowing animals. *Hydrobiologia*, **426**, 1–24.

KRISTENSEN, E. & ANDERSEN, F. Ø. 1987. Determination of organic carbon in marine sediments: a comparison of two CHN-analyzer methods. *Journal of Experimental Marine Biology and Ecology*, **109**, 15–23.

KRISTENSEN, E. & BLACKBURN, T. H. 1987. The fate of organic carbon and nitrogen in experimental marine sediment systems: influence of bioturbation and anoxia. *Journal of Marine Research*, **45**, 231–257.

KRISTENSEN, E. & HOLMER, M. 2001. Decomposition of plant materials in marine sediment exposed to different electron acceptors (O_2, NO_3^-, and SO_4^{2-}), with emphasis on substrate origin, degradation kinetics, and the role of bioturbation. *Geochimica et Cosmochimica Acta*, **65**, 419–433.

KRISTENSEN, E., AHMED, S. I. & DEVOL, A. H. 1995. Aerobic and anaerobic decomposition of organic matter in marine sediment: which is fastest? *Limnology and Oceanography*, **40**, 1430–1437.

Leftley, J. W. & MacDougall, N. 1991. *The Dunstaffnage Sedimentation Trap and its Moorings*. DML Internal Report No. **174**, Oban.

Loh, P. S. 2005. *An Assessment of the Contribution of Terrestrial Organic Matter to Total Organic Matter in Sediments in Scottish Sea Lochs*. PhD thesis. Open University, UHI Millennium Institute, Oban.

Loh, P. S., Reeves, A. D.*, Overnell, J., Harvey, S. M. & Miller, A. E. J. 2002. Assessment of terrigenous organic carbon input to the total organic carbon in sediments from Scottish transitional waters (sea lochs): methodology and preliminary results. *Hydrology and Earth System Sciences*, **6**, 959–970.

Loh, P. S., Miller, A. E. J., Reeves, A. D., Harvey, S. M. & Overnell, J. 2008*a*. Assessing the biodegradability of terrestrially-derived organic matter in Scottish sea loch sediments. *Hydrology and Earth System Sciences*, **12**, 811–823.

Loh, P. S., Miller, A. E. J., Reeves, A. D., Harvey, S. M. & Overnell, J. 2008*b*. Optimised recovery of lignin-derived phenols in a Scottish fjord by the cuo oxidation method. *Journal of Environmental Monitoring*, **10**, 1187–1194.

Loh, P. S., Reeves, A. D., Harvey, S. M., Overnell, J. & Miller, A. E. J. 2008*c*. The fate of terrestrial organic matter in two Scottish sea lochs. *Estuarine, Coastal and Shelf Science*, **76**, 566–579.

Louchouarn, P., Lucotte, M. & Farella, N. 1999. Historical and geographical variations of sources and transport of terrigenous organic matter within a large-scale coastal environment. *Organic Geochemistry*, **30**, 675–699.

Ludwig, W., Probst, J. L. & Kempe, S. 1996. Predicting the oceanic input of organic carbon by continental erosion. *Global Biogeochemical Cycles*, **10**, 23–41.

Lund-Hansen, L. C., Valeur, J., Pejrup, M. & Jensen, A. 1997. Sediment fluxes, re-suspension and accumulation rates at two wind-exposed coastal sites and in a sheltered bay. *Estuarine, Coastal and Shelf Science*, **44**, 521–531.

MacKenzie, F. T., Bewers, J. M. et al. 1991. Group report: what is the importance of ocean margin processes in global change? *In*: Mantoura, R. F. C., Martin, J.-M. & Wollast, R. (eds) *Dahlem Workshop Reports. Ocean Margin Processes in Global Change. Physical, Chemical, and Earth Sciences Research Report 9*. John Wiley & Sons, Chichester, 433–454.

Mantoura, R. F. C., Martin, J.-M. & Wollast, R. 1991. Introduction. *In*: Mantoura, R. F. C., Martin, J.-M. & Wollast, R. (eds) *Dahlem Workshop Reports. Ocean Margin Processes in Global Change. Physical, Chemical, and Earth Sciences Research Report 9*. John Wiley & Sons, Chichester, 1–3.

Martin, J. P. & Haider, K. 1980. Microbial degradation and stabilization of 14C-labeled lignins, phenols, and phenolic polymers in relation to soil humus formation. *In*: Kirk, T. K., Higuchi, T. & Chang, H.-M. (eds) *Lignin Biodegradation: Microbiology, Chemistry, and Potential Applications Volume I*. CRC Press, Bota Racon, Fl, 77–100.

Meyers-Schulte, K. J. & Hedges, J. I. 1986. Molecular evidence for a terrestrial component of organic matter dissolved in ocean water. *Nature*, **321**, 61–63.

Nickell, L. A., Black, K. D. et al. 2003. Bioturbation, sediment fluxes and benthic community structure around a salmon cage farm in Loch Creran, Scotland. *Journal of Experimental Marine Biology and Ecology*, **285–286**, 221–233.

Nuwer, J. M. & Keil, R. G. 2005. Sedimentary organic matter geochemistry of Clayoquot Sound, Vancouver Island, British Columbia. *Limnology and Oceanograph*, **50**, 1119–1128.

Opsahl, S. & Benner, R. 1997. Distribution and cycling of terrigeneous dissolved organic matter in the ocean. *Nature*, **386**, 480–482.

Overnell, J. & Young, S. 1995. Sedimentation and carbon flux in a Scottish Sea Loch, Loch Linnhe. *Estuarine, Coastal and Shelf Science*, **41**, 361–376.

Overnell, J., Edwards, A., Grantham, B. E., Harvey, S. M., Jones, K. J., Leftley, J. W. & Smallman, D. J. 1995*a*. Sediment-water column coupling and the fate of the spring phytoplankton bloom in Loch Linnhe, a Scottish Fjordic Sea-loch. Sediment processes and sediment–water fluxes. *Estuarine, Coastal and Shelf Science*, **41**, 1–19.

Overnell, J., Harvey, S. M. & Parkes, R. J. 1995*b*. A biogeochemical comparison of sea loch sediments. Manganese and iron contents, sulphate reduction and oxygen uptake rate. *Oceanologica Acta*, **19**, 41–55.

Parkes, R. J. & Buckingham, W. J. 1986. The flow of organic carbon through aerobic respiration and sulphate-reduction in inshore marine sediments. *In*: Megusar, F. & Gantar, G. (eds) *Perspective in Microbial Ecology. Proceedings of the 4th International Symposium on Microbial Ecology*, 617–624.

Pearson, T. H. 2001. Functional group ecology in soft-sediment marine benthos: the role of bioturbation. *Oceanography and Marine Biology: An Annual Review*, **39**, 233–267.

Pejrup, M., Valeur, J. & Jensen, A. 1996. Vertical fluxes of particulate matter in Aarhus Bight, Denmark. *Continental Shelf research*, **16**, 1047–1064.

Postma, H. 1980. Sediment transport and sedimentation. *In*: Olausson, E. & Cato, I. (eds) *Chemistry and Biogeochemistry of Estuaries*. John Wiley & Sons, Chichester, 153–186.

Raymond, P. A. & Bauer, J. E. 2000. Riverine export of aged terrestrial organic matter to the North Atlantic Ocean. *Nature*, **409**, 497–500.

Readman, J. W., Mantoura, R. F. C., Llewellyn, C. A., Preston, M. R. & Reeves, A. D. 1986. The use of pollutant and biogenic markers as source discriminants of organic inputs to estuarine sediments. *International Journal of Environmental Analytical Chemistry*, **27**, 29–54.

Rowe, G. T. & Gardner, W. D. 1979. Sedimentation rates in the slope water of the northwest Atlantic Ocean measured directly with sediment traps. *Journal of Marine Research*, **37**, 581–600.

Rowe, G. T., Smith, S., Falkowski, P., Whitledge, T., Theroux, R., Phoel, W. & Ducklow, H. 1986. Do continental shelves export organic matter? *Nature*, **324**, 559–561.

SCHLUNZ, B. & SCHNEIDER, R. R. 2000. Transport of terrestrial organic carbon to the oceans by rivers: re-estimating flux- and burial rates. *International Journal of Earth Science*, **88**, 599–606.

SKOOG, D. A., WEST, D. M., HOLLER, F. J. & CROUCH, S. R. 2004. *Fundamentals of Analytical Chemistry*. 8th edn. Thomson Brooks/Cole, Belmont, USA.

SPITZY, A. & ITTEKKOT, V. 1991. Dissolved and particulate organic matter in rivers. *In*: MANTOURA, R. F. C., MARTIN, J.-M. & WOLLAST, R. (eds) *Dahlem Workshop Reports. Ocean Margin Processes in Global Change. Physical, Chemical, and Earth Sciences Research Report 9*. John Wiley & Sons, Chichester, 5–17.

STAHL, H., TENGBERG, A. *ET AL*. 2004. Factors influencing organic carbon cycling and burial in Skagerrk sediments. *Journal of Marine Research*, **62**, 867–907.

SUTHERLAND, R. A. 1998. Loss-on-ignition estimates of organic matter and relationships to organic carbon in fluvial bed sediments. *Hydrobiologia*, **389**, 153–167.

VANGRIESHEIM, A. & KHRIPOUNOFF, A. 1990. Near-bottom particle concentration and flux: temporal variations observed with sediment traps and nepholometer on the Meriadzek Terrace, Bay of Biscay. *Progress in Oceanography*, **24**, 103–116.

WALSH, E. M., INGALLS, A. E. & KEIL, R. G. 2008. Sources and transport of terrestrial organic matter in Vancover Island fjords and the Vancouver-Washington Margin: a multiproxy approach using $\delta^{13}C_{org}$, lignin phenols, and the ether lipid BIT index. *Limnology and Oceanography*, **53**, 1054–1063.

WASSMAN, P. 1984. Sedimentation and benthic mineralization of organic detritus in a Norwegian fjord. *Marine Biology*, **83**, 83–94.

WESTRICH, J. T. & BERNER, R. A. 1984. The role of sedimentary organic matter in bacterial sulfate reduction: the G model tested. *Limnology and Oceanography*, **29**, 236–249.

WHEATCROFT, R. A. & SOMMERFIELD, C. K. 2005. River sediment flux and shelf sediment accumulation rates on the Pacific Northwest margin. *Continental Shelf Research*, **25**, 311–332.

WILSON, J. O., VALIELA, I. & SWAIN, T. 1985. Sources and concentrations of vascular plant material in sediments of Buzzards Bay, Massachusetts, USA. *Marine Biology*, **90**, 129–137.

WOLLAST, R. 1991. The coastal organic carbon cycle: fluxes, sources, and sinks. *In*: MANTOURA, R. F. C., MARTIN, J.-M. & WOLLAST, R. (eds) *Dahlem Workshop Reports. Ocean Margin Processes in Global Change. Physical, Chemical, and Earth Sciences Research Report 9*. John Wiley & Sons, Chichester, 365–381.

ZEIKUS, J. G., WELLSTEIN, A. L. & KIRK, T. K. 1982. Molecular basis for the biodegradative recalcitrance of lignin in anaerobic environments. *FEMS Microbiology Letters*, **15**, 193–197.

Experimental exploration of the stratigraphy of fjords fed by glaciofluvial systems

IRINA OVEREEM* & JAMES P. M. SYVITSKI

Community Surface Dynamics Modeling System, INSTAAR, University of Colorado at Boulder, CO, USA

**Corresponding author (e-mail: irina.overeem@colorado.edu)*

Abstract: Whereas most Late Quaternary sedimentary systems experienced only sea-level rise, fjords record unique sequences because rapid uplift after the unloading of the Last Glacial Maximum (LGM) ice sheets outpaced global eustatic sea-level rise. This study aims to disentangle how rapid initial uplift and high variability of eustatic sea-level change affects fjord sedimentary records. Two numerical models are coupled, ICE-5G and SedFlux, and show that timing and duration of deglaciation and total uplift strongly affect fjord stratigraphy. The ICE-5G model predicts a number of distinct time intervals during which many fjords deglaciate, independent of latitude and short-term climate. Deglaciation of the entire fjord system takes significantly longer (*c.* 6 ka) for fjords that deglaciate early (17–15 ka BP) than for fjords deglaciating after 9 ka BP (*c.* 1 ka). Exponential uplift curves totalled *c.* 220–280 m, and have half-lives of 1–1.4 ka.

High uplift rates consistently cause rapid progradation of the rivermouth over tens of kilometres. Thick packages of glaciomarine, and glaciofluvial sediments emerged above sea level and are subsequently incised. Sensitivity tests predict high frequency of submarine mass movements. Fjords that deglaciated early additionally show deposition to be strongly dominated by rapid sea-level rise; signs of drowning are pronounced and subsequent thick fine-grained sequences aggrade. We conclude that recently deglaciated fjords record solely deposition under falling sea-level and thus provide the best modern analogues of forced-regressive systems.

A glance at any world map confirms that fjords dominate mountainous coasts at latitudes higher than *c.* 42–43° both in the Northern and Southern Hemisphere. Fjords are deep, narrow valleys carved by advancing and retreating ice streams at the margins of the Quaternary Ice Sheets (Johnson 1915; Syvitski *et al.* 1986; Anderson *et al.* 2006; Kessler *et al.* 2008). Advancing ice streams are relatively efficient in eroding existing sedimentary deposits, which were deposited during a previous deglacial cycle. Present-day fjords, which are freed from ice occupation after the decay of the large land-ice sheets after the Last Glacial Maximum (LGM) (*c.* 21 ka BP), store records mostly of the last deglacial period.

It is widely acknowledged that our understanding of sedimentary dynamics in general is biased towards recent records built up during the Latest Pleistocene eustatic sea-level rise and especially the Holocene period of relatively stable climate and sea level. If we put this time period into the perspective of the entire Quaternary, it is evident that highstand conditions as experienced during the Late Holocene are more of an exception than a rule (Fig. 1a). Fjord sedimentary systems are unique in this aspect; sediments may be deposited over the same time period but, when land-based ice sheets retreated, the Earth's crust experienced unloading and started rebounding. These glacio-isostatic adjustments most often resulted in relative uplift of the land and, as a consequence, many fjord valley fills have been deposited under conditions of relative sea-level fall. Fjord valley fills are considered unique modern analogues of a forced-regressive system. But are all fjord deposits dominated by relative sea-level fall? Is there a systematic difference between fjords that started filling in during the post-LGM periods of rapid eustatic sea-level rise and fjords that only started filling in when eustatic sea level already stabilized during the late Holocene? (Fig. 1c).

Stratigraphic surveys show the complexity of fjord deposits, documenting interfingering of tills, glaciomarine sediments, submarine slides, moraines of small readvances of the ice and glaciofluvial sediments (Syvitski *et al.* 1986; Powell & Elverhøi 1989; Stravers & Syvitski 1991; Dowdeswell & Andrews 1995; Syvitski & Shaw 1995; Sejrup *et al.* 1996; Aarseth 1997; Hjelstuen *et al.* 2009). Careful mapping and unravelling of the local depositional history of many specific fjords made it possible to formulate theoretical depositional models (Boulton 1990; Syvitski 1991; Powell & Cooper 2002; Corner 2007).

In this paper, we eliminate many complexities of fjord sedimentation in an attempt to isolate the

From: Howe, J. A., Austin, W. E. N., Forwick, M. & Paetzel, M. (eds) *Fjord Systems and Archives*. Geological Society, London, Special Publications, **344**, 125–142.
DOI: 10.1144/SP344.11 0305-8719/10/$15.00

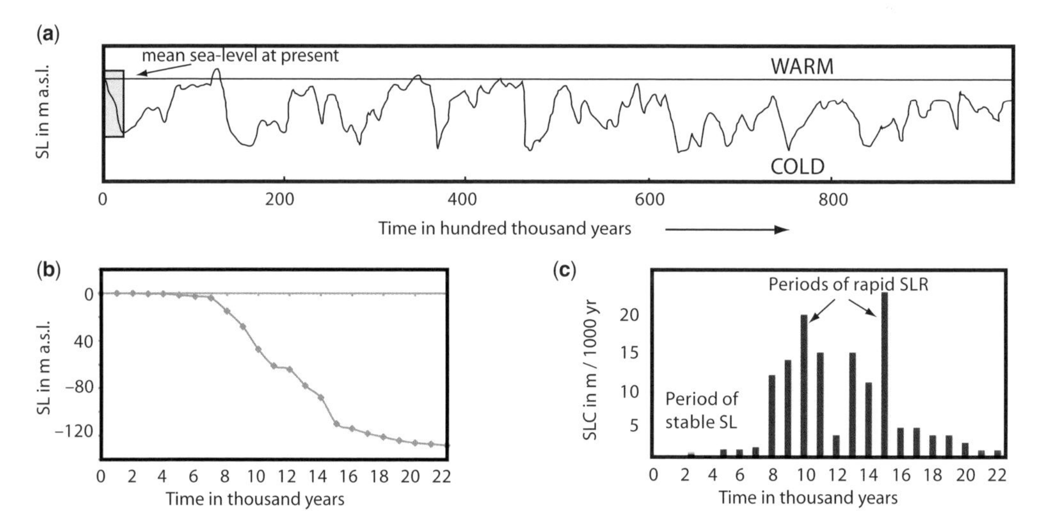

Fig. 1. (**a**) Global sea-level fluctuations over the last million years illustrate the uniqueness of the present mean sea level. (**b**) Eustatic sea-level rise over the last deglacial cycle (modified after Milne & Mitrovica 2008). (**c**) Eustatic sea-level change rates over the last deglacial cycle show a number of time intervals of rapid sea-level rise, that is, >20 m in 1 ka, and stable sea level towards the present day (based on reanalysis of sea level curve of Milne & Mitrovica 2008).

effects of the interaction of eustatic sea-level rise and glacio-isostatic uplift on fjord stratigraphy. We focus on fjord systems, which receive sediments dominantly from glaciofluvial systems in the head-waters, since the sedimentary dynamics for fjords dominated by tidewater glaciers are so distinctly different.

We analyse ICE-5G model predictions (Peltier 2004) on glacio-isostatic uplift and relative sea-level change for a representative suite of over 30 fjord regions in the Northern Hemisphere. The ICE-5G model analysis is coupled to a stratigraphic numerical model to simulate the depositional history of a suite of generic fjords.

Analysis of deglacial history, glacio-isostatic adjustment and relative sea-level change with ICE-5G

We selected 34 fjords located in the Northern Hemisphere to form a representative coverage ranging from Spitsbergen and the West Coast of Norway to Greenland, to Ellesmere Island and Baffin Island in the Canadian Archipelago, to Alaska and to lower latitude sites in British Columbia and Quebec, Canada (Fig. 2; Table 1).

We mapped the location of the present-day fjordhead and the outer fjord region for each of these selected fjords. They include relatively short fjords, such as Kongsfjorden on Spitsbergen (27 km) and Yakutat Bay in Alaska (58 km), long, narrow fjords such as Søndre Strømfjord in Western Greenland (172 km) and long, wide bays such as Cumberland Sound on Baffin Island (310 km) and Cook Inlet in Alaska (272 km).

The ICE-5G (VM2) model (Peltier 2004) mathematically analyses global glacio-isostatic adjustment processes and provides data on global ice-sheet coverage, ice thickness and paleotopography at 10 min spatial resolution for 21 ka and 0 ka, and at 1 degree spatial resolution for 500-year intervals between these two snapshots. The model inversely predicts the ice sheet and topography parameters from available relative sea-level curves over the entire world (Peltier 2002). An earlier version of the model was tuned to 392 relative sea-level curves (Tushingham & Peltier 1992) and it has been gradually refined for specific areas (Peltier 2002; Tarasov & Peltier 2002).

We use ICE-5G to determine (1) timing of deglaciation (2) paleotopography, and (3) ice thickness for each of the selected outer fjord and fjordhead regions over the last 21 ka. Whereas the two snapshots at 21 ka BP and 0 ka BP have a higher resolution, and thus resolve the locations of individual fjords relatively well, it is evident that at 1 degree spatial resolution (which is about 100 × 100 km at the equator) each grid cell in the ICE-5G model is representative of *regional* uplift and *regional* ice thickness only. The timing of the onset of deglaciation is determined by the moment that the large-scale, 1-degree gridcell overlaying the location of a specific outer fjord becomes ice-free. The duration of the active deglaciation is defined as the time period from when the outer

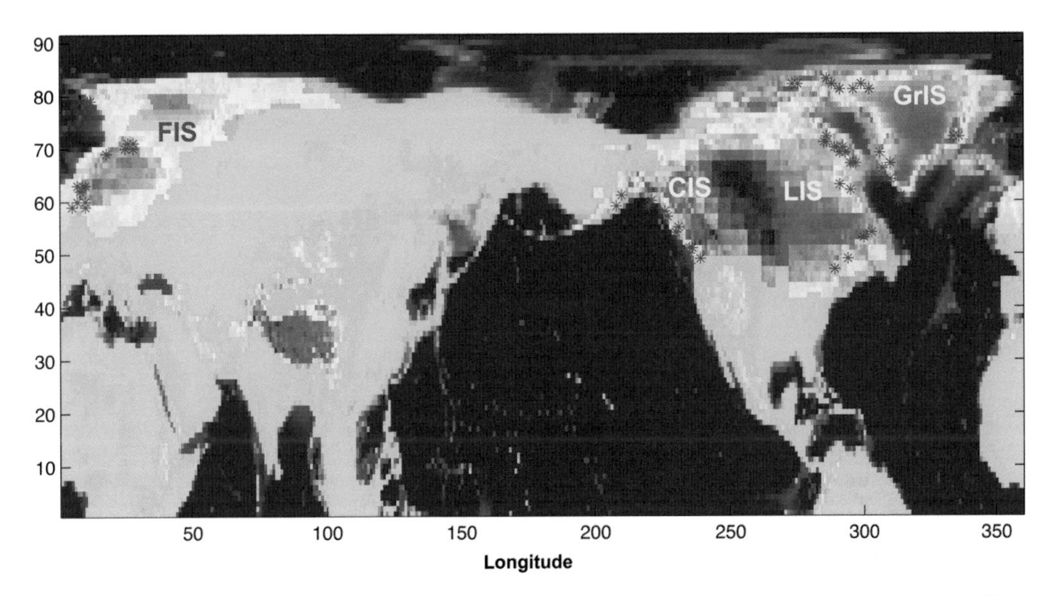

Fig. 2. Northern Hemisphere topography and ice-sheet thickness at 21 ka BP on 10-min resolution grid as predicted by ICE-5G (Peltier 2004). Blue stars indicate an arbitrary selection of fjords.

fjord zone becomes ice-free until the fjordhead zone becomes ice-free.

It is clear that this low spatial resolution oversimplifies a fjord landscape, where in deep fjord valleys erosive ice streams show fast advances and retreats (Briner *et al.* 2007; Kleman & Glasser 2007; Hjelstuen *et al.* 2009) and in the low-relief highlands non-erosive plateau glaciers remain stagnant for extensive periods of time (Briner *et al.* 2006). Similarly, the temporal resolution of the ICE-5G model predictions is limited. The predictions have 500 a time intervals, except for the millennia directly after the LGM (21–18 ka BP), which has 1 ka time intervals. In summary, the low spatial and temporal resolution may hamper very precise local reconstructions, but ICE-5G makes consistent predictions for arbitrary fjord regions and allows coherent comparisons to set up generalized boundary conditions for stratigraphic modelling.

Timing of deglaciation and glacio-isostatic adjustment

The timing of deglaciation and the concurrent onset of sediment supply to the fjord are key parameters controlling fjord stratigraphy.

ICE-5G data reveals the major phases of Northern Hemisphere deglaciation over the last 21 ka (Fig. 3). During early phases of deglaciation (21–15 ka BP), the Fennoscandian Ice Sheet disintegrated more rapidly than the Laurentide Ice Sheet. The Cordilleran Ice Sheet, covering the western parts of North America, retreated rapidly over the interval 15–11 ka BP, onsetting sedimentation in the fjords of Alaska and British Columbia. Remaining eastern domes of the Laurentide Ice Sheet retreated relatively late during the deglacial period (11–6 ka BP), and some remnants of this ice mass remain today as the Barnes and Penny Ice Caps on Baffin Island, Canada. During the Late Holocene, ice margins of the remaining Greenland Ice Sheet were relatively stable and only minor fluctuations occurred.

Figure 4a shows the relative distribution of outer fjord deglaciation ages, that is, how many of the 34 selected outer fjords became free of ice at what time. Even at the maximum LGM ice extent, some outer fjord areas were not glaciated either because the fjord outlets are located near deep water or they are situated in very dry areas that were not glaciated during the LGM. Similarly, some fjords are not yet ice-free today such as fjords on Ellesmere Island or some of the fjords on Baffin Island which are still influenced by the remnants of the Laurentide Ice Sheet (LIS). The number of fjord regions becoming ice-free does not directly reflect high-resolution climate fluctuations, but rather integrates warming climate conditions over longer time periods (Fig. 4a). The 'average' outer fjord was deglaciated by 11 265 $\pm$ 6861 BP, and the 'average' fjordhead is ice-free by 7794 $\pm$ 5772 BP. Outer fjords that became ice-free at 17, 15, 13, 11, 9 and 5 ka BP are used to explore patterns of sedimentary infill.

After the process of ice retreat has started, an extremely rapid retreat of the major ice stream

Table 1. *The discussed parameters for 34 selected fjords: location of the outer fjord (Lat OF and Long OF); the length of the fjords in km (Length); ICE-5G prediction of timing of deglaciation in ka BP (T-deglac); ICE-5G prediction of duration of deglaciation in years (D-deglac); total ice melt at both the outer fjord and fjordhead in m (Ice melt OF and FH); and total uplift at both the outer fjord and fjordhead in m (Uplift OF and FH)*

Fjord ID	Lat OF	Long OF	Length	T-deglac	D-deglac	Ice melt OF	Ice melt FH	Uplift OF	Uplift FH
Kongsfjorden, Nor	78.54	12.29	27	9000	10 000	1471	1475	151	141
Porsanger fjord, Nor	70.03	25.02	120	11 000	2000	1606	1600	426	397
Tana Fjord, Nor	70.25	28.16	60	11 000	2000	1456	1506	408	401
Malangen Fjord, Nor	69.13	18.29	50	10 000	0	1948	881	325	277
Trondhjeim fjord, Nor	63.17	10.15	80	IR	1000	1237	1293	460	413
Sundalls Fjord, Nor	62.4	8.33	75	IR	4000	1137	1178	394	336
Hardangerfjord, Nor	60.29	7.1	164	IR	6000	1477	1156	370	273
Oslo fjord, Nor	59.54	10.41	95	8000	3000	2191	1639	464	424
Cambridge Fjord, Baf	71.11	−75.06	86	8000	−7500	1408	689	278	215
Sam Ford Fjord, Baf	70.01	−71.32	116	IR	7500	662	1062	256	286
Clyde Fjord, Baf	69.51	−70.26	120	8000	0	1155	675	260	132
Itirbilung Fjord, Baf	69.15	−69.15	85	8000	0	1088	0	246	192
Maktak Fjord, Baf	67.21	−65.04	77	12 000	5500	1210	1309	198	139
N-Pangnirtung Fjord, Baf	67	−64.42	70	11 000	5500	1309	1309	185	139
Cumberland Sound, Baf	63.44	−68.56	310	11 000	−3000	1326	1309	276	139
Frobisher Bay, Baf	63.43	−68.56	265	12 000	0	1326	0	276	128
Disraeli Fjord, Ell	82.33	−72.32	80	13 000	0	1301.5	0	213	140
Phillips Inlet, Ell	81.59	−85	42	11 000	0	2146	0	234	173
Baird Inlet, Ell	81.12	−69.19	96	11 000	7500	1124	1485	234	203
Hall Land, Grl	81.31	−58.47	55	15 000	7500	773	1019	161	180
Disco Bay, Grl	69.06	−54.04	116	13 000	0	0	0	166	169
Sondre Stromfjord, Grl	67	−50.41	172	7000	1000	1727.5	998.5	258	160
Kapisillit, Grl	64.26	−50.12	113	IR	13 000	1303	0	200	128
JulianeHaab, Grl	61.01	−46.42	55	5000	13 000	1001	0	128	94
Alpe Fjord, Grl	72.03	−26.23	98	8000	7000	−14	890.7	172	162
Cook Inlet, AK	61.16	−150.37	272	8000	−1000	408	630	112	102
Yakutat Bay, AK	60	−139.29	58	7000	0	1351	0	239	177
Stephens Passage, AK	58.17	−134.02	133	9000	0	1386	1389	311	267
Observatory Inlet, Can	55.4	−129.48	122	21 000	10 000	1517	0	324	262
Knight Inlet, Can	51.05	−125.35	120	11 000	10 000	1506	1210	350	233
Bute Inlet, Can	50.55	−124.49	106	11 000	4000	1199	1199	323	245
Frazer Lowlands, Can	49.14	−121.44	272	21 000	9000	1000	127	227	88
Hamilton Inlet, Can	53.2	−60.1	188	21 000	1000	2732	2044	567	298
St Lawrence, Can	46.52	−71.34	454	13 000	0	1530	0	445	278

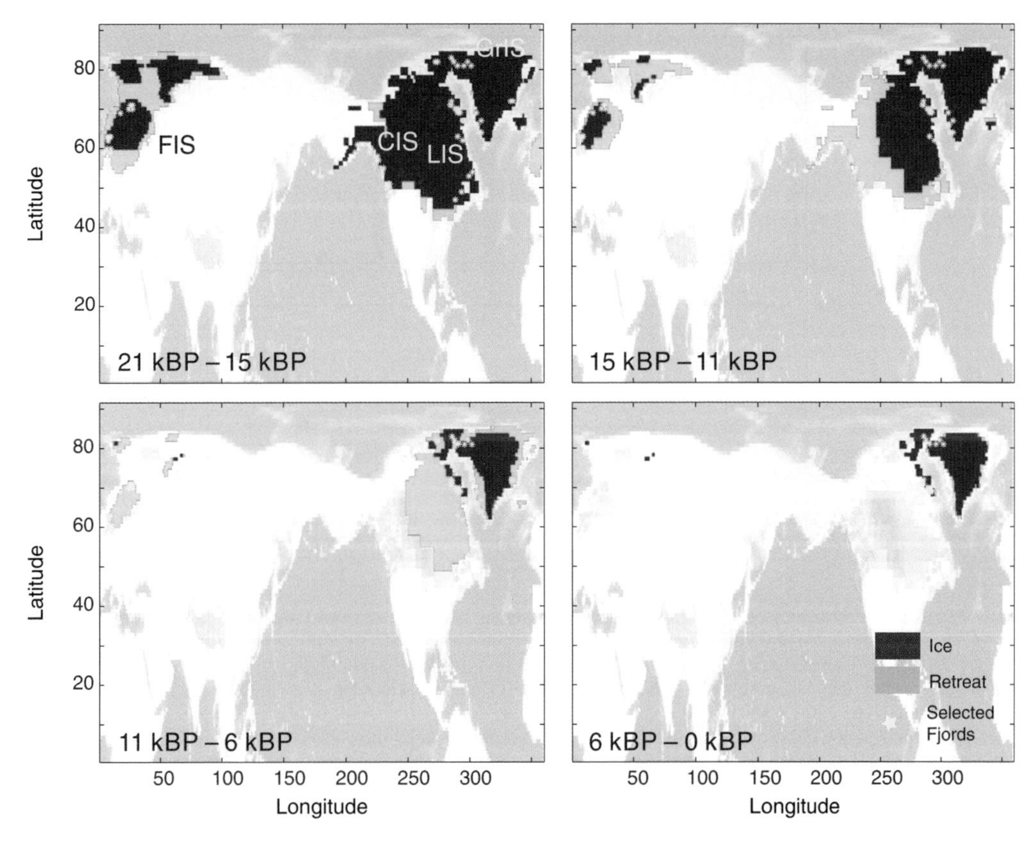

Fig. 3. Difference grids calculated from 1-degree resolution ICE-5G ice-cover grids at 1 ka intervals. The grids are centred on the Pacific Ocean. Yellow stars indicate the selected fjords. Between 21–15 ka BP, most ice retreat occurs in the area of the Fennoscandinavian Ice Sheet (FIS). Between 15–11 ka BP, many fjords along the Cordilleran Ice Sheet (CIS) become ice-free. Between 11–6 ka BP, remaining ice of the Eastern Laurentide (LIS) and Foxe Dome disintegrates and between 6–0 ka BP there are only a few changes along the outer edges of the remaining ice in Baffin Island, Ellesmere Island and Greenland (GIS).

occupying a fjord can occur as the ice is retreating from the outer fjord; this is mostly because the ice retreat is dominated by tidewater glacier dynamics during that phase. However, it is postulated that the ice sheet influences water and sediment supply as long as it lingers in the fjordhead region. Timing of ice retreat from the fjordhead region is considered the controlling parameter that captures this switch in sediment supply regime. Figure 4b plots deglaciation age of the outer fjord against the duration of the ice retreat from the entire fjord. Although the plot shows a wide variety of durations for different deglacial ages, there is a weak trend ($R^2 = 0.32$) towards a decreasing duration of ice retreat for the fjords, which are deglaciated later during the deglacial cycle. This can be explained by the fact that the ice sheets are in an advanced state of disintegration and are thinner at later stages, thus retreating at relatively higher rates.

Total uplift for each of the fjord locations has been derived from the detailed grids of topography and ice thickness at 10 min resolution at both 21 ka BP and 0 ka BP. Outer fjords generally have less ice load at LGM (843 $\pm$ 641 m) than fjordheads (1351 $\pm$ 487 m). This difference is also reflected in the average total uplift, which is 217 $\pm$ 96 m at the outer fjords compared to 283 $\pm$ 109 m at the fjordhead regions. There is no significant trend between age of deglaciation for an individual fjord and total uplift. Instead, total uplift reflects both the thickness and extent of the ice sheet and whether fjords are located closer to the margins or further towards the centre of the ice sheet.

The ICE-5G data provides insight in the rate of the uplift over time. The majority of uplift curves show exponentially decreasing uplift rates, with relatively rapid uplift directly after deglaciation and much slower rates towards the latest Holocene.

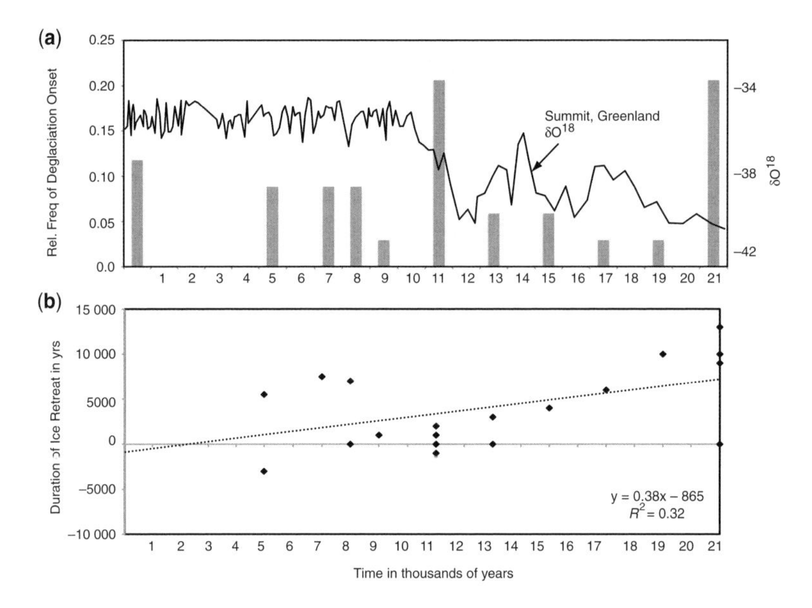

Fig. 4. (**a**) Relative frequency of calculated deglaciation onset for 34 fjordheads. Fjords that were already deglaciated at 21 ka BP, or still glaciated at 0 ka BP, have been excluded from further analysis. δO^{18} reconstruction from Summit Station, Greenland (Johnsen *et al.* 1995) shows that deglaciation predictions do not directly follow temperature fluctuations; rather, they integrate warm periods over time. (**b**) Duration of ice retreat as calculated from the difference of deglaciation onset in the outer fjord and the potentially later deglaciation onset in the fjordhead region. A trend to decreasing deglaciation duration over the deglacial cycle is represented by the dotted line.

Some fjords are located close to the LGM ice margins, and are affected by forebulge migration and collapse, especially in their outer fjord regions. Exponentially decreasing uplift is evident from the analysis of relative sea-level curves, resulting from field mapping and radiocarbon dating which summarize the net result of both glacio-isostatic adjustment and eustatic sea-level change (Dyke & Peltier 2000). Dyke & Peltier (2000) showed that relative sea-level curves for sites towards the margins of the ice sheets have curve 'half-lives' of only *c.* 1–1.4 ka, whereas further towards the centre of the ice sheet half-lives are *c.* 2 ka. The selected fjords are mostly located in the marginal region and are thus dominated by rapid rebound rates.

Numerical modelling methodology

Model description

Sedflux 2.0 is a process-based forward model, which produces synthetic stratigraphy for sedimentary basins (Syvitski & Hutton 2001; Hutton & Syvitski 2008). The model is capable of modelling in either two or three spatial dimensions over time. The model simulates sediment transport for multiple grain-size classes. For this specific study SedFlux, is used to simulate a 2D longitudinal profile. This is thought to be a valid simplification since fjord basins are relatively narrow and constrained and they dominantly receive sediment input from a single fjordhead river.

Relevant process modules used include river avulsion, river braidplain sedimentation, river erosion and generation of hypopycnal plumes upon entering the marine fjord basin. Sedflux presently incorporates no process components to deal with glacial processes such as the formation of moraines or sedimentation from tidewater glaciers. This means that the simulations are restricted to mimicking the glaciofluvial processes and the glaciomarine suspended sediment transport processes. The entire model grid is subject to user-defined subsidence or uplift to account for glacio-isostasy. For detailed process descriptions for river processes and hypopycnal plume dynamics, refer to Syvitski & Hutton (2001), Overeem *et al.* (2005), Hutton & Syvitski (2008) and Peckham (2008).

A predecessor of Sedflux was originally developed for and tested against fjord sedimentation (Syvitski & Daughney 1992). Recently, SedFlux has successfully predicted stratigraphic patterns for a number of lower latitude margins over a deglacial cycle (Overeem *et al.* 2005; Kubo *et al.* 2006; Jouet *et al.* 2008).

The user controls the boundary conditions of the simulated basin (i.e. the paleotopography and bathymetry) and the parameters of each sedimentary process through input files (Syvitski & Hutton 2001; Hutton & Syvitski 2008).

Every fjord has a unique set of boundary conditions ranging from different conditions in its hinterland (such as lithology and topography, vegetation cover or occurrence of mountain glaciers and fingerlakes and possibly multiple river systems feeding the fjord) to varied conditions in the fjord itself (in its aspect and bathymetry, its oceanography such as tidal currents and sensitivity to storm events). Our experiments are highly generalized and focused on the effects of the timing of deglaciation, the duration of ice retreat and the interaction of eustatic sea-level change and glacio-isostatic movements. Table 2 summarizes the input conditions described in this section.

Initial fjord geometry. All simulations use a simplified concept of a fjord basin. A fjord is parameterized as hundreds of metres deep with a narrow (i.e. 4–8 km wide) U-shaped valley. Fjords have a complex longitudinal profile with local depressions due to the increased scouring of the glacial system when tributary valley glaciers combine with the main ice stream (MacGregor *et al.* 2000; Kessler *et al.* 2008). These scoured depressions function as local depocentres and are considered to be such an important constraint on the fjord infill that the generic initial fjord topography and bathymetry include several local *c.* 100 m deep depressions. Many fjords shallow towards their outer region and have a 'sill', originating from the divergence of the eroding ice streams which reduced their scouring capacity; this does not mean that fjords are completely closed basins, however. In our model simulations we maintain an open boundary,

Table 2. *Input parameters for SedFlux experiments*

SedFlux input description	Parameter	Value
Bathymetry	Fjord length in km	120
Stepped profile, multiple tributary valleys	Fjord width in km	7
	Fjord depth in m	815
Grid dimensions	y-resolution in m	50
2D longitudinal profile	z-resolution in m	0.05
Time dimensions	Total simulation time in years	17 000, 15 000, 13 000, 11 000, 9000, 5000
Buffer	Duration of deglaciation in years	6000, 4000, 2000, 1000, 1000, 0
	Simulation time step	Daily
Water discharge		
Arctic river stable SS	Q daily variation, summer in m^3/a	200–1125
	Q daily variation, winter in m^3/a	5
Arctic river non-uniform SS-HIGH	Q daily variation, summer in m^3/a	200–1500
	Q daily variation, winter in m^3/a	5
Arctic river non-uniform SS-LOW	Q daily variation, summer in m^3/a	75–500
	Q daily variation, winter in m^3/a	5
Sediment discharge		
Arctic river stable SS	Qb daily variation, summer in kg/s	8–12
	Qb daily variation, winter in kg/s	0
	Qs daily variation, summer in kg/m^3	7–0.6
	Qs daily variation, winter in kg/m^3	0.004
Arctic river non-uniform SS-HIGH	Qb daily variation, summer in kg/s	10–20
	Qb daily variation, winter in kg/s	0
	Qs daily variation, summer in kg/m^3	8–0.8
	Qs daily variation, winter in kg/m^3	0.004
Arctic river non-uniform SS-LOW	Qb daily variation, summer in kg/s	3–6
	Qb daily variation, winter in kg/s	0
	Qs daily variation, summer in kg/m^3	2.7–0.27
	Qs daily variation, winter in kg/m^3	0.004
Sediment grain size	Number of grain-size classes	5
	Grain size in micron	1200, 200, 120, 30, 2
Glacio-isostasy	Total uplift in m	275
Exponential curve	Half-time in years	1000

representing the potential escape of sediment towards the open shelf and ocean.

Sediment supply signal. Present-day monitored glaciated watersheds have relatively high sediment loads (Guymon 1974; Hallet *et al.* 1996; Hasholt *et al.* 2005) and readily available stored sediment in the form of moraines and glaciofluvial floodplains; this may even result in high sediment loads well beyond the disappearance of the ice (Church & Ryder 1972; Church & Slaymaker 1989). Stratigraphic reconstructions of deglacial systems point to a phase of high sediment supply during deglaciation and a strong reduction of sediment supply after the ice has retreated completely from a basin (Andrews 1987; Helle 2004). Field studies aiming for a complete on- and offshore sediment budget reconstruction corroborate this twofold pattern (e.g. Hansen 2004; Eilertsen *et al.* 2007), although there are unique deviations in every one of these case studies related to short-term climate fluctuations and possibly ice regime changes. Corner (2007) used these studies to guide a sequence stratigraphical model for fjords. We use his conceptual model here, which distinguishes between a *deglacial* phase and a *postglacial* phase. High sediment supply during active deglaciation (i.e. the deglacial phase) is thought to be caused by the drainage of enormous amounts of meltwater from the decaying ice sheet. When the decaying ice sheet has retreated from the hinterland of the fjord the postglacial phase starts; small local river systems are generally the sole source of sediment to the fjord basin.

A first set of simulations focuses exclusively on the interaction of eustatic sea-level change and glacio-isostatic adjustments; we therefore use a uniform, stable sediment supply over the time of simulation. A second set of experiments simplifies the complexity of changing sediment loads during deglacial and postglacial phases with a step function. The imposed step function dictates that total annual sediment load is relatively high during the deglacial phase, whereas total annual loads abruptly decrease to one-third of the deglacial loads when the fjordhead region and local drainage basin become ice-free (Table 2).

Cold-climate rivers are highly dynamic with short 2–3 month discharge seasons, incorporating snowmelt periods and runoff peaks due to extreme events (e.g. Overeem & Syvitski 2008). Sedflux incorporates such daily variations in the water and sediment load. We run a typical flashy river with a short discharge season of 3 months incorporating a snowmelt season and additional high daily variation and 9 months of a very small base flow.

The sediment load is distributed over 5 grain-size classes ranging from 1200 μm, for bedload material, to 2 μm for the finest suspended sediment. This sediment distribution is kept constant over the time span of the entire simulation implying that, even if there is less energy for transport during postglacial times, all sediment grain-size classes are still assumed to be potentially available. This assumption is justified further if local mountain glaciers persist in the fjordhead region postglacially; in this case, there is still a significant new source for both coarse sand and large amounts of fine glacial flour.

Eustatic sea level and uplift. Forward modelling experiments require disentangling of the moving boundaries of eustatic sea-level change and glacio-isostatic movement (i.e. uplift) which are controlled by separate model process modules. An emergence curve of a fjord is defined as the sum of the glacio-isostatic uplift and sea-level rise.

Sedflux uses a eustatic sea-level curve from LGM to the present (Milne & Mitrovica 2008). For our first-order experiments, we use a simple spatially uniform uplift curve. It has a total uplift of 275 m with the greatest rates directly after deglaciation and a short half-life of 1 ka, as derived from the ICE-5G data. The interaction of sea-level change rates, timing of deglaciation and uplift result in distinctly different emergence curves, as illustrated in Figure 5. Fjord deglaciating around 17 ka BP experience rapid emergence for only *c.* 2 ka; subsequently, there are two distinct periods during which sea-level rise outpaces the uplift. Fjords deglaciating at *c.* 15 ka BP experience rapid emergence for *c.* 3 ka and then have a continuous period of *c.* 4 ka when sea-level rise outpaces uplift. The fjords that deglaciated from 11 ka BP onwards have a more simple emergence pattern with rapid emergence during the first 3 ka and subsequent stable conditions, with balance between Late Holocene eustatic sea-level rise and ongoing slow uplift (Fig. 5).

Predicted fjord stratigraphy

Stable sediment supply scenarios

Stratigraphic profiles for six different simulations are shown in Figure 6. The 2D longitudinal profiles show the dominant deposited sediment grain size over the 120 km length of the fjord at the end of the respective simulation.

Fjords that were deglaciated relatively recently, that is, at *c.* 5 ka BP, show a simple emergence record. The deposited sedimentary wedge is stretched out over a large distance of tens of kilometres due to migration of the river mouth with the rapidly falling relative sea level. River braidplain sands and deltaic sediments fill the local overdeepened valley sections and are uplifted. Marine sediments, even the finest marine muds deposited

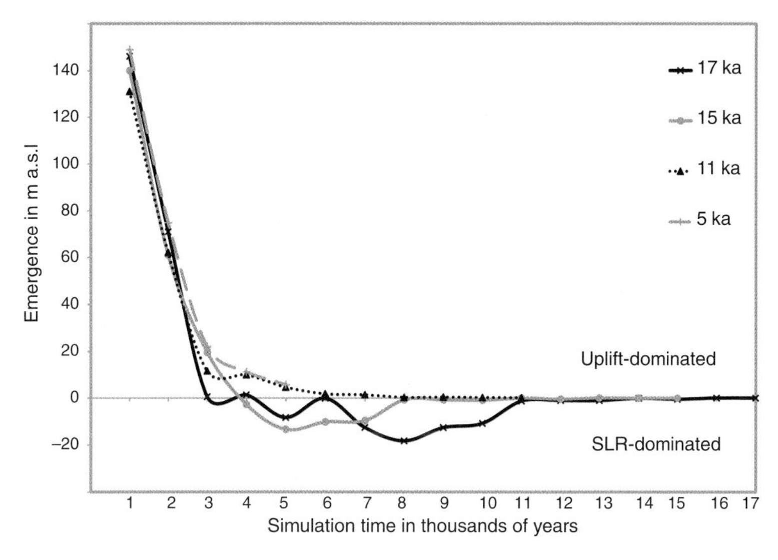

Fig. 5. Emergence curves as defined by global sea-level change and an average uplift curve of 275 m with a half-life of 1 ka. Emergence over the duration of a model simulation is defined for a set of generic fjords, which deglaciated at 17, 15, 11, 5 ka BP. All fjords experience initial rapid emergence and are then uplift-dominated. Fjords which deglaciated relatively early, that is, at 17 ka BP and 15 ka BP, show distinct sea-level rise dominated periods for a few thousand years which taper out to the present stable situation.

from suspension plumes, outcrop as thick packages on the present-day onshore (Fig. 6).

This unit deposited during rapid uplift directly after deglaciation is indeed typical at many recently deglaciated fjordheads. Figure 7a shows a more detailed view of the simulated stratigraphic profile. Two 'pseudo-cores' derived from the 5 ka BP simulation at 30 and 41 km along the profile (Fig. 7c, e) show the deposited marine muds and delta foreset and topset sands that are all uplifted above present-day sea level. Both mud-dominated packages and deltaic sedimentary units are typically encountered in fjordheads on Baffin Island, as depicted in two photographs of the fjordheads of McBeth Fjord (Fig. 7b) and Clyde Fjord (Fig. 7d).

Fjord deglaciated at the Early Holocene, at *c.* 9 ka BP and 11 ka BP, show a forced-regressive unit onshore and at the base of the offshore deposits. This unit consists of a widespread and relatively thin sedimentary wedge which has been incised during rapid uplift. Similar river braidplain sands and deltaic sediments fill the locally overdeepened valley sections and the oldest marine sediments outcrop on the present-day onshore. In addition, fjords of these ages show the first signs of an aggrading and even prograding sedimentary unit on top of the forced-regression unit. The second unit consists of >100 m of coarsening upward sediments with a thin layer of coarse sandy topsets (Fig. 6). The foreset slope of the prograding wedge becomes steeper over time; it is much more gentle in the 9 ka BP scenario than in the 11 and 13 ka BP simulations. This implies that, assuming all other conditions are similar, the fjords that deglaciated earlier ought to be more prone to slope failures along the present-day foreset slopes because they have had more time to build up a steep bathymetric profile. Hutton & Syvitski (2004) modelled high frequency of slope failures associated with rapid continental margin slope steepening, occurring in that case during periods of sea-level lowstand.

Fjords that deglaciated relatively early after the LGM, that is, with onset of deglaciation at 17 and 15 ka BP, still show a forced-regressive unit at the base of the deposited sedimentary sequence (Fig. 6). This unit consistently records the tremendously fast uplift rates due to glacio-isostatic adjustments, which initially easily outpace eustatic sea-level rise. However, the half-time of the uplift curves is only of the order *c.* 1 ka so these fjords experienced distinct periods when global sea-level rise outpaced the waning uplift rates (Fig. 5). The deposits show subsequent rapid drowning and aggradation and only a progradational unit during the late Holocene.

A more detailed view of the modelled sequence is shown in Figure 8a. It can be seen that the unit deposited under forced regression has brought glaciofluvial braidplain or sandur sediment out to about −50 m present-day water depth. An example

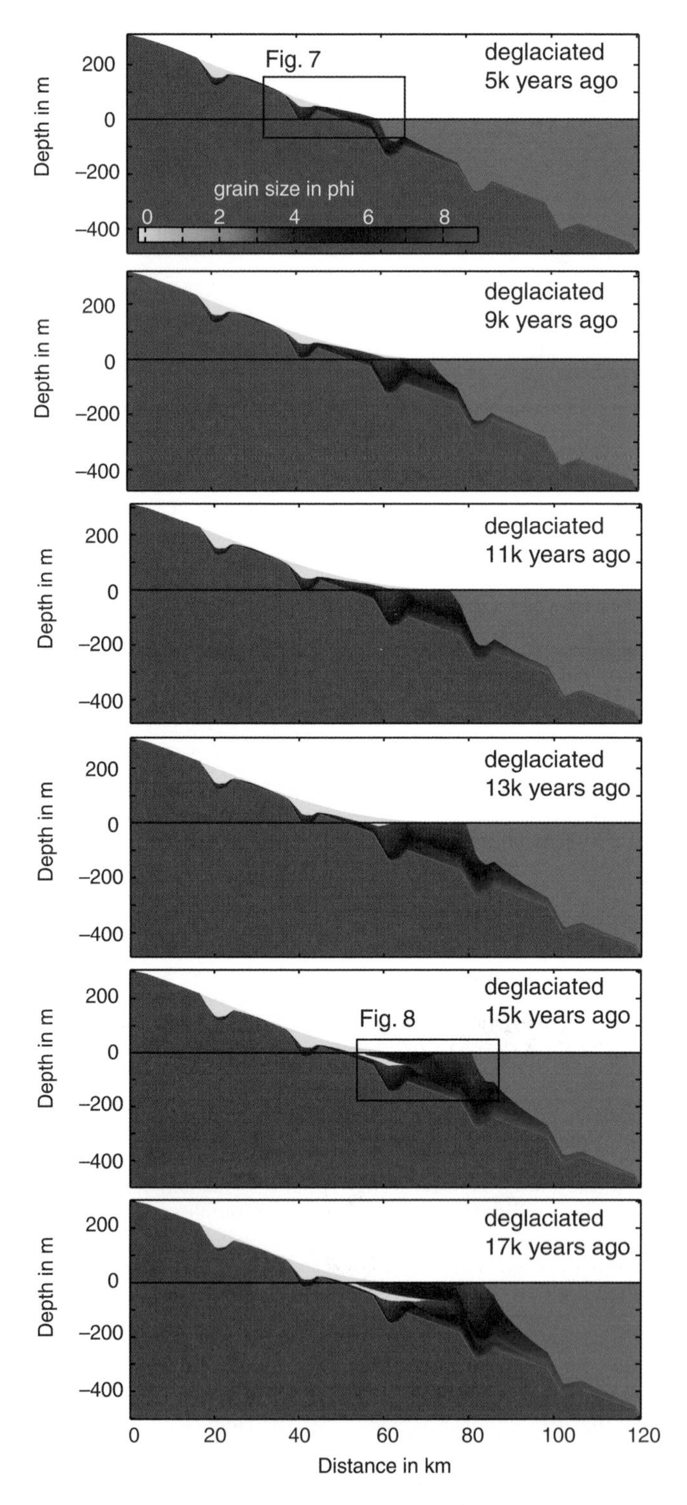

Fig. 6. SedFlux model predictions of longitudinal stratigraphic fjord profiles for the different age scenarios (recently deglaciated fjords at 5 ka BP to early deglaciated fjords at 17 ka BP) show the dominant deposited sediment grain size over the 120 km length of the fjord at the end of the respective simulation. Insets point to subsequent figures.

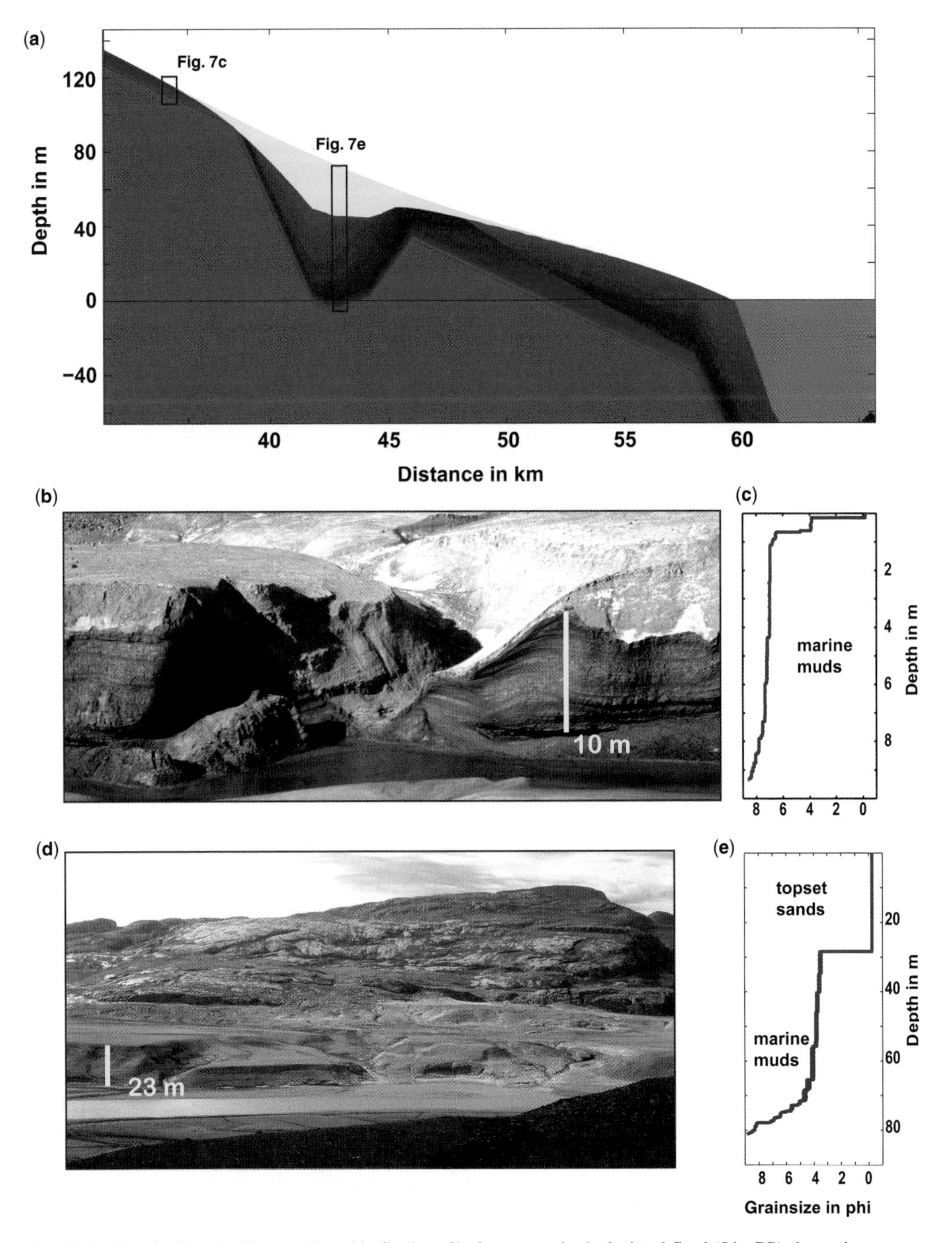

Fig. 7. (**a**) Detail of longitudinal stratigraphic fjord profile for a recently deglaciated fjord (5 ka BP) shows the sedimentary architecture of the onshore deposits, which are part of the forced-regressive unit. (**b**) Emerged marine muds in McBeth Fjord, Baffin Island. (**c**) SedFlux model prediction of a pseudo-core penetrating *c.* 10 m thick marine mud deposits emerged to *c.* 105 m a.s.l. (**d**) Topset sands packages of >20 m form part of the present-day valley fill of Clyde Fjord; marine muds cannot be seen in the photo but were uncovered during sedimentary logging. (**e**) SedFlux model prediction of a pseudo-core penetrating *c.* 80 m thick coarsening upwards sequence of glaciomarine muds and glaciodeltaic to fluvial sands emerged to *c.* 75 m a.s.l.

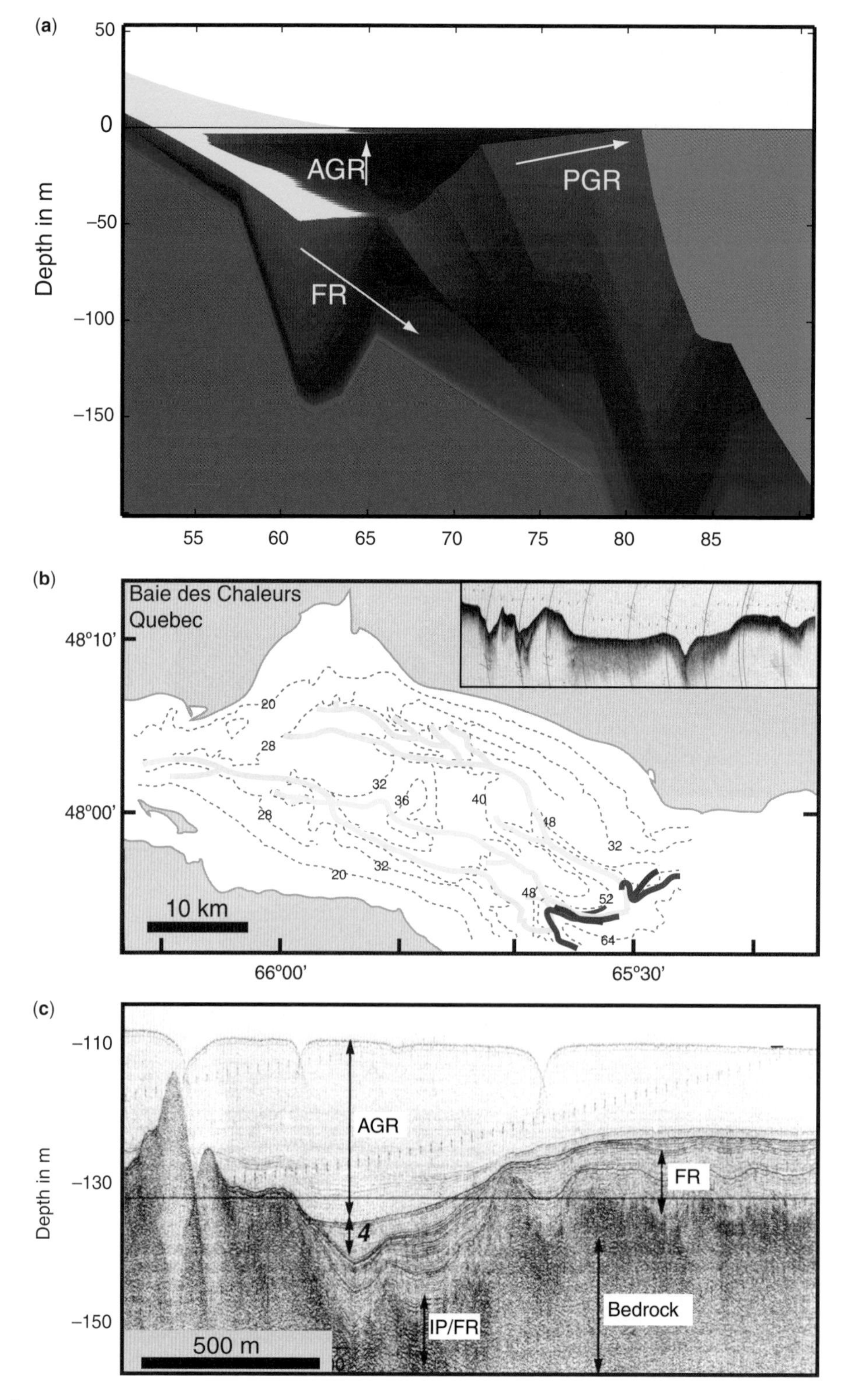
(a)
50
0
-50
-100
-150
Depth in m
AGR
PGR
FR
55
60
65
70
75
80
85
(b)
Baie des Chaleurs
Quebec
48°10'
48°00'
20
28
32
36
40
48
32
28
20
32
48
52
64
10 km
66°00'
65°30'
(c)
-110
-130
-150
Depth in m
AGR
FR
4
IP/FR
Bedrock
500 m

Fig. 8.

of an analogous situation has been mapped from seismic data in the Baie des Chaleur, Quebec (Syvitski 1992) (Fig. 8b). An extensive incised fluvial channel network cuts into previous glacio-marine deposits and extends to about −56 m water depth, indicating that the sea-level lowstand must have been at least as deep as *c.* 60 m. These incised channels are often filled in with Holocene fine sediments but remain unfilled in some cases due to tidal current activity, as can be seen in the hydrographic profile (inset Fig. 8b). The typical phases in the sequence can also be recognized in somewhat deeper water (*c.* 110 m) of the Baie des Chaleurs. HUNTEC high-resolution seismic data, collected in 1987 (Praeg *et al.* 1987), shows a number of phases of marine muds with glacial influence at the base attributed to the forced-regressive unit and a number of more aggradational wedges stacked thickly on top (Fig. 8c).

Non-uniform sediment supply scenarios

It was found that a non-uniform sediment supply has a profound effect on the simulated stratigraphic profiles. The drastic reduction of available sediment during the postglacial phase causes an overall thickness reduction of the deposits and lower progradation rates at the end of the simulation. All simulations show this reduction of deposited sediment thickness in deeper water and slower progradation rates. ICE-5G predictions of duration of deglaciation had a large range, but generally durations were found to be longer for fjords that deglaciated early after LGM (i.e. *c.* 6 ka for fjords that deglaciated about 17 ka BP), varying to only *c.* 1 ka for fjords that deglaciated around 9 ka BP (Fig. 4b). We discuss the effects of a variable sediment supply pulse on the stratigraphy of the 17 ka simulation.

Figure 9 compares the stratigraphic profiles for a simulation of 17 ka duration of the stable sediment supply scenario discussed in the previous paragraph (the lowest panel in Fig. 6) with the non-uniform sediment supply scenario. It is immediately clear that the system prograded less far into the fjord basin. There is also less sediment deposited overall; even in the present-day floodplain there has been more incision and thus less preservation of the oldest deposits onshore. Both simulations have a comparable basal unit that was built up during the forced regression. The high sediment supply during the first 6 ka of the simulation causes the forced-regression unit at the base of the sequence to develop even faster than in the uniform supply scenario. Available accommodation space at a depression at 62 km gets filled in *c.* 3.4 ka, and then the system progrades further out into the basin. At 62 km, only a condensed surface of thin braidplain sediments is maintained up to *c.* 6.3 ka. In the case with constant sediment supply, the system instead starts aggrading which explains a much thicker topset layer. After 6 ka the sedimentation regime switches; there is much less available sediment and a fining upwards sequence is deposited when the system retrogrades under relative sea-level rise until sea level finally stabilizes and aggradation occurs. These last two phases are much less pronounced when sediment supply remains constant and the system is able to keep up with eustatic sea-level rise. Pseudo cores at 72 km (Fig. 9c) enforce the notion that the imposed switches in the sediment supply regime may strongly influence the stratigraphic architecture. A marine flooding surface develops as a consequence of sea-level rise, simply because the system now has little sediment to keep up with sea-level rise. The remaining accommodation space is filled up slowly by progradation over the rest of the Holocene.

Fjords, which become ice-free during the later part of the deglacial cycle, have a shorter period than the pulse of high sediment supply associated with an active retreating ice sheet in the drainage basin, affecting deposition. Those fjords with large river drainage basins draining towards the fjordhead will have a higher likelihood of a significant period of high sediment supply, either because the ice retreat from a large basin takes more time or because river sediment load and basin area are directly correlated.

Fig. 8. (*Continued*) (**a**) Detail of longitudinal stratigraphic fjord profile for a 15 ka simulation shows the sedimentary architecture of the offshore deposits. A thick unit is deposited under forced-regression conditions (FR) with a distinct fluvial topset unit that extends tens of kilometres into the present offshore. Sea level subsequently rose relatively fast and isostatic rebound tapers off, resulting in a finer grained aggradational unit (AGR). When sea level stabilizes, the sedimentary system switches into progradational mode (PGR). (**b**) Paleo-drainage network of incised fluvial channels mapped from HUNTEC and airgun seismic data in the Baie des Chaleurs, Canada (cruise data described by Praeg *et al.* 1987). Inset shows a Kelvin-Hughes 14.25 kHz bathymetric profile of incised channels at 30 m water depth, which are partly filled in with Holocene muddy sediments. (**c**) HUNTEC seismic profile at 110 m water depth from Baie des Chaleurs, Canada (Syvitski 1992) illustrates a similar sequence of ice proximal (IP) and several phases of FR sediments, grading to a much more transparent AGR.

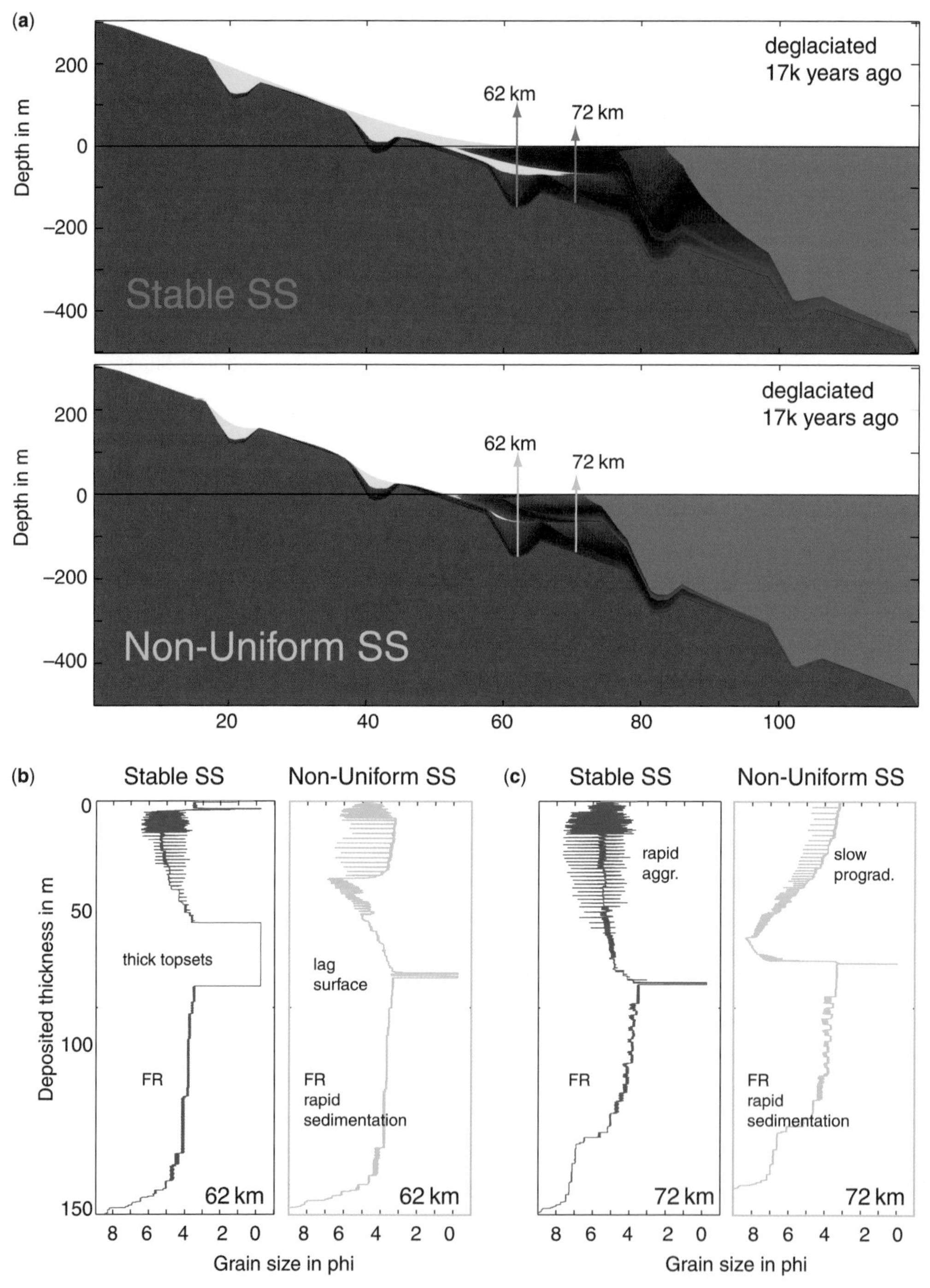

Fig. 9. (**a**) Longitudinal stratigraphic fjord profile for a 17 ka simulation with stable sediment supply and non-uniform sediment supply. High deglacial sediment supply lasted for 6 ka; the remaining 11 ka see postglacial reduced sediment supply. (**b**) Pseudo-cores at 62 km for stable sediment supply (in blue) compared to non-uniform sediment supply scenarios (in green). (**c**) Pseudo-cores at 72 km for stable sediment supply (in blue) compared to non-uniform sediment supply scenarios (in green).

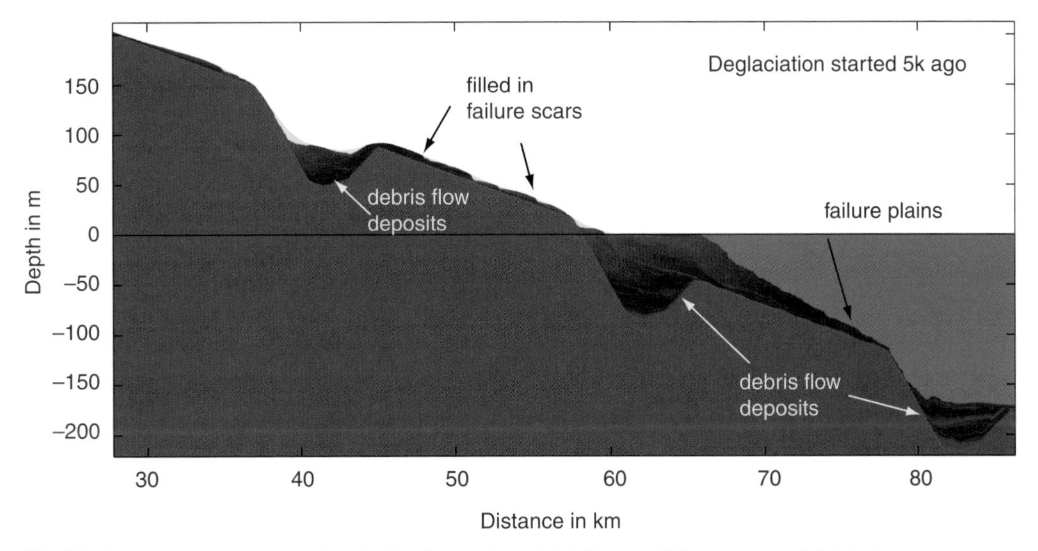

Fig. 10. Sensitivity test simulates 5 ka fjord sedimentation with failure, turbidity current and debris flow processes active. Many slope failures occur and transport coarse sediment down into the fjord depressions, causing coarser layers to occur in a finer matrix.

Discussion

The simulated stratigraphy may be found to be highly dependent on the timing of deglaciation and dominated by rapid uplift. Every simulation result will inherently be dependent on (1) boundary conditions and (2) model simplifications, however.

We ran a number of sensitivity tests to explore how robust the predicted sequences are under a variation of those boundary conditions. The ICE-5G model indicated that the selected fjords do not experience uniform uplift. Outer fjords experienced less ice load and were located a larger distance from the centre of the ice sheets; consequently, there is about 25% less uplift in those areas as compared to the fjordheads. Simulations of the endmember scenarios of 17 ka and 5 ka duration, with a slower uplift rate in the outer fjord (210 m) and rapid uplift in the fjordhead (275 m), show no apparent effect on the simulated stratigraphy. This can be explained by the fact that the uplift dominates the depositional patterns early after deglaciation, when the sedimentary wedge is mainly built up in the fjordhead region and hardly any sediment reaches the outer fjord regions. We can envision natural systems that receive sediment from several rivers along the length of a fjord and, in those cases, non-uniform uplift would affect the stacking. In our simplified model world, however, the predicted depositional sequences are relatively robust. Similarly, uplift rates for different fjords have a large range. Simulations with 25% less or 25% more uplift have been run, and stratal patterns remain comparable to the original scenarios. Eustatic sea-level rise will not outpace the rapid uplift even for the fjords that experience 'slow' uplift rates (as low as 94 m; see Table 1) and forced-regression units will develop in any case.

We emphasize that fjord stratigraphy is impacted by a range of processes previously neglected. One of the most important processes is the generation of failures and subsequent turbidity currents of debris flows. First-order sensitivity tests of the endmember scenarios, including these mass movement processes, do show extensive occurrence of failures and debris flow deposits (Fig. 10).

Other important but previously neglected processes include sediment remobilization in push moraines and sediment transport and deposition mechanisms linked to tidewater glaciers (Syvitski 1989; Boulton 1996; Ottesen & Dowdeswell 2006). We consider these processes as important components to be included in future fjord sedimentation models to make our approach applicable in tidewater glacier settings.

Conclusions

The analysis of glacio-isostatic adjustments on the scale of the ICE-5G model predictions (i.e. 10 min or even 1 degree) provides a regional trend of ice cover and topographic changes for Northern Hemisphere fjords over the last 21 ka. Whereas ICE-5G is well validated against *c.* 400 specific relative sea-level curves and locally improved when necessary, it is still an oversimplification if interested in precise glacio-isostatic adjustments on a fjord-to-fjord basis. Our numerical modelling of fjord sedimentary processes is equally simplified. We ignore tidewater glacier sedimentation and simplify complex sediment supply signals to isolate the effects of sea-level change and rebound on fjord stratigraphy. However, this generalized approach allows for the establishment of a number of overarching relations which help to identify what fjords are the most prominent modern analogues of forced-regressive systems.

ICE-5G predicts the age of deglaciation of a fjord region; if fjords are long enough, the outer fjord and fjordhead were separately considered. Fjords deglaciate in clusters at 17, 15, 13, 11, 9 and 5 ka BP. There is not necessarily a trend between deglaciation age and latitude; rather, the Fennoscandian Ice Sheet melt onset is relatively early and the Laurentide Ice Sheet remains stable for a longer time. On the large spatial and temporal scale of this study, the number of fjord regions becoming ice-free integrates warming climate conditions over longer time periods and cannot be directly correlated to shorter duration warm and cold periods. The 'average' outer fjord was deglaciated by 11 265 $\pm$ 6861 BP and the 'average' fjordhead is ice-free by 7794 $\pm$ 5772 BP.

The ICE-5G analysis showed a systematic difference in ice load at LGM for outer fjords and fjordheads. This difference is reflected in the average total uplift, which is 217 $\pm$ 96 m at the outer fjords compared to 283 $\pm$ 109 m at the fjordhead regions. Most uplift curves show exponential patterns with a short half-life of only *c.* 1–1.4 ka.

The interaction of eustatic sea-level change rates and the timing of deglaciation and uplift result in distinctly different fjord emergence curves and stratigraphy. Fjords that have been deglaciated since 11 ka have a simple emergence pattern: a rapid emergence during the first 3 ka and subsequent stable conditions, with balance between Late Holocene stable sea level and ongoing slow uplift. These are the fjords that are dominated by the initial uplift signal, and thus host the most unique 'forced-regression' records. Simulations show that the typical deposits are stretched out over a large distance of tens of kilometres, due to migration of the river mouth with the rapidly falling relative sea-level. River braidplain sands and deltaic sediments fill overdeepened valley sections and are uplifted and incised. Marine sediments are found as thick packages on the present-day onshore as well as in the fjord basin. This unit is potentially strongly affected by frequent failures and resulting turbidity currents and debris flows due to rapid sedimentation rates. Typical examples of such fjords are McBeth Fjord, Cambridge Fjord and Clyde Fjord on Baffin Island, Cook Inlet and Yakutat Bay in Alaska and Søndre Strømfjord in West Greenland.

Fjord deglaciating around 17 ka BP experience rapid emergence for only *c.* 2 ka; subsequently, there are two distinct periods during which sea-level rise outpaces the uplift. Fjords deglaciating at *c.* 15 ka BP experience rapid emergence for *c.* 3 ka and then have a continuous period of *c.* 4 ka when sea-level rise outpaces uplift. Simulated deposits show a forced-regression unit at their base resulting from the fast uplift rates due to glacio-isostatic adjustments. However, uplift dominates only for a short time and later eustatic sea-level rise outpaced the waning uplift rates. As a result, the simulated deposits show signs of rapid drowning, aggradation of sediments and, eventually, a progradational unit during the late Holocene. Some examples of these systems exist: St Lawrence Bay and Frosbisher Bay, Baffin Island. We used seismic data from the Baie des Chaleur, Canada, to illustrate the preservation of an incised river network that reflects the final phases of the forced-regressive unit. The incised surface is covered in most of the fjord with younger muddy deposits. These fjords have a much thicker, muddy and more complex fill and are less appropriate as a model of forced-regression sequences *per se*.

We found that there is a trend towards decreasing duration of active ice retreat for fjords, which are deglaciated later during the deglacial cycle. We explain this trend by the general thinning of the ice caps with time and the more rapid disintegration of the land-based ice towards the early Holocene. The trend implies that fjords that deglaciated early after the LGM are affected by ice in their hinterlands for a much more significant period of time than fjords that are deglaciated relatively late during the deglaciation. This implies that fjords with early onset of deglaciation ought to have a longer period of high sediment supply associated with the deglacial period. If forced-regression units may be buried with later depositional sequences, they are still relatively well developed and thick. Fjords that deglaciate relatively late are more likely to be solely influenced by postglacial conditions and thus lower sediment supply. In these younger fjords the best-developed forced-regression system tracts are predicted to occur in those with a relatively large drainage basin with local mountain glaciers drains towards the fjordhead, hence ensuring persisting high sediment fluxes.

All systems have a completely different fill signature than a simple Late Quaternary drowning system. Fjord systems are uniquely influenced by their initial syndepositional uplift, and provide us with an opportunity to study forced-regressive systems up close and with good process control.

We thank reviewers, A. Jennings and J. Dowdeswell, for constructive comments and the editor J. Howe for his support. This research project was supported by ConocoPhilips; we thank J. Suter and M. Hoffmann for their continued interest in the uniqueness of high-latitude sedimentary systems.

References

Aarseth, I. 1997. Western Norwegian fjord sediments: age, volume, stratigraphy, and role as temporary depository during glacial cycles. *Marine Geology*, **143**, 39–53.

Anderson, R. S., Molnar, P. & Kessler, M. A. 2006. Features of glacial valley profiles simply explained. *Journal of Geophysicsical Research*, **111**, F01004, doi: 10.1029/2005JF000344.

Andrews, J. T. 1987. Late Quaternary marine sediment accumulation in fiord-shelf-deep-sea transects, Baffin Island to Baffin Bay. *Quaternary Science Reviews*, **6**, 231–243.

Boulton, G. S. 1990. Sedimentary and sea level changes during glacial cycles and their control on glacimarine facies architecture. *In*: Dowdeswell, J. A. & Scourse, J. D. (eds) *Glacimarine Environments: Processes and Sediments*. Geological Society, Special Publications, **53**, 15–52.

Boulton, G. S. 1996. Theory of glacial erosion, transport and deposition as a consequence of subglacial sediment deformation. *Journal of Glaciology*, **42**, 43–62.

Briner, J. P., Miller, G. M., Davis, T. P. & Finkel, R. C. 2006. Cosmogenic radionuclides from fiord landscapes support differential erosion by overriding ice sheets. *Geological Society of America Bulletin*, **118**, 406–420.

Briner, J. P., Overeem, I., Miller, G. M. & Finkel, R. C. 2007. The deglaciation of Clyde Inlet, northeastern Baffin Island, Arctic Canada. *Journal of Quaternary Science*, **22**, 223–232.

Church, M. & Ryder, J. M. 1972. Paraglacial sedimentation: a consideration of fluvial processes conditioned by glaciation. *Geological Society of America Bulletin*, **83**, 3059–3071.

Church, M. & Slaymaker, O. 1989. Disequilibrium of Holocene sediment yield in glaciated British Columbia. *Nature*, **337**, 452–454.

Corner, G. 2007. A transgressive–regressive model of fjord-valley fill: stratigraphy, facies and depositional controls. *In*: Dalrymple, R. W., Dale, A. E. & Tillman, R. W. (eds) *Incised Valleys in Time and Space*. SEPM Special Publications, **85**, 161–178.

Dowdeswell, E. K. & Andrews, J. T. 1995. The fjords of Baffin Island: description and classification. *In*: Andrews, J. T. (ed.) *Eastern Canadian Arctic, Baffin Bay and West Greenland, Quaternary Environments*. Allen and Unwin, Boston, 93–123.

Dyke, A. S. & Peltier, W. R. 2000. Forms, response times and variability of relative sea-level curves, glaciated North America. *Geomorphology*, **32**, 315–333, doi: 10.1016/S0169-555X(99)00102-6.

Eilertsen, R., Corner, G. D., Aasheim, O., Andreassen, K., Kristoffersen, Y. & Ygstborg, H. 2007. Valley-fill stratigraphy and evolution of the Malselv fjord-valley, northern Norway. *In*: Dalrymple, R. W., Dale, A. E. & Tillman, R. W. (eds) *Incised Valleys in Time and Space*. SEPM Special Publications, **85**, 179–195.

Guymon, G. L. 1974. Regional sediment analysis of Alaskan Streams. *Journal of Hydraulics Division, Proceedings of the American Society of Civil Engineers*, **100**, 41–51.

Hallet, B., Hunter, L. & Bogen, J. 1996. Rates of erosion and sediment evacuation by glaciers: a review of field data and their implications. *Global and Planetary Change*, **12**, 213–235.

Hansen, L. 2004. Deltaic infill of a deglaciated arctic fjord, East Greenland: sedimentary facies and sequence stratigraphy. *Journal of Sedimentary Research*, **74**, 422–437.

Hasholt, B., Bobrovitskaya, N. N., Bogen, J., McNamara, J., Mernild, S. H., Milburn, D. & Walling, D. 2005. Sediment transport to the Arctic Ocean and adjoining cold oceans. *Nordic Hydrology*, **37**, 413–432.

Helle, S. K. 2004. Sequence stratigraphy in a marine moraine at the head of Hardangerfjorden, western Norway: evidence for a high-frequency relative sea-level cycle. *Sedimentary Geology*, **164**, 251–281.

Hjelstuen, B. O., Haflidason, H., Sejrup, H. P. & Lysa, A. 2009. Sedimentary processes and depositional environments in glaciated fjord systems – evidence from Nordfjord, Norway. *Marine Geology*, **258**, 88–99, doi: 10.1016/j.margeo.2008.11.0.

Hutton, E. W. H. & Syvitski, J. P. M. 2004. Advances in the numerical modeling of sediment failure during the development of a continental margin. *Marine Geology*, **203**, 367–380, doi: 10.1016/S0025-3227(03)00316-5.

Hutton, E. W. H. & Syvitski, J. P. M. 2008. Sedflux 2.0: an advanced process-response model that generates three-dimensional stratigraphy. *Computers & Geosciences*, **34**, 1319–1337, doi: 10.1016/j.cageo.2008.02.013.

Johnsen, S. J., Dahl-Jensen, D., Dansgaard, W. & Gundestrup, N. 1995. Greenland palaeotemperatures derived from GRIP bore hole temperature and ice isotope profiles. *Tellus*, **47B**, 624–629.

Johnson, D. W. 1915. The nature and origin of fjords. *Science*, **41**, 537–543.

Jouet, G., Hutton, E. W. H., Syvitski, J. P. M. & Berne, S. 2008. Response of the Rhone deltaic margin to loading and subsidence during the last climatic cycle. *Computers & Geosciences*, **34**, 1338–1357, doi: 10.1016/j.cageo.2008.02.003.

Kessler, M. A., Anderson, R. S. & Briner, J. P. 2008. Fjord insertion into continental margins driven by topographic steering o ice. *Nature Geoscience*, 2008–06, **1**, 365–369. Available online at: http://dx.doi.org/10.1038/ngeo201.

Kleman, J. & Glasser, N. F. 2007. Subglacial thermal organization (STO) of ice sheets. *Quaternary Science Reviews*, **26**, 585–597.

Kubo, Y., Syvitski, J. P. M., Hutton, E. W. & Kettner, A. J. 2006. Inverse modeling of post Last Glacial Maximum transgressive sedimentation using 2D-sedflux: application to the northern Adriatic Sea. *Marine Geology*, **234**, 233–243, doi: 10.1016/j.margeo.2006.09.011.

MacGregor, K. R., Anderson, R. S., Anderson, S. P. & Waddington, E. D. 2000. Numerical simulations of glacial-valley longitudinal profile evolution. *Geology*, **28**, 1031–1034.

Milne, G. A. & Mitrovica, J. X. 2008. Searching for eustasy in deglacial sea-level histories. *Quaternary Science Reviews*, doi: 10.1016/j.quascirev.2008.08.018.

Ottesen, D. & Dowdeswell, J. A. 2006. Assemblages of submarine landforms produced by tidewater glaciers in Svalbard. *Journal of Geophysical Research*, **111**, F01016, doi: 10.1029/2005JF000330.

Overeem, I. & Syvitski, J. P. M. 2008. *Proceedings of Symposium Changing Sediment Supply in Arctic Rivers. Sediment Dynamics in Changing Environments*, Christchurch, New Zealand, December 2008. IAHS Press, Wallingford, **325**.

Overeem, I., Syvitski, J. P. M., Hutton, E. W. H. & Kettner, A. J. 2005. Stratigraphic variability due to uncertainty in model boundary conditions: a case-study of the New Jersey Shelf over the last 40,000 years. *Marine Geology*, **224**, 23–41, doi: 10.1016/j.margeo.2005.06.044.

Peckham, S. D. 2008. A new method for estimating suspended sediment concentrations and deposition rates from satellite imagery based on the physics of plumes. *Computers & Geosciences*, **34**, 1198–1222.

Peltier, W. R. 2002. Global glacial isostatic adjustment: paleogeodetic and space-geodetic tests of the ICE-4G (VM2) model. *Journal of Quaternary Science*, **17**, 491–510.

Peltier, W. R. 2004. Global glacial isostasy and the surface of the ice-age Earth: the ICE-5G model and GRACE. *Annual Review of Earth and Planetary Science*, **32**, 111–149.

Powell, R. D. & Elverhøi, A. 1989. Modern glacimarine environments: glacial and marine controls of modern lithofacies and biofacies. *Marine Geology*, **85**, 101–418.

Powell, R. D. & Cooper, J. M. 2002. A glacial sequence stratigraphic model for temperate, glaciated continental shelves. *In*: Dowdeswell, J. A. & Cofaigh, C. O. (eds) *Glacier Influenced Sedimentation on High-latitude Continental Margins*. Geological Society, London, Special Publications, **203**, 215–244.

Praeg, D. B., Syvitski, J. P. M., Schafer, C. T., Johnston, B. L. & Hackett, D. W. 1987. CSS Dawson 86-016 Cruise Report. Geological Survey of Canada, Open File Report 1412.

Sejrup, H. P., King, E., Aarseth, I., Haflidason, H. & Elverhøi, A. 1996. Quaternary erosion and depositional processes: Western Norwegian fjords, Norwegian Channel and North Sea Fan. Geological Society, London, Special Publications, **117**, 187–202.

Stravers, J. A. & Syvitski, J. P. M. 1991. Early Holocene land-sea correlations and deglacial evolution of the Cambridge Fjord basin, Northern Baffin Island. *Quaternary Research*, **35**, 72–90.

Syvitski, J. P. M. 1989. On the deposition of sediment within glacier-influenced fjords: oceanographic controls. *Marine Geology*, **85**, 301–329.

Syvitski, J. P. M. 1991. Towards an understanding of sediment deposition on glaciated continental shelves. *Continental Shelf Research*, **11**, 897–937.

Syvitski, J. P. M. 1992. Marine geology of Baie des Chaleurs. *Geographie physique et Quaternaire*, **46**, 331–348.

Syvitski, J. P. M. & Daughney, S. 1992. Delta2: delta progradation and basin filling. *Computers & Geosciences*, **18**, 839–897, doi: 10.1016/0098-3004(92)90028-P.

Syvitski, J. P. M. & Shaw, J. 1995. Sedimentology and geomorphology of fjords. *In*: Perillo, G. M. E. (ed.) *Geomorphology and Sedimentology of Estuaries*. Developments in Sedimentology, **53**, 113–178.

Syvitski, J. P. M. & Hutton, E. W. H. 2001. 2D SEDFLUX 1.0C: an advanced process-response numerical model for the fill of marine sedimentary basins. *Computers & Geosciences*, **27**, 731–753, doi: 10.1016/S0098-3004(00)00139-4.

Syvitski, J. P. M., Burrell, D. C. & Skei, J. M. 1986. *Fjords, Processes and Products*. Springer-Verlag, New York Inc.

Tarasov, L. & Peltier, W. R. 2002. Greenland Glacial history and local geodynamic consequences. *Geophysical Journal International*, **150**, 198–229.

Tushingham, A. M. & Peltier, W. R. 1992. Validation of the ICE-3G model of Wurm – Wisconsin deglaciation using a global database of relative sea level histories. *Journal of Geophysical Research*, **97**, 3285–3304.

Canadian west coast fjords and inlets of the NE Pacific Ocean as depositional archives

AUDREY DALLIMORE[1,2]* & DANIELLE G. JMIEFF[1]

[1]*School of Environment and Sustainability, Royal Roads University, Victoria, BC, Canada V9B 5Y2*

[2]*Geological Survey of Canada-Pacific, Institute of Ocean Sciences, Sidney, BC, Canada V8L 4B2*

**Corresponding author (e-mail: Audrey.dallimore@royalroads.ca)*

Abstract: The west coast of Canada is one of the major fjord coastlines of the world, hosting about 150 fjords which are locally known as inlets. Much of the coastline remains remote without extensive research, and this paper summarizes what is known of the Canadian west coast fjord environments. Two fjord regimes are recognized along the British Columbia coastline. Mainland fjords drain high mountains and ice fields with sediment input from snowmelt and glacier runoff in spring and summer. By contrast, the inlets on Vancouver Island are in a milder marine climate, and sediment input occurs mostly during heavy rains of autumn and winter. Due to unique oceanographic conditions and shallow sills at the mouth of some Vancouver Island inlets, anoxic bottom waters exist which allow the preservation of annually laminated sediments. Studies of these annually laminated sediment archives over the past decade, including two international drill ship investigations, have characterized the deglacial, sea-level and palaeoenvironmental history of the British Columbia coastal area. These coastal depositional archives have also advanced our knowledge of the cyclical nature of the NE Pacific ocean and climate system, as well as given evidence of infrequent, yet significant abrupt changes that characterize it.

This paper summarizes the palaeoenvironmental history of the northeastern Pacific Ocean fjords in the province of British Columbia, Canada, and gives an overview of the marine geological and oceanographic research to date which characterizes these fjord environments. The Pacific coast of British Columbia (BC) has been glaciated many times, and is indented everywhere by fjords which are referred to locally as sounds, inlets or channels but are most commonly known in the scientific literature as inlets; they will therefore be referred to as such in this paper (Fig. 1) (Thomson 1981).

Late Pleistocene/Holocene postglacial sedimentation and sea-level changes in BC coastal inlets is of interest to shore zone engineers and planners involved in future coastal and offshore development (Barrie & Conway 2002*a*, *b*) as well as ecologists and archaeologists with interests in coastal fisheries, ecosystems and cultures (Josenhans *et al.* 1995, 1997; Fedje & Josenhans 2000; Hetherington *et al.* 2003; McKechnie 2005; Wright *et al.* 2005). An interpretation of palaeoenvironments of BC coastal inlets, particularly with respect to rapidly changing sea level after deglaciation, also enhances our understanding of the early migration routes of early humans into North America (Mandryk *et al.* 2001). Precise relative sea-level curves for the Holocene have been interpreted from inlet sedimentary records. They yield valuable information for geophysical modelling of crustal response and mantle rheology in this seismically active area of the Cascadia subduction zone, since geodetic observations measuring crustal strain leading to earthquakes must be corrected for isostatic adjustments of the crust (Wang *et al.* 2001; James *et al.* 2002, 2005, 2009).

In recent years, research in BC inlets has focused on the proxy records of past terrestrial, oceanic and atmospheric conditions of the northeastern Pacific Ocean, preserved in annually laminated marine sediments which are found in several anoxic inlets of the BC coast. These natural archives contain key information on the causes and dynamics, including rates and critical thresholds, of climate change in timescales of human interest (Chang *et al.* 2003; Hay *et al.* 2003, 2007, 2009; Chang & Patterson 2005; Dallimore *et al.* 2005, 2009*a*; Wake *et al.* 2006; Ivanochko *et al.* 2008*a*, *b*). Defining past climate variability is essential to put modern climate change detection in a proper long-term context. Understanding the forcing mechanisms of these changes is equally essential to help anticipate their occurrences and consequences, thus guiding present and future climate change adaptation strategies in

From: Howe, J. A., Austin, W. E. N., Forwick, M. & Paetzel, M. (eds) *Fjord Systems and Archives.* Geological Society, London, Special Publications, **344**, 143–162.
DOI: 10.1144/SP344.12 0305-8719/10/$15.00

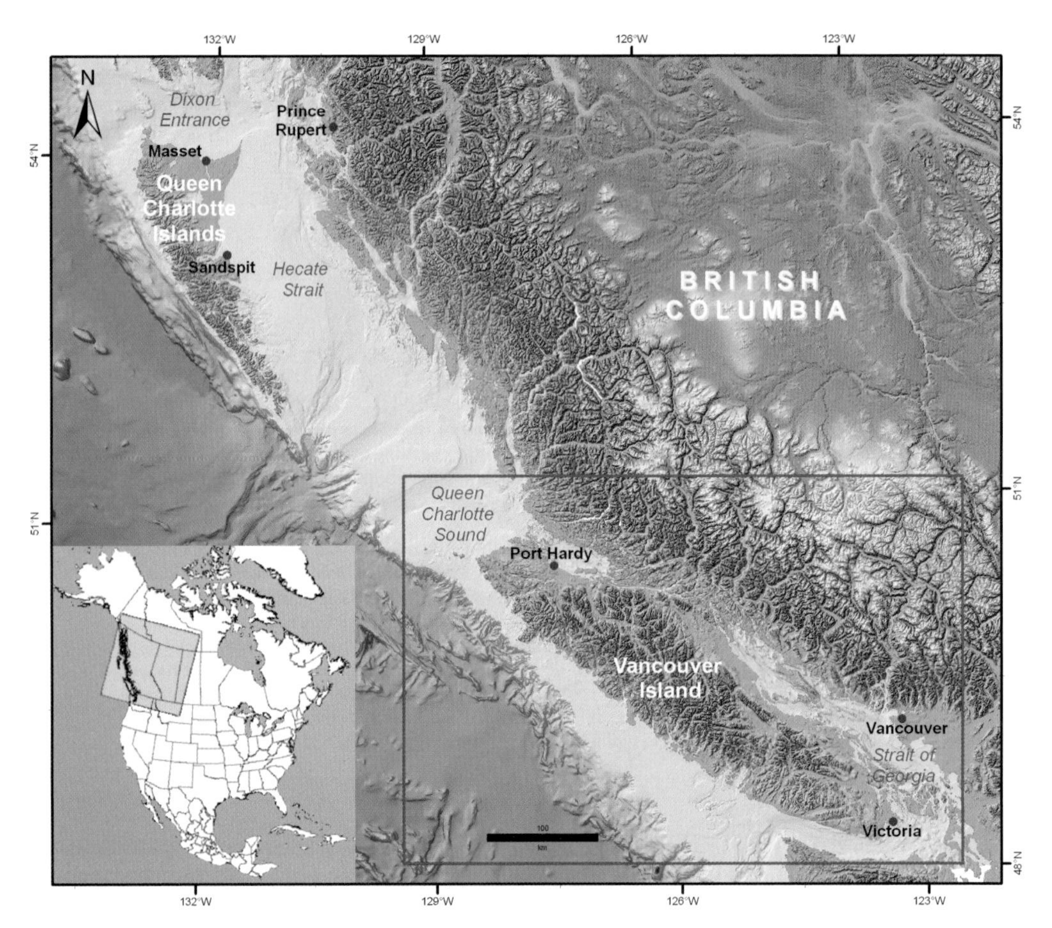

Fig. 1. Map of coastal British Columbia showing mainland coastline and the west coast of Vancouver Island deeply incised with inlets. Bathymetry of offshore areas shows remnant glaciofluvial channels from the Fraser Glaciation.

Canada (Natural Resources Canada 2007*a*; Lemmen *et al.* 2008).

Inlets of British Columbia

British Columbia, Canada has one of the longest fjord coastlines in the world which consists of about 150 inlets. The inlets are most extensive on the mainland coast of BC (Figs 1 & 2) with mountainous relief up to 3300 m. Most of the inlets in the south of the province have been studied to some degree, but little is comprehensively known of these inlets on the wild and mostly unpopulated more northern reaches of the BC coast. Typically, the inlets are U-shaped with a river at the head, have a high tidal range of >4 m and host high primary productivity which is related to nutrient delivery from river input as well as upwelled ocean waters during the summer months. Grey glacial clay is the dominant sediment of the mainland inlets, with soft muddy brown organic sediments most common in the inlets of the west coast of Vancouver Island (Pickard 1961, 1963, 1975). Some of the inlets of BC contain remnant alpine glaciers from the last glaciation (classified as Stage 3 following Syvitski & Shaw 1995) and many are now completely deglaciated (Stage 4).

Sills or a series of sills are important features of many of the inlets and represent the terminus of glaciers (Fig. 2). At the mouth of the central coast Seymour-Belize inlet complex, for example, a shallow tidal rapid of about 30 m depth (the Nakwakto Rapids) is passable only at slack tide and runs a spectacular 16 knots (8 m s^{-1}) at flood tide, making it the swiftest tidal channel in the world. Once through this treacherous channel, the landscape opens into several hundred kilometres of mainland inlets up to 650 m deep, behind a series of sills as shallow as 7 m (Fig. 2) (Thomson 1981; Nicklen 2006).

In general, BC inlets are considered to be partially mixed with river runoff influencing the

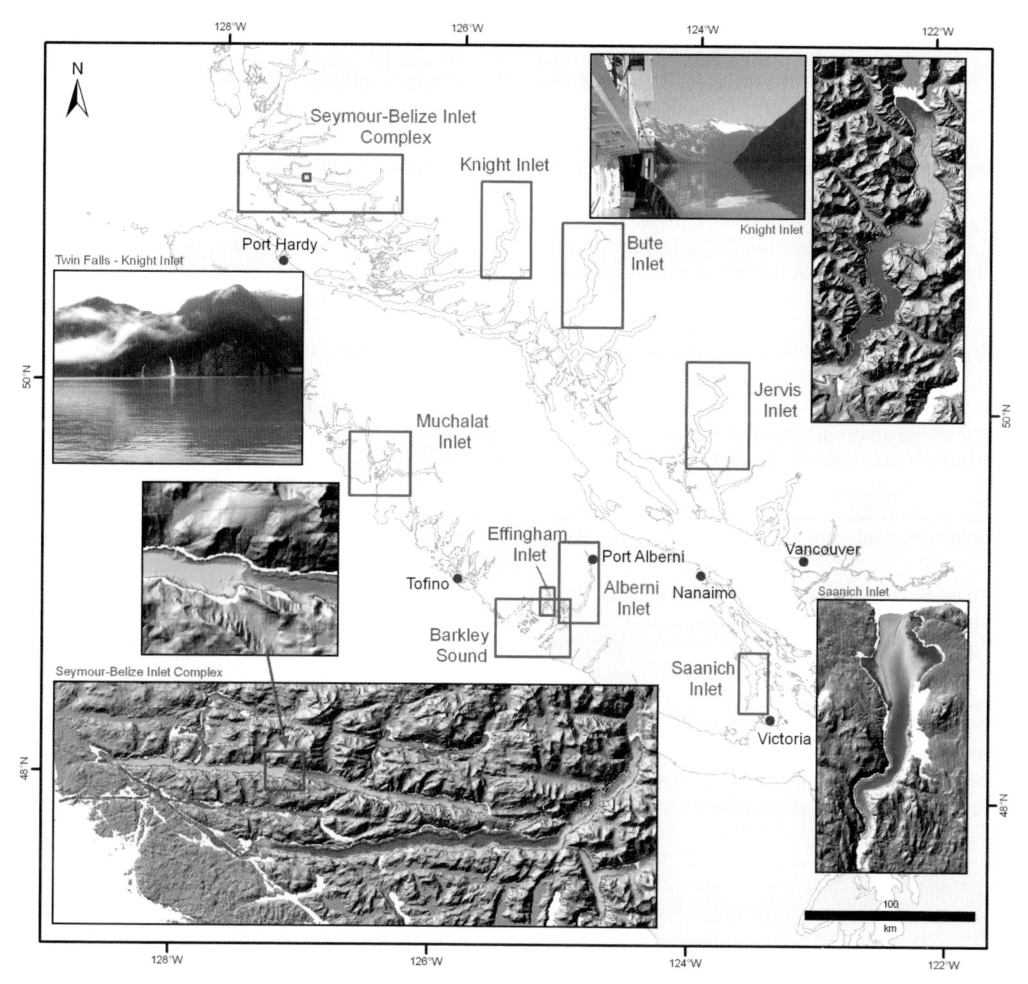

Fig. 2. Map of southern coastal British Columbia showing locations and inlets referred to in the text, with some multibeam bathymetry of selected inlets. Colour ramp shows: shallow depths <50 m in the red end of the scale to depths +200 m at the blue end. Turbidite channels are shown on the bottom of Knight Inlet and an inset of the Seymour-Belize inlet complex shows a glacial moraine indicating a temporary stillstand of ice retreat of the Fraser Glaciation, an example of one of many ice-retreat and sea-level lowstand features now submarine along the BC coast and inlets. The photo insets are twin falls from Knight Inlet, and the Klinaklini Glacier at the head of Knight Inlet. Bathymetry for Sasnich Inlet shows a typical sill at the mouth of 80 m depth.

strength and timing of the estuarine flow and therefore the deep water characteristics (Thomson 1981). The humid, mesothermic maritime climate of BC is the warmest in Canada. Abundant winter precipitation falls mostly as rain on the coastal area, feeding river runoff and contributing to snow accumulation at high altitudes. Mean annual precipitation in the BC coastal area ranges from 1000 to 4000 mm a^{-1} and can be as much as 10 000 mm in some areas (Pojar *et al.* 1991; Hebda 1997; Lemmen *et al.* 2008).

In his pioneering studies of BC inlets, Pickard (1961, 1963) noted differences in the character of the mainland and Vancouver Island west coast inlets based on two general hydrographic regimes. In the inlets of Vancouver Island, peak runoff is in the winter from November to January and is related to high levels of precipitation at that time of year. In contrast, on the mainland where many rivers drain mountainous snow capped peaks, the maximum runoff occurs in the summer from June to September due to snowmelt. The peak summer discharge in the mainland inlets can be up to ten times that of fluvial discharge during the rainy winter season (Pickard 1961, 1963; Thomson 1981).

These differences in hydrographic regimes and therefore estuarine flow characteristics are the expressions of inlet location and physiography. The inlets on the outer coast of Vancouver Island have small drainage basins with no ice fields in their upper mountainous reaches, receive little or no snowfall, have high primary productivity and have terrigenous sediment input related to rainstorm events. Sills in the Vancouver Island inlets are generally shallower than on the mainland. These inlets are less lengthy and have a tide range of 4–5 m. Low values of dissolved oxygen are more common in Vancouver Island inlets than on the mainland because of comparatively reduced river runoff. A few known anoxic inlets exist on Vancouver Island where bottom waters are seasonally or permanently anoxic behind shallow sills which restrict estuarine circulation and where high rates of decomposition of organic matter consumes the available oxygen (Pickard 1961, 1963, 1975; Syvitski & Shaw 1995).

In all BC inlets, density, salinity and temperature are horizontally stratified but oxygen distributions are not; oxygen distribution is attributed to factors such as sill depth, photosynthetic production in the surface layers and the level of oxygen consumption at depth (Pickard 1961; Dallimore *et al.* 2005). The deep water characteristics of Vancouver Island inlets with the shallowest sills show the greatest differences in water properties from the outside open ocean. In general, the Vancouver Island inlet sediments are brown organic muds which contain shells and wood. In contrast, the mainland inlets are longer and deeper. At the head of most mainland inlets, rivers exhibit a large variation year to year in monthly mean fluvial runoff since they drain areas of mountain glaciers. In general, the sediments in mainland inlets are grey in colour due to glacial flour; on Vancouver Island the sediments are predominantly brown (Pickard 1961, 1963; Syvitski & Shaw 1995).

History of BC inlet research

A thriving subsidence marine-based culture of First Nations peoples has populated the BC coast and inlets for at least the past 5000 years and probably longer (McMillan 1999; McMillan & Yellowhorn 2003). British Columbia's present population is about 4.25 million, with 75% of the population living in the metropolitan Vancouver–Lower Mainland region and 10% in the city of Victoria on Vancouver Island (Fig. 1). The remaining 15% of the population lives in rural and remote areas of the province in smaller towns, with First Nations communities spread along the coast and in the interior of the province (Lemmen *et al.* 2008).

The first documented marine research carried out on the BC coast was by the early European explorers, including Juan Perez in 1774, James Cook in 1778 and George Vancouver in 1792, who charted the coastline, sampled the seafloor and conducted gravity and magnetic surveys. Accurate charts of the coast were made soon after in the years 1827–1860 by British hydrographers (Thomson 1981). Oceanographic research along the BC coast began in earnest in 1908 with the establishment of the Fisheries Research Board at the Marine Biological Station in Nanaimo on Vancouver Island. In the central coast region, the oceanography of the Seymour-Belize inlet complex was investigated in 1938 (Thompson & Barkey 1938). Dr J. P. Tully conducted oceanographic work in several BC inlets from the 1930s. His classic BC inlet study of Alberni Inlet on the west coast of Vancouver Island was published in 1949 (Levson 1992).

In 1951, in collaboration with Dr Tully of the Marine Biological Station, Dr George Pickard of the Institute of Oceanography at the University of British Columbia initiated a series of inlet cruises. For almost 50 years, students obtained and analysed water samples for temperature, salinity and dissolved oxygen in many inlets, straits, passages and channels on the entire BC coast. The dataset stands today as the longest oceanographic instrument record that spans the entire length of the mainland coast and Vancouver Island. Pickard's early work was summarized in 1961 in a publication characterizing the mainland coast inlet. It was soon followed by a companion publication characterizing the inlets of Vancouver Island (Pickard 1961, 1963).

In 1978, Pacific marine oceanographic and geosciences research was given a home at the new federal Institute of Ocean Sciences (IOS) in Sidney, BC, on the shores of Saanich Inlet on Vancouver Island. This institute houses oceanographers under Fisheries and Oceans, Canada; marine geoscientists and seismologists with the Geological Survey of Canada-Pacific under Natural Resources Canada; hydrographers under the Canadian Hydrographic Service; and science research vessels operated by the Canadian Coast Guard Service. Inlet research launched from IOS over the years has provided extensive datasets of oceanographic observations from several inlets, most particularly Knight, Bute and Jervis Inlets on the mainland coast and Saanich, Alberni and Effingham Inlets on Vancouver Island (Figs 1 & 2). Modern inlet oceanographic transects have been maintained in the Vancouver Island inlets of Saanich and Muchalat since 1950 and in Effingham Inlet on the outer coast since 1995. On the mainland coast, the Vancouver harbour inlet areas of Indian Arm and Howe Sound as well as the more northern Jervis, Bute and Knight Inlets off the Strait of Georgia have been monitored since 1950. On the central coast, an oceanographic transect in the Seymour–Belize Inlet complex has been monitored sporadically since 2000, while most of the distant northern inlets are beyond the reach of present annual research cruises. All of the IOS datasets and most of the early inlet and coastal water surveys are now preserved in digital form in the IOS data archive (Fisheries and Oceans Canada-Pacific Region 2008*a*, *b*).

Multibeam bathymetry data have been collected on the BC coast by the Canadian Hydrographic Service in co-operation with the Geological Survey of Canada-Pacific (Natural Resources Canada) since the first shallow water survey in 1999. Since then, 12 000 km^2 of multibeam swath bathymetry coverage of the Pacific coast of Canada in the province of BC have been obtained. Surveys were undertaken from the CCGS *Otter Bay* for shallow water depths <50 m using a hull-mounted Kongsberg-Simrad EM3002 system, which operates at a frequency of 300 kHz utilizing 121–135 beams. Surveys were also undertaken from the CCGS *Vector* in water depths >50 m, using a hull-mounted Kongsberg-Simrad EM1002 system operating at a frequency of 95 kHz with 127 beams.

Multibeam surveying takes place continuously in the summer months during relatively calm sea conditions and survey coverage of the BC coast and shelf area has progressed on a priority basis, depending on navigation hazards and research objectives. To date, the Straits of Georgia and Juan de Fuca between Vancouver Island and the mainland have been fully surveyed and many of the mainland coast inlets now have full multibeam coverage maps. The Vancouver Island inlets of Effingham and Muchalat as well as Clayoquote Sound have also been surveyed with multibeam swath bathymetry. Much of the multibeam imagery is available to the public once it has been processed, via the Fisheries and Oceans Canada Geoportal website (Fisheries and Oceans Canada 2008*c*).

A permanent cabled subsea observatory called VENUS (the Victoria Experimental Network Under the Sea) run by the University of Victoria on Vancouver Island was installed in Saanich Inlet in 2006, just a few hundred metres off the main dock of the Institute of Ocean Sciences. In the following years more nodes of the cable will be installed and presently the VENUS cable has two additional research nodes installed in the strait of Georgia between Vancouver Island and the mainland. VENUS provides real-time, long-term ecosystem science capability utilizing remotely operated water property sensors (current, temperature and depth profilers or CTDs), acoustic profilers, cameras and sediment traps and has the capability for interactive user controlled experiments. Real-time data over the internet is available from fibre-optics cables connecting the remote research nodes to the shore station. VENUS research directions in Saanich Inlet include water mass behaviour, organism responses to hypoxia, ocean change, zooplankton dynamics, microturbulence and nutrient resuspension (University of Victoria 2009).

Geologic setting of BC coastal inlets

The BC coast lies along a convergent continental margin flanked on the east by rugged mountain ranges of more than 3000 m in elevation (Fig. 1). The continental shelf of BC consists of two sedimentary basins: the enclosed Georgia Basin to the south and the semi-enclosed Queen Charlotte Basin to the north. Many areas of the present nearshore and shelf were subaerial in the late Pleistocene and early Holocene. Vivid channel, delta and other bedform features are preserved underwater today in many areas and can be identified in recently acquired multibeam bathymetric imaging (Fig. 2) (Barrie *et al.* in press).

The present-day Canadian Cordillera which comprises the BC coast area was formed during Mesozoic and Cenozoic time as multiple terranes were accreted to the North American craton during an oblique convergence of the North American and Pacific Ocean lithosphere. The strong NW–SE trend of the present-day Canadian Cordillera is a result of this accretion (Clague 1989*c*; Yorath 1990; Wheeler & McFeely 1991; Wheeler *et al.* 1991; Gabrielse & Yorath 1992). Over the last 40 million years, the Juan de Fuca plate has been subducting beneath the North American plate at the rate of about 45 mm a^{-1}, which produces considerable seismic shaking in the coastal and offshore regions of the province (Riddihough & Hyndman 1991). The northern part of the province is also seismically active as the Pacific and North American plates slide past each other along the active Queen Charlotte fault at a rate of about 58 mm a^{-1} (Fig. 3) (Rogers 1988; Atwater *et al.* 1995*a*, *b*; Hyndman 1995). Most of the inlets of BC exhibit subaqueous slope failures which are a consequence of high sedimentation rates and seismic events related to tectonic activity (Syvitski & Shaw 1995).

Hundreds of small earthquakes are recorded each year in BC and earthquakes up to magnitude 7 are frequent in the modern instrument record (Clague *et al.* 2006). The epicentre of most of the earthquakes is shallow and they occur along the plate boundaries; however, prehistoric strong earthquakes of magnitude 8+ in the region have also been identified from the palaeoseismic record (Clague *et al.* 1982; Rogers 1988; Atwater *et al.*

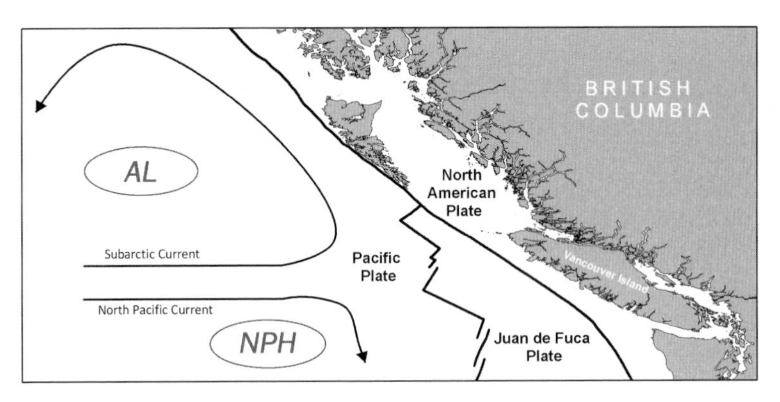

Fig. 3. Location map showing the tectonic regime of coastal BC, major current gyres as well as the location of the AL and NPH atmospheric gyres. These atmospheric gyres drive the coastal BC ocean dynamics and the prevailing spring and summer upwelling of deep ocean waters onto the continental shelf.

1995*a*, *b*; Hyndman 1995; Atwater & Hemphill-Haley 1996, 1997; Nelson *et al.* 2000, 2003; Kelsey *et al.* 2002, 2005; Goldfinger *et al.* 2003*a*, *b*). The last historical great crustal earthquake in Canada of $M = 8.1$ occurred along a 500 km length of the Queen Charlotte fault on 22 August 1946. It is known from the palaeoseismic record, and also from historical records in Japan of the resulting pan-Pacific tsunamis wave, that the most recent great subduction zone earthquake occurred in 1700 AD (Clague *et al.* 2000, 2006). The frequency of these major subduction zone earthquake events in the geologic record seems to be *c.* 1 every 300–500 years, with the maximum time between major events about 1000 years (Blais-Stevens & Clague 2001; Blais-Stevens *et al.* 2001; Dallimore *et al.* 2005, 2009*b*; Blais-Stevens *et al.* 2009).

According to interpretations of the palaeoseismic record, large tsunamis are known to have struck the BC coast *c.* 7 times in the past 3500 years, although there has only been one major historical tsunami strike on the BC coast (Clague *et al.* 2006). Many smaller harmless tsunamis are regularly registered on tide gauges along the BC coast with waves less than 1 m in height. These seismogenic tsunamis can originate along the Cascadia subduction zone, or from large earthquakes originating elsewhere in the Pacific 'Ring of Fire'. This was the case on 27 March 1964, when the Vancouver Island town of Port Alberni at the head of Alberni Inlet on the west coast of Vancouver Island was inundated by three wave trains in the early morning hours. This tsunami wave was the result of the M 9.2 earthquake which originated in Alaska. The wave train travelled 1700 km to the BC coast in 4 hours. It then underwent wave amplification in Alberni Inlet, where wave height grew to 7 m at the inlet head compared to 0.6 m at the mouth (Clague *et al.* 2000, 2006). Although extensive damage affected the shoreline areas of the town, there were fortunately no deaths.

Tsunami waves remain very real risks along the BC coast and emergency preparedness measures are progressively being implemented by the BC Provincial Emergency Preparedness Program. Awareness is increasing through the outreach activities of researchers at the Geological Survey of Canada (Natural Resources Canada 2009*a*). Computer simulations of tsunamis waves reaching the BC coastline and palaeoseismic evidence show that tsunamis have a run-up potential of up to 15 m in height on the west coast Vancouver Island inlets and along the Queen Charlotte Islands to the north (Clague *et al.* 2006; Fisheries and Oceans Canada 2008*d*). The first wave from a Cascadia subduction zone earthquake is modelled to reach the inlets of the outer coast of Vancouver Island in less than 30 min. However, the major population centre of BC in the Vancouver–Lower Mainland area will not be at such risk, since a tsunami wave would attenuate and lose height as it travelled up the Strait of Juan de Fuca and into the Strait of Georgia (Fisheries and Oceans Canada 2008*d*). There have been other smaller tsunamis, both historically and in the palaeo record, generated locally from crustal earthquakes, landslides and rockfalls in inlets along the BC coast. These potential waves are, and will continue to be, a threat to coastal infrastructure and communities (Prior *et al.* 1982; Bornhold *et al.* 2007).

Glacial history

Canada's Pacific coast has been glaciated repeatedly, with valley glaciers flowing seaward carving out the many inlets along the coastline. Geological evidence of the most recent Fraser Glaciation (*c.* 30 000–11 000 a BP) is found everywhere along the coastline. The Cordilleran Ice Sheet covered British Columbia, the southern part of Yukon Territory and southern Alaska during the Fraser Glaciation. At the glacial maximum (*c.* 15 ^{14}C ka BP; 18 cal ka BP), the ice sheet was up to 2000 m thick and extended to the shelf edge off the west coast of Vancouver Island (Clague 1981, 1983, 1989*a*, *b*, 1994; Josenhans *et al.* 1995; James *et al.* 2000; Clague & James 2002; Booth *et al.* 2004). Ice retreat progressed sequentially along the BC coast from north to south beginning first in the Queen Charlotte Basin to the north *c.* 16 ^{14}C ka BP (*c.* 18.5 cal ka BP). Ice disappeared from the west coast *c.* 14 ^{14}C ka BP (16.75 cal ka BP) (Ward *et al.* 2003) and subsequently from the east coast of the island and the Georgia Basin after 12.4 ^{14}C ka BP (14.25 ka BP) (Clague 1989*a*, 1994; Barrie & Conway 1999, 2002*a*, *b*) (Fig. 1).

Open marine conditions were then soon established in the Georgia Basin *c.* 12.4 ^{14}C ka BP (14.25 cal ka BP) signalling marine circulation around Vancouver Island about that time (Howes 1983; Barrie & Conway 2002*a*, *b*). Open marine circulation to the north in Dixon Entrance was also established about this time at 13 ^{14}C ka BP (15.5 cal ka BP) (Guilbault *et al.* 1997) indicating that modern oceanographic conditions and coastal ocean dynamics were established by the earliest Holocene (Dallimore *et al.* 2008). By *c.* 10 ^{14}C ka BP, ice cover in BC had about the same areal extent as today (Clague & James 2002).

Rapid changes in sea level of up to 200 m over several centuries occurred on the BC coast immediately following deglaciation, caused by a complex interaction between glacio-isostatic crustal adjustments and changes in eustatic sea level throughout the Late Pleistocene and Holocene. This was further complicated by tectonic activity and a variable character of the mantle along the Cascadia subduction zone (Clague 1994; James *et al.* 2005, 2009). Consequently, the local expression of deglaciation combined with these factors make the history of sea-level change in the BC coastal region highly complex (Clague *et al.* 1982; Clague 1983, 1989*a*, *b*; Linden & Schurer 1988; Luternauer *et al.* 1989*a*, *b*; Hutchinson 1992; Friele & Hutchinson 1993; Josenhans *et al.* 1995, 1997; Barrie & Conway 1999, 2002*a*, *b*; Fedje & Josenhans 2000; Hetherington & Barrie 2004; Hetherington *et al.* 2004; Hutchinson *et al.* 2004*a*, *b*; Mosher & Hewitt 2004; James *et al.* 2005, 2009; Dallimore *et al.* 2008).

Oceanographic setting

The ocean dynamics of the BC coast are strongly seasonally dependent (Thomson 1981; Thomson & Gower 1998; Ware & Thomson 2000) and influenced by the relative positions of the Aleutian Low (AL) and the North Pacific High (NPH) atmospheric pressure systems (Fig. 3), the Jet Stream and El Niño/La Niña events (Roden 1989; Patterson *et al.* 2000). The position, timing and intensity of these atmospheric systems determine wind fields and surface currents of the Pacific, which in turn influence mixed layer depth, coastal upwelling and offshore advection. These factors all affect phytoplankton abundance and distribution in the coastal ocean, and ultimately impact coastal BC fisheries (Mackas & de Young 2001; Hay *et al.* 2003, 2007, 2009).

Wind-driven currents that flow over the coastal shelf are strongly seasonally dependent and are of particular significance to the summer upwelling onto the continental shelf of deep (150 m) slope waters (Thomson 1981; Thomson & Gower 1998; Patterson *et al.* 2000; Hay *et al.* 2007). Strong winds associated with these systems are a major contributor to the three oceanographic domains that characterize the west coast of North America: the Coastal Upwelling Domain, extending from Baja California to northern Vancouver Island; the Coastal Transition Domain, extending from northern Vancouver Island to Dixon Entrance; and the Coastal Downwelling Domain of the Alaska Coast and the Panhandle (Ware & Thomson 2000). These nutrient-rich upwelled ocean waters are the source of a dramatic increase in primary productivity in the upper layer of the coastal water column and lead to diatom blooms, common in spring and summer in the BC coastal ocean and inlets. Upwelling of deep shelf waters also occurs in the winter months but, because of low light and water temperatures, conditions are not conducive to diatom blooms (Thomson 1981; Patterson *et al.* 2000; Chang *et al.* 2003; Hay *et al.* 2003, 2007, 2009).

The expressions of a strong El Niño event in BC coastal waters are warm nearshore waters and a reduced upwelling tendency along the BC shelf (Ware & Thomson 2000). At times, a marked 'transition' to moderate to strong La Niñas (or cold phases) of the ENSO (El Niño-Southern Oscillation) cycle follows a major El Niño event. La Niña conditions appear to be linked to colder than normal BC coastal waters and enhanced upwelling tendency (Castro *et al.* 2002; Chavez *et al.* 2002; Dallimore *et al.* 2005).

Traditional ecological knowledge (TEK) of inlet history

There is a rich archaeological (McMillan 1999; McMillan & Yellowhorn 2003), traditional ecological knowledge (TEK) (Turner *et al.* 2000) and local ecological knowledge (LEK) (Morton & Proctor 2001; Proctor & Maximchuk 2003) history of First Nations occupations of the inlets of BC for many millennia, and at least for the past 5 ka.

Archaeologists of the NE Pacific coast need to understand regional sea-level history to interpret pre-history sites of human occupation. Evidence that the broad banks of Hecate Strait between the Queen Charlotte Islands and the mainland were subaerially exposed and possibly occupied in earliest Holocene times, during the sea-level lowstand following isostatic crustal rebound, has modified theories on human migration and occupation along the BC coast (Fig. 1). Stone tools were recovered from a drowned delta and palaeo-beach sites in the nearshore areas of the Queen Charlotte Islands, giving positive proof that the drowned landscapes of the Queen Charlotte Basin hosted early humans as early as 9.3 ^{14}C ka BP (Josenhans *et al.* 1995, 1997; Fedje & Josenhans 2000; Mandryk *et al.* 2001; Barrie & Conway 2002*a*, *b*; Hetherington *et al.* 2003; Mackie & Sumpter 2005).

Near the palaeoenvironmental study area of Effingham Inlet in Barkley Sound on Vancouver Island (Fig. 2), archaeologists have recently identified the temporal trends of nearshore fish populations over several millennia from cultural midden deposits in prehistory First Nations' village sites. When this knowledge of the changes in the marine-based food source of early peoples is combined with the physical record of the palaeoenvironment interpreted from laminated sediments preserved in anoxic Effingham Inlet as described in this paper, interesting insights are emerging with respect to the context and challenges of sustainable resource use in pre-history. These studies also shed light on the apparent lag or contemporaneous nature of abrupt changes in ecosystems as they relate to similarly abrupt climate and ocean changes in the Late Holocene (Dallimore *et al.* 2005; McKechnie 2005, 2007; McKechnie *et al.* 2008).

First Nations oral history is replete with the TEK of tsunamis waves as well as earthquakes. Scientists have successfully combined the palaeoseismic record of great ($M = 8$) earthquakes and tsunamis of the northern Cascadia subduction zone region with the archaeological record of village abandonment and native oral traditions, allowing pre-historic late Holocene seismic events to be identified (Hutchinson & McMillan 1997; McMillan & Hutchinson 1999, 2002; Clague *et al.* 2000). A unique study in central mainland coast Knight Inlet, utilizing ethnographic, archaeological and geological investigations, characterized a devastating landslide-generated tsunami that destroyed the First Nations village of Kwalate about 600 years ago. This event was recorded in the oral history of the A'wa'etlala and Da'naxda'xw First Nations peoples, and a study of the resulting rock slide into the inlet informs the current models of the physics of tsunamis wave generation and behaviour on the Pacific coast (Bornhold *et al.* 2007).

Vancouver Island anoxic inlets: implications for palaeoenvironmental studies

It has been known for some years that it may be possible to resolve the depositional history as well as palaeoceanographic, palaeoclimate and palaeoproductivity signals from the preservation of finely laminated sediments in the anoxic (oxygen-free) and dysoxic (low oxygen) bottom waters of inlets along the coast of BC (Gross *et al.* 1963; Gucluer & Gross 1964). Many inlets have highly stratified water columns due to the high rainfall in the region, and a few unique inlets on Vancouver Island have a combination of shallow sills and weak freshwater recharge from small rivers at the head of the inlets. Consequently, the normal estuarine inflow of marine water at depth can be interrupted and bottom waters become stagnant (Tully 1949; Syvitski & Shaw 1995). The resulting dysoxic or anoxic bottom waters aid in the preservation of annually laminated sediments since a thriving benthos cannot become established and bioturbation, which destroys the laminated structure of sediments, is absent. In highly productive fjords such as Effingham Inlet and Saanich Inlet on Vancouver Island, a rapid consumption of oxygen occurs during decomposition of organic matter, which returns the inlet bottom waters within weeks or months to an anoxic or dysoxic state even after periodic recharge by oxygenated oceanic waters (Dallimore *et al.* 2005).

Annually laminated sediments are therefore preserved undisturbed in these rare inlets and are valuable high temporal resolution (annual) archives which record past terrestrial climate, ecosystem and environmental change on a number of timescales (Kennett & Ingram 1995; Kemp 1996, 2003; Pike & Kemp 1996; Sancetta 1996). The dark 'terrigenous' winter laminae within the annual varve couplet record variations in precipitation while the light 'diatomaceous' laminae are indicative of annual productivity in the coastal ocean, specifically the abundance of diatom remains from the spring and summer blooms. Diatomaceous variability in the sedimentary record can infer pre-historic oceanographic and atmospheric conditions conducive to coastal upwelling of nutrient rich water which, in turn, create favourable conditions for spring and summer diatom blooms. Conversely, the absence of diatomaceous laminae and the presence of thick terrigenous laminae represent a failure of the annual diatom bloom and periods of high precipitation, respectively (Kemp 1996; Sancetta 1996; McQuoid & Hobson 2001; Chang *et al.* 2003; Hay *et al.* 2003, 2007, 2009). The variability in preservation and deposition of laminae archive the abrupt crossing of depositional thresholds that are related to climate changes (Chang & Patterson 2005; Dallimore *et al.* 2005; Hay *et al.* 2009).

Saanich Inlet, southeastern coast Vancouver Island

Studies of laminated marine sediments in British Columbia anoxic inlets were initiated in Saanich Inlet, on the SE coast of Vancouver Island (Gucluer 1962; Gross *et al.* 1963; Gucluer & Gross 1964). This work was expanded upon in studies of an anoxic ocean basin off Santa Barbara, California, investigated during ODP Leg 146 (Kennett & Ingram 1995; Bull & Kemp 1996; Pike & Kemp 1996; Schimmelman & Lange 1996). Subsequently, the high-resolution sediment record spanning the entire Holocene and Late Pleistocene in Saanich Inlet was studied in 1996 during ODP Leg 169S to investigate the palaeoenvironment and palaeoseismic record of the inlet (Blais-Stevens *et al.* 1997, 2001; Bornhold *et al.* 1998; Blais-Stevens & Clague 2001). The body of work resulting from ODP Leg 169S is contained in a Special Issue (volume 174) of Marine Geology (2001). The comprehensive results of the Saanich Inlet research give a highly detailed record for the entire Holocene and Late Pleistocene of: sedimentary deposition (Blais-Stevens *et al.* 2001); palaeoseismicity (Blais Stevens & Clague 2001); deglacial history (Cowan 2001; Mosher & Moran 2001); palaeo-fish populations (O'Connell & Tunnicliffe 2001; Tunnicliffe *et al.* 2001); palaeoproductivity (Dean *et al.* 2001; McQuoid & Hobson 2001) and palaeoclimate (Pellatt *et al.* 2001).

The single basin fjord is narrow (0.4 km width −7.6 km, 24 km length) with gently sloping terrain to the north and steep bedrock walls at its southern end. The communities of North Saanich, Victoria and Mill Bay surround this mostly urban watershed (Fig. 2) (Blais-Stevens *et al.* 2001). The inlet has a single basin which is 238 m at it deepest and is located behind a sill at 70 m depth that separates the waters of the inlet from Satellite Channel and the protected ocean waters of Haro Strait. Anoxic conditions exist in the bottom waters below *c.* 145 m depth, due to a stable stratified water column with a thick freshwater wedge at the surface combined with high primary productivity and sluggish estuarine circulation (due to the minor fluvial input in the winter months from the Goldstream River at the head of the inlet). Saanich is unique since deepwater renewal events lasting about 10 days occur regularly in late summer and early autumn; anoxia quickly re-develops, however (Anderson & Devol 1973; Stucchi 2008). These persistent anoxic water conditions allow the preservation of annually laminated sediments due to the absence of a bioturbating benthos (Blais-Stevens *et al.* 2001).

The varved sediments consist of laminated couplets of dark terrigenous and light organic-rich sediments with varve thicknesses ranging from 1 mm to 1 cm. The dark silt and clay laminae are deposited from autumn and winter runoff and the light-coloured diatom-rich laminae are deposited in spring and summer blooms (Blais-Stevens *et al.* 2001). Primary production is elevated in Saanich Inlet (Timothy & Soon 2001) and intermittent upwelling of nutrient-rich waters throughout the fjord is caused by processes occurring in the Strait of Georgia and the passages leading to Saanich Inlet (Gargett *et al.* 2003).

One of the main goals of ODP Leg 169S was to understand the earthquake frequency in this region of Cascadia by looking for sediment disruption in Saanich Inlet that may be related to seismic events. One of the advantages of a laminated sediment column is the accurate geochronology which can be established on an almost annual resolution. This allows accurate dating of events recorded in the sediment record, such as disturbance by earthquakes. The palaeoseismic record in Saanich Inlet is interpreted by the presence of unlaminated lithofacies which are intercalated within the laminated sediments. These massive units contain benthic foraminifera which are interpreted to have been transported from the shallow, oxygenated waters on the shoulders of the inlet. Most units are interpreted to be emplaced by debris flow events related to seismic shaking. Interpretation of the frequency of occurrence of these debris flow units indicates that in Saanich Inlet the average recurrence interval for moderate to large earthquakes which trigger debris flows has been *c.* 150 a during the past 6 ka (Blais-Stevens *et al.* 1997; Blais-Stevens & Clague 2001).

The Saanich Inlet work revealed a Late Quaternary palaeoenvironmental history of deglaciation. This was followed by progressive isolation of the inlet as isostatic rebound from the weight of ice of the Fraser Glaciation occurred. Full anoxic conditions were reached *c.* 7 ^{14}C ka BP (Blais-Stevens *et al.* 2001) allowing the preservation of annually laminated sediments since that time. Changing Holocene oceanographic and climatic conditions were also interpreted from diatom and pollen assemblages, showing a warmer early Holocene climate and modern conditions becoming established *c.* 2.8 cal ka BP (Blais-Stevens *et al.* 2001; Pellatt *et al.* 2001). Palaeoclimate work on sediment flux (using digital images, scanning electron microscopy and spectral analysis techniques on sediments deposited in the mid to late Holocene) have revealed a climatic cyclicity in this region of the Pacific, associated with the Pacific Decadal Oscillation

(PDO) and the ENSO. These results illustrate the use of these high-resolution laminated records for quantitative interpretations of the functioning of the large-scale north Pacific climate system (Nederbragt & Thurow 2001; Dean & Kemp 2004).

Data from the IOS time series of deep-water temperatures in Saanich Inlet since the 1930s indicate that global climate change is being expressed on a local level. The IOS time series of deep-water temperatures in Saanich Inlet show an increase of 1 °C since the 1930s and an increase of 0.5 °C since the mid-1970s. The weaker oxygenation of deeper waters of the basin observed in recent years indicates that there is a decline in oxygen levels in offshore waters, which are the source of the oxygen for the deep waters of Saanich Inlet. This may be an indicator of warmer ocean waters due to global warming; if this trend continues in the future, it will intensify bottom water anoxia (Stucchi 2008).

Effingham Inlet, west coast Vancouver Island

To investigate the possibility of studying laminated sediments in an open marine setting, a team from IOS began a search for an anoxic inlet on Vancouver Island's west coast in 1995. They identified Effingham Inlet in Barkley Sound, part of the Pacific Rim National Park. Coastal ocean processes influencing Effingham Inlet sedimentation are likely modified by climate-scale ocean variability making Effingham Inlet an ideal site for palaeoenvironmental studies over the past ten years, shedding light on oceanic variability during the Holocene (Patterson *et al.* 2000; Dallimore 2001; Chang *et al.* 2003; Hay *et al.* 2003, 2007, 2009; Chang & Patterson 2005; Dallimore *et al.* 2005; Natural Resources Canada 2007*a*). Effingham Inlet was also the site of an international drill ship expedition in 2002. The MONA (Marges Ouest Nord Américaines) campaign studied the palaeoenvironmental record of the entire Holocene (Core MD02-2494) (Beaufort 2002; Ivanochko *et al.* 2008*a*, *b*; Dallimore *et al.* 2009*a*).

Effingham Inlet is a multiple-silled inlet located on the southwestern coast of Vancouver Island, which opens to the Pacific Ocean through Barkley Sound. The inlet is rimmed by steep bedrock walls and has an inner basin (120 m depth) and an outer basin (205 m depth), each isolated from the open ocean waters by shallow bedrock sills of 65 m depth in the outer basin and 46 m depth in the inner basin (Fig. 4). Water property distributions in the modern inlet are characteristic of a weakly mixed estuary (Fig. 4) (Thomson 1981) with small (<2 m) tidal variation and negligible bottom currents. The bottom waters of the inner basin are usually anoxic due to the sills, combined with low fluvial input from the Effingham River at the head and high levels of productivity (Fig. 4) (Patterson *et al.* 2000). Winter temperatures usually remain above freezing at lower elevations, preventing significant snow accumulation and thereby limiting the spring freshet. Peak freshwater input into Effingham Inlet occurs during the heavy rains of late autumn and early winter (October to January). Bottom water renewals are rare, with only one having ever been recorded in the winter of 1999.

Interpretations of the upper sediment column in Effingham Inlet indicate that this single oxygenation event was the only one since 1946, when an earthquake disrupted the sediment column and masked the oceanographic signal in the sediment record for the following years until about 1997 (Chang *et al.* 2003). The 1999 oxygenation event was associated with the transition from an exceptionally strong El Niño to a strong La Niña in 1999, which in turn was associated with enhanced coastal upwelling which may be indicative of a recent profound change in the BC coastal ocean dynamics (Dallimore *et al.* 2005). There is some pan-Pacific indication that other profound disruptions of ocean dynamics and ecosystems at about the turn of this century are also indicative of an abrupt 'regime change' in the Pacific Ocean, apparent in records of recent oceanographic and ecological conditions (Chavez *et al.* 2003)

In Effingham Inlet, the variability in preservation and deposition of the annual winter laminae, which record variation in precipitation, and the spring to summer diatomaceous laminae, which are indicative of annual productivity in the coastal ocean, archive the abrupt crossing of depositional thresholds that are related to climate changes (Dallimore 2001; Chang *et al.* 2003; Hay *et al.* 2003, 2007, 2009; Chang & Patterson 2005; Dallimore *et al.* 2005). Varve thicknesses in Effingham Inlet are about 0.75 to 3.5 mm, somewhat thinner than in Saanich Inlet; this indicates higher productivity and fluvial input in Saanich. Varves can contain multiple micro-laminae within a seasonal varve which represents the seasonal succession of diatom blooms in any given year, giving not only annual but inter-annual resolution of ocean and climate processes (Chang & Patterson 2005). To date in Effingham Inlet, an interpreted abrupt climatic and oceanic change which occurred *c.* 4 ka ago (also found in other palaeorecords of the Pacific Ocean) and the abrupt coastal oceanographic changes of 1999 have been recorded. These observations could possibly indicate 'regime changes' of the Pacific or 'thresholds' of environmental conditions which may help to inform our understanding and modelling of future climate changes (Fig. 5) (Wake *et al.* 2006).

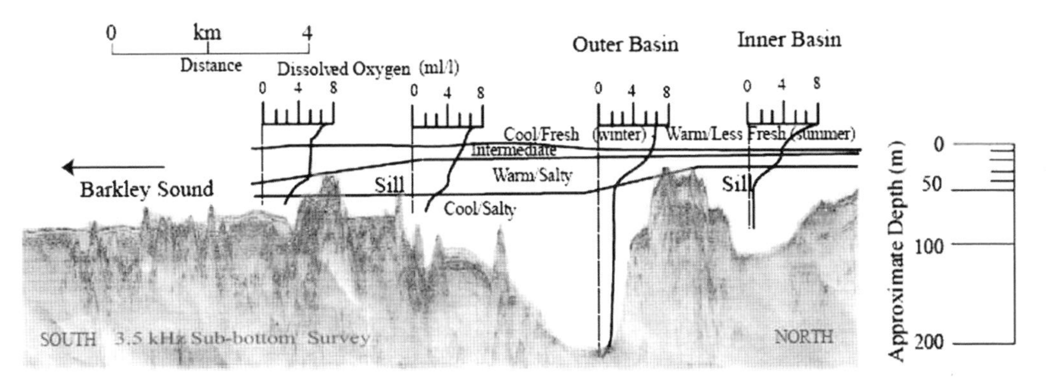

Fig. 4. 3.5 kHz sub-bottom survey profile of the Effingham Inlet study site showing draping of Holocene sediments over bedrock. The water property structure shows the weakly mixed estuarine stratified water column and the anoxic bottom waters of the inner basin where annually laminated sediments are preserved (from Dallimore *et al.* 2005).

Similarly the occurrence of a type of unlaminated mud units (termed homogenites), which are sporadically intercalated among the laminated sediments, can be interpreted in terms of rare climatically related, wind-driven disruptions of the coastal ocean dynamics (Fig. 5). These units can be identified geochemically and by physical properties as being identical to the surrounding laminated sediments. They are therefore interpreted as being the result of the re-working of previously laminated sediments by dense bottom currents triggered by strong episodes of upwelling, prompted by extreme storm conditions. A 200-year running average of the occurrence of massive homogenite units shows that these stormy, extreme-weather conditions have become more frequent in the coastal ocean throughout the Holocene towards the present (Dallimore *et al.* 2005, 2008; Hay *et al.* 2007, 2009).

Unlaminated mud units in the Effingham sediment record can be identified by physical property measurements as having an allocthonous source, and thus are interpreted to be associated with seismic shaking and/or sediment slumping (Dallimore *et al.* 2008). These lithofacies are termed seismite and debris flow units, respectively, and testify to the occurrence of large earthquake events at the northern Cascadia subduction zone (Fig. 3). The frequency of earthquake events in the Effingham sediments is less than in Saanich Inlet, possibly due to the greater frequency of crustal earthquakes in the Saanich Inlet area (Blais-Stevens *et al.* 2009; Dallimore *et al.* 2009*b*). However, the timing of five of the seismites from Effingham Inlet correlate approximately to some of the seismites identified from Saanich Inlet and those probably attest to regional great earthquake events originating along the Cascadia subduction zone. These events occurred at about 1840 ^{14}C a BP (1900 cal a BP), 3890 ^{14}C a BP (4280 cal a BP), 7020 ^{14}C a BP (7900 cal a BP), 8060 ^{14}C a BP (9090 cal a BP) and 9710 ^{14}C a BP (11.02 cal ka BP). Volcanic ash layers identified in the core may be products of unrecorded eruptions from Cascadia vents at about 5650 ^{14}C a BP (6310 cal a BP); 9190 ^{14}C a BP (10.4 cal ka BP) and 9220 ^{14}C a BP (10.44 cal ka BP (Dallimore *et al.* 2009*a*, *b*).

Stratigraphic analysis of the 40 m, 12 ka sediment record from Effingham Inlet core MD02-2494 from the 2002 MONA campaign shows that the inlet experienced a sea-level fall and then a rise of almost 50 m due to eustatic and isostatic crustal adjustments that persisted from *c.* 11 to 4.5 ^{14}C ka BP. This crustal adjustment, combined with rising global eustatic sea-level change, resulted in the inner basin of the inlet becoming a freshwater lake for about 850 years at *c.* 11.6 ^{14}C ka BP (13.5 cal ka BP), when the sill which today is at 46 m depth became briefly subaerially exposed. The timing of the sea-level lowstand when Effingham Inlet was an isolation basin provides the two index points needed for a new sea-level curve for the outer coast of Vancouver Island. These indicate a variable mantle response to deglaciation composed of an early response controlled by the low viscosity upper mantle and a later response controlled by a more viscous lower mantle, and contributes to our very recent understanding of the nature of the mantle in this part of the Cascadia subduction zone (James *et al.* 2000, 2002, 2005, 2009; Dallimore *et al.* 2009*a*, *b*).

The collective work on Effingham Inlet and most particularly the MD02-2494 long core, which is to date our longest and best dated palaeoenvironmental record of the NE Pacific, has refined and developed interpretive techniques for working with laminated sediment records. The results of the work have shed light on: changes in fisheries ecosystems over time (Wright *et al.* 2005), earthquake frequency in

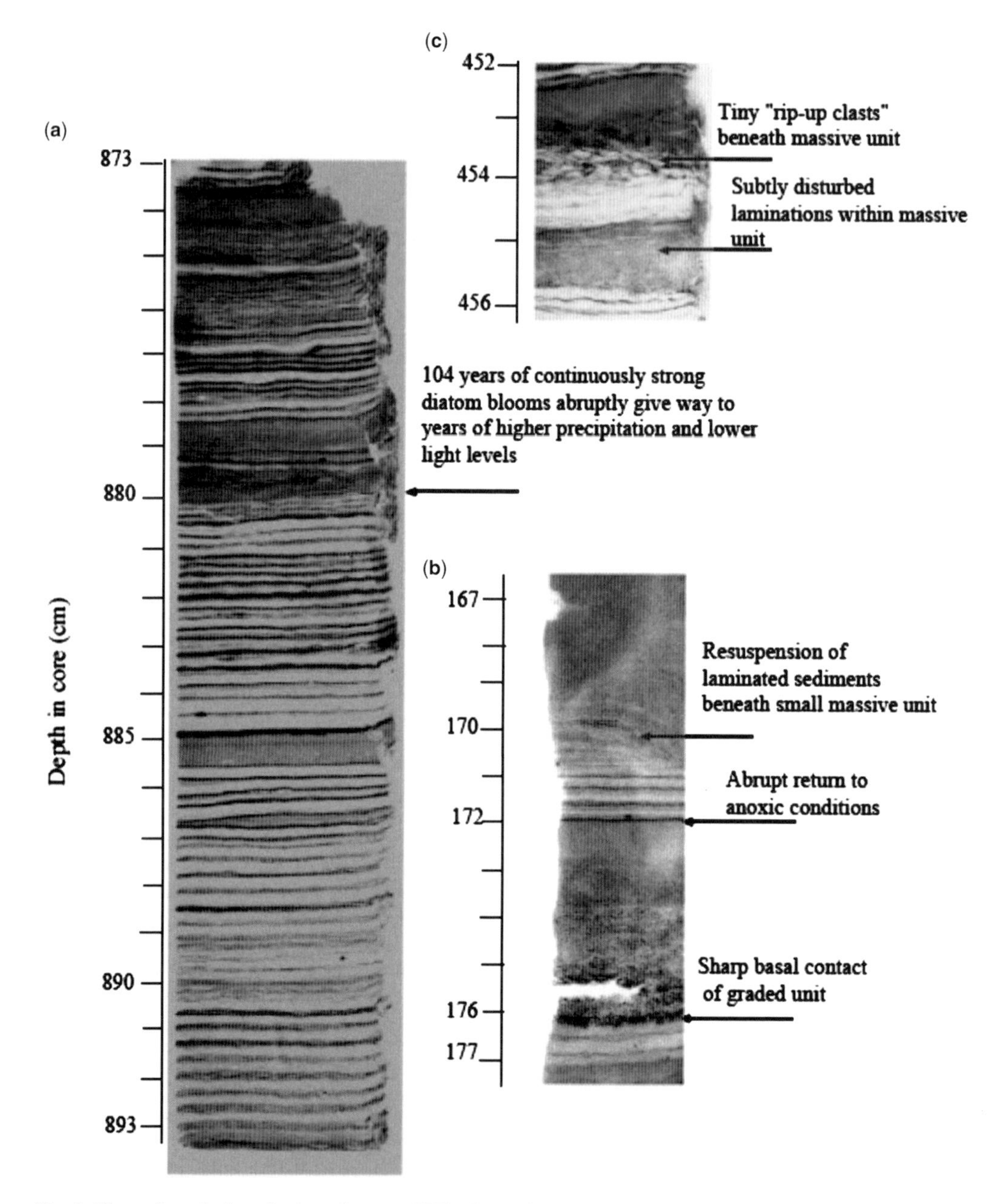

Fig. 5. X-rays from the inner basin sediments of Effingham Inlet showing typically well-laminated sediments. At 880 cm in (**a**) an abrupt change in climate and ocean conditions dated *c.* 4 ka BP is evident. Unlaminated lithofacies in (**b**) and (**c**) show 'homogenites' which are re-mixed laminated sediments associated with incoming density currents and climatic conditions. In lower (b) x-ray, the graded unlaminated unit is interpreted as a 'seismite' which represents sediment disturbance by seismic shaking (from Dallimore *et al.* 2005).

this region of the Cascadia subduction zone (Dallimore *et al.* 2001; Dallimore *et al.* 2005, 2009*b*); and the nature of deglaciation and mantle response to isostatic loading (Dallimore *et al.* 2008), which in turn contributes to our understanding of the geodynamics of the Cascadia subduction zone. The palaeoenvironmental studies from Effingham Inlet have shed light on the coupled functioning of our ocean and climate system (Patterson *et al.* 2000; Dallimore 2001;

Chang *et al.* 2003; Hay *et al.* 2003, 2007, 2009; Dallimore *et al.* 2005, 2008; Ivanochko *et al.* 2008*a*, *b*) which in turn helps us to understand what our future climate functioning may be in a changing climate on the west coast of Canada (Wake *et al.* 2006; Lemmen *et al.* 2008). As in Saanich Inlet, it appears that the sediment record of Effingham Inlet is indicating that the natural climate system in the region experiences thresholds and abrupt changes. It is also indicating that the north Pacific climate is largely driven now (and was throughout the Holocene) by the bidecadal and pentadecadal periods of the Pacific Decadal Oscillation (PDO), which is related to the strength and position of the atmospheric Aleutian low pressure system (Ivanochko *et al.* 2008*a*).

Research in mainland coast inlets: Knight, Bute and Jervis inlets

Studies in three large and well-oxygenated inlets of the mainland coast, Knight Inlet, Bute Inlet and Jervis Inlet, have described the oceanographic processes influencing these environments. Sediment transport and modelling research has defined the depositional environment of these representative inlets (Figs 1 & 2). On the mainland coast, fluvial runoff peaks in July due to snowmelt of snowfields and glaciers in the interior watershed. In Knight Inlet, for example, the glacier terminus of the Klinaklini Glacier (Fig. 2) in the basin is retreating at a rate of between 5 and 32 m a^{-1}. Consequently, 25% of the annual river flow is ice melt in the summer and 50% is snowmelt during spring and early summer (Morehead *et al.* 2001). This increased fluvial input on the mainland inlets as compared to inlets of Vancouver Island also translates to much higher sedimentation rates, for example about 300 mm a^{-1} in Bute Inlet (Syvitski & Shaw 1995) compared to about 2.5 mm a^{-1} in Effingham Inlet (Chang & Patterson 2005). In contrast to the Vancouver Island inlets, the lowest river discharge occurs in winter on the mainland because most of the precipitation is stored as snow in higher elevations of the watershed. Relatively high (*c.* 3 ml l^{-1}) deep water dissolved oxygen contents are common in these inlets, since bottom water renewals are frequent with a large portion of the deep waters exchanged (Bornhold *et al.* 1994).

Multiyear sediment trap studies in deep-silled (380 m) and well-oxygenated Jervis Inlet (780 m depth), where deepwater renewals are less frequent than for Saanich Inlet, give insights into inlet estuarine dynamics which determine whether bottom waters are anoxic. In Jervis Inlet, moderate freshwater input at the head drives weak estuarine circulation (Pickard 1961). Deepwater renewals occur primarily during summer or autumn as a result of wind-driven upwelling. Even although there are fewer deep water renewal events in Jervis than in Saanich Inlet, anoxia is not known to occur in Jervis. Since the inlet is so deep, sinking organic matter decomposes before stagnant water is reached which decreases the chances of oxygen depletion in the bottom water (Timothy & Soon 2001; Timothy *et al.* 2003).

Extensive channel systems are observed on the seafloor in both deep Knight (540 m depth) (Fig. 2) and Bute inlets (660 m depth). It is thought that powerful and frequent sand transporting turbidity currents, fed by high volumes of sediment transport associated with annual glacier and snowmelt, are responsible for creating these dramatic features (Ren *et al.* 1996). In Knight Inlet, many channels emanate from a fjord-head delta and are up to 40 m deep and 30 m wide and extend 41 km down-inlet (Fig. 2, inset). Average flow velocities of turbidity currents in Knight Inlet are as high as 7 m s^{-1}, and they occur principally during the summer and early autumn (June to October). Turbidity currents are related to river discharge peaks of the Klinaklini River at the head of the inlet and are initiated by slope failure on the oversteepened slopes of the steep delta front (Bornhold *et al.* 1994). In Bute Inlet, a year-long monitoring program detected sand-transporting turbidity currents with maximum velocities of 3.3 m s^{-1} and turbidity flow thicknesses of more than 30 m. The Bute Inlet turbidity currents flow a minimum distance of 26 km, but possibly as far as 40–50 km (Prior *et al.* 1987). These deep channel features are not known in the inlets of Vancouver Island, probably due to the lack of a high-volume summer peak fluvial flow carrying sand-size sediments originating from snow and ice melt.

Unique and rough 'blocky' terrain was also described from the floor of Bute and Knight Inlets from sidescan sonar and seismic profiling surveys. Blocks are large, 150 m by 60 m and 1–3 m in height, with calcareous concretions. Although their genesis is somewhat a mystery, it is hypothesized that the blocky terrain may be linked to uplift related to the formation of gas hydrates; alternatively, they may represent sediment slumping as a result of extensive liquefaction and subsidence caused by seismic shaking (Bornhold & Prior 1989).

Numerical modelling of sediment transport in a fjord depositional environment was carried out in Knight Inlet (Morehead *et al.* 2001). This model can build sedimentary sequences and generate 'synthetic cores' to address issues of preservation of abrupt climate signals in the sedimentary record over millennia. These simulated studies can inform interpretations of the sensitivity of the complex sedimentological inlet systems to evolving boundary conditions that would be expressed in the sedimentary record during sea-level and climate changes. In Knight Inlet, the model successfully predicted sediment deposition and disturbance sequences related to submarine slope failures. This indicates that, although the basin does not host laminated sediments, it may also be useful to interpret its preserved lithofacies in terms of a palaeoenvironmental record.

Future directions

Marine science funding for research in Canada is at the moment closely tied to resource development. Natural resources extraction continues to dominate economic activity along the BC inlet coastline (mainly wood, fish and minerals), although these traditional sectors are now declining in response to various environmental, social and economic factors both locally and globally (Lemmen *et al.* 2008). The offshore region of BC has been under a federal government moratorium for oil and gas exploration as well as oil tanker traffic through BC waters since 1972. As a result, little extensive marine research has been carried out in BC coastal waters with modern survey techniques. However, the moratorium has been revisited a number of times in response to pressure from the province for economic development. The public consensus, which was confirmed in a federal government series of public hearings on the topic in 2003, determined that 75% of participants and all 70 BC First Nations groups continue to support the moratorium. The federal government's present position is that 5–7 years of research needs to be done to fill in gaps in research before reconsidering lifting the moratorium. If this research were to be undertaken, more activity in inlet research would no doubt proceed to characterize the modern environments which would provide environmental impact assessments (Natural Resources Canada 2009*b*).

Inlets research mounted from IOS is continuing with a focus towards the central coast Seymour-Belize inlet complex. When complete, this central coast study will provide some indication of the functioning of the coastal ocean dynamics in the more northern Coastal Transition Upwelling Domain (Hay *et al.* 2009). The central coast region inlet sediment record will also contain information about the presence or absence of sediment disturbance resulting from seismic activity along the Queen Charlotte-Fairweather transform fault at the northern limit of the Cascadia subduction zone. The tectonic regime at this transitional area north of Vancouver Island, between the Cascadia subduction zone and the Queen Charlotte fault, is largely unknown. This is partially because the sea-level history, which informs models of mantle viscosity, is also unknown in the central coast region. An on-going study of perched coastal lakes, combined with multibeam imagery of sea-level lowstand bedform features in the Seymour-Belize inlet complex and central coast shelf area, is presently underway. Within the next few years, it is anticipated that it will add the missing pieces necessary to interpret the palaeoenvironmental record of the central coast region.

An IODP proposal, should it come to fruition, may be the opportunity for another long core in a central or north coast anoxic inlet at some point in the near future. A north coast study site will be particularly valuable, not only to improve the robustness of our regional interpretations of NE Pacific present and past climate and tectonic functioning, but also to bring the palaeomarine records closer to the palaeo-atmospheric record of the Mount Logan ice core. This core was sampled in 2002 from the top of Mount Logan, Canada's highest mountain, by a team from the Geological Survey of Canada (Natural Resources Canada 2007*b*). Ongoing work on this core gives an annually resolved indication of the atmospheric circulation of the north Pacific Ocean throughout the Holocene and beyond (Fisher *et al.* 2004; Wake *et al.* 2006; Osterberg *et al.* 2008). New techniques of spectral analysis developed on the laminated sedimentary record (Ivanochko *et al.* 2008*a*) will improve our capability to quantitatively compare various palaeorecords in the near future to solidify our understanding of the functioning of past, present and future north Pacific climate (Wake *et al.* 2006).

The IOS time series of inlet oceanographic properties continues on an opportunity basis, and has to date proven the need for ongoing monitoring studies. The time series of deep-water temperature anomalies from seven of the better sampled BC inlets (Bute, Jervis, Knight, Muchalat, Howe Sound and Indian Arm of the southern mainland coast and Saanich on Vancouver Island) show an overall increase of about 0.5 to 1 °C since the mid 1970s. This rise in temperature is more consistent in the last 20–25 years than in the previous 25 years. The warmest deep waters in Knight Inlet ever observed since 1950 occurred in 1999 (Fisheries & Oceans 2008*b*). The rare deep-water oxygenation event, unexpectedly captured in a routine oceanographic transect of Effingham Inlet in 1999 (Dallimore *et al.* 2005), also attests to the importance of monitoring work to understand the intricacies of our climate system. This need grows more important with our rapidly changing climatic conditions under global climate change (Lemmen *et al.* 2008; Thomson *et al.* 2008).

The future may also hold many new discoveries and research capabilities for inlet sedimentary and oceanographic studies we have yet to imagine, with the recent installation of the VENUS fibre-optic cable in Saanich Inlet. The NEPTUNE cable became fully operational in March, 2010 (Neptune Canada 2009). The fibre-optic cable is 800 km long and was laid down in Alberni Inlet and out into Barkley Sound on the west coast of Vancouver Island. Perhaps fittingly, the onshore terminus of this futuristic cabled observatory is located in Port Alberni and the cable runs along the bottom of Alberni Inlet; this was the first inlet of the BC coast which was comprehensively investigated with a modern scientific approach by Dr Tully in 1949. Presently, the research funding picture in Canada for coastal regions is more promising on the east and Arctic coasts than the Pacific, due to the Pacific oil and gas exploration moratorium. However, the future for inlet research on the BC coast is positive with the current international focus on coastal, environmental and climate change issues. Multidisciplinary researchers and students from Canadian universities, and around the world, are continuing and initiating collaborative research to enhance our understanding of one of the last-remaining extensive, largely pristine and unpopulated fjord coastlines of the world.

The authors would like to acknowledge funding and support from the Geological Survey of Canada-Pacific (Natural Resources Canada), Fisheries and Oceans Canada, the Canadian Hydrographic Service, the Natural Sciences and Engineering Research Council of Canada and Royal Roads University. This support contributed not only to some of the research reported in this paper but also to the research and writing of this manuscript. Marine work along the west coast of Canada would not be possible without the dedication and professionalism of the officers and crew of the Canadian Coast Guard who operate Canadian research vessels, and we acknowledge their contribution to coastal and marine science in Canada. The manuscript was greatly improved by the critical revision of two anonymous reviewers to whom we extend our thanks.

References

Anderson, J. J. & Devol, A. H. 1973. Deep water renewal in Saanich Inlet, an intermittently anoxic basin. *Estuarine and Coastal Marine Science*, **1**, 1–10.

Atwater, B. F. & Hemphill-Haley, E. 1996. *Preliminary estimates of recurrence intervals for great earthquakes of the past 3,500 years at North Eastern Willapa Bay, Washington*. U.S. Geological Survey Open File Report 96-001.

Atwater, B. F. & Hemphill-Haley, E. 1997. Recurrence intervals for great earthquakes in coastal Washington. *Geological Society of America Annual Meeting, 1997 Abstracts with Programs*, **29**.

Atwater, B. F., Nelson, A. R. & Clague, J. J. 1995*a*. Summary of coastal geologic evidence for past great earthquakes at the Cascadia subduction zone. *Earthquake Spectra*, **11**, 1–18.

Atwater, B. F., Nelson, A. R. *et al.* 1995*b*. Summary of coastal geologic evidence for past great earthquakes at the Cascadia subduction zone. *Earthquake Spectra*, **11**, 1–18.

Barrie, J. V. & Conway, K. W. 1999. Late Quaternary glaciation and postglacial stratigraphy of the northern Pacific margin of Canada. *Quaternary Research*, **51**, 113–123.

Barrie, J. V. & Conway, K. W. 2002*a*. Rapid sea-level change and coastal evolution on the Pacific margin of Canada. *Sedimentary Geology*, **150**, 171–183.

Barrie, J. V. & Conway, K. W. 2002*b*. Contrasting glacial sedimentation processes and sea-level changes in two adjacent basins on the Pacific margin of Canada. *In*: Dowdeswell, J. A. & O'Cofaigh, C. (eds) *Glacier-Influenced Sedimentation on High-Latitude Continental Margins*. Geological Society, London, Special Publications, **203**, 181–194.

Barrie, J. V. & Conway, K. W. in press. *Sea level terraces, sedimentary processes and a paleogeographic reconstruction of offshore British Columbia, Canada, using multibeam imagery*. International Association of Sedimentologists Special Publication.

Barrie, J. V., Conway, K. W., Picard, K. & Greene, H. G. 2009. Large-scale sedimentary bedforms and sediment dynamics on a glaciated tectonic continental shelf: examples from the Pacific margin of Canada. *Continental Shelf Research*, **11**, 701–715.

Beaufort, L. 2002. Les rapports de campagnes à la mer MD126/MONA Marges Ouest Nord Américaines IMAGES VIII à bord du Amrion Dufresne. Réf: OCE/2002/03, Insitute Polaires Français, Plouzané, France. Available online at: www.images-pages.org, accessed June 2010.

Blais-Stevens, A. & Clague, J. J. 2001. Paleoseismic signature in late Holocene sediment cores from Saanich Inlet, British Columbia. *Marine Geology*, **175**, 131–148.

Blais-Stevens, A., Clague, J. J., Bobrowsky, P. T. & Patterson, R. T. 1997. Late Holocene sedimentation in Saanich Inlet, British Columbia, and its paleoseismic implications. *Canadian Journal of Earth Sciences*, **34**, 1345–1357.

Blais-Stevens, A., Bornhold, B. D., Kemp, A. E. S., Dean, J. M. & Vaan, A. A. 2001. Overview of Late Quaternary stratigraphy in Saanich Inlet, British Columbia: results of Ocean Drilling Program Leg 169S. *Marine Geology*, **174**, 3–26.

Blais-Stevens, A., Rogers, G. C. & Clague, J. J. 2009. A 4000 year record of earthquakes in the late Holocene sediments from Saanich Inlet, an anoxic fjord near Victoria, Brisith Columbia. *Seismological Society of America*, Annual Meeting, Monterey California, April 2009.

Booth, D. B., Troost, K. G., Clague, J. J. & Waitt, R. B. 2004. The Cordilleran ice sheet. *In*: Gillespie, A. R., Porter, S. C. & Atwater, B. F. (eds) *The Quaternary Period in the United States*. Developments in Quaternary Science, **1**, 17–43.

Bornhold, B. D. & Prior, D. B. 1989. Sediment blocks on the sea floor in British Columbia fjords. *Geo-marine Letters*, **9**, 135–144.

Bornhold, B. D., Ren, P. & Prior, D. B. 1994. High-frequency turbidity currents in British Columbia fjords. *Geo-Marine Letters*, **14**, 238–243.

Bornhold, B. D., Firth, J. V. et al. 1998. *Proceedings of the Ocean Drilling Program Sites 1033 and 1034.* Initial Reports, Saanich Inlet, Victoria, BC, **169S**.

Bornhold, B. D., Harper, J. R., McLaren, D. & Thomson, R. E. 2007. Destruction of the First Nations Village of Kwalate by a rock avalanche-generated tsunami. *Atmosphere-Ocean*, **45**, 123–128.

Bull, D. & Kemp, A. E. S. 1996. Composition and origins in laminae in late Quaternary and Holocene sediments from the Santa Barbara basin. *In*: Kemp, A. E. S. (ed.) *Paleoceanography and Paleoclimatology in Laminated Sediments.* Geological Society, London, Special Publications, **116**, 143–146.

Castro, G. C., Baumgartner, T. R. et al. 2002. Introduction to the 1997–98 El Niño Atlas of oceanographic conditions along the west coast of North America (23°N–50°N). *Progress in Oceanography*, **54**, 503–511.

Chang, A. S. & Patterson, R. T. 2005. Climate shift at 4400 years BP: evidence from high-resolution diatom stratigraphy, Effingham Inlet, British Columbia, Canada. *Palaeogeography, Palaeoclimatology, Palaeoecology*, **226**, 72–92.

Chang, A. S., Patterson, R. T. & McNeely, R. 2003. Seasonal sediment and diatom record from the late Holocene laminated sediments, Effingham Inlet, British Columbia, Canada. *Palaios*, **18**, 477–494.

Chavez, F. P., Collins, C. A., Huyer, A. & Mackas, D. L. 2002. El Niño along the west coast of North America. *Progress in Oceanography*, **54**, 1–5.

Chavez, J. R., Salvador, E. L.-C. & Ñiquen, C. 2003. From anchovies to sardines and back: multi-decadal changes in the Pacific Ocean. *Science*, **299**, 217–221.

Clague, J. J. 1981. Late Quaternary geology and geochronology of British Columbia. Part 2: summary and discussion of radiocarbon-dated Quaternary history. *Geological Survey of Canada*, paper 80-35.

Clague, J. J. 1983. Glacio-isostatic effects of the Cordilleran ice sheet, British Columbia, Canada. *In*: Smith, D. E. & Dawson, A. G. (eds) *Shorelines and Isostacy.* Institute of British Geographers, Special Publications, **16**, 321–343.

Clague, J. J. 1989*a*. Cordilleran ice sheet. *In*: Fulton, R. J. (ed.) *Quaternary Geology of Canada and Greenland.* Geological Survey of Canada, **1**, 40–42.

Clague, J. J. 1989*b*. Quaternary geology of the Canadian Cordillera. *In*: Fulton, R. J. (ed.) *Quaternary Geology of Canada and Greenland.* Geological Survey of Canada, **1**, 17–22.

Clague, J. J. 1989*c*. Bedrock geology (Canadian Cordillera). *In*: Fulton, R. J. (ed.) *Quaternary Geology of Canada and Greenland.* Geological Survey of Canada, **1**, 2–24.

Clague, J. J. 1994. Quaternary stratigraphy and history of south-coastal British Columbia. *In*: Monger, J. W. (ed.) *Geology and Geological Hazards of the Vancouver Region, Southwestern British Columbia.* Geological Survey of Canada Bulletin, **481**, 181–192.

Clague, J. J. & James, T. S. 2002. History and isostatic effects of the last ice sheet in southern British Columbia. *Quaternary Science Reviews*, **21**, 71–87.

Clague, J. J., Harper, J. R., Hebda, R. J. & Howes, D. E. 1982. Late Quaternary sea-levels and crustal movements, coastal British Columbia. *Canadian Journal of Earth Sciences*, **19**, 597–618.

Clague, J. J., Bobrowsky, P. T. & Hutchinson, I. 2000. A review of geological records of large tsunamis at Vancouver Island, British Columbia, and implications for hazard. *Quaternary Science Reviews*, **19**, 849–863.

Clague, J. J., Yorath, C., Franklin, R. & Turner, R. J. W. 2006. *At Risk: Earthquakes and Tsunamis on the West Coast.* Tricouni Press, Vancouver.

Cowan, E. A. 2001. Late Pleistocene glacimarine record in Saanich Inlet, British Columbia, Canada. *Marine Geology*, **174**, 43–57.

Dallimore, A. 2001. *Late Holocene geologic, oceanographic and climate history of an anoxic fjord: Effingham Inlet, west coast, Vancouver Island.* PhD thesis, Carleton University, Ottawa, Ontario, Canada.

Dallimore, A., Patterson, R. T. & Thomson, R. E. 2001. *Discrimination of oxygenation episodes from earthquake events in late Holocene sediments of anoxic Effingham Inlet, Vancouver Island, British Columbia.* 3rd International Conference IGCP Project 437, Durham, UK.

Dallimore, A., Thomson, R. E. & Bertram, M. A. 2005. Modern to Late Holocene deposition in an anoxic fjord on the west coast of Canada: implications for regional oceanography, climate and paleoseismic history. *Marine Geology*, **219**, 47–69.

Dallimore, A., Enkin, R. J. et al. 2008. Postglacial evolution of a Pacific coastal fjord in British Columbia, Canada: interactions of sea-level change, crustal response, and environmental fluctuations – results from MONA core MD02-2494. *Canadian Journal of Earth Sciences*, **45**, 1345–1362.

Dallimore, A., Enkin, R. J., Baker, J. & Pienitz, R. 2009*a*. *Stratigraphy and late Pleistocene-Holocene history of Effingham Inlet, B.C.: results from MONA Core MD02-2494 and GSC freeze cores.* Geological Survey of Canada, Open File Report 5930.

Dallimore, A., Enkin, R. J., Baker, J. & Rogers, G. 2009*b*. Identification and frequency of Holocene earthquakes from the laminated sedimentary record of Effingham Inlet, Pacific coast, Vancouver Island, British Columbia. *Seismological Society of America*, Annual Meeting, Monterey California, April 2009.

Dean, J. M. & Kemp, A. E. S. 2004. A 2100 year BP record of Pacific Decadal Oscillation, El Niño Soithern Oscillation in marine production and fluvial input from Saanich Inlet, British Columbia. *Palaeoceanography, Palaeoclimatology, Palaeoecology*, **213**, 207–229.

Dean, J. M., Kemp, A. E. S. & Pearce, R. B. 2001. Palaeoflux records from electron microscope studies of Holocene laminated sediments in Saanich Inlet, British Columbia. *Marine Geology*, **174**, 139–158.

Fedje, D. & Josenhans, H. W. 2000. Drowned forests and archaeology on the continental shelf of British Columbia, Canada. *Geology*, **28**, 99–102.

Fisher, D. A., Wake, C. et al. 2004. Stable isotope records from Mount Logan, Eclipse ice cores and nearby Jelly Bean Lake. Water cycle of the North Pacific over 2,000 years and over five vertical kilometers: sudden shifts and tropical connections. *Géographie physique et Quaternaire*, **58**, 337–352.

FISHERIES AND OCEANS CANADA – PACIFIC REGION 2008*a*. Monitoring Southern BC Coastal Waters. Available online at: http://www-sci.pac.dfo-mpo.gc.ca/osap/projects/straitofgeorgia/default_e.htm, accessed June 2010.

FISHERIES AND OCEANS CANADA – PACIFIC REGION 2008*b*. Long term trends in deep water properties of BC Inlets. Available online at: http://www.pac.dfo-mpo.gc.ca/science/oceans/BCinlets/index_eng.htm.

FISHERIES AND OCEANS CANADA 2008*c*. Fisheries and Oceans Canada, Maps and Services. Available online at: http://geoportail-geoportal.gc.ca/en/services.html, accessed June 2010.

FISHERIES AND OCEANS CANADA 2008*d*. Tsunamis and tsunamis research. Available online at: http://www.pac.dfo-mpo.gc.ca/sci/OSAP/projects/tsunami/default_e.htm, accessed June 2010.

FRIELE, P. A. & HUTCHINSON, I. 1993. Holocene sea-level change on the central west coast of Vancouver Island, British Columbia. *Canadian Journal of Earth Sciences*, **30**, 832–840.

GABRIELSE, H. & YORATH, C. J. (eds) 1992. *Geology of the Cordilleran Orogen in Canada*. The Geology of Canada. **4**, Geological Survey of Canada.

GARGETT, A. E., STUCCHI, D. & WHITNEY, F. 2003. Physical processes associated with high primary production in Saanich Inlet, British Columbia. *Estuarine, Coastal and Shelf Science*, **56**, 1141–1156.

GOLDFINGER, C., NELSON, C. H. & JOHNSON, J. 2003*a*. Holocene Earthquake Records from the Cascadia Subduction Zone and Northern San Andreas Fault based on Precise dating of Offshore Turbidites. *Annual Reviews of Geophysics*, **31**, 555–577.

GOLDFINGER, C., NELSON, C. H. & JOHNSON, J. E. 2003*b*. Deep-water turbidites as Holocene earthquake proxies: the Cascadia subduction zone and northern San Andreas fault systems. *Annali Geofisica*, **46**, 1169–1194.

GROSS, M. G., GUCLUER, S. M., CREAGER, J. S. & DAWSON, W. A. 1963. Varved marine sediments in a stagnant fjord. *Science*, **141**, 918–919.

GUCLUER, S. M. 1962. *Recent sediments in Saanich Inlet, British Columbia*. MSc thesis, University of Washington.

GUCLUER, S. M. & GROSS, M. G. 1964. Recent marine sediments in Saanich Inlet, a stagnant marine basin. *Limology and Oceanography*, **9**, 359–376.

GUILBAULT, J.-P., PATTERSON, R. T., THOMSON, R. E., BARRIE, J. V. & CONWAY, K. W. 1997. Late Quaternary paleoceanographic changes in Dixon Entrance, northwest British Columbia, Canada: evidence from the foraminiferal faunal succession. *Journal of Foraminiferal Research*, **27**, 151–174.

HAY, M. B., PIENITZ, R. & THOMSON, R. E. 2003. Distribution of diatom surface sediment assemblages within Effingham Inlet, a temperate fjord on the west coast of Vancouver Island (Canada). *Marine Micropaleontology*, **48**, 291–320.

HAY, M. B., DALLIMORE, A., THOMSON, R. E., CALVERT, S. & PIENITZ, R. 2007. Siliceous microfossil record of late Holocene oceanography and climate along the west coast of Vancouver Island, British Columbia (Canada). *Quaternary Research*, **67**, 33–49.

HAY, M. B., CALVERT, S. E., PIENITZ, R., DALLIMORE, A., THOMSON, R. E. & BAUMGARTNER, T. R. 2009. Geochemical and diatom signatures of bottom-water renewal events in Effingham Inlet, British Columbia (Canada). *Marine Geology*, **262**, 50–61.

HEBDA, R. J. 1997. Impact of climate change on biogeoclimatic zones of British Columbia and Yukon. *In*: TAYLOR, E. & TAYLOR, B. (eds) *Responding to Global Climate Change in British Columbia and Yukon*. Environment Canada, 13.1–13.15.

HETHERINGTON, R. & BARRIE, J. V. 2004. Interaction between local tectonics and glacial unloading on the Pacific margin of Canada. *Quaternary International*, **120**, 65–77.

HETHERINGTON, R., BARRIE, J. V., REID, R. G. B., MACLEOD, R., SMITH, D. J., JAMES, T. S. & KUNG, R. 2003. Late Pleistocene coastal paleogeography of the Queen Charlotte Islands, British Columbia, Canada, and its implications for terrestrial biogeography and early postglacial human occupation. *Canadian Journal of Earth Sciences*, **40**, 1755–1766.

HETHERINGTON, R., BARRIE, J. V., REID, R. G. B., MACLEOD, R. & SMITH, D. J. 2004. Paleogeography, glacially induced crustal displacement, and Late Quaternary coastlines on the continental shelf of British Columbia, Canada. *Quaternary Science Reviews*, **23**, 295–318.

HOWES, D. E. 1983. Late Quaternary sediments and geomorphic history of northern Vancouver Island, British Columbia, Canada. *Canadian Journal of Earth Science*, **20**, 57–65.

HUTCHINSON, I. 1992. *Holocene sea-level change in the Pacific Northwest: a catalogue of radiocarbon dates and an atlas of regional sea-level curves*. Simon Fraser University, Institute of Quaternary Research Occasional Paper 1.

HUTCHINSON, I. & MCMILLAN, A. D. 1997. Archaeological evidence for village abandonment associated with Late Holocene earthquakes at the Northern Cascadia subdiction zone. *Quaternary Research*, **48**, 79–87.

HUTCHINSON, I., JAMES, T. S., CLAGUE, J. J., BARRIE, J. V. & CONWAY, K. W. 2004*a*. Reconstruction of late Quaternary sea-level change in southwestern British Columbia from sediments in isolation basins. *Boreas*, **33**, 183–194.

HUTCHINSON, I., JAMES, T., REIMER, P. J., BORNHOLD, B. D. & CLAGUE, J. J. 2004*b*. Marine and limnic radiocarbon reservoir corrections for studies of late- and postglacial environments in Georgia Basin and Puget Lowland, British Columbia, Canada and Washington, USA. *Quaternary Research*, **61**, 193–203.

HYNDMAN, R. D. 1995. Giant earthquakes of the Pacific Northwest. *Scientific American*, **276**, 1621–1623.

IVANOCHKO, T. S., CALVERT, S. E., SOUTHON, J. R., ENKIN, R. J., BAKER, J., DALLIMORE, A. & PEDERSEN, T. F. 2008*a*. Determining the post-glacial evolution of a northeast Pacific coastal fjord using a multiproxy geochemical approach. *Canadian Journal of Earth Sciences*, **45**, 1–14.

IVANOCHKO, T. S., CALVERT, S. E., THOMSON, R. E. & PEDERSEN, T. F. 2008*b*. Geochemical reconstruction of Pacific decadal variability from the eastern North Pacific during the Holocene. *Canadian Journal of Earth Sciences*, **45**, 1317–1329.

James, T. S., Clague, J. J., Wang, K. & Hutchinson, I. 2000. Postglacial rebound at the northern Cascadia subduction zone. *Quaternary Science Reviews*, **19**, 1527–1541.

James, T. S., Hutchinson, I. & Clague, J. J. 2002. *Improved relative sea-level histories for Victoria and Vancouver, British Columbia, from isolation-basin coring*. Geological Survey of Canada, Ottawa, Geological Survey of Canada, Current Research 2002-A16.

James, T. S., Hutchinson, I., Barrie, J. V., Conway, K. C. & Mathews, D. 2005. Relative sea-level change in the northern Strait of Georgia, British Columbia. *Geographie Physique et Quaternaire*, **59**, 113–127.

James, T., Gowan, E. J., Hutchinson, I., Clague, J. J., Barrie, J. V. & Conway, K. W. 2009. Sea-level change and paleogoegraphic reconstructions, southern Vancouver Island, British Columbia, Canada. *Quaternary Science Reviews*, **28**, 1200–1216.

Josenhans, H. W., Fedje, D. W., Conway, K. W. & Barrie, J. V. 1995. Post glacial sea-levels on the western Canadian continental shelf: evidence for rapid change, extensive subaerial exposure, and early human habitation. *Marine Geology*, **125**, 73–94.

Josenhans, H. W., Fedje, D., Pienitz, R. & Southon, J. 1997. Early humans and rapidly changing Holocene sea-levels in the Queen Charlotte Islands, Hecate Strait, British Columbia, Canada. *Science*, **227**, 71–74.

Kelsey, H. M., Nelson, A. R., Hemphill-Haley, E. & Witter, R. C. 2005. Tsunami history of an Oregon coastal lake reveals a 4,600 year record of great earthquakes on the Cascadia subduction zone. *Geological Society of America Bulletin*, **117**, 1009–1032.

Kelsey, H. M., Witter, R. C. & Hemphill-Haley, E. 2002. Plate-boundary earthquakes and tsunamis of the past 5,500 yr, Sixes River estuary, southern Oregon. *Geological Society of American Bulletin*, **114**, 298–314.

Kemp, A. E. S. 1996. Laminated sediments as paleo-indicators. *In*: Kemp, A. E. S. (ed.) *Palaeoclimatology and Palaeoceanography from Laminated Sediments*. Geological Society, London, Special Publications, **116**, vii–xii.

Kemp, A. E. S. 2003. Evidence for abrupt climate changes in annually laminated marine sediments. *In*: *Abrupt Climate Change; Evidence, Mechanisms and Implications*; Papers of a Discussion Meeting. *Philosophical Transactions-Royal Society*. Mathematical, Physical and Engineering Sciences, **361**, 1851–1870.

Kennett, J. P. & Ingram, B. L. 1995. A 20,000-year record of ocean circulation and climate change from the Santa Barbara Basin. *Nature*, **377**, 510–514.

Lemmen, D. S., Warren, F. J., Lacroix, J. & Bush, E. (eds) 2008. *From Impacts to Adaptation: Canada in a Changing Climate 2007*. Government of Canada, Ottawa.

Levson, V. 1992. Pioneering Geology in the Canadian Cordillera. *Proceedings of 'The Earth Before Us, Pioneering Geology in the Canadian Cordillera'*, March 1991. B.C Geological Survey Open File OF92-19. Available online at: http://www.empr.gov.bc.ca/Mining/geoscience/PublicationsCatalogue/OpenFiles/1992/1992-19/Pages/default.aspx.

Linden, R. H. & Schurer, P. J. 1988. Sediment characteristics and sea-level history of Royal Roads Anchorage, Victoria, British Columbia. *Canadian Journal of Earth Sciences*, **25**, 1800–1810.

Luternauer, J. L., Clague, J. J., Conway, K. W., Barrie, J. V., Blaise, B. & Mathewes, R. W. 1989*a*. Late Pleistocene terrestrial deposits on the continental shelf of western Canada: evidence for rapid sea-level change at the end of the last glaciation. *Geology*, **17**, 357–360.

Luternauer, J. L., Conway, K. W., Clague, J. J. & Blaise, B. 1989*b*. Late Quaternary geology and geochronology of the central continental shelf of western Canada. *Marine Geology*, **89**, 57–68.

Mackas, D. L. & de Young, B. 2001. GLOBEC Canada: 1996–2000: a sampler. *Response of marine ecosystems to environmental variability. Canadian Journal of Fisheries and Aquatic Sciences*, **58**, 685–702.

Mackie, A. P. & Sumpter, I. D. 2005. Shoreline settlement patterns in Gwaii Hannas during the early and late Holocene. *In*: Fedje, D. W. & Mathewes, R. W. (eds) *Haida Gwaii: Human History and Environment from the Time of Loon to the Time of the Iron People Since the Time of Raven*. UBC Press, University of British Columbia, Vancouver, Canada, 337–371.

Mandryk, C. A. S., Josenhans, H. W., Fedje, D. W. & Mathewes, R. W. 2001. Late Quaternary paleoenvironments of northwestern North America: implications for inland v. Coastal migration routes. *Quaternary Science Reviews*, **20**, 301–314.

McKechnie, I. 2005. *Five thousand years of fishing at a shell midden in the Broken Group Islands, Barkley Sound, British Columbia*. MSc Thesis, Department of Archaeology, Simon Fraser University, Burnaby, British Columbia.

McKechnie, I. 2007. Investigating the complexities of sustainable fishing at a prehistoric village on western Vancouver Island, British Columbia, Canada. *Journal for Nature Conservation*, **15**, 208–222.

McKechnie, I., Dallimore, A., Frederrick, G. & Smith, N. 2008. Marine fisheries and marine climate before and during the Little Ice Age on western Vancouver Island. *Society of America Archaeologists*, Annual Meeting, Vancouver, BC, March 2008.

McMillan, A. D. 1999. *Since the time of the transformers: the Ancient heritage of the Nuu-chah-nulth, Ditidaht, and Makah*. UBC Press, Vancouver.

McMillan, A. D. & Hutchinson, I. 1999. Archaeological evidence of village abandonment associated with late Holocene earthquakes at the northern Cascadia subduction zone. *Quaternary Research*, **48**, 79–87.

McMillan, A. D. & Hutchinson, I. 2002. Aboriginal traditions of paleoseismic events along the Cascadia subduction zone of western North America. *Ethnohistory*, **40**, 1.

McMillan, A. D. & Yellowhorn, E. 2003. *First Peoples in Canada*. 3rd edn. Douglas and McIntyre, Vancouver.

McQuoid, M. R. & Hobson, L. A. 2001. A Holocene record of diatom and silicoflagellate microfossils in sediments of Saanich Inlet, ODP Leg 169S. *Marine Geology*, **174**, 111–124.

MOREHEAD, M. D., SYVITSKI, J. P. & HUTTON, E. W. H. 2001. The link between abrupt climate change and basin stratigraphy: a numerical approach. *Global and Planetary Change*, **28**, 107–127.

MORTON, A. & PROCTOR, B. 2001. *Heart of the Raincoast: A Life Story*, Touchwood Editions, Victoria, Canada.

MOSHER, D. C. & MORAN, K. 2001. Post-glacial evolution of Saanich Inlet, British Columbia: results of physical property and seismic reflection stratigraphic analysis. *Marine Geology*, **174**, 59–77.

MOSHER, D. C. & HEWITT, A. T. 2004. Late Quaternary deglaciation and sea-level history of eastern Juan de Fuca Strait, Cascadia. *Quaternary International*, **121**, 23–39.

NATURAL RESOURCES CANADA 2007*a*. Paleoenvironmental Perspectives on Climate Change, Rapid Climate Changes and Extreme Weather Events on the Pacific Coast. http://ess.nrcan.gc.ca/ercc-rrcc/proj4/theme1/act4_e.php.

NATURAL RESOURCES CANADA 2007*b*. Paleoclimatic and paleo-environmental reconstructions of the Northwest Pacific region from Mt. Logan ice cores. Available online at: http://ess.nrcan.gc.ca/2002_2006/rcvcc/j27/2_1_e.php?p=1, accessed June 2010.

NATURAL RESOURCES CANADA 2009*a*. Preparing for earthquakes. Available online at: http://earthquakescanada.nrcan.gc.ca/info-gen/prepare-preparer/index-eng.php, accessed June 2010.

NATURAL RESOURCES CANADA 2009*b*. Offshore British Columbia, Review of the Federal Moratorium on Oil and Gas Activities Offshore British Columbia. Available online at: http://www.nrcan.gc.ca/eneene/sources/offext/offcbextcb-eng.php, accessed June 2010.

NEDERBRAGT, A. J. & THUROW, J. W. 2001. A 600 year record of Holocene climate in Saanich Inlet, British Columbia, from digital sediment color analysis of ODP Leg 169S cores. *Marine Geology*, **174**, 95–110.

NELSON, C. H., GOLDFINGER, C., JOHNSON, J. E. & DUNHILL, G. 2000. Variation of modern turbidite systems along the subduction zone margin of Cascadia Basin and implications for turbidite reservoir beds. *In*: WEIMER, P. W. & NELSON, C. H. (eds) *Deep-water Reservoirs of the World*. Gulf Coast Section Society of Economic Paleontologists and Mineralogists Foundation 20, Annual Research Conference.

NELSON, C. H., GOLDFINGER, C., JOHNSON, J., DUNHILL, G., VALLIER, T. L., KASHGARIN, M. & MCGUNN, R. 2003. *Turbidite event history, methods and implications for Holocene paleoseismicityof the Cascadia subduction zone*. U.S. Geological Survey Open File and Professional Paper.

NEPTUNE CANADA 2009. Transforming ocean science. Available online at: http://www.neptunecanada.ca, accessed June 2010.

NICKLEN, P. 2006. Where currents collide: in wild tides surging through the straits of Vancouver Island off British Columbia, marine life grows up strong and beautiful. *National Geographic*, **210**, 120–135.

O'CONNELL, J. M. & TUNNICLIFFE, V. 2001. The use of sedimentary fish remains for interpretation of long term fish population fluctuations. *Marine Geology*, **174**, 177–196.

OSTERBERG, E., MAYEWSKI, P. *ET AL*. 2008. Ice core record of rising lead pollution in the north Pacific atmosphere. *Geophysical Research Letters*, **35**, LO5810.

PATTERSON, R. T., GUILBAULT, J.-P. & THOMSON, R. E. 2000. Oxygen level control of foraminiferal distribution in Effingham Inlet, Vancouver Island, British Columbia. *Journal of Foraminiferal Research*, **30**, 321–335.

PELLATT, M. G., HEBDA, R. J. & MATHEWES, R. W. 2001. High-resolution Holocene vegetation history and climate from Hole 1034B, ODP leg 169S, Saanich Inlet, Canada. *Marine Geology*, **174**, 211–226.

PICKARD, G. L. 1975. Annual and longer term variations of deepwater properties in the coastal waters of southern British Columbia. *Journal of the Fisheries Research Board of Canada*, **32**, 1561–1587.

PICKARD, G. L. 1961. Oceanographic features of inlets in the British Columbia mainland coast. *Journal of the Fisheries Research Board of Canada*, **18**, 907–999.

PICKARD, G. L. 1963. Oceanographic characteristics of inlets of Vancouver Island, British Columbia. *Journal of Fisheries Research Board of Canada*, **20**, 1109–1144.

PIKE, J. & KEMP, A. E. S. 1996. Records of seasonal flux in Holocene laminated sediments, Gulf of California. *In*: KEMP, A. E. S. (ed.) *Palaeoclimatology and Palaeoceanography from Laminated Sediments*. Geological Society, London, Special Publications, **116**, 157–170.

POJAR, J., KLINKA, K. & DEMARCHI, D. A. 1991. Coastal western hemlock zone. *In*: MEIDINGER, D. & POJAR, J. (eds) *Ecosystems of British Columbia*. Victoria B.C. Research Branch, Ministry of Forests, 95–112.

PRIOR, D. B., BORNHOLD, B. D., COLEMAN, J. M. & BRYANT, W. R. 1982. Morphology of a submarine slide, Kitimat Arm, British Columbia. *Geology*, **10**, 588–592.

PRIOR, D. B., BORNHOLD, B. D., WISEMAN, W. J. & LOWE, D. R. 1987. Turbidity current activity in a British Columbia fjord. *Science*, **237**, 1330–1333.

PROCTOR, B. & MAXIMCHUK, Y. 2003. *Full Moon, Flood Tide: Bill Proctor's Raincoast*. Harbour Publishers, British Columbia, Canada.

REN, P., BORNHOLD, B. D. & PRIOR, D. B. 1996. Seafloor morphology and sedimentary processes, Knight Inlet, British Columbia. *Sedimentary Geology*, **1030**, 201–228.

RIDDIHOUGH, R. P. & HYNDMAN, R. D. 1991. Modern plate tectonic regime of the continental margin of western Canada. *In*: GABRIELSE, H. & YORATH, C. J. (eds) *Geology of the Cordilleran Orogen in Canada*. Geological Survey of Canada, Geology of Canada No. 4, Chapter 13, 435–455.

RODEN, G. I. 1989. Analysis and interpretation of long-term climate variability along the west coast of North America. *Geophysical Monographs*, **1**, 93–111.

ROGERS, G. C. 1988. An assessment of the megathrust earthquake potential of the Cascadia subduction zone. *Canadian Journal of Earth Sciences*, **25**, 844–852.

SANCETTA, C. 1996. Laminated diatomaceous sediments: controls on formation and strategies for analysis. *In*: KEMP, A. E. S. (ed.) *Palaeoclimatology and Palaeoceanography from Laminated Sediments*. Geological Society, London, Special Publications, **116**, 17–23.

SCHIMMELMAN, A. & LANGE, C. V. 1996. Tales of 1001 varves: a review of Santa Barbara Basin sediment studies. *In*: KEMP, A. E. S. (ed.) *Palaeoclimatology and Palaeoceanography from Laminated Sediments*. Geological Society, London, Special Publications, **116**, 121–142.

STUCCHI, D. 2008. Long term trends in deep water properties of B.C. inlets. Available online at: http://www.pac.dfo-mpo.gc.ca/SCI/osap/projects/bcinlets/default_e.htm, accessed June 2010.

SYVITSKI, J. P. M. & SHAW, J. 1995. Sedimentology and geomorphology of fjords. Chapter 5. *In*: PERILLO, G. E. M. (ed.) *Geomorphology and Sedimentology of Estuaires*. Elsevier, Developments in Sedimentology, **53**.

THOMSON, R. E. 1981. *Oceanography of the British Columbia Coast*. Special Publications of Fisheries and Aquatic Sciences, The Government of Canada, Ottawa, Ontario, **56**, 291.

THOMSON, R. E., BORNHOLD, B. D. & MAZZOTTI, S. 2008. *An examination of the factors affecting relative and absolute sea level in coastal British Columbia*. Canadian Technical Report of Hydrography and Ocean Sciences, Fisheries and Oceans Canada, **260**.

THOMSON, R. E. & GOWER, J. F. R. 1998. A basin-scale instability event in the Gulf of Alaska. *Journal of Geophysical Research*, **103**, 3033–3040.

THOMPSON, T. G. & BARKEY, K. T. 1938. Observations on fjord-waters. *Transactions, American Geophysical Union: Reports and Papers Oceanography*, 254–260.

TIMOTHY, D. A. & SOON, M. Y. S. 2001. Primary production and deep-water oxygen content of two British Columbian fjords. *Marine Chemistry*, **73**, 37–51.

TIMOTHY, D. A., SOON, M. Y. S. & CALVERT, S. E. 2003. Settling fluxes in Saanich and Jervis Inlets, British Columbia, Canada: sources and seasonal patterns. *Progress in Oceanography*, **59**, 31–73.

TULLY, J. P. 1949. *Oceanography and prediction of pulp mill pollution in Alberni Inlet*. PhD Thesis, University of Washington.

TURNER, N., IGNACE, M. & IGNACE, R. 2000. Traditional ecological knowledge and wisdom of Aboriginal Peoples in B.C. *Ecological Applications*, **10**, 1275–1287.

TUNNICLIFFE, V., O'CONNELL, J. M. & MCQUOID, M. R. 2001. A Holocene record of fish remains from the Northeastern Pacific. *Marine Geology*, **174**, 197–210.

UNIVERSITY OF VICTORIA 2009. VENUS: Victoria Experimental Network Under The Sea. Available online at: http://www.venus.uvic.ca/, accessed June 2010.

WAKE, C. P., DALLIMORE, A. & FISHER, D. A. 2006. *North Pacific Climate Workshop Final Report*. Available online at: http://ess.nrcan.gc.ca/extranet/npw/index_e.php (user name: npw password: npw123**).

WANG, K., HE, J., DRAGERT, H. & JAMES, T. S. 2001. Three-dimensional viscoelastic interseismic deformation model for the Cascadia subduction zone. *Earth Planets Space*, **53**, 295–306.

WARD, B. C., WILSON, M. C., NAGORSEN, D. W. & DRIVER, J. 2003. Port Eliza Cave: North American West Coast interstadial environment and implications for human migrations. *Quaternary Science Reviews*, **22**, 1383–1388.

WARE, D. M. & THOMSON, R. E. 2000. Interannual to multidecadal timescale climate variations in the Northeast Pacific. *Journal of Climate*, **13**, 3209–3220.

WHEELER, J. O. & MCFEELY, P. 1991. *Tectonic Assemblage Map of the Canadian Cordillera and Adjacent Parts of the United States of America*. Geological Survey of Canada, Map1712A, Scale 1:2 000 000.

WHEELER, J. O., BROOKFIELD, A. J., GABRIELSE, H., MONGER, J. W. H. & WOODSWORTH, G. J. 1991. *Terrane Map of the Canadian Cordillera*. Geological Survey of Canada, Map 1713A, Scale 1:2 000 000.

WRIGHT, C. A., DALLIMORE, A., THOMSON, R. E., PATTERSON, R. T. & WARE, D. M. 2005. Late Holocene paleofish populations in Effingham Inlet, British Columbia, Canada. *Palaeogeography, Palaeoclimatology, Palaeoecology*, **224**, 367–384.

YORATH, C. J. 1990. *Where Terranes Collide*. Orca Book Publishers, British Columbia, Canada.

Spatial and temporal influence of glaciers and rivers on the sedimentary environment in Sassenfjorden and Tempelfjorden, Spitsbergen

MATTHIAS FORWICK[1]*, TORE O. VORREN[1], MORTEN HALD[1], SERGEI KORSUN[2], YUL ROH[3], CHRISTOPH VOGT[4] & KYU-CHEUL YOO[5]

[1]*Department of Geology, University of Tromsø, N-9037 Tromsø, Norway*

[2]*Shirshov Institute of Oceanology, Nakhimovsky Prospect 36, Moscow 117995, Russia*

[3]*Department of Earth and Environmental Sciences, Chonnam National University, 300 Yongbong-Dong, Buk-Gu, Gwangju, 500-757, Republic of Korea*

[4]*Central Laboratory for Crystallography and Applied Material Sciences (ZEKAM), Geosciences, University of Bremen, Klagenfurter Str. 2, D-28359 Bremen, Germany*

[5]*Korea Polar Research Institute, Korea Ocean Research & Development Institute, Songdo TP, Incheon 406-840, Republic of Korea*

**Corresponding author (e-mail: Matthias.Forwick@uit.no)*

Abstract: Multiproxy analyses including hydrographical, geochemical, foraminferal, lithological and geophysical data reveal variable influences of the glaciers Tunabreen and von Postbreen as well as the river Sassenelva on the sedimentary environment in two Spitsbergen fjords during the Late Weichselian and the Holocene. Grounded ice covered the study area during the last glacial. The glacier fronts retreated stepwise during the latest Weichselian/earliest Holocene, and the glaciers were probably small during the early Holocene. A growth of Tunabreen occurred between 6 and 4 cal ka BP. Reduced input from Tunabreen from *c.* 3.7 cal ka BP was probably a result of suppressed iceberg rafting related to the enhanced formation of sea ice and/or reduced meltwater runoff. During the past two millennia, the glacier fronts advanced and retreated several times. The maximum Holocene glacier extent was reached at the end of a surge of von Postbreen in AD 1870. Characteristics of the modern glaciomarine environment include: (1) different colours and bulk-mineral assemblages of the turbid waters emanating from the main sediment sources; (2) variable locations of the turbid-water plumes as a consequence of wind forcing and the Coriolis effect; (3) stratified water masses during summers with interannual variations; (4) increasing productivity with increasing distance from the glacier fronts; (5) foraminifera-faunal assemblages typical for glacierproximal settings; and (6) periodical mass-transport activity.

Fjords are regarded as miniature oceans (Skei 1983), hence providing the opportunity to study a variety of sedimentary processes within a comparatively small area. Due to their vicinity to sediment sources, fjords are characterized by high sedimentation rates so that they act as climate archives with high temporal resolution. Because sea ice is absent during the summers, fjords on Spitsbergen are of particular interest because they allow glacial and non-glacial sedimentary processes to be studied relatively easily.

Spitsbergen fjords acted as pathways for fast-flowing ice streams during the Late Weichselian Glaciation (e.g. Landvik *et al.* 1998, 2005; Ottesen *et al.* 2005). The final glacier retreat to the inner fjords occurred stepwise and terminated *c.* 11.3/11.2 ka BP (e.g. Mangerud *et al.* 1992, 1998; Elverhøi *et al.* 1995; Svendsen *et al.* 1996; Lønne 2005; Baeten *et al.* 2010; Forwick & Vorren 2009). Tidewater glaciers apparently existed in Spitsbergen fjords throughout the entire Holocene (Hald *et al.* 2004; Baeten *et al.* 2010; Forwick & Vorren 2009). They were, however, relatively small during the early Holocene. A regional cooling starting at *c.* 9 to 8.8 cal ka BP led to the asynchronous growth of glaciers on Spitsbergen that reached their maximum Holocene extents at different times (Werner 1993; Huddart & Hambrey 1996; Svendsen & Mangerud 1997; Hald *et al.* 2004; Plassen *et al.* 2004; Forwick & Vorren 2007; Baeten *et al.* 2010; Forwick & Vorren 2009). Apart from surges, the glacier fronts generally retreated after *c.* AD 1900

From: Howe, J. A., Austin, W. E. N., Forwick, M. & Paetzel, M. (eds) *Fjord Systems and Archives.* Geological Society, London, Special Publications, **344**, 163–193.
DOI: 10.1144/SP344.13 0305-8719/10/$15.00

(Hagen *et al.* 1993; De Geer in Plassen *et al.* 2004; Ottesen & Dowdeswell 2006; Ottesen *et al.* 2008; Baeten *et al.* 2010).

Until now, the systematic study of the seafloor morphology in the Isfjorden area included the study of Late Weichselian glacial linear features in Isfjorden and Billefjorden (Ottesen *et al.* 2005; Baeten *et al.* 2010), glaciogenic landforms and deposits related to Neoglacial glacier re-advances and surges (Elverhøi *et al.* 1995; Boulton *et al.* 1996; Plassen *et al.* 2004; Ottesen & Dowdeswell 2006), Holocene mass-transport activity (Prior *et al.* 1981; Forwick & Vorren 2007) and pockmarks (Forwick *et al.* 2009).

In this paper, we present multiproxy analyses including hydrographic, geochemical, foraminiferal, lithological and acoustic data from the fjords Sassenfjorden and Tempelfjorden which receive sediments from rivers and glaciers. We provide an overview of the influence of the main sediment sources in the modern environment. Based on this, we reconstruct the influence of these sources on the sedimentary environment during the past *c.* 12 ka. We focus particularly on glacier growths and retreats and how they correlate with palaeoenvironmental changes in the region of Svalbard and the Barents Sea.

Physiographic setting

Sassenfjorden and Tempelfjorden are the easternmost tributaries of Isfjorden on Spitsbergen (Fig. 1). Sassenfjorden is 13 km long, up to 12 km wide and *c.* 130 km^2 large. The greatest water depth of about 150 m occurs at the transition to Isfjorden. A transverse and hummocky threshold crosses the fjord in the NW–SE direction. Tempelfjorden is *c.* 14 km long, up to 5 km wide and has an area of *c.* 57 km^2. It comprises two basins: a sinuous-shaped basin with a maximum water depth of 110 m in the central and outer fjord (referred to as 'main basin') and a smaller 'glacier-proximal basin' with up to 70 m water depth in the inner fjord. A NNW–SSE oriented terminal moraine, deposited during a surge of von Postbreen in AD 1870 (Plassen *et al.* 2004), separates these basins.

Two drainage basins surround the study area. The northern drainage basin is 785 km^2 large and has a glacier coverage of 58.4% (Hagen *et al.* 1993). It comprises the glaciers Tunabreen (tidewater glacier) as well as von Postbreen and Bogebreen (land-based glaciers) at the head of Tempelfjorden (Fig. 1c). Tunabreen and von Postbreen are the major sediment sources from this

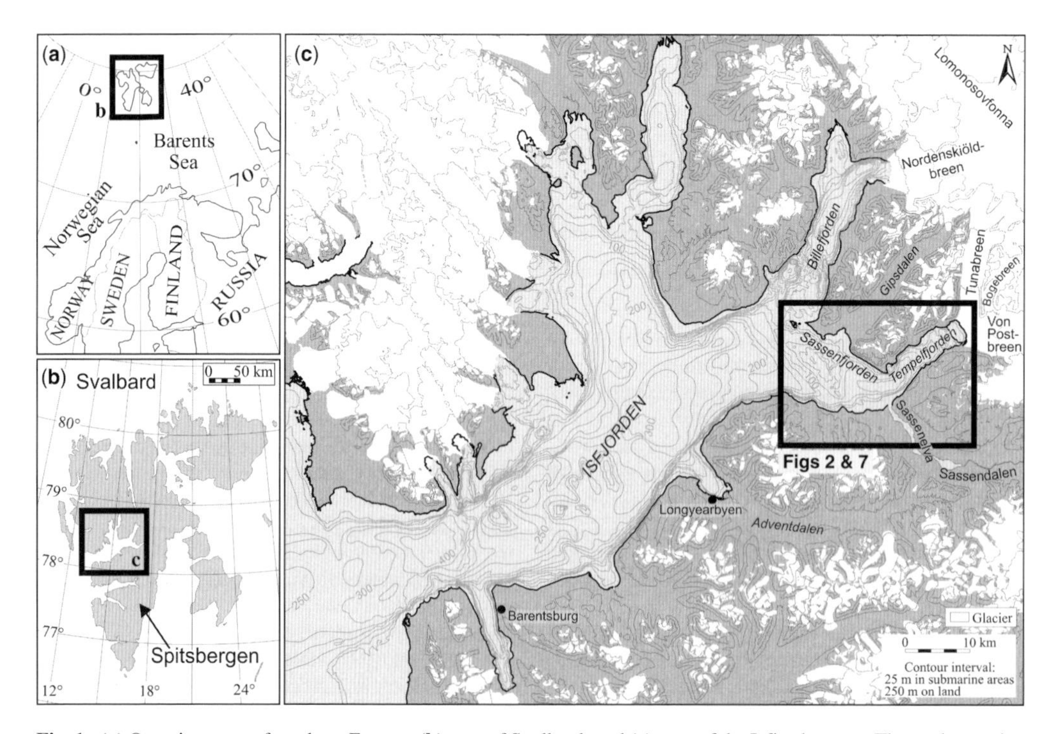

Fig. 1. (**a**) Overview map of northern Europe; (**b**) map of Svalbard; and (**c**) map of the Isfjorden area. The study area is marked with the frame. Locations mentioned in the text are indicated.

drainage area. The southern basin has a size of 1085 km^2 and a glacier coverage of 21.5% (Hagen *et al.* 1993). The river Sassenelva at the transition from Tempelfjorden to Sassenfjorden is the main sediment source from this drainage area (Fig. 1c).

Mesozoic and Late Palaeozoic sandstones and shales dominate the bedrock geology in the southern drainage basin (Dallmann *et al.* 2002). Silicified carbonates, carbonates, evaporitic sedimentary rocks and dolerite occur subordinately. Carbonate and evaporitic rocks dominate in the northern basin. However, the distribution of silicified carbonate and clastic sedimentary rocks can also be significant. A patch of dolerite occurs at the mouth of Sassenfjorden (Dallmann *et al.* 2002).

Sea ice usually forms in west Spitsbergen fjords from November and starts to break up between April and July (Węsławski *et al.* 1995; Svendsen *et al.* 2002; Nilsen *et al.* 2008). However, large interannual variations of the location of the ice edge as well as the timing of melting and break-up can occur. The formation of sea ice can occasionally be significantly reduced (Cottier *et al.* 2007; Gerland & Renner 2007).

Glacier surges in the study area occurred in AD 1870 (von Postbreen), AD 1930 and AD 1970 (Tunabreen), AD 1980 (Bogebreen; Hagen *et al.* 1993) and between *c.* AD 2002 and 2005 (Tunabreen; own observations).

Material and methods

This study is based on multiple analyses including the physical properties of the water column, the bulk-mineral assemblages of turbid waters and acoustic, lithological, geochemical and foraminifera-fauna analyses of the seafloor and the sub-seafloor. All material was collected onboard RV *Jan Mayen* of the University of Tromsø, Norway.

The physical properties of the water masses were investigated along a transect including 7 locations overnight on 8/9 July 2004 (stations 1 to 7) and at 8 locations during the afternoon and evening of 4 August 2006 (stations 1–8; Fig. 2a; Table 1). The measurements were carried out with a Seabird 911 current-temperature-depth profiler (CTD). A Seatech transmissometer with a path length of 25 cm was attached to the CTD for turbidity measurements. The transmissometer had a voltage (V) unit of 0–5 V that corresponds to 0–100% transmittance (Bishop 1986).

Geochemical analyses were carried out on surface samples (0–1 cm) from 7 box cores (50 × 50 × 60 cm^3 volume) and one gravity core retrieved on 4 August 2006 (stations 1 to 8; Fig. 2a; Table 1). Total carbon (TC) and total nitrogen (TN) contents were measured on powdered samples using a Carlo Erba NA-1500 elemental analyser. Total inorganic carbon (TIC) content was determined using a UIC carbon dioxide coulometer after dissolving the powdered samples using phosphoric acid. Total organic carbon (TOC) and the contents of carbonate minerals were calculated as the difference between total and inorganic carbon (Stein 1991).

Approximately 80–200 cm^3 of sediment from the uppermost centimetre of 8 box cores retrieved on 4 August 2006 (stations 1–8; Fig. 2a; Table 1) were mixed with 96% alcohol stained with rose bengal in order to distinguish living from dead benthic foraminifera species. The samples were subsequently wet sieved with mesh sizes of 125 μm and 1 mm and dried at 80 °C. Foraminifera tests were examined under a dissecting microscope and identified to the lowest possible taxon. Specimens with more than one brightly stained chamber were considered to have been alive at the time of collection. Both living and dead specimens were counted in the 0.125–1 mm and >1 mm fractions.

Ten gravity cores (GC) and one piston core (PC) were analysed for lithology, colour, bulk-mineral assemblages (only PC) and sedimentation rates. They were retrieved in 2002, 2003 and 2006 (Fig. 2b; Table 1). Both corers comprised a lead bomb of 1600 kg weight. Their core barrels were 6 m (GC) and 12 m (PC) long, respectively. The inner diameter of the plastic liners was 10 cm.

Prior to opening, the physical properties of the cores were measured with a GEOTEK multi-sensor core logger (MSCL). The parameters included bulk density, magnetic susceptibility (loop sensor) and fractional porosity. After opening, the magnetic susceptibility was measured with a point sensor (Bartington MS2EI). Furthermore, colour images of the sediment surfaces were acquired with the line-scan camera attached to the MSCL. Radiographs of half-core sections were acquired and described. A visual description was recorded as well as colour information based on the Munsell Soil colour charts.

Grain-size analyses were carried out on 1-cm thick slices from selected depths. The fractions <63 μm, 63–2000 μm and >2000 μm were separated.

The chronology for this study is partly based on 12 Accelerator Mass Spectrometer (AMS)-radiocarbon dates of molluscs (Table 2). The targets were produced at the Radiological Dating Laboratory in Trondheim, Norway, and the measurements were carried out at the Ångström Laboratory in Uppsala, Sweden. A marine reservoir effect of 440 years was applied (Mangerud & Gulliksen 1975). The radiocarbon ages were calibrated with the software programme Calib Rev 5.1 (Stuiver &

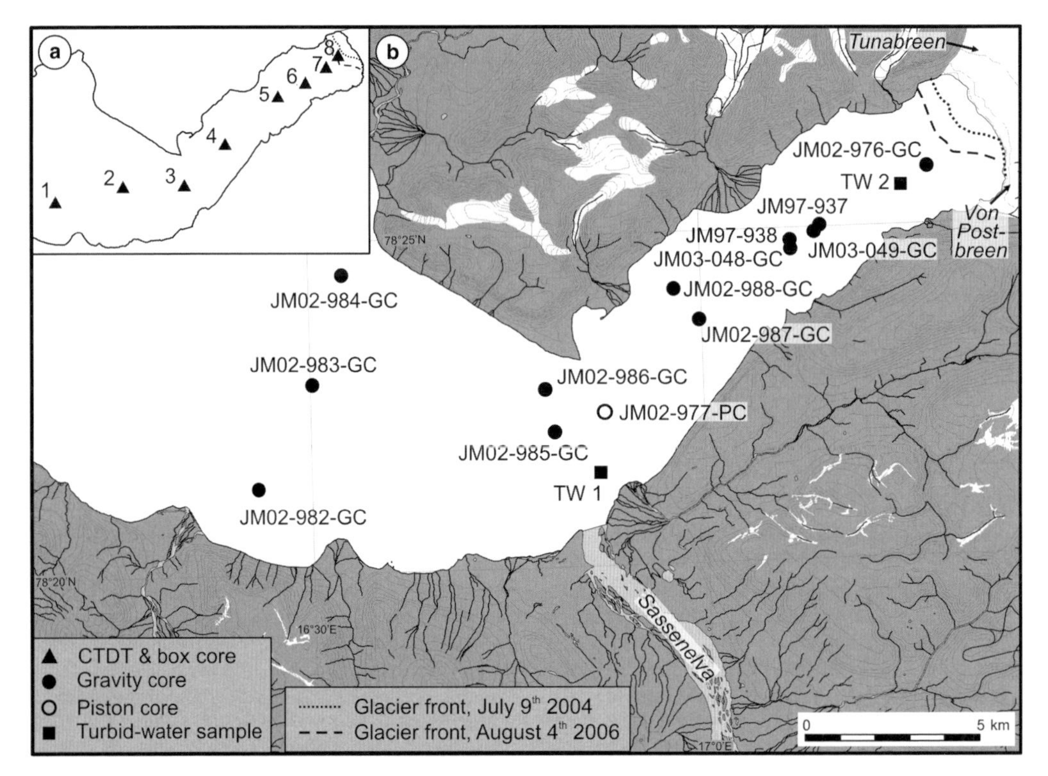

Fig. 2. Sampling stations and locations of the glacier fronts on 9 July 2004 and 4 August 2006. (**a**) locations of CTDT (conductivity-temperature-depth-transmissivity) and box-core stations. Note that these stations were sampled in 2004 and 2006. For exact positions see Table 1. In 2004, only the stations 1 to 7 were sampled because station 8 was covered by Tunabreen. (**b**) Gravity-core, piston-core and 'turbid-water' (TW) stations. The cores JM97-937 and JM97-938 were investigated by Plassen *et al.* 2004.

Reimer 1993) using $\Delta R = 99 \pm 39$ (regional average for Spitsbergen; http://radiocarbon.pa.qub.ac.uk/marine/). Results were obtained from the calibration data set marine04.14C of Hughen *et al.* (2004). Other chronological information was obtained from correlation with published data and varve counting.

Powder X-ray diffraction (XRD) analyses were carried out on bulk samples from the cores JM02-977-PC (43 samples), JM03-048-GC (46 samples) and on two samples of the turbid waters off the glaciers and the river (the turbid-water samples were retrieved on the early morning of 23 July 2003). The XRD measurements were performed with a Philips X'Pert Pro MD, Cu-radiation and X'Celerator detector system. A $1/4°$ 2θ fixed divergence slit with a first angle of $3°$ 2θ and a last angle of $85°$ 2θ was used, with a step size of $0.016°$ 2θ and a calculated time per step of 100 s. Semi-quantitative interpretations of the mineral content were carried out with Quantitative Phase-Analysis with X-ray Powder Diffraction (QUAX) (for further information see Vogt *et al.* 2002).

Swath bathymetry data were collected on 31 October and 1 November 2008 with a Kongsberg Maritime EM 300 multibeam echosounder. A sound–velocity profile through the water column for calibrating the equipment was recorded from a CTD cast prior to the data acquisition. The data were processed and corrected for tidewater variations using the software programme Neptune version 6.0. They are presented with the programme software ArcMap.

Sparker data were acquired in the summer of 1997 using a 700 J Bennex multi-electrode sparker. The digital interpretation of the seismic data was performed using Kingdom Software from Seismic Micro-Technology Inc. (versions 7.4, 7.5 and 8.2).

P-wave velocities in marine sediments can range between less than 1250 m s^{-1} and up to 2000 m s^{-1} depending on, for example, gas content, porosity and mineralogical composition (Hamilton 1985; MacKay *et al.* 1994). We applied a p-wave velocity of 1600 m s^{-1} to estimate thicknesses (see Elverhøi *et al.* 1995).

Table 1. *Sampling stations*

Station no.	Latitude (N)	Longitude (E)	Sampling date	Water depth (m)	Gear (Recovery (m))	Analysed parameters
Station 1						
JM04-236-CTDT	78°21.56′	016°27.98′	07/08/2004	65	CTDT	Hydrography
JM04-236-BC	78°21.56′	016°27.98′	07/08/2004	65	BC	Seafloor colour
JM06-175-CTDT	78°21.59′	016°27.73′	08/04/2006	64	CTDT	Hydrography
JM06-176-GC	78°21.54′	016°28.52′	08/04/2006	60	GC (*c.* 2.20)	Foraminifera
JM06-191-BC	78°21.68′	016°28.15′	08/04/2006	57	BC	Geochemistry
Station 2						
JM04-237-CTDT	78°22.05′	016°39.72′	07/08/2004	87	CTDT	Hydrography
JM04-237-BC	78°22.05′	016°39.72′	07/08/2004	87	BC	Seafloor colour
JM06-177-CTDT	78°22.07′	016°39.92′	08/04/2006	93	CTDT	Hydrography
JM06-178-BC	78°22.11′	016°40.56′	08/04/2006	95	BC	Geochemistry; foraminifera
Station 3						
JM04-238-CTDT	78°22.09′	016°51.14′	07/08/2004	100	CTDT	Hydrography
JM04-238-BC	78°22.09′	016°51.14′	07/08/2004	100	BC	Seafloor colour
JM06-179-CTDT	78°22.02′	016°50.94′	08/04/2006	101	CTDT	Hydrography
JM06-180-BC	78°22.02′	016°50.94′	08/04/2006	101	BC	Geochemistry; foraminifera
Station 4						
JM04-239-CTDT	78°23.54′	016°58.23′	07/08/2004	101	CTDT	Hydrography
JM04-239-BC	78°23.54′	016°58.23′	07/08/2004	101	BC	Seafloor colour
JM06-181-CTDT	78°23.34′	016°58.57′	08/04/2006	93	CTDT	Hydrography
JM06-182-BC	78°23.34′	016°58.97′	08/04/2006	88	BC	Geochemistry; foraminifera
Station 5						
JM04-240-CTDT	78°25.05′	017°08.19′	07/08/2004	71	CTDT	Hydrography
JM04-240-BC	78°25.05′	017°08.19′	07/08/2004	71	BC	Seafloor colour
JM06-183-CTDT	78°25.01′	017°08.11′	08/04/2006	73	CTDT	Hydrography
JM06-184-BC	78°24.96′	017°08.51′	08/04/2006	71	BC	Geochemistry; foraminifera
Station 6						
JM04-241-CTDT	78°25.47′	017°12.80′	07/09/2004	30	CTDT	Hydrography
JM04-241-BC	78°25.47′	017°12.80′	07/09/2004	30	BC	Seafloor colour
JM06-189-CTDT	78°25.47′	017°13.57′	08/04/2006	31	CTDT	Hydrography
JM06-190-BC	78°25.51′	017°13.81′	08/04/2006	38	BC	Geochemistry; foraminifera

(*Continued*)

Table 1. *Continued*

Station no.	Latitude (N)	Longitude (E)	Sampling date	Water depth (m)	Gear (Recovery (m))	Analysed parameters
Station 7						
JM04-242-CTDT	78°25.98′	017°17.10′	07/09/2004	37	CTDT	Hydrography
JM04-242-BC	78°25.98′	017°17.10′	07/09/2004	37	BC	Seafloor colour
JM06-185-CTDT	78°25.99′	017°16.93′	08/04/2006	40	CTDT	Hydrography
JM06-186-BC	78°25.96′	017°17.27′	08/04/2006	36	BC	Geochemistry; foraminifera
Station 8						
JM06-187-CTDT	78°26.35′	017°19.31′	08/04/2006	41	CTDT	Hydrography
JM06-188-BC	78°26.45′	017°18.97′	08/04/2006	40	BC	Geochemistry; foraminifera
Turbid waters						
TW 1	78°21.40′	016°52.00′	07/23/2003	Water surface	Bucket	Bulk-mineral assemblages, colour
TW 2	78°25.58′	017°15.71′	07/23/2003	Water surface	Bucket	Bulk-mineral assemblages, colour
Sediment cores						
JM02-976-GC	78°25.85′	017°17.78′	08/15/2002	28	GC (0.57)	Lithology, colour, chronology
JM02-982-GC	78°21.30′	016°25.80′	08/16/2002	82	GC (3.37)	Lithology, colour, chronology
JM02-983-GC	78°22.85′	016°30.16′	08/16/2002	38	GC (2.02)	Lithology, colour, chronology
JM02-984-GC	78°24.49′	016°32.61′	08/16/2002	57	GC (2.49)	Lithology, colour, chronology
JM02-985-GC	78°22.04′	016°48.65′	08/16/2002	102	GC (2.50)	Lithology, colour, chronology
JM02-986-GC	78°22.68′	016°47.99′	08/16/2002	78	GC (3.37)	Lithology, colour, chronology
JM02-987-GC	78°23.67′	016°59.83′	08/16/2002	89	GC (2.46)	Lithology, colour, chronology
JM02-988-GC	78°24.14′	016°57.97′	08/16/2002	98	GC (3.36)	Lithology, colour, chronology
JM03-048-GC	78°24.68′	017°07.02′	07/23/2003	77	GC (2.58)	Lithology, colour, chronology, bulk-mineral assemblages
JM03-049-GC	78°24.93′	017°08.87′	07/23/2003	71	GC (2.43)	Lithology, colour, chronology
JM02-977-PC	78°22.32′	016°52.45′	08/15/2002	96	PC (8.49)	Lithology, colour, chronology, bulk-mineral assemblages
JM97-937	78°25.01′	017°09.40′	09/26/1997	66	GC (3.18)	Correlation with core JM03-049-GC
JM97-938	78°24.80′	017°07.05′	09/26/1997	76	GC (4.15)	Correlation with core JM03-048-GC

Note: The cores JM97-937 and JM97-938 were described by Plassen *et al.* (2004). The following gear was used: GC, gravity corer; PC, piston corer; BC, box corer; CTDT, conductivity-temperature-depth-transmissivity; TW, turbid-water sample retrieved with a plastic bucket.

Table 2. *Overview of radiocarbon ages and calibrated ages*

Core	Depth (cm)	Material	Lab. no.	^{14}C age ($\pm 2\sigma$)	Calibrated age BP (1σ)	Calibrated age BP (2σ)
JM02-982-GC	305	Gastropoda indet. (fragment)	TUa-5136	8125 ± 45	8995–9155	8955–9269
JM02-983-GC	151	*Margarites cinereus* (well preserved)	TUa-5137	7685 ± 50	8409–8544	8356–8627
JM02-984-GC	248.5	Bivalvia indet. (fragment)	TUa-5138	8230 ± 50	9116–9313	9033–9392
JM02-985-GC	238.5	*Thyasira equalis* (fragments of paired shell)	TUa-5139	2075 ± 40	1975–2127	1895–2220 (99.4%) 2225–2235 (0.06%)
JM02-986-GC	305.5	*Thyasira equalis* (paired) *Portlandia Lenticula* (one valve)	TUa-5140	4300 ± 40	4808–4936	4714–4756 (2.9%) 4764–5030 (97.1%)
JM02-987-GC	239.5	*Nucula tenuis* (paired)	TUa-5141	2435 ± 35	2417–2610 (95.7%) 2632–2643 (4.3%)	2348–2674
JM02-988-GC	325.5	*Cyclichna albe* (well preserved)	TUa-5142	2380 ± 40	2331–2502	2306–2615 (97.2%) 2623–2650 (2.8%)
JM02-977-PC	187.5	*Yoldiella lenticula* (paired shell)	TUa-7370	1775 ± 35	1624–1773	1553–1832
JM02-977-PC	436	*Yoldiella lenticula* (paired shell)	TUa-7371	3675 ± 40	3950–4120	3868–4206
JM02-977-PC	517.5–519.5	*Buccinidae* sp. (partly damaged)	TUa-7372	4635 ± 40	5274–5415	5131–5159 (1.6%) 5168–5470 (98.4%)
JM02-977-PC	568–569	*Yoldiella lenticula* (paired shell)	TUa-7373	5430 ± 45	6133–6267	6010–6299
JM02-977-PC	774	*Nuculana pernula* (one valve)	TUa-7374	8160 ± 45	9031–9209	8982–9295

Note: The radiocarbon ages are corrected for a marine reservoir age of 440 years (Mangerud & Gulliksen 1975).

Results

Hydrography

The CTD measurements reveal stratified water masses in Sassenfjorden and Tempelfjorden during the summers of 2004 and 2006, respectively (Fig. 3). In 2004, the water temperatures ranged from *c.* +4 °C at the surface to *c.* −1.5 °C in the deepest part of the basin. They varied between +8 °C at the surface and *c.* 0 °C in the deepest part of the basin in 2006. A thermocline occurred in both years at around 10 m water depth.

The salinities varied in 2004 between 26.62 psu at the surface and 34.24 psu in the basin. In 2006, they ranged from 10.82 psu at the surface to 34.49 psu in the basin, and a pronounced halocline developed in the uppermost 15 m of the water column.

Lower transmissivity values generally indicate higher turbidity in the entire water column during the measurements in 2004 compared to 2006. In both years, the most turbid water occurred off Tunabreen and von Postbreen and a layer of more turbid water in the uppermost 5–10 m extended over the lengths of these transects. In 2004, comparatively high turbidities were measured at the water surface at station 2 and in the deepest part of the main basin (Fig. 3).

Sediment sources: colours and bulk-mineral assemblages

Tunabreen and von Postbreen, as well as Sassenelva, are the main sediment sources in the study area. Both sources supply turbid waters (Fig. 4a, b). However, Tunabreen also releases icebergs that carry fine-grained and coarser sediments (Fig. 4d, e).

The turbid water off Sassenelva is brown in colour (Fig. 4a). It comprises *c.* 70% phyllosilicates (mainly illite and mica, but also chlorite, kaolinite and 'mixed-layered clays'; Fig. 4c). Other characteristic minerals are quartz and plagioclase. The turbid waters emanating from Tunabreen and von Postbreen are reddish-beige in colour (Fig. 4b). Their mineral assemblage is characterized by *c.* 30% ankerite/dolomite, 18% calcite and 10% epidote, as well as almost 30% phyllosilicates (mainly illite and mica; Fig. 4c).

Seafloor-sediment colour

The surfaces of the box cores retrieved from stations 1 to 7 in 2004 were photographed in order to obtain information about the colours of the fjord floor. At the westernmost three stations, the sediment colour was brownish (Fig. 4f–h). Reddish patches

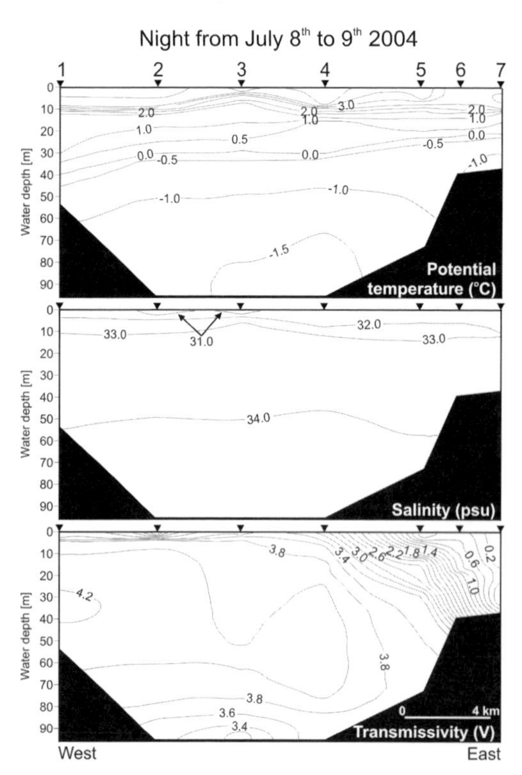

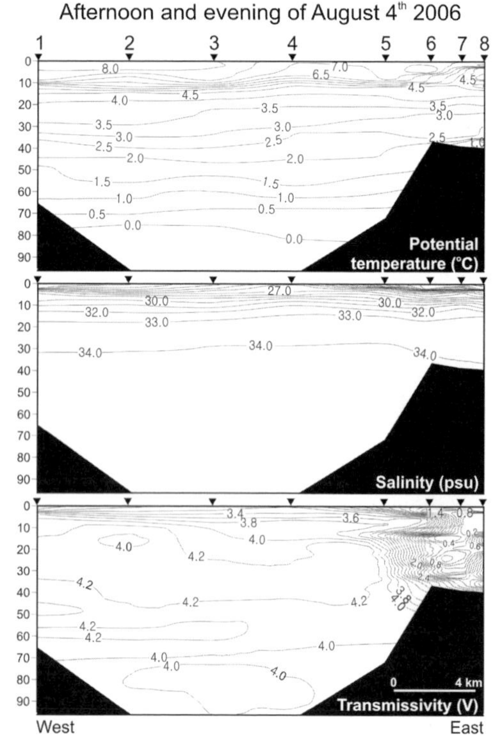

Fig. 3. CTDT transects (for locations see Fig. 2a; Table 1).

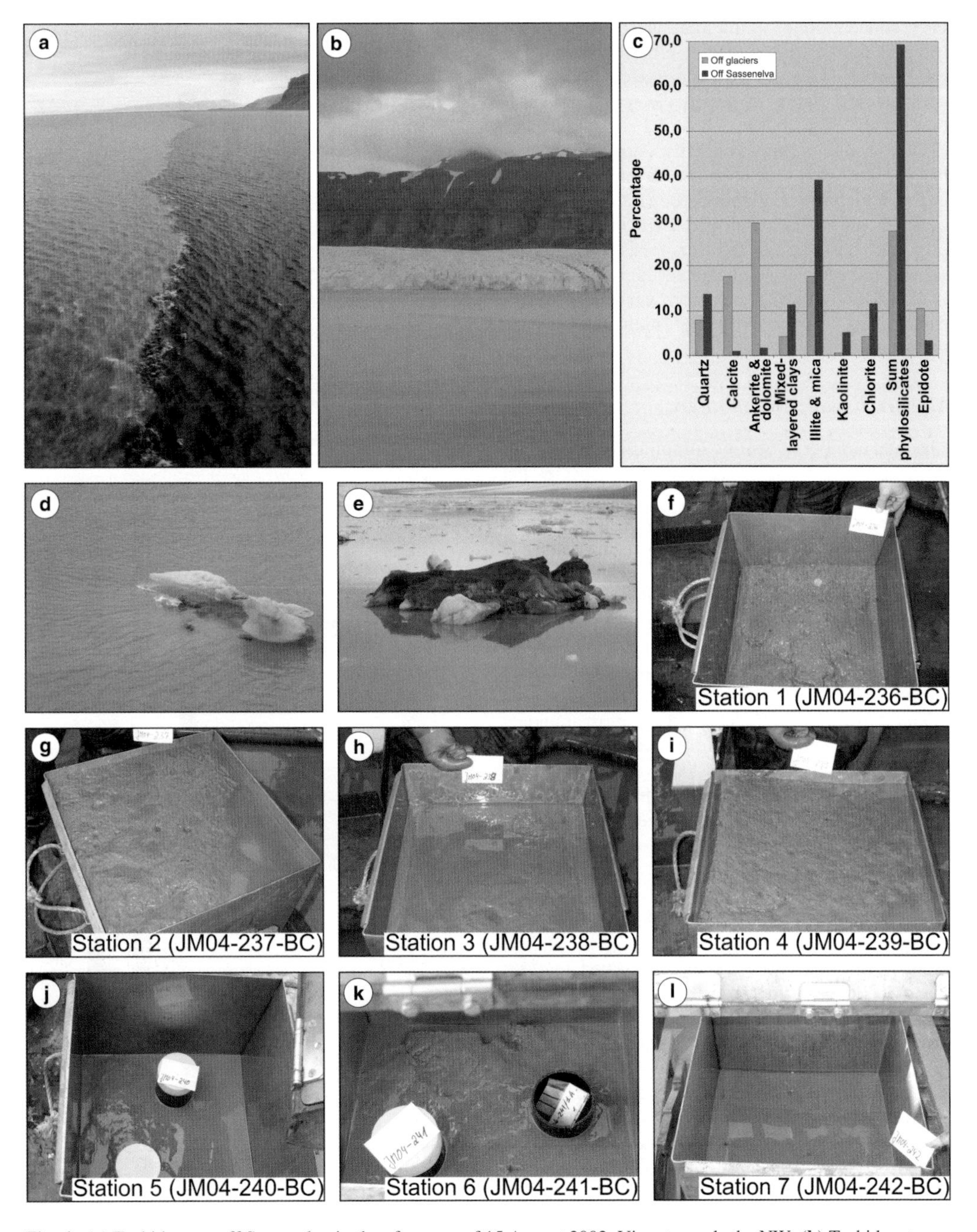

Fig. 4. (**a**) Turbid water off Sassenelva in the afternoon of 15 August 2002. View towards the NW. (**b**) Turbid water off Tunabreen and von Postbreen on 31 July 2005. (**c**) Bulk-mineral assemblages of the turbid waters off Sassenelva and Tunabreen/von Postbreen. (**d**), (**e**) Icebergs released during a surge of Tunabreen (photographs taken on 9 July 2004). (**f**)–(**l**) Surfaces of the box cores JM04-236-BC (station 1) to JM04-242-BC (station 7; for location see Fig. 2a; Table 1).

occurred at station 4 (Fig. 4i) and, at the three glacier-proximal stations, the sediment surface was mainly red with a minor component of brown (Fig. 4j–l).

Seafloor-sediment geochemistry

The geochemistry of the surface sediments shows marked lateral variations (Fig. 5). Total nitrogen

(TN) concentrations in the main basin are generally twice as high as in the glacier-proximal basin. The total carbon (TC) and total inorganic carbon (TIC) contents are higher in the glacier-proximal basin than in the main basin. The total organic carbon (TOC) contents are generally similar in both basins, but reach a peak at the easternmost station of the main basin. Calcium carbonate ($CaCO_3$) percentages are at least twice as high in the glacier-proximal basin as in the main basin. The C/N ratio is more or less constant in the outer basin. It is significantly higher and more variable from the easternmost station of the main basin and further east.

Modern benthic foraminifera

The total number of benthic foraminifera per 10 cm^3 of sediment, as well as the number of species per sample, increase from the glacier front towards the west (Fig. 6; Table 3). Simultaneously, the percentage of calcareous and living foraminifera decreases. *Elphidium excavatum f. clavata* dominates the assemblage in the glacier-proximal basin (stations 6–8). Its percentage drops markedly to the west of the moraine separating the basins in Tempelfjorden. *Cassidulina reniforme*, *Quinqueloculina stalkeri*, *Nonionellina labradorica* and *Islandiella norcrossi/helenae* are the most abundant species in the central part of the study area (stations 3–5). At the two westernmost stations (1 and 2), the foraminifera fauna is significantly influenced by the agglutinating species *Labrospira crassimargo*, *Recurvoides turbinatus* and *Reophax curtus/scorpius* (Fig. 6; Table 3).

Acoustic data

Glaciogenic landforms. Linear to curved features occur in Sassenfjorden and the southern slope of central Tempelfjorden (Fig. 7). They are generally oriented parallel to the fjord axis. Sparker data reveal that these features can be related to landforms with a relatively rough morphology, deposited immediately above bedrock (Fig. 8). Their internal reflection pattern is chaotic. We interpret these features as glacial lineations that formed beneath fast-flowing ice (compare with Ottesen *et al.* 2005; Baeten *et al.* 2010). Their curvature is most likely related to changes in the direction of the fjord axis or to bedrock highs.

One bent sinuous-shaped ridge, overlying the glacial lineations, occurs off Sassenelva (Fig. 7). On the sparker profile, it appears as a multicrest feature with absent to chaotic acoustic signature covering glacial lineations (Fig. 8b). We interpret the ridge as an esker that was formed by sedimentary infilling of a subglacial conduit (compare with

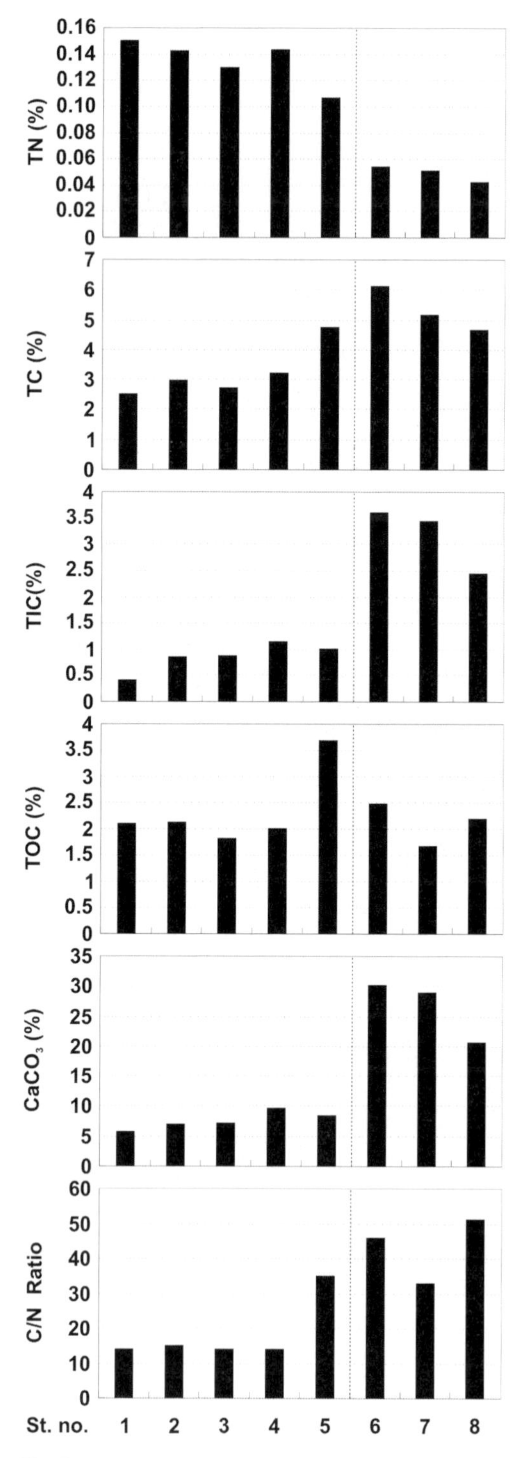

Fig. 5. Geochemistry of the seafloor on 4 August 2006 (for locations see Fig. 2a; Table 1). The dotted vertical lines separate the main basin (stations 1 to 5) and the proximal basin (stations 6 to 8).

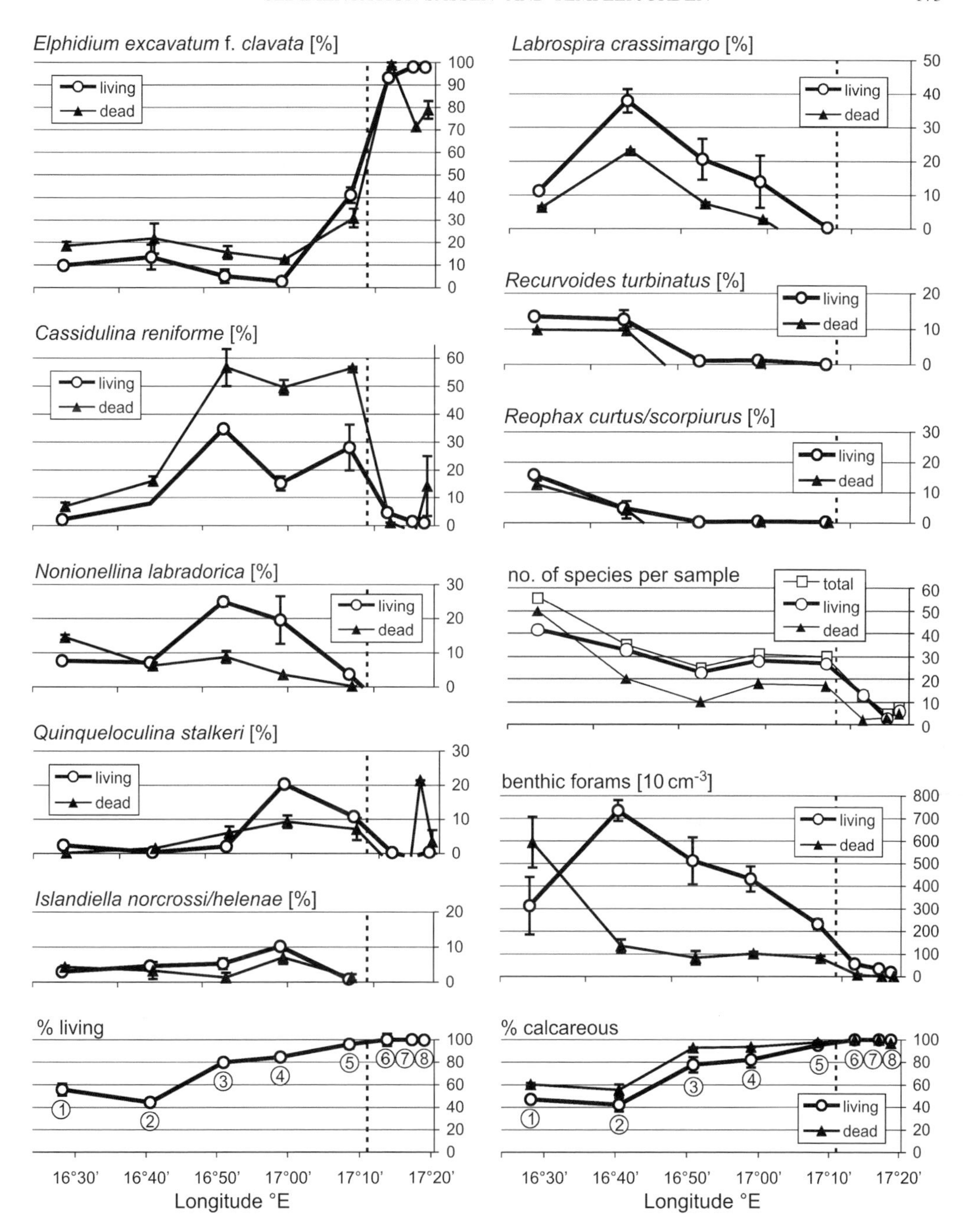

Fig. 6. Benthic-foraminifera-fauna assemblage at the seafloor on 4 August 2006. Exclusively species over 10% abundance are shown. For more detailed information about the assemblages, see Table 3. Station numbers are indicated in the lowermost figure in each column (for locations see Fig. 2a; Table 1). The dotted vertical lines separate the main basin (stations 1–5) and the proximal basin (stations 6–8).

Ottesen *et al.* 2008). For additional support of this suggestion, we refer to the discussion.

Several transverse ridge-like features are located in the central and eastern parts of the main basin (Fig. 7). These are smoothed expressions of positive depositional landforms with absent to chaotic internal signature occurring above the acoustic basement (Fig. 8b, c). Because of their relatively large number

Table 3. *Overview of dead and living benthic foraminifera from sediment-surface samples retrieved on 4 August 2006*

Station no.	Living foraminifera (%)								Dead foraminifera (%)							
	1	2	3	4	5	6	7	8	1	2	3	4	5	6	7	8
Original station no. (JM-XXX-BC)	191	178	180	182	184	190	186	188	191	178	180	182	184	190	186	188
Longitude °E	16.5	16.7	16.8	17.0	17.1	17.2	17.3	17.3								
Water depth (m)	57	95	101	88	71	38	36	40								
Hyperammina subnodosa	1.4	0.5						0.2	0.5	0.1						
Armorella sp.									0.2			0.4	0.7			
Hippocrepina indivisa	1.4			0.3					0.7	0.4						
Ammodiscus sp.	0.1	0.1		0.1	0.1				0.2	0.7			0.2			
Reophax curtus/scorpiurus	15.9	4.8	0.3	0.5	0.3				12.7	4.3		0.4	0.2			
Reophacidae spp.	0.2				0.1				0.9							
Cuneata arctica									0.0							
Ammotium cassis	1.0									1.3						
Labrospira crassimargo	11.3	37.9	20.6	14.0	0.3				6.4	23.2	7.5	2.8				
Recurvoides turbinatus	13.6	12.7	1.0	1.2	0.1				9.8	9.6		0.4				
Adercotryma glomerata	2.9	0.1							2.1							
Portatrochammina karica	0.2								1.4							1.7
Trochamminella atlantica	0.4	0.1							0.5							
Trochamminella bullata									0.0							
Spiroplectammina biformis	0.6	0.4	1.1	0.5	3.5	0.1			1.2			2.3	0.9			
Textularia torquata		0.1							0.0							
Egerella advena									0.1							
Agglutinated indefinite			0.3		0.3											
Silicosigmoilina groenlandica	3.4	0.5							2.9	3.6			0.2			
Cornuspira sp.	0.2	0.1	0.1		0.2			0.3								
Sigmoilina sp.				0.3												
Siphonaperta agglutinata	2.5	0.6		0.3					0.3	0.4		0.4				
Quinqueloculina seminula	0.1	0.1	0.4			0.8										
Quinqueloculina stalkeri	2.4	0.4	2.1	20.3	10.8	0.3		0.4	0.2	1.6	6.1	9.3	7.1		21.4	3.4
Quinqueloculina sp1		0.5	0.3	0.1	0.7						0.7					
Triloculina trihedra	0.1			0.4	0.2				0.0			4.8	0.5			
Pyrgo williamsoni	2.3	3.6	0.7	4.4	1.3	0.1			0.2	0.9		0.9			7.1	
Miliolidae spp.	0.1	1.0	0.1	3.8	1.9	0.3	0.5					0.4				

Cibicides lobatulus	1.5		0.3						2.4							
Rosalina sp.	2.9								0.0							
Buccella tenerrima	0.1								2.1							
Buccella frigida		0.3	0.1	2.8	0.2				0.5	3.7		0.9	0.2			
Epistominella vitrea	0.6			0.3					0.5							
Astrononion gallowayi	1.4	0.3		0.1	0.6				2.3	0.4						
Nonionellina labradorica	7.7	7.1	24.9	19.6	3.8				14.5	6.2	8.8	3.6	0.2			
Haynesina orbiculare						0.2										
Elphidium incertum	0.1	0.1	0.1	0.1	1.1	0.2			0.2			0.7	0.2			
Elphidium frigidum									0.2							
Elphidium subarcticum									0.2							
Elphidium bartletti	2.0	0.1	0.3		0.1				1.8	0.4		1.5				
Elphidium excavatum f. clavata	9.8	13.5	5.0	2.8	41.1	93.2	98.0	98.0	18.4	21.8	15.5	12.4	30.9	98.9	71.4	78.9
Cassidulina reniforme	2.1	7.9	34.8	15.2	28.0	4.7	1.5	1.1	6.9	16.0	56.7	49.6	56.5	1.1		14.2
Cassidulina terretis									0.1							
Islandiella norcrossi/helenae	2.9	4.6	5.3	10.2	0.8				4.4	3.3	1.3	7.0	1.4			
Islandiella islandica									0.2		0.7		0.2			
Angulogerina fluens	2.2								1.1							
Bulimina marginata									0.0							
Stainforthia loeblichi	1.3	0.3		0.5	0.2				2.2	0.9						
Globobulimina auriculata + turgida	0.2	0.2	0.5						0.3	0.9	2.0	2.0				
Bolivina pseudopunctata											0.7					
Robertina arctica	4.8	1.6	1.3	2.0	4.1											
Polymorphinidae spp.	0.1	0.3	0.1		0.1	0.1			0.6				0.2			
Astacolus hyalacrulus				0.1					0.0							
Dentalina spp.	0.1		0.2		0.2	0.1			0.2				0.1			
Lagena spp.	0.1								0.2							1.7
Oolina sp.									0.0							
Fissurina spp.	0.1															
Calcareous indefinite		0.1						0.1								
Benthic foramanifera (10 cm^{-3})	313	735	512	431	232	57	35	19	594	135	83	102	82	8	1	1
Benthic foraminifera counted	1132	1074	682	750	1235	901	403	731	2011	193	109	184	427	110	14	33
No. of species per sample	42	33	23	28	27	13	3	6	50	20	10	18	17	2	3	5
Living/total (%)	55.8	44.4	79.9	84.4	95.8	99.9	100	99.6								
% Calcareous	47.0	42.3	77.8	82.3	95.2	99.9	100	99.7	60.4	55.4	92.7	93.6	97.6	100	100	97.0
% Specimens >1 mm	1.49	0.57	0.04	0.01	0	0	0	0.01	0.58	0.31	0	0	0.06	0	0	0

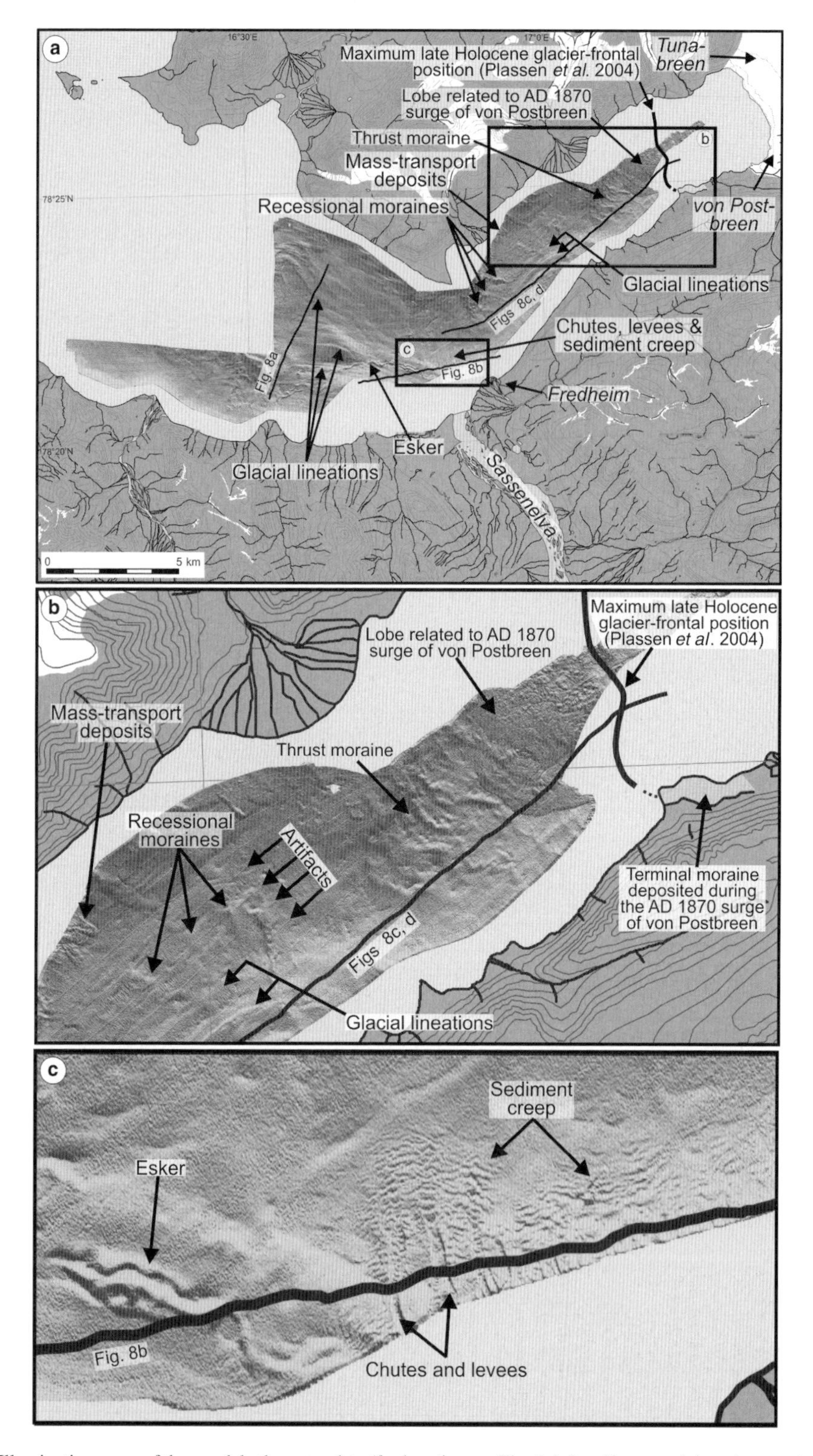

Fig. 7. Illumination maps of the swath bathymetry data (for location see Fig. 1c). Landforms and deposits mentioned in the text are indicated. Black lines with reference to Figure 8 mark the locations of sparker profiles. (**a**) bathymetry map of the study area; (**b**) bathymetry in the inner parts of the main basin; (**c**) bathymetry off Sassenelva.

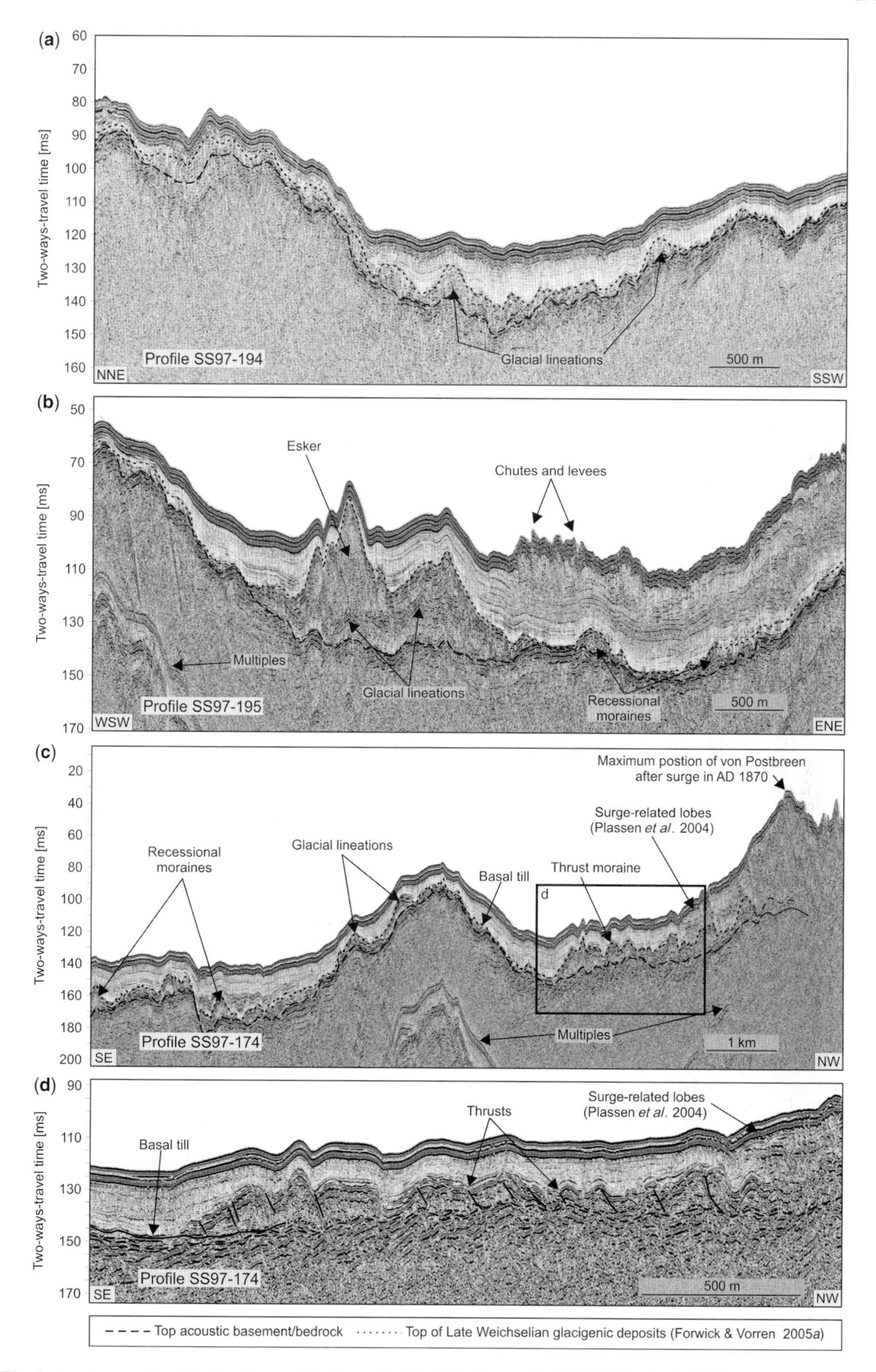

Fig. 8. Sparker profiles (for location see Fig. 7): (**a**) SS97-194; (**b**) profile SS97-195; (**c**), (**d**) SS97-174.

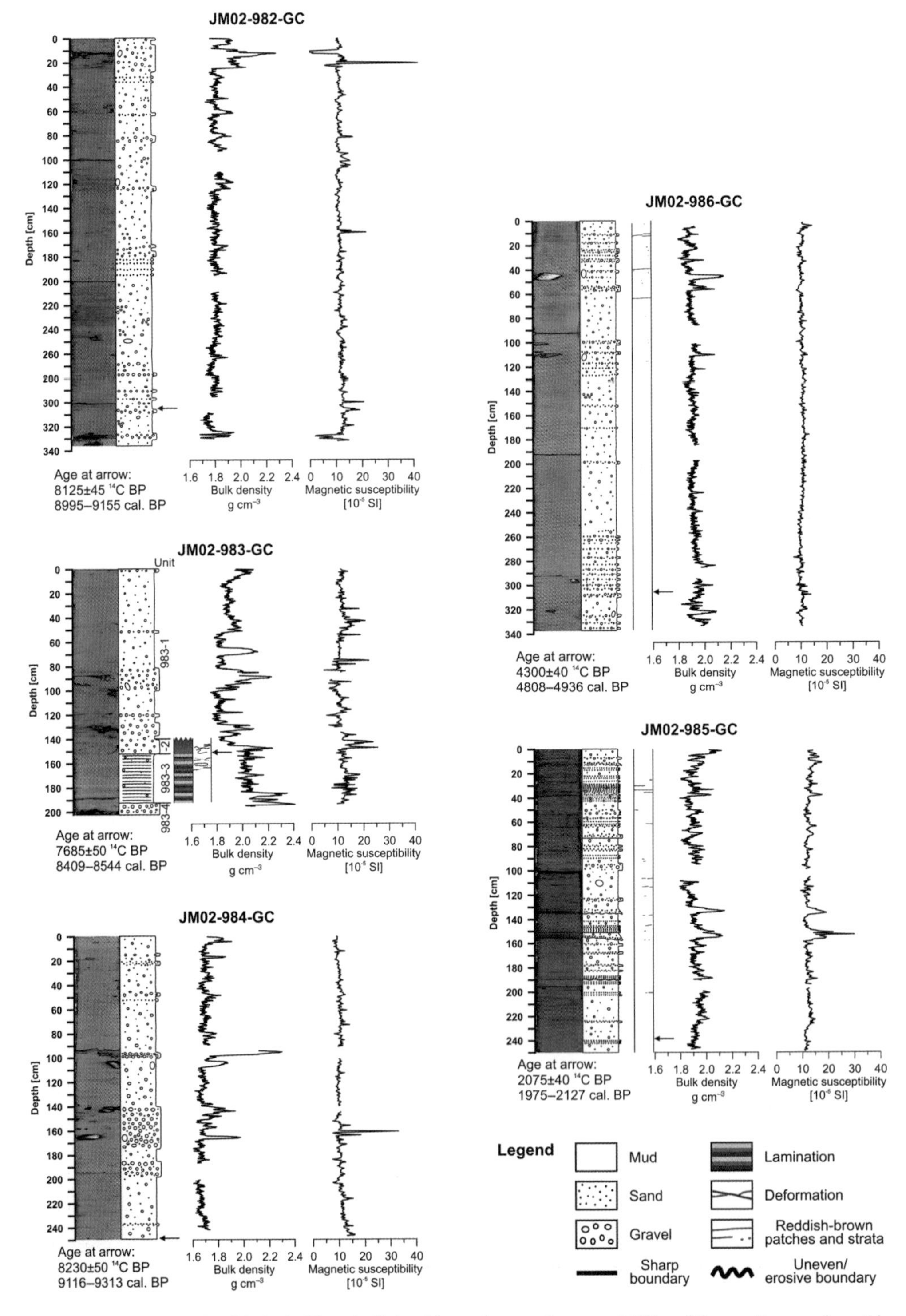

Fig. 9. Colour photographs, lithological logs, bulk densities and magnetic susceptibilities of the gravity cores from this study. Arrows indicate the sampling depths of the dating material. Reservoir-corrected radiocarbon and calendar ages are shown.

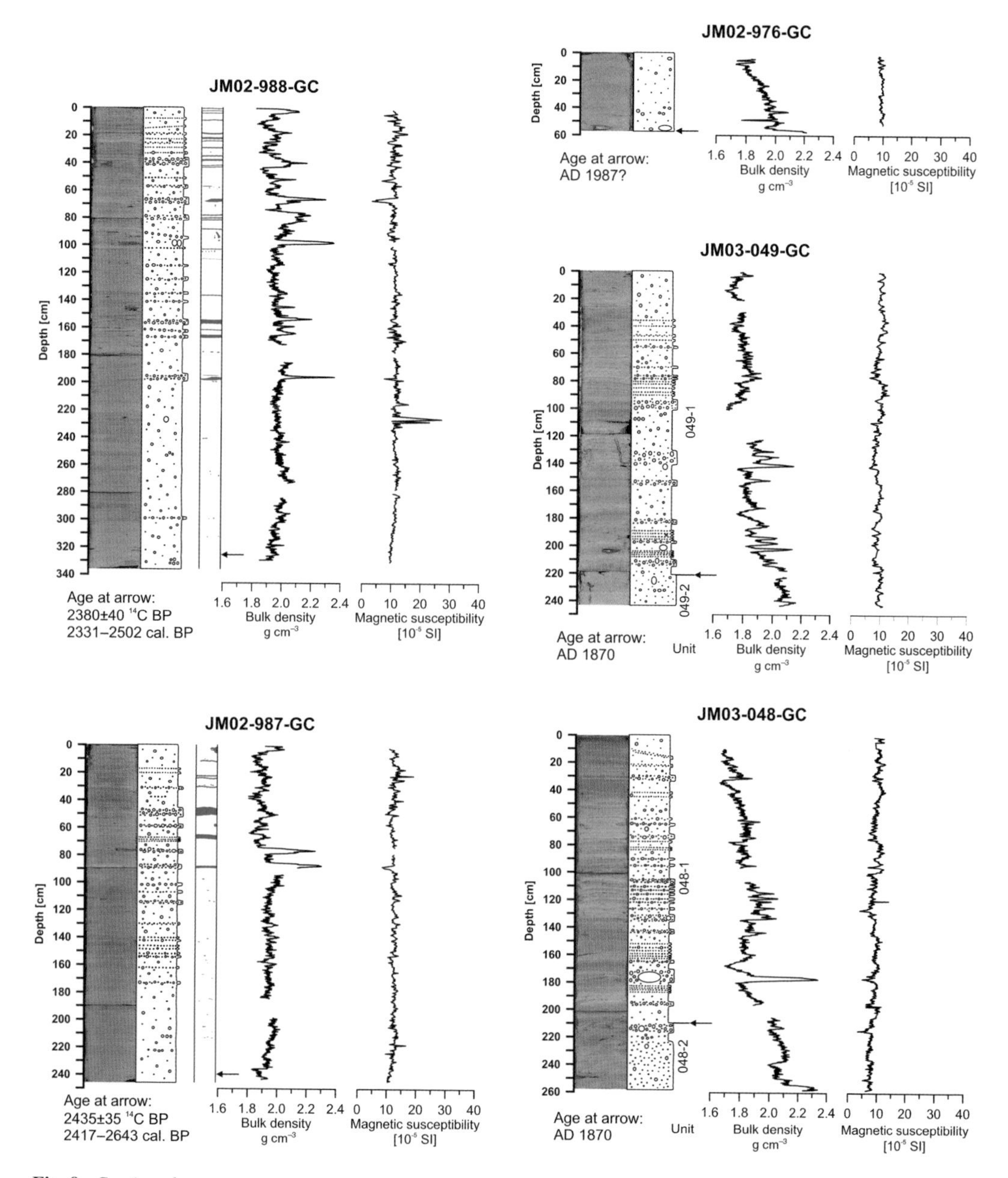

Fig. 9. *Continued.*

and regular spacing, as well as the similarity to other transverse ridges from Spitsbergen (Ottesen *et al.* 2008; Baeten *et al.* 2010), we suggest that they are recessional moraines that were deposited during halts or minor re-advances of the glacier fronts during a period of general glacier retreat.

Plassen *et al.* (2004) described a glaciogenic debris lobe that was deposited beyond the limit of a surge of Paulabreen in AD 1870. This lobe has a 'fresh' appearance with clearly defined limits and more subtle surface morphology (e.g. creep and block-like features; Fig. 7).

Another lobate feature occurs beneath this lobe and extends further to the west. Its morphology is less clearly outlined. Several curved and irregular ridges occur on top of this lobe. The sparker data reveal a multicrested, (at least) 3 km long, 1.5 km wide and up to 13 m high deposit with internal thrusts that covers basal till (Fig. 8c, d). The comparatively dense spacing of the crests, as well as the greater thickness of the deposit, may indicate a different depositional process compared to the recessional moraines. We assume that this feature is a thrust moraine (compare with Bennett 2001).

Non-glaciogenic landforms. Several incisions and ridges occur off the mouth of Sassenelva (Figs 7 & 8b). These are interpreted to be chutes and levees that are the result of down-slope processes (e.g. mass wasting, hyperpycnal flows) related to the progradation of the delta of Sassenelva. A wavy seafloor at the lower slope indicates sediment creep.

Lobate features on the northern slope of Tempelfjorden reflect mass-transport activity along the fjord walls (Fig. 7b). The 'fresh' appearance of some features indicates that they were deposited relatively recently.

Sediment cores: lithology

Ten gravity cores and one piston core were analysed (Figs 2, 9 & 10; Table 1). Unless otherwise stated, all cores comprise one lithological unit and have the following properties. They are composed of bioturbated massive mud with up to 7 cm large clasts (Figs 9–11; Table 4). The clasts occur irregularly as single grains, in layers or in random accumulations. The sediment colour was generally very dark grey (Munsell Code 2.5Y 3/1) to black (2.5Y 2.5/1) immediately after opening. When exposed to air, it changed to very dark greyish brown (2.5Y 3/2) with a trace of reddish colour. An increase in the amount of reddish sediments occurs from the west to the east. Marked increases in the bulk density usually reflect intervals of coarser lithological composition (Figs 9 & 10). The magnetic susceptibility generally decreases towards the head of Tempelfjorden.

The cores JM02-982-GC, JM02-983-GC (units 983-1 to 983-4) and JM02-984-GC were retrieved from Sassenfjorden (Figs 2b & 9; Table 1). Unit 983-2 (140–152 cm) consists of a massive diamicton with a sharp lower and a gradational upper boundary. It is slightly more reddish in its lower parts. Unit 983-3 (152–192 cm) comprises laminated mud with comparatively low clast content. The colours of the laminae vary between dark reddish brown to dark brown (5YR 3/2 to 7.5 YR 3/2) and dark greyish brown (2.5 YR 3/2). Some folding and faulting occurs in the lower parts of unit 983-2 and the upper parts of unit 983-3. Unit 983-4 (192–202 cm) comprises a massive, slightly consolidated diamicton of dark reddish brown/dark brown (5YR 3/2/7.5 YR 3/2) to dark greyish brown (2.5 YR 3/2) colour. Bioturbation is absent below 110 cm in core JM02-983-GC.

The cores JM02-985-GC to JM02-988-GC and JM02-977-PC were retrieved from the transition between Sassen- and Tempelfjorden and in outer Tempelfjorden (Figs 2b, 9 & 10; Table 1). In comparison to the cores from Sassenfjorden, the sediment colour in the cores JM02-986-GC to JM02-988-GC is slightly more reddish and the colour in the cores JM02-977-PC and JM02-985-GC is slightly more brownish. Patches (usually <5 mm) of reddish-brown sediments (2.5 YR 4/4 to 5YR 4/4) occur occasionally. Reddish-brown strata of up to 3 cm thickness (mostly <1 cm) are generally limited to the upper parts of the cores from this area. They occur generally more frequently and are more pronounced in the cores JM02-987-GC and JM02-988-GC. In some occasions, the reddish-brown strata coincide with higher amounts of clasts and lower magnetic susceptibility (e.g. JM02-987-GC, *c.* 89 cm; JM02-988-GC, 68 cm). Several sandy strata with sharp or erosive lower boundaries occur in the cores JM02-977-PC and JM02-985-GC.

The cores JM03-048-GC and JM03-049-GC were retrieved from central Tempelfjorden (Figs 2b & 9; Table 1). They are both divided into two units. The units 048-1 (0–210 cm) and 049-1 (0–220 cm) comprise couplets of brown to dark greyish brown (7.5YR 4/2 to 10YR 3/2) and reddish brown (5YR 4/3) strata. More beige-coloured intervals occur occasionally, mostly within the reddish strata. Individual strata can be correlated between the two cores. We suggest that these couplets are glaciomarine varves and that the brown strata were deposited during summer when Sassenelva is not frozen (compare with Plassen *et al.* 2004). The units 048-2 (210–258 cm) and 049-2 (220–243 cm) are massive. They are generally reddish-brown in colour with some influence of greyish brown.

Plassen *et al.* (2004) described the two cores JM97-937 and JM97-938 containing the units 937-1, 937-2, 938-1 and 938-2, respectively (for location see Fig. 2b; Table 1). The cores from our study can be correlated to the cores JM97-937 and JM97-938 (JM03-049-GC and JM97-937; JM03-048-GC and JM97-938). We therefore refer to Plassen *et al.* (2004) for a more detailed description of the lithology and the origin of the glaciomarine varves.

Core JM02-976-GC was retrieved from the glacier-proximal basin. The sediment colour is generally reddish brown (5YR 4/3). However, some beige and more brownish intervals occur. It might be reasonable to assume that the brownish sediments derive from Sassenfjorden, hence reflecting summer deposition because the river is frozen during winter (compare with Plassen *et al.* 2004). Based on this, we suggest that core JM02-976-GC comprises glaciomarine varves.

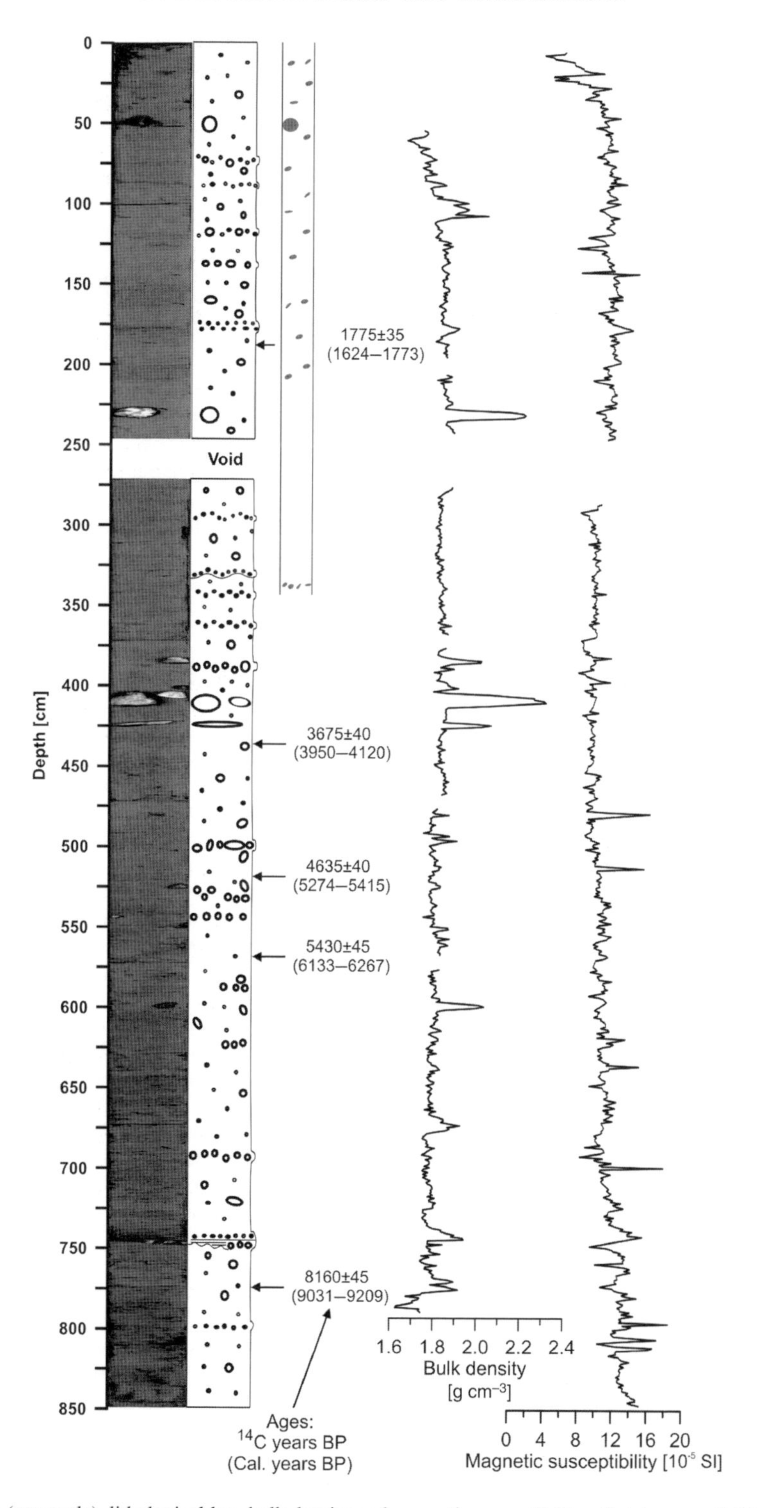

Fig. 10. Photo (greyscale), lithological log, bulk density and magnetic susceptibility of piston core JM02-977-PC. Arrows indicate the sampling depths of the dating material. Reservoir-corrected radiocarbon and calendar ages are shown. The column with the black dots to the right of the lithological log indicates the occurrence of reddish-brown patches. (For legend, see Fig. 9.)

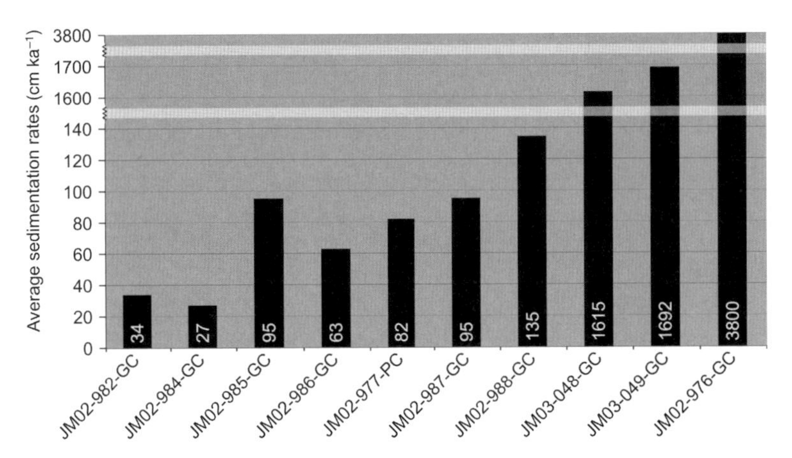

Fig. 11. Sedimentation rates. The rates for the cores JM02-982-GC to JM02-988-GC and JM02-977-PC are based on radiocarbon dates. The value for core JM02-977-PC shows the average sedimentation rate from the oldest radiocarbon data to the present. Note that the sedimentation rates in this core varied through time (see Fig. 12b). The sedimentation rates for the cores JM03-048-GC and JM03-049-GC are based on the duration of deposition of the units 048-1 and 049-1, respectively, starting at *c.* AD 1870. The value for core JM02-976-GC is based on varve counting.

Sediment cores: chronology and sedimentation rates

We sampled material for radiocarbon dating from the lowest possible level in the cores JM02-982-GC to JM02-988-GC, with the purpose of obtaining maximum ages for the oldest sediments in these cores as well as to estimate linear sedimentation rates. Furthermore, we took five samples from core JM02-977-PC with the purpose of dating marked changes in the bulk-mineral assemblages in order to estimate sedimentation rates. The sampling depths and ages are shown on Figures 9 & 10 and in Table 2, respectively. All dates provided Holocene ages and are regarded to be correct, with exception of the date from core JM02-983-GC.

The material in this core was found in deformed and more reddish sediments close to the base of unit 983-2 (Fig. 9; Table 2). It provided a radiocarbon age of 7685 $\pm$ 50 years (8477 cal a BP). If this date was correct the average sedimentation rate at the core site would have been *c.* 18 cm ka^{-1}, that is, it would have been significantly lower than at the sites of cores JM02-982-GC and JM02-984-GC that were also retrieved in Sassenfjorden. The location of core JM02-983-GC at shallower water depth than the other two cores (see Table 1) would in fact support a lower sedimentation rate because the sedimentation rates in the Isfjorden area are generally lower in shallower areas (see Forwick & Vorren 2005*a*). However, the deformation most probably indicates that the dated material has been reworked (see 'Sedimentary processes', below). We therefore exclude this result from further discussion.

Unit 983-3 was most probably deposited during the latest Weichselian or the earliest Holocene because the stratification of differently coloured sediments is assumed to require the proximity to several sediment sources (see 'Sedimentary processes', below). Its deposition occurred most likely during a late stage of the last glaciation when one or several glacier fronts were located close to the core site. This must have taken place after the retreat of the glacier fronts from the Younger Dryas maximum glacier extent in central Isfjorden (Forwick & Vorren 2005*b*) and prior to the termination of this retreat in the inner parts of Tempelfjorden. Our material does not provide the opportunity to date the end of this retreat but, based on studies from other Spitsbergen fjords, we assume that it took place c. 11.3–11.2 cal ka BP (Mangerud *et al.* 1992, 1998; Elverhøi *et al.* 1995; Svendsen *et al.* 1996; Hald *et al.* 2004; Lønne 2005; Baeten *et al.* 2010; Forwick & Vorren 2009). We suggest that unit 983-4 was deposited during the last glacial or earlier, because it directly underlies the stratified glaciomarine sediments.

The deposition of units 048-1 and 049-1 commenced around AD 1870, that is, after the termination of the surge of von Postbreen (based on correlations to the cores JM97-938 and JM97-937 described by Plassen *et al.* 2004 – see 'sediment cores: lithology'). Because hiatuses are absent in both cores, we suggest that the units 048-2 and 049-2 were deposited during a final phase of this surge or shortly after. Core JM02-976-GC comprises 15 couplets of glaciomarine varves, suggesting that base of the core dates to *c.* AD 1987.

Table 4. *Overview of the granulometric compositions at selected depths in the gravity cores and the piston core*

Core and sampling depth (cm)	Weight % <63 μm	Weight % 63–2000 μm	Weight % >2000 μm
JM02-976-GC			
9–10	99.35	0.51	0.14
49–50	99.88	0.12	0.00
JM02-982-GC			
9–10	89.80	10.16	0.04
109–110	95.51	4.34	0.15
209–210	97.23	2.63	0.14
309–310	96.43	3.08	0.48
JM02-983-GC			
9–10	96.50	3.46	0.04
109–110	88.27	8.45	3.28
159–160	97.49	2.03	0.49
189–190	96.05	3.84	0.11
JM02-984-GC			
9–10	98.02	1.88	0.11
109–110	96.55	3.45	0.00
209–210	96.55	3.30	0.16
JM02-985-GC			
9–10	97.02	2.81	0.18
109–110	92.89	5.12	1.99
209–210	97.73	2.27	0.00
JM02-986-GC			
9–10	96.50	3.43	0.07
104–105	96.81	2.79	0.39
209–210	96.56	3.30	0.14
319–320	85.44	5.49	9.07
JM02-987-GC			
9–10	98.43	0.98	0.58
109–110	93.86	1.51	4.63
209–210	98.17	1.76	0.06
JM02-988-GC			
9–10	98.73	1.27	0.00
109–110	99.35	0.65	0.00
209–210	99.18	0.82	0.00
309–310	98.47	1.35	0.18
JM03–048-GC			
9–10	99.31	0.53	0.16
109–110	98.08	0.81	1.11
219–220	97.83	1.35	0.82
JM03-049-GC			
9–10	99.46	0.43	0.11
109–110	98.60	1.37	0.03
209–210	79.08	5.69	15.23
224–225	93.54	5.05	1.41
JM02-977-PC			
19.5–20.5	93.25	5.92	0.83
69.5–70.5	94.51	4.97	0.52
119.5–120.5	96.86	2.25	0.89
169.5–170.5	91.55	1.68	6.78
219.5–220.5	98.29	1.71	0.00
293.5–294.5	98.61	1.39	0.00
343.5–344.5	97.99	1.85	0.16
393.5–394.5	97.99	2.01	0.00
443.5–444.5	97.48	1.68	0.84
493.5–494.5	98.42	1.58	0.00
543.5–544.5	97.14	2.25	0.62
593.5–594.5	97.60	2.37	0.03
643.5–644.5	98.53	1.47	0.00
693.5–694.5	96.57	3.02	0.41
743.5–744.5	82.29	17.71	0.00
794.5–794.5	99.18	0.82	0.00
843.5–844.5	98.65	1.18	0.17

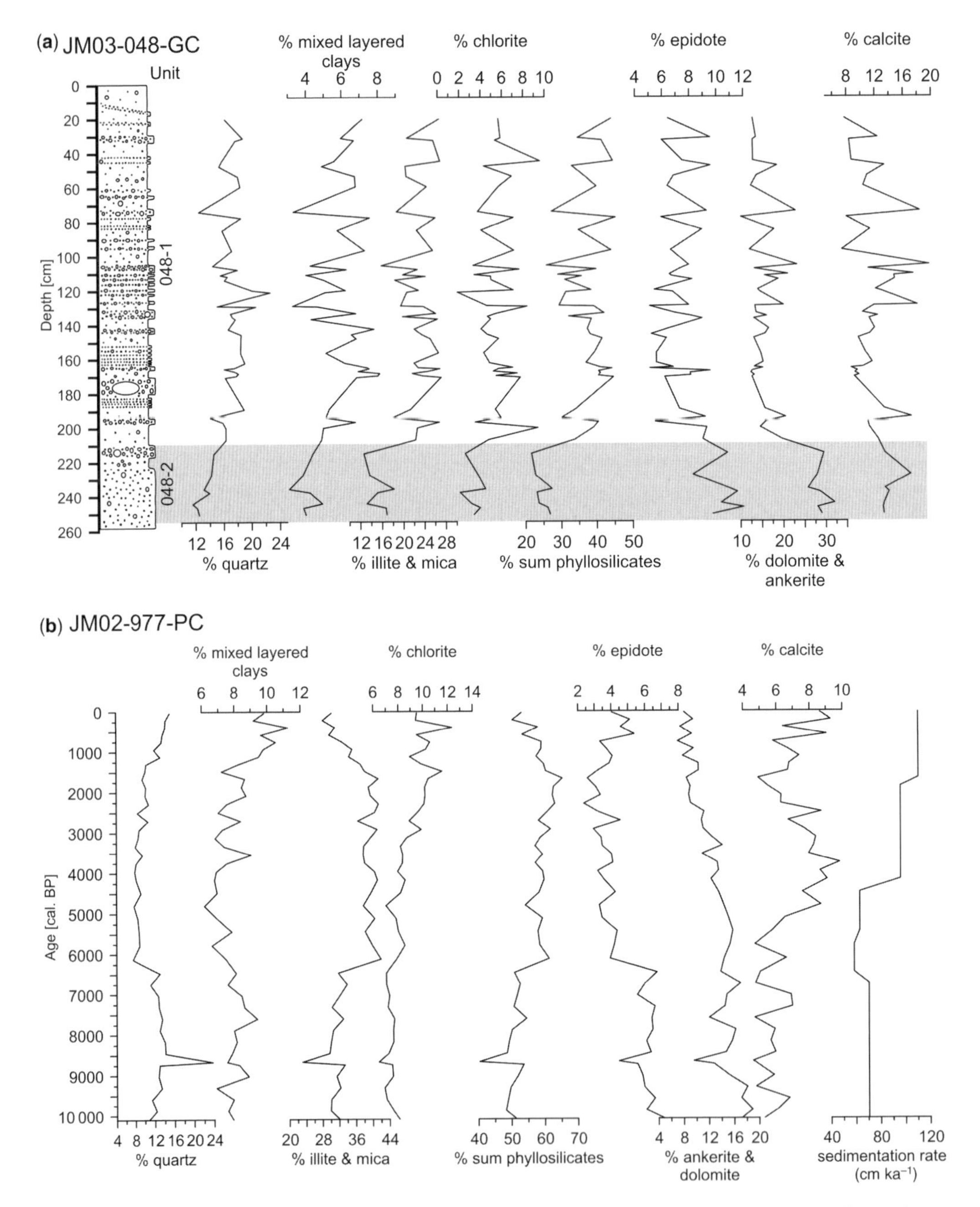

Fig. 12. Bulk-mineral assemblages (**a**) in core JM03-048-GC v. depth, and (**b**) JM02-977-PC v. age. Sedimentation rates for the latter core are shown in the right column. Note that the age model below 9120 cal a BP is tentative, because it is based on a continued sedimentation rate from the overlying interval.

The sedimentation rates vary between 27 cm ka^{-1} in the outer parts of the study area and 3800 cm ka^{-1} in the vicinity of the present glacier front (Fig. 11). They increase in the proximity of the main sediment sources, that is, towards the glaciers and off Sassenelva (cores JM02-977-PC, JM02-985-GC). The chronology in core JM02-977-PC reveals that sedimentation rate in outer Tempelfjorden varied over time, reaching a minimum between *c.* 6.5 and 4.5 cal ka BP (Fig. 12b).

It subsequently increased, reaching the highest values at the present.

Sediment cores: bulk-mineral assemblages

We investigated the bulk-mineral assemblages for the cores JM02-977-PC and JM03-048-GC with the purpose of (1) identifying temporal variations in the sediment input from the glaciers and the river, and (2) distinguishing sediments supplied from Tunabreen and von Postbreen.

The most abundant minerals in core JM03-048-GC are phyllosilicates (mainly illite and mica), quartz, dolomite, calcite and epidote (Fig. 12a). The glaciomarine varves comprising unit 048-1 are characterized by relatively high but fluctuating percentages of phyllosilicates and quartz, but comparatively low concentrations of ankerite/dolomite and epidote. Higher percentages of phyllosilicates and quartz generally correlate with brownish strata; higher calcite and ankerite/dolomite percentages with reddish strata. The fluctuations are less pronounced in unit 048-2 and the percentages of ankerite/dolomite and epidote are relatively high, whereas the concentrations of clay minerals and quartz are generally low. The percentages of calcite fluctuate throughout the core, but no trend changes occur (Fig. 12a).

The sediments in core JM02-977-PC generally comprise more than 50% phyllosilicates: mainly illite and mica, mixed-layered clays and chlorite (Fig. 12b). Other abundant minerals are ankerite/dolomite, calcite, quartz and epidote. The marked peak in most logs at *c.* 8650 cal a BP shows the composition of a mass-transport deposit. Five levels with marked changes in the mineralogical composition should be noted: (1) a marked increase of phyllosilicates (due to an increase in illite & mica) and a decrease in quartz and epidote occurred around 6.5 cal ka BP; (2) an increase in the calcite percentages starting at *c.* 5.6 cal ka BP; (3) an onset of decreasing calcite percentages around 3.7 cal ka BP; (4) the onset of decreasing phyllosilicate percentages (due to illite and mica decrease), as well as increasing calcite and epidote percentages at *c.* 1750 cal a BP; and (5) another increase in calcite and epidote around 650 cal BP.

Discussion

Sedimentary processes

We suggest that most of the analysed sediments were deposited in a glaciomarine environment where sediment supply occurred from glaciers, rivers and sea ice. The preservation of landforms from the last glacial (Fig. 7) and the draping character of the overlying sediments (Fig. 8) indicate that the bottom-current activity in the study area is low. We therefore regard coarser grains – apart from a few exceptions (see below) – as ice-rafted debris (IRD) that was transported by icebergs and sea ice. Fine grains derived from the turbid waters emanating from Tunabreen, von Postbreen and Sassenelva, from ice rafting by sea ice and icebergs (Fig. 4a, b & d) and, to a very limited degree, from smaller rivers along the fjord sides.

The sandy strata with sharp or erosive lower boundaries in the cores JM02-977-PC and JM02-985-GC are interpreted to be mass-transport deposits (e.g. debris-flow deposits or turbidites). They most probably originated from slope failure at the mouth of Sassenelva, as a consequence of high sediment supply. The coarse strata with non-erosive boundaries in these two cores might be mass-transport deposits that moved with a non-erosive behaviour due to hydroplaning (Elverhøi *et al.* 2000), or they were deposited by dumping or slumping from icebergs (Vorren *et al.* 1983) or ice floes.

Reddish sediments presently originate exclusively from the glaciers (see Fig. 4b). It is therefore reasonable to assume that the reddish strata in the sediment cores reflects periods of increased sediment supply from these sources. Reddish strata without higher clast content may have formed: (1) during winter, when Sassenelva is frozen and sediment input from this source is absent (Plassen *et al.* 2004); (2) during periods of increased meltwater runoff in relation to surges (compare with Kamb *et al.* 1985; Fleischer *et al.* 1998; Murray *et al.* 2003); (3) during periods of increased meltwater runoff related to glacier retreat as a consequence of climatic warming; and (4) meltout of exclusively fine-grained IRD (compare with Fig. 4d).

Reddish strata with a higher clast content are suggested to be the result of: (1) random dumping or slumping from icebergs (Vorren *et al.* 1983); (2) increased iceberg rafting during and after glacier surges (see also Fig. 4d, e); and (3) enhanced iceberg rafting due to climatically induced glacier growth and retreat. We favour icebergs rather than sea ice as transporting agents of the coarse reddish material, because (1) icebergs calving off the front of Tunabreen during its surge between *c.* AD 2002 and AD 2005 comprised significant amounts of reddish sediments (Fig. 4d, e); and (2) red bedrock material occurs exclusively in the terminal moraine deposited during the surge of von Postbreen in AD 1870 (for location see Fig. 7), that is, in a laterally restricted area implying that its inclusion into sea ice is rather limited.

The units 983-1 to 983-4, 048-2 and 049-2 (Fig. 9) most probably reflect different sedimentary environments and/or they show signs of reworking. Unit 983-4 was deposited during the last glacial or earlier (see above). Because of its massive composition, high bulk density (Fig. 9) and increased stiffness, we suggest that it is a basal till that was deposited beneath an ice stream draining through the study area during the last glacial. The occurrence of differently coloured stratification in unit 983-3 indicates that the sediment supply occurred from multiple sources that were located relatively close to the core site. We suggest that unit 983-3 was deposited in a glacier-proximal environment where several glacier fronts were located in Sassenfjorden and Tempelfjorden (compare with Forwick & Vorren 2009). The deformation of units 983-3 and 983-2 could have occurred synchronously or asynchronously. Because the core was retrieved from a relatively shallow water depth, the deformation in unit 983-3 might have been caused by grounding icebergs that were released during a final phase of the last glacial. The dated shell from unit 983-2 (Fig. 9; Table 2) was well preserved, even though it was found in deformed sediments. Because the core was retrieved from an area with relatively irregular bathymetry (compare with Fig. 1c), it is reasonable to assume that the deformation is the result of mass wasting. In the case that the top of unit 983-3 was not deformed by ploughing icebergs, it could have been possible that the mass-wasting event leading to the deposition of unit 983-2 also deformed the top of unit 983-3. The general coarsening and the absence of bioturbation below 110 cm could be attributed to winnowing (Vorren *et al.* 1984) as a consequence of the shallow water depth at the core site, resulting in increased bottom-current speeds.

Unit 049-2 contains the top of the debris lobe that was deposited in relation to the surge of von Postbreen in AD 1870 (based on correlation with Plassen *et al.* 2004). Unit 048-2 was interpreted to be deposited during a final phase of this surge, or shortly after. Because the core site is located beyond the extent of the lobe related to this surge, we assume that the general coarsening and increased bulk density in unit 048-2 reflect a stronger current regime in front of the glacier. This was probably caused by meltwater outbursts related to the surge (compare with Kamb *et al.* 1985; Fleischer *et al.* 1998; Murray *et al.* 2003).

Our results reveal an asymmetric sedimentation pattern in outer Tempelfjorden, for example reflected by higher sedimentation rates in core JM02-988-GC (north) compared to lower rates in core JM02-987-GC (south). Furthermore, even though the deposition of reddish strata and reddish spots commenced almost synchronously in both cores (Fig. 13), their amount is higher in core JM02-988-GC. Based on these observations, we suggest that the Coriolis effect significantly influences the distribution of sediments in the study area, by deflecting turbid water plumes to the right (compare with Dowdeswell 1987).

Modern environments

The benthic foraminiferal fauna reflects a typical glacier-proximal setting (Fig. 6; Table 3; cf. Hald & Korsun 1997; Korsun & Hald 1998, 2000). The dominance of *Elphidium excavatum f. clavata* in the inner parts of the fjord (stations 6–8) indicates that this species thrives in cold, turbid waters and probably scant nutrients. The relatively high abundances of *Cassidulina reniforme* and *Spiroplectammina biformis* (Fig. 6; Table 3) in combination with the proximity of the turbid-water plume, reflected by low transmissivity (Fig. 3), suggest still glacier-proximal conditions in the eastern part of the central basin (stations 3–5; e.g. Elverhøi *et al.* 1980; Hald & Korsun 1997). However, the occurrence of *Cassidulina reniforme* may also be attributed to an increase in nutrients, probably resulting from a reduced glacial input in comparison to the three easternmost stations (compare with Korsun & Hald 1998).

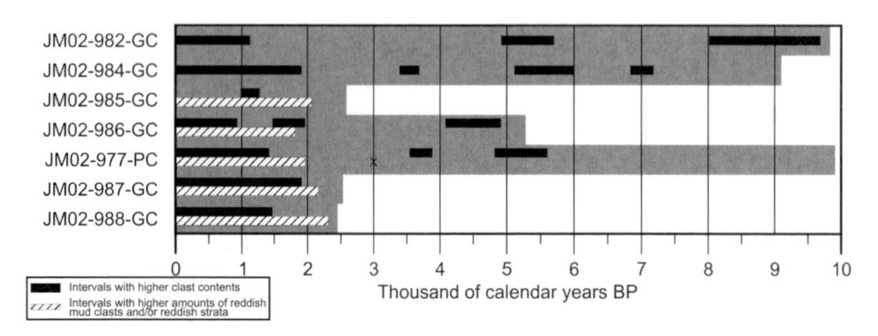

Fig. 13. Graph showing intervals with higher clast content, as well as intervals with larger amounts of reddish-brown strata and patches. The 'x' around 3000 indicates one level comprising several reddish patches in core JM02-977-PC. The time spans archived in the individual cores are indicated with grey colour.

The increasing numbers of benthic foraminifera indicate that living conditions generally improve with increasing distance from the glacier fronts (Fig. 6; Table 3). This is supported by an increase in TN percentages as well as decreasing C/N ratios (Fig. 5). The low numbers of calcareous benthic foraminifera, low TOC and high TIC, in addition to high amounts of suspended particulate matter, indicate that most of the $CaCO_3$ in the proximal basin is of detrital origin and derived from the turbid water emanating from Tunabreen and von Postbreen (Figs 5 & 6; compare also with high $CaCO_3$ contents on Fig. 4c).

It is striking that the number of dead benthic foraminifera at the westernmost station (station 1) is much higher than the number of living benthic foraminifera (Fig. 6). The largest differences are registered for the species *Elphidium excavatum f. clavata*, *Cassidulina reniforme* and *Nonionellina labradorica*. This is most probably related to the variations in the sedimentation rates along the transect (compare with Fig. 12). The high rates in the inner parts of the fjord mask the dead organisms by quickly removing them from the upper parts of the sediment column. On the other hand, low sedimentation rates can result in larger numbers of dead organisms at or close to the sediment surface.

The well-developed stratification in the hydrographic data from 2004 and 2006 (Fig. 3) indicates that the meltwaters emanating from Tunabreen rise to the water surface and are distributed as a turbid overflow plume. Low transmissivity values show that the meltwater runoff from the glaciers and/or suspension settling affected the entire water column in the glacier-proximal basin and in the easternmost parts of the main basin.

The generally lower transmissivity (i.e. higher turbidity) in 2004 compared to 2006 could be attributed to the collection of the data during an earlier stage of the melting season, that is, during a time when larger amounts of suspended matter were supplied from the sediment sources. However, it could also be related to increased amounts of turbid waters released during the surge of Tunabreen between *c.* AD 2002 and 2005.

Varying surface transmissivities in the western parts of the transect (station 2; Fig. 3) are suggested to reflect different locations of the turbid-water plume off Sassenelva. The low transmissivity values from the evening of 8 July 2004 reveal that the plume was directed towards the NW. In the afternoon of 4 August 2006, it was deflected to the NE and located at the southern shore of Tempelfjorden, that is, it did not reach central parts of Sassenfjorden and Tempelfjorden (personal observation). This explains the absence of an interval of reduced transmissivity as observed in 2004. Weather observations from Adventdalen (see Fig. 1c for location) show that the wind direction on 8 July 2004 was generally from the south, whereas the wind direction on 4 August 2006 was from the north (obtained from weather observation dataset from the University Centre in Svalbard; http://www.unis.no/20_RESEARCH/2060_Online_Env_Data/weatherstations.htm). This indicates that wind can significantly influence the locations of the turbid-water plumes and, in consequence, the sediment distribution in the study area, in addition to the Coriolis effect.

The reduced transmissivity in the central parts of the main basin (station 3; Fig. 3) may reflect a temporarily higher turbidity during the measurements in 2004. Because the sampling station is located off the delta of the river Sassenelva comprising chutes and levees, as well as sediment creep (Figs 7 & 8), it is reasonable to assume that mass wasting may account for this. It might have been caused, for example, by the tail of a turbidity flow or by a hyperpycnal flow.

The data show large interannual temperature and salinity variations. These are most probably the result of different sampling times during the summer season and/or significant meteorological differences during preceding winters. Whereas low temperatures and a well-developed sea-ice cover occurred during winter 2003/2004, winter 2005/2006 was comparatively mild and sea ice was limited (for temperatures see weather-observation dataset from the University Centre in Svalbard; http://www.unis.no/20_RESEARCH/2060_Online_Env_Data/weatherstations.htm). Additionally, winter 2005/2006 was characterized by a relatively strong inflow of Atlantic Water into west Spitsbergen fjords (Cottier *et al.* 2007), most probably causing the higher temperatures and salinities in 2006.

The reddish colour of the turbid water emanating from the glaciers is assumed to reflect the subcropping of reddish bedrock beneath the glaciers. These might be Devonian ('Old Red') sandstones and shales that, to a larger extent, occur in the northwestern parts of Spitsbergen (Dallmann *et al.* 2002). There, they generally underlie the Late Palaeozoic rocks that predominate in the drainage area of Tunabreen and von Postbreen. It is therefore reasonable to assume that they strike out beneath the glaciers.

The bulk-mineral assemblages indicate that the phyllosilicates and quartz in the turbid waters off Sassenelva originate from the Mesozoic and Late Palaeozoic sandstones and shales in the southern drainage basin (Fig. 4c; Dallmann *et al.* 2002). Ankerite/dolomite and calcite in the turbid waters off Tunabreen and von Postbreen derive from carbonate and evaporitic rocks in the northern drainage basin. Quartz and phyllosilicates in these waters most probably have their origin in the 'Old Red' sandstones and shales sub-cropping underneath the glaciers.

We suggested earlier that the deposition of unit 048-2 was significantly influenced by increased meltwater runoff during the surge of von Postbreen. Hence, we assume that the bulk-mineral assemblage of this unit carried a pronounced imprint of the turbid waters emanated from this glacier. Because it comprises relatively high percentages of ankerite/dolomite and epidote (Fig. 12a), we suggest that material deriving from von Postbreen contains comparatively high percentages of epidote and ankerite/dolomite. This may, on the other hand, indicate that higher calcite contents might be ascribed to material originating in the drainage area of Tunabreen.

Past environments

The Late Weichselian. The acoustic and lithological data suggest that glaciers dominated the sedimentary environment during the Late Weichselian.

The glacial lineations, the esker, the recessional moraines and the thrust moraine (Figs 7 & 8), as well as the basal till comprising unit 983-4 (Fig. 9), reflect the presence of grounded ice in Sassenfjorden and Tempelfjorden. We suggest that these deposits were formed during the last glacial, because: (1) they occur on top of, or slightly above, bedrock; (2) they are located beyond the Holocene maximum glacier-frontal position as interpreted by Plassen *et al.* (2004; Fig. 2a); and (3) they are draped with up to 30 m thick deposits (Fig. 8).

The glacial lineations were apparently formed beneath ice streams draining the Svalbard and Barents Sea ice sheet through Spitsbergen fjords (compare with Landvik *et al.* 1998, 2005; Ottesen *et al.* 2005). The esker in Sassenfjorden is suggested to represent the location of a subglacial conduit that was filled with sediments. It was most probably deposited after the termination of rapid ice flow, because it covers the glacial lineations. Its multi-crest morphology reflects lateral variations in the position of the conduit (compare with Ottesen *et al.* 2008). A close correlation between eskers/subglacial conduits and medial moraines has been observed in modern settings (Ottesen *et al.* 2008; Benn *et al.* 2009). If the esker is extended along a straight line in a southeasterly direction, it terminates at Fredheim (for location see Fig. 7). It might therefore reflect the confluence of the ice streams flowing through Tempelfjorden and Sassendalen.

Recessional moraines indicate that minor halts and/or re-advances of the glacier fronts occurred during the period of general glacier retreat at the end of the last glacial (compare with Baeten *et al.* 2010). Whether they are annual recessional moraines as observed in other areas on Spitsbergen (e.g. Boulton 1986; Ottesen & Dowdeswell 2006; Ottesen *et al.* 2008; Baeten *et al.* 2010) cannot be inferred from our data. The deposition of the thrust moraine in the inner parts of Tempelfjorden is suggested to reflect a re-advance of the glacier front close to the end of the last glacial. This re-advance could have been caused by a surge or could have been climatically induced.

Based on studies from other fjords on Spitsbergen (Mangerud *et al.* 1992, 1998; Elverhøi *et al.* 1995; Svendsen *et al.* 1996; Hald *et al.* 2004; Lønne 2005; Baeten *et al.* 2010; Forwick & Vorren 2009) we suggest that the retreat of the glacier fronts in the study area terminated *c.* 11.3/11.2 cal ka BP.

The Holocene. We assume that Tunabreen and von Postbreen existed during the entire Holocene because of the continuous occurrence of IRD in core JM02-977-PC (Fig. 10) and because other studies infer that tidewater glaciers in central Spitsbergen existed during this period (Hald *et al.* 2004; Baeten *et al.* 2010; Forwick & Vorren 2009).

Variations in the IRD content can be caused by a variety of glacial, oceanographic and climatic fluctuations. High amounts of IRD can, for example, reflect intensive iceberg calving related to the presence of ice streams, glacier advances or glacier retreats (Powell 1984; Dowdeswell *et al.* 1994, 1998; Vorren & Plassen 2002). However, high amounts of IRD can also be the result of low IRD flux combined with very limited supply of fine-grained material (compare with Plassen *et al.* 2004). Furthermore, variations in IRD content may reflect changes in surface water conditions. Whereas relatively warm surface waters can cause a rapid meltout of the transported material in glacier-proximal areas, low water temperatures can provide conditions that allow the material to be transported over larger distances. Based on the above-mentioned reasons, one should be cautious when using IRD as a single proxy to reconstruct climate variations. We therefore combine the variations in IRD contents in our cores with fluctuations in the bulk-mineralogy assemblages, colour variations, as well as with published data to reconstruct the glacial history in the study area during the Holocene.

The brown and red sediments (Figs 9 & 10) indicate that Tunabreen and von Postbreen, as well as Sassenelva, influenced the sedimentary environment during the Holocene. However, their influence varied in space and time. High percentages of phyllosilicates in core JM02-977-PC (Fig. 12b) are suggested to reflect the proximity and a generally higher influence from Sassenelva. Marked fluctuations in the mineral assemblages can be caused by short-lasting changes in the sedimentary processes at the river mouth (e.g. fluctuation related to mass-transport activity at 8650 cal a BP). Variations in the percentages of ankerite/dolomite, calcite and epidote may, on the other hand, be used as indicators for the activities/sizes of the glaciers.

Prolonged intervals with increased clast contents occur in several cores between 6 and 4 cal ka BP (Fig. 13). They might reflect increased ice rafting related to a regional cooling (Birks 1991; Wohlfarth *et al.* 1995; Sarnthein *et al.* 2003; Hald *et al.* 2004; Rasmussen *et al.* 2007; Baeten *et al.* 2010; Forwick & Vorren 2009; Skirbekk *et al.* 2010). We suggest that the marked increase in the calcite content around 5.6 cal ka BP (Fig. 12b) reflects the growth of Tunabreen rather than of von Postbreen; sediments supplied from the latter glacier are characterized by higher ankerite/dolomite contents (see above). This growth of Tunabreen is almost synchronous with the proposed growth of the glacier Nordenskiöldbreen in Billefjorden around 5470 cal a BP (for location see Fig. 1c; Baeten *et al.* 2010). Both glaciers are outlets of the ice cap Lomonosovfonna. It might therefore be reasonable to assume that these almost synchronous increases in size reflect the growth of Lomonosovfonna.

The late Holocene is characterized by the still ongoing oceanographic cooling in fjords on Spitsbergen and on the continental margin off western Spitsbergen (e.g. Hald *et al.* 2004; Rasmussen *et al.* 2007), as well as by a climatic cooling on Svalbard (Birks 1991; Svendsen & Mangerud 1997). The calcite content decreases, nevertheless, from *c.* 3.7 cal ka BP (Fig. 12b). This onset of decrease occurs almost synchronously with a decrease in iceberg rafting in central Isfjorden (Forwick & Vorren 2009). Forwick & Vorren (2009) suggest that a decrease in ice rafting could be related to the enhanced formation of shorefast sea ice and/or more permanent sea-ice cover; ice rafting was generally reduced and icebergs were trapped and forced to release their debris close to the calving fronts. This argument could also apply to our study area. Another reason might be that the meltwater runoff from the glaciers decreased because more ice and water were retained in the growing glacier. However, the occurrence of red mud clasts at *c.* 335 cm depth in core JM02-977-GC (*c.* 3 cal ka BP) indicates that ice rafting must have occurred temporarily. Our suggestion of increased sea-ice formation precedes observations from Billefjorden by *c.* 500 a (Baeten *et al.* 2010). These differences might result from local differences in the onset of enhanced formation of shorefast sea ice or incorrect chronologies. Further research is needed to explain these discrepancies.

A marked increase in the influence of the glaciers on the sedimentary environment occured between *c.* 2 and 1.75 cal ka BP. This is reflected by: (1) increased amounts of reddish patches and strata from *c.* 2 cal ka BP (Fig. 13); (2) larger amounts of IRD since *c.* 2 cal ka BP (Fig. 13); and (3) the increase in the percentages of epidote and calcite, as well as decreasing percentages of clay minerals around 1.75 cal ka BP (Fig. 12).

Higher numbers of IRD-rich intervals may reflect increased ice rafting as a consequence of climatically induced glacier growth or glacier retreat during this period, characterized by marked climatic fluctuations on the Northern Hemisphere (e.g. Moberg *et al.* 2005, 2006; Seidenkrantz *et al.* 2008). An increase in IRD content around 2 cal ka BP might reflect increased ice rafting due to glacier retreat, related to warming during the Roman Warm Period, the increases in epidote and calcite percentages around 650 cal a BP might be the result of glacier growth related to the Little Ice Age. However, because Tunabreen and von Postbreen are surge glaciers (Hagen *et al.* 1993), surging should not be excluded for the input of larger amounts of IRD (compare with Hald *et al.* 2001, 2004).

Summary

During the last glacial, Sassenfjorden and Tempelfjorden were covered with grounded ice depositing a basal till and leading to the formation of glacial lineations (Fig. 14a). After the end of rapid ice flow the glacier fronts retreated stepwise, resulting in the deposition of recessional moraines (Fig. 14b). Stratified glaciomarine sediments (unit 983-3) were deposited in front of several tidewater glaciers in the study area during a late phase of the last glacial. The infill of a subglacial conduit led to the formation of an esker after the termination of rapid ice flow. The esker most probably reflects the position of the medial moraine between ice streams draining through Tempelfjorden and Sassendalen (Fig. 14a, b). A glacier re-advance close to the termination of the last glacial led to the formation of a thrust moraine in central Tempelfjorden (Fig. 14c).

Sediment supply during the Holocene occurred mainly from the glaciers Tunabreen and von Postbreen at the head of Tempelfjorden, as well as from Sassenelva at the transition between the fjords (Fig. 14d–h). The glaciers were most probably small during the early Holocene (Fig. 14d). An increase in ice rafting and a growth of the glaciers, in particular Tunabreen, occurred between 6 and 4 cal ka BP (Fig. 14e). From 3.7 cal ka BP, the supply from Tunabreen decreased. This was probably due to the enhanced formation of shorefast and/or more permanent sea-ice cover, suppressing ice rafting (Fig. 14f) and/or reduced meltwater runoff from Tunabreen, hence retaining more water in the growing glacier. However, occasional ice rafting apparently occurred. The glacier fronts

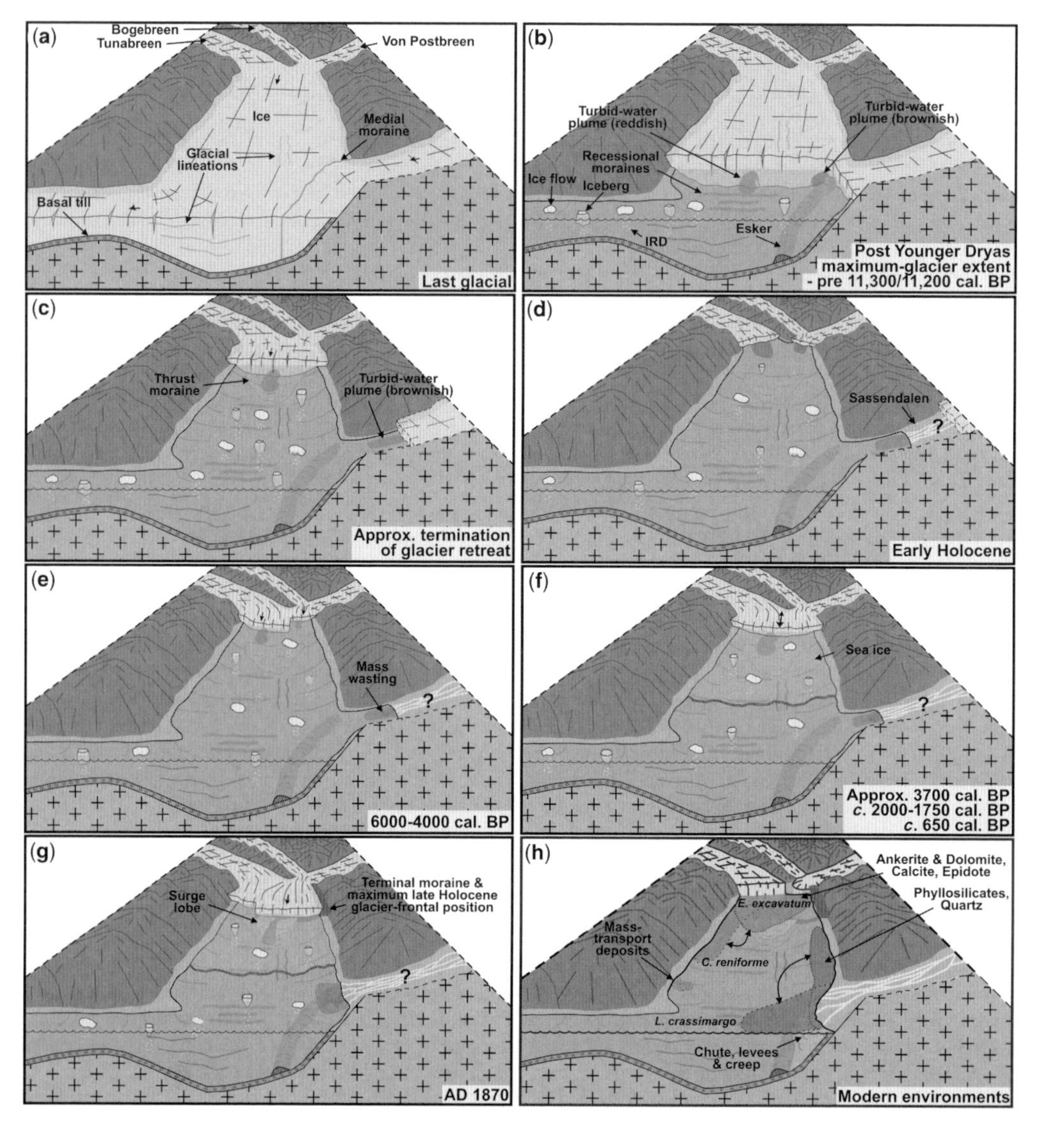

Fig. 14. Conceptual model showing the influence of Tunabreen and von Postbreen, as well as of Sassenelva on the sedimentary environment in Sassenfjorden and Tempelfjorden (**a**)–(**c**) during the last glacial, (**d**) the early Holocene, (**e**)–(**g**) the mid and late Holocene, and (**h**) in modern times. The '?' in Sassendalen (d)–(g) indicates the uncertainty of the extents of glaciers and the river Sassenelva. (f) The double arrow indicates general glacier growth, as well as some periods of glacier retreat. (h) The double arrow indicates switching directions of the water plumes due to wind and/or the Coriolis effect.

probably retreated around 2 cal ka BP due to warming related to the Roman Warm Period, and they advanced at *c.* 650 a as a consequence of cooling during the Little Ice Age (Fig. 14f). The maximum Holocene glacier extent was reached at the end of a surge of von Postbreen in AD 1870 (Fig. 14g). A debris lobe was deposited in relation to this surge. Apart from advances related to surges (Hagen *et al.* 1993), the glacier fronts have retreated since *c.* AD 1900 (De Geer in Plassen *et al.* 2004).

Characteristics of the modern environment include (Fig. 14h): (1) different colours and mineral assemblages of the turbid waters emanating from the glaciers and Sassenelva; (2) variable locations of the turbid-water plumes as a consequence of wind forcing and the Coriolis effect; (3) stratified water masses during summers with interannual differences; (4) increasing productivity with increasing distance from the glacier fronts; (5) foraminifera-faunal assemblages typical for glacier-proximal settings; and (6) periodical mass-transport activity off the river mouth and along the fjord slopes.

This study was performed as part of the Strategic University Programme SPONCOM (Sedimentary Processes and Palaeoenvironment on Northern Continental Margins), financed by the Research Council of Norway. The masters and crews of RV *Jan Mayen* supported the data collection. S. Iversen and K. M. Sandaker provided technical help during and after the cruises. J. P. Holm contributed to the creation of the figures. E. Ellingsen and T. Dahl supported the laboratory analyses. The *Roald Amundsen Centre for Arctic Research* at the University of Tromsø provided funding for some radiocarbon dates. The mineral-assemblage analyses were carried out at the University of Bremen as part of the EU-funded programme *Paleostudies*. J. A. Howe and an anonymous reviewer provided constructive comments to improve an earlier version of this manuscript. We extend our most sincere thanks to these persons and institutions.

References

BAETEN, N. J., FORWICK, M., VOGT, C. & VORREN, T. O. 2010. Late Weichselian and Holocene sedimentary environments and glacial activity in Billefjorden, Svalbard. *In*: HOWE, J. A., AUSTIN, W. E. N., FORWICK, M. & PAETZEL, M. (eds) *Fjord Systems and Archives*. Geological Society, London, Special Publications, **344**, 207–223.

BENN, D. I., KRISTENSEN, L. & GULLEY, J. D. 2009. Surge propagation constrained by a persistent subglacial conduit, Bakaninbreen-Paulabreen, Svalbard. *Annals of Glaciology*, **50**, 81–86.

BENNETT, M. R. 2001. The morphology, structural evolution and significance of push moraines. *Earth-Science Reviews*, **53**, 197–236.

BIRKS, H. 1991. Holocene vegetation history and climatic change in west Spitsbergen – plant macrofossils from Skardtjørna, an Arctic lake. *The Holocene*, **1**, 209–218.

BISHOP, J. K. B. 1986. Instrument and methods: the correction and suspended particulate matter calibration of Sea Tech transmissometer data. *Deep-Sea Research*, **33**, 121–134.

BOULTON, G. S. 1986. Push-moraines and glacier-contact fans in marine and terrestrial environments. *Sedimentology*, **33**, 677–698.

BOULTON, G. S., VAN DER MEER, J. J. M., HART, J., BEETS, D., RUEGG, G. H. J., VAN DER WATEREN, F. M. & JARVIS, J. 1996. Till and moraine emplacement in a deforming bed surge – an example from a marine environment. *Quaternary Science Reviews*, **15**, 961–987.

COTTIER, F. R., NILSEN, F., INALL, M. E., GERLAND, S., TVERBERG, V. & SVENDSEN, H. 2007. Wintertime warming of an Arctic shelf in response to large-scale atmospheric circulation. *Geophysical Research Letters*, **34**, doi: 10.1029/2007/GL029948.

DALLMANN, W. K., OHTA, Y., ELVEVOLD, S. & BLOMEIER, D. 2002. *Bedrock map of Svalbard and Jan Mayen*. Norsk Polarinstitutt Temakart No. 33.

DOWDESWELL, J. A. 1987. Processes of glacimarine sedimentation. *Progress in Physical Geography*, **11**, 52–90.

DOWDESWELL, J. A., WHITTINGTON, R. J. & MARIENFELD, P. 1994. The origin of massive diamicton facies by iceberg rafting and scouring, Scoresby Sund, East Greenland. *Sedimentology*, **41**, 21–35.

DOWDESWELL, J. A., ELVERHØI, A. & SPIELHAGEN, R. 1998. Glacimarine sedimentary processes and facies on the polar North Atlantic margins. *Quaternary Science Reviews*, **17**, 243–272.

ELVERHØI, A., LIESTØL, O. & NAGY, J. 1980. Glacial erosion, sedimentation and microfauna in the inner part of Kongsfjord, Spitsbergen. *Norsk Polarinstitutt Skrifter*, **172**, 33–61.

ELVERHØI, A., SVENDSEN, J. I., SOLHEIM, A., ANDERSEN, E. S., MILLIMAN, J., MANGERUD, J. & HOOKE, LEB. R. 1995. Late Quaternary Sediment Yield from the High Arctic Svalbard Area. *Journal of Geology*, **103**, 1–17.

ELVERHØI, A., HARBITZ, C. B., DIMAKIS, P., MOHRIG, D. & MARR, J. 2000. On the dynamics of subaqueous debris flows. *Oceanography*, **13**, 109–117.

FLEISCHER, P. J., CADWELL, D. H. & MULLER, E. H. 1998. Tsivat Basin Conduit System persists through two surges, Bering Piedmont Glacier, Alaska. *Geological Society of America Bulletin*, **110**, 877–887.

FORWICK, M. & VORREN, T. O. 2005*a*. Late Weichselian and Holocene sedimentation and environments in the Isfjorden area. *In*: FORWICK, M. (ed.) *Sedimentary Processes and Palaeoenvironments in Spitsbergen Fjords*. A dissertation for the degree of *Doctor Scientiarum*, University of Tromsø.

FORWICK, M. & VORREN, T. O. 2005*b*. Late Weichselian deglaciation history of the Isfjorden area, Spitsbergen. *In*: FORWICK, M. (ed.) *Sedimentary Processes and Palaeoenvironments in Spitsbergen Fjords*. A dissertation for the degree of *Doctor Scientiarum*, University of Tromsø.

FORWICK, M. & VORREN, T. O. 2007. Holocene mass-transport activity and climate in outer Isfjorden, Spitsbergen: marine and subsurface evidence. *The Holocene*, **17**, 707–716.

FORWICK, M. & VORREN, T. O. 2009. Late Weichselian and Holocene sedimentary environments in Isfjorden, Spitsbergen. *Palaeogeography, Palaeoclimatology, Palaeoecology*, **280**, 258–274.

FORWICK, M., BAETEN, N. J. & VORREN, T. O. 2009. Pockmarks in Spitsbergen fjords. *Norwegian Journal of Geology*, **89**, 65–77.

GERLAND, S. & RENNER, A. H. H. 2007. Sea-ice mass-balance monitoring in an Arctic fjord. *Annals of Glaciology*, **46**, 435–424.

HAGEN, J. O., LIESTØL, O., ROLAND, E. & JØRGENSEN, T. 1993. Glacier Atlas of Svalbard and Jan Mayen. *Meddelelser*, Norsk Polarinstitutt, **129**, 5–41.

HALD, M. & KORSUN, S. 1997. Distribution of modern benthic foraminifera from fjords of Svalbard, European Arctic. *Journal of Foraminiferal Research*, **27**, 101–122.

HALD, M., DAHLGREN, T., OLSEN, T.-E. & LEBESBYE, E. 2001. Late Holocene palaeoceanography in Van Mijenfjorden, Svalbard. *Polar Research*, **20**, 23–35.

HALD, M., EBBESEN, H. *ET AL.* 2004. Holocene paleoceanography and glacial history of the West Spitsbergen area, Euro-Arctic margin. *Quaternary Science Reviews*, **23**, 2075–2088.

HAMILTON, E. L. 1985. Sound velocity as a function of depth in marine sediments. *Journal of the Acoustic Society of America*, **78**, 1348–1355.

HUDDART, D. & HAMBREY, M. J. 1996. Sedimentary and tectonic development of a High-Arctic, thrust-moraine complex: comfortlessbreen, Svalbard. *Boreas*, **25**, 227–243.

HUGHEN, K. A., BAILLIE, M. G. L. *ET AL.* 2004. Marine04, Marine radiocarbon age calibration, 26–0 ka BP. *Radiocarbon*, **46**, 1059–1086.

KAMB, B., RAYMOND, C. F. *ET AL.* 1985. Glacier surge mechanism: 1982–1983 surge of Variegated Glacier, Alaska. *Science*, **227**, 469–479.

KORSUN, S. & HALD, M. 1998. Modern benthic foraminfera off Novaya Zamlya tidewater glaciers, Russian Arctic. *Arctic and Alpine Research*, **30**, 61–77.

KORSUN, S. & HALD, M. 2000. Seasonal dynamics of benthic foraminifera in a glacially fed fjord of Svalbard, European Arctic. *Journal of Foraminiferal Research*, **30**, 251–271.

LANDVIK, J. Y., BONDEVIK, S. *ET AL.* 1998. The last glacial maximum of Svalbard and the Barents Sea area: ice sheet extent and configuration. *Quaternary Science Reviews*, **17**, 43–75.

LANDVIK, J. Y., INGÓLFSSON, Ó., MIENERT, J., LEHMAN, S. J., SOLHEIM, A., ELVERHØI, A. & OTTESEN, D. 2005. Rethinking Late Weichselian ice sheet dynamics in coastal NW Svalbard. *Boreas*, **34**, 7–24.

LØNNE, I. 2005. Faint traces of high Arctic glaciations: an early Holocene ice-front fluctuation in Bolterdalen, Svalbard. *Boreas*, **34**, 308–323.

MACKAY, M. E., JARRARD, R. D., WESTBROOK, G. K., HYNDMAN, R. D. & SHIPBOARD SCIENTIFIC PARTY OF OCEAN DRILLING PROGRAM LEG 146 1994. Origin of bottom-simulating reflectors: geophysical evidence from the Cascadia accretionary prism. *Geology*, **22**, 459–462.

MANGERUD, J. & GULLIKSEN, S. 1975. Apparent radiocarbon ages of recent marine shells from Norway, Spitsbergen, and Arctic Canada. *Quaternary Research*, **5**, 263–273.

MANGERUD, J., BOLSTAD, M. *ET AL.* 1992. The Last Glacial Maximum on Spitsbergen. *Quaternary Research*, **38**, 1–31.

MANGERUD, J., DOKKEN, T. *ET AL.* 1998. Fluctuations of the Svalbard–Barents Sea ice sheet during the last 150 000 years. *Quaternary Science Reviews*, **17**, 11–42.

MOBERG, A., SONECHKIN, D. M., HOLMGREN, K., DATSENKO, N. M. & KARLÉN, W. 2005. Highly variable Northern Hemisphere temperatures reconstructed from low- and high-resolution proxy data. *Nature*, **433**, 613–617.

MOBERG, A., SONECHKIN, D. M., HOLMGREN, K., DATSENKO, N. M., KARLÉN, W. & LAURITZEN, S.-E. 2006. Corrigendum: highly variable Northern Hemisphere temperatures reconstructed from low- and high-resolution proxy data. *Nature*, **439**, 1014.

MURRAY, T., STROZZI, T., LUCKMAN, A., JISKOOT, H. & CHRISTAKOS, P. 2003. Is there a single surge mechanism? Contrasts in dynamics between glacier surges in Svalbard and other regions. *Journal of Geophysical Research*, **108**, No. B5 2237, doi: 10.1029/2002JB001906.

NILSEN, F., COTTIER, F., SKOGSETH, R. & MATTSSON, S. 2008. Fjord-shelf exchanges controlled by ice and brine production: the interannual variation of Atlantic Water in Isfjorden, Svalbard. *Continental Shelf Research*, **28**, 1838–1853.

OTTESEN, D. & DOWDESWELL, J. A. 2006. Assemblages of submarine landforms produced by tidewater glaciers in Svalbard. *Journal of Geophysical Research*, **111**, 1–16.

OTTESEN, D., DOWDESWELL, J. A. & RISE, L. 2005. Submarine landforms and the reconstruction of fast-flowing ice streams within a large Quaternary ice sheet: the 2500-km-long Norwegian Svalbard margin (57°–80°N). *Geological Society of America Bulletin*, **117**, 1033–1050.

OTTESEN, D., DOWDESWELL, J. A. *ET AL.* 2008. Submarine landforms characteristic of glacier surges in two Spitsbergen fjords. *Quaternary Science Reviews*, **27**, 1583–1599.

PLASSEN, L., VORREN, T. O. & FORWICK, M. 2004. Integrated acoustic and coring investigation of glacigenic deposits in Spitsbergen fjords. *Polar Research*, **23**, 89–110.

POWELL, R. D. 1984. Glacimarine processes and inductive lithofacies modelling of ice shelf and tidewater glacier sediments based on Quaternary examples. *Marine Geology*, **57**, 1–52.

PRIOR, D. B., WISEMAN, W. J. JR, & BRYANT, W. R. 1981. Submarine chutes on the slopes of fjord deltas. *Nature*, **290**, 326–328.

RASMUSSEN, T. L., THOMSEN, E., ŚLUBOWSKA, M. A., JESSEN, S., SOLHEIM, A. & KOÇ, N. 2007. Paleoceanographic evolution of the SW Svalbard margin (76°N) since 20,000 ^{14}C yr BP. *Quaternary Research*, **67**, 100–114.

SARNTHEIN, M., VAN KREVELD, S., ERLENKEUSER, H., GROOTES, P. M., KUCERA, M., PFLAUMANN, U. & SCHULZ, M. 2003. Centennial-to-millennial-scale periodicities of Holocene climate and sediment injections off the western Barents shelf, 75°n. *Boreas*, **32**, 447–461.

SEIDENKRANTZ, M.-S., RONCAGLIA, L., FISCHEL, A., HEILMANN-CLAUSEN, C., KUIJPERS, A. & MOROS, M. 2008. Variable North-Atlantic climate seasaw patterns documented by a late Holocene marine record from Disko Bugt, West Greenland. *Marine Micropaleontology*, **68**, 66–83.

SKEI, J. 1983. Why sedimentologists are interested in fjords. *Sedimentary Geology*, **36**, 75–80.

SKIRBEKK, K., KRISTENSEN, D. K., RASMUSSEN, T., KOÇ, N. & FORWICK, M. 2010. Holocene climate variations at the entrance to a warm Arctic fjord: evidence from Kongsfjorden trough, Svalbard. *In*: HOWE, J. A., AUSTIN, W. E. N., FORWICK, M. & PAETZEL, M. (eds) *Fjord Systems and Archives*. Geological Society, London, Special Publications, **344**, 289–304.

STEIN, R. 1991. *Accumulation of Organic Carbon in Marine Sediments*. Lecture Notes in Earth Sciences, Springer-Verlag, New York, **34**.

STUIVER, M. & REIMER, P. J. 1993. Extended ^{14}C database and revised CALIB radiocarbon calibration program. *Radiocarbon*, **35**, 215–230.

SVENDSEN, J. I. & MANGERUD, J. 1997. Holocene glacial and climatic variations on Spitsbergen, Svalbard. *The Holocene*, **7**, 45–57.

SVENDSEN, J. I., ELVERHØI, E. & MANGERUD, J. 1996. The retreat of the Barents Sea Ice Sheet on the western Svalbard margin. *Boreas*, **25**, 244–256.

SVENDSEN, H., BESZCZYNSKA-MØLLER, A. *ET AL.* 2002. The physical environment of Kongsfjorden–Krossfjorden, an Arctic fjord system in Svalbard. *Polar Research*, **21**, 133–166.

VOGT, C., LAUTERJUNG, J. & FISCHER, R. X. 2002. Investigation of the clay fraction (<2 μm) of the clay mineral society reference clays. *Clays and Clay Minerals*, **50**, 388–400.

VORREN, T. O. & PLASSEN, L. 2002. Deglaciation and palaeoclimate of the Andfjord–Vågsfjord area, north Norway. *Boreas*, **31**, 97–125.

VORREN, T. O., HALD, M., EDVARDSEN, M. & LIND-HANSEN, O.-W. 1983. Glacigenic sediments and sedimentary environments on continental shelves: general principles with a case study from the Norwegian shelf. *In*: EHLERS, J. (ed.) *Glacial Deposits in North-West Europe*. Balkema, Rotterdam, 61–73.

VORREN, T. O., HALD, M. & THOMSEN, E. 1984. Quaternary sediments and environments on the continental shelf of northern Norway. *Marine Geology*, **57**, 229–257.

WERNER, A. 1993. Holocene moraine chronology, Spitsbergen, Svalbard: lichenometric evidence for multiple Neoglacial advances in the Arctic. *The Holocene*, **3**, 128–137.

WĘSŁAWSKI, J. M., KOSZTEYN, J., ZĄJACZKOWSKI, M., WIKTOR, J. & KWAŚNIEWSKI, S. 1995. Fresh water in Svalbard fjord ecosystems. *In*: SKJOLDAL, H. R., HOPKINS, C., ERIKSTAD, K. E. & LEINAA, H. P. (eds) *Ecology of Fjords and Coastal Waters*. Elsevier, Amsterdam, 229–241.

WOHLFARTH, B., LEMDAHL, G., OLSSEN, S., PERSSON, T., SNOWBALL, I., ISING, J. & JONES, V. 1995. Early Holocene environment on Bjørnøya (Svalbard) inferred from multidisciplinary lake sediment studies. *Polar Research*, **14**, 253–275.

Morphodynamic evolution of Kongsfjorden–Krossfjorden, Svalbard, during the Late Weichselian and Holocene

S. E. MACLACHLAN[1]*, J. A. HOWE[1] & M. E. VARDY[2]

[1]*The Scottish Association for Marine Science, Scottish Marine Institute, Oban, Argyll, PA37 1QA*

[2]*National Oceanography Centre, Southampton, University of Southampton, European Way, Southampton, SO30 3ZH*

**Corresponding author (e-mail: sucl@noc.soton.ac.uk)*

Abstract: We present a combination of fjord bathymetry and shallow seismic data from Kongsfjorden and Krossfjorden, Svalbard, to characterize and analyse change in the fjord coastal environment physiography and the glaciosedimentary processes since the Last Glacial Maximum. Swath bathymetry reveals a series of several styles of landform, frequently superimposed upon each other, permitting the reconstruction of the relative timings of deposition of each landform with the oldest successively overlain and cross-cut by younger landforms and erosional processes. Large transverse ridges interpreted as recessional moraines are overlain by streamlined lineations formed subglacially during a subsequent ice advance. A complex of recessional morainal ridges occurring within the central fjord are incised by glacial lineations and meltwater channels from younger glacial events.

The larger fjords of western Spitsbergen form gateways to the cross-shelf troughs for fast-flowing palaeo-ice streams that delivered a high flux of ice and sediment to the ice-sheet margin (e.g. Vorren & Laberg 1997; Elverhøi *et al.* 1998; Ottesen *et al.* 2007; Vanneste *et al.* 2007). Tidewater glaciers, together with ice streams, play a key role in controlling the stability and dynamics of the ice caps on Svalbard.

The morphology of the western Svalbard continental margin was primarily formed by glacial activity during the Plio-Pleistocene, when the glaciers repeatedly extended to the shelf break (Solheim *et al.* 1996; 1998; Vorren *et al.* 1998). Ottesen *et al.* (2005) studied the seafloor morphology of the entire western margin of the Scandinavian, Barents Sea and Svalbard ice sheets and demonstrated that *c.* 20 fast-flowing ice streams were directed through fjords and cross-shelf troughs. Previous workers have suggested that the Late Weichselian ice sheet was the most recent to reach the western and northern continental shelf edge of Svalbard (Svendsen *et al.* 1992, 1996, 2004; Andersen *et al.* 1996; Landvik *et al.* 1998; Mangerud *et al.* 2002). However, recent investigations by Landvik *et al.* (2005) suggest that the cross-shelf troughs were filled by fast-flowing ice streams with sharp boundaries to dynamically less active ice on the adjacent shelves and strandflats.

The study by Landvik *et al.* (2005) suggests a complicated topographically controlled configuration of Late Weichselian glacier ice along the west coast of Svalbard. The ice stream in Kongsfjordrenna was fed by ice draining through the deep fjord systems of Kongsfjorden and Krossfjorden, which must have drained a large portion of the ice culmination over NW Spitsbergen. It is proposed that fast-flowing ice streams filled Kongsfjorden and Isfjorden troughs, whereas the glacial ice in the area between was dynamically less active. It is also suggested that deglaciation of the ice stream was rapid: the ice margin reached the coast of Kongsfjorden by 13 ka and the fjord was completely deglaciated by *c.* 9 ka during a two-step deglaciation (Lehman & Forman 1992). At present there are no detailed studies of the deglaciation for northwestern Spitsbergen.

The deep cross-shelf troughs extending to the shelf break are the most prominent features of the west Spitsbergen shelf (Fig. 1 inset). Together with Bellsund and Isfjorden, Kongsfjorden is one of the largest outlets for palaeo-ice streams in western Svalbard. Our aim in this paper is to investigate the former Late Weichselian ice stream of the Kongsfjorden–Krossfjorden system, which was an important drainage outlet of the northwestern Spitsbergen ice culmination. The geometry of subglacial landforms may reveal new insight into the

From: Howe, J. A., Austin, W. E. N., Forwick, M. & Paetzel, M. (eds) *Fjord Systems and Archives*. Geological Society, London, Special Publications, **344**, 195–205.
DOI: 10.1144/SP344.14 0305-8719/10/$15.00

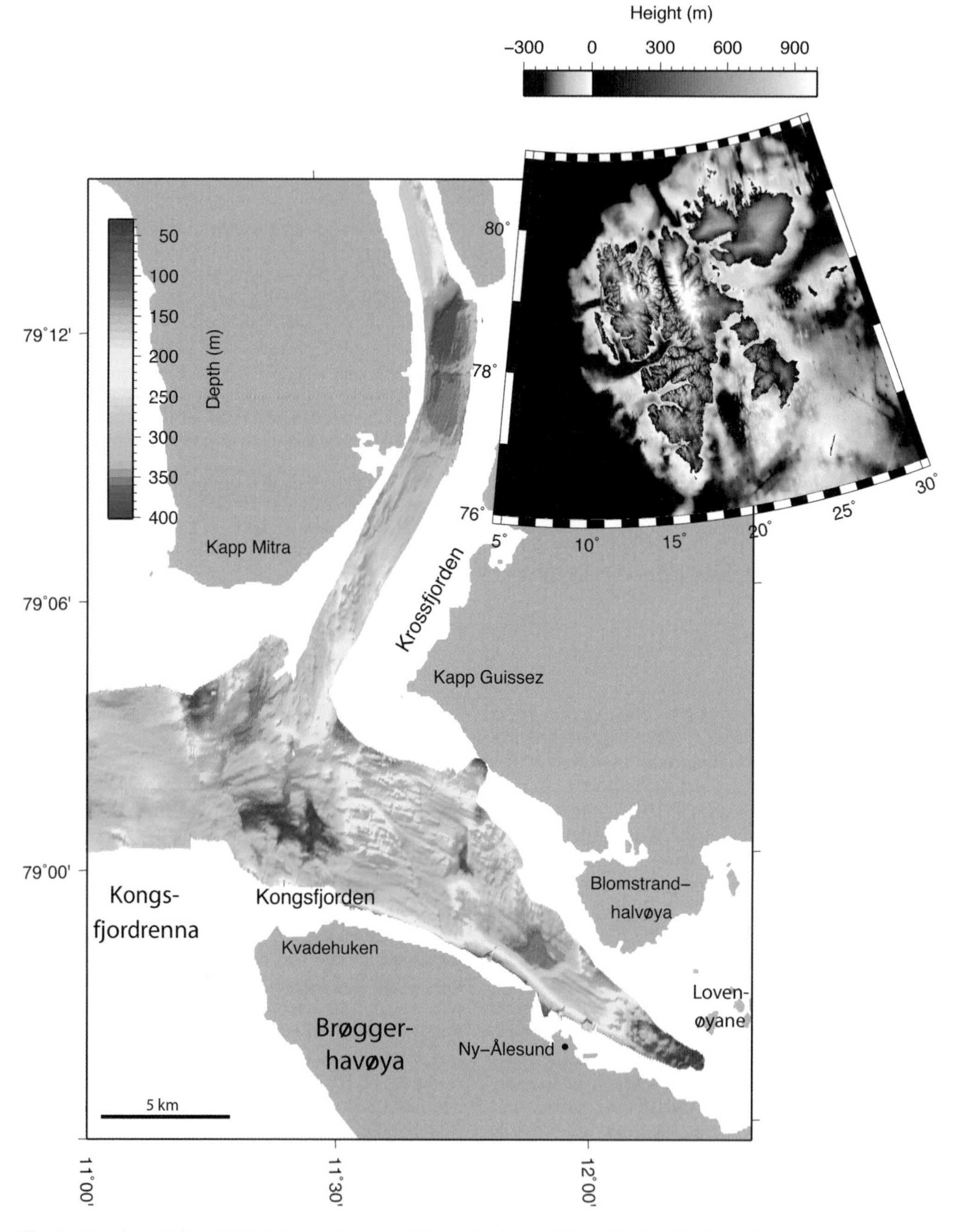

Fig. 1. Illuminated (from NW) bathymetric map of Kongsfjorden and Krossfjorden obtained with EM120 multibeam system. Inset map showing the archipelago of Svalbard and the surrounding continental margin. The red box highlights the study area within northwestern Spitsbergen.

processes acting at the ice-bed–sediment interface, thus characterizing the dynamics of the ice stream. This work builds on the initial geomorphological descriptions of the relict subglacial, ice-scoured seabed topography and the acoustic character classification of depositional environments within the fjord system outlined by Howe *et al.* (2003).

Study area and geological setting

Study area

The fjordic system comprising Kongsfjorden and Krossfjorden opens onto the western Svalbard shelf by a submarine glacial trough (Kongsfjordrenna, Fig. 1). The southern arm of the fjord system, Kongsfjorden, is 20 km long with a maximum depth of 394 m and a width of 4–10 km (Fig. 1). To the north of Kongsfjorden, Krossfjorden is *c.* 30 km long and 3–6 km in width. The deepest point of Krossfjorden is 374 m, where Kongs Haakon Halvøya divides the inner fjord region of Krossfjorden into Lilliehöökfjorden and Möllerfjorden (Fig. 2). The large icefields of Isachsenfonna and Holtedalfonna drain the adjacent landmass, feeding into the fjords through large glacier complexes at the head of the fjords; in total, they drain an area of *c.* 3074 km^2 (Svendsen *et al.* 2002). The Kongsfjorden–Krossfjorden system is strongly influenced by the presence of the tidewater glaciers, namely Lilliehöökbreen at the head of Krossfjorden and Kronebreen, Kongvegen, Kongsbreen, Conwaybreen and Blomstrandbreen at the head of Kongsfjorden.

Geological setting

The Kongsfjorden–Krossfjorden system is located at a major tectonic boundary which separates the Cenozoic fold- and thrust belt of western Spitsbergen to the SW and the Northwestern Basement Province, which lies to the NE (Bergh *et al.* 2000). The region north of the fault zone is a unified province comprising pre-Devonian metasediments and related igneous rocks; the general trend of lithological boundaries and structural fabrics is NNW–SSE. The island of Blomstrandhalvøya within Kongsfjorden (Fig. 1) consists predominantly of Devonian red conglomerates and sandstones, interlayered with the Generalfjella Formation marbles which are also exposed to the SE on some of the islands of Lovénøyane. Brøggerhalvøya (Fig. 1), south of the fault zone, comprises Late Palaeozoic sedimentary strata within the Cenozoic fold- and thrust belt made up of carbonates, calcareous sandstones and conglomerates. The Kings Bay Coalfield, comprising Cenozoic strata including conglomerates, sandstones and shales with numerous interbedded coal seams, is situated in the area surrounding Ny-Ålesund (Harland 1997).

Methods

EM120 multibeam swath bathymetry and TOpographic PArametric Sonar (TOPAS) sub-bottom profiler (0.5–5 kHz) data were collected during cruises JR75 and JR127 on the RRS *James Clark Ross*. The surveys extended across the central and outer basins of the Kongsfjorden–Krossfjorden system and into a limited area of the adjacent Kongsfjordrenna trough on the shelf, west of Svalbard (Fig. 2). The hull-mounted Kongsberg-Simrad EM120 multibeam swath bathymetry system emits 191 beams, each with a frequency of 12 kHz and a maximum port- and starboard-side angle of 75°. This gives, for example, a total swath width of approximately three times the water depth. Swath data were processed and gridded at cell sizes of 30–50 m. TOPAS uses the parametric interference between primary waves to produce a secondary acoustic beam of narrow width and frequency range of 0.5–5 kHz. Navigation data were aquired using differential GPS. The total area covered by the fjord surveys was *c.* 300 km^2.

General pattern of submarine landform deposition

The full swath bathymetry survey indicating the seafloor morphology of Kongsfjorden–Krossfjorden system is presented in Figure 1. The seabed morphology of Kongsfjorden reveals a series of various landform types, many of which are superimposed upon each other. Initial interpretations of the primary survey (JR75) were presented by Howe *et al.* (2003). Here, we have built on these initial observations and present further interpretation of the different landform types. We then discuss these landforms with reference to the historical record of deglaciation of the Late Weichselian ice sheet within the fjords and relate their formation to the processes of glacial activity.

Transverse ridges: recessional moraines

Description. Three large, slightly sinuous transverse ridges (TR1–3) dominate the seafloor topography of Kongsfjorden and the outer basin of Krossfjorden (Fig. 3). The outermost submerged ridge (TR1) partially crosses the northern area of the fjord mouth. The ridge is at least 4 km long; its northerly extent is unknown as it has not been fully surveyed. However, the surveyed area indicates a height of *c.* 240 m with a maximum width of *c.* 2 km (Fig. 3). The large central ridge (TR2) is *c.* 7 km in length, trends from north to south and has a maximum height of *c.* 200 m. The ridge straddles the outer basins of both fjords. The southern edge of the ridge occurs *c.* 3 km north of the Kvadehuken shoreline and is *c.* 3 km wide. The northern extent of the ridge has not been fully mapped; however, it appears to become a low relief, hummocky seabed feature. A third transverse ridge (TR3, *c.* 1.5 km in

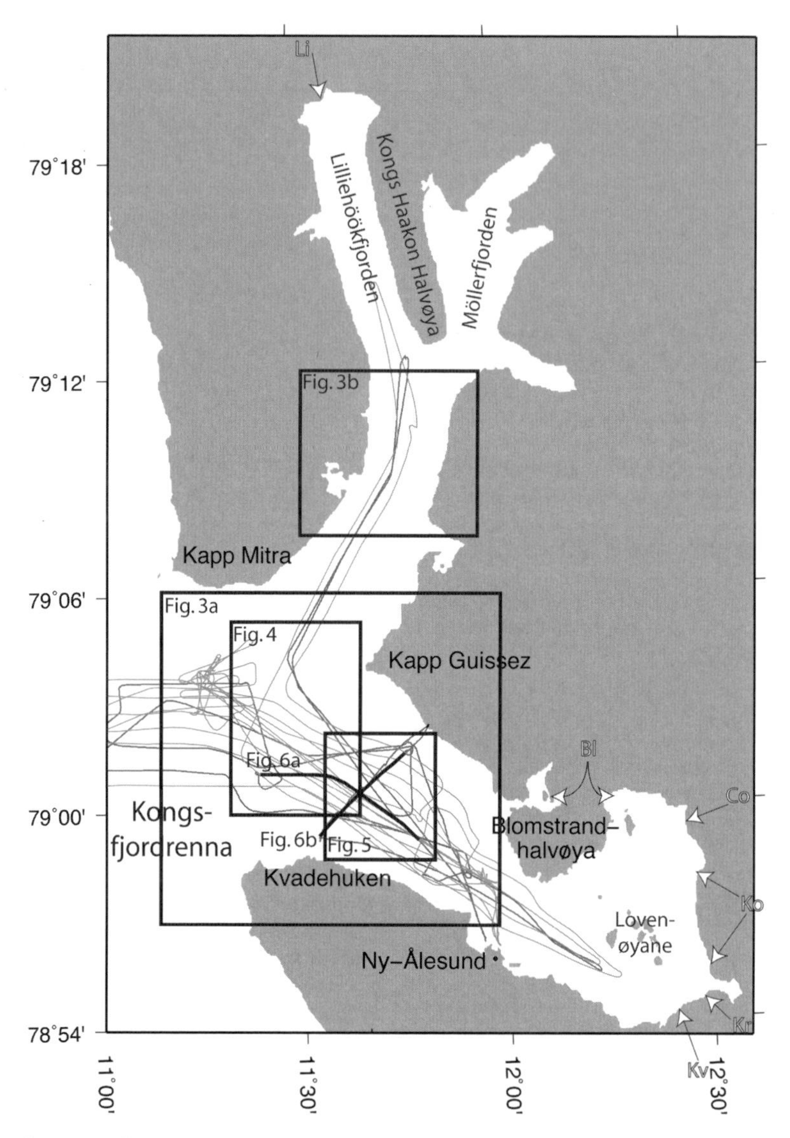

Fig. 2. Map of the Kongsfjorden–Krossfjorden system, with major place names and tidewater glaciers marked. Thin grey line shows extent of JR75 survey track; thin black line indicates coverage of JR127 survey track; thick black lines indicate selected TOPAS sub-bottom profiles from JR127 for Figure 6; rectangles indicate areas discussed in Figures 3–5. Tidewater glaciers: Bl, Blomstrandbreen; Co, Conwaybreen; Ko, Kongsbreen; Kr, Kronebreen; Kv, Kongsvegen; Li, Lilliehöökbreen.

length) also trending from north to south occurs within the central area of the fjord and is *c.* 140 m high and *c.* 1.5 km wide (Fig. 3). A deep hollow with a maximum water depth of 370 m breaks the central area of the ridge (Fig. 3).

Two slightly sinuous transverse ridges are present within the central region of Krossfjorden. These ridges are much less distinct than those in Kongsfjorden. The outer ridge (TR4) is *c.* 10 m high, between 150 and 300 m wide and *c.* 1.8 km long. The inner ridge (TR5) has approximately the same dimensions (see Fig. 3).

Interpretation. A previously documented transverse ridge at the shelf break in the southern part of Kongsfjordrenna was interpreted to mark the maximum position of the Late Weichselian ice stream relating to the Kongsfjorden–Krossfjorden system (Ottesen *et al.* 2007). We therefore propose that the transverse ridges (TR1–5) within the

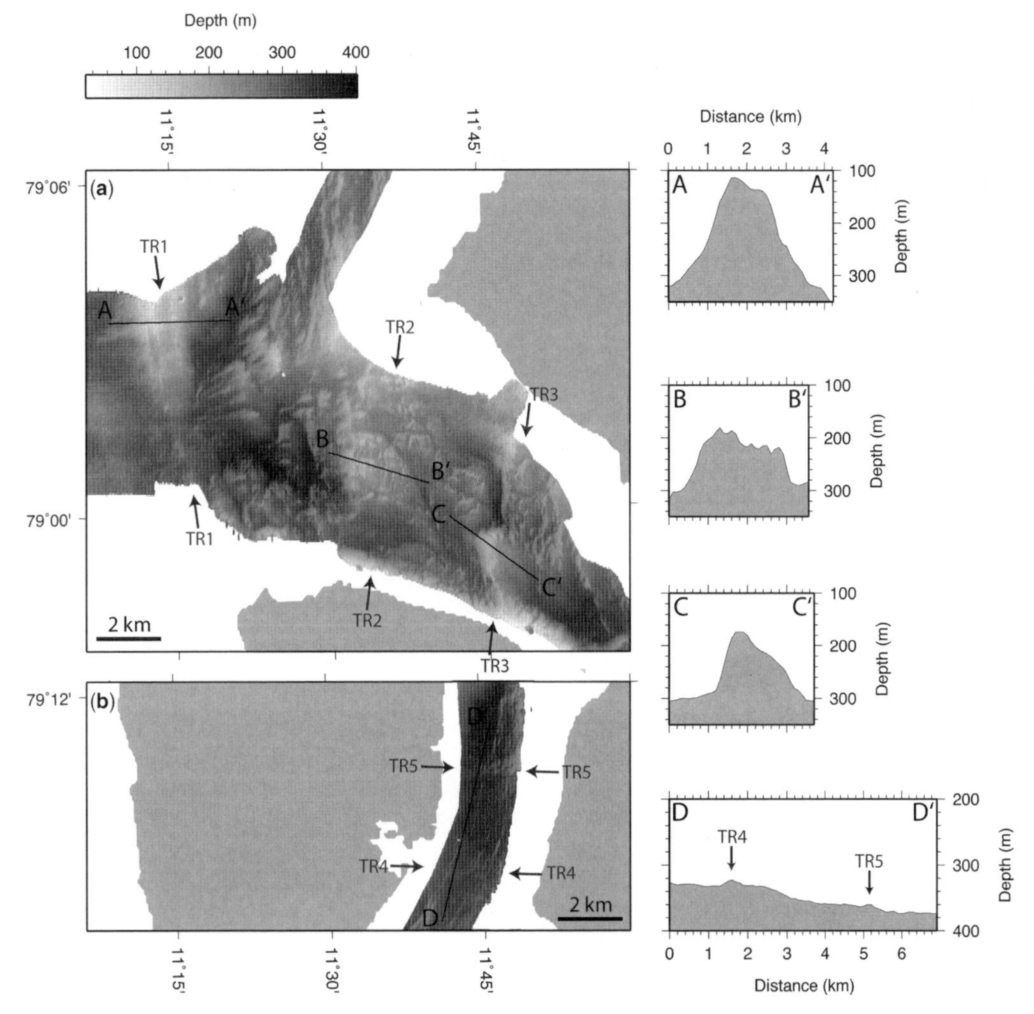

Fig. 3. Interpreted recessional moraine distribution: (**a**) transverse ridge structures (TR1–3) within the central and outer basins of Kongsfjorden; (**b**) transverse ridge structures (TR4, 5) within central Krossfjorden (see Fig. 2 for location). Topographic profiles across the ridge axis demonstrate the classic recessional moraine morphology as described by Hambrey (1994). Note that the surface of TR2 has been heavily modified during a Late Weichselian re-advance (see text for details).

Kongsfjorden as terminal or recessional moraine ridges forming part of a series of moraines occurring landward of the terminal moraine. Furthermore, it is probable that these transverse ridges were deposited during deglaciation, during a halt or re-advance within the general phase of ice recession (Hambrey *et al.* 1997) of the Late Weichselian ice sheet.

At the fjord mouth, the outermost ridge (TR1) is asymmetric in cross-section with a hummocky inner (proximal) slope and steeper distal slope (Fig. 3). The inner ridge (TR3) displays a similar asymmetric cross-sectional shape. The observed shape of these features is analogous to the description of terminal and recessional moraines outlined by Hambrey (1994); the outer slope of the moraine is steep, whereas the inner slope is generally irregular and hummocky and of a relatively low angle (10–20°). The central fjord moraine (TR2) is a more complex feature composed of interconnected lobate morainal segments, breached in places by low-relief seafloor. It is interpreted that the morainal ridge has been modified by further glacial processes after formation. In the acoustic surveys, there was no evidence to define the substructure of these features due to penetration being problematic as a result of the likely heterogeneous grain size of the deposits (Sexton *et al.* 1992). However, the possibility of bedrock being present cannot be discounted.

The transverse ridges (TR4, 5) within Krossfjorden were previously interpreted as channels (Howe *et al.* 2003). However, this interpretation is inaccurate due to the convex morphology of the features (Fig. 3). It is more likely that these features are sedimentary deposits, possibly small recessional moraines that record brief stillstands during the general retreat phase of the ice stream (Ottesen & Dowdeswell 2006). These ridges comprise volumes of debris smaller by an order of magnitude than the larger recessional moraines (TR1–3), which suggests that they are recording shorter lived events during deglaciation (Ottesen *et al.* 2007).

Drumlins

Description. Streamlined submarine landforms have been observed within both fjords. In particular, several elongated hills are present within the central and outer part of Kongsfjorden, landward of the outermost transverse ridge (TR1) and within the outer basin of Krossfjorden. Individual features average 500 m to 1 km in width, 1–2 km in length and are tens of metres high (Fig. 4). The features exhibit a strongly coherent set of alignments ranging between NE–SW and ENE–WSW. The composition and internal stratigraphy is imaged poorly on the TOPAS records, but is generally characterized by chaotic reflection patterns.

Interpretation. These features were previously documented as relict subglacial drumlins within the central area of Kongsfjorden, south of Kapp Guissez (Fig. 1) by Howe *et al.* (2003). Our investigations show that these features are more widespread than previously noted. The 'smooth teardrop-shaped lineations' described by Howe *et al.* (2003) are the most common form; however, a subgroup of these features has also been identified to be crag-and-tail ridges. Figure 4 illustrates the bedrock peak of the

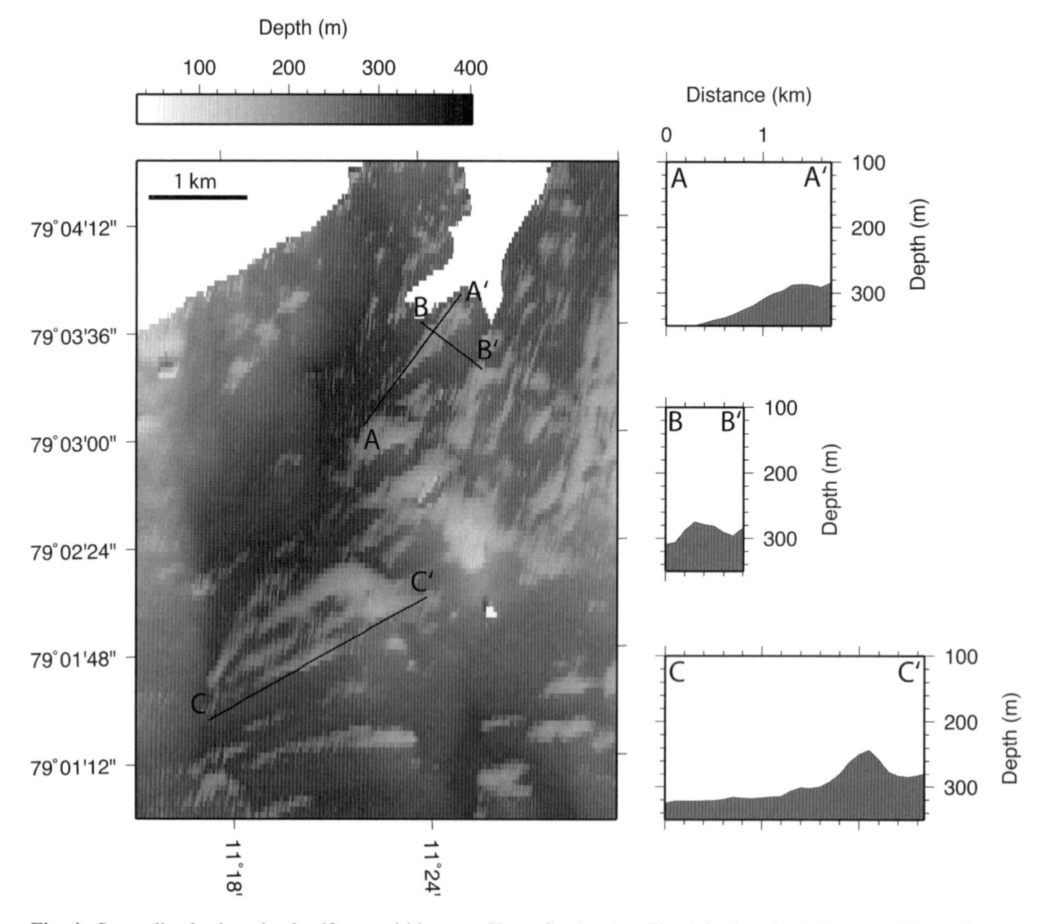

Fig. 4. Streamlined submarine landforms within outer Kongsfjorden (see Fig. 2 for location). Topographic profiles A–A′ and B–B′ show classic cross-sectional morphology of subglacial drumlin. Topographic profile C–C′ illustrates the bedrock peak of the 'crag' connected to the low relief sediment 'tail'.

'crag' connected to the low-relief slightly irregular linear sediment 'tail' that forms this feature. Both types of subglacial landforms develop parallel to the ice-flow direction.

Glacial lineations

Description. The seabed of the central and outer areas of Kongsfjorden exhibit linear ridge/groove features that trend NW–SE (Fig. 1). These features are *c.* 100–200 m wide, up to 4 km long and have a relief of the order 5–15 m (Fig. 5). They are orientated roughly parallel to the fjord axis, and occur in water depths of 200–300 m deep.

Interpretation. The lineations are interpreted to be glacial lineations (e.g. Clark 1993), formed from soft sediment deformation at the base of the fast-flowing ice stream (Tulaczyk *et al.* 2001; Ó Cofaigh *et al.* 2005). They represent the direction of past ice flow and are widely regarded as diagnostic of the presence of fast-flowing ice streams (Clark 1993). Landvik *et al.* (2005) suggested that the lineations in the outer part of the Kongsfjorden trough were probably formed at the Late Weichselian glaciation, whereas the lineations in the fjord were probably formed during deglaciation.

Glacial channels

Description. The basin floor of inner and central Kongsfjorden is characterized by a distinctive channelized morphology. The channels are 50–100 m wide and display predominantly low sinuosity and straight planform morphology (Fig. 5). Channel sides vary in relief (20–50 m) with some very steep slopes (45° or more) (Figs 5 & 6). The inner fjord channel is bound with 10–20 m high levees.

Interpretation. The convergent channels of the central fjord area converge and cross-cut the transverse ridge (TR2). The slightly sinuous planform morphology and the variable depth of their thalweg with common overdeepened depressions suggest that the channels were formed by fluvioglacial processes. The inner fjord channel was interpreted as a subglacial meltwater channel supplying sediment to the deep basin SW of Blomstrandhalvøya (Howe *et al.* 2003). Considering the channel aligns parallel to the direction of ice flow, it is possible that the channel was formed by subglacial erosion. The occurrence of such features indicates that significant volumes of meltwater and sediment were transferred out onto the shelf during deglaciation.

The succession of bedforms

The stratigraphic arrangement of the observed landforms permits the reconstruction of the relative timings of deposition of each landform; the oldest are successively overlain and cross-cut by younger

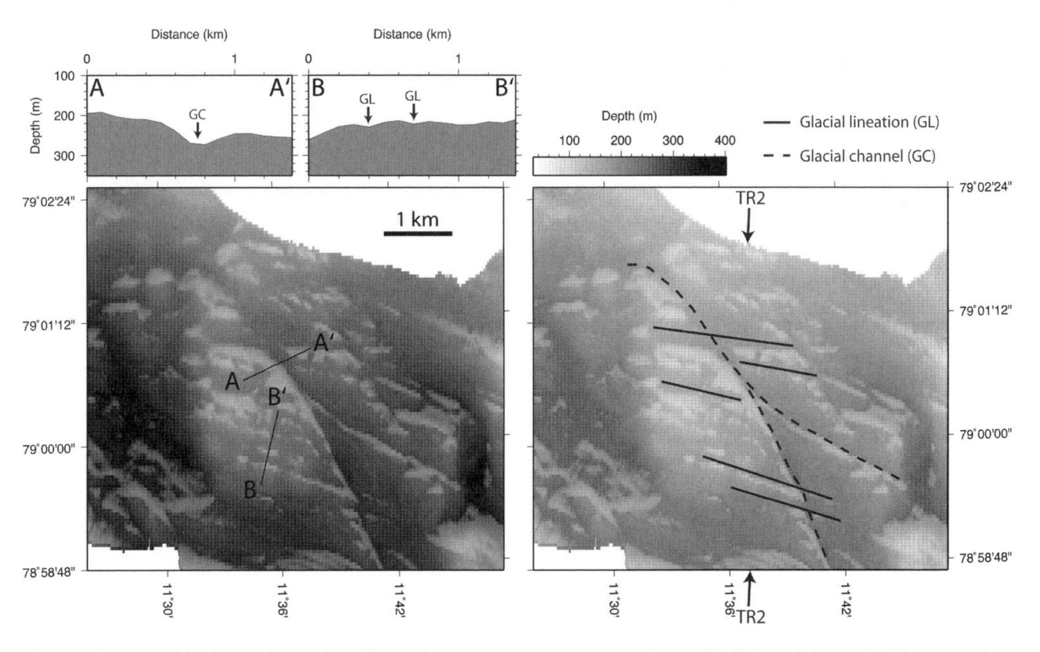

Fig. 5. Illuminated bathymetric section illustrating glacial lineations (trending NW–SE) and channel within central Kongsfjorden. Topographic profile A–A′ shows the deeply scoured U-shaped channel, B–B′ the shallowly incised parallel lineations.

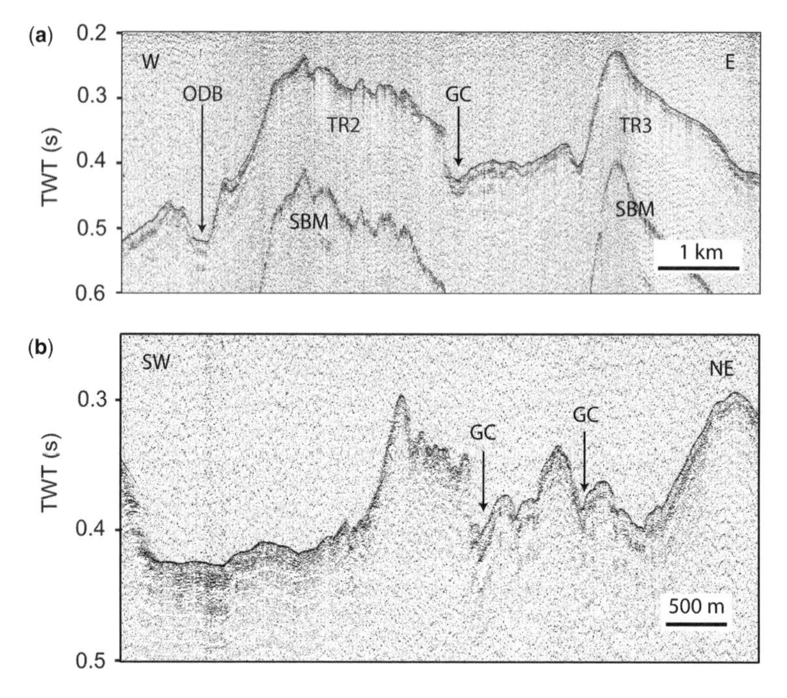

Fig. 6. Topas sub-bottom profiles of: (**a**) cross-section along the fjord axis cross-cutting the central and inner Kongsfjorden recessional moraines; (**b**) cross-fjord view from SW to NE through the central fjord basin and over inner slope of the central fjord recessional moraine towards the northern fjord flank. Abbreviated geographical features include: SBM, seabed multiple; GC, subglacial meltwater channel; ODB, over-deepened basin.

landforms and erosional processes (compare with Ottesen & Dowdeswell 2006).

The oldest seabed features within the fjord system are the set of transverse ridges, interpreted as a series of recessional moraines. The central and outer basin ridges are cross-cut by streamlined linear bedforms oriented parallel to the fjord axis. The streamlined bedforms have modified the ridge surface and have, in parts, remobilized material from the TR2 ridge below (Fig. 7). Similar streamlined bedforms are also observed within the basins between the ridges. These are likely to have formed through a combination of processes including bedrock plucking/gouging and moulding of sediment from meltwater discharge. Within the depressions of central Kongsfjorden, the lineations are overprinted by cross-cutting meltwater channels. This suggests that the meltwater channels of the inner and central basins are possibly one of the youngest seabed features in the succession. The succession of retreat moraines identified by Howe *et al.* (2003) probably represents the deglaciation of the Late Weichselian ice sheet within inner Kongsfjorden (Fig. 7).

Kongsfjorden is also characterized by extensive over-deepened basins, probably located in the Late Palaeozoic sedimentary strata or Quaternary sediments. The observed deep hollows are interpreted to have been carved by glacial erosion, which preferentially exploited less resistant bedrock (Fig. 6). In contrast, the crystalline basement rocks north of the fault have been more resistant to erosion and the basin within central Krossfjorden is much shallower (see Fig. 1).

Discussion

Formation of streamlined bedforms and ice-stream dynamics

Streamlined lineations have been reported to form subglacially during recent ice advances where the movement of ice over the sedimentary substrate creates streamlined bedforms aligned parallel to the direction of ice flow across a range of scales (e.g. Clark 1994; Ottesen & Dowdeswell 2006; Ottesen *et al.* 2008). A study of the mid-Norwegian shelf by Ottesen *et al.* (2002) inferred that the presence of streamlined subglacial bedforms is evidence for fast-flowing ice streams. Additionally, it is noted by Stokes & Clark (2001) that it is not the presence of single features but rather the arrangement of several features that provides the strongest

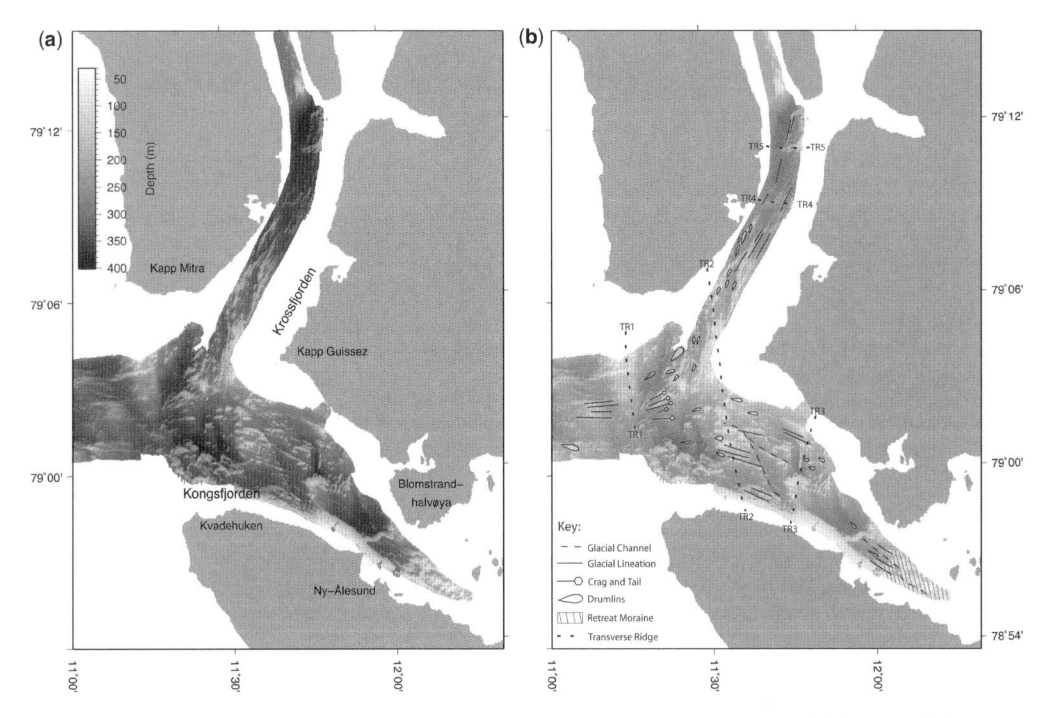

Fig. 7. Geomorphological map of the study area showing the generalized distribution of glacial features within the Kongsfjorden–Krossfjorden system. Five distinct recessional moraines are identified, along with numerous subglacial features that suggest a converging ice flow at the mouth of the fjord system.

indication of palaeo-ice stream locations. Therefore, the mega-scale glacial lineations and drumlins within the Kongsfjorden–Krossfjorden system may have formed by: (1) rapid ice flow; (2) slow ice flow over a long period; or (3) developed from variations in sediment characteristics beneath the ice sheet. The parallel nature of the lineated bedforms and the convergent flow pattern at the mouth of the system implies that there is a possible topographic control present (Fig. 7). However, the uneven distribution of these features suggests that a change in the bedrock composition from south to north may explain the predominant absence of these features from Krossfjorden. Furthermore, the irregular distribution of the streamlined bedforms suggests they are unlikely to have developed under a sustained fast-flowing ice stream and an episodic flow is more probable. For instance, Ottesen *et al.* (2007) suggests where grounding-zone wedges are present within Svalbard cross-shelf troughs and fjords, episodic ice retreat occurs between points of relative stability where the ice front remained long enough to build the sediment forming the wedge. Landvik *et al.* (2005) mapped two ridges from seismic profiles near the mouth of Kongsfjorden, proposing these were interruptions to the ice retreat.

The Kongsfjorden–Krossfjorden system was one of the major ice streams flowing in western Svalbard (Ottesen *et al.* 2005, 2007). Ice masses from the NW (Krossfjorden) and SW (Kongsfjorden) joined to form a major ice stream following Kongsfjordrenna to the shelf break. Figures 3 and 7 depict the glacial lineations and drumlins indicating ice flow along the fjord axis out into the trough. These well-developed flow-parallel features indicate that this ice stream had a substantial velocity when the tributaries coalesced at the fjord mouth. The flow lines illustrate how the tributaries assimilate into the major trunk of the Kongsfjordrenna ice stream and are deflected out onto the shelf.

The sequence of deglaciation

Geomorphic features indicative of ice retreat, such as recessional moraines and annual retreat moraines (Howe *et al.* 2003), are a prominent characteristic of the seabed in Kongsfjorden (Fig. 3). The serial nature of the recessional moraines are interpreted to indicate a staggered retreat, recording stillstands or ice-front oscillations during the retreat of a grounded ice margin. The presence of such moraines has also been used to infer the occurrence of an ice margin that was not fast flowing (Dowdeswell

& Elverhøi 2002; Landvik *et al.* 2005). The sub-bottom profiles show distinct units of acoustically well-laminated sediments throughout outer and central Kongsfjorden, predominantly within bedrock basins in steep irregular topography (up to 20 m thick). It is interpreted that these sediments accumulated during deglaciation with bedrock highs marking the stepped regression of the ice front retreat. The bedrock highs act as 'pinning points', providing a position of temporary stability for the ice front during which ice-proximal sedimentation into the basins is promoted. These distinct isolated sediment units are mainly separated by expanses of thin sediment cover and/or bedrock highs, signifying periods of rapid retreat between pinning points (Ó Cofaigh *et al.* 2001).

In contrast, the retreat features within Krossfjorden are much less prominent. This may indicate decoupling of the ice stream at the mouth of the fjord or a lack of available sediment to form such features (Fig. 1). If the lack of retreat features is indicative of fast retreat, then a potentially rapid retreat occurred in Krossfjorden where it is characterized by crystalline substrate.

Both streamlined glacial lineations and small retreat moraines are observed on the top of the inner two moraine ridges (Fig. 3). The majority of the central recessional moraine ridge appears to have been modified by the streamlined bedforms that overlie the ridge; this modification must have occurred during a small ice advance during the Late Weichselian deglaciation (Ottesen & Dowdeswell 2006). The fact that the channels cross-cut the recessional moraines is also evidence of ice stagnation and the reworking of the surface sediment by meltwater. In summary, the complex cross-cutting relationship between bedforms observed within Kongsfjorden suggests that there was a series of stillstands and a re-advance of the ice front during the deglaciation following the Late Weichselian glacial phase.

Conclusion

The data we present suggests that the ice retreat within Kongsfjorden was probably less rapid than the previously assumed two-step deglaciation proposed by Lehman & Forman (1992). We propose that there were several stillstands (TR1-3) and possible re-advances. A series of smaller annual retreat moraines are found within central Kongsfjorden; retreat of the ice front is inferred to have been slow but quasi-continuous relative to the outer fjord where the larger grounding-zone wedges were formed. These smaller volume ridges were probably the product of the sediments pushing up during minor re-advances or halts (e.g. Boulton 1986; Ottesen & Dowdeswell 2006; Ottesen *et al.* 2008). Contrastingly, the morphology of Krossfjorden would suggest a more rapid retreat of the ice front to the tributaries at the head of the fjord. The decoupling of the ice stream at the mouth of the fjord is attributed to the regional geology and a complex topographic control of the ice stream.

We thank the officers and crew of RSS *James Clark Ross* during the 2002 and 2005 cruises. D. Benn and M. Forwick are thanked for their comments and constructive reviews which improved the manuscript considerably.

References

Andersen, E. S., Dokken, T. M., Elverhøi, A., Solheim, A. & Fossen, I. 1996. Late Quaternary sedimentation and glacial history of the western Svalbard continental margin. *Marine Geology*, **133**, 123–156.

Bergh, S. G., Maher, H. D. & Braathen, A. 2000. Tertiary divergent thrust directions from partitioned transpression, Brøggerhalvøya, Spitsbergen. *Norsk Geologisk Tidsskrift*, **80**, 63–82.

Boulton, G. S. 1986. Push-moraines and glacier-contact fans in marine and terrestrial environments. *Sedimentology*, **33**, 677–698.

Clark, C. D. 1993. Mega-scale glacial lineations and cross-cutting ice-flow landforms. *Earth Surface Processes and Landforms*, **18**, 1–29.

Clark, C. D. 1994. Large-scale ice-moulding: a discussion of genesis and glaciological significance. *Sedimentary Geology*, **91**, 253–268.

Dowdeswell, J. A. & Elverhøi, A. 2002. The timing of initiation of fast-flowing ice streams during a glacial cycle inferred from glacimarine sedimentation. *Marine Geology*, **188**, 3–14.

Elverhøi, A., Hooke, R. L. & Solheim, A. 1998. Late Cenozoic erosion and sediment yield from the Svalbard-Barents Sea region: implications for understanding erosion of glacierized basins. *Quaternary Science Reviews*, **17**, 209–241.

Hambrey, M. J. 1994. *Glacial Environments*. UCL Press Limited, London, 145–146.

Hambrey, M. J., Huddart, D., Bennett, M. R. & Glasser, N. F. 1997. Genesis of 'hummocky moraines' by thrusting in glacier ice: evidence from Svalbard and Britain. *Journal of the Geological Society*, **154**, 623–632.

Harland, R. B. 1997. *The Geology of Svalbard*. Geological Society Memoir No. 17. The Geological Society, London.

Howe, J. A., Moreton, S. G., Morri, C. & Morris, P. 2003. Multibeam bathymetry and the depositional environments of Kongsfjorden and Krossfjorden, western Spitsbergen, Svalbard. *Polar Research*, **22**, 301–316.

Landvik, J. Y., Bondevik, S. *et al.* 1998. The last glacial maximum of the Barents Sea and Svalbard area: ice sheet extent and configuration. *Quaternary Science Reviews*, **17**, 43–75.

Landvik, J. Y., Ingólfsson, Ó., Mienert, J., Lehman, S. J., Solheim, A., Elverhøi, A. & Ottesen, D. 2005. Rethinking Late Weichselian ice-sheet dynamics in coastal NW Svalbard. *Boreas*, **34**, 7–24.

LEHMAN, S. J. & FORMAN, S. L. 1992. Late Weichselian glacier retreat in Kongsfjorden, West Spitsbergen, Svalbard. *Quaternary Research*, **37**, 139–154.

MANGERUD, J., ASTAKOV, V. & SVENDSEN, J. I. 2002. The extent of the Barents-Kara ice sheet during the Last Glacial Maximum. *Quaternary Science Reviews*, **21**, 111–119.

Ó COFAIGH, C., DOWDESWELL, J. A. & GROBE, H. 2001. Holocene glacimarine sedimentation, inner Scoresby Sund, East Greenland: the influence of fast-flowing ice sheet outlet glaciers. *Marine Geology*, **175**, 103–129.

Ó COFAIGH, C., DOWDESWELL, J. A., ALLEN, C. S., HIEMSTRA, J. F., PUDSEY, C. J. & EVANS, J. 2005. Flow dynamics and till genesis associated with a marine-based Antarctic palaeo-ice stream. *Quaternary Science Reviews*, **24**, 709–740.

OTTESEN, D. & DOWDESWELL, J. A. 2006. Assemblages of submarine landforms produced by tidewater glaciers in Svalbard. *Journal of Geophysical Research*, **111**, 1–16.

OTTESEN, D., DOWDESWELL, J. A., RISE, L., ROKOENGEN, K. & HENRIKSEN, S. 2002. Large-scale morphological evidence for past ice-stream flow on the mid-Norwegian continental margin. *In*: DOWDESWELL, J. A. & COFAIGH, C. Ó. (eds) *Glacier-Influenced Sedimentation on High-Latitude Continental Margins*. Geological Society, London, Special Publications, **203**, 245–258.

OTTESEN, D., DOWDESWELL, J. A. & RISE, L. 2005. Submarine landforms and the reconstruction of fast-flowing ice streams within a large Quaternary ice sheet: the 2500-km-long Norwegian-Svalbard margin (57 degrees–80 degrees N). *Geological Society of America Bulletin*, **117**, 1033–1050.

OTTESEN, D., DOWDESWELL, J. A., LANDVIK, J. Y. & MIENERT, J. 2007. Dynamics of the Late Weichselian ice sheet on Svalbard inferred from high-resolution sea-floor morphology. *Boreas*, **36**, 286–306.

OTTESEN, D., DOWDESWELL, J. A. ET AL. 2008. Submarine landforms characteristic of glacier surges in two Spitsbergen fjords. *Quaternary Science Reviews*, **27**, 1583–1599.

SEXTON, D. J., DOWDESWELL, J. A., SOLHEIM, A. & ELVERHOI, A. 1992. Seismic architecture and sedimentation in northwest Spitsbergen fjords. *Marine Geology*, **103**, 53–68.

SOLHEIM, A., ANDERSEN, E. S., ELVERHOI, A. & FIEDLER, A. 1996. Late Cenozoic depositional history of the western Svalbard continental shelf, controlled by subsidence and climate. *Global and Planetary Change*, **12**, 135–148.

SOLHEIM, A., FALEIDE, J. I. ET AL. 1998. Late Cenozoic seismic stratigraphy and glacial geological development of the East Greenland and Svalbard-Barents Sea continental margins. *Quaternary Science Reviews*, **17**, 155–184.

STOKES, C. R. & CLARK, C. D. 2001. Palaeo-ice streams. *Quaternary Science Reviews*, **20**, 1437–1457.

SVENDSEN, J. I., MANGERUD, J., ELVERHØI, A., SOLHEIM, A. & SCHÜTTENHELM, R. T. E. 1992. The Late Weichselian glacial maximum on western Spitsbergen inferred from offshore sediment cores. *Marine Geology*, **104**, 1–17.

SVENDSEN, J. I., ELVERHØI, A. & MANGERUD, J. 1996. The retreat of the Barents Ice Sheet on the western Svalbard margin. *Boreas*, **25**, 244–256.

SVENDSEN, H., BESZCZYNSKA-MOLLER, A. ET AL. 2002. The physical environment of Kongsfjorden–Krossfjorden, an Arctic fjord system in Svalbard. *Polar Research*, **21**, 133–166.

SVENDSEN, J. I., ALEXANDERSON, H. ET AL. 2004. Late Quaternary ice-sheet history of northern Eurasia. *Quaternary Science Reviews*, **23**, 1229–1271.

TULACZYK, S. M., SCHERER, R. P. & CLARK, C. D. 2001. A ploughing model for the origin of weak tills beneath ice streams: a qualitative treatment. *Quaternary International*, **86**, 59–70.

VANNESTE, M., BERNDT, C., LABERG, J. S. & MIENERT, J. 2007. On the origin of large shelf embayments on glaciated margins-effects of lateral ice flux variations and glacio-dynamics west of Svalbard. *Quaternary Science Reviews*, **26**, 2406–2419.

VORREN, T. O. & LABERG, J. S. 1997. Trough Mouth Fans-Paleoclimate and ice-sheet monitors. *Quaternary Science Reviews*, **16**, 865–881.

VORREN, T. O., LABERG, J. S. ET AL. 1998. The Norwegian-Greenland Sea continental margins: morphology and late quaternary sedimentary processes and environment. *Quaternary Science Reviews*, **17**, 273–302.

Late Weichselian and Holocene sedimentary environments and glacial activity in Billefjorden, Svalbard

NICOLE J. BAETEN[1]*, MATTHIAS FORWICK[1], CHRISTOPH VOGT[2] & TORE O. VORREN[1]

[1]*Department of Geology, University of Tromsø, N-9037 Tromsø, Norway*

[2]*Central Laboratory for Crystallography and Applied Material Sciences (ZEKAM), Geosciences, University of Bremen, Klagenfurter Str. 2, D-28359 Bremen, Germany*

**Corresponding author (e-mail: Nicole.Baeten@ig.uit.no)*

Abstract: Swath bathymetry data and one sediment core were used to improve the understanding of the Late Weichselian and Holocene glacier activity in Billefjorden, Svalbard. Grounded ice existed in Billefjorden prior to 11.23 cal ka BP (calendar years before present), depositing a basal till and producing glacial lineations. The glacier front retreated from the central parts to the inner parts of the fjord between *c.* 11.23 and 11.2 cal ka BP. Annual recessional moraines suggest that this retreat occurred at a rate of up to 170 m a^{-1}. During the early Holocene, the glacier Nordenskiöldbreen was comparatively small and sediment supply to central Billefjorden occurred mainly from the fjord sides. An increase in ice rafting around 7930 cal a BP is ascribed to enhanced sea-ice formation. The activity of Nordenskiöldbreen increased around 5470 cal a BP. Ice rafting was generally low during the past *c.* 3230 a. This was most likely related to the formation of a more permanent sea-ice cover. Nordenskiöldbreen reached its maximum Holocene extent around AD 1900, generating glacial lineations and depositing a terminal moraine in the inner fjord. Annual recessional moraines were formed during its subsequent retreat. Icebergs from Nordenskiöldbreen generated iceberg ploughmarks during the late Holocene.

During the Late Weichselian, fjords on Spitsbergen acted as pathways for fast-flowing ice streams draining the Svalbard and Barents Sea ice sheet (e.g. Landvik *et al.* 1998, 2005; Ottesen *et al.* 2005). The deglaciation of the mouth of Isfjorden occurred about 14.1 cal ka BP (12.3 ^{14}C ka BP radiocarbon years before the present, Mangerud *et al.* 1992). Based on a single radiocarbon date, Boulton (1979) interpreted a till at Kapp Ekholm (for location see Fig. 1) in Billefjorden to indicate a major glacial re-advance between *c.* 12.58 and 11.1 cal ka BP (11.0–9.7 ^{14}C ka BP). Mangerud & Svendsen (1990) could not, however, reproduce this date, and question Boulton's interpretations. In addition, Forwick & Vorren (2005) could not find any evidence for a Late Weichselian glacier re-advance in Billefjorden.

According to Forwick & Vorren (2005), the deglaciation of Billefjorden took place during the late Younger Dryas and the early Preboreal. Recessional moraines in the fjord may indicate a stepwise retreat of the glaciers. In the Isfjorden area, the main retreat of the glacier front terminated at *c.* 11.3 cal ka BP (*c.* 10 ^{14}C ka BP) in the inner fjords (Mangerud *et al.* 1992, 1998; Elverhøi *et al.* 1995; Svendsen *et al.* 1996; Lønne 2005; Forwick & Vorren 2009).

Based on constant low sedimentation rates during the early and mid Holocene, Svendsen & Mangerud (1997) suggested that glaciers in Billefjorden were small or absent during this interval. Possible glacier expansion, in which the glacier Nordenskiöldbreen would reach the fjord after 2.8 cal ka BP, is indicated by a pronounced increase in the sedimentation rate. Recent studies on sea-surface temperature and ice-rafted debris (IRD), however, indicate that central parts of Spitsbergen have never been ice-free during the Holocene and that the glacial activity increased asynchronously after *c.* 8.8 cal ka BP (Hald *et al.* 2004; Hald & Korsun 2008; Forwick & Vorren 2009).

In this paper, one sediment core and swath bathymetry data are used to investigate sedimentary environments and landforms with the purpose of reconstructing the glacial history in Billefjorden during the Late Weichselian and the Holocene. We improve the existing lithostratigraphy and the chronology of ice-rafting. Bulk mineral assemblage data are used to identify sediment sources and to infer their influence on the sedimentary environment in central Billefjorden.

Physiographic setting

Billefjorden is a tributary of Isfjorden, the largest fjord on western Spitsbergen (Fig. 1). It is located between 78°27′ and 78°45′N and 16°00′ and

From: Howe, J. A., Austin, W. E. N., Forwick, M. & Paetzel, M. (eds) *Fjord Systems and Archives*. Geological Society, London, Special Publications, **344**, 207–223.
DOI: 10.1144/SP344.15 0305-8719/10/$15.00

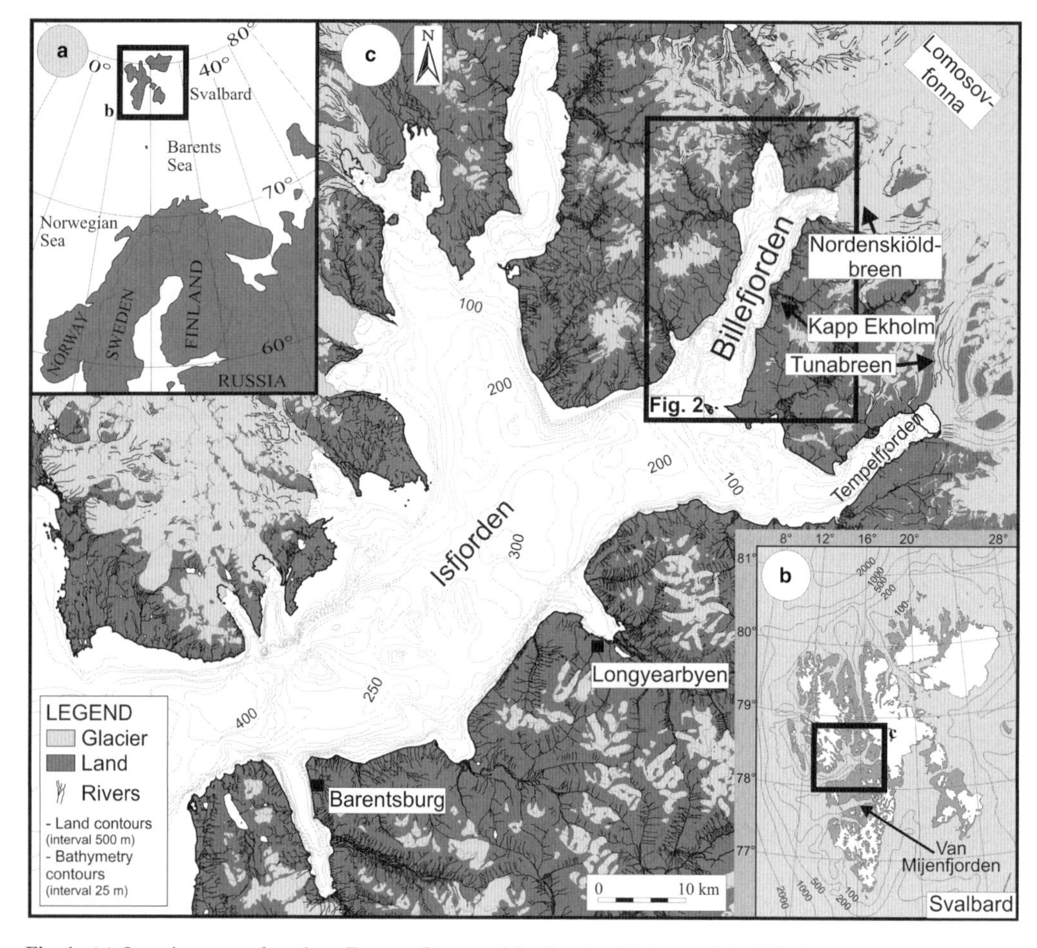

Fig. 1. (**a**) Overview map of northern Europe; (**b**) map of Svalbard and (**c**) map of the Isfjorden area.

17°00′E. It is 30 km long, between 5 and 8 km wide and has a maximum water depth of 226 m. The area of the fjord is *c*. 200 km^2. The catchment area of Billefjorden is 907 km^2 and it has a glacier coverage of 43.8% (Hagen *et al.* 1993).

Sediment supply is from several rivers and the tidewater glacier Nordenskiöldbreen in Adolfbukta (Figs 1 & 2). Nordenskiöldbreen is an outlet of the ice cap Lomonosovfonna (Fig. 1).

The approximately north–south oriented Billefjorden Fault Zone dominates the geology in the fjord (Fig. 2; Dallmann *et al.* 2004). A relatively large area of Precambrian basement rocks occurs in the drainage area of Nordenskiöldbreen. Smaller areas occur north of Petuniabukta and west of Mimerbukta. These mainly crystalline rocks are very resistant and form a terrace in front of the present glacier front. Along the fjord sides the rocks mainly consist of Billefjorden Trough strata of late Carboniferous age, including limestones, dolomites, conglomerates, sandstones, gypsum/anhydrite, mudstone and shale (Dallmann *et al.* 2004).

Oceanographic conditions in western Spitsbergen fjords are strongly determined by the West Spitsbergen Current, a northern extension of the Norwegian Atlantic Current. It transports relatively warm and saline water along the continental slope of west Spitsbergen. Because of this current, the area west of Spitsbergen is essentially ice-free (Aagaard *et al.* 1987; Gascard *et al.* 1995). However, Spitsbergen fjords are generally covered by sea ice during winter. The sea ice starts to break up between April and July (Węsławski *et al.* 1995; Svendsen *et al.* 2002; Nilsen *et al.* 2008).

Material and methods

The swath bathymetry data was collected with RV *Jan Mayen* in the summers of 2005 and 2006 using a Kongsberg Maritime Simrad EM 300 multibeam

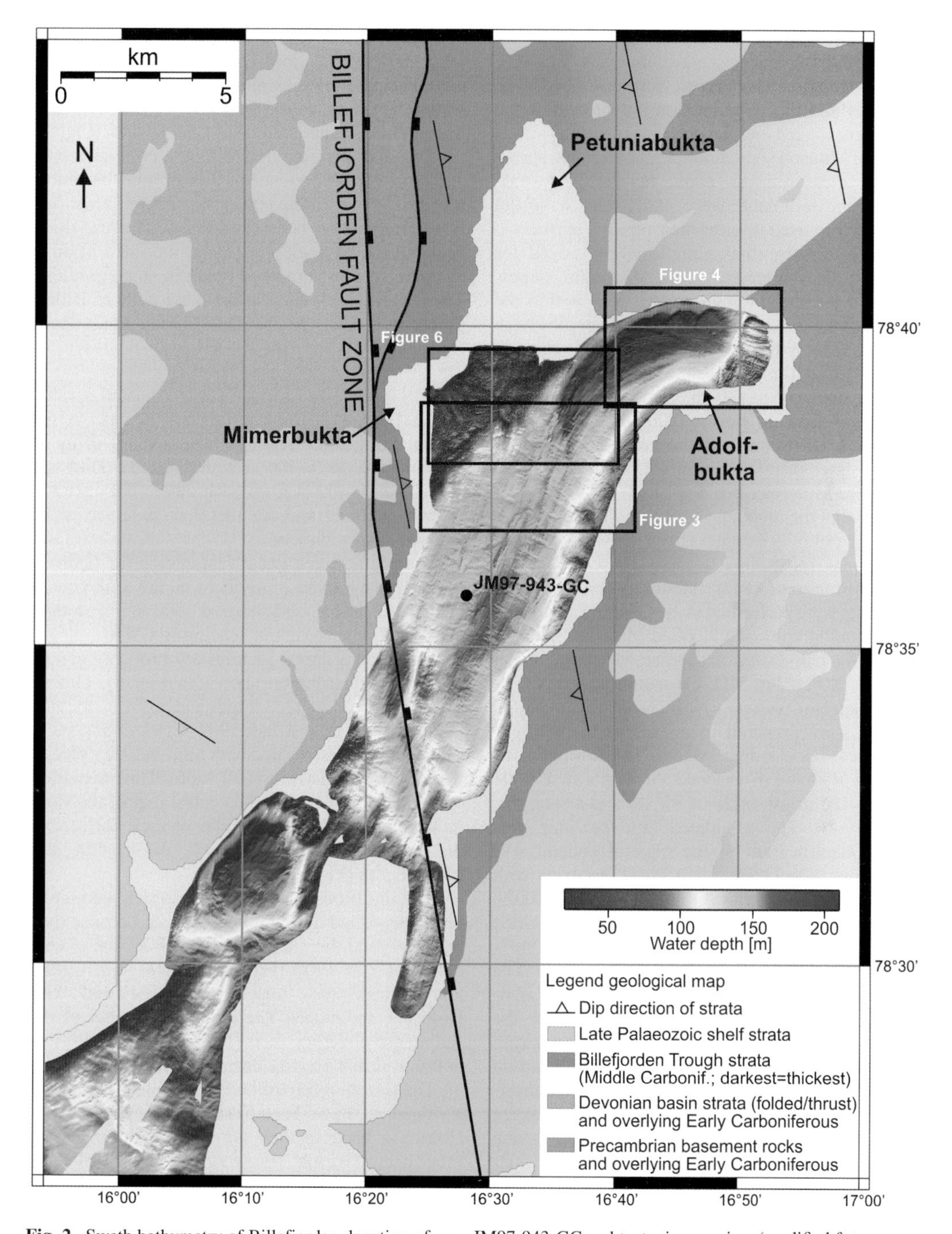

Fig. 2. Swath bathymetry of Billefjorden, location of core JM97-943-GC and tectonic overview (modified from Dallmann *et al.* 2004).

echosounder. Data cleaning and processing was performed using the post-processing software Neptune version 4.1.2. Gridding and visualization of the results have been carried out in The General Mapping Tools (GMT) (Wessel & Smith 1998) and Fledermaus 6.4.

Gravity core JM97-943-GC was retrieved with RV *Jan Mayen* in 1997 in the central part of the fjord (Fig. 2; 78°35.7′N/16°29.1′E; 193 m water depth). The physical properties of the sediments were measured with a GEOTEK multisensor core logger (MSCL) prior to opening. This included bulk density and magnetic susceptibility (loop sensor).

Radiographs of half cores were described. The structures and the sediments on the surface of the cores were logged and the colours were determined using the Munsell soil colour charts. Reliable shear-strength measurements could not be performed because the sediments were partly dry. A needle was therefore used to infer their relative stiffness.

One-centimetre-thick samples were sieved for grain-size distribution analysis. The sampling depths depended on the lithological variations within the core but were *c.* 20 cm intervals. The samples were wet sieved in order to separate the finest fraction (<0.063 mm) from the rest of the sediments. The material courser than 0.063 mm was dry sieved using the following sieves: 0.063 mm, 0.125 mm, 0.250 mm, 0.50 mm, 1.0 mm and 2.0 mm. The material finer than 0.063 mm was analysed with a sedigraph (Micromeritics Sedigraph 5100) in order to determine the amount of silt and clay.

Powder of bulk samples (from the same depth as the samples for grain-size analysis) was used for X-ray diffraction (XRD) measurements. The measurements were performed at the Central Laboratory for Crystallography and Applied Material Sciences (ZEKAM), University of Bremen, Germany, using a Philips X'Pert Pro MD, Cu-radiation and X'Celerator detector system. Near randomly oriented samples were prepared by means of the Philips/Panalytical backloading system. The settings used were a $1/4^\circ$ fixed divergence slit with a first angle of 3° 2θ and a last angle of 85° 2θ and a step size of 0.016° 2θ. The calculated time per step was 100 s. Quantification of the mineral content was carried out with Quantitative Phase-Analysis with X-ray Powder Diffraction (QUAX) (Vogt *et al.* 2002).

Seven Accelerator Mass Spectrometry (AMS) radiocarbon dates on shells from core JM97-943-GC provide the chronology for this study. The targets were prepared at the Laboratory of Radiometric Dating in Trondheim, Norway and the measurements were carried out at the Ångstrøm Laboratory in Uppsala, Sweden. All radiocarbon ages in this study were corrected for a marine reservoir effect of 440 years (Mangerud & Gulliksen 1975). The radiocarbon ages were calibrated using the software Calib Rev 5.0.2 (Stuiver & Reimer 1993) with a ΔR of 99 ± 39 (http://calib.qub.ac.uk/marine) and the results were obtained from the calibration data set *marine 04.14C* (Hughen *et al.* 2004).

Results

Morphology

Billefjorden has a shallow sill at its mouth and is *c.* 12 km wide (Fig. 2). A basin of up to 226 m depth is located within this sill. The central and inner parts of the fjord comprise a *c.* 15 km long main basin with a maximum water depth of 211 m. Water depths are generally shallower than 100 m in Mimerbukta, Petuniabukta and Adolfbukta (Fig. 2).

The shallow area in Adolfbukta correlates well to the onshore occurrence of Precambrian basement rocks (Fig. 2; Dallmann *et al.* 2004). This part is referred to as bedrock terrace. The transition between the bedrock terrace and the main basin is characterized by a marked break in slope gradient. The rough morphology in the outer parts of Billefjorden is related to the Billefjorden Fault Zone (Fig. 2).

Glacial lineations. Linear features oriented parallel to the fjord axis occur at several places. In the central fjord, these features have a relief of up to 8 m and are up to 300 m wide (Fig. 3). They are spaced up to 400 m from each other. They are not sharply outlined and are therefore assumed to be covered with sediment.

Sharply outlined linear features on the bedrock terrace have a relief of up to 14 m, are a maximum of 300 m wide and exceed 500 m in length (Fig. 4). Their maximum spacing is 125 m.

Both sets of linear features are probably glacial lineations (distribution shown in Fig. 5). Glacial lineations are elongated ridges and grooves oriented parallel to ice flow (Clark 1994; Clark *et al.* 2003). They are often associated with high-velocity glacier flow (Ottesen & Dowdeswell 2006). The orientation of the linear features in the central part of the fjord indicates that, during the time of their formation, the ice was flowing through Adolfbukta and Mimerbukta (Fig. 5).

The lineations in central Billefjorden were most likely deposited during the last glacial because they occur beyond the maximum Late Holocene glacial extent at AD 1900 (Fig. 5; Plassen *et al.* 2004) and because they appear to be covered with Holocene sediments. The draping character of the sediments covering these lineations suggests low bottom-current activity during the Holocene.

The lineations on the bedrock terrace (Fig. 4) are assumed to have been formed during the latest Holocene because they are sharply outlined and they occur within the area of Late Holocene maximum extent (Fig. 5).

Morainal ridges. Numerous ridges are oriented almost perpendicular to the fjord axis (Figs 3–5). They superimpose the glacial lineations and are therefore assumed to be younger.

In the central fjord, the ridges are up to 9 m high, up to 250 m wide and have a maximum spacing of 600 m (Fig. 3). They are not sharply outlined, indicating that they are draped by sediment.

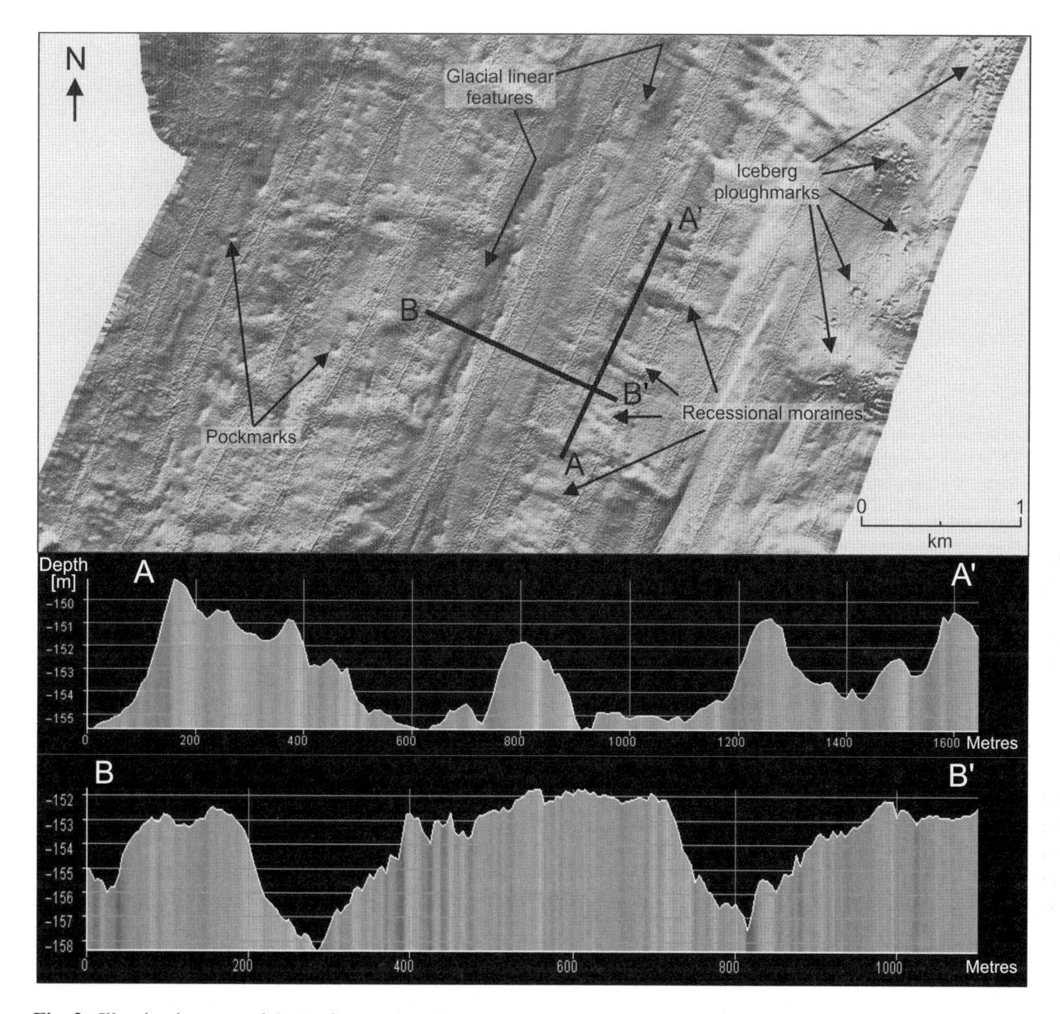

Fig. 3. Illumination map of the seafloor and profiles of the central part of Billefjorden (see Fig. 2 for location). Profile A–A′ crosses morainal ridges; profile B–B′ crosses glacial linear features. Iceberg ploughmarks and pockmarks (Forwick *et al.* 2009) are indicated.

Ridges on the bedrock terrace range between 2 and 8 m in height, are up to 100 m wide and are spaced up to 125 m from each other (Fig. 4).

Both sets of ridges are interpreted as morainal ridges. Their undulating and discontinuous morphology (Fig. 5) might indicate that they were deposited as crevasse squeeze ridges during surges of Nordenskiöldbreen (compare with Solheim & Pfirman 1985; Solheim 1991; Boulton *et al.* 1996; Ottesen & Dowdeswell 2006; Ottesen *et al.* 2008). However, we assume that the morainal ridges are recessional moraines because: (1) even although being statistically possible (Jiskoot *et al.* 2000), no surges for Nordenskiöldbreen have been recorded; (2) the morainal ridges are similar in appearance to small morainal ridges observed beyond tidewater glaciers on Svalbard (Liestøl 1976; Boulton 1986; Whittington *et al.* 1997; Ottesen *et al.* 2005; Ottesen & Dowdeswell 2006; Ottesen *et al.* 2008); and (3) the ridges are concordant with the ice margin and are quite regularly spaced (see also Ottesen *et al.* 2008).

The morainal ridges in central Billefjorden were interpreted to have been formed during the deglaciation of the fjord at the end of the last glacial because they are not sharply outlined and occur beyond the Late Holocene maximum glacier extent reached around 1900 AD (Plassen *et al.* 2004). The sharply outlined morainal ridges on the bedrock terraces were most likely deposited during the retreat after the Late Holocene maximum extent of Nordenskiöldbreen.

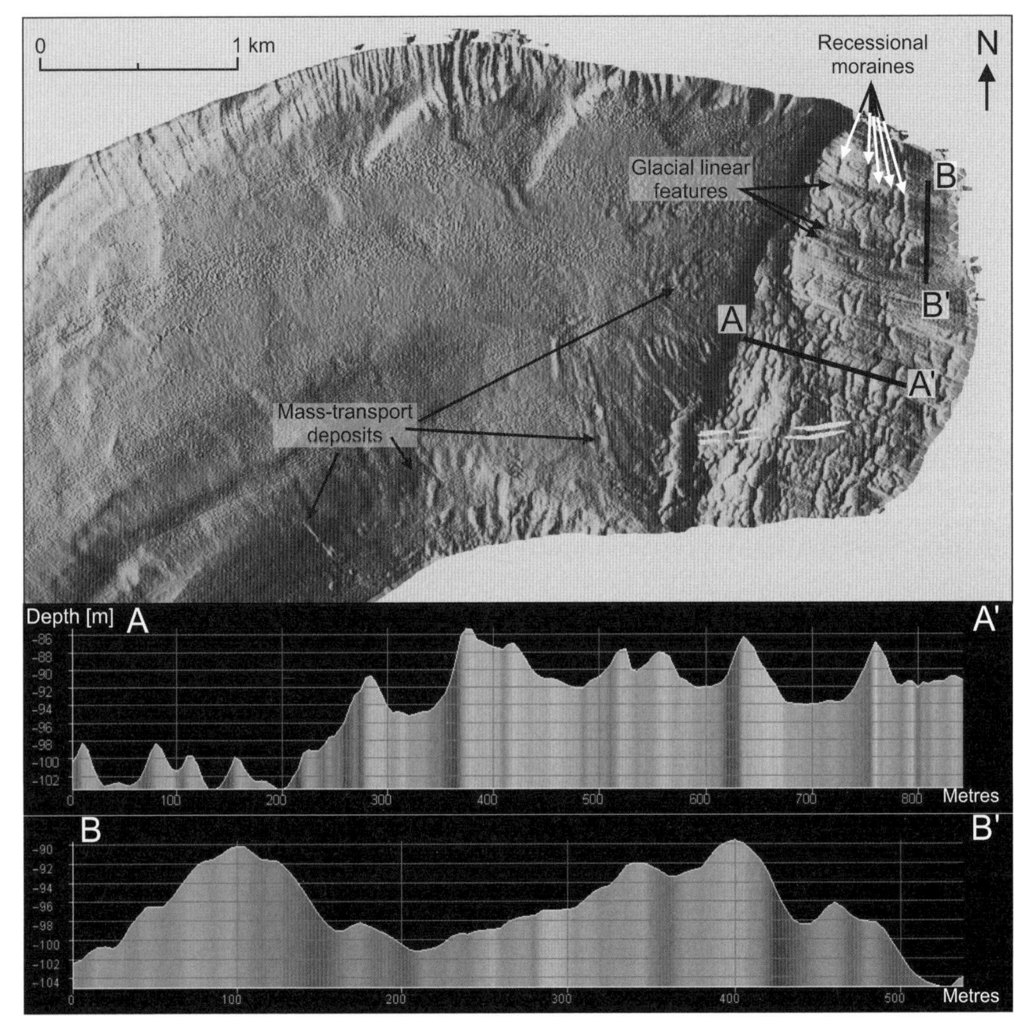

Fig. 4. Illumination map of the seafloor and profiles of the inner parts of Billefjorden (Adolfbukta; see Fig. 2). Profile A–A′ crosses morainal ridges; profile B–B′ crosses glacial linear features.

Iceberg ploughmarks. Slightly elongated depressions in a chaotic pattern occur in most areas shallower than 120 m (Figs 5 & 6). Their depths range from 2 to 5 m, their widths vary between 30 and 50 m and they are on average 30 to 80 m long. These depressions are interpreted as iceberg ploughmarks because of their presence within a specific depth range in several parts of the fjord (Fig. 5).

Their fresh appearance indicates that they are relatively young. The fact that they only occur at shallow depths suggests that they were generated by relatively small icebergs, most likely not from the last glacial. It is suggested that these ploughmarks were formed by icebergs calving off Nordenskiöldbreen during glacier growth and retreat in the Late Holocene.

Sediment core JM97-943-GC

Core JM97-943-GC is 460.5 cm long and comprises the units 943-A to 943-E (Fig. 7).

Unit 943-A (460.5–440 cm). The lowermost unit comprises a dark greyish-brown, massive, clast-rich diamicton with a muddy matrix (Fig. 7; Table 1). Its upper boundary is sharp and defined by a marked decrease in clast content (Figs 7 & 8). The magnetic susceptibility is relatively high and the sediments are comparatively stiff. The stiffness might be the result of compaction. It is suggested that this unit is composed of a basal till because of the massive texture, high amounts of clasts and the absence of shells and bioturbation. This till was correlated to

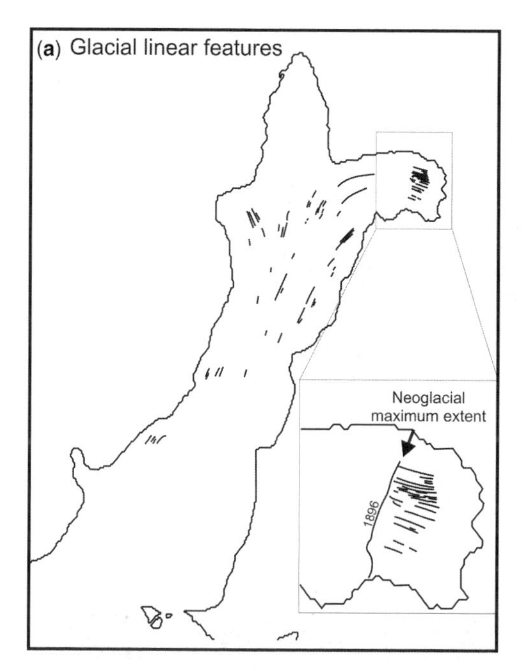

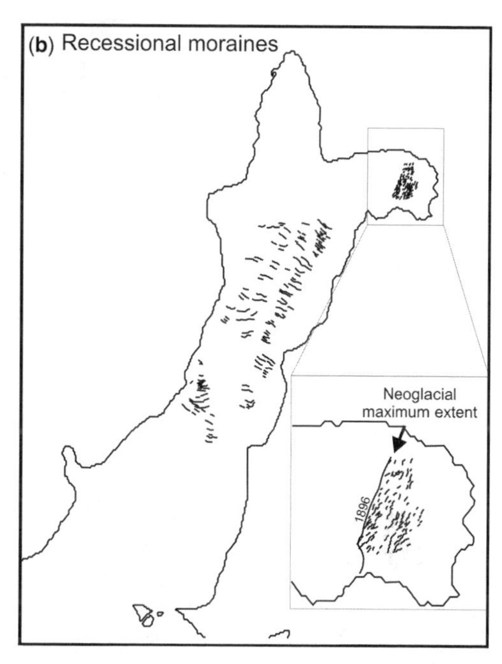

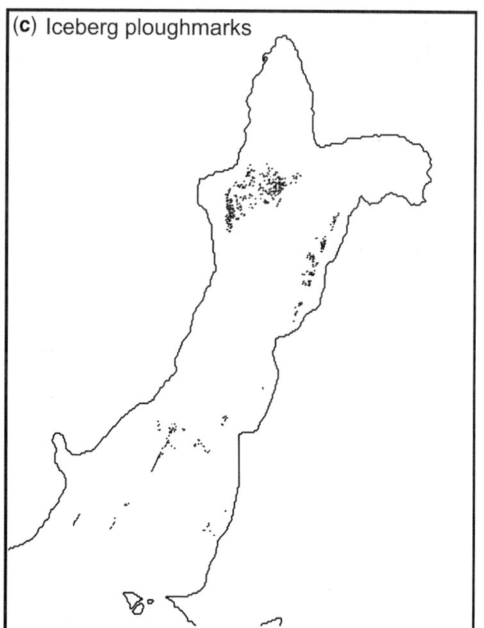

Fig. 5. Maps showing the distribution of morphological features: (**a**) glacial lineations; (**b**) recessional morains; and (**c**) iceberg ploughmarks. Maximum Neoglacial extent suggested by De Geer (1910) in Plassen *et al.* (2004) is indicated in a and b.

a slightly overconsolidated till from Billefjorden (Svendsen *et al.* 1996).

Unit 943-B (440–415 cm). Unit 943-B contains a dark greyish-brown, faintly laminated, clast-rich diamicton with a muddy matrix (Fig. 7; Table 1). Its lower boundary is sharp and defined by a marked upcore decrease in clast content. The upper boundary is gradational and defined by a colour change, increasing shell content and decreasing

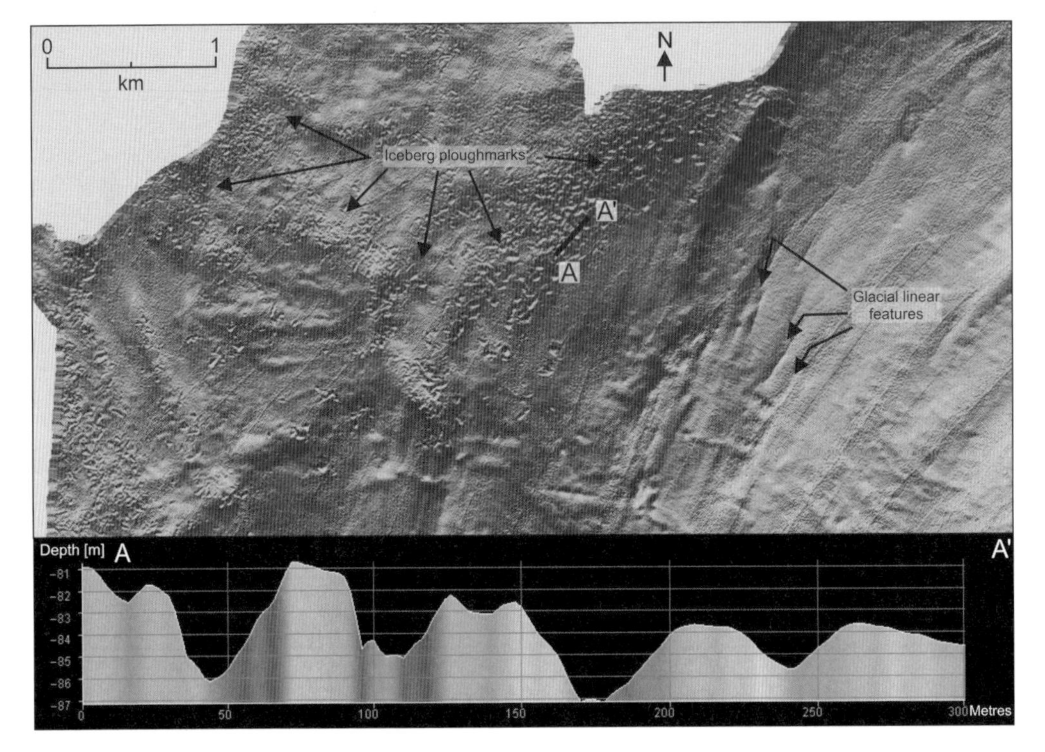

Fig. 6. Illumination map of the inner parts of Billefjorden (Mimerbukta; see Fig. 2). Profile A–A′ crosses several iceberg ploughmarks.

clast content (Fig. 8). The magnetic susceptibility is lower than in the underlying unit, but higher than in the overlying sediments (Fig. 7). Because of the most likely low bottom-current activity in Billefjorden, clasts coarser than 1 mm are regarded as IRD (compare with Hald *et al.* 2004; Forwick & Vorren 2009). Based on the relatively high IRD and low shell content compared to the overlying unit, unit 943-B is interpreted as a glaciomarine diamicton, deposited proximal to the glacier front. This unit was correlated to a laminated glaciomarine mud that was found in cores from Billefjorden, interpreted as a transitional zone between the till and overlying the Holocene mud (Svendsen *et al.* 1996).

Unit 943-C (415–255 cm). Unit 943-C comprises a dark greyish-brown, bioturbated, massive mud with scattered IRD (Fig. 7; Table 1). It has relatively high amounts of shells, in particular between 305 and 325 cm. Its lower boundary is gradational and it is defined by a decrease in the amount of IRD, increase in the amount of shells and a change in colour. A gradual increase in the amount of IRD and a decrease in the amount of shells characterize its upper boundary (Fig. 8). The silt content is comparatively high (Fig. 7).

The low amount of IRD is assumed to reflect a glaciomarine environment with limited ice rafting. Furthermore, a high concentration of shells could indicate warmer water with more favourable living conditions. Based on this and the relatively low amounts of IRD, it is suggested that the climate was relatively warm during the deposition of unit 943-C.

Unit 943-D (255–125 cm). Unit 943-D comprises a dark greyish-brown, massive mud with IRD (Fig. 7; Table 1). It contains more IRD and fewer shells than the underlying unit. Its lower boundary is gradational and defined by marked changes in these parameters. A gradual decrease in the IRD content defines its upper boundary. Four IRD-rich layers occur. Between these layers, the sediments are bioturbated.

These sediments were probably deposited in a glaciomarine environment. The increase in IRD content and the decrease in shell content are suggested to reflect a climatic cooling resulting in an increase in glacial activity in the form of glacier growth and/or the increased formation of sea ice.

Unit 943-E (125–0 cm). Unit 943-E comprises a dark greyish-brown, massive mud with scattered

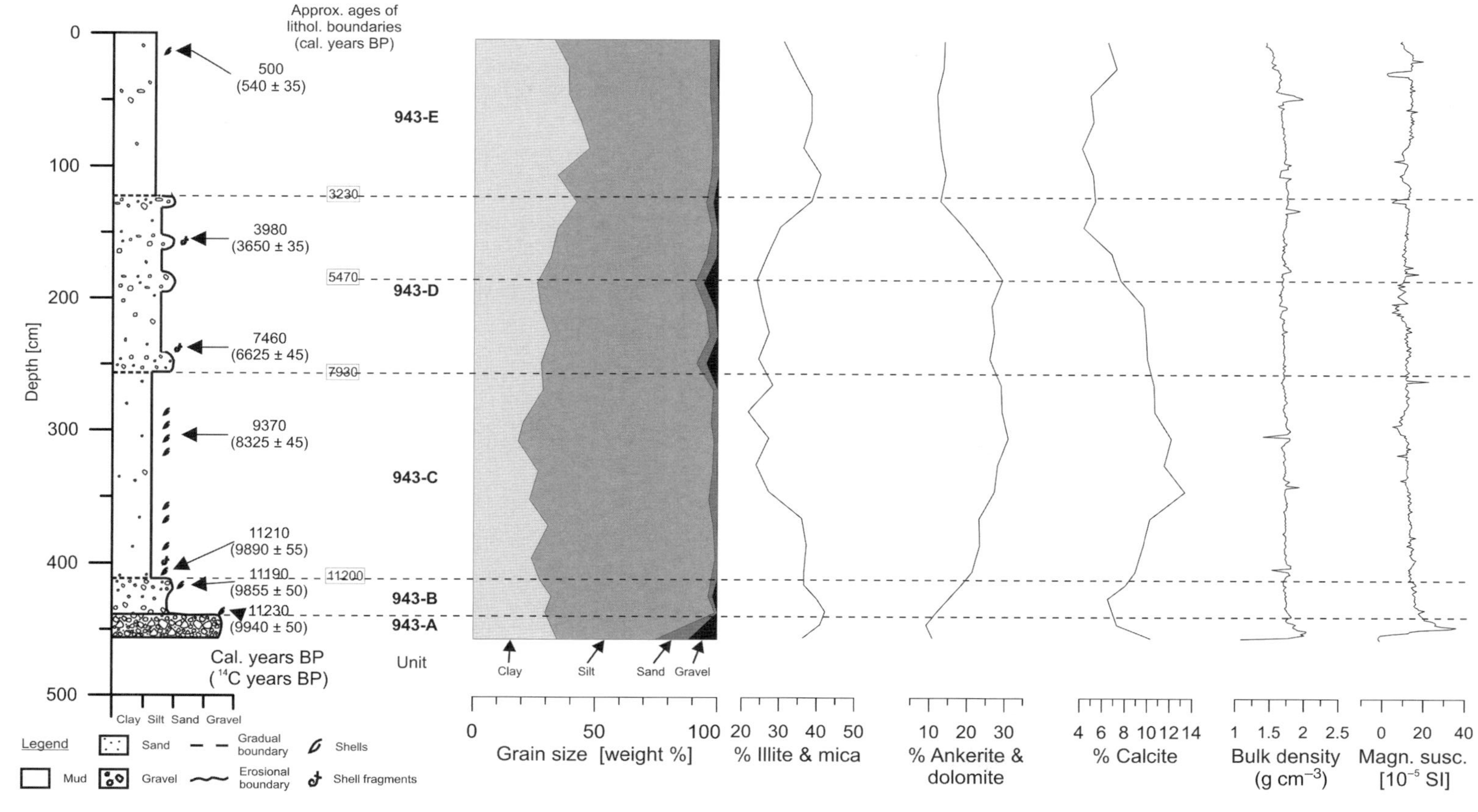

Fig. 7. Lithological log of core JM97-943-GC, together with radiocarbon ages and calibrated ages, estimated ages of lithological boundaries, grain-size distribution and selected minerals and physical properties.

Table 1. *Properties of lithological units in core JM97-943-GC*

Unit Parameter	943-A (460.5–440 cm)	943-B (440–415 cm)	943-C (415–255 cm)	943-D (255–125 cm)	943-E (125–0 cm)
Lithology	Massive clast-rich diamicton with muddy matrix	IRD-rich diamicton with muddy matrix	Massive mud with scattered IRD	Massive mud with IRD	Massive mud with scattered IRD
Clasts (IRD)	High amounts <3 cm	Rel. high amounts (less 943-A) <2 cm	Rel. low amounts <3 cm	Rel. high amounts <3 cm	Rel. low amounts <2 cm
Clasts (IRD) and maximum sizes	High amounts <3 cm	Rel. high amounts (less than 943-A) <2 cm	Rel. low amounts <3 cm	Rel. high amounts <3 cm	Rel. low amounts <2 cm
Colour (Munsell code)	Dark greyish brown (10 YR 4/2)	Dark greyish brown (2.5 Y 4/2)	Dark greyish brown (2.5 Y 4/2)	Dark greyish brown (10 YR 4/2 to 2.5 Y 4/2)	Dark greyish brown (10 YR 4/2 to 2.5 Y 4/2)
Lower boundary	Not recovered	Sharp, clasts ↓	Gradational, shells ↑, clasts ↓	Gradational clasts ↑	Gradational shells ↓, clasts ↑
Upper boundary	Sharp, change in colour, clasts ↓	Gradational, shells ↑, clasts ↓	Gradational clasts ↑	Gradational ↑ shells ↓, clasts	Seafloor
Bulk density [g/cm^3]	1.75–2	*c.* 1.75	*c.* 1.75	1.75–2	1.5–2
Mag. sus. [10^{-5} SI]	20–40	15–20	10–20	10–20	10–30
Bioturbation	Absent	Very little	Intense	Moderate	Intense
Shells (from X-radiographs)	Absent	2 in living position	High amounts, in particular between 305–325 cm, <5 cm	2 shell fragments	1 shell in living position
Mineralogy				(lower/ upper)	
% illite/mica	36–42	36–43	22.37	25–28/25–40	30–40
% ankerite/dolomite	10–12	12–20	20–23	25–40/13–40	14–16
% calcite	7–10	6.5–8.5	8.5–13.5	8.5–11/4–8.5	4.5–7
Other	Stiff	Faint lamination	Black mottles on fresh surfaces	4 clasts rich layers (5–15 cm thick)	Black mottles on fresh surfaces, polychaeta tube at 77–79 cm
Approx. times of Deposition (cal ka BP)	>11.23	±11.23–11.2	±11.2–7.93	±7.93–3.23	<3.23
Interpretation and genesis	Late Weichselian basal till	Late Weichselian glaciomarine sediments, proximal	Early Holocene glaciomarine sediments; low glacial activity	Mid-Holocene glaciomarine sediments; increasing sea ice and iceberg rafting	Late Holocene glaciomarine sediments; possible presence of multi-annual shorefast sea-ice

Note: Arrows in the descriptions of the boundaries indicate an upcore increase (↑) or decrease (↓) of the mentioned parameters. The explanation 'lower/upper' for the mineralogy of unit 943-D refers to the lower and upper parts of the unit, respectively.

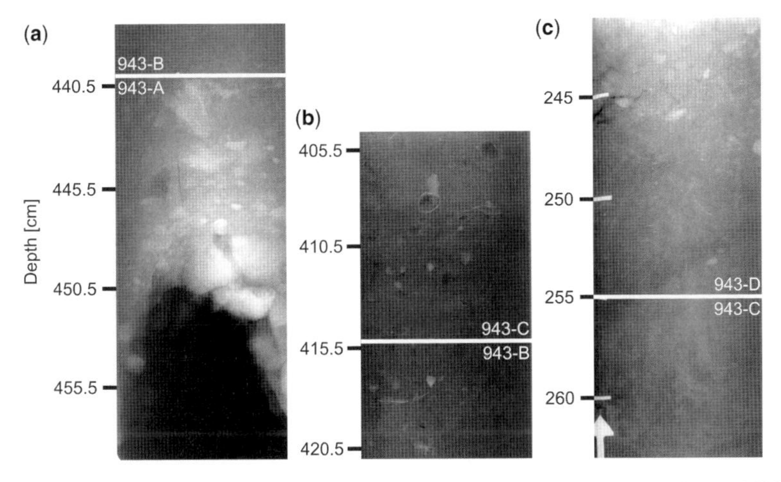

Fig. 8. X-radiographs from core JM97-943-GC showing sections of the lithological units 943-A to 943-D and their boundaries.

IRD (Fig. 7; Table 1). The unit contains comparatively low amounts of IRD and shells, but has a relatively high clay content. We suggest this unit was deposited in a glaciomarine environment. The relatively low IRD content could indicate a decrease in ice rafting or it could indicate the presence of a more permanent sea-ice cover, blocking the way for icebergs or sea ice transporting IRD to the core site (compare with Ó Cofaigh & Dowdeswell 2001; Hald *et al.* 2004; Forwick & Vorren 2009).

Mineral assemblage and sediment sources

The dominating minerals in core JM97-943-GC are illite/mica, ankerite/dolomite and calcite (Fig. 7; Table 1).

Illite/mica percentages are high in units 943-A, 943-B and 943-E, comparatively low in 943-C and increase in 943-D. The percentages of ankerite/dolomite and calcite mirror this pattern. Unit 943-D was divided into two subunits based on the clear increase in illite/mica and corresponding decrease in ankerite/dolomite in the mineral assemblage record approximately halfway along this unit (Fig. 7).

It is suggested that illite/mica derives mainly from Nordenskiöldbreen because mica occurs in almost all the metamorphic rocks in the drainage area of Nordenskiöldbreen and, to more limited extents, in Mimerbukta and Petuniabukta (Fig. 2; Dallmann *et al.* 2004). Illite does not indicate a specific source area because it is a weathering product of muscovite and feldspar and occurs in sedimentary rocks as well as in metamorphic rocks. It dominates the clay fraction in surface sediments in western Svalbard fjords (Winkelmann & Knies 2005; Vogt & Knies 2009). The sides of the fjords are dominated by limestones and dolomite (Fig. 2; Dallmann *et al.* 2004). It is therefore suggested that ankerite/dolomite and calcites have their origin predominantly along the fjord sides.

Chronology

Seven shells from core JM97-943-GC were radiocarbon dated (Fig. 7; Table 2). The results show generally decreasing ages upcore. However, the two dates at the upper boundary of unit 943-B show an age inversion. A possible explanation for this could be reworking of the material providing the upper date (11.21 cal ka BP). According to the age range of the calendar ages, the upper date could in fact be younger than the lower date (Table 2). Hence, this date is regarded as correct.

The ages for the lithological boundaries were defined based on linear sedimentation rates between the dates (Fig. 7). The basal till comprising unit 943-A was deposited before *c.* 11.23 cal ka BP. The overlying glaciomarine diamicton (unit 943-B) was deposited between *c.* 11.23 and 11.2 cal ka BP. Unit 943-C was deposited between *c.* 11.2 and 7.93 cal ka BP. Unit 943-D was divided into two subunits deposited approximately between 7.93 and 5.47 cal ka BP, and between 5.47 and 3.23 cal ka BP, respectively. Unit 943-E was deposited during the last *c.* 3.23 ka.

Discussion

Subglacial environment, >11.23 cal ka BP (Fig. 9-1)

The radiocarbon date obtained immediately above the basal till (unit 943-A; Fig. 7; Table 2) indicates

Table 2. *Radiocarbon ages and calibrated ages from core JM97-943-GC*

Lab. ref.	Depth (cm)	Species	^{14}C years BP	^{14}C years BP incl. res. effect	Calibrated age (BP) 1σ	Peak distr. 1σ (BP)	Calibrated age (BP) 2σ
TUa-6327	19	*Enucula tenuis*	980 ± 35	540 ± 35	459–535	500	418–608
TUa-6329	152–155	*Malcoma calcarea*	4090 ± 35	3650 ± 35	3917–4074	3980	3840–4145
TUa-6665	238	*Nuculana pernula*	7065 ± 45	6625 ± 45	7416–7519	7460	7360–7575
TUa-6329	305–309	*Yoldia hyperborea*	8765 ± 45	8325 ± 45	9275–9419	9370	9172–9470
TUa-6666	408	*Enucula tenuis*	10 330 ± 55	9890 ± 55	11 156–11 265	11 210	11 106–11 377
TUa-6667	422	*Enucula tenuis*	10 295 ± 50	9855 ± 50	11 144–11 230	11 190	11 096–11 313
TUa-6330	439	*Enucula tenuis*	10 380 ± 50	9940 ± 50	11 191–11 323	11 230	11 143–11 423

Note: The radiocarbon ages were corrected for a marine reservoir effect of 440 years (Mangerud & Gulliksen 1975). They were calibrated with the program Calib Rev. 5.0.2 (Stuiver & Reimer 1993) with a ΔR of 99 ± 39 (calib.qub.ac.uk/marine) and the results were obtained from the calibration data set marine 04.14C (Hughen *et al.* 2004).

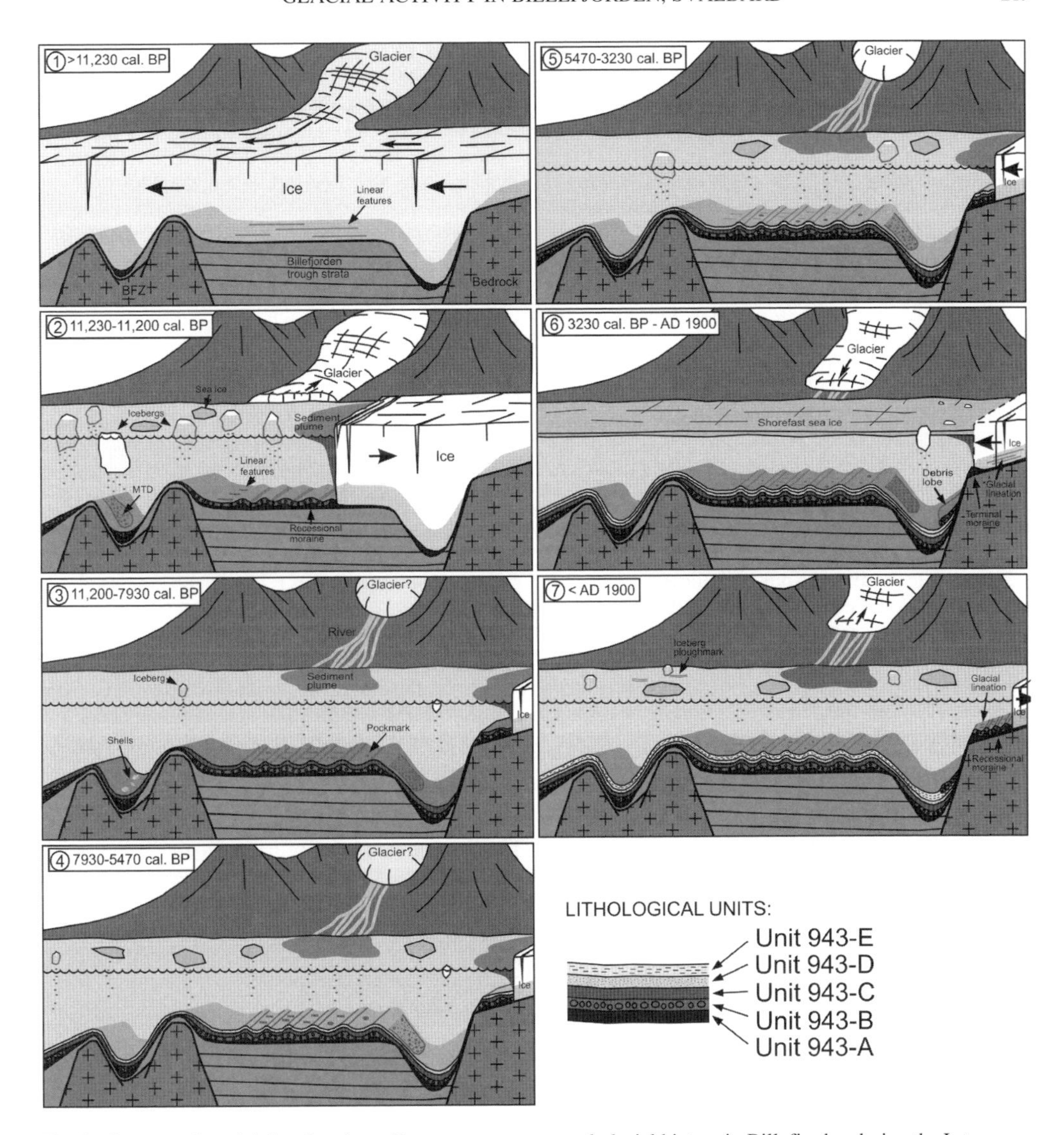

Fig. 9. Conceptual model showing the sedimentary processes and glacial history in Billefjorden during the Late Weichselian and the Holocene. The Billefjorden Fault Zone (BFZ) pockmarks and a debris lobe (Plassen *et al.* 2004; Forwick *et al.* 2009) are indicated.

that Billefjorden was covered with a grounded glacier prior to 11.23 cal ka BP. The glacial lineations in the central parts of the fjord were most likely formed beneath this glacier. The comparatively high amounts of illite/mica in the basal till indicate that sediment supply was largely from Nordenskiöldbreen. However, minor amounts of material could also have been supplied from glaciers draining through Mimerbukta and Petuniabukta. The till may be correlated to the till found by Boulton (1979) on sparker profiles at the mouth of Billefjorden, as well as to unit 6 of the Kapp Ekholm sections that was deposited prior to 11.29 cal ka BP (*c.* 10 ^{14}C ka BP; Mangerud & Svendsen 1992).

Deglaciation, c. *11.23–11.2 cal ka BP (Fig. 9-2)*

Unit 943-B comprises glaciomarine sediments that were deposited between *c.* 11.23 and 11.2 cal ka BP (Fig. 7; Table 2). It is correlated to the glaciomarine mud found by Svendsen *et al.* (1996) in Billefjorden dated to *c.* 11.2 cal ka BP (9730 ± 110 ^{14}C a BP). It is suggested that this unit was deposited during the rapid deglaciation of the fjord at the end of the last glacial.

This suggestion is supported by decreasing percentages of illite/mica and generally increasing percentages of calcite and ankerite/dolomite, reflecting a gradually lower sediment supply from Nordenskiöldbreen and a larger input from the fjord sides.

Even though this period is characterized by a pronounced ^{14}C plateau, that is, exact age estimates may appear speculative (see also age reversal, Fig. 7; Table 2), it is assumed that the retreat of the glacier front from the core site to the inner parts of the fjord occurred within a few decades.

The moraines in the central part of Billefjorden are regarded as annual recessional moraines formed during halts and/or minor re-advances in winters during a period of general glacier retreat (Figs 3 & 5b; compare with Boulton 1986; Ottesen & Dowdeswell 2006; Ottesen *et al.* 2008). The occurrence of *c.* 30 recessional moraines between the core site and Adolfbukta indicates that the glacier fronts retreated *c.* 5 km in *c.* 30 years. This reflects an annual retreat rate of about 170 m a^{-1}. Furthermore, it has been concluded that the remaining ice tongues in coastal areas of Svalbard calved away quickly due to a sudden climatic warming (Lehman & Forman 1992; Mangerud *et al.* 1992). Retreat rates of several hundred metres per year during the final phase of the deglaciation have also been estimated in northern Norway (Corner 1980; Vorren & Plassen 2002).

Low glacial activity, c. *11.2–7.93 cal ka BP (Fig. 9-3)*

The high number of shells and low amounts of IRD in unit 943-C (Figs 7 & 8) most probably reflect warmer climatic conditions during the early Holocene. This coincides well with the results from other studies on Spitsbergen and in the Barents Sea region (Birks 1991; Salvigsen *et al.* 1992; Wohlfarth *et al.* 1995; Sarnthein *et al.* 2003; Hald *et al.* 2004; Rasmussen *et al.* 2007; Forwick & Vorren 2009).

High percentages of ankerite/dolomite (Fig. 7) show that the sediment supply into the fjord mainly occurred from the fjord sides. A low percentage of illite/mica in the sediments indicates that some material most likely originated from Nordenskiöldbreen.

Because of the occurrence of clasts and the moderate amounts of illite/mica it is suggested that Nordenskiöldbreen existed as a tidewater glacier during the early Holocene. Its size was most probably significantly reduced in comparison to earlier periods. This clarifies the suggestion by Svendsen & Mangerud (1997) that the glaciers in Billefjorden were small or absent during this interval. It supports continuous glacial activity on Spitsbergen, as suggested by Hald *et al.* (2004) and Forwick & Vorren (2009).

The warmer climate would most likely have suppressed the formation of sea ice. We therefore assume that ice rafting was dominated by scattered icebergs (see also Forwick & Vorren 2009).

Increased glacial activity, c. *7930–5470 cal a BP (Fig. 9-4)*

Higher IRD contents in the lower part of unit 943-D indicate increased ice rafting between 7930 and 5470 cal a BP (Figs 7 & 8). This is most probably related to the onset of a regional climatic cooling (Birks 1991; Wohlfarth *et al.* 1995; Sarnthein *et al.* 2003; Hald *et al.* 2004; Rasmussen *et al.* 2007; Hald & Korsun 2008).

The illite/mica percentages remained low (Fig. 7), indicating that the influence of Nordenskiöldbreen was limited. This may indicate that enhanced ice rafting was related to an increased formation of sea ice before 5470 cal a BP. However, the possibility that Nordenskiöldbreen started to grow during this period cannot be excluded.

These results differ from investigations from central Isfjorden indicating generally enhanced ice rafting after 9 cal ka BP (Forwick & Vorren 2009). However, an increase in IRD flux in van Mijenfjorden also occurred around 8 cal ka BP (Hald *et al.* 2004; Hald & Korsun 2008; for location see Fig. 1). The delayed increases of ice rafting in Billefjorden and van Mijenfjorden could be attributed to the occurrence of sills at their mouths, restricting the water exchange with the open ocean. They might also be related to the lower surface-water temperatures, restricting iceberg meltout in the inner fjords. More investigation is needed to explain these differences.

Increased glacial activity, c. *5470–3230 cal a BP (Fig. 9-5)*

The period between *c.* 5470 and 3230 cal a BP (upper part of unit 943-D) is also characterized by relatively high IRD contents, reflecting higher glacial activity compared to the early Holocene (Fig. 7). It is suggested that the increase in the percentage of illite/mica and the decrease of ankerite/dolomite at *c.* 5470 cal a BP reflect the growth of Nordenskiöldbreen, resulting in increased iceberg rafting. This is almost synchronous with the growth of Tunabreen in Tempelfjorden, another outlet of the ice cap Lomonosovfonna (for location see Fig. 1; Forwick *et al.* 2010).

The increase in the activity of Nordenskiöldbreen inferred from this study is at least 1500 years earlier than suggested by Svendsen & Mangerud (1997). This could be due to the fact that these authors did not regard sand grains as

IRD. However, because indications for reworking in core JM97-943-GC were not found (Fig. 7) and because bottom-current activity has most likely been low, it is suggested that sand grains were transported to the central parts of the fjord by ice.

Holocene glacial maximum, c. *3230 cal a BP to* c. *AD 1900 (Fig. 9-6)*

The IRD content in unit 943-E is very low (Fig. 7). The concentration of illite/mica decreases towards the top of the unit. A possible explanation for the lack of IRD and decrease of material originating from Nordenskiöldbreen is the presence of a more permanent sea-ice cover (compare with Dowdeswell *et al.* 2000; Ó Cofaigh & Dowdeswell 2001; Forwick & Vorren 2009). In this scenario, most icebergs were trapped within the sea ice so that most of their debris melted out in the vicinity of the glacier front. Furthermore, material incorporated in the sea ice would rarely reach the central fjord if the sea ice broke up occasionally.

The formation of a more permanent sea-ice cover would support the suggestion of an ongoing climatic cooling in the Svalbard Barents Sea region during the late Holocene (Birks 1991; Svendsen & Mangerud 1997; Hald *et al.* 2004; Rasmussen *et al.* 2007).

The glacial lineations on the bedrock terrace were formed during the late Holocene re-advance of Nordenskiöldbreen and/or its maximum extent at the edge of the bedrock terrace. A terminal moraine was deposited during its maximum extent (see Plassen *et al.* 2004).

Post Holocene glacial maximum, c. *AD 1900 to present (Fig. 9-7)*

The late Holocene maximum glacier extent of Nordenskiöldbreen culminated at 1900 AD (Plassen *et al.* 2004); at this time, Nordenskiöldbreen covered the bedrock terrace. The recessional moraines on the bedrock terrace were deposited during the subsequent retreat (Figs 4 & 5). It is suggested that these moraines are annual recessional moraines (compare with Ottesen & Dowdeswell 2006; Ottesen *et al.* 2008).

Iceberg ploughmarks in the shallow areas (Figs 3, 5 & 6) were most likely generated by icebergs calving from Nordenskiöldbreen during the retreat and/or the maximum glacier extent.

Conclusions

- A grounded glacier occupied Billefjorden prior to *c.* 11.23 cal ka BP. It deposited a basal till and formed glacial lineations in central parts of the fjord.
- Annual recessional moraines and radiocarbon dates indicate a rapid (*c.* 170 m a^{-1}) retreat of the glacier front from central to inner parts of Billefjorden between *c.* 11.23 and 11.2 cal ka BP. The sediment supply from Nordenskiöldbreen decreased gradually, whereas the delivery from the fjord sides increased.
- Nordenskiöldbreen existed as a tidewater glacier throughout the entire Holocene.
- The glacial activity in the fjord was significantly reduced from *c.* 11.2 to 7.93 cal ka BP. Nordenskiöldbreen reached a minimum size and sediment supply to central Billefjorden occurred mainly from the fjord sides.
- An increase in sea ice rafting at *c.* 7930 cal a BP was followed by an increase in iceberg rafting related to the growth of Nordenskiöldbreen around 5470 cal a BP.
- Ice rafting in central parts of the fjord was reduced during the last *c.* 3230 a. The formation of a more permanent sea-ice cover most likely suppressed sea ice and icebergs from rafting in Billefjorden.
- Nordenskiöldbreen advanced to the edge of the bedrock terrace in Adolfbukta and deposited a terminal moraine around AD 1900 (Plassen *et al.* 2004). Glacial lineations formed during the advance and/or during the maximum glacier extent.
- Annual recessional moraines were deposited at the front of Nordenskiöldbreen during its subsequent retreat.
- Icebergs calving from Nordenskiöldbreen during and/or after the late Holocene maximum glacier extent grounded and led to the formation of iceberg ploughmarks at water depths shallower than 120 m.

This study was a part of the strategic university program *Sedimentary Processes and Palaeoenvironment on Northern Continental Margins* (SPONCOM), funded by the Research Council of Norway. The masters and the crews of *Jan Mayen* supported the data collection. Teachers and course participants of AG-301 (2005) collected some multibeam data in Billefjorden. S. Iversen provided technical help during and after the cruises. J. P. Holm contributed significantly to the figures. M. Vanneste and J. S. Laberg helped with GMT and provided constructive comments. T. Dahl and C. Davids supported the work in the laboratory. J. Juhlin, M. Baeten and H. Mourmets supported NB with the writing process. We extend our most sincere thanks to these persons and institutions.

References

Aagaard, K., Foldvik, K. & Hillman, S. R. 1987. The west Spitsbergen current: disposition and water mass transformation. *Journal of Geophysical Research*, **92**, 6729–6740.

Birks, H. 1991. Holocene vegetation history and climatic change in west Spitsbergen – plant macrofossils from Skardtjørna, an Arctic lake. *The Holocene*, **1**, 209–218.

Boulton, G. S. 1979. Glacial history of the Spitsbergen archipelago and the problem of the Barents Shelf ice sheet. *Boreas*, **8**, 31–57.

Boulton, G. S. 1986. Push-moraines and glacier-contact fans in marine and terrestrial environments. *Sedimentology*, **33**, 677–698.

Boulton, G. S., Van der Meer, J. J. M., Hart, J., Beets, D., Ruegg, G. H. J., Van der Wateren, F. M. & Jarvis, J. 1996. Till and moraine emplacement in a deforming bed surge – an example from a marine environment. *Quaternary Science Reviews*, **15**, 961–987.

Clark, C. D. 1994. Large-scale ice-moulding: a discussion of genesis and glaciological significance. *Sedimentary Geology*, **91**, 253–268.

Clark, C. D., Tulaczyk, M., Stokes, C. R. & Canals, M. 2003. A groove-ploughing theory for the production of mega-scale glacial lineations, and implications for ice-stream mechanisms. *Journal of Glaciology*, **49**, 240–256.

Corner, G. D. 1980. Preboreal deglaciation chronology and marine limits of the Lyngen-Storfjord area, Troms, North Norway. *Boreas*, **9**, 239–249.

Dallmann, W. K., Piepjohn, K. & Blomeier, D. 2004. *Geological map of Billefjorden, Central Spitsbergen, Svalbard, with geological excursion guide*. Norsk Polarinstitutt Temakart Nr. 36.

Dowdeswell, J. A., Whittington, R. J., Jennings, A. E., Andrews, J. T., Mackensen, A. & Marienfeld, P. 2000. An origin for laminated glacimarine sediments through sea-ice build-up and suppressed iceberg rafting. *Sedimentology*, **47**, 557–576.

Elverhøi, A., Andersen, E. S. *et al.* 1995. Late Quaternary Sediment Yield from the High Arctic Svalbard Area. *Journal of Geology*, **103**, 1–17.

Forwick, M. & Vorren, T. O. 2005. Late Weichselian deglaciation history of the Isfjorden area, Spitsbergen. *In*: Forwick, M. (ed.) *Sedimentary Processes and Palaeoenvironments in Spitsbergen Fjords*. Dr. Scient. thesis, University of Tromsø.

Forwick, M. & Vorren, T. O. 2009. Late Weichselian and Holocene sedimentary environments and ice rafting in Isfjorden, Spitsbergen. Submitted to *Palaeogeography, Palaeoclimatology, Palaeoecology*, **280**, 258–274.

Forwick, M., Baeten, N. J. & Vorren, T. O. 2009. Pockmarks in Spitsbergen fjords. *Norwegian Journal of Geology*, **89**, 65–77.

Forwick, M., Vorren, T. O. *et al.* 2010. Spatial and temporal influence of glaciers and rivers on the sedimentary environment in Sassenfjorden and Tempelfjorden, Spitsbergen. *In*: Howe, J. A., Austin, W. E. N., Forwick, M. & Paetzel, M. (eds) *Fjord Systems and Archives*. Geological Society, London, Special Publications, **344**, 163–193.

Gascard, J. C., Richez, C. & Rouault, C. 1995. New insights from large-scale oceanography in Fram Strait: the West Spitsbergen Current. *In*: Smith, W. & Grebmeier, J. (eds) *Oceanography of the Arctic: Marginal Ice Zones and Continental Shelves*. Coastal and Estuarine Studies Series, American Geophysical Union, Washington, DC, 131–182.

Hagen, J. O., Liestøl, O., Roland, E. & Jørgensen, T. 1993. Glacier atlas of Svalbard and Jan Mayen. *Meddelelser*, Norsk Polarinstitutt, **129**, 5–41.

Hald, M. & Korsun, S. 2008. The 8200 cal. yr BP event reflected in the Arctic fjord, Van Mijenfjorden, Svalbard. *The Holocene*, **18**, 981–990.

Hald, M., Ebbesen, H. *et al.* 2004. Holocene paleoceanography and glacial history of the West Spitsbergen area, Euro-Arctic margin. *Quaternary Science Reviews*, **23**, 2075–2088.

Hughen, K. A., Baillie, M. G. L. *et al.* 2004. Marine04, Marine radiocarbon age calibration, 26–0 ka BP. *Radiocarbon*, **46**, 1059–1086.

Jiskoot, H., Murray, T. & Boyle, P. 2000. Controls on the distribution of surge-type glaciers in Svalbard. *Journal of Glaciology*, **46**, 412–422.

Landvik, J. Y., Bondevik, S. *et al.* 1998. The last glacial maximum of Svalbard and the Barents Sea area: ice sheet extent and configuration. *Quaternary Science Reviews*, **17**, 43–75.

Landvik, J. Y., Ingólfsson, Ó., Mienert, J., Lehman, S. J., Solheim, A., Elverhøi, A. & Ottesen, D. 2005. Rethinking Late Weichselian ice sheet dynamics in coastal NW Svalbard. *Boreas*, **34**, 7–24.

Lehman, S. J. & Forman, S. L. 1992. Late Weichselian Glacier Retreat in Kongsfjorden, West Spitsbergen, Svalbard. *Quaternary Research*, **37**, 139–154.

Liestøl, O. 1976. Årsmorener foran Nathorstbreen? *In Norsk Polarinstitutt Årbok 1976*, Norsk Polarinstitutt, 361–363.

Lønne, I. 2005. Faint traces of high Arctic glaciations: an early Holocene ice-front fluctuation in Bolterdalen, Svalbard. *Boreas*, **34**, 308–323.

Mangerud, J. & Gulliksen, S. 1975. Apparent radiocarbon ages of recent marine shells from Norway, Spitsbergen, and Arctic Canada. *Quaternary Research*, **5**, 73–263.

Mangerud, J. & Svendsen, J. I. 1990. Deglaciation chronology inferred from marine sediments in a proglacial lake basin, western Spitsbergen, Svalbard. *Boreas*, **19**, 249–272.

Mangerud, J. & Svendsen, J. I. 1992. The last interglacial–glacial period on Spitsbergen, Svalbard. *Quaternary Science Reviews*, **11**, 633–664.

Mangerud, J., Bolstad, M. *et al.* 1992. The Last Glacial Maximum on Spitsbergen. *Quaternary Research*, **38**, 1–31.

Mangerud, J., Dokken, T. *et al.* 1998. Fluctuations of the Svalbard–Barents Sea ice sheet during the last 150 000 years. *Quaternary Science Reviews*, **17**, 11–42.

Nilsen, F., Cottier, F., Skogseth, R. & Mattsson, S. 2008. Fjord-shelf exchanges controlled by ice and brine production: the interannual variation of Atlantic Water in Isfjorden, Svalbard. *Continental Shelf Research*, **28**, 1838–1853.

Ó Cofaigh, C. & Dowdeswell, J. A. 2001. Laminated sediments in glacimarine environments: diagnostic criteria for their interpretation. *Quaternary Science Reviews*, **20**, 1411–1436.

Ottesen, D. & Dowdeswell, J. A. 2006. Assemblages of submarine landforms produced by tidewater glaciers in Svalbard. *Journal of Geophysical Research*, **111**, 1–16.

OTTESEN, D., DOWDESWELL, J. A. & RISE, L. 2005. Submarine landforms and the reconstruction of fast-flowing ice streams within a large Quaternary ice sheet: the 2500-km-long Norwegian Svalbard margin (57°–80°N). *Geological Society of America Bulletin*, **117**, 1033–1050.

OTTESEN, D., DOWDESWELL, J. A. *ET AL.* 2008. Submarine landforms characteristic of glacier surges in two Spitsbergen fjords. *Quaternary Science Reviews*, **27**, 1583–1599.

PLASSEN, L., VORREN, T. O. & FORWICK, M. 2004. Integrated acoustic and coring investigation of glacigenic deposits in Spitsbergen fjords. *Polar Research*, **23**, 89–110.

RASMUSSEN, T. L., THOMSEN, E., ŚLUBOWSKA, M. A., JESSEN, S., SOLHEIM, A. & KOÇ, N. 2007. Paleoceanographic evolution of the SW Svalbard margin (76°N) since 20 000 ^{14}C yr BP. *Quaternary Research*, **67**, 100–114.

SALVIGSEN, O., FORMAN, S. L. & MILLER, G. H. 1992. Thermophilous molluscs on Svalbard during the Holocene and their paleoclimatic implications. *Polar Research*, **11**, 1–10.

SARNTHEIN, M., VAN KREVELD, S., ERLENKEUSER, H., GROOTES, P. M., KUCERA, M., PFLAUMANN, U. & SCHULZ, M. 2003. Centennial-to-millennial-scale periodicities of Holocene climate and sediment injections off the western Barents shelf, 75°n. *Boreas*, **32**, 447–461.

SOLHEIM, A. 1991. The depositional environment of surging sub-polar tidewater glaciers: a case study of the morphology, sedimentation and sediment properties in a surge affected marine basin outside Nordaustlandet, the Northern Barents Sea. *Norsk Polarinstitutt Skrifter*, **194**, 1–97.

SOLHEIM, A. & PFIRMAN, S. L. 1985. Sea-floor morphology outside a grounded, surging glacier; Bråsvellbreen, Svalbard. *Marine Geology*, **65**, 127–143.

STUIVER, M. & REIMER, P. J. 1993. Extended ^{14}C database and revised CALIB radiocarbon calibration program. *Radiocarbon*, **35**, 215–230.

SVENDSEN, H., BESZCZYNSKA-MØLLER, A. *ET AL.* 2002. The physical environment of Kongsfjorden-Krossfjorden, an Arctic fjord system in Svalbard. *Polar Research*, **21**, 133–166.

SVENDSEN, J. I. & MANGERUD, J. 1997. Holocene glacial and climatic variations on Spitsbergen, Svalbard. *The Holocene*, **7**, 45–57.

SVENDSEN, J. I., ELVERHØI, E. & MANGERUD, J. 1996. The retreat of the Barents Sea Ice Sheet on the western Svalbard margin. *Boreas*, **25**, 244–256.

VOGT, C. & KNIES, J. 2009. Sediment pathways in the Western Barents Sea inferred from clay mineral assemblages in surface sediments. *Norwegian Journal of Geology*, **89**, 41–55.

VOGT, C., LAUTERJUNG, J. & FISCHER, R. X. 2002. Investigation of the clay fraction (<2 μm) of the clay mineral society reference clays. *Clays and Clay Minerals*, **50**, 388–400.

VORREN, T. & PLASSEN, L. 2002. Deglaciation and palaeoclimate of the Andfjord-Vågsfjord area, North Norway. *Boreas*, **31**, 97–125.

WĘSŁAWSKI, J. M., KOSZTEYN, J., ZĄJACZKOWSKI, M., WIKTOR, J. & KWAŚNIEWSKI, S. 1995. Fresh water in Svalbard fjord ecosystems. *In*: SKJOLDAL, H. R., HOPKINS, C., ERIKSTAD, K. E. & LEINAA, H. P. (eds) *Ecology of Fjords and Coastal Waters*. Elsevier, Amsterdam, 229–241.

WESSEL, P. & SMITH, W. H. F. 1998. New, improved version of Generic Mapping Tools released. *EOS*, **79**, 579.

WHITTINGTON, R. J., FORSBERG, C. F. & DOWDESWELL, J. A. 1997. Seismic and side-scan sonar investigations of recent sedimentation in an ice-proximal glacimarine setting, Kongsfjorden, north-west Spitsbergen. *In*: DAVIES, T. A., BELL, T., COOPER, A. K., JOSENHANS, H., POLYAK, L., SOLHEIM, A., STOKER, M. S. & STRAVERS, J. A. (eds) *Glaciated Continental Margins: An Atlas of Acoustic Images*. Chapman & Hall, London, 175–178.

WINKELMANN, D. & KNIES, J. 2005. Recent distribution and accumulation of organic carbon on the continental margin west off Spitsbergen. *Geochemistry Geophysics Geosystems*, **6**, doi: 10.1029/2005GC000916.

WOHLFARTH, B., LEMDAHL, G., OLSSEN, S., PERSSON, T., SNOWBALL, I., ISING, J. & JONES, V. 1995. Early Holocene environment on Bjørnøya (Svalbard) inferred from multidisciplinary lake sediment studies. *Polar Research*, **14**, 253–275.

Paraglacial slope instability in Scottish fjords: examples from Little Loch Broom, NW Scotland

MARTYN S. STOKER[1]*, CHARLES R. WILSON[2,3], JOHN A. HOWE[2], TOM BRADWELL[1] & DAVID LONG[1]

[1]*British Geological Survey, West Mains Road, Edinburgh, EH9 3LA*

[2]*Scottish Association for Marine Science, Scottish Marine Institute, Oban, Argyll, PA37 1QA, UK*

[3]*Department of Earth Sciences, Durham University, Durham, DH1 3LE, UK*

**Corresponding author (e-mail: mss@bgs.ac.uk)*

Abstract: Lateglacial–Holocene fjord sediments in Little Loch Broom preserve evidence of extensive slope instability. The major area of reworking is in the outer loch and mid-loch sill region where ice-contact/ice-proximal deposits of the Lateglacial Assynt Glaciogenic Formation have been disrupted by sliding and mass-flow processes linked to the Little Loch Broom Slide Complex and the adjacent Badcaul Slide. Mass failure was instigated about 14–13 ka BP, and is probably the response of the landscape to deglaciation immediately following the removal of ice support during glacial retreat. An initial phase of translational sliding was followed by rotational sliding, as revealed by the superimposition of scallop-shaped slumps on a larger-scale rectilinear pattern of failure. Paraglacial landscape readjustment may also have been enhanced by episodic seismic activity linked to glacio-isostatic unloading. In the inner fjord, evidence of Holocene mass failure includes the Ardessie debris lobe and a discrete intact slide block preserved within the postglacial basinal deposits. The former is a localized accumulation linked to a fluvial catchment on the adjacent An Teallach massif. These mass-transport deposits may represent an ongoing response to paraglacial processes, albeit much reduced (relative to the major slides) in terms of sediment supply to the fjord.

It is well established that fjords are areas of major landslide activity due to their steep lateral slopes, characteristically high rates of sedimentation and commonly exceptional rates of isostatic uplift (Syvitski & Shaw 1995; Hampton *et al.* 1996). These factors make them ideal environments for all kinds of sediment deformation, either due to gravitational reworking of the rapidly accumulating deposits, fluid flow or to external stimuli such as earthquakes. Examples of sediment slides, slumps and extensional and compressional deformation structures are to be found in fjords worldwide, including Canada (Syvitski & Hein 1991), East Greenland (Whittington & Niessen 1997; Niessen & Whittington 1997) and Norway (Aarseth *et al.* 1989; Hjelstuen *et al.* 2009).

Scotland's fjords are no exception. Recent work in the Summer Isles region of NW Scotland (Fig. 1) has identified an up to 100 m thick sequence of fjord sediments which was rapidly deposited during the landward retreat of the ice margin from the Summer Isles into the present-day sea lochs of Loch Broom and Little Loch Broom (Stoker *et al.* 2009). Swath bathymetric imagery, high-resolution seismic reflection profiles and sediment core data have shown that mass failure is pervasive throughout the Summer Isles region. Regional mapping has demonstrated that early post-depositional deformation in and around Loch Broom, including the Cadail Slide (Fig. 1), occurred during the Lateglacial interval (Stoker *et al.* 2006, 2009; Stoker & Bradwell 2009). In this paper we focus on one of the largest areas of submarine mass failure yet to be identified in nearshore UK waters – herein referred to as the Little Loch Broom Slide Complex (Fig. 1). We also describe more localized features of instability in the inner part of Little Loch Broom, including the Badcaul Slide and the Ardessie debris lobe; the latter is most probably a Holocene feature. We use geophysical and geological data to describe the style of the instability and the nature of the resulting sediments, as well as to provide constraints on the timing of failure. Considered together with the regional pattern of neotectonic deformation, the development of the Little Loch Broom Slide Complex and adjacent mass failures may provide clues as to the stability of the deglaciating Scottish hinterland during the Lateglacial–Holocene interval.

From: Howe, J. A., Austin, W. E. N., Forwick, M. & Paetzel, M. (eds) *Fjord Systems and Archives*. Geological Society, London, Special Publications, **344**, 225–242.
DOI: 10.1144/SP344.16 0305-8719/10/$15.00

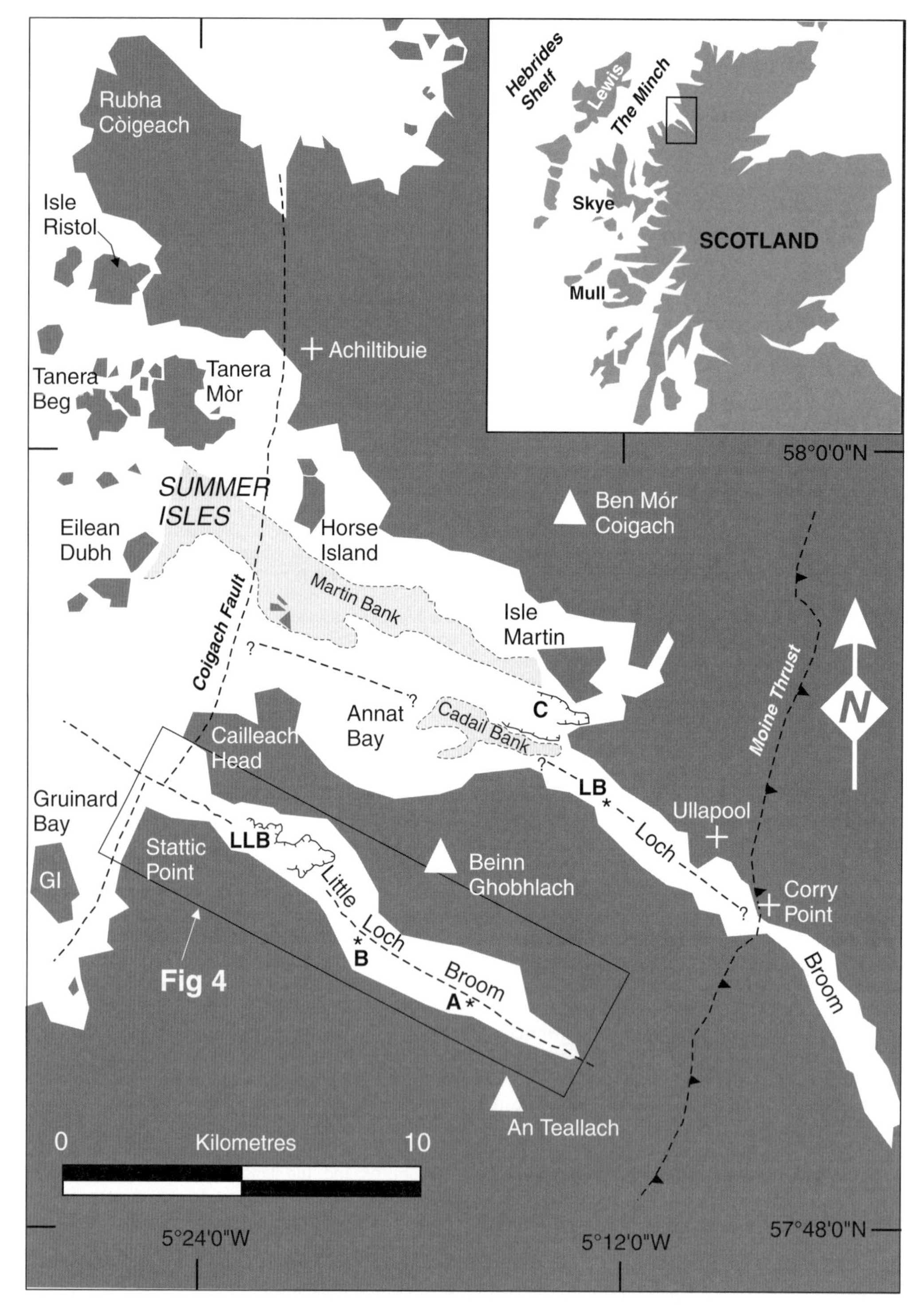

Fig. 1. Location of study area (expanded in Figure 4) in relation to the regional structural grain. Occurrences of all areas of mass failure cited in text are shown. A: Ardessie debris lobe; B: Badcaul Slide; C: Cadail Slide; LB: rotational slumping in Loch Broom; LLB: Little Loch Broom Slide Complex; GI, Gruinard Island.

Regional setting

Little Loch Broom is a NW trending sea loch situated approximately 10 km west of Ullapool (Fig. 1). The flanks of the loch are characterized by rugged headlands backed by mountains such as An Teallach to the south and Beinn Ghobhlach to the north. The loch is 12 km long, 0.5 to 2.0 km wide and is divided more or less at its mid-point into inner and outer basins by a mid-loch sill (*c.* 25 m below sea level). The deepest part of the loch is the inner basin (119 m), while the outer basin reaches a maximum depth of 75 m (Stoker *et al.* 2006). Slope angles are also greater in the inner loch (locally >50°) compared to the outer loch (locally up to 25°). No detailed hydrographic survey from Little Loch Broom has been published; however, the Scottish Environmental Protection Agency site data report for the loch (SEPA 2006) states that it is protected from SW winds but exposed to winds blowing from the NW. Heath *et al.* (2000) used data from Lee and Ramster's *Atlas of the Sea* (1981) to estimate a mean spring tidal current of 1–2 m s^{-1} in both Loch Broom and Little Loch Broom. The approximate average flushing time for the whole loch is 7 days although the deeper basins take longer; freshwater input is generally low (SEPA 2006).

The depth to bedrock in Little Loch Broom is locally up to 160 m (maximum) below OD in the inner loch. The bedrock geology is dominated by Neoproterozoic Torridonian sandstone with sporadic inliers of Archaean gneiss near the mouth of the loch. The orientation of the loch is structurally controlled by a fault that trends along the length of Little Loch Broom (Fig. 1). Significant NE trending structures in the area include the Moine Thrust and the Coigach fault.

The regional glacial geology of the Summer Isles area is summarized in Figure 2 and Table 1; for details see Stoker *et al.* 2009. All dates have been calibrated (Fairbanks *et al.* 2005) and are expressed as calendar years (ka BP). The major part of the succession records a time-transgressive landward retreat of the Lateglacial ice-sheet margin from The Minch, across the Summer Isles and back to the sea lochs of Loch Broom and Little Loch Broom. The deposition of ice-contact to ice-proximal glaciomarine and ice-distal glaciomarine facies, assigned to the Assynt Glaciogenic and Annat Bay Formations, respectively, comprise the bulk of the sediment infilling the fjord region. Cosmogenic-isotope surface-exposure ages of boulders from onshore moraines within the Assynt Glaciogenic Formation, combined with Accelerator Mass Spectrometry (AMS) radiocarbon dating of

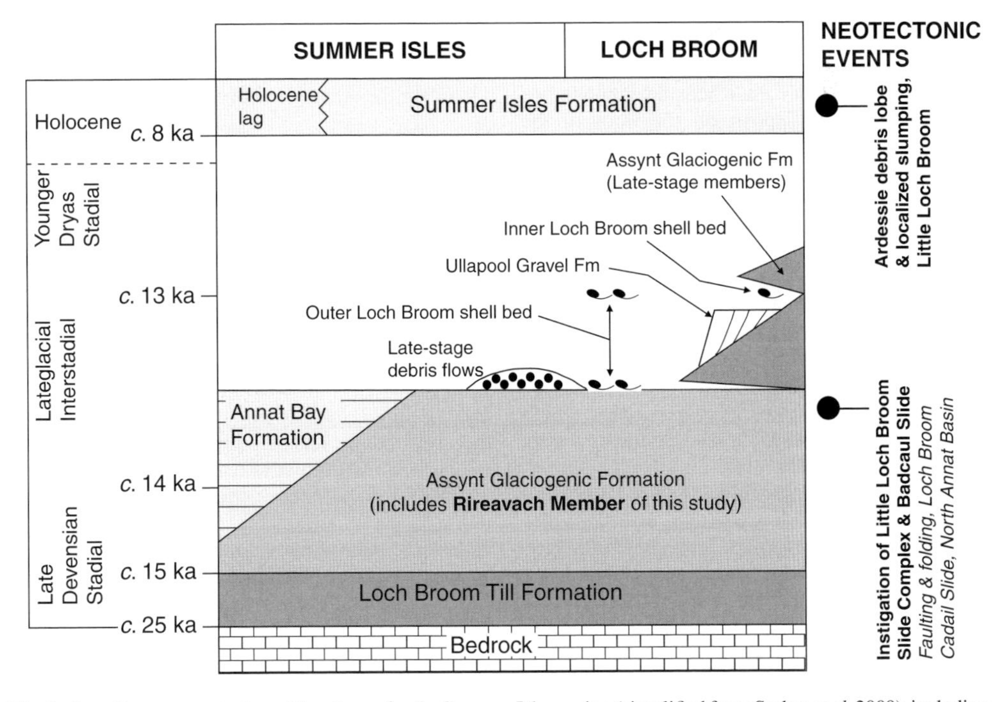

Fig. 2. Late Quaternary stratigraphic scheme for the Summer Isles region (simplified from Stoker *et al.* 2009), including inferred relative timing of neotectonic events.

Table 1. *Interpretation of Late Quaternary stratigraphic units in the Summer Isles region (after Stoker et al. 2009)*

Stratigraphic unit	Depositional setting
Summer Isles Fm	Marine deposits strongly influenced by bottom currents. Localized mass failure
Ullapool Gravel Fm	Fluvioglacial outwash fan-deltas
Inner and Outer Loch Broom shell beds	Time-transgressive condensed section in Loch Broom
Late-stage debris flows	Discrete, localized debris-flow deposits
Annat Bay Fm	Distal glaciomarine facies, diachronous with Assynt Glaciogenic Fm
Assynt Glaciogenic Fm (including Rireavach Member in Little Loch Broom and other Late-stage members in Loch Broom)	Recessional, oscillating, ice-contact and proximal glaciomarine facies. Contemporaneous mass failure, for example: Little Loch Broom slide complex; Cadail slide (pre-Annat Bay Fm); neotectonic deformation in Loch Broom
Loch Broom Till Fm	Subglacial lodgement till

marine shells from (and micropalaeontological analysis of) both the Assynt Glaciogenic and Annat Bay Formations in offshore sediment cores all suggest that these units were deposited largely between *c.* 14 and 13 ka BP, that is, during the Lateglacial Interstadial (Bradwell *et al.* 2008; Stoker *et al.* 2009). In Loch Broom, there is evidence of Late-stage oscillation of outlet glacier lobes back into the fjord; this correlates with several discrete Late-stage members of the Assynt Glaciogenic Formation. An associated series of large fan-deltas comprise the Ullapool Gravel Formation, which is sandwiched between these Late-stage members. As the fjord gradually became ice free, the Outer and Inner Loch Broom shell beds accumulated as a time-transgressive deposit on the floor of the fjord. The Inner Loch Broom shell bed is overlain by glacial diamicton belonging to one of the Late-stage members of the Assynt Glaciogenic Formation in the inner loch. This relationship provides an age constraint of <13 ka BP for the late-stage ice-margin oscillation within the inner fjord. A discrete lithogenetic unit, informally named the 'Late-stage debris flows', occurs sporadically throughout the Summer Isles region. This unit post-dates the Assynt Glaciogenic and Annat Bay Formations, but pre-dates the Summer Isles Formation which forms a cover of Holocene marine sediments deposited after about 8 ka BP.

The Assynt Glaciogenic, Annat Bay and Summer Isles Formations, together with the Late-stage debris flow lithogenetic unit, have all been mapped in and around the inner part of Little Loch Broom (Fig. 3). In contrast, the shallower, outer part of the loch is dominated almost completely by the Assynt Glaciogenic Formation and its reworked upper component assigned to the Rireavach Member; these units are overlain by a seismically unresolvable Holocene veneer (Stoker *et al.* 2009) (Figs 2 & 3). The mid-loch sill that separates the inner and outer lochs is formed by a coincidence of a bedrock high and a major moraine ridge associated with the Assynt Glaciogenic Formation. Other prominent moraine ridges also belonging to this unit are preserved on sills at the mouth (outer loch moraine) and near the head (inner loch moraine) of the loch. From a stratigraphic perspective, the Rireavach Member defines the reworked extent of the Assynt Glaciogenic Formation within the confines of the Little Loch Broom Slide Complex. Additional mass-transport deposits are associated with the Late-stage debris flow unit linked to the Badcaul Slide and with the Summer Isles Formation including the Ardessie debris lobe.

Methods

This study combines geophysical and geological data collected by the British Geological Survey (BGS) and the Scottish Association for Marine Science (SAMS) in the Summer Isles region between 2005 and 2007. A marine geophysical survey of the Summer Isles region, including Little Loch Broom, was undertaken in July 2005 and acquired multibeam swath bathymetry and high-resolution seismic reflection data (Stoker *et al.* 2006). Bathymetric data were acquired using a GeoSwath system operating at 125 kHz, mounted on a retractable bow pole on the RV *Calanus*. Swath survey lines were traversed at a spacing of 200 m, thereby enabling swath overlap and full coverage bathymetry across an area of 225 km^2. The data were collected on a GeoSwath computer with post-acquisition processing carried out on a separate workstation. Output was in the form of xyz data with a typical grid spacing of 3 m. The grid was converted into a depth-coloured

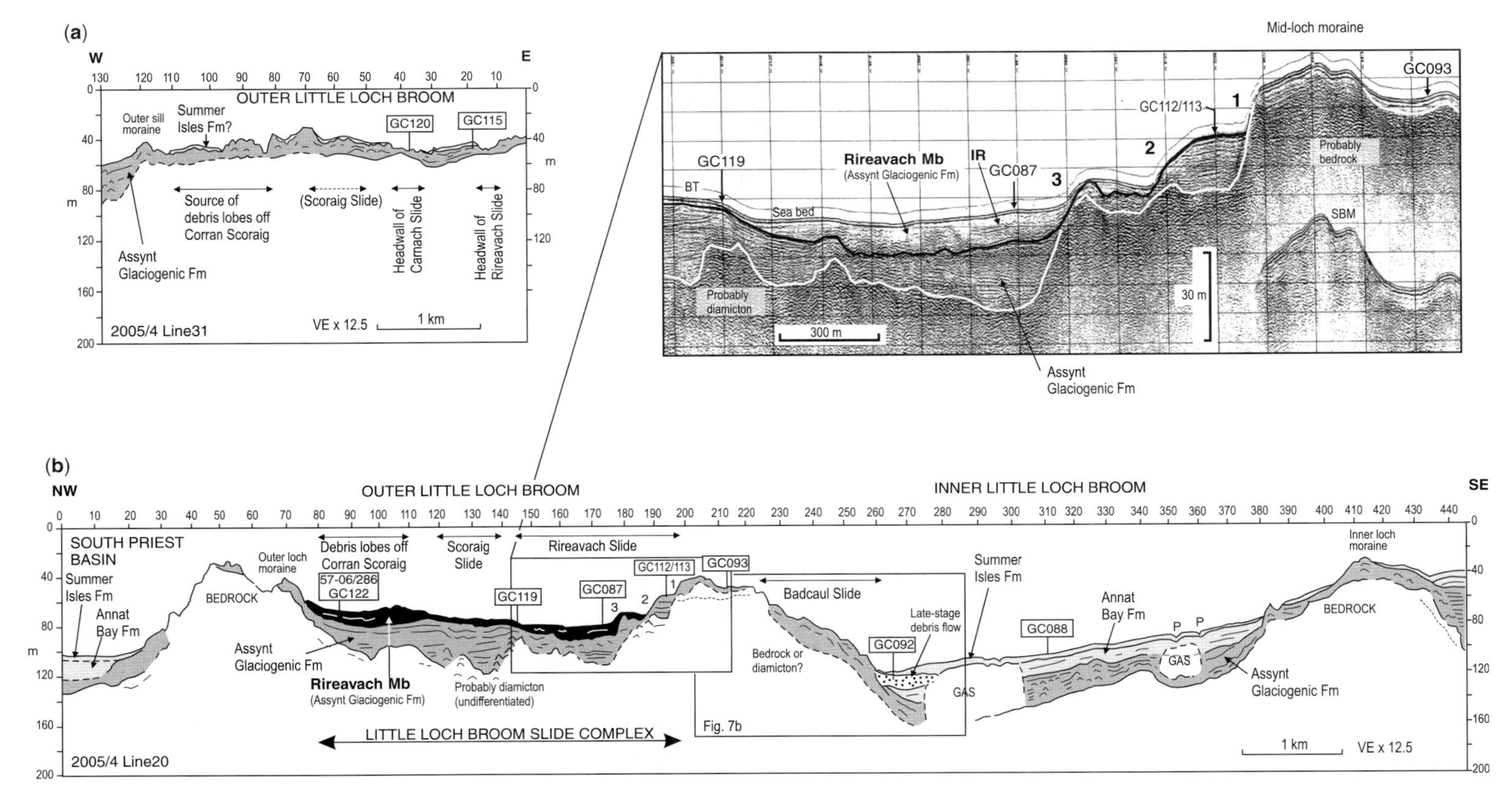

Fig. 3. Geoseismic profiles showing distribution of Quaternary units, major zones of sliding and location of sediment cores (used in this study): (**a**) slope-parallel profile on northern flank of outer Little Loch Broom; (**b**) axial profile along length of Little Loch Broom, with seismic inset showing sub-bottom detail of the Rireavach Slide and relationship to pre-slide stratigraphy. Profiles are located in Figure 4. 1–3, main slide scars associated with Rireavach Slide; BT, bottom tracking indicator; IR, internal reflector in disturbed section; P, pockmark; SBM, sea-bed multiple.

shaded-relief image using Fledermaus (processing and visualization software). The shaded-relief image of the study area is shown in Figure 4. The seismic reflection data were acquired using a BGS-owned Applied Acoustics surface-towed boomer and hydrophone. Fifty-seven boomer profiles (a total length of about 235 km) were collected across the region, including twelve profiles specifically acquired in Little Loch Broom. The data were recorded and processed (Time Varied Gain, Bandpass Filter 800–200 Hz) on a CODA DA200 seismic acquisition system and output as SEG-Y and TIFF format. Further technical details of the geophysical data collection are outlined in Stoker *et al.* (2006).

On the basis of regional measurements of superficial sediments offshore Scotland, sound velocities in the fjord sediment fill are taken to be in the range of 1500–2000 m s^{-1} depending upon their composition and degree of induration (McQuillin & Ardus 1977; Stoker *et al.* 1994). In this paper, the conversion of sub-bottom depths from milliseconds to metres has been generally taken as a maximum estimate (e.g. 20 ms two-way travel time (TWTT) $\leq$20 m) of sediment thickness. The relief of features with expression at the sea bed is based on the sound velocity in water of 1450 m s^{-1} (Hamilton 1985).

Geological calibration of the geophysical data was established using SAMS gravity cores GC087, 088, 092, 093, 112, 113, 115, 119, 120 and 122 and BGS vibrocore 57-06/286 (a re-occupation of site GC122) (Fig. 4). The SAMS cores were collected from the RV *Calanus* in August 2006, whereas the BGS core was collected in September 2007 using the RV *James Cook*. Stratigraphic correlation of these cores is based on a regional study of

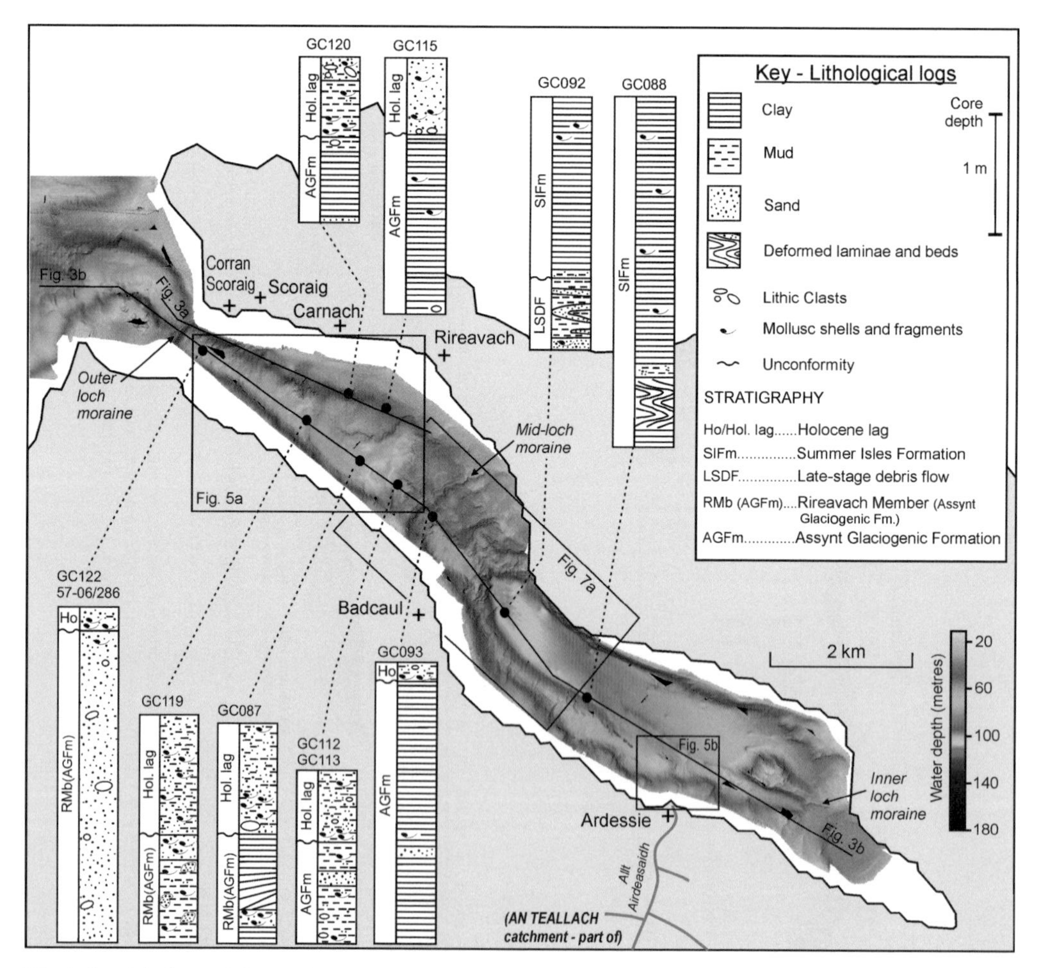

Fig. 4. Swath bathymetric image of Little Loch Broom showing: (1) the location of the enlarged panels in Figures 5 & 7; (2) the location of the geoseismic profiles in Figure 3; and (3) the location and summary lithology logs of the SAMS and BGS cores used in this study.

all cores collected in the Summer Isles region (a total of 50 sample stations), which is detailed in Stoker *et al.* (2009). The lithology of the cores is summarized in Table 2.

Indicators of instability in outer Little Loch Broom

Little Loch Broom Slide Complex

The Little Loch Broom Slide Complex extends between the mid- and outer loch moraine/bedrock sills (Figs 3–5a). In this area, the maximum water depth is *c.* 75 m. The sea bed in the centre of the outer loch is flat, whereas much of the rest of the basin is characterized by undulating bathymetry and an irregular scalloped margin (Fig. 5a). On this basis three main areas of sliding have been identified designated the Rireavach, Carnach and Scoraig slides, with a further area of debris lobes offshore Corran Scoraig near the outer sill. The morphology of these features is detailed below, together with the resultant deposits which collectively comprise the Rireavach Member (Assynt Glaciogenic Formation).

Rireavach Slide. This slide is the largest feature identified in Little Loch Broom, and has affected an area of the sea bed about 1–2 km^2 (Fig. 5a).

Table 2. *Summary of stratigraphy and lithofacies proved in SAMS and BGS cores in Little Loch Broom. SAMS cores prefixed by GC; BGS core prefixed by 57-06*

Stratigraphy	Cores	Lithofacies description
HOLOCENE		
Holocene lag (Outer Little Loch Broom)	GC087 GC093 GC112/113 GC115 GC119 GC120 GC122/57-06/286	Predominantly grey, dark grey and olive grey, very poorly sorted muddy, very fine-grained sand and sandy mud, with gravel clasts and shells/shell fragments, including *Turritella* sp. and paired bivalves, commonly concentrated at the base of the unit. In core GC120, muddy sandy gravel bed crops out at sea bed. In core GC122, fine- to coarse-grained muddy sand is predominant with abundant shells at the base of the unit.
Summer Isles Formation (Inner Little Loch Broom)	GC092 GC088	Dark to very dark greenish grey, homogeneous, massive, mottled (bioturbated), soft and sticky, organic-rich silty to slightly silty clay, with sporadic fine- to medium-grained sand grains and shells/shell fragments. Core GC088 preserves a discrete slumped bed, 0.43 m thick, of greenish-grey laminated clay bounded by homogeneous clay. In core GC092, base of core is slightly sandy and mottled through bioturbation.
LATEGLACIAL		
Late-stage debris flow (Lithogenetic unit)	GC092	Interbedded dark olive-grey homogeneous silty clay and greyish brown, muddy, very fine- to fine-grained sand, with scattered shells/shell fragments. Sandy beds are 1.5–4.0 cm thick and display tight isoclinal folding.
Rireavach Member (Assynt Glaciogenic Fm)	GC122 57-06/286	Reddish brown, massive, compact, medium- to coarse-grained, gravelly (granule grade) sand dominated by quartz and lithics. Very poorly sorted, with sporadic shell fragments. Top of sand is reworked by Holocene lag.
	GC119	0.2 m-thick bed of dark grey to grey, very fine- to fine-grained sand, moderately sorted with shells/shell fragments; on 0.69 m of dark grey to grey, slightly sandy clay with abundant shells/shell fragments and patches of lithic grains (intraclasts?).
	GC087	0.62 m-thick bed of colour laminated (grey, dark greenish-grey, pale red and dark greyish-brown) silty clay, with inclined and disrupted lamination; on 0.14 m-thick bed of grey sandy mud that includes abundant intact and comminuted shells and subangular lithic clasts; on homogeneous pale grey clay, becoming dark grey-brown and slightly silty towards the base.
Assynt Glaciogenic Formation (Pre-slide deposits)	GC093 GC112 GC113 GC115 GC120	Homogeneous to colour laminated (dark grey to brown, greyish-brown and reddish-grey), soft and buttery, clay and silty clay. Laminae range from 0.5–1.5 cm. Bioturbation and reduction spots locally observed. Common pebbles (up to 2 cm) and shells/shell fragments. Sporadic very thin to thin beds (2–12 cm) of reddish grey, fine- to coarse-grained muddy sand.

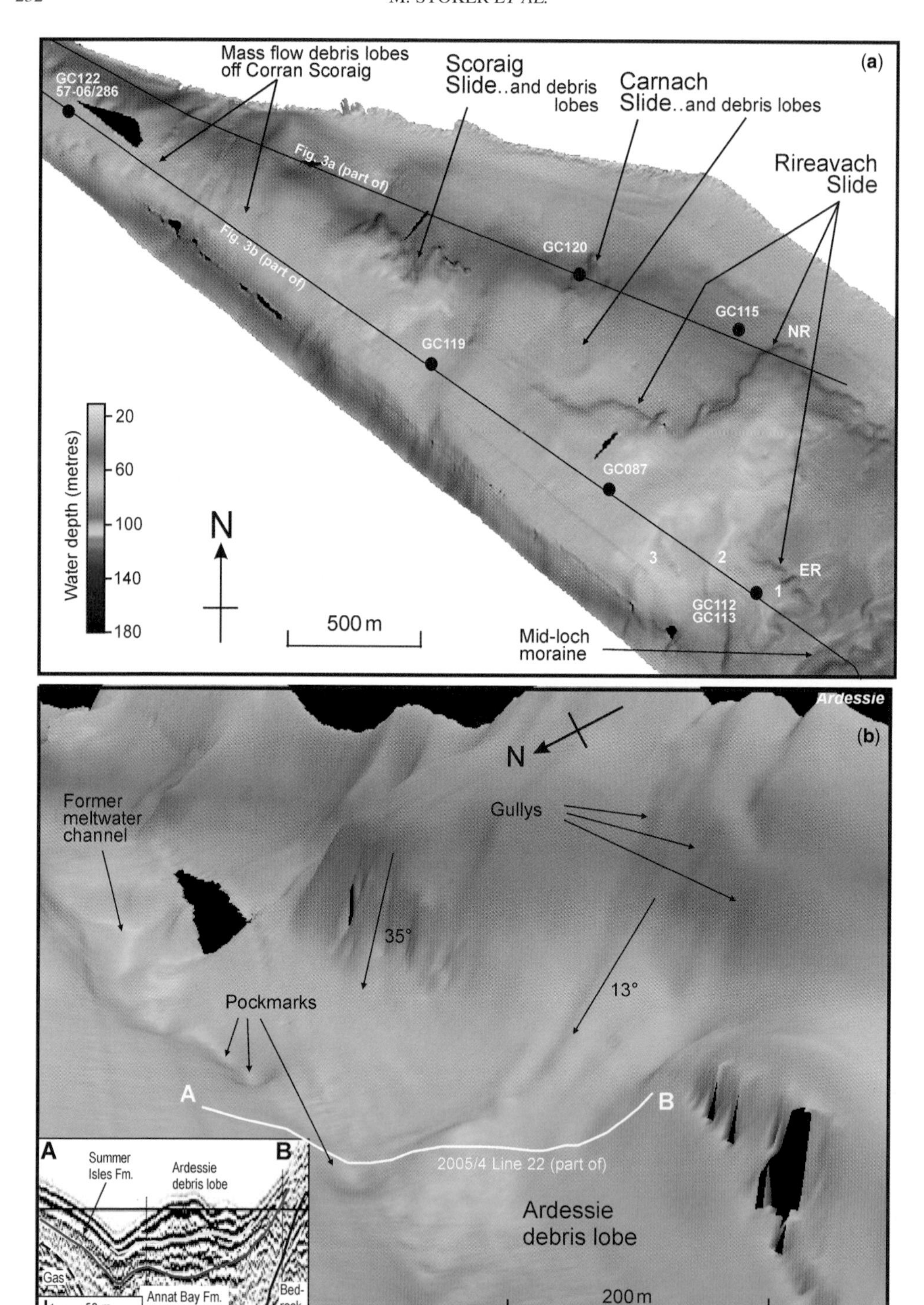

Fig. 5. (**a**) Detailed swath bathymetric image of the Little Loch Broom Slide Complex, showing the distribution of the component slides and mass-transport deposits (see Fig. 4 for location). ER, eastern re-entrant; NR, northern re-entrant; 1–3, slide scars of Rireavach Slide. (**b**) Perspective view of Ardessie debris lobe, looking SE within inner Little Loch Broom (see Fig. 4 for location). Seismic inset shows sub-bottom detail of the debris lobe (base = yellow reflector) and relationship to Summer Isles Formation (base = red reflector).

Three distinct scarp surfaces (1–3) are observed on the swath image, confirmed by their correlation with the boomer profile that transects the slide (Fig. 3b). The most distinctive backscarp (1) displays a generally rectilinear margin at a water depth of about 35–40 m. This can be traced around the entire area of the slide a distance of about 4 km, and defines two main erosional hollows or re-entrants: a northern re-entrant (NR) constrained by the side of the loch and an eastern re-entrant (ER) constrained by the morainic rampart that forms the mid-loch sill. These re-entrants indicate that material has been displaced into the basin in both southerly (NR) and westerly (ER) directions. However, the rectilinear shape of the scarp appears to have been modified by smaller scale ($\leq$200 m) curved or scalloped indentations that impart an irregularity to the backscarp. The southern wall of the loch appears to have remained linear, steep and less obviously affected by sliding. Scarps 1–3 (in the eastern re-entrant) range from 7 to 15 m high with slope angles between 7° and 15°. They display subplanar to predominantly curved concave-up profiles. Distinct terraces occur between scarps 1–2 and 2–3 at water depths of about 55 m and 65 m, respectively. The present-day basin floor at the foot of scarp 3 is at about 75 m water depth. Along the western edge of the northern re-entrant (NR), the delineation of the separate scarps is less clear. A single scarp is present with an upper headwall limit at 30–35 m water depth, a vertical relief of 35 m and a slope angle up to 10°. In Figure 3b, the seismic profile shows the basin floor to be slightly undulatory, a characteristic also observed on the swath image (Fig. 5a).

Carnach Slide. The Carnach Slide forms a much smaller sea bed hollow on the northern slope of the loch, between about 35 and 55 m water depth (Fig. 5a). It occupies an area of about 0.05 km^2, has displaced sediment to a depth of about 5 m below sea bed, displays a curved, concave-up, profile, with a maximum backscarp slope angle up to about 13°. Downslope of the scar, an equivalent area of seabed is raised *c.* 1–2m (convex-up) above the adjacent sea floor for a distance of up to 300 m from the foot of the scar. At least two separate lobes are identified from the swath image, which represent material derived from the slide scar.

Scoraig Slide. The Scoraig Slide is expressed as a discrete crenulate scar that broadly parallels the northern slope of the loch for about 500 m (Fig. 5a). The slide scar covers an area of about 0.25 km^2 and is cut into the slope between about 30 and 65 m water depth. It displays a curved, concave-up, slide surface profile with a maximum backscarp angle of 18°. Several debris lobes are visible at the base of the scar raised *c.* 1–2 m above the sea bed, extending 250–300 m into the basin and up to 600 m along the strike of the basin.

Debris lobes offshore Corran Scoraig. Further NW along the northern flank of the outer loch a series of less well-defined crenulations and hollows are visible on the swath image. Some of these have a gully-like appearance and are up to 10 m deep and several tens of metres wide (Fig. 5a). A series of overlapping debris lobes, up to 4 m in relief and up to several hundred metres wide, are found towards the base of the slope forming a package that extends for up to 1.3 km along the axis of the basin. There may also be some input from the southern slope, although the swath image is increasingly restricted (due to operational constraints) in extent near the mouth of the loch. The basin floor topography is hummocky on both the swath image and the seismic profile data.

Rireavach Member (slide complex deposits). The Rireavach Member of the Assynt Glaciogenic Formation forms a discrete fjord slope to basin-floor package of mass-transport deposits derived from multiple slide sources. It has accumulated below about 30 m water depth, the approximate upper bounding limit of the headwalls of the slides. On seismic profiles, it displays an irregular, hummocky, sheet-like geometry up to 12 m thick. In Figure 3, the base of the Rireavach Member (the base of the slide complex) is clearly depicted by the truncation of acoustically layered strata in the underlying, undisturbed deposits of the Assynt Glaciogenic Formation. Internally, the mass-transport deposits display a predominantly chaotic internal reflection configuration; however, sporadic subhorizontal reflections are locally observed, particularly in the area of the Rireavach Slide and within the pile of debris lobes off Corran Scoraig near the mouth of the outer loch.

The deposits of the Rireavach Member have been sampled at three sites: GC087, GC119 and GC122 (57-06/286). All three cores are from the basin floor (Figs 4 & 5), and all recovered different lithologies (Table 2). Deposits of the Rireavach Slide were tested by core GC087, which recovered 0.62 m of colour laminated silty clay overlying a 0.14 m thick bed of shelly, gravelly, sandy mud, overlying homogeneous clay and silty clay. The contacts between all three beds are sharp and the laminated clay displays angular discordance with the underlying bed. In Figure 6a, the laminations are clearly observed to be inclined relative to the underlying bed of sandy mud and partially disrupted and offset along small faults. The laminated clay is in sharp contact with the overlying Holocene unit, the base of which is marked by a gravelly and shelly lag deposit (Fig. 4). Core GC119 is located on the edge of the debris lobes derived from the Scoraig Slide, where the Rireavach Member contained a 0.2 m thick sandy bed overlying 0.69 m of slightly sandy clay. Discrete patches of lithic grains are scattered throughout the clay and possibly represent coarser, matrix-supported intraclasts. There is a sharp contact with the overlying Holocene sandy mud. Cores GC122 and 57-06/286 tested the debris lobes off Corran Scoraig and recovered 2.57 m of massive, compact, reddish-brown gravelly sand, the top of which is reworked and overlain by a veneer of Holocene muddy sand.

Pre-slide deposits (Assynt Glaciogenic Formation). Cores GC112/113 and 120 are located in the slide scar region of the Rireavach and Carnach slides, respectively (Fig. 4). They both recovered homogeneous to colour-laminated mud and clay with sporadic pebbles, shelly material and thin beds of muddy sand. The same lithofacies was also present in cores GC093 and GC115, which penetrated the Assynt Glaciogenic Formation outside of the slide complex; GC093 is located adjacent to the major moraine on the mid-loch sill. In all of these cores, this lithofacies is sharply overlain by a Holocene shelly and gravelly lag deposit. The colour-laminated mud and clay is distinctive of the Assynt Glaciogenic Formation throughout the Summer Isles region (Stoker *et al.* 2009); in Little Loch Broom the sediments in these cores are regarded as being part of the undisturbed pre-sliding fjord section.

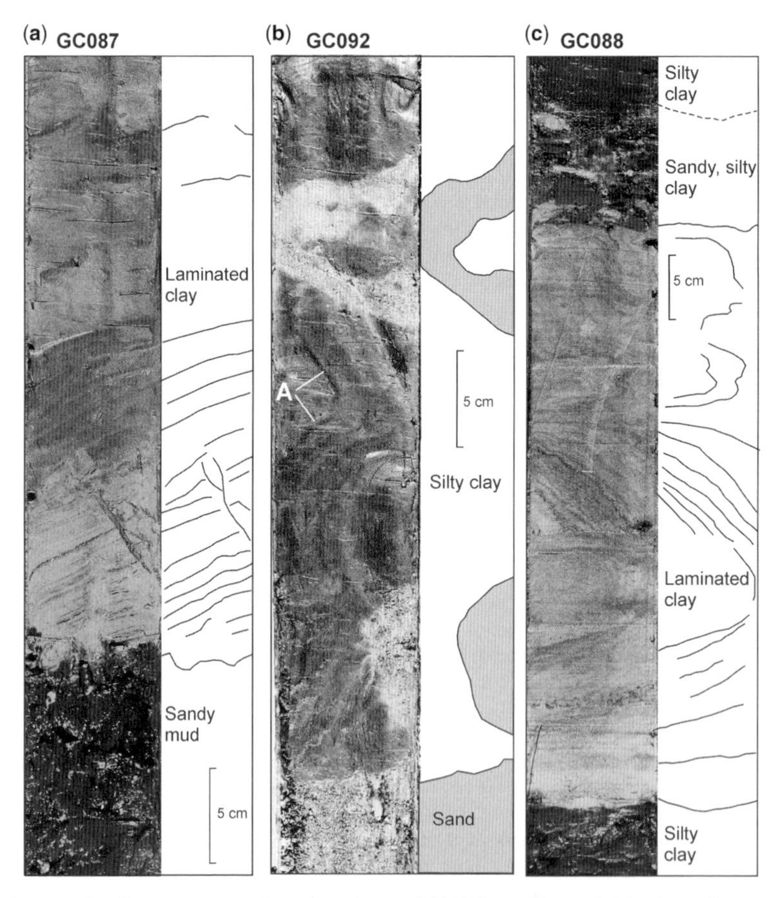

Fig. 6. Core photographs of mass-transport deposits: (**a**) core GC087 from Rireavach Member (Rireavach Slide) showing sharp, discordant contact between sandy mud and dipping laminated clay (1.24–1.68 m); (**b**) core GC092 from Late-stage debris flow unit (Badcaul Slide) showing recumbent, isoclinal folding of sand beds (1.61–2.05 m) (A: artefact, caused by sweep of osmotic knife during cleaning of surface of core); and (**c**) core GC088 from Summer Isles Formation showing recumbent, isoclinally folded laminated clay sharply bounded above and below by basinal silty clay (2.16–2.8 m).

Indicators of instability in inner Little Loch Broom

The fjord infill succession in inner Little Loch Broom is far less disturbed in comparison to the outer loch; this is most probably a function of the steeper sides of the inner loch precluding the accumulation of significant sidewall deposits. However, there are several indicators of instability, specifically the Badcaul Slide and associated Late-stage debris flow unit, together with the Ardessie debris lobe and a discrete slide deposit in core GC088. These latter two deposits are both associated with the Holocene Summer Isles Formation.

Badcaul Slide

The term Badcaul Slide is used here to refer to a 1-km^2 diffuse zone of disturbance on the SE side of the mid-loch sill that has affected the Assynt Glaciogenic Formation (Figs 3 & 7). The swath image displays a number of terraces with irregular, scalloped, backscarps stepping down into the inner loch. However, the most distinctive terrace (which occurs on the main part of the sill at *c.* 60 m water depth) displays a rectilinear shape up to 800 m wide and is backed by a headwall with a slope angle up to 15° which is traced for 1.5 – 2 km. A series of narrower (*c.* 50 m wide) terraces occur between about 70 and 90 m water depth (Fig. 7). The narrower terraces coincide with what appear to be several slide and/or slump blocks observed on the seismic profile. The internal seismic reflection configuration of the Assynt Glaciogenic Formation on this part of the fjord wall is mostly structureless to irregular and chaotic with discontinuous subplanar reflecting surfaces dipping into the basin; the latter indicate the presence of a general large-scale tabular structure within the depositional package (Fig. 7b). Gently curved, smaller scale, concave-up, slide surfaces are also observed within the upper part of the slide package and, where bedding is observed, some rotation of reflections (including the sea bed) into the slope is locally evident (Fig. 7b, upper inset). At the base of the slope, the sea bed occurs at 110 m water depth and is characterized by a hummocky morphology that becomes smoother into the basin. The seismic profile shows that this reflects a large debris lobe that has accumulated at the base of the slope and which becomes progressively buried beneath younger sediments into the basin (Figs 3 & 7b). The debris lobe is up to 10–12 m thick, is lensoid in shape with a chaotic internal reflection configuration and can be traced for about 800 m along the line of the profile (Fig. 7b, lower inset). This deposit forms part of the Late-stage debris flow lithogenetic unit of Stoker *et al.* (2009). The overlying deposits belong to the Holocene Summer Isles Formation.

The debris lobe has been sampled by core GC092 which contained 1.5 m of homogeneous silty clay of the Summer Isles Formation, with a slightly sandy base, sharply overlying 0.6 m of interbedded silty clay and muddy sand of the Late-stage debris flow unit. Most of the latter deposit is folded with recumbent isoclinal folds depicted by the paler sandy beds in Figure 6b. It is unclear whether the basal sandy bed in the core is part of the deformed section; its contact with the overlying interbedded section is sharp.

Ardessie debris lobe

The Ardessie debris lobe has been identified on the southern flank of inner Little Loch Broom, immediately offshore Ardessie, at a water depth of 75–80 m (Figs 4 & 5b). The swath image shows several gullies 2–5 m deep and 20–50 m wide, eroded into the side of the fjord on a slope angle of 13°. A debris lobe is clearly observed at the base of the slope with a convex-up relief of up to 2 m above the surrounding sea bed and covering an area of *c.* 0.2 km^2. Traced landward, the gullies trend back towards the coastline at the point where the Allt Airdeasaidh drains into the loch. This stream and its tributaries drain part of the An Teallach massif (Fig. 4) and have a catchment area of *c.* 10 km^2.

The debris lobe is similarly well imaged on seismic profile data which show a discrete lobe at the sea bed with a double hump that may be indicative of several smaller component lobes. The hummocky nature of the sea bed associated with the debris lobe contrasts with the generally smoother morphology of the basin floor (Fig. 5b). The seismic profile also reveals an acoustically structureless internal reflection configuration and that the debris lobe overlies the bulk of the basinal deposits associated with the Summer Isles Formation. In contrast to the Badcaul Slide, there is no indication that the debris lobe is buried beneath any younger sediment (within seismic resolution *c.* 0.5 m). A linear trail of pockmarks, which is associated with shallow gas in the basin, appears to follow the course of a buried, former meltwater channel that retains expression at the sea bed (Fig. 5b); however, it is unclear whether the debris lobe pre- or post-dates pockmark formation. Although there are no core data available, the seismic stratigraphy indicates that this is a Holocene deposit.

Core GC088 (Summer Isles Formation)

Core GC088 is located close to the southern slope of the inner loch at a water depth of 102 m (Fig. 4). The core penetrated 2.88 m into the Summer Isles

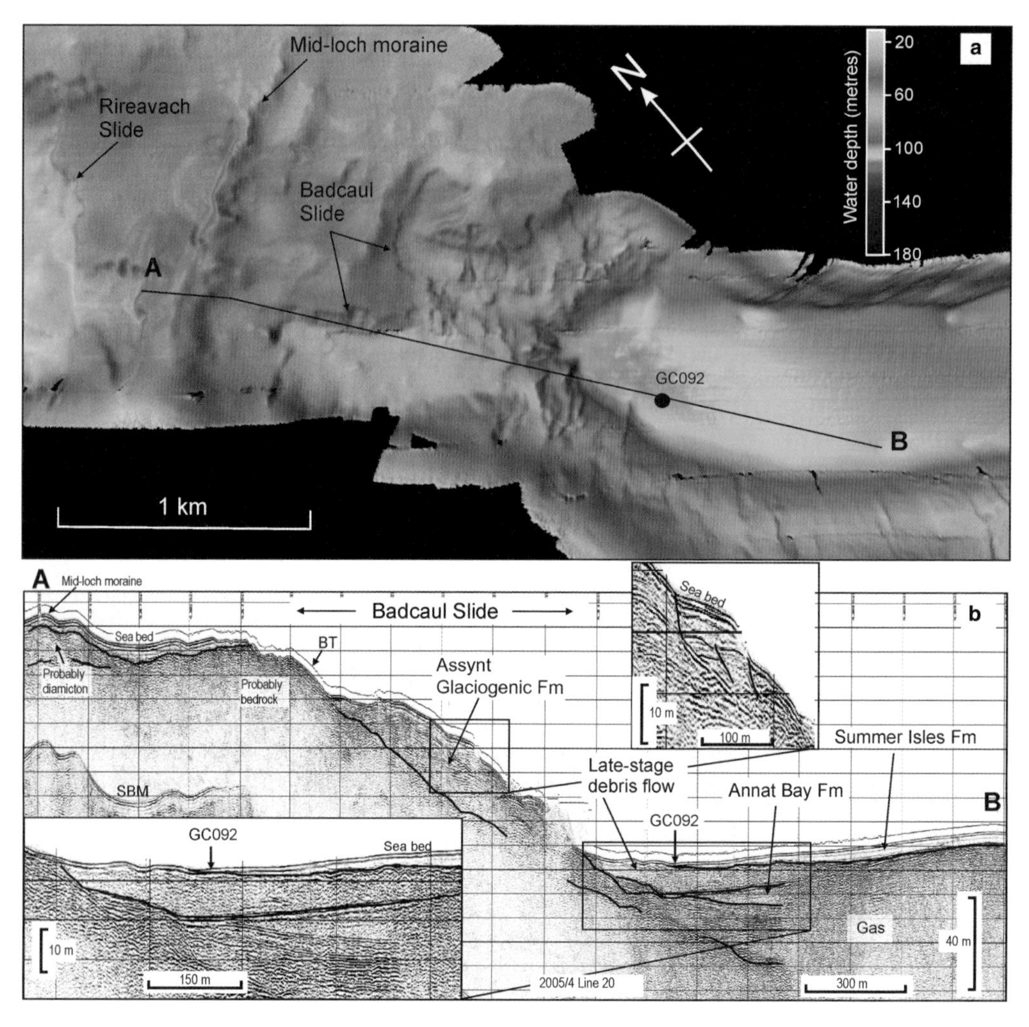

Fig. 7. (**a**) Detailed swath bathymetric image and (**b**) seismic profile showing the Badcaul Slide on the inner part of the mid-loch sill. The seismic profile shows the disposition of the Late Quaternary units, in particular the disturbed Assynt Glaciogenic Formation and the Late-stage debris flow unit at the base of the slope, sandwiched between the Annat Bay and the Summer Isles formations. Lower seismic inset in (b) shows detail of Late-stage debris flow unit; upper seismic inset shows evidence of rotational sliding (slumping) in the Assynt Glaciogenic Formation. BT, bottom tracking indicator; SBM, sea-bed multiple.

Formation and recovered predominantly soft, sticky, homogeneous, mottled silty clay. The seismic profile at the core site shows an undisturbed acoustically bedded character; however, the core revealed a discrete bed of folded laminated clay between 2.32 and 2.75 m (Fig. 6c) in contrast to the enclosing sediment. The deformed lamination reveals recumbent folds showing varying degrees of complexity in fold pattern from symmetrical to asymmetrical isoclinal folds. The contact of the deformed bed with the underlying homogeneous silty clay is sharp; the contact with the overlying bed also appears to be sharp, although there may be some disruption of the top of the bed due to subsequent bioturbation or erosion. It seems probable that this is a discrete bed of deformed clay incorporated within the more typical basinal silty clay of the Summer Isles Formation.

Interpretation and discussion

Types of mass transport

The swath bathymetric, seismic reflection and core data provide unambiguous evidence for widespread slope instability within Little Loch Broom. Two main types of mass-transport process can be identified: (1) sliding; and (2) mass flow (Table 3). These processes are not mutually exclusive as the mass-flow deposits are commonly sourced from the adjacent slides (e.g. Carnach Slide, Fig. 5a) although the Ardessie debris lobe is a more discrete feature linked to a series of slope gullies. The main characteristics of these two types of mass-transport process are described in the following and summarized in Table 3.

Slides. A major characteristic of the slide failure surfaces found within the Little Loch Broom Slide Complex, as well as the Badcaul Slide, is their curved concave-up profile; the morphology of this is enhanced on the swath bathymetric image by the scallop-shaped nature of the scarps (Figs 5 & 7). Tilting of bedding is locally observed in the upper part of the Badcaul Slide mass-transport deposit, where bedding is rotated into the failure surfaces (Fig. 7b, upper inset). However, the swath bathymetry also reveals that this irregular pattern of scarps is superimposed on a larger-scale rectilinear pattern of failure, as indicated by both the northern and eastern re-entrants of the Rireavach Slide as well as the shallow part of the Badcaul Slide on the mid-loch sill. Slides on curved surfaces are classified as slumps, whereby failure is accompanied by rotation; in contrast, planar slides are classified as glides with failure facilitated by translation (Nardin *et al.* 1979; Cook *et al.* 1982; Mulder & Cochonat 1996). Our data suggest that both translational and rotational sliding have occurred in Little Loch Broom, the implications of which – in terms of release mechanisms and timing of mass movement – are considered elsewhere in this section (p. 239).

Probable slide deposits were recovered in cores GC087 and GC088 from the Rireavach Member and Summer Isles Formation, respectively, which proved beds of deformed laminated clay in sharp contact with undeformed beds (Figs 4 & 6a, c). The deformed laminations range from gently inclined (core GC087) to recumbent and isoclinally folded (core GC088). The abrupt contact with undeformed beds suggests that these beds represent discrete blocks that have moved downslope. Despite some internal deformation, they have retained an internal coherency in that continuous laminations are still preserved. Deformation of the sediment may have begun as creep which, as gravitational stresses increased, may have ultimately failed as an intact block (Syvitski & Shaw 1995). Comparable structures have been described from slide deposits cored on the continental slope offshore Nova Scotia and elsewhere (Cook *et al.* 1982; Jenner *et al.* 2007) and are commonly attributed to elastic mechanical behaviour of the sediment during submarine slope failure (Nardin *et al.* 1979). As core GC087 did not penetrate the entire mass-transport deposit associated with the Rireavach Slide, the possibility that this bed represents a rafted block within a debris flow cannot be discounted.

Mass flows. Mounded, lensoid and lobate packages of sediment form the predominant basin floor deposit associated with the Little Loch Broom Slide Complex and the Badcaul Slide, as well as the more discrete Ardessie debris lobe. This depositional morphology, combined with a general lack of internal reflectors, is characteristic of mass-flow deposits where the absence of internal structure is commonly related to deformational homogenization of the sediment mass during submarine slope failure (Nardin *et al.* 1979). The swath bathymetry shows clearly that the Scoraig and Carnach debris lobes are derived from the adjacent slides, whereas several failure surfaces have contributed to the accumulation of the basinal mass-flow package associated with the Rireavach Slide (Fig. 5a). The Late-stage debris flow unit at the foot of the Badcaul Slide is sourced from the Badcaul Slide, the deposit gradually thinning into the basin (Fig. 7b). By way of contrast, the Ardessie debris lobe is localized at the base of a series of erosional gulleys that transect the adjacent slope (Fig. 5b). The hummocky surface of many of these mass-flow packages has been imaged on swath bathymetry or seismic profiles. The images suggest that each of the main packages, which may extend >1 km across the basin floor, consists of an amalgamation of smaller stacked flows (e.g. the Scoraig and Ardessie debris lobes) which may be up to a maximum of a few hundred metres in width.

The core data suggest that a number of different mass-flow processes may have operated during slope failure. The occurrence of a clay bed with matrix-supported intraclasts in core GC119, from a mass-flow lobe at the base of the Scoraig Slide (Fig. 4), is consistent with muddy debris-flow deposition. This is associated with plastic mechanical behaviour where the strength of the flow is principally a result of cohesion due to the clay content (Nardin *et al.* 1979; Mulder & Cochonat 1996). In contrast, the massive gravelly sand recovered in cores GC122 and 57-06/286, from the area of debris lobes offshore Corran Scoraig (Fig. 4), may be more characteristic of a grain flow (Lowe 1982) or a sandy debris flow (Middleton 1967; Shanmugan 1996). In this case, the less cohesive sandy material is supported by dispersive pressure, thus exhibiting pseudo-plastic flow, and deposited by ‘freezing’. Although rapid mass deposition from a high-concentration turbidity current – the Bouma A division – cannot be discounted, there is no evidence in the cores for grading which is commonly a defining

Table 3. *Summary of characteristics of mass failure and mass-transport deposits in Little Loch Broom*

Stratigraphy	Indicator of instability		Morphology – swath and seismic characteristics	Sedimentary structures	Mass-transport process
HOLOCENE					
Summer Isles Formation	Ardessie debris lobe		Base of slope lobe(s); hummocky sea bed; structureless internal reflection pattern; sourced by several slope gullies	No data	Mass flow
	Core GC088		Seismically unresolvable bed within parallel bedded sequence	Recumbent, isoclinally folded clay bed; sharp upper and lower bed contacts	Slide
LATEGLACIAL					
Late-stage debris flow (Lithogenetic unit)	Badcaul Slide & core GC092		Irregular, hummocky sea bed; series of terraces backed by scalloped headwall scarps; subplanar and curved, concave-up slide surfaces; bedding locally rotated into slide surface; mounded, lensoid lobe at base of slope with structureless to irregular internal reflection pattern	Recumbent, isoclinally folded interbeds of sand and silty clay; sharp bounding bed contacts	Slide and mass flow
Rireavach Member (Assynt Glaciogenic Fm)	Little Loch Broom Slide Complex	Rireavach Slide & core GC087 Carnach Slide & core GC119 Scoraig Slide Corran Scoraig debris lobes & cores GC122 & 57-06/286	Irregular, hummocky sea bed bounded by several discrete scarps, including multiple scarps and terraces of Rireavach Slide; scalloped to rectilinear slide scars; subplanar and curved concave-up slide surfaces; stacked, lensoid lobes on lower slope and basin floor; mainly structureless to chaotic internal reflection pattern	Highly variable lithofacies, including dipping laminated clay, massive gravelly sand, mud with matrix-supported clasts and thin-bedded shelly gravelly sand; bed contacts are sharp	Slide and mass flow

structure in such deposits (Pickering *et al.* 1989; Shanmugan 1996). Whereas debris flows can be initiated and moved along low-angle slopes, grain flows usually require steep slopes for initiation and sustained downslope movement (Nardin *et al.* 1979). All of these sediment types have been described in association with mass-flow processes from continental slopes and fjords (Cook *et al.* 1982; Syvitski & Hein 1991; Jenner *et al.* 2007). In core GC092, from the Late-stage debris flow unit, a deformed, folded bed was present at the top of the mass-flow lobe, underlain by a sandy bed. The deformed bed is more typical of a slide block as described above; in this setting it may represent a rafted block incorporated within the mass flow. Unfortunately, no core data are available at this time from the Ardessie debris lobe. However, its association with erosional gullies implies some kind of sediment gravity-flow process linked to the formation of the gullies. On continental slopes, such gullies are commonly associated with turbidity currents (Pickering *et al.* 1989). In Little Loch Broom, the Ardessie debris lobe and associated gullies appear to represent the sink for part of the fluvial catchment area on the northern slope of An Teallach drained by the Allt Airdeasaidh (Fig. 4); this potential linkage is further discussed below.

Timing of mass failure

A significant phase of mass failure is reported from elsewhere within the Summer Isles region between about 14 and 13 ka BP, including the Cadail Slide in the North Annat Basin and slumping of basinal sediments in outer Loch Broom (Stoker & Bradwell 2009; Stoker *et al.* 2009) (Fig. 1). Seismic-stratigraphy indicates that the Cadail Slide, which also deformed sediments of the Assynt Glaciogenic Formation, occurred immediately prior to the deposition of the glaciomarine Annat Bay Formation in the North Annat Basin. A similar relationship is observed in this study from the Badcaul Slide, where the main package of mass-transport deposits (of the Assynt Glaciogenic Formation) is onlapped by the Annat Bay Formation (Figs 3b & 7b). However, the Late-stage debris flow unit, also linked to the Badcaul Slide, overlies the Annat Bay Formation, but is itself onlapped by the Summer Isles Formation. This stratigraphical relationship strongly supports the idea that mass failure in Little Loch Broom (including the Little Loch Broom Slide Complex) was initiated during the Lateglacial interval. It also suggests that large-scale mass-flow processes persisted for some time after the deposition of the Annat Bay Formation, but are no younger than *c.* 8 ka BP, the onset of deposition of the Summer Isles Formation (Stoker *et al.* 2009). A Holocene veneer also overlies the slide and mass-flow deposits of the Rireavach Member in outer Little Loch Broom. In contrast, the Ardessie debris lobe and the slide deposit in core GC088 are both part of the Summer Isles Formation, and thus are of Holocene age. This suggests that discrete areas of mass failure have continued to develop in Little Loch Broom within the postglacial interval.

Triggering mechanism for mass failure and implications for Lateglacial–Holocene instability in the Summer Isles region

When glacier ice occupies a fjord, subglacial sediment and morainic debris are deposited on sidewalls which are commonly very steep. However, during glacier retreat, such sediment-mantled slopes become inherently unstable as they lose support during glacial downwasting (Church & Ryder 1972; Ballantyne 2002; Powell 2005). Thus, sidewall sediment is prone to failure by gravitational processes soon after the removal of glacier ice (Syvitski 1989; Syvitski & Shaw 1995; Ballantyne 2002). The walls of Little Loch Broom are no exception; they are locally very steep, exceeding 50° on the northern slope of the inner loch close to the mid-loch sill. The triggering of mass failure in Little Loch Broom due to the removal of ice support is consistent with a timing of failure during the Lateglacial interval as the depositional environment changed from an ice-contact/ice-proximal setting to an ice-distal setting. Arguably, the initial failure on both the Rireavach and Badcaul slides was as rectilinear glide blocks that slid downslope. The subplanar basinward-dipping reflections in the mass-transport package of the Badcaul Slide imply a tabular geometry to this sediment package. However, the rectilinear shape of these two slides has been subsequently modified by numerous smaller scale scallop-shaped failures whose curved surfaces and evidence of bed rotation indicate rotational sliding (slumps). This is also the dominant style of the smaller Carnach and Scoraig slides.

The time lag between large-scale slide initiation and subsequent modification by smaller scale slumps is unknown, although the available stratigraphic evidence that has been presented suggests that it had mostly occurred before 8 ka BP. This scenario is consistent with the paraglacial concept of Church & Ryder (1972), who emphasized the relatively rapid adjustment of deglaciated landscapes to non-glacial conditions through the enhanced operation of a wide range of processes including slope failure and mass transport. This essentially involves the progressive relaxation of unstable or metastable elements of the formerly glaciated landscape to a new more stable state (Ballantyne 2002). Although the reason for the change in the style of sliding is unclear, we propose the following two-stage process: (1) the planar rupture surfaces associated with the rectilinear slide blocks most likely follow weak bedding layers within the poorly consolidated sediment pile, rendering them highly susceptible to gravity sliding in the early stage of deglaciation of the fjord; (2) rotational sliding is less influenced by bedding; instead, it may relate more to subsequent rupturing of the infill through fractures generated by stress relaxation in the later stage of deglaciation.

The identification of mass-transport deposits within the Summer Isles Formation in inner Little Loch Broom indicates that paraglacial processes probably continued into the Holocene. The relaxation of the landscape following a widespread glaciation can occur over timescales of $10-10^4$ years, although the rate of sediment transfer is greatest immediately after deglaciation and probably declines exponentially with time (Ballantyne 2002). On this basis, we infer that this Little Loch Broom fjord region may not yet have fully adjusted (in terms of sediment transfer) to non-glacial conditions. In particular, the Ardessie debris lobe appears to be a localized anomalous Holocene seafloor sediment accumulation linked via a series of slope gullies to part of the An Teallach drainage basin. We suggest that the source of the mass-flow material is reworked glacial deposits on the northern flank of the An Teallach massif, through fluvial incision by the Allt Airdeassaidh and its tributaries (Fig. 4).

Paraglacial landscape readjustment may also have been enhanced by episodic seismic activity. Mass failure linked to glacio-isostatic rebound is a well-established phenomenon along the Atlantic continental margin of NW Europe. On the SW Norwegian margin, a detailed study of the giant Storrega Slide concluded that a major seismic pulse most likely accompanied deglaciation (Evans *et al.* 2002; Bryn *et al.* 2003; Haflidason *et al.* 2004). Differential rebound following ice unloading is also known to reactivate pre-existing structural lineaments and bedrock weaknesses as the new stress regime is accommodated, and enhanced neotectonic seismicity along the coastal areas of northern, western and southeastern Norway is an established fact (Olesen *et al.* 2008). In the UK, the earliest postglacial reactivation of pre-existing Caledonian and older lineaments is known to have generated normal faulting with metre-scale displacement in the southern Sperrin Mountains, Northern Ireland (Knight 1999). Differential rebound and seismicity may also have resulted in movement on faults such as the Kinloch Hourn Fault, western Scotland (Stewart *et al.* 2001), and possibly caused liquefaction of lake sediments at Glen Roy, Scotland (Ringrose 1989). Consequently, it seems probable that palaeoseismic activity was also occurring in the Summer Isles region during Lateglacial–Holocene time. This hypothesis is strengthened when it is noted that the west coast of Scotland, from Ullapool to Arran, continues to the present day to be a major focus for earthquakes (Musson 2003). Indeed, Stoker & Bradwell (2009) concluded that earthquake activity was the most likely trigger of slumping and deformation of the basin-floor fjord sediments in outer Loch Broom. It may be no coincidence that all three areas of large-scale sediment deformation in the Summer Isles region (North Annat Basin, Loch Broom and Little Loch Broom) are located along lines of NW trending faults (Fig. 1).

Conclusions

Swath bathymetry, seismic reflection profiles and sediment core data have revealed evidence of extensive slope instability within Lateglacial and Holocene sediments in Little Loch Broom. The major area of reworking is in the outer loch where ice-contact/ice-proximal fjord infill deposits of the Lateglacial Assynt Glaciogenic Formation have been extensively reworked by sliding and mass-flow processes linked to the Little Loch Broom Slide Complex. Collectively, this consists of the Rireavach, Carnach and Scoraig slides and associated mass-transport deposits – assigned to the Rireavach Member of the Assynt Glaciogenic Formation. In the inner loch, the Badcaul Slide has reworked the Assynt Glaciogenic Formation on the eastern flank of the mid-loch sill. A major mass-flow deposit at the foot of the slope is assigned to the Late-stage debris flow lithogenetic unit. Elsewhere in the inner loch, localized sliding and mass-flow deposition, including the Ardessie debris lobe and an intact slide block in core GC088, are preserved in the Holocene Summer Isles Formation.

Regional stratigraphic and isotopic dating evidence suggest that the Little Loch Broom Slide Complex and the Badcaul Slide were instigated between 14 and 13 ka BP, and that the bulk of the mass failure activity had occurred prior to 8 ka BP. The Ardessie debris lobe and GC088 slide-block deposit are both younger than 8 ka BP.

The sea-bed morphology and sub-bottom profiles suggest that both translational and rotational sliding mechanisms were active in the generation of the Rireavach and Badcaul slides; the superimposition of scallop-shaped slumps on a larger scale rectilinear pattern of failure implies that an initial glide phase was superseded by rotational backwall failure. A variety of associated mass-transport deposits include intact blocks of laminated clay and mounded mass-flow deposits (muddy debris flow and sandy debris flow or grain-flow deposits) preserved on the floor of the fjord. In contrast, the Holocene Ardessie debris lobe is a localized anomalous accumulation in the inner loch that appears to be fed by a series of discrete slope gullies; these are downslope continuations of the Allt Airdeasaidh, which drains part of the An Teallach massif. The intact Holocene slide block in core GC088 implies continuing (albeit sporadic) postglacial slide activity in the fjord.

On the basis that the bulk of the mass failure in Little Loch Broom probably occurred between about 14 and 8 ka BP, we infer that the major trigger of instability is probably the response of the landscape to deglaciation immediately following the retreat of the last ice sheet. We further suggest that paraglacial landscape readjustment may have been enhanced in this particular region by episodic seismic activity linked to glacio-isostatic unloading along pre-existing geological faults. By way of contrast, the Holocene Ardessie debris lobe may relate to erosion of the drift-mantled northern slopes of the An Teallach massif. This, together with the slide block in core GC088, may represent an ongoing (albeit much reduced) response to paraglacial processes in the fjords of NW Scotland.

The authors would like to thank the masters and crew of the RV *Calanus* and RV *James Cook* for their skill and assistance during the collection of the geophysical and geological datasets in 2005 and 2006. We thank R. Gatliff for comments on an earlier version of this manuscript, which was further improved by the reviews of R. Duck and C. O'Cofaigh. Published with the permission of the Executive Director, BGS (NERC).

References

Aarseth, I. A., Lønne, O. & Giskeødegaard, O. 1989. Submarine slides in glaciomarine sediments in some western Norwegian fjords. *Marine Geology*, **88**, 1–21.

Ballantyne, C. K. 2002. Paraglacial geomorphology. *Quaternary Science Reviews*, **21**, 1935–2017.

Bradwell, T., Fabel, D., Stoker, M. S., Mathers, H., McHargue, L. & Howe, J. A. 2008. Ice caps existed throughout the Lateglacial Interstadial in northern Scotland. *Journal of Quaternary Science*, **23**, 401–407.

Bryn, O., Solheim, A. *et al.* 2003. The Storegga Slide complex; repeated large scale sliding in response to climate cyclicity. *In*: Locat, J. & Mienert, J. (eds) *Submarine Mass Movements and their Consequences*. Kluwer Academic Publishing, Netherlands, 215–222.

Church, M. & Ryder, J. M. 1972. Paraglacial sedimentation: a consideration of fluvial processes conditioned by glaciation. *Geological Society of America, Bulletin*, **83**, 3059–3071.

Cook, H. E., Field, M. E. & Gardner, J. V. 1982. Characteristics of sediments on modern and ancient continental slopes. *In*: Scholle, P. A. & Spearing, D. (eds) *Sandstone Depositional Environments*. American Association of Petroleum Geologists, Tulsa, Oklahoma, 329–364.

Evans, D., McGiveron, S., Harrison, Z., Bryn, P. & Berg, K. 2002. Along-slope variation in the late Neogene evolution of the mid-Norwegian margin in response to uplift and tectonism. *In*: Doré, A. G., Cartwright, J. A., Stoker, M. S., Turner, J. P. & White, N. (eds) *Exhumation of the North Atlantic Margin: Timing, Mechanisms and Implications for Petroleum Exploration*. Geological Society, London, Special Publications, **196**, 139–151.

Fairbanks, R. G., Mortlock, R. A., Chiu, T. C., Kaplan, A., Guilderson, T. P., Fairbanks, T. W. & Bloom, A. L. 2005. Marine radiocarbon calibration curve spanning 0 to 50,000years B.P. based on paired $^{230}Th/^{234}U/^{238}U$ and ^{14}C dates on pristine corals. *Quaternary Science Reviews*, **24**, 1781–1796.

Haflidason, H., Sejrup, H. P. *et al.* 2004. The Storegga Slide: architecture, geometry and slide development. *Marine Geology*, **213**, 201–234.

Hamilton, E. L. 1985. Sound velocity as a function of depth in marine sediments. *Journal of the Acoustic Society of America*, **78**, 1348–1355.

Hampton, M. A., Lee, H. J. & Locat, J. 1996. Submarine landslides. *Revue of Geophysics*, **34**, 33–59.

Heath, S., Hastings, T. & Rae, G. (eds). 2000. *Final Report of the Joint Government/Industry Working Group on Infectious Salmon Anaemia (ISA) in Scotland*. Scottish Executive.

Hjelstuen, B. O., Haflidason, H., Sejrup, H. P. & Lyså, A. 2009. Sedimentary processes and depositional environments in glaciated fjord systems – Evidence for Nordfjord, Norway. *Marine Geology*, doi: 10.1016/j.margeo.2008.11.010.

Jenner, K. A., Piper, D. J. W., Campbell, D. C. & Mosher, D. C. 2007. Lithofacies and origin of late Quaternary mass transport deposits in submarine canyons, central Scotian Slope, Canada. *Sedimentology*, **54**, 19–38.

Knight, J. 1999. Geological evidence for neotectonic activity during deglaciation of the southern Sperrin Mountains, Northern Ireland. *Journal of Quaternary Science*, **14**, 45–57.

Lee, A. J. & Ramster, J. W. 1981. *Atlas of the Seas around the British Isles*. Ministry of Agriculture, Fisheries and Food (MAFF).

Lowe, D. R. 1982. Sediment gravity flows, II. Depositional models with special reference to the deposits of high-density turbidity currents. *Journal of Sedimentary Petrology*, **52**, 279–297.

McQuillin, R. & Ardus, D. A. 1977. *Exploring the Geology of Shelf Seas*. Graham & Trotman, London.

Middleton, G. V. 1967. Experiments on density and turbidity currents, III. Deposition of sediment. *Canadian Journal of Earth Science*, **4**, 475–505.

Mulder, T. & Cochonat, P. 1996. Classification of offshore mass movements. *Journal of Sedimentary Research*, **66**, 43–57.

Musson, R. 2003. *Seismicity and Earthquake Hazard in the UK*. British Geological Survey: http://www.quakes.bgs.ac.uk/hazard/Hazard_UK.htm.

Nardin, T. R., Hein, F. J., Gorsline, D. S. & Edwards, B. D. 1979. A review of mass movement processes, sediment and acoustic characteristics, and contrasts in slope and base-of-slope systems v. Canyon-fan-basin floor systems. *Society of Economic Palaeontologists and Mineralogists Special Publication*, **27**, 61–73.

Niessen, F. & Whittington, R. J. 1997. Synsedimentary faulting in an East Greenland Fjord. *In*: Davies, T. A., Bell, T., Cooper, A. K., Josenhans, H., Polyak, L., Solheim, A., Stoker, M. S. & Stravers, J. A. (eds) *Glaciated Continental Margins: An Atlas of Acoustic Images*. Chapman & Hall, London, 130–131.

Olesen, O., Bungum, H., Dehls, J., Lindholm, C., Pascal, C. & Roberts, D. 2008. Neotectonics in Norway – mechanisms and implications. *Abstract: International Geological Congress*, Oslo, 2008. Available at: http://www.cprm.gov.br/33IGC/1398408.html.

Pickering, K., Hiscott, R. & Hein, F. 1989. *Deep-marine Environments: Clastic Sedimentation and Tectonics*. Unwin Hyman Ltd, London.

Powell, R. D. 2005. Subaquatic landsystems: fjords. *In*: Evans, D. J. A. (ed.) *Glacial Landsystems*. Hodder Arnold, London, 313–347.

Ringrose, P. S. 1989. Palaeoseismic (?) Liquefaction event in late Quaternary lake sediment at Glen Roy, Scotland. *Terra Nova*, **1**, 57–62.

SEPA. 2006. *Designated Shellfish Waters in Scotland: Site Data. 61 Little Loch Broom*. Scottish Environmental Protection Agency.

Shanmugan, G. 1996. The Bouma Sequence and the turbidite mind set. *Earth Science Reviews*, **42**, 201–229.

Stewart, I. S., Firth, C. R., Rust, D. J., Collins, P. E. F. & Firth, J. A. 2001. Postglacial fault movement and palaeoseismicity in western Scotland: a reappraisal of the Kinloch Hourn fault, Kintail. *Journal of Seismology*, **5**, 307–328.

Stoker, M. S. & Bradwell, T. 2009. Neotectonic deformation in a Scottish fjord, Loch Broom, NW Scotland. *Scottish Journal of Geology*, **45**, 107–116.

Stoker, M. S., Leslie, A. B. *et al.* 1994. A record of late Cenozoic stratigraphy, sedimentation and climate change from the Hebrides Slope, NE Atlantic Ocean. *Journal of the Geological Society, London*, **151**, 235–249.

Stoker, M. S., Bradwell, T., Wilson, C. K., Harper, C., Smith, D. & Brett, D. 2006. Pristine fjord landsystem revealed on the sea bed in the Summer Isles region, NW Scotland. *Scottish Journal of Geology*, **42**, 89–99.

Stoker, M. S., Bradwell, T., Howe, J. A., Wilkinson, I. P. & McIntyre, K. 2009. Lateglacial ice-cap dynamics in NW Scotland: evidence from the fjords of the Summer Isles region. *Quaternary Science Reviews*, **28**, 3161–3184.

Syvitski, J. P. M. 1989. On the deposition of sediment within glacier-influenced fjords: oceanographic controls. *Marine Geology*, **85**, 301–329.

Syvitski, J. P. M. & Hein, F. J. 1991. *Sedimentology of an Arctic Basin: Itirbilung Fiord, Baffin Island, Northwest Territories*. Geological Survey of Canada Paper 91–11.

Syvitski, J. P. M. & Shaw, J. 1995. Sedimentology and Geomorphology of Fjords. *In*: Perillo, G. M. E. (ed.) *Geomorphology and Sedimentology of Estuaries. Developments in Sedimentology 53*. Elsevier Science BV, Amsterdam, 113–178.

Whittington, R. J. & Niessen, F. 1997. Staircase rotational slides in an ice-proximal fjord setting, East Greenland. *In*: Davies, T. A., Bell, T., Cooper, A. K., Josenhans, H., Polyak, L., Solheim, A., Stoker, M. S. & Stravers, J. A. (eds) *Glaciated Continental Margins: An Atlas of Acoustic Images*. Chapman & Hall, London, 132–133.

Climate fluctuations during the past two millennia as recorded in sediments from Maxwell Bay, South Shetland Islands, West Antarctica

H. C. HASS[1]*, G. KUHN[2], P. MONIEN[3], H.-J. BRUMSACK[3] & M. FORWICK[4]

[1]*Alfred Wegener Institute for Polar and Marine Research, Wadden Sea Research Station, Hafenstrasse 43, D-25992 List, Germany*

[2]*Alfred Wegener Institute for Polar and Marine Research, Am Alten Hafen 26, D-27568 Bremerhaven, Germany*

[3]*ICBM, Oldenburg University, P.O. Box 2503, D-26111 Oldenburg, Germany*

[4]*Department of Geology, University of Tromsø, N-9037 Tromsø, Norway*

**Corresponding author (e-mail: christian.hass@awi.de)*

Abstract: The climate evolution of the South Shetland Islands during the last *c.* 2000 years is inferred from the multiproxy analyses of a long (928 cm) sediment core retrieved from Maxwell Bay off King George Island. The vertical sediment flux at the core location is controlled by summer melting processes that cause sediment-laden meltwater plumes to form. These leave a characteristic signature in the sediments of NE Maxwell Bay. We use this signature to distinguish summer and winter-dominated periods. During the Medieval Warm Period, sediments are generally finer which indicates summer-type conditions. In contrast, during the Little Ice Age (LIA) sediments are generally coarser and are indicative of winter-dominated conditions. Comparison with Northern and Southern Hemisphere, Antarctic, and global temperature reconstructions reveals that the mean grain-size curve from Maxwell Bay closely resembles the curve of the global temperature reconstruction. We show that the medieval warming occurred earlier in the Southern than in the Northern Hemisphere, which might indicate that the warming was driven by processes occurring in the south. The beginning of the LIA appears to be almost synchronous in both hemispheres. The warming after the LIA closely resembles the Northern Hemisphere record which might indicate this phase of cooling was driven by processes occurring in the north. Although the recent rapid regional warming is clearly visible, the Maxwell Bay record does not show the dominance of summer-type sediments until the 1970s. Continued warming in this area will likely affect the marine ecosystem through meltwater induced turbidity of the surface waters as well as an extension of the vegetation period due to the predicted decrease of sea ice in this area.

Changes in the climate of Antarctica in general and of West Antarctica in particular are likely signposting more radical environmental change to come on decadal as well as on millennial timescales (Steig *et al.* 1998; Blunier & Brook 2001; Weaver *et al.* 2003; Thomas *et al.* 2004; Cook *et al.* 2005; Vaughan 2005). The recent rapid regional (RRR) warming of atmospheric temperatures on the western Antarctic Peninsula and adjacent islands (abbreviated as WAP in the following) observed over the last 50 years is exceptional and unprecedented within the past 500 years, according to ice-core data (Vaughan *et al.* 2001). At Vernadsky Station (former Faraday Beascochea Bay) the atmosphere has warmed by 0.56 °C per decade since the 1950s (Turner *et al.* 2005). Research results show that the Antarctic system, and specifically the Antarctic Peninsula (AP), has been highly sensitive to such short-term climate change (e.g. Kreutz *et al.* 1997; Domack *et al.* 2003). Results from Palmer Deep (WAP area), a high-resolution marine record extending back *c.* 13 ka suggest that Holocene climatic changes were not driven by thermohaline reorganizations but the dynamics of the Southern Hemisphere westerly wind stream (Shevenell & Kennett 2002) and solar variability (Warner & Domack 2002; Bentley *et al.* 2009). Denton & Broecker (2008) suggested a wobbly ocean conveyor to be instrumental in driving late Holocene millennial-scale climate fluctuations (see also Clarke *et al.* 2007). Vaughan *et al.* (2003) emphasize the importance of regional climate-change processes in understanding the RRR warming trend that affects the AP and

From: Howe, J. A., Austin, W. E. N., Forwick, M. & Paetzel, M. (eds) *Fjord Systems and Archives*. Geological Society, London, Special Publications, **344**, 243–260.
DOI: 10.1144/SP344.17 0305-8719/10/$15.00

other high-latitude areas. They suggest any combination of three candidate mechanisms (oceanic and atmospheric circulations, as well as local greenhouse warming) but point out that it is not yet understood which of the mechanisms initiates and sustains RRR warming and finally causes the regional amplification of the global temperature trend (see also King *et al.* 2004; Bentley *et al.* 2009).

Whatever the ultimate cause, the WAP region has undergone considerable changes over the past 50 years. The disintegration of the Larsen A and B ice shelves in 1995 and 2002, respectively (Rott *et al.* 1996; Shepherd *et al.* 2003) demonstrates the sensitivity of the WAP; it is possible that such changes will serve as an analogue for future changes in the West Antarctic Ice Sheet (Rignot *et al.* 2004; Scambos *et al.* 2004). In addition, the glaciers of the AP and King George Island have responded directly to climatic change through retreat of ice fronts and increased meltwater production (Park *et al.* 1998; Chung *et al.* 2004; Cook *et al.* 2005). For example, the glacier fronts in Marian Cove retreated *c.* 250 m between 1956 and 1986 (6 and 36 a AP, AP = after present) and 270 m between 1989 and 1994 (39 and 44 a AP) as a result of the recent global warming (Park *et al.* 1998).

A direct consequence of glacier retreat on sub-Antarctic islands has been the emergence of newly ice-free areas for the colonization of terrestrial and intertidal plant vegetation and animals, which has resulted in increased erosion and denudation (Pichlmaier *et al.* 2004). The increased production of meltwater plumes has imposed further pressure on marine phytoplankton and thus the food web through increasing turbidity and associated light attenuation of the surface waters (Schloss & Ferreyra 2002; Schloss *et al.* 2002). Loss of sea ice has also been observed in several places since the 1970s (Parkinson 2002) and this correlates with the loss of krill stock density, likely to result in further impacts on the coastal Antarctic food webs (Loeb *et al.* 1997; Atkinson *et al.* 2004).

Here we present results of multiproxy analyses of a sediment record from Maxwell Bay that contains a clear climate signal of the past *c.* 1800 years. Granulometric and geochemical data allow us to distinguish local, regional and global climate signals. The present study has three main goals:

- to explore the potential of coast-proximal fjordic sediments of the sub-Antarctic region for high-resolution climate reconstructions of the Late Holocene;
- to reconstruct and evaluate the impact of historical climate change; and
- to place climatic changes from sub-Antarctica within a global framework of change.

Environmental setting of Maxwell Bay

Maxwell Bay is located off King George Island, South Shetland Islands, Antarctica (Fig. 1). It is *c.* 14 km long and 6–14 km wide, eroded into volcanoclastic rocks (Yoon *et al.* 1998). The bathymetry is characterized by one main basin more than 500 m deep. Maxwell Bay forms a fjordic environment surrounded by the tributary fjords Potter Cove, Marian Cove and Collins Harbor in the northeast and shallower, plateau-like areas in the north, west and southwest. The main water exchange proceeds through the mouth of the bay into Bransfield Strait. Minor water exchange with the Drake Passage takes place through the narrow Fildes Strait between King George and Nelson islands. Sea ice in the South Shetland bays generally occurs between July and October (Domack & Ishman 1993; Yoon *et al.* 1997).

The supply of freshwater and terrigenous sediments to Maxwell Bay is controlled by glaciofluvial rivers and tidewater glaciers (Griffith & Anderson 1989; Yoon *et al.* 1998). Significant amounts of meltwater runoff can cause stratified water masses comprising a surface layer that is characterized by low salinity and high terrigenous suspended particulate matter (Domack & Ishman 1993; Yoon *et al.* 1998). The location of the surface layer can vary significantly in space and time due to variations in meltwater runoff or wind activity, for example (Domack & Ishman 1993; Brandini & Rebello 1994).

Sediment supply to the bays on the South Shetland Islands occurs mainly through suspension settling from meltwater emanating from glaciofluvial rivers and tidewater glaciers, as well as by ice rafting of icebergs calving off the surrounding glaciers during the summer months (Griffith & Anderson 1989; Domack & Ishman 1993; Yoon *et al.* 1997, 1998). The highest concentrations of suspended particulate matter occur in Marian Cove and it has been assumed that this tributary fjord supplies more sediment to Maxwell Bay than the glaciers at the head of the bay (Yoon *et al.* 1998).

Sediment and ice cores from the WAP usually include tephra layers from volcanic sources close by (Matthies *et al.* 1990; Björck *et al.* 1991; Lee *et al.* 2007). Thus, coarser material may also originate from episodic tephra fall out, ice rafting and reworking (Yoon *et al.* 1997; (Shallow drilling on the Antarctic continental margin) SHALDRIL 2005).

The acoustic basement in Maxwell Bay is covered with a more than 120 ms (two-way travel time) thick wedge of acoustically stratified deposits that thin towards the bay mouth and mask the underlying basement topography (Griffith & Anderson 1989). The longest sedimentary record from Maxwell Bay is *c.* 108 m long and is predominantly Holocene in age (SHALDRIL 2005).

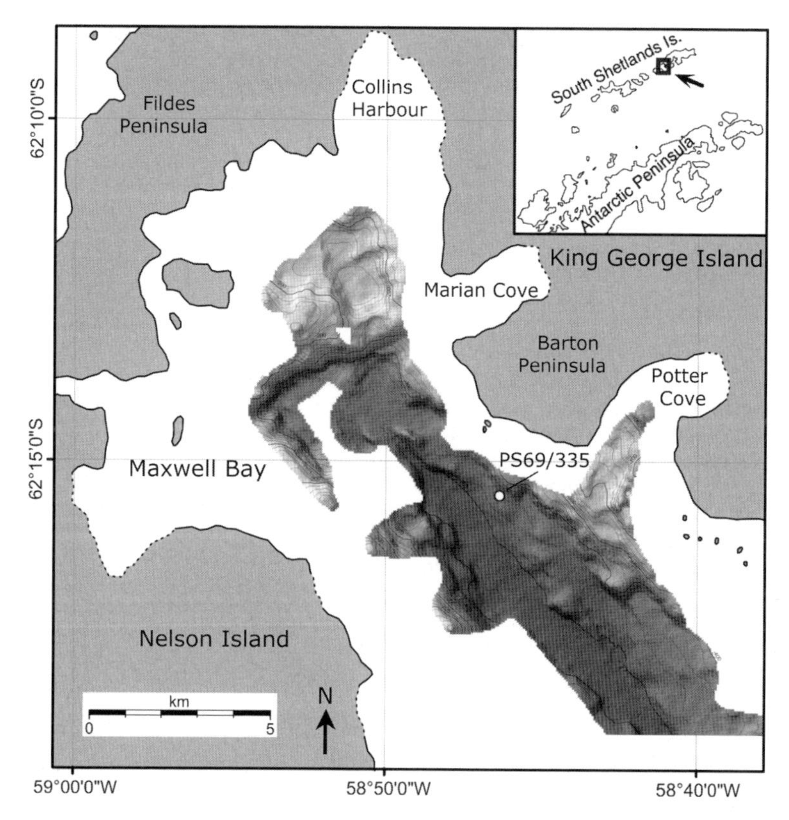

Fig. 1. Map of the working area including Hydrosweep bathymetry recorded during Expedition ANTXXIII/4 and the location of cores PS69/335-1 and -2.

Materials and methods

Gravity core PS69/335-2 and giant box core PS69/335-1 were recovered during RV *Polarstern* Expedition ANTXXIII/4 in April 2006 (Gohl 2007). The core location is 62°15.50′S and 58°46.34′W, at 446 m of water depth in the NE part of Maxwell Bay between Potter and Marian coves off King George Island (Fig. 1). The box core was principally taken to retrieve an undisturbed surface sample but it also retrieved 60 cm of the uppermost sediment column. PS69/335-2 recovered a 928 cm long sediment core consisting largely of muddy sediments. The cores were cut into 1 m pieces (PS69/335-2) and subsequently stored at 4 °C.

Physical properties were measured at 1 cm steps aboard RV *Polarstern* with a multisensor core logger (MSCL, GEOTEK Ltd., UK) on the un-split 1-m long core sections. Analyses included non-destructive determinations of core temperature and diameter, wet-bulk density via gamma-ray attenuation, P-wave velocity and magnetic susceptibility.

The cores were split, visually described and x-rayed at Alfred Wegener Institute for Polar and Marine Research (AWI) laboratories in Bremerhaven and List. Magnetic susceptibility was then measured in high resolution (1 cm steps) using an F-sensor (HRMS, volume-corrected).

For granulometry analyses the gravity core was subsampled every 1 cm. Carbonates were removed with acetic acid and organic matter was removed using hydrogen peroxide. Both chemicals were allowed to react overnight then washed using deionized water (twice after each treatment). Sodium polyphosphate was then added to each sample for dispersion and they were left overnight on a shaking table. In order to avoid adverse effects on the bulk samples that may result from chemical removal (e.g. destruction of clay minerals), biogenic opal was not removed. The mean content of bio-opal is around 7% (see results section below).

Grain-size measurements were carried out using a Cilas 1180L laser-diffraction particle sizer (range 0.04–2500 μm). Data processing and statistical calculations are based on self-programmed routines and the software GRADISTAT (Blott & Pye 2001). The Cilas particle sizer measures grain volume rather than grain mass. Thus, grain-size percentage given in the following text and in the figures relates to volume percent. Statistical parameters always refer to geometric (modified) Folk & Ward (1957) graphical measures (cf. Blott & Pye 2001).

Our core chronology is based on 5 Accelerator Mass Spectrometry (AMS) ^{14}C data that were measured at the 'Leibniz Labor für Altersbestimmung und Isotopenforschung' at Kiel University (Table 1). In addition, ^{210}Pb analyses were carried out for 8 samples obtained from both the gravity and the giant box corers at the Institute for Chemistry and Biology of the Marine Environment (ICBM) at Oldenburg University in order to accurately date the core top. Radionuclide concentrations were determined by gamma spectrometry using a Ge detector (GWC 2522-7500 SL, Canberra Industries Inc., Meriden, USA) and evaluated with the software GENIE 2000 3.0 (Canberra Industries Inc., Meriden, USA). Dry bulk density data, calculated from pycnometer and water-content data, was used to correct ^{210}Pb excess activity for compactional effects. Sediment ages t_{sed} were then calculated according to the constant rate of supply (CRS) model (Appleby & Oldfield 1978; Durham & Oliver 1983) with slight modifications, using:

$$t_{sed} = \frac{-1}{\lambda_{210_{Pb}}} \cdot \ln\left(1 - \frac{\sum A_m}{\sum A_\infty}\right) \quad (1)$$

where $\sum A_m$ represents the integrated and corrected ^{210}Pb activity from the surface to the depth m, $\sum A_\infty$ is the total integrated concentration of unsupported ^{210}Pb expressed in Bq cm^{-3} and λ is the radioactive decay constant of ^{210}Pb (0.3114 a^{-1}).

For geochemical analyses, subsamples were taken every 5 cm. Total organic carbon (TOC) was measured with a LECO CS-analyser after the inorganic carbon had been removed from the samples with concentrated HCl acid.

An automated leaching method was applied for the determination of the opal content, providing a relative analytical precision between 4% and 10% (Müller & Schneider 1993). The opal concentration was calculated with 10% water in the opal.

XRF analyses were carried out using a conventional wavelength dispersive X-ray fluorescence spectrometer (Philips PW 2400) equipped with a rhodium tube. Reproducibility of the results was checked by octuple measurements of in-house (DR-BS, Loess) and international standard samples (BIR-1, JA-2, JB-1, PACS-1, JG-1a) revealing a pooled estimated precision of $<1\%$ for major components. The mean relative error of this method ranges from -1.6% to 1%. Both TOC and elemental concentrations were corrected for salt contents, based on the water content of each sample and an estimated salt concentration of the pore water of 35‰. Furthermore, high-resolution colour scans (150 px cm^{-1}) were carried out with a 3-CCD line scan camera (CV-L105, JAI) and evaluated with the Software LineScan V 1.1 (den Burg, Texel, Netherlands). Statistics were carried out using MatLab 7.0 (The Math Works Inc., Natick, MA, USA).

Results

Characteristics of core PS69/335-2

Core PS69/335-2 consists of silty clay with comparably few scattered ice-refted debris (IRD). (Average values for the 63–2500 μm fraction are 1.1% with only 4 samples exceeding 5%. One of these samples – 612 cmbsf (centimetres below sea floor) – includes a tephra layer.) There are traces of moderate bioturbation throughout the core. The core is mainly homogenous with the exception of the lowermost 74 cm that shows flaser bedding (Fig. 2).

Magnetic-susceptibility values vary around a mean of 9700 ($\times 10^{-6}$ SI units) with minima at *c.* 7000 and maxima at *c.* 11 000 (Fig. 2). A cyclic pattern with a periodicity of 37 cm is visible. At 612 cm the HRMS values rise to 16 800 due to a volcanic ash layer.

Stratigraphy

Carbonate for AMS radiocarbon analyses was picked from five 1-cm thick samples from core

Table 1. *Details of the AMS ^{14}C age determinations. Year AD shows the age value after correction for the local reservoir effect (1100 years) and after calibration (see text). These data form the basis for the age model used in this study*

Lab Code	Core number	Sample depth (cm)	Conventional age (a BP)	Year AD	Years BP
KIA 33872	PS 69/335-2	109–111	1660 ± 25	1751	199
KIA 33873	PS 69/335-2	258–259	1800 ± 25	1603	347
KIA 33874	PS 69/335-2	499–500	2195 ± 25	1294	656
KIA 33875	PS 69/335-2	603–604	2280 ± 25	1233	717
KIA 33876	PS 69/335-2	733.5–734.5	2645 ± 25	847	1103

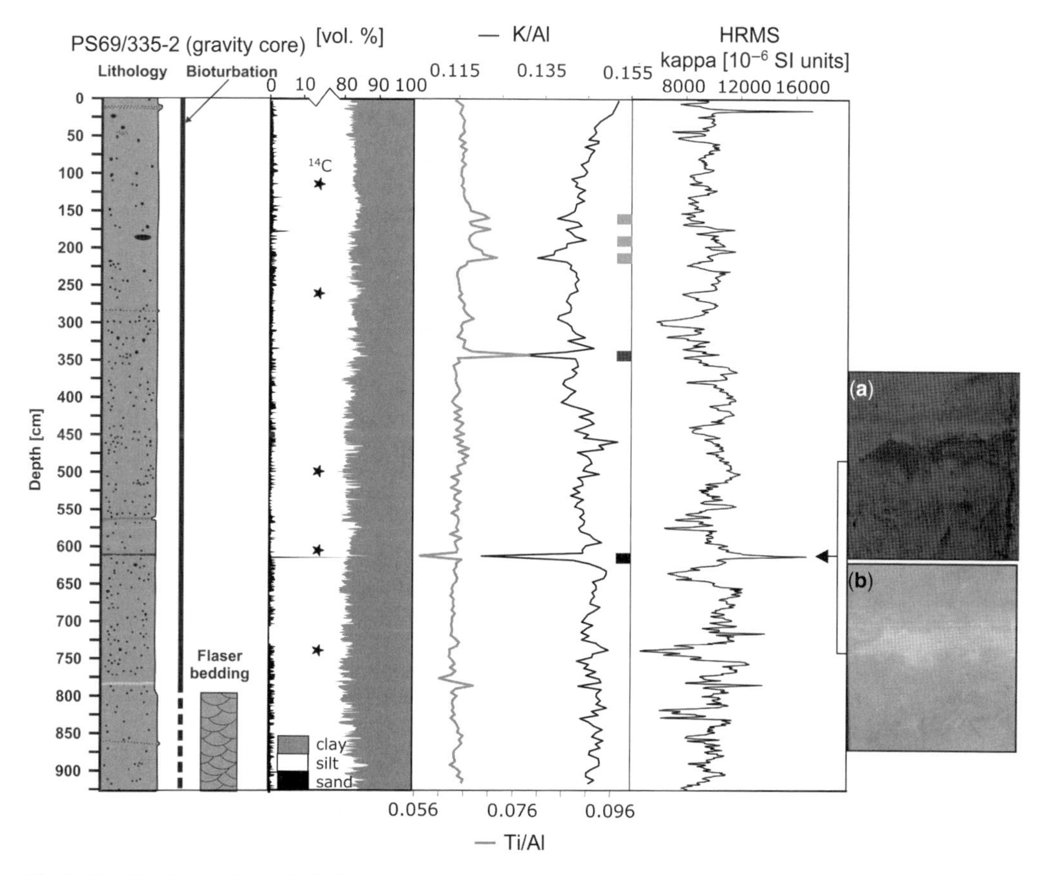

Fig. 2. Details of core PS69/335-2: lithlogy and structure; clay-silt-sand proportions and location of the AMS radiocarbon dated samples (stars); K/Al and Ti/Al records revealing positions of tephra-rich layers (light-grey squares, Cluster 2; dark-grey squares, Cluster 3; black squares, Cluster 4; no square, Cluster 1, see text); high-resolution record of the volume-corrected magnetic susceptibility (HRMS). (**a**) Photo- and (**b**) radiograph of the tephra layer at 612 cm core depth.

PS69/335-2. Benthic and planktic foraminifers were not present in suitable numbers, which indicates a high degree of dilution. Even shells of mussels or other organisms were very sparsely distributed in the core. Thus, the upper four AMS radiocarbon dates are based on shell fractions of small mussels; the lowermost sample is a mixture of benthic and planktic foraminifers and mussel-shell parts.

Due to the unusually low radiocarbon concentrations in Antarctic waters and the non-uniform distribution of water masses with certain radiocarbon concentrations, the local reservoir varies (e.g. Domack 1992; Gordon & Harkness 1992; Domack *et al.* 2005, 2001). Based on previous work in this area and our own dating of the gravity core (AMS and ^{210}Pb), we estimate the local reservoir to be 1100 years. The age data were first corrected using the local reservoir and subsequently calibrated using the MARINE04 dataset in the Calib 5.0.2html software (http://calib.qub.ac.uk/calib/) (Stuiver & Reimer 1993; Stuiver *et al.* 1998; Reimer *et al.* 2004).

The AMS data were then used to calculate linear sedimentation rates (LSR, Bruns & Hass 1999) and to construct an age model (Fig. 3). The core spans the past 1750 years with an average linear sedimentation rate of 0.74 cm a^{-1} (range is between 0.34 and 1.70 cm a^{-1}). Although the sedimentation rate varies, there are no age reversals that would indicate significant reworking.

The results of the radiocarbon analyses are confirmed by ^{210}Pb-analysis results obtained from selected samples of the cores PS 69/335-1 and PS 69/335-2. In order to avoid dating errors due to losses and disturbances of surface sediments from the gravity core, a depth correction based on ^{210}Pb excess and high-resolution reflectance data

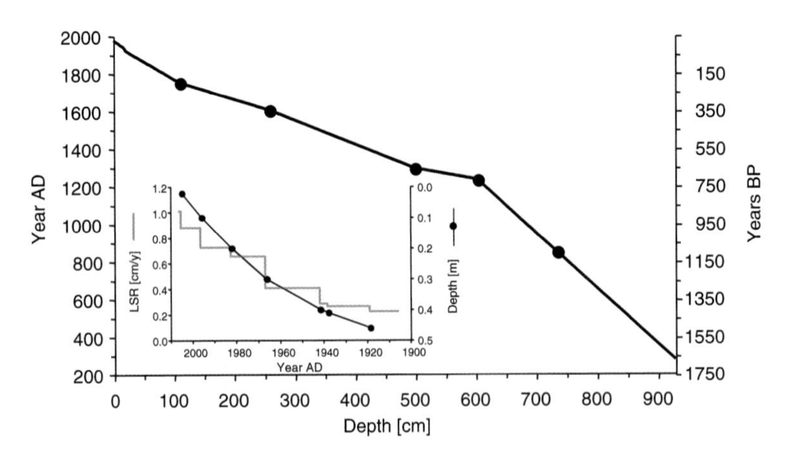

Fig. 3. Age–depth diagram of core PS69/335-2 (black dots mark radiocarbon-dated samples, see also Table 1). Inset: depth v. CRS age; sedimentation rate v. year AD of cores PS 69/335-1 and 2 according to ^{210}Pb dating method (see also Fig. 4).

(b*, linescan) from both cores was conducted. Core correlation reveals a loss of the upper *c.* 24 cm during the coring process and further disturbances within the following *c.* 20 cm, which was incorporated into the CRS age and sedimentation rate determination (Fig. 4).

Since *c.* AD 1900 (50 a BP) a gradual increase in the sedimentation rate from *c.* 0.22 cm a^{-1} to 1.00 cm a^{-1} has been observed. Highest growth rates are calculated between AD 1967 and AD 1982 (17 and 32 a AP) (0.017 cm a^{-1}), AD 1996 and AD 2005 (46 and 55 a AP) (0.016 cm a^{-1}) and especially during 2005 (0.13 cm a^{-1}). Data addressing the period after AD 1973 (23 a AP) are derived from the box corer.

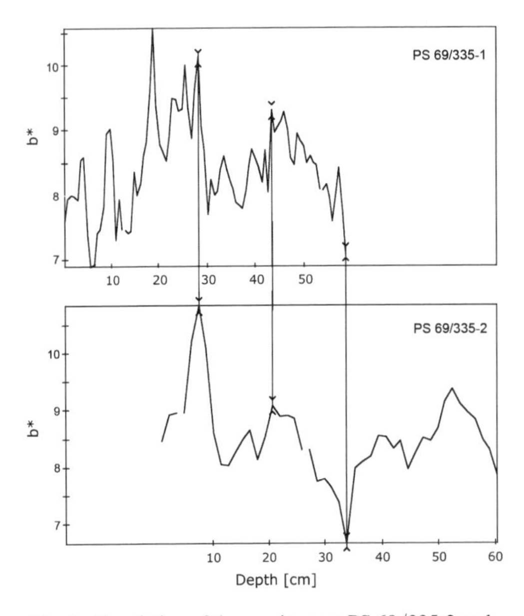

Fig. 4. Correlation of the gravity core PS 69/335-2 and the giant box core PS 69/335-1 using high-resolution reflectance (b*) data. Note that in the disturbed gravity core a constant offset of 24 cm is reached after *c.* 20 cm.

Granulometry

The principal mode of the sediments of core PS69/335-2 occurs in the coarse silt at 28 μm (minimum = 15 μm; maximum = 69 μm), whereas the total mean grain size is at 8.6 μm (minimum = 6 μm; maximum = 17 μm). This indicates that the grain-size distributions are significantly fine-skewed (mean = 0.94; minimum = 0.84; maximum = 1.05) and thus poorly sorted (mean = 4.19; minimum = 3.48; maximum = 5.33). About 85% of the samples are bi- or polymodal. Polymodality is indicative of different depositional processes that occur simultaneously. They form a key diagnostic tool to reconstruct conditions and energy of the depositional environment (Folk & Ward 1957; Beierle *et al.* 2002; Sun *et al.* 2002). Weltje & Prins (2003) suggest it is necessary to decompose polymodal and not-lognormal distributed grain-size distributions into their endmembers because the statistics calculated from the composite samples may be misleading. Since there is an indefinite number of combinations to produce a given polymodal sample and since it is not clear how the different processes interact, we use the composite samples for the study presented here and discriminate different sediment classes instead.

We identify two general classes of sediments that clearly represent two distinct depositional environments. The conditions seem to switch

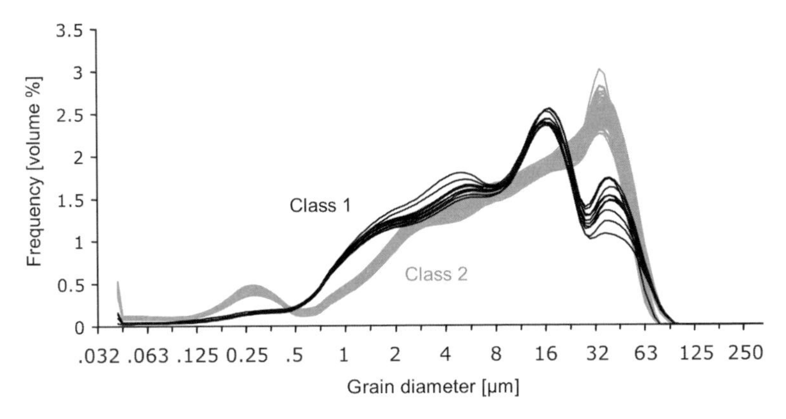

Fig. 5. Example of Class 1 and 2 samples: grain-size frequency distributions of samples 0 to 100 cm core depth.

between those that produce these sediment classes, but there is no clear or consistent transition between them. As an example, Figure 5 shows the frequency distributions of all samples from 0 to 100 cm core depth. The two classes are easily distinguishable. Class 1 accounts for 43% of the total number of samples and is generally characterized by fine-skewed samples with the principal mode between 14 and 18 μm. Most of these samples show two more modes: one between 5 and 6 μm and one slightly below 40 μm. If either one of the latter values forms the second mode, the other value forms the third mode. The coarser mode is usually clearer shaped whereas the finer mode is less pronounced. In this class, the coarsest grain size does not exceed 110 μm. The mean grain size is 7.68 μm.

Class 2 accounts for 57% of the total number of samples. All samples in this class are fine skewed, but are generally coarser than Class 1 sediments. The principal mode in this class ranges between 28 and 37 μm with a mean grain size of 36.35 μm. Although further modes could be calculated mathematically, the general characteristic of this class of samples is that they are largely unimodal. The coarsest grain size measured in this class is at 116 μm, thus slightly coarser than in Class 1. The mean grain size amounts to 9.28 μm, which is also slightly coarser than in Class 1. In detail, the first mode of Class 2 samples corresponds largely to the second mode of Class 1, whereas the first mode of Class 1 has no counterpart in Class 2. All modal values are largely homogenous and show only minor deviations within their classes.

Figure 6 (top panel) shows the mean grain sizes of both classes plotted separately in the same graph. Although both classes occur all over the core, it is evident that the finer Class 1 is more frequent between AD 600 and 1300 (1350 and 650 a BP) while Class 2 dominates the upper 700 and the lower 300 years of the core. When compared to the bulk set of samples, both curves influence the profile of the grain-size record in different aspects. For example the period from AD 680–1200 (1270–750 a BP) shows a slight fining trend in Class 2 samples whereas there is a slight coarsening trend in Class 1 samples. Furthermore, the mean grain-size parameter proves to be a meaningful parameter even for the polymodal samples; coarsening and fining trends are well matched in both classes (the more or less unimodal Class 2 and the largely polymodal Class 1).

Figure 7 ('MaxB') shows the mean grain size v. calendar years AD. From the base of the core until AD 630 (1320 a BP) the sediment shows a fining by about 3 μm. Until about AD 1400 (550 a BP), the overall mean grain size remains at a finer level with some finer or coarser intervals. Following this, a stepwise coarsening occurs with coarsest values around AD 1700 (250 a BP). A strong fining between AD 1800 and AD 1830 (150 and 120 a BP) is then followed by a weaker fining trend until the core top.

Geochemistry

In general, the geochemical proxies in the Maxwell Bay sediment record are mostly homogenous and are characterized by low variability. However, the K/Al and Ti/Al ratios (Fig. 2) show some notable variations (e.g. at 137, 142, 197, 332, 612 and 792 cm). Using a hierarchical cluster analysis (single-linkage clustering) yields four principal clusters (Fig. 8). To avoid the constant sum constraint of compositional data (Aitchison 1982), we normalized the elements using titanium which shows the highest correlation coefficient (0.7632). Cluster 1 characterizes 'average' conditions and represents most of the samples (>95%). Cluster 2 shows increased Ti and decreased K values;

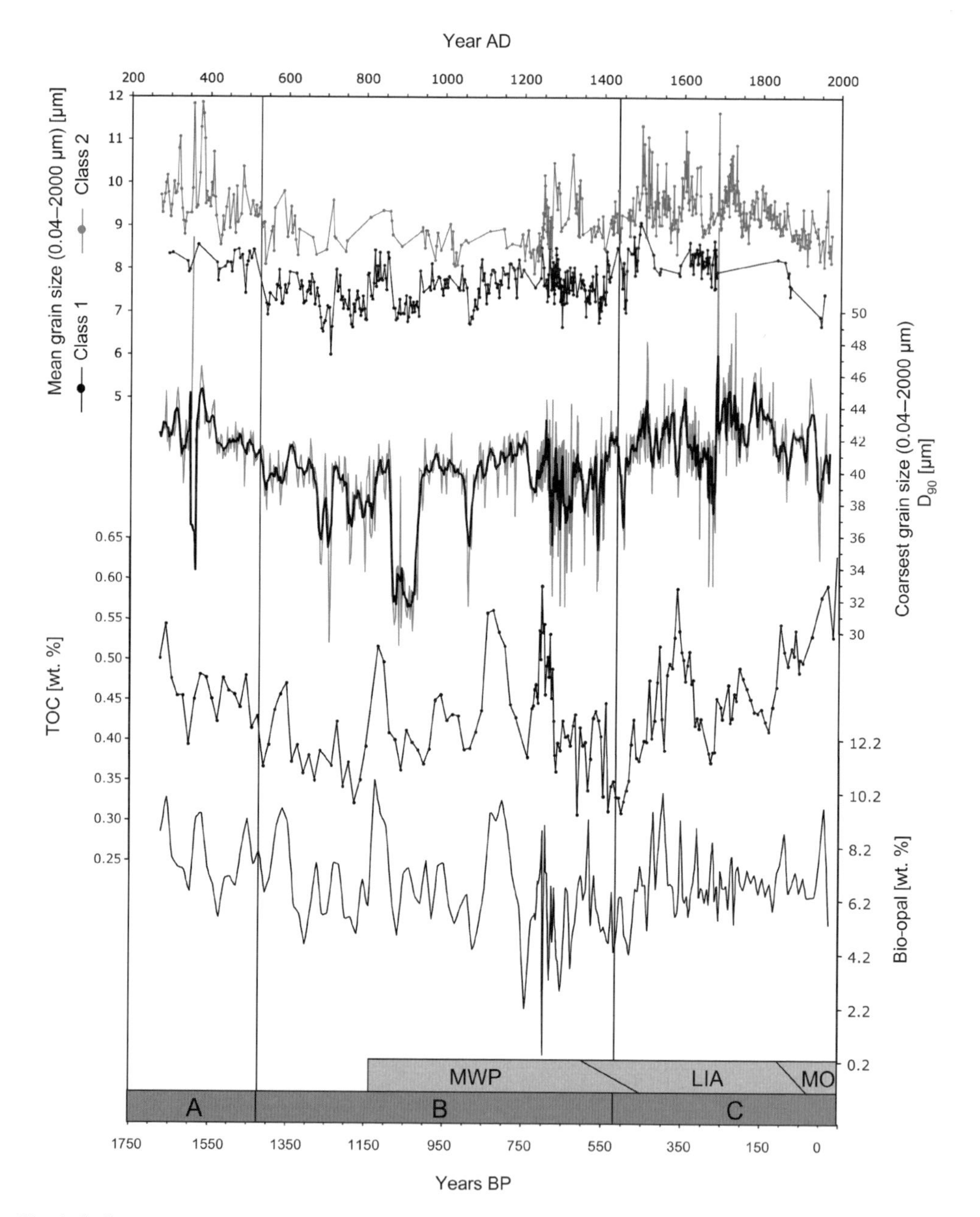

Fig. 6. Sediment-core parameters v. year AD and years BP. From top: Mean grain size of Class 1 and 2 samples depicted separately in the same diagram; coarsest grain size expressed as D90 (black line: 5 pt. running mean); TOC and bio-opal. Granulometric units A-C are shown at the bottom of the figure along with frequently used limits of the MWP, LIA and MO. Tephra layers were removed from the records to avoid bias.

Cluster 3 shows the same relationship as Cluster 2 but with much higher values and Cluster 4 indicates significantly decreased K and Ti values. Clusters 2 to 4 represent a significant change in depositional processes that we interpret here as tephra layers. These include the tephra layer identified visually at 612 cm as well as four other potential ash-bearing horizons (see Fig. 2) which are not detected in the radiographs.

TOC concentrations range between 0.26% and 0.58% with an average of 0.39% (Fig. 6). To investigate the downcore palaeo-productivity trend using

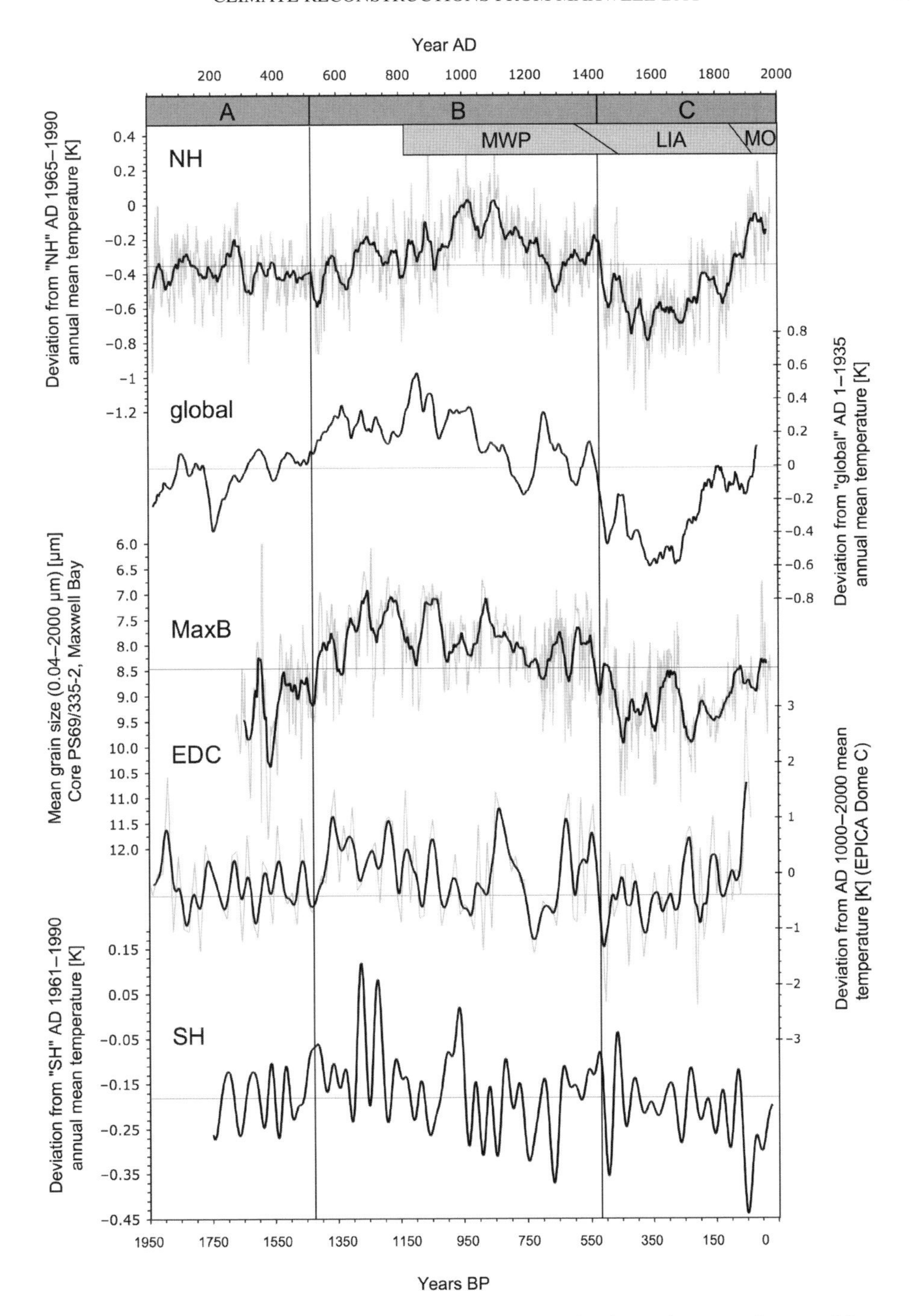

Fig. 7. Temperature reconstructions expressed as deviations from a calculated mean (see respective *y* axes) from different areas compared to the grain-size record from Maxwell Bay. Thin horizontal lines mark the bi-millennial average value. From top: Northern Hemisphere (NH, Moberg *et al.* 2005), global (Loehle 2007; Loehle & McCulloch 2008); mean grain size record from Maxwell Bay (MaxB, this study, tephra layers were removed, note direction of *y* axis), Antarctica: EPICA Dome C (EDC, Jouzel *et al.* 2007) and Southern Hemisphere (SH, Mann & Jones 2003). Black lines are 31-year (NH, gloabal, MaxB, EDC) and 40-year (SH) running means. Locations of granulometric units A–C and frequently used limits of the MWP, LIA and MO are shown on top.

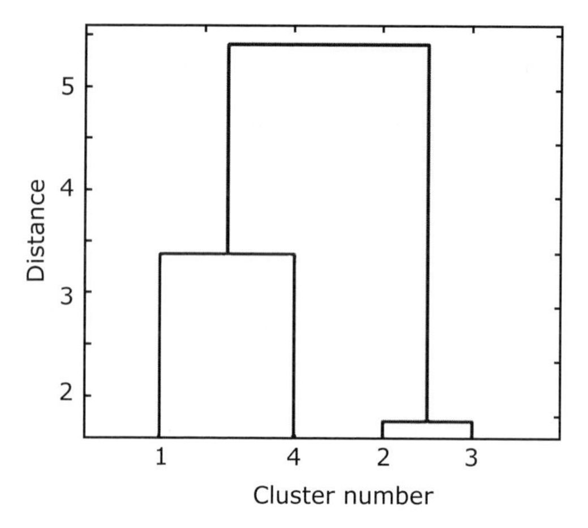

Fig. 8. Cluster tree of hierarchical cluster analysis using Ti-normalized data (corr. coeff. = 0.7632).

TOC content, we first removed the tephra layers (clusters 2–4) in order to prevent misinterpretations. Periods with higher amounts of TOC are centred at <AD 550 (1400 a BP), *c.* AD 850 (1100 a BP), AD 1100 to 1300 (850–650 a BP), AD 1600 (1350 a BP), and from AD 1880 (70 a BP), respectively.

The origin of the granulometric signal

Near-shore glaciomarine sediments reflect the complex interplay between glaciogenic, oceanographic, meteorologic and physiographic controls. Hence it can be assumed that grain-size characteristics carry important information on the dominant sedimentary processes. It is therefore critical to discriminate the predominant processes and interpret the most likely combination of controls to explain the suite of signals preserved in the sediment. It is essential that we understand where the sediment derives from, how and why it was transported to the core location and which processes and controls were involved.

Unfortunately, there exists no detailed information on the hydrography of Maxwell Bay and its tributaries. However, on the basis of limited monitoring and observational evidence, it has been assumed that there is a predominant clockwise circulation of the surface waters since inflowing surficial water from either east or west is deflected to the left. This is supported by satellite images which show that sediment-rich water (coloured brown) is deflected to the left when exiting tributary fjords such as Potter and Marian coves (e.g. Vogt & Braun 2004). These turbid waters are formed within the tributary fjords through meltwater discharged by the tidewater glaciers, through sediment resuspension (Klöser *et al.* 1994; Yoon *et al.* 2000; Khim *et al.* 2007) and through surficial snowmelt runoff (Yoon *et al.* 1998). Flowing east within Maxwell Bay these turbid waters remain close to the coast where they are responsible for high sediment accumulation within the upper 200 m of water depth. Here, the slope is still gentle before it steepens down to the bottom of Maxwell Bay (Yoon *et al.* 2000). Due to the location of core PS69/335-2 southwest off the mouth of Potter Cove in combination with the Coriolis-controlled clockwise rotation of the surface water masses within Maxwell Bay, we assume the sediments at the core location are principally derived from Marian Cove and Collins Harbor and, to a lesser degree, from other areas further upstream and Potter Cove slightly downstream. Due to the high discharge of meltwaters via the tributary fjords into Maxwell Bay, Domack & Ishman (1993) postulated an estuarine circulation pattern for this area. Hence, at 460 m water depth there may also be material that is transported from the SE via deeper waters flowing in from Bransfield Strait.

The granulometric data suggest that there are two basic modes of sediment transport and/or deposition (Class 1 and 2). Figure 6 shows both classes (here shown as the mean grain size) v. time. It is clear that the scatter of data of the mean grain size (standard deviation STDV = 1.05) is the result of switching between the classes rather than the result of unsteady conditions (Class 1 STDV = 0.48; Class 2 STDV = 0.62). The grain-size fraction <10 μm is regarded as the permanently suspended (i.e. largely not bedload-transported) fraction, although part of this fraction might be transported in the form of aggregates due to the cohesive behaviour of the fraction <10 μm (cf. McCave *et al.* 1995). Because of the shape of the overall grain-size frequency distributions, the importance of the fine-fraction <10 μm for this study and the fact that detailed current-speed reconstructions are not the focus of this study, we emphasize the bulk grain-size distributions (at least down to 0.5 μm) rather than focusing on the sortable silt fraction (10–63 μm, McCave *et al.* 1995). Initiation and transport of the fraction <10 μm is predominantly controlled by the shear velocity of the current (Middleton 1976). The subpopulations centred at 16 μm and 40 μm modal size represent the usual bedload fraction (the rolling and saltating fractions).

Class 1 and Class 2 sediments have modes centred at 35 μm and, together with variations in the coarsest grain size (e.g. D90, Fig. 6), reflect the energy of the transporting medium. For Class 2 the principal mode is around 35 μm. In Class 1, however, the principal mode is centred at 16 μm with the secondary mode at the 35 μm. A general deceleration of the sediment-transporting currents does not explain the occurrence of a finer mode and the significant weakening of the coarser mode, however. If currents decelerated there would be a lateral signal (i.e. a shift to fine) in the coarser modus, which is not the case. Hence, it is inferred that Class 2 characterizes the basic conditions and that Class 1 occurs in the form of rapid discharge events (i.e. turbid meltwater plumes) whenever finer materials are released near to the core location.

It follows therefore that between AD 600 and 1400 (1350 and 550 a BP) the number of finer-material release events was higher than before and after this period. However, both classes also show some internal variability. Between AD 600 and 1400 (1350 and 550 a BP), Class 1 sediments were not only predominant but both Class 1 and 2 samples also exhibit a finer mean grain size than both before and after this period (Fig. 6). This could indicate that current speeds were slightly lower during this period, thus allowing a slightly finer fraction to settle. The coarsest grain size (here expressed as the D90 value) which shows a drop by *c.* 2 μm during the same time intervals supports this interpretation.

Since the tributary fjords of Maxwell Bay form the source area of most of the sediments of core PS69/335-2, it is also very likely that the episodically discharged fine-grained sediments derive from there. Summer temperatures are as high as +2 °C (Arctowsky station) in this part of the South Shetland Islands which causes strong meltwater discharge during the austral summer season (Domack & Ishman 1993). Sea ice begins to rapidly break up in late October and Maxwell Bay remains ice-free from November to June (Yoon *et al.* 2000). Pack-ice formation starts in July and Maxwell Bay is usually covered with up to 60 cm thick sea ice until November (Yoon *et al.* 2000). Large meltwater streams transporting significant amounts of terrigenous sediment into Maxwell Bay via the tributary fjords tend to establish in late December (Yoon *et al.* 1998). We therefore suggest that the finer-grained samples of Class 1, in particular the mode centred at 16 μm, can be related to the summer discharge of sediment-laden meltwaters. The coarser mode reflects the largely uniform motion of the bottom currents at the core location. Class 1 samples thus represent summer-dominated conditions or conditions that are influenced by strong meltwater production. In contrast, Class 2 samples characterize winter-dominated periods, that is, periods with reduced summer meltwater production.

Variability of summer- and winter-type sedimentation

According to the data shown in Figure 6 (top) the following units can be defined. Unit A consists of the end of the core to AD 516 (1434 a BP) and is dominated by Class 2 sediments (Class 1/2 frequency ratio = 0.4) and increased mean grain-size values which indicates reduced summer melting and slightly faster bottom currents. Unit B (AD 516–1426) (1434–524 a BP) shows finer mean grain sizes and a dominance of Class 1 sediments (Class 1/2 frequency ratio = 1.9). Accordingly, Unit B represents a period in which summer melting processes were strong and the average bottom currents were slightly slower. However, some episodes of stronger and weaker bottom currents are clearly discriminated within Unit B, for example: slower than average currents between AD 868–932 (1082–1018 a BP); and faster than average currents between AD 803–856 (1147–1094 a BP). Unit C (AD 1426 until present) (524 a BP until today) is again dominated by Class 2 sediments (Class 1/2 frequency ratio = 0.3) indicating reduced summer meltwater discharge and stronger bottom currents. There are periods within Unit C in which Class 1 samples are very rare or even absent, a pattern which is not visible anywhere else except in the lower part of the core.

Climate change and sedimentation in Maxwell Bay

The sediment record from Maxwell Bay contains a detailed environmental record that reflects the interplay between glacier melting current speed, sea-ice and wind stress.

To investigate whether this signal reflects local, Antarctic, Southern Hemisphere, global or even a Northern Hemisphere signal, we compare the Maxwell Bay grain-size record to temperature reconstructions from Antarctica (EPICA (European Project for Ice Coring in Antarctic) Dome C temperature reconstruction, 'EDC': Jouzel *et al.* 2007), the Southern Hemisphere ('SH', Mann & Jones 2003), a global temperature stack ('global', Loehle 2007; Loehle & McCulloch 2008) and the Northern Hemisphere ('NH', Moberg *et al.* 2005) (Fig. 7). There have been many local and regional marine and lacustrine studies that include a large amount of discussion mainly on the regional climate development. However, this chapter aims at placing the Maxwell Bay results within a large climatic frame. We therefore utilize temperature reconstructions that represent larger areas (see e.g. Bentley *et al.* 2009 for a comparison of regional climate records and a discussion on potential forcings).

The Maxwell Bay record extends back to AD 275 (1675 a BP) and covers climate phases commonly identified in the Northern Hemisphere records such as the Dark Ages Cold Period (DACP, *c.* AD 200–900 or 1750–1050 a BP), the Medieval

Warm Period (MWP, *c.* AD 900–1350 or 1050–600 a BP), the Little Ice Age (LIA, *c.* AD 1350–1900 or 600–50 a BP) and the Modern (Climate) Optimum (MO, from AD 1900 or 50 a BP). The triggers for these climate oscillations are complex.

Usually these climate phases are related to a 'modern' mean temperature and are therefore defined as periods above or below this value (e.g. Bradley *et al.* 2003; Hunt 2006). However, the timing and magnitude of these late Holocene climate fluctuations (in particular the MWP and LIA) are controversial since the available temperature reconstructions appear to be heterogeneous, lacking temporal and spatial correspondence, and they seem dependent on the proxies used and on the number of records from different places that are considered (e.g. IPCC 2001; Bradley *et al.* 2003; Mann & Jones 2003; Jones & Mann 2004; Hunt 2006; IPCC 2007; Loehle 2007; Esper & Frank 2008; Loehle & McCulloch 2008). Since there is no agreement on the dates for the climate phases, we use the mean value calculated over the time shown in Figure 7 as the reference line that indicates whether the data are above or below average. For better comparability, the records shown in Figure 7 were recalculated to a 31-year running mean with the exception of the SH reconstruction of Mann & Jones (2003); this was left in the original 40-year running mean. Thus, maxima and minima might be slightly less pronounced when compared to the unsmoothed data.

The Medieval warm period and before

The Maxwell Bay sediment core shows the coarsest sediments of the entire record between AD 300 and 400 (1650–1550 a BP), thus indicating colder conditions. At the same time the global temperature reconstruction shows a minor but clearly visible temperature drop. Whether or not these signals are related to the DACP is uncertain since the record starts only after AD 200 thus providing no comparison to earlier conditions.

The timing of the MWP is usually related to the Viking era (*c.* AD 800–1100 or 1150–850 a BP) and one or two centuries beyond (Lamb 1965, 1995; Bradley *et al.* 2003).

Although Moberg *et al.* (2005) do not explicitly name the MWP they do present evidence for exceptional warm conditions between AD 1000 and 1100 (950 and 850 a BP) (Fig. 7). While the peak warmth occurs around AD 1000 and 1100 (950 and 850 a BP), the onset of the warming trend starts as early as *c.* AD 550 (1400 a BP) when related to the mean temperature of the past two millennia. The temperature record then remains largely above the mean until *c.* AD 1430 (520 a BP). There are a few cold spells (the most prominent centred at AD 640 and 1300); however, a longer term temperature decrease below the mean does not occur before AD 1430 (520 a BP). For this study, it seems appropriate to set the boundaries for the warmer phase that includes the peak MWP to *c.* AD 550 and 1430 (1400 and 520 a BP) (Unit B, see above). Within these boundary dates, all the other records shown in Figure 7 indicate warmer conditions.

The global record matches these dates almost exactly, revealing temperatures that are almost exclusively above the 2000-year mean. Two cold spells occur in this record at AD 1200 and 1360 (750 and 590 a BP). The period centred at AD 1200 (750 a BP) is the coldest of the preceding millennium.

The Maxwell Bay sediment record suggests summer-dominated conditions exactly within the boundary dates. With only two exceptions the (mean) data are finer than the average values; the preceding and following centuries are all coarser than this average. The two exceptions are almost synchronous with those that occur in the global record at AD 1200 and 1360 (750 and 590 a BP).

The Antarctic ice-core record is more complicated. There is a clear shift to higher temperatures at AD 550 (1400 a BP) after more than four centuries of lower mean temperatures. Following this, there is an overall decreasing trend towards an ultimate minimum of the 2000-year record after AD 1430 (520 a BP). The two cold episodes at AD 1200 and 1360 (750 and 590 a BP) that are clearly visible in the global and the Maxwell Bay records have synchronous counterparts in the EDC record. The cold phase centred at AD 1200 (750 a BP) is the coldest of the preceding 1200 years.

The most problematic in this context appears to be the Southern Hemisphere temperature reconstruction. The scatter of data around the mean value is relatively large; there appears to be no clear signal in this record.

Figure 7 shows that the time period from AD 550 to 1430 (1400 to 520 a BP), sedimentary Unit B, covers the highest average temperatures of the past two millennia with the likely exception of the post-1980s (see discussion on this issue, e.g. IPCC 2001; Bradley *et al.* 2003; Mann & Jones 2003; Jones & Mann 2004; IPCC 2007; Loehle 2007; Loehle & McCulloch 2008; Mann *et al.* 2008). The Northern Hemisphere record reveals that temperatures peak during the middle of the MWP phase. All the other records reveal curves that are 'left-skewed,' meaning the peak medieval warmth occurred earlier in these records. This is most pronounced in the EDC record which is the southernmost of the records, but the signal is also visible in the SH reconstruction. The global and the Maxwell Bay records are clearly but only weakly 'left-skewed'. Since the 'skewness' appears to increase from north to south, it is likely that Antarctica and the Southern Hemisphere are the sources of the early warming signals during the medieval times. Indeed, the NH reconstruction is based on proxies that mostly originate from the higher NH; only 2 of 18 proxy records come from slightly south of the tropic of Cancer. The global reconstruction is based on 18 proxy records spread over the globe. The majority of data originate from north of the tropic of Cancer; however, there are more data from the tropics and even 3 records from the SH (but still north of the tropic of Capricorn). It is questionable whether the observed offset is caused by: a more southern data source; the avoidance of tree-ring data (Loehle 2007; Loehle & McCulloch 2008); or an unknown bipolar or hemispherical see-saw effect that operates not only in glacial/deglacial times (Broecker 1998; Stocker 1998) but also in the interglacial (Holocene) at a pace of about 400 years. Ljung & Björck (2007) have proposed that such a see-saw process was in operation, but simple adjustments of atmospheric heat transport might lead to similar results (Steig 2006; Wunsch 2006). In this context it is interesting that both the sediment record from Maxwell Bay and the EDC record clearly indicate warming around AD 1100 (850 a BP) when the NH record shows peak warming. It is uncertain whether the NH warming has influence on Antarctica or whether this is simply a coincidence.

After the peak warmth of the MWP, temperatures drop to a minimum around AD 1300 (650 a BP) in the NH. However, all the other records indicate a strong cooling event around AD 1200 (750 a BP) while the NH only shows a minor temperature fluctuation. We suggest that this cooling signal originated in Antarctica where it had the strongest impact. It probably then propagated north producing a clear signal in the sediment core from Maxwell Bay, causing an even stronger signal in the global temperature record and probably initiating the cooling trend in the NH record.

The sedimentary record of Maxwell Bay and the global temperature curve show greatest similarities, although slight shifts due to possible errors in the age model of the Maxwell Bay record can be expected. Evidently, our interpretations of summer- and winter-dominated (i.e. warmer and colder) periods fit to the global temperature record. They also fit into the frame of regional climate reconstructions (e.g. James Ross Island, Björck *et al.* 1996; Palmer Deep/Andvord Bay, Domack *et al.* 2003), however they are slightly different to results from other places (e.g. South Orkney Islands, Jones *et al.* 2000; Signy Island, Noon *et al.* 2003; see also Bentley *et al.* 2009). The MWP on King George Island (Unit B) can therefore be characterized as a phase of intense summer warmth and associated intense meltwater production (reflected by the dominance of Class 1 sediments). The TOC and bio-opal records (Fig. 6) that are assumed to be indicative of bioproductivity do not, however, follow this interpretation. Ice-free waters are essential for algae to grow. Summer-dominated periods should therefore reveal an increase in the TOC/bio-opal values. Although inorganic nutrients are sufficiently present, the amount of TOC appears to be quite low throughout the core suggesting restricted bioproductivity. This is likely to result from light attenuation through the formation of turbid waters by ice-melting processes (Schloss *et al.* 2002 referring to Potter and Marian coves; Hernando & Ferreyra 2005). The Maxwell Bay TOC and bio-opal records show a general decrease at the onset of the MWP, synchronous with the beginning of the summer-dominated period. Most of the samples remain below average until the 16th century AD (4th century BP). We interpret here that the strong summer discharge of turbid waters limits the light available for plants to grow. We suggest that this effect is likely to be even more adverse than a prolonged sea-ice season, since the light limitation occurs during the most productive phase of the year. As a result, bioproductivity is likely to have decreased during warm phases with strong meltwater input from land. Turbid water of either meltwater or resuspension origin is reported from Potter Cove to occur under strong wind conditions (causing resuspension) in the summer season (meltwater) (e.g. Klöser *et al.* 1994; Schloss *et al.* 2002). Pakhomov *et al.* (2003) show that even higher organisms (here, tunicates) are significantly affected by suspended matter. In this case, the mass death of salps was caused by dense suspension load in Potter Cove. We therefore argue that TOC content in the Maxwell Bay record is indicative of the turbidity of the coastal waters rather than of the potential bioproductivity. Consequently, the grain-size and TOC records show opposing trends until *c.* AD 1400 (550 a BP) with TOC maxima likely occurring during episodes of reduced summer meltwater production. Significantly decreased TOC values occur between AD 1290 and 1530 (660 and 420 a BP) during a phase of high sediment accumulation, thereby further diluting the already weak signal.

The Little Ice Age until the present

Similar to the MWP and other climate phases mentioned, the existence and timing of the LIA is controversial (e.g. Lamb 1995; Bradley 2000; IPCC 2001; Nesje & Dahl 2003; IPCC 2007; Bentley *et al.* 2009). There is neither an agreed beginning nor an agreed end date of the LIA. The beginning is usually set between AD 1350 and 1500 (600 and 450 a BP) and the end is set between AD 1850 and 1920 (100 and 30 a BP). Based on the 2000-year averages used here, we place the beginning of the LIA to slightly after AD 1400 (550 a BP) (beginning of Unit C) and the ending of the LIA to slightly before AD 1900 (50 a BP).

Within this time period the NH temperature curve, the global temperature reconstructions and the Maxwell Bay sedimentary record plot below the average value. The EDC record remains largely below the average; however, there are two warmer episodes around AD 1710 and 1790 (240 and 160 a BP). The SH record shows no signal of the LIA. The global record suggests that the end of the LIA occurred at *c.* AD 1770 (180 a BP), which is the earliest termination signal of the records shown in Figure 7. EDC, the Maxwell Bay and the NH records all show a synchronous termination around AD 1860 (90 a BP). The Maxwell Bay record reveals a clear dominance of winter-type conditions during the LIA. This is in accordance with marine studies from the region (e.g. Palmer Deep, Domack *et al.* 2001; Bransfield Basin, Fabrés *et al.* 2000; Khim *et al.* 2002) but contrasts with results from lakes of the region (Bentley *et al.* 2009; Lee *et al.* 2009).

Until AD 1490 (460 a BP), the situation is balanced although the overall coarsening of the sediment is already heralding changing conditions. Between AD 1490 and 1610 (460 and 340 a BP), summer type sediments are very sparse. In the following until AD 1680 (270 a BP), a warmer episode is indicated by the sudden dominance of summer-type sediments. However, this episode is sharply terminated and summer-type sediments are very rare until the top of the core. We expected that the Maxwell Bay record would have shifted into the summer-dominated mode shortly after the peak cold of the LIA. However, the grain-size mean becomes finer possibly in response to climate warming and associated increasing melting processes. However, the sediments are still almost exclusively of the winter type, that is, they lack the finer fraction that is assumed to be linked to meltwater discharge even after AD 1870 (80 a BP). A similar signal of delayed warming after the LIA (compared to the MWP) occurs in a magnetic-susceptibility record from Bransfield Basin (Khim *et al.* 2002).

The mean grain size reaches a level that is comparable to MWP/LIA transition and there is a fining trend visible towards the top of the core, indicating that the bottom-current speed decreased. The TOC record (Fig. 6) which reflects the limiting effects of turbid waters on bioproduction rather than the potential bioproductivity responding directly to climate, reaches a maximum parallel to the bio-opal around AD 1600 (350 a BP) (i.e. well within the LIA cold phase). Meltwater flux and thus the turbidity of the coastal system was likely reduced at this time; so phytoplankton could take better advantage of shorter periods of ice-free waters than during the presumably longer periods of open waters and the presumably warmer conditions during the MWP. The similarity of the TOC and bio-opal records ceases after *c.* AD 1700 (250 a BP). Specifically, during the last 100 to 150 years increasing TOC contents are not matched by the bio-opal data. This is possibly the result of labile organic matter, which is absent in older sediments and which has not yet decomposed in the younger sediments. Conversely, increasing TOC values may reflect an enhanced primary production due to the recent warming trend in this region, which is not solely based on siliceous phytoplankton. Between 1991 and 1995 Moline *et al.* (2004) observed a consistent transition from opal-building diatoms to cryptophytes in near-shore waters of the WAP region. Claustre *et al.* (1997) and Lewitus & Caron (1990) relate the dominance of the smaller cryptophytes that are characterized by an opal-free plasma membrane and a much higher salt tolerance to increasing glacial meltwater input and the formation of low-salinity surface layers. Further micropaleontological and microbiological investigations off King George Island may confirm this interpretation.

Interestingly, the samples of the early 20th century belong to the finest Class 1 samples of the entire record. One possible explanation is that although warming occurred prior to AD 1970 (top of core PS69/335-2), it did not have significant impact on the summer melting processes. This explanation is supported by increasing TOC values that would be over-printed under meltwater conditions comparable to those of the MWP. For example, the glacier fronts in Marian Cove retreated far more than 500 m since 1956 (Park *et al.* 1998). It is therefore possible that meltwater discharge was reduced due to smaller glaciers or that the meltwater was routed differently (possibly closer to the coastline) due to changing hydrography and/or wind systems. A shoreward retreating estuarine front due to decreasing meltwater production (because of the beginning winter) has been described by Domack *et al.* (2003).

The global temperature record shows that from the end of the last LIA cold phase (i.e. from AD 1765, 185 a BP) until AD 1920 (30 a BP) temperatures remained at a certain level. Strong, possibly MO-related, warming commenced sharply at AD 1920 (30 a BP). This contrasts with the NH record that shows a strong warming trend until exactly AD 1920 (30 a BP) followed by a stagnation or even a cooling trend until the mid-1970s. The EDC record reveals a strong warming similar to the NH record whereas the SH record drops to the lowest temperatures in 2000 years. The climate records between AD 1850 and 1970 show major differences suggesting regional factors were important during this period. In this context, the Maxwell Bay record reveals more aspects of the NH record. The development of the late LIA from *c.* AD 1700 (250 a BP) until shortly before AD 1900 (50 a BP) is similar to the NH record, whereas the global record indicates post-LIA conditions from AD 1760 (190 a BP). The Maxwell Bay record (gravity core) ends at AD 1973 (23 a AP) after about 100 years of minor variations.

Summary and conclusions

We have investigated a 928 cm long gravity core (plus associated box core) from Maxwell Bay off King George Island, Antarctica. The core site is dominated by sediments from the tributary fjords entering Maxwell Bay, namely Potter and Marian coves and Collins Harbor. We have grouped the sediment of the core into two classes. Class 1 is characterized by two grain-size subpopulations. The coarser subpopulation represents the bedload fraction, whereas the finer subpopulation is interpreted to represent suspension load that was introduced into the system via meltwater. Since there are only significant amounts of meltwater in summer, it is suggested that Class 1 sediments characterize periods of intense summer meltwater production and thus warmer climate phases. Class 2 samples show the same coarse component but they lack the fine subpopulation. We suggest that these sediments indicate that summer meltwater production is less intense and thus colder climatic conditions prevail. Class 1 sediments are generally finer (mean grain size) than Class 2 samples, but the mean grain size (independent of this classification) also indicates the average bottom-current speeds were slightly higher during colder climate phases than during the warmer phases. Potential bioproduction at the core location and in the sediment source areas is only indirectly linked to bio-productivity proxies (TOC, bio-opal) since warm-phase meltwater discharge adversely affects bioproduction through light attenuation by turbid waters. Due to increased deposition of terrigenous fine sediment, the TOC signal becomes diluted during warmer climate phases and more focused during colder climate phases.

Comparisons with Antarctic, hemispherical and global temperature reconstructions reveal a clear signal of the MWP, the LIA and the post-LIA climate recovery. Class 1 sediments dominate the MWP and Class 2 sediments dominate the LIA. The Maxwell Bay record appears to moderate between the SH and NH records. It carries climate signals that are partly unique to either one of the hemispheres. Consequently, it best resembles the global temperature reconstruction. MWP conditions occur earlier in the SH than in the NH, suggesting that the source for the warming might also be in the SH. The timing of the beginning and the ending of the LIA appears to be largely synchronous between the hemispheres. LIA-type sediments are deposited at the core location until at least the 1970s; the sedimentary conditions therefore did not change at the end of the LIA. Generally finer sediments clearly mark the end of the LIA; however, no stronger meltwater influence can be detected. The conditions in the 1970s resemble those at the beginning of the MWP. According to the results presented here, the RRR warming cannot be detected in the Maxwell Bay record until AD 1975 (25 a AP).

If the present warming trend continues, two opposing effects can be expected that will have a significant impact on the ecosystem in this part of Antarctica. Negatively, there will be intense discharge of turbid meltwater into fjords and bays, adversely affecting the euphotic zone and life that depends on that. The positive effect for marine plants (but negative for e.g. large ice-associated predators; Clarke *et al.* 2007) of the complete disappearance of sea ice in the working area may occur within 80 years from now (winter temperatures need to rise by *c.* 3 °C; see Ferron *et al.* 2004). Consequently, the vegetation period will extend which is likely to lead to further changes of the ecosystem.

Supplementary material is available at http://www.pangaea.de. The authors are especially indebted to B. Austin and one anonymous reviewer for enthusiastic and critical reviews. We would like to thank the master and crew of RV *Polarstern* for support during Expedition ANT XXIII/4. HCH is grateful to J.P. Klages and J. Sobiech for their help with the grain-size analyses. S. Gauger processed the Hydrosweep data and J.P. Holm helped to produce Figure 1.

References

Aitchison, J. 1982. The statistical-analysis of compositional data. *Journal of the Royal Statistical Society Series B-Methodological*, **44**, 139–177.

Appleby, P. G. & Oldfield, F. 1978. The calculation of lead-210 dates assuming a constant rate of supply of unsupported 210 Pb to the sediment. *Catena*, **5**, 1–8.

Atkinson, A., Siegel, V., Pakhomov, E. & Rothery, P. 2004. Long-term decline in krill stock and increase in salps within the Southern Ocean. *Nature*, **432**, 100–103.

Beierle, B. D., Lamoureux, S. F., Cockburn, J. M. H. & Spooner, I. 2002. A new method for visualizing sediment particule size distributions. *Journal of Paleolimnology*, **27**, 279–283.

Bentley, M. J., Hodgson, D. A. *et al.* 2009. Mechanisms of Holocene palaeoenvironmental change in the Antarctic Peninsula region. *The Holocene*, **19**, 51–69.

Björck, S., Hakansson, H., Zale, R., Karlén, W. & Jonsson, B. L. 1991. A late Holocene lake sediment sequence from Livingston Island, South Shetland Islands (Antarctica), with paleoclimatic implications. *Antarctic Science*, **3**, 61–72.

Björck, S., Olsson, S., Ellis-evans, J. C., Håkansson, H., Humlum, O. & De lirio, J. M. 1996. Late Holocene palaeoclimatic records from lake sediments on James Ross Island, Antarctica. *Palaeogeography, Palaeoclimatology, Palaeoecology*, **121**, 195–220.

Blott, S. J. & Pye, K. 2001. Gradistat: a grain size distribution and statistics package for the analysis of unconsolidated sediments. *Earth Surface Processes and Landforms*, **26**, 1237–1248.

Blunier, T. & Brook, E. J. 2001. Timing of millennial-scale climate change in Antarctica and Greenland during the last glacial period. *Science*, **291**, 109–112.

Bradley, R. S. 2000. Climate paradigms for the last millennium. *Pages Newsletter*, **8**, 2–3.

Bradley, R. S., Hughes, M. K. & Diaz, H. F. 2003. Climate in medieval time. *Science*, **302**, 404–405.

Brandini, F. P. & Rebello, J. 1994. Wind field effect on hydrography and chlorophyll dynamics in the coastal pelagial of Admiralty Bay, King George Island, Antarctica. *Antarctic Science*, **6**, 433–442.

Broecker, W. S. 1998. Paleocean circulation during the last deglaciation: a bipolar seasaw. *Paleoceanography*, **13**, 119–121.

Bruns, P. & Hass, H. C. 1999. On sediment accumulation rates and their determination. *In*: Bruns, P. & Hass, H. C. (eds) *On the Determination of Sediment Accumulation Rates*. TransTech Publications, Zürich, **5**, 1–14.

Chung, H., Lee, B. Y., Chang, S.-K., Kim, J. H. & Kim, Y. 2004. Ice cliff retreat and sea-ice formation observed around King Sejong Station in King George Island, West Antarctica. *Ocean and Polar Research*, **26**, 1–10.

Clarke, A., Murphy, E. J., Meredith, M. P., King, J. C., Peck, L. S., Barnes, D. K. A. & Smith, R. C. 2007. Climate change and the marine ecosystem of the western Antarctic Peninsula. *Philosophical Transactions of the Royal Society B*, **362**, 149–166.

Claustre, H., Moline, M. A. & Prezelin, B. B. 1997. Sources of variability in the column photosynthetic cross section for Antarctic coastal waters. *Journal of Geophysical Research – Oceans*, **102**, 25047–25060.

Cook, A. J., Fox, A. J., Vaughan, D. G. & Ferrigno, J. G. 2005. Retreating glacier fronts on the Antarctic Peninsula over the past half-century. *Science*, **308**, 541–544.

Denton, G. H. & Broecker, W. S. 2008. Wobbly ocean conveyor circulation during the Holocene? *Quaternary Science Reviews*, **27**, 1939–1950.

Domack, E. W. 1992. Modern carbon-14 ages and reservoir corrections for the Antarctic Peninsula and Gerlache Strait area. *Antarctic Journal of the United States*, **27**, 63–64.

Domack, E. W. & Ishman, S. 1993. Oceanographic and physiographic controls on modern sedimentation within Antarctic fjords. *Geological Society of America Bulletin*, **105**, 1175–1189.

Domack, E., Leventer, A., Dunbar, R., Taylor, F., Brachfeld, S., Sjunneskog, C. & Party, O. L. S. 2001. Chronology of the Palmer Deep site, Antarctic Peninsula: a Holocene palaeoenvironmental reference for the circum-Antarctic. *The Holocene*, **11**, 1–9.

Domack, E., Leventer, A. *et al.* 2003. Marine sedimentary record of natural environmental variability and recent warming in the Antarctic Peninsula. *Antarctic Research Series*, **79**, 205–224.

Domack, E., Duran, D. *et al.* 2005. Stability of the Larsen B ice shelf on the Antarctic Peninsula during the Holocene epoch. *Nature*, **436**, 681–685.

Durham, R. W. & Oliver, B. G. 1983. History of Lake Ontario contamination from the Niagara river by sediment radiodating and chlorinated hydrocarbon analysis. *Journal of Great Lakes Research*, **9**, 160–183.

Esper, J. & Frank, D. C. 2008. The IPCC on a heterogeneous Medieval Warm Period. *Climatic Change*, **94**, 267–273.

Fabrés, J., Calafat, A., Canals, M., Barcena, M. A. & Flores, J. A. 2000. Bransfield Basin fine-grained sediments: late-Holocene sedimentary processes and Antarctic oceanographic condition. *Holocene*, **10**, 703–718.

Ferron, F. A., Simões, J. C., Aquino, F. E. & Setzer, A. W. 2004. Air temperature time series for King George Island, Antarctica. *Pesquisa Antártica Brasileira*, **4**, 155–169.

Folk, R. L. & Ward, W. C. 1957. Brazos River bar, a study in the significance of grain-size parameters. *Journal of Sedimentary Petrology*, **27**, 3–27.

Gohl, K. 2007. The Expedition ANTARKTIS-XXIII/4 of the research vessel "Polarstern" in 2006. *Reports on Polar and Marine Research*, **557**.

Gordon, J. E. & Harkness, D. D. 1992. Magnitude and geographic variation of the radiocarbon content in Antarctic marine life: implications for reservoir corrections in radiocarbon dating. *Quaternary Science Reviews*, **11**, 697–708.

Griffith, T. W. & Anderson, J. B. 1989. Climatic control of sedimentation in bays and fjords of the northern Antarctic Peninsula. *Marine Geology*, **85**, 181–204.

Hernando, P. M. & Ferreyra, G. A. 2005. The effects of UV radiation on photosynthesis in an Antarctic diatom (*Thalassiosira*, sp.): does vertical mixing matter? *Journal of Experimental Marine Biology and Ecology*, **325**, 35–45.

Hunt, B. G. 2006. The Medieval Warm Period, the Little Ice Age and simulated climate variability. *Climate Dynamics*, **27**, 677–694.

IPCC 2001. *Climate Change 2001: The Scientific Basis. Contribution of Working Group I to the Third Assessment Report of the Intergovernmental Panel on Climate Change*. Cambridge University Press, Cambridge.

IPCC 2007. *Climate Change 2007: The Physical Science Basis. Contribution of Working Group I to the Fourth Assessment Report of the Intergovernmental Panel on Climate Change*. Cambridge University Press, Cambridge.

Jones, P. D. & Mann, M. E. 2004. Climate over past millennia. *Reviews of Geophysics*, **42**, doi: 10.1029/2003RG000143.

JONES, V., HODGSON, D. & CHEPSTOW-LUSTY, A. 2000. Palaeolimnological evidence for marked Holocene environmental changes on Signy Island, Antarctica. *The Holocene*, **10**, 43–60.

JOUZEL, J., MASSON-DELMOTTE, V. ET AL. 2007. Orbital and millennial Antarctic climate variability over the past 800,000 Years. *Science*, **317**, 793–796.

KHIM, B. K., YOON, H. I., KANG, C. Y. & BAHK, J. J. 2002. Unstable Climate Oscillations during the Late Holocene in the Eastern Bransfield Basin, Antarctic Peninsula. *Quaternary Research*, **58**, 234–245.

KHIM, B. K., SHIM, J., YOON, H. I., KANG, Y. C. & JANG, Y. H. 2007. Lithogenic and biogenic particle deposition in an Antarctic coastal environment (Marian Cove, King George Island): seasonal patterns from a sediment trap study. *Estuarine, Coastal and Shelf Science*, **73**, 111–122.

KING, J. C., TURNER, J., MARSHALL, G. J., CONNOLLEY, W. M. & LACHLAN-COPE, T. A. 2004. Antarctic peninsula climate variability and its causes as revealed by analysis of instrumental record. *Antarctic Research Series*, **79**, 205–224.

KLÖSER, H., FERREYRA, G., SCHLOSS, I., MERCURI, G., LATURNUS, F. & CURTOSI, A. 1994. Hydrography of Potter Cove, a small fjord-like inlet on King George Island (South Shetland). *Estuarine, Coastal and Shelf Science*, **38**, 523–537.

KREUTZ, K. J., MAYEWSKI, P. A., MEEKER, L. D., TWICKLER, M. S., WHITLOW, S. I. & PITTALWALA, I. I. 1997. Bipolar changes in atmospheric circulation during the Little Ice Age. *Science*, **277**, 1294–1296.

LAMB, H. H. 1965. The early medieval warm epoch and its sequel. *Palaeogeography, Palaeoclimatology, Palaeoecology*, **1**, 13–37.

LAMB, H. H. 1995. *Climate, History and the Modern World*. 2nd edn. Routledge, London.

LEE, Y. I., LIM, H. S., YOON, H. I. & TATUR, A. 2007. Characteristics of tephra in Holocene lake sediments on King George Island, West Antarctica: implications for deglaciation and paleoenvironment. *Quaternary Science Reviews*, **26**, 3167–3178.

LEE, K., YOON, S.-K. & YOON, H. I. 2009. Holocene paleoclimate changes determined using diatom assemblages from Lake Long, King George island, Antarctica. *Journal of Paleoclimatology*, **42**, 1–10.

LEWITUS, A. J. & CARON, D. A. 1990. Relative effects of nitrogen or phosphorus depletion and light-intensity on the pigmentation, chemical-composition, and volume of *Pyrenomonas salina* (Cryptophyceae). *Marine Ecology-Progress Series*, **61**, 171–181.

LJUNG, K. & BJÖRCK, S. 2007. Holocene climate and vegetation dynamics on Nightingale Island, South Atlantic—an apparent interglacial bipolar seesaw in action? *Quaternary Science Reviews*, **26**, 3150–3166.

LOEB, V., SIEGEL, V., HOLM-HANSEN, O., HEWITT, R., FRASER, W., TRIVELPIECE, W. & TRIVELPIECE, S. 1997. Effects of sea-ice extent and krill or salp dominance on the Antarctic food web. *Nature*, **387**, 897–900.

LOEHLE, C. 2007. A 2000 year global temperature reconstruction based on non-tree ring proxy data. *Energy & Environment*, **18**, 1049–1058.

LOEHLE, C. & MCCULLOCH, J. H. 2008. Correction to: a 2000 year global temperature reconstruction based on non-tree ring proxy data. *Energy & Environment*, **19**, 93–100.

MANN, M. E. & JONES, P. D. 2003. Global surface temperatures over the past two millennia. *Geophysical Research Letters*, **30**, doi: 10.1029/2003GL017814.

MANN, M. E., ZHANG, Z., HUGHES, M. K., BRADLEY, R. S., MILLER, S. K., RUTHERFORD, S. & NI, F. 2008. Proxy-based reconstructions of hemispheric and global surface temperature variations over the past two millennia. *Proceedings of the National Academy of Sciences*, **105**, 13252–13257.

MATTHIES, D., MÄUSBACHER, R. & STORZER, D. 1990. Deception Island tephra: a stratigraphical marker for limnic and marine sediments in Bransfield Strait area. *Zentralblatt für Geologie und Paläontologie, Teil 1*, **1/2**, 153–165.

MCCAVE, I. N., MANIGHETTI, B. & ROBINSON, S. G. 1995. Sortable silt and fine sediment size/composition slicing: parameters for palaeocurrent speed and palaeoceanography. *Paleoceanography*, **10**, 593–610.

MIDDLETON, G. V. 1976. Hydraulic interpretation of sand size distributions. *Journal of Geology*, **84**, 405–426.

MOBERG, A., SONECHKIN, D. M., HOLMGREN, K., DATSENKO, N. M. & KARLÉN, W. 2005. Highly variable Northern Hemisphere temperatures reconstructed from low- and high-resolution proxy data. *Nature*, **433**, 613–617.

MOLINE, M. A., CLAUSTRE, H., FRAZER, T. K., SCHOFIELD, O. & VERNET, M. 2004. Alteration of the food web along the Antarctic Peninsula in response to a regional warming trend. *Global Change Biology*, **10**, 1973–1980.

MÜLLER, P. J. & SCHNEIDER, R. 1993. An automated leaching method for the determination of opal in sediments and particulate matter. *Deep-Sea Research*, **40**, 425–444.

NESJE, A. & DAHL, S. O. 2003. The 'Little Ice Age' – only temperature? *The Holocene*, **13**, 171–177.

NOON, P. E., LENG, M. J. & JONES, V. J. 2003. Oxygen-isotope ($\delta^{18}O$) evidence of Holocene hydrological changes at Signy Island, maritime Antarctica. *The Holocene*, **13**, 251–263.

PAKHOMOV, E. A., FUENTES, V., SCHLOSS, I., ATENCIO, A. & ESNAL, G. B. 2003. Beaching of the tunicate *Salpa thompsoni* at high levels of suspended particulate matter in the Southern Ocean. *Polar Biology*, **26**, 427–431.

PARK, B.-K., CHANG, S.-K., YOON, H. I. & CHUNG, H. S. 1998. Recent retreat of ice cliffs, King George Island, South Shetland Islands, Antarctic Peninsula. *Annals of Glaciology*, **27**, 633–635.

PARKINSON, C. L. 2002. Trends in the length of the Southern Ocean sea-ice season, 1979–99. *Annals of Glaciology*, **34**, 435–440.

PICHLMAIER, M., AQUINO, F. E., SANTOS DA SILVA, C. & BRAUN, M. 2004. Suspended sediments in Admiralty Bay, King George Island (Antarctica). *Pesquisa Antartica Brasileira*, **4**, 77–85.

REIMER, P. J., BAILLIE, M. G. L. ET AL. 2004. Intcal04 Terrestrial radiocarbon age calibration, 26–0 ka BP. *Radiocarbon*, **46**, 1029–1058.

RIGNOT, E., CASASSA, G., GOGINEN, P., KRABILL, W., RIVERA, A. & THOMAS, R. 2004. Accelerated ice discharge from the Antarctic Peninsula following the collapse of Larsen B ice shelf. *Geophysical Research Letters*, **31**, doi: 10.1029/2004GL020697.

Rott, H., Skvarca, P. & Nagler, T. 1996. Rapid Collapse of Northern Larsen Ice Shelf, Antarctica. *Science*, **271**, 788–792.

Scambos, T. A., Bohlander, J. A., Shuman, C. A. & Skvarca, P. 2004. Glacier acceleration and thinning after ice shelf collapse in the Larsen-B embayment, Antarctica. *Geophysical Research Letters*, **31**, 1–4.

Schloss, I. R. & Ferreyra, G. A. 2002. Primary production, light and vertical mixing in Potter Cove, a shallow coastal Antarctic environment. *Polar Biology*, **25**, 41–48.

Schloss, I. R., Ferreyra, G. A. & Ruiz-pino, D. 2002. Phytoplankton Biomass in Antarctic Shelf Zones: a conceptual model based on Potter Cove, King George Island. *Journal of Marine Systems*, **36**, 129–143.

SHALDRIL-SHIPBOARD-SCIENTIFIC-PARTY 2005. SHALDRIL 2005 Cruise Report.

Shepherd, A., Wingham, D. J., Payne, T. & Skvarca, P. 2003. Larsen ice shelf has progressively thinned. *Science*, **203**, 856–859.

Shevenell, A. E. & Kennett, J. P. 2002. Antarctic Holocene Climate Change: a Benthic Foraminiferal Stable Isotopic Record from Palmer Deep. *Paleoceanography*, **17**, doi: 10.1029/2000PA000596.

Steig, E. J. 2006. The south-north connection. *Nature*, **444**, 152–153.

Steig, E. J., Brook, E. J. *et al.* 1998. Synchronous climate changes in Antarctica and the North Atlantic. *Science*, **282**, 92–95.

Stocker, T. F. 1998. The seesaw effect. *Science*, **282**, 61–62.

Stuiver, M. & Reimer, P. J. 1993. Extended 14C database and revised CALIB radiocarbon calibrating program. *Radiocarbon*, **35**, 215–230.

Stuiver, M., Reimer, P. J. *et al.* 1998. INTCAL 98 Radiocarbon age calibration, 24,000–0 cal BP. *Radiocarbon*, **40**, 1041–1083.

Sun, D., Bloemendal, J., Rea, D. K., Vandenberge, J., Jiang, F., An, Z. & Su, R. 2002. Grain-size distribution function of polymodal sediments in hydraulic and aeolian environments, and numerical partitioning of the sedimentary components. *Sedimentary Geology*, **152**, 263–277.

Thomas, R., Rignot, E. *et al.* 2004. Accelarated sea-level rise from west Antarctica. *Science*, **306**, 255–258.

Turner, J., Colwell, S. R. *et al.* 2005. Antarctic climate change during the last 50 years. *International Journal of Climatology*, **25**, 279–294.

Vaughan, D. G. 2005. How does the Antarctic ice sheet affect sea level rise? *Science*, **308**, 1877–1878.

Vaughan, D. G., Marshall, G. J., Connolley, W. M., King, J. C. & Mulvaney, R. 2001. Devil in the detail. *Science*, **293**, 1777–1779.

Vaughan, D. G., Marshall, G. J. *et al.* 2003. Recent rapid regional climate warming on the Antarctic Peninsula. *Climate Change*, **60**, 243–274.

Vogt, S. & Braun, M. 2004. Influence of glaciers and snow cover on terrestrial and marine ecosystems as revealed by remotely-sensed data. *Pesquisa Antartica Brasileira*, **4**, 105–118.

Warner, N. R. & Domack, E. W. 2002. Millennial- to decadal-scale paleoenvironmental change during the Holocene in the Palmer Deep, Antarctica, as recorded by particle size analysis. *Paleoceanography*, **17**, doi: 10.1029/2000PA000602.

Weaver, A. J., Saenko, O. A., Clark, P. U. & Mitrovica, J. X. 2003. Meltwater pulse 1A from Antarctica as a trigger of the Bolling-Allerod warm period. *Science*, **299**, 1709–1713.

Weltje, G. J. & Prins, M. A. 2003. Muddled or mixed? Inferring palaeoclimate from size distributions of deep-sea clastics. *Sedimentary Geology*, **162**, 39–62.

Wunsch, C. 2006. Abrupt climate change: an alternative view. *Quaternary Research*, **65**, 191–203.

Yoon, H. I., Han, M. W., Park, B.-K., Oh, J.-K. & Chang, S.-K. 1997. Glaciomarine sedimentation and paleo-glacial setting of Maxwell Bay and its tributary embayment, Marian Cove, South Shetland Islands. *Marine Geology*, **140**, 265–282.

Yoon, H. I., Park, B.-K., Domack, E. W. & Kim, Y. 1998. Distribution and dispersal pattern of suspended particulate matter in Maxwell Bay and its tributary, Marian Cove in the South Shetland Islands, West Antarctica. *Marine Geology*, **152**, 261–275.

Yoon, H. I., Park, B. K., Kim, Y. & Kim, D. 2000. Glaciomarine sedimentation and its paleoceanographic implications along the fjord margins in the South Shetland Islands, Antarctica during the last 6000 years. *Palaeogeography, Palaeoclimatology, Palaeoecology*, **157**, 189–211.

Variations in organic carbon flux and stagnation periods during the last 2400 years in a Skagerrak fjord basin, inferred from benthic foraminiferal $\delta^{13}C$

HELENA L. FILIPSSON[1,3]* & KJELL NORDBERG[2]

[1]*Europrox, Universität Bremen, Leobener Str, 28359 Bremen, Germany*

[2]*Department of Earth Sciences, University of Gothenburg, PO Box 460, SE 405 30, Sweden*

[3]*Present address: GeoBiosphere Science Centre, Quaternary Sciences, Lund University, Sölvegatan 12, SE-223 62 Lund, Sweden*

**Corresponding author (e-mail: Helena.filipsson@geol.lu.se)*

Abstract: A well-dated high-resolution $\delta^{13}C$ record of the last 2400 a, based on the benthic foraminifer *Cassidulina laevigata*, is presented for Gullmar Fjord, Sweden. The time interval covers the Roman Warm Period (RWP), the Viking Age/Medieval Warm Period (VA/MWP), the Little Ice Age (LIA) and the most recent warming. There is little variation in the $\delta^{13}C$ record until the early Viking Age (AD 800), when the $\delta^{13}C$ signal becomes significantly more negative and continues to decrease throughout the VA/MWP. The $\delta^{13}C$ signal increases both at the beginning and at the end of the LIA but is marked by more negative values during the larger part of the period. Since about 1970, the $\delta^{13}C$ values are more negative than the long-term average. This general negativity of the record may result from a higher flux of organic matter, possibly of terrestrial origin due to land-use changes together with moderate changes in stagnation periods since the VA/MWP. In most recent times, the oceanic Suess effect together with increased number of extended stagnation periods are probably the main causes of the shift towards more negative $\delta^{13}C$ values.

Northern Europe has experienced several significant climatic variations during the last 2500 a: the Roman Warm Period (RWP, *c*. 300 BC to AD 400); Medieval Warm Period (MWP, *c*. AD 800–1300); the Little Ice Age (LIA, *c*. AD 1300–1850) and the most recent times. The global significance of these climate changes has also been debated (e.g. Bradley *et al.* 2001; Broecker 2001); however, the impact in this region is well documented (e.g. Lamb 1995; Wefer *et al.* 2002) and the causes behind the climate variations have been studied in detail (see recent review by Wanner *et al.* 2008). The climate classifications are, however, informal and the boundaries between them are not strictly defined.

Over the last years there have also been several paleoceanographic studies in the Skagerrak through the Holsmeer and Millenium EU projects (e.g. Brückner & Mackensen 2006; Gil *et al.* 2006; Hebbeln *et al.* 2006; Brückner & Mackensen 2008). However, the outcome and interpretations are not straightforward, partly depending on difficulties in obtaining a reliable age model. Here we present a well-dated, high-resolution record of stable carbon isotopes representing the last 2400 a based on the benthic foraminifer *Cassidulina laevigata* from Gullmar Fjord, a fjord which debouches into the Skagerrak (Fig. 1). We describe in detail both how the duration of the deep water's stagnation periods and the flux of organic matter to the fjords have varied over last two millennia and compare these results with those from the Skagerrak. Since it is well documented (e.g. Lamb 1995; Wefer *et al.* 2002) that the region has experienced sustainable climate changes during the latest part of the Holocene, it is particularly interesting to study how the marine environment has responded to these climate changes. The fjord is a particularly good high-resolution environmental archive since it experiences a continuous large net-deposition of fine-grained sediments and negligible tidal activity. Modern sedimentation rates in the deepest part of the fjord are exceptionally high, of the order 0.7–1.4 cm a^{-1} (Filipsson & Nordberg 2004). The fjord acts as a large sediment trap, and any material reaching the more stagnant basin below sill depth has little possibility of being transported out of the fjord. In addition, the area is well situated to record the climate in NE Europe since it lies within the westerlies and is governed by the hydrography of the Skagerrak. To complement this, there is a long time-series of hydrographic and nutrient

From: HOWE, J. A., AUSTIN, W. E. N., FORWICK, M. & PAETZEL, M. (eds) *Fjord Systems and Archives*. Geological Society, London, Special Publications, **344**, 261–270.
DOI: 10.1144/SP344.18 0305-8719/10/$15.00

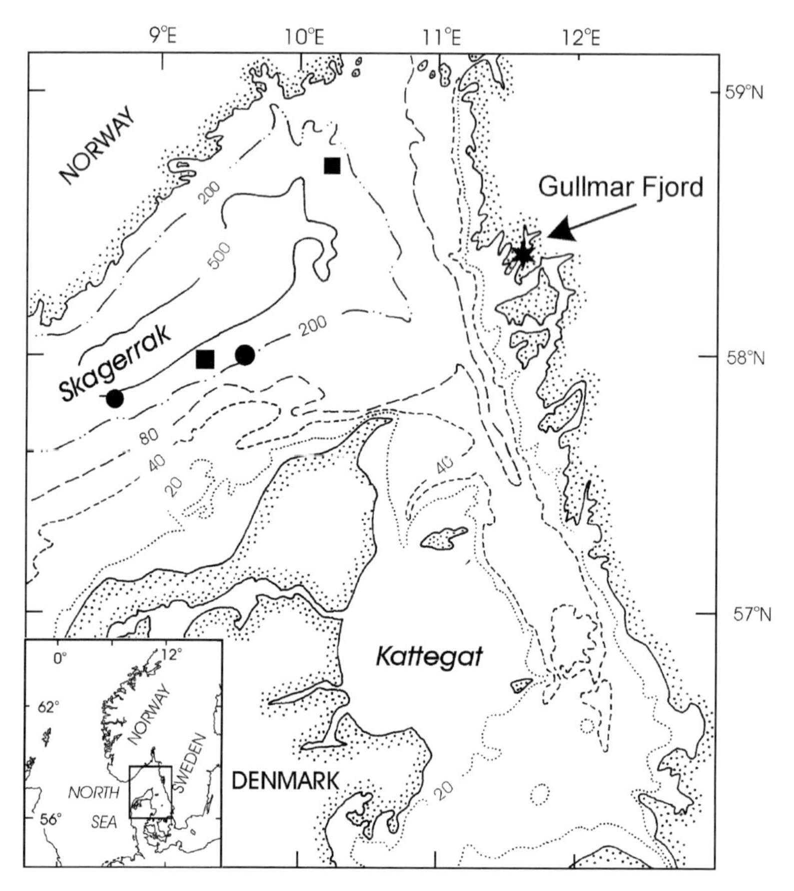

Fig. 1. Location map over the Skagerrak and the Gullmar Fjord. Cores 9004 and GA113-2Aa are indicated with a star (✱) (position: 58°19′10N/11°32′84E). The coring sites from Hebbeln *et al.* (2006) are marked with filled square (■) and from Brückner & Mackensen (2008) with a filled circle (●).

measurements from the fjord from as early as 1869 by Ekman (1870).

Modern hydrographic setting and primary productivity

Gullmar Fjord, a sill fjord on the Swedish west coast, is oriented southwest–northeast and is 28 km long and 1–2 km wide (Fig. 1). The sill depth is 42 m, and the maximum fjord depth is 120 m. The fjord's deep basin, below 100 m, is *c.* 5 km long and 1 km wide. The adjoining Skagerrak largely determines the hydrography of the fjord (Rydberg 1977; Arneborg 2004) with the water being stratified due to salinity differences into 3–4 layers. The first is a thin layer of river water from the Örekilsälven, which varies both spatially and temporally due to wind mixing and variations in the river outflow. The average river discharge from the Örekilsälven between the years 1909 and 1990 was about 22 $m^3 s^{-1}$ (SMHI 1994). A second layer is brackish ($S < 28$), extends to 15 m and has a residence time of between 20–38 days (Arneborg 2004, salinity measured using the Practical Salinity Scale). This layer is primarily composed of water from the Baltic current, which carries brackish water from the Baltic northwards along the Swedish west coast. A third layer of more saline water ($S > 28$) occurs between water depths of *c.* 15 and 50 m and is derived from the Skagerrak; its mean residence time is between 29 and 62 days (Arneborg 2004). The fourth and deepest layer (>50 m) is more stagnant with little seasonal variation; an average salinity of 34.4 and temperature *c.* 6 °C was measured between the years 1980 and 2001 (Filipsson 2003). The stratification of the water column is strengthened during the summer by the development of a strong thermocline. The fjord's deep water originates from the Skagerrak and the North Sea and is usually exchanged every year, typically between January and March. The oxygen levels in the deep water are usually <2 ml l^{-1} between October and March (Filipsson 2003). In sill fjords, more prolonged low-oxygen conditions can evolve for several reasons. The most striking example is when deep-water exchange fails to occur, as it did in Gullmar Fjord in 1997. A second reason is when an inflow occurs later in the year than average and a third when only partial exchange occurs (as happened in 1979/80 and 1988/89). However, even after long periods of low-oxygen conditions in Gullmar Fjord, no H_2S has ever been measured in the water. The general hydrography of the fjord is more thoroughly discussed in Svansson (1984) and the water renewal processes in Arneborg *et al.* (2004) and Erlandsson *et al.* (2006). The hydrography of the adjoining Skagerrak is reviewed by Rodhe (1998). The primary productivity in the mouth of the fjord is of the order 220 g C m^{-2}, averaged between 2000 and 2005 (Lindahl 2007) and based on ^{14}C incubations.

Material and methods

Sediment sampling

Two sediment cores were analysed: a 731 cm long piston core (core 9004) and a 60 cm long Gemini core (core GA1132Aa) from the deepest part of the fjord (Fig. 1). Core 9004 was sampled in July 1990 and was stored unopened in a cold room until November 2003; core GA1132A was sampled and sectioned in June 1999. Core GA1132A (Ø 80 mm) was sectioned into 1-cm samples and split in half. One half (GA1132Aa) was used for foraminiferal fauna and isotopic analyses (Filipsson 2003; Filipsson & Nordberg 2004) and the other (GA1132Ab) for dinoflagellate cysts analysis (Harland *et al.* 2006; Harland *et al.* 2010). Core 9004 (Ø 70 mm) was subsampled into 1-cm samples down to 28 cm and thereafter into 2-cm samples. The samples for foraminiferal analyses were freeze dried, washed over a 63 μm sieve, treated with sodiumdiphosphate ($Na_4P_2O_7$) during sieving to disintegrate sediment aggregates and dried at 40 °C. Samples for organic carbon (C. org) were freeze dried, homogenized and treated with hydrochloric acid before being analysed. C. org samples were analysed at all levels for core GA1132Aa, whereas every second interval was run for core 9004. All organic carbon samples for core GA1132Aa were analysed at the Department of Earth Sciences, Göteborg University, Sweden using a Carlo Erba NA 1500 CHN analyser. Samples from core 9004 were run at the Department of Geosciences, Bremen University, Germany using a Vario EL III CHN analyser. The sand-sized fraction was estimated by weighing the material larger than 63 μm after sample preparation.

Age model

The sedimentation rate for GA113 2A is 0.7 cm a^{-1} (Filipsson & Nordberg 2004) and the core represents the time between 1999 and approximately 1915. The age model for core GA113 2A was established by a combination of ^{210}Pb methodology using the constant rate of supply (CRS) model as described by Appleby & Oldfield (1978) and biostratigraphy (Filipsson & Nordberg 2004). The ^{210}Pb analyses were performed at the Nuclear Safety Research Department, Risø National Laboratory, Denmark.

Eleven mollusc shells from core 9004 were ^{14}C dated at the Ångström laboratory, Uppsala University, Sweden (Fig. 2, Table 1) using the marine model calibration curve (Reimer *et al.* 2004; Bronk Ramsey 2005) together with a global reservoir age of 400 a ($\Delta R = 0$). The chronology was constructed by visual line fitting through the calibrated radiocarbon dates, giving an average sedimentation rate of 0.5 cm a^{-1} in the uppermost part of the core and 0.2 cm a^{-1} in the lowermost part (Fig. 2). Two dates gave ages deviating from the average (Fig. 2, Table 1) and were ignored; these shells were probably reworked. It is suspected that *c.* 100 cm of the core top of core 9004 is missing (an effect from the combination of the heavy piston-corer and the very soft, unconsolidated sediments). Estimating an assumed continuous sedimentation rate of *c.* 0.5–0.7 cm a^{-1}, 100 cm has been added to the record to compensate for the lost interval.

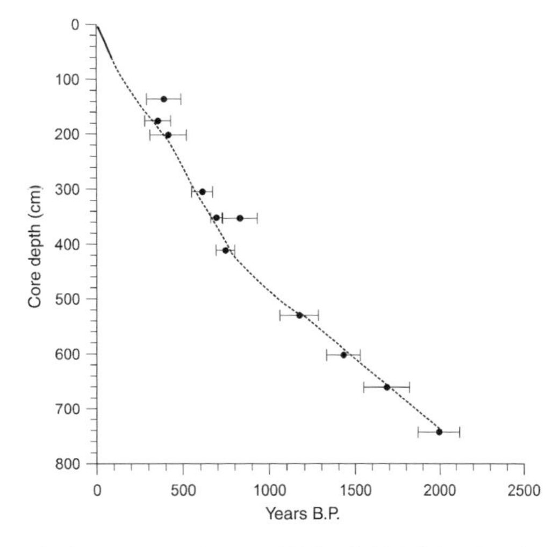

Fig. 2. Age model for core 9004 and GA113-2Aa based on ^{210}Pb dating (solid black line) and AMS ^{14}C dating (dotted line); see text for details.

Table 1. *AMS ^{14}C dates for core 9004 from Gullmar Fjord*

Core depth (cm)	Lab. ID	Species	^{14}C and reservoir corrected ages (BP)	Error (±)	Calibrated ages (cal a BP)	Error (±)	Years AD
136	Ua-24043	*Nuculana minuta*	310	40	390	100	1560
176	Ua-35966	*Nuculana pernula*	275	40	355	75	1595
202	Ua-23075	*Yoldiella lenticula*	400	40	415	105	1535
305	Ua-35967	*Nucula* sp.	625	30	610	60	1340
352	Ua-35968	*Clamys septemradiatus*	745	25	690	35	1260
353	Ua-23000	*Abra nitida*	905	45	830	100	1120
411	Ua-35969	*Nucula tenuis*	845	25	745	55	1205
530	Ua-23001	*Abra nitida*	1240	45	1170	110	780
602	Ua-23002	*Nuculana minuta*	1525	40	1430	100	520
661	Ua-23003	*Thyasira flexuosa*	1755	45	1685	135	265
743	Ua-23004	*Thyasira flexuosa*	2015	45	1995	125	−45

Stable carbon isotope analysis

All samples for $\delta^{13}C$ were analysed at the Department of Geosciences, University of Bremen using a Finnigan Mat 251 mass spectrometer equipped with an automatic carbonate preparation device. Isotope composition is given in the usual δ-notation and is calibrated to Vienna Pee Dee Belemnite (V-PDB) standard. The analytical standard deviation is <0.05‰ for $\delta^{13}C$ based on the long-term standard deviation of an internal standard (Solnhofen limestone). On average, 20 specimens of *Cassidulina laevigata* were necessary for each analysis. The samples were run in random order and outliners were run again when enough specimens were present. All data including duplicates are presented in Figure 3.

In the short core (GA113 2Aa) every centimetre between 5 and 60 cm was analysed. In the uppermost 4 cm, there were not enough specimens of *C. laevigata* for analysis. Samples had to be combined at some levels, but never more than 2 cm. For core 9004, every sample was analysed where possible (a total of 353 samples). There are, however, a few short gaps in the 9004 record when only a few specimens of *C. laevigata* were found; these gaps correspond to between 20 and 40 a of time only.

Results

$\delta^{13}C$ record for the time period between 430 BC and AD 1850

Core 9004 approximately represents the time between 430 BC and AD 1850. The stable carbon isotope record of *Cassidulina laevigata* is negative throughout the record, varying between −1.38 and −0.23‰ (Fig. 3). We focus on the major climatic variations during this period: the RWP (300 BC to AD 400), the VA/MWP (AD 800–1300) and the LIA (AD 1300–1850).

The $\delta^{13}C$ record displays relatively little variability during the RWP (Fig. 3). The average $\delta^{13}C$ value is −0.56‰ and the variability is between −0.36 and −0.96‰. There are slightly more negative values at the beginning of the RWP compared to the rest of this period, but not compared to the long-term average for the entire record. The most marked change in the record from core 9004 occurred during the VA/MWP, with the $\delta^{13}C$ values displaying a steady but fast decrease from more or less the onset and throughout the period. The $\delta^{13}C$ values are −0.70‰ around AD 800 and decrease to −1.38‰ around AD 1200. There is a gap in the record, that is, there are very few tests of *C. laevigata* around 400 cm corresponding to *c*. AD 1100. The values both before and after this gap are considerably more negative than the long-term average (*c*. −1‰). By the end of the VA/MWP the $\delta^{13}C$ starts to increase and reaches the same levels as before its onset. During the VA/MWP there are other short gaps in the record when there was not enough material for isotopic analyses; these gaps occur during periods of both low and high isotopic signals. The LIA is also marked with relatively large variations in the carbon isotopic signal. The $\delta^{13}C$ values are −0.6‰ at the beginning of the LIA, display a negative trend and reach a minimum value of −1.15‰ around AD 1500. This decrease is slightly less than that during the VA/MWP. There is a positive trend from around AD 1770 to the beginning of the 19th century; the $\delta^{13}C$ values are between −0.23 and −0.31‰ which are as high as they were at *c*. 440 BC.

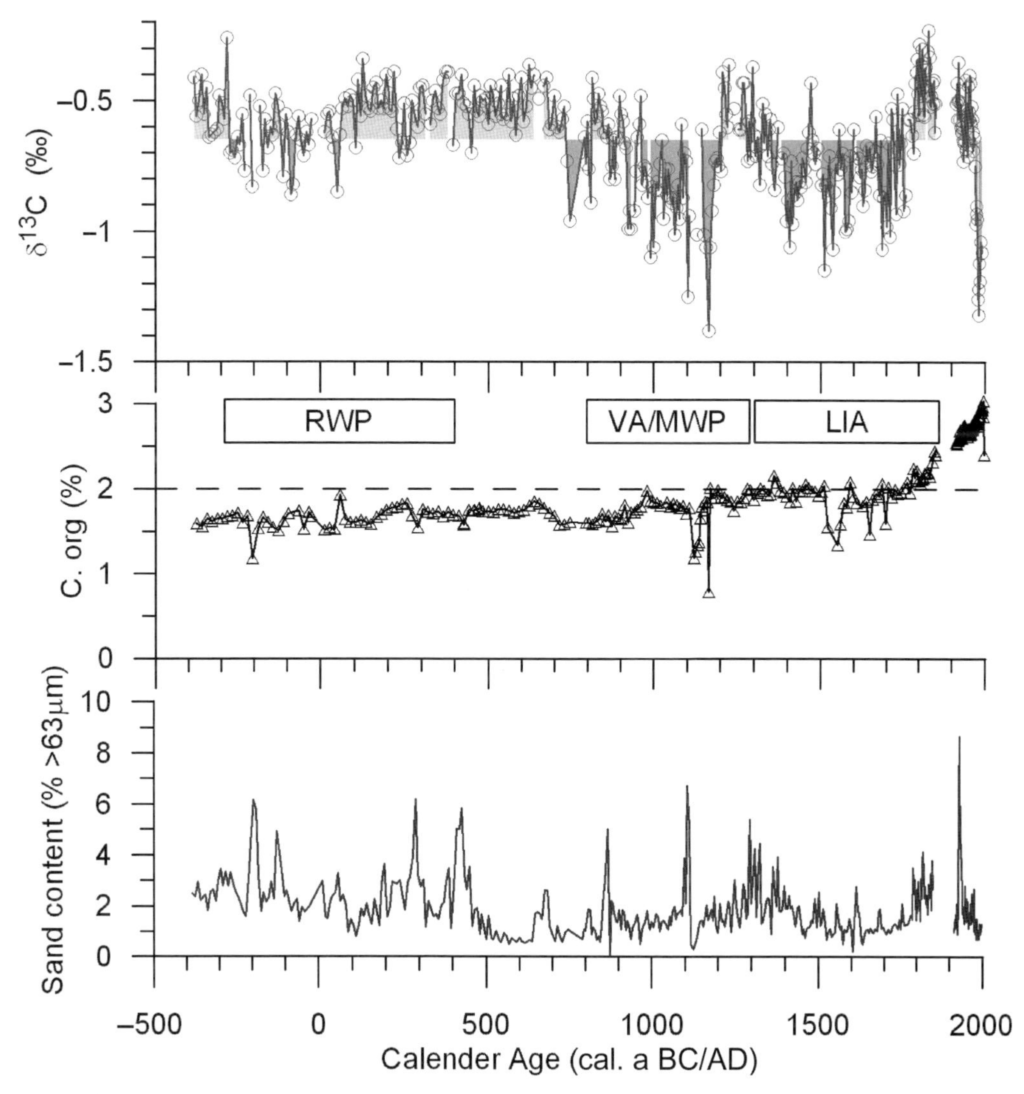

Fig. 3. $\delta^{13}C$ values from the benthic foraminifer *Cassidulina laevigata* (upper panel), C. org content (middle panel) and sand-sized fraction (lower panel) from cores 9004 and GA113-2Aa v. time. The long-term average of the $\delta^{13}C$ record is −0.65‰ and is indicated with a line; the long-term average for the C. org content is 2.0% and is indicated with a dotted line. The boxes indicate the climate periods Roman Warm Period (RWP), Viking Age/Medieval Warm Period (VA/MWP) and Little Ice Age (LIA).

Organic carbon and sand-sized fraction between

The organic carbon (C. org) shows little similarity to the $\delta^{13}C$ record (Fig. 3); instead, it displays an increasing trend towards the top. The average C. org content for the entire record is 1.79%; maximum is 2.4% and minimum 0.79%. A larger deviation with low C. org values occurs at AD 1100; the same interval where there is a gap in the $\delta^{13}C$ record. Organic carbon is subject to degradation in the sediment after deposition hence any trend in the record must be analysed with this phenomenon in mind.

The sand-sized fraction (material >63 μm) has a long-term average of 1.89% with a maximum of 6.7% at around AD 1100: the period where *C. laevigata* is sparse and the C. org content is very low (Fig. 3). It is possible that this peak of coarser material represents a slumping or storm event.

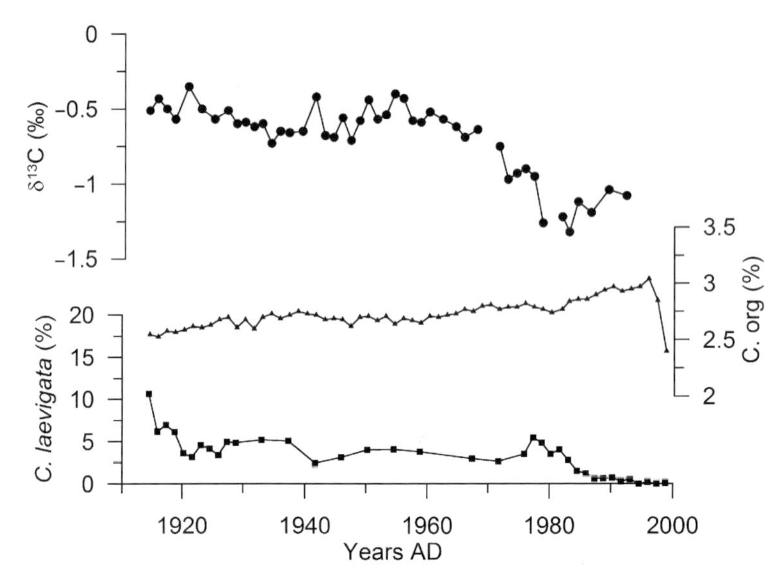

Fig. 4. Stable carbon isotope data from *C. laevigata*, C. org content and relative abundance of *C. laevigata* from core GA113-2Aa.

$\delta^{13}C$ record and organic carbon content for the time period 1915 and 1999

The carbon isotope record from core GA113-2Aa covers the years 1915 until 1992 when the abundances of *C. laevigata* become too low for analysis. There is little variability in the record until approximately 1970; the $\delta^{13}C$ values fluctuate between -0.73 and -0.35‰ with an average of -0.57‰ and there is no trend in the data set (Fig. 4). Around 1970 there is a marked decrease in $\delta^{13}C$; the average after 1970 is -1.06‰ with a minimum value at -1.32‰ (Fig. 4). The relative abundance of *C. laevigata* is around 5% until the beginning of the 1980s when the abundance decreases and the species disappears from the record (Filipsson & Nordberg 2004; Fig. 4). The organic carbon content is on average 2.7% between 1915 and 1999 and displays the typical increasing trend towards the top (except for the uppermost samples where the organic carbon content is lower; Fig. 4). There is no marked increase in organic carbon content around 1970.

Discussion

Stable carbon isotopes

The $\delta^{13}C$ signal in benthic foraminifera is complex and can be related to processes involving the content of dissolved inorganic carbon (DIC) in seawater/pore water, the flux of organic carbon to the sea floor, the sea water's alkalinity and biomineralization processes when the foraminiferal test is formed (e.g. Rohling & Cooke 1999). Some of the more important processes affecting the benthic foraminiferal $\delta^{13}C$ record in a fjord environment are summarized below. An increase in primary productivity in the surface water and thereby a larger flux of organic matter to the sea floor will result in a more negative $\delta^{13}C$ signal in the bottom water, since phytoplankton heavily discriminate towards ^{13}C. This is reflected in the DIC content of the bottom water due to remineralization of organic matter; this signal will be incorporated in the foraminiferal tests during calcification.

However, changes in the residence time or origin of both surface and deep water would also affect the $\delta^{13}C$ signal. If the residence time of the fjord's surface water became longer, the proportion of locally produced organic matter would increase. In addition, a longer residence time of the surface water would affect the $\delta^{13}C$ signal in two opposing directions. More organic matter would have time to settle from the shallower zone, resulting in a more negative $\delta^{13}C$ signal, while remineralization processes occurring in the surface water would be more effective, turning the deep-water $\delta^{13}C$ signal in the opposite direction. The exposure time of organic matter in the bottom water will also influence the $\delta^{13}C$ values such that longer stagnation periods will result in greater remineralization of organic matter, assuming enough oxygen is available, resulting in a more negative $\delta^{13}C$ signal. To complicate matters further, general circulation changes in the adjoining Skagerrak could also affect the carbon content of the water entering the fjord.

In addition, the carbon isotopic composition of the pore water and of the foraminiferal tests does not necessarily correspond due to kinetic, ontogenetic and metabolic effects and to the living depth of the foraminifera (the so-called microhabitat effect) (McCorkle *et al.* 1997; Rohling & Cooke 1999; Fontanier *et al.* 2006).

Changes in $\delta^{13}C$ and organic carbon and their possible causes during the VA/MWP and LIA

The $\delta^{13}C$ record in Gullmar Fjord displays little variation before the onset of the Viking Age (Fig. 3). The focus here is on large and significant environmental changes that evidently occurred in the fjord and its surroundings during the VA/MWP and LIA. A regime shift seems to have taken place around AD 800; the fjord environment shifted from a stable environment with little variability to a more unstable environment where the flux and sources of organic carbon varies greatly. The record is marked by two periods of more negative $\delta^{13}C$ values, one during the VA/MWP and a second one during the LIA. The causes behind these two periods could be a combination of several processes and they could also be different between the VA/MWP and the LIA. Potential processes involved could be a larger flux of organic matter from the surface water, surrounding land and/or from the inflowing water from the Skagerrak, changes in residence time of both surface and deep water or indeed changes in the hydrography of the Skagerrak.

A higher primary productivity signal and a changed flora have also been documented in the Skagerrak and Kattegat during the VA/MWP (Thorsen *et al.* 1995; Hebbeln *et al.* 2006) which would support the theory of increased surface water productivity. However, core 9004 has recently been analysed for its dinoflagellate cyst content (Harland & Nordberg, pers. comm., 2009). These results indicate no significant changes in cyst concentration during the last 2000 years, suggesting no large variations in local primary productivity in the fjord in as much as cyst concentrations can be an indication of primary productivity. In addition, the organic carbon content in the sediment does not suggest a large increase of organic matter during the VA/MWP or LIA (Fig. 3). If there was a higher input of organic material, it is likely that the increase was in the form of easily degradable labile carbon with little potential for preservation in the sediment record.

A second option involves changes in the residence time of the water. The surface-water exchange is driven by density differences along the west coast (Arneborg *et al.* 2004) and there is little evidence to suggest that the hydrography would have been substantially different during VA/MWP and the LIA. The stagnation periods of the deep water could have been slightly longer during VA/MWP and LIA compared to before AD 800, but not considerably longer. This would have caused even lower oxygen conditions and *C. laevigata* would have disappeared; in addition, *Cassidulina laevigata* is not particularly tolerant towards low-oxygen conditions (Nordberg *et al.* 2000; Gustafsson & Nordberg 2001; Filipsson & Nordberg 2004). There are short periods where the abundance of *C. laevigata* is too low to allow for isotopic analysis but these periods occur throughout the record and are not more frequent during the VA/MWP or the LIA.

A substantial difference in regional circulation and flux in organic matter would have affected the Skagerrak; no similarities are found between the $\delta^{13}C$ record from the fjord and the four published $\delta^{13}C$ records from the Skagerrak (Hebbeln *et al.* 2006; Brückner & Mackensen 2008). However, few similarities can be noted between the Skagerrak records. Hebbeln *et al.* (2006) investigated two cores comprising records of the last 2700 a. One core came from southern Skagerrak (312 mbsl (metres below sea level)) and one from the NE (225 mbsl) (Fig. 1). One of the investigated variables on both cores was $\delta^{13}C$, measured on the deep infaunal foraminiferal species *Melonis barleeanus*. The $\delta^{13}C$ record from the NE part of Skagerrak displays some variability but no trend, suggesting that the organic flux did not vary in concert with climate changes. The southern core exhibits slightly more negative benthic $\delta^{13}C$ values between AD 200 and 900 compared to the entire record (Hebbeln *et al.* 2006). The values increased after AD 900; however, in the youngest part of the record the $\delta^{13}C$ values decrease again. The authors (Hebbeln *et al.* 2006) interpreted the change around AD 900 as a decrease in productivity. In contrast, the surface water displays an opposite signal: the $\delta^{13}C$ data measured from the planktonic foraminifer *Globigerinita uvula* shows a weak increase from AD 700, hence indicating greater productivity (Hebbeln *et al.* 2006). These contrasting results between surface and bottom water suggest that there is no direct link between the two. Brückner & Mackensen (2008) also studied two cores, one from the western part of Skagerrak (420 mbsl) and one from the southern Skagerrak (285 mbsl). They presented a time slice between AD 1500 and the present day. No trend in the benthic carbon isotope records was found between AD 1500 and

1950 for any of the investigated foraminiferal species (including *C. laevigata*). It is clear that the $\delta^{13}C$ signal is challenging to interpret when analyses fail to match neighbouring cores.

It is however possible that the composition of particulate organic carbon (POC) could have varied with the amount of terrestrial carbon increasing in the fjord during the VA/MWP and LIA, due to land-use changes as both farming and logging activities became more extensive. There are documented large-scale land-use changes at the onset of the VA/MWP from several areas in Scandinavia (e.g. Bradshaw *et al.* 2005; Zillen *et al.* 2008). The end of the VA/MWP and the LIA were also marked by several productive herring periods (Alheit & Hagen 1997); these affected both the organic carbon flux, had large socio-economic impacts and caused further land-use changes.

The $\delta^{13}C$ signal could also be affected by microhabitat or ontogenetic effects; however, we think these are of minor importance for this record. We analysed the same species, *C. laevigata*, throughout the record. *C. laevigata* is known to be a shallow infaunal species which does not tolerate low-oxygen conditions well (Gustafsson & Nordberg 2001). It is unlikely that the species would change its microhabitat preference to such a large extent that it affects the $\delta^{13}C$ record. We were also careful in choosing specimens of the similar size for stable isotopic analysis, limiting any ontogenetic effect (McCorkle *et al.* 2008).

Time period between 1915 and 1992

The $\delta^{13}C$ values in core GA113-2Aa (representing the period from 1915 to 1992 in the fjord) show little variability until 1970 when there is a considerable decrease from $-0.57‰$ on average before 1970 to $-1.06‰$ afterwards (Fig. 4). The organic carbon content does not indicate any large increase during this time (Fig. 4). There are, however, floral changes in the fjord's surface water occurring around the same time (Harland *et al.* 2006; Harland *et al.* 2010). The number of *Lingulodinium polyedrum* cysts decreases to very low values at the beginning of the 1970s; the dinoflagellate cyst record from the fjord is from the other half of core GA113-2A (Harland *et al.* 2006). The total number of cysts does not, however, change significantly. *Lingulodinium polyedrum* has been used as a proxy for eutrophication (e.g. Dale 2000), but what the ecology of the species indicates is still under discussion (Godhe & McQuoid 2003; Harland *et al.* 2006; Harland *et al.* 2010). Harland *et al.* (2006) also discuss the local pollution history and conclude that, beyond the mid-1970s, there were no significant ongoing pollution sources in the fjord. The benthic foraminiferal composition reveals a significant shift in the fauna in the early 1980s (Nordberg *et al.* 2000; Filipsson & Nordberg 2004), most likely related to a low-oxygen event taking place around this time. Since 1979 there have been an increased number of low oxygen events, probably related to changes in climate (Nordberg *et al.* 2000; Filipsson & Nordberg 2004), where the oxygen concentration has been lower than 1 ml l^{-1} for at least 3 months (Filipsson & Nordberg 2004). Hence the decrease in $\delta^{13}C$ does not occur in connection with changes in abundance of *C. laevigata*, but it is possible that the two are related even with a time lag.

Erlandsson *et al.* (2006) suggested an increasing trend in oxygen consumption in the fjord since the 1950s due to both abiotic and biotic factors. This is based on long-term measurements of dissolved oxygen content from the fjord's deep water. Erlandsson *et al.* (2006) proposed that the increase in biological production is not local in the fjord but originates from the Skagerrak. However, there are large differences in sampling frequencies in the time series and the effect on oxygen consumption caused by increased bottom water temperatures that have characterized the fjord since the 1980s, these were not taken account in the analysis.

Brückner & Mackensen (2008) also noted a weak but significant decreasing trend from the 1950s until present day in the carbon isotope records from several benthic foraminiferal species (including *C. laevigata*). They interpret this as a result of a larger organic input to the sea floor due to changes in climate. This would agree with the findings of Erlandsson *et al.* (2006), but it could also be caused by a negative shift in $\delta^{13}C_{DIC}$ caused by burning of fossil fuels. Human activities, such as burning of fossil fuels and land-use changes, have emitted anthropogenic CO_2 which is strongly depleted in ^{13}C into the atmosphere. This increase of ^{13}C is also noted in the oceans' surface water, the so-called oceanic Suess effect (Keeling 1979).

Recent estimates suggest that the effect for the upper 1000 m of the North Atlantic is of the order $-0.026 \pm 0.002‰$ per year (Körtzinger *et al.* 2003); a global estimate is -0.018% per year (Gruber *et al.* 1999). It is possible that this is partly the cause of the decrease noted in the record from core GA113-2Aa as the deep water in the Gullmar Fjord partly originates from surface water in the North Sea. If it is assumed that the decrease recorded in the North Atlantic is of the same order in the North Sea, the Suess-related decrease between 1970 and 1992 would then be *c.* 0.6%, compensating for a large part of the decrease recorded in the Gullmar Fjord $\delta^{13}C$ record. The data from Hebbeln *et al.* (2006) also show a decrease in $\delta^{13}C$ record in the uppermost samples; however, the resolution is not high enough to determine anything conclusively. Brückner & Mackensen (2008) also noted a decrease in their $\delta^{13}C$ records. A similar decrease in benthic foraminiferal $\delta^{13}C$, also related to oceanic Suess effect, has been shown in Tagus Prodelta, Portugal (Lebreiro *et al.* 2006) and off the Moroccan coast, NW Africa (McGregor *et al.* 2007).

Conclusions

This paper present a well-dated high-resolution $\delta^{13}C$ record from the Skagerrak area covering the last 2400 a. Significant changes in the $\delta^{13}C$ record have occurred since the Viking Age (AD 800) and therefore in the system's carbon cycling. The causes of the changes might have had similar or dissimilar origins and different explanations are explored. The period of more negative $\delta^{13}C$ values both during the Viking Age/Medieval Warm Period and the Little Ice Age may result from a large influx of terrestrial carbon due to land-use changes, higher primary productivity or moderate changes in stagnation periods of the fjord's deep water. There are, however, no indications of considerably longer stagnation periods in the fjord or of anoxic periods over longer time periods until modern times. A rapid decrease in $\delta^{13}C$ occurred during the early 1970s, possibly related to the oceanic Suess effect and to the more frequent deep-water low-oxygen events that have taken place since the late 1970s. At this point it is not possible to explain all the processes behind the $\delta^{13}C$ changes, but it is evident that the flux and the sources of organic matter to the fjord have varied significantly over last millenium.

The authors would like to thank the crews of the RV *Arne Tiselius* and RV *Skagerak* for their assistance in collecting the cores. M. Segl is thanked for running the mass spectrometer at the University of Bremen, Germany. R. Harland checked the English of the text. The two reviewers, E. Alve and an anonymous reviewer, are thanked for their insightful comments on the manuscript. The study was financed by the Swedish Research Council (VR grants no 621-2004-5320, 621-2007-4369 (KN) and 621-2005-4265 (HLF)), Lamm Foundation, Göteborg University Marine Research Centre (GMF) and EUROPROX (European Graduate College – Proxies in Earth History). The PALEOSTUDIES program, University of Bremen, Germany covered the costs for stable O and C isotope analyses.

References

Alheit, J. & Hagen, E. 1997. Long-term climate forcing of European herring and sardine populations. *Fisheries Oceanography*, **6**, 130–139.

Appleby, P. G. & Oldfield, F. 1978. The calculation of Lead-210 dates assuming a constant rate of supply of unsupported ^{210}Pb to the sediment. *Catena*, **5**, 1–8.

Arneborg, L. 2004. Turnover times for the waters above sill level in Gullmar Fjord. *Continental and Shelf Research*, **24**, 443–460.

Arneborg, L., Erlandsson, C. P., Liljebladh, B. & Stigebrandt, A. 2004. The rate of inflow and mixing during deep-water renewal in a sill fjord. *Limnology and Oceanography*, **49**, 768–777.

Bradley, R. S., Briffa, K. R., Crowley, T. J., Hughes, M. K., Jones, P. D. & Mann, M. E 2001. The scope of medieval warming. *Science*, **292**, 2011–2012.

Bradshaw, E. G., Rasmussen, P. & Odgaard, B. V. 2005. Mid- to late-Holocene land-use change and lake development at Dallund So, Denmark: synthesis of multiproxy data, linking land and lake. *The Holocene*, **15**, 1152–1162.

Broecker, W. S. 2001. Paleoclimate – was the medieval warm period global? *Science*, **291**, 1497–1499.

Bronk Ramsey, C. 2005. Improving the resolution of radiocarbon dating by statistical analysis. *In*: Levy, T. E. & Higham, T. F. G. (eds) *The Bible and Radiocarbon Dating: Archaeology, Text and Science*. Equinox, London, 57–64.

Brückner, S. & Mackensen, A. 2006. Deep-water renewal in the Skagerrak during the last 1200 years triggered by the North Atlantic Oscillation: evidence from benthic foraminiferal delta O-18. *The Holocene*, **16**, 331–340.

Brückner, S. & Mackensen, A. 2008. Organic matter rain rates, oxygen availability, and vital effects from benthic foraminiferal delta C-13 in the historic Skagerrak, North Sea. *Marine Micropaleontology*, **66**, 192–207.

Dale, B. 2000. Dinoflagellate cysts as indicators of cultural eutrophication and industrial pollution in coastal sediments. *In*: Martin, R. E. (ed.) *Environmental Micropaleontology: the Application to Environmental Geology*. Topics in Geobiology. Kluwer Academic/Plenum publishers, New York, 305–321.

Ekman, F. L. 1870. Om hafsvattnet utmed Bohuslänska kusten. *Kongl. Svenska Vetenskaps-akademiens handlingar*, **9**, 16–21.

Erlandsson, C. P., Stigebrandt, A. & Arneborg, L. 2006. The sensitivity of minimum oxygen concentrations in a fjord to changes in biotic and abiotic external forcing. *Limnology and Oceanography*, **51**, 631–638.

Filipsson, H. L. 2003. *Recent changes in the marine environment in relation to climate, hydrography and human impact – with special reference to fjords on the Swedish west coast*. PhD thesis, Göteborg University, Göteborg, Sweden.

Filipsson, H. L. & Nordberg, K. 2004. Climate variations, an overlooked factor influencing the recent marine environment. An example from Gullmar Fjord, Sweden, illustrated by Benthic Foraminifera and Hydrographic Data. *Estuaries*, **27**, 867–881.

Fontanier, C., Mackensen, A., Jorissen, F. J., Anschutz, P., Licari, L. & Griveaud, C. 2006. Stable oxygen and carbon isotopes of live benthic foraminifera from the Bay of Biscay: microhabitat impact and seasonal variability. *Marine Micropaleontology*, **58**, 159–183.

Gil, I. M., Abrantes, F. & Hebbeln, D. 2006. The North Atlantic Oscillation forcing through the last 2000 years: Spatial variability as revealed by high-resolution marine diatom records from N and SW Europe. *Marine Micropaleontology*, **60**, 113–129.

Godhe, A. & McQuoid, M. R. 2003. Influence of benthic and pelagic environmental factors on the distribution of dinoflagellate cysts in surface sediments along the Swedish west coast. *Aquatic Microbial Ecology*, **32**, 185–201.

Gruber, N., Keeling, C. D. *et al.* 1999. Spatiotemporal patterns of carbon-13 in the global surface oceans and the oceanic Suess effect. *Global Biogeochemical Cycles*, **13**, 307–335.

Gustafsson, M. & Nordberg, K. 2001. Living (stained) benthic foraminiferal response to primary production and hydrography in the deepest part of the Gullmar Fjord, Swedish west coast, with comparisons to Höglund's 1927 material. *Journal of Foraminiferal Research*, **31**, 2–11.

Harland, R., Nordberg, K. & Filipsson, H. L. 2006. Dinoflagellate cysts and hydrographical change in Gullmar Fjord, west coast of Sweden. *Science of the Total Environment*, **355**, 204–231.

Harland, R., Nordberg, K. & Filipsson, H. L. 2010. A major change in the dinoflagellate cyst flora of Gullmar Fjord, Sweden, at around 1969/1970 and its possible explanation. *In*: Howe, J. A., Austin, W. E. N., Forwick, M. & Paetzel, M. (eds) *Fjord Systems and Archives*. Geological Society, London, Special Publications, **344**, 75–82.

Hebbeln, D., Knudsen, K. L. *et al.* 2006. Late Holocene coastal hydrographic and climate changes in the eastern North Sea. *The Holocene*, **16**, 987–1001.

Keeling, C. D. 1979. The Suess effect: 13Carbon-14 Carbon interrelations. *Environment International*, **2**, 229–300.

Körtzinger, A., Quay, P. D. & Sonnerup, R. E. 2003. Relationship between anthropogenic CO_2 and the C-13 suess effect in the North Atlantic Ocean. *Global Biogeochemical Cycles*, **17**, 1005, doi: 10.1029/2001gb001427.

Lamb, H. H. 1995. *Climate, History and the Modern World*. Routledge, London.

Lebreiro, S. M., Frances, G. *et al.* 2006. Climate change and coastal hydrographic response along the Atlantic Iberian margin (Tagus Prodelta and Muros Ria) during the last two millennia. *The Holocene*, **16**, 1003–1015.

Lindahl, O. 2007. Primary production in the Gullmar Fjord. *In*: Håkansson, B. (ed.) *Swedish National Report on Eutrophication Status in the Kattegat and the Skagerrak*. OSPAR ASSESSMENT 2007, 41–43.

McCorkle, D. C., Corliss, B. H. & Farnham, C. A. 1997. Vertical distributions and stable isotopic compositions of live (stained) benthic foraminifera from the North Carolina and California continental margins. *Deep-Sea Research Part I–Oceanographic Research Papers*, **44**, 983–1024.

McCorkle, D. C., Bernhard, J. M., Hintz, C. J., Blanks, J. K., Chandler, G. T. & Shaw, T. J. 2008. The Carbon and oxygen stable isotopic composition of cultured benthic foraminifera. *In*: Austin, W. E. N. & James, R. H. (eds) *Biogeochemical Controls on Paleoceanographic Environmental Proxies*. Geological Society, London, Special Publications, **303**, 135–154.

McGregor, H. V., Dima, M., Fischer, H. W. & Mulitza, S. 2007. Rapid 20th-century increase in coastal upwelling off Northwest Africa. *Science*, **315**, 637–639.

Nordberg, K., Gustafsson, M. & Krantz, A.-L. 2000. Decreasing oxygen concentrations in the Gullmar Fjord, Sweden, as confirmed by benthic foraminifera, and the possible association with NAO. *Journal of Marine Systems*, **23**, 303–316.

Reimer, P. J., Baillie, M. G. L. *et al.* 2004. Intcal04 terrestrial radiocarbon age calibration, 0–26 cal kyr BP. *Radiocarbon*, **46**, 1029–1058.

Rodhe, J. 1998. The Baltic and North Seas: a process-oriented review of the physical oceanography. *In*: Robinson, A. R. & Brink, K. H. (eds) *The Sea*. John Wiley & Sons, New York, 699–732.

Rohling, E. J. & Cooke, S. 1999. Stable oxygen and carbon isotopes in foraminiferal carbonate shells. *In*: Sen Gupta, B. K. (ed.) *Modern Foraminifera*. Kluwer Academic Publishers, Dordrecht/Boston/London, **297**, 239–258.

Rydberg, L. 1977. *Circulation in the Gullmaren – a sill fjord with externally maintained stratification*. Department of Oceanography, Göteborg University, Report no **23**.

SMHI 1994. Vattenföring i Sverige, Del 4 Vattendrag till Västerhavet. no 43, SMHI Svenskt Vattenarkiv.

Svansson, A. 1984. Hydrography of the Gullmar fjord. *Meddelande från Havsfiskelaboratoriet*, Lysekil, Sweden, **297**, 1–22.

Thorsen, T. A., Dale, B. & Nordberg, K. 1995. 'Blooms' of the toxic dinoflagellate Gymnodinium catenatum as evidence of climatic fluctuations in the late Holocene of southwestern Scandinavia. *The Holocene*, **5**, 435–446.

Wanner, H., Beer, J. *et al.* 2008. Mid- to Late Holocene climate change: an overview. *Quaternary Science Reviews*, **27**, 1791–1828.

Wefer, G., Berger, W. H., Behre, K.-E. & Jansen, E. (eds) 2002. *Climate Development and History of the North Atlantic Realm*. Springer-Verlag, Berlin Heidelberg, New York.

Zillen, L., Conley, D. J., Andren, T., Andren, E. & Björck, S. 2008. Past occurrences of hypoxia in the Baltic Sea and the role of climate variability, environmental change and human impact. *Earth-Science Reviews*, **91**, 77–92.

Climate proxies for recent fjord sediments in the inner Sognefjord region, western Norway

MATTHIAS PAETZEL* & TORBJØRN DALE

Sogn og Fjordane University College, The Faculty of Science, P.O.Box 133, N 6851 Sogndal, Norway

** Corresponding author (e-mail: matthias.paetzel@hisf.no)*

Abstract: Three high-resolution sediment cores from the periodically anoxic Inner Barsnesfjord and the oxic Outer Barsnesfjord and Sogndalsfjord, Western Norway, are analysed for signals of climate variation. Sedimentation rates are 0.85, 0.75 and 0.45 cm a^{-1}, respectively. Sediment slices are taken of 0.3–0.5 cm thickness revealing annual resolution. Corresponding peaks of mineral clay particles, total organic matter and freshwater diatoms correlate well with maxima in regional precipitation, temperature and cloud cover over the last 20 years. This regional climate record is also correlated to the wider NAO (North Atlantic Oscillation) winter index, northern hemisphere temperature and solar activity (cosmic rays and open solar flux) based on 60 years of continuous observation. There is a strong indication that the fjord sediment record contains a climate archive spanning the last 20–100 a; it is suggested that these proxies might also work on longer timescales. A simple box model is presented describing the use of the sediment record for the interpretation of climate variability on a regional scale and at a longer timescale on a millennium scale. It is predicted that this model can be transferred to sediment settings of similar regional climate influence.

The impact of climate change has been widely discussed over the last 20 years. The interpretation of this impact is based on two major premises: (1) the reliable documentation of the geological record of climate change, and (2) the modelling of climate change using recent, historical and geological climate proxies. For a comprehensive summary see IPCC (2007) and Jones *et al.* (2009).

Here we contribute to the understanding on how climate variation has influenced sediment deposition during the last 20 years. The periodically anoxic Inner Barsnesfjord, the oxic Outer Barsnesfjord and the oxic Sogndalsfjord, Western Norway (Fig. 1), reveal high-resolution sediments with sedimentation rates of 0.5–1.0 cm a^{-1} (Paetzel & Schrader 1991). These fjord sediments provide the opportunity to identify sediment signals of annually documented environmental change (Paetzel & Schrader 1992) and climate variation. The recent sediment evidence is converted into climate proxies. The ultimate goal will be to use these proxies for reconstructing the regional climate during the last 1000 a, including the build up and retreat of the Little Ice Age.

The major objectives of this study are to answer the following questions.

(1) Do recent fjord sediments hold signals that can be related to documented regional climate variation over the last 20 years?

(2) Are there indications of possible links between regional fjord deposition processes and hemispherical climate change observations?

The last 100 years of Barsnesfjord sediment deposition are constrained using fallout ^{137}Cs from the Chernobyl radioactive event (1986) and from the 1963 maximum release from the world-wide nuclear weapon testing. A supporting chronology is gained from recognizing documented historical slide events in the sediment (Paetzel & Schrader 1991). On the other hand, dating of the sediments between 100 and 1000 a BP (before present) remains uncertain. Seasonal lamination is inconsistent and cannot be counted continuously; the ^{14}C-dating method does not produce reliable dating of marine sediments younger than 335–801 a BP (Hughen *et al.* 2004).

While the dating of fjord sediments seems difficult for the last 1000 a, it is also possible to correlate the sediment signals with meteorological variations over the last 100 a. These are represented by annual and seasonal averages, including winter precipitation and summer temperature as suggested by Nesje *et al.* (2001). This regional variation can be extended to established timescales of palaeoclimatological and even cosmic records, as this paper will show. Millennium to Holocene palaeoclimatological timescales, among others, are evident from ice cores or the northern hemisphere climate

From: HOWE, J. A., AUSTIN, W. E. N., FORWICK, M. & PAETZEL, M. (eds) *Fjord Systems and Archives*. Geological Society, London, Special Publications, **344**, 271–288.
DOI: 10.1144/SP344.19 0305-8719/10/$15.00

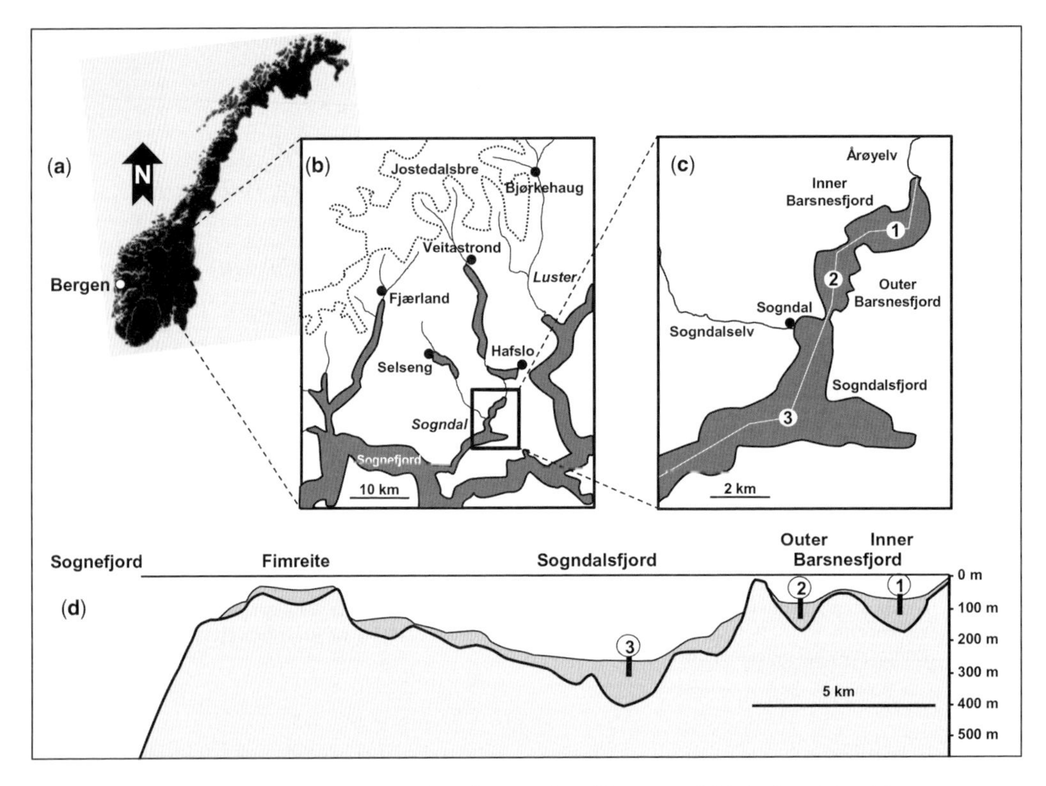

Fig. 1. Locations maps of (**a**) western Norway including the city of Bergen; and (**b**) the Sogndal and Luster municipalities. The dashed line indicates the glacier Jostedalsbre. Meteorological stations are Luster: Bjørkehaug, Hafslo and Sogndal: Selseng, Fjærland. (**c**) The Inner and Outer Barsnesfjord and the Sogndalsfjord. Numbers indicate the 2007 location 1 (=BC-74-6 of 1988), location 2 (=BC-74-2 of 1988) and location 3; the white line illustrates the transect line of the bathymetrical profile. (**d**) The bathymetrical profile including water depth and sediment thickness (Dale & Hovgaard 1993); numbers correspond to locations as shown in map (c).

variability as documented for example in the North Atlantic Oscillation (NAO) index, summarized in the recent reviews of Wanner *et al.* (2008) and Jones *et al.* (2009). Cosmic forcing has been reconstructed and related to temperature variability for the last 1000 years and throughout the Holocene Period. This includes observations of the atmospheric production rate of the ^{14}C and ^{10}Be sunspot number index, the open solar magnetic flux and cosmic rays (Usoskin *et al.* 2005; Bard & Frank 2006; Usoskin & Kovaltsov 2008).

This paper presents the first step towards a regional 1000 year timescale reconstructed from Norwegian fjord sediment signals. The relationship between regional climatic indicators and solar forcing parameters is established from literature values for the last 20 years of direct and continuous measurement. Climate variation is correlated with fjord sediment parameters of annual resolution during the same time period.

Setting

The Barsnesfjord and the Sogndalsfjord are two NE–SW oriented tributaries of the Sognefjord, Western Norway (Fig. 1a, b). The Barsnesfjord consists of an inner, periodically anoxic, basin of maximum water depth 66 m and of an outer, mostly oxic, basin of maximum water depth 80 m. The Inner Barsnesfjord is connected with the Outer Barsnesfjord (Fig. 1c) by a 29 m deep sill. Another 7.5 m deep sill marks the boundary between the Outer Barsnesfjord and the oxic basin of the Sogndalsfjord, which has a maximum depth of 260 m (Fig. 1c). The Sogndalsfjord connects across a 25 m deep sill at Fimreite to the >900 m deep Sognefjord. This combination of shallow sills is restricting water circulation in the Barsnesfjord (Paetzel & Schrader 1991; Dale & Hovgaard 1993). Maximum sediment thickness exceeds 50–100 m of Holocene deposition in all three investigated fjord basins (Fig. 1d).

The river Årøyelv (Fig. 1c) represents the major freshwater inlet into the northern end of the Inner Barsnesfjord basin. The Årøyelv is connected with the lake Hafslovatnet (169 m above sea level), located SW of Hafslo (Fig. 1b). The Hafslovatnet is connected across a shallow sill to the lake Veitastrondvatnet (172 m above sea level), located south of Veitastrond (Fig. 1b). The two lakes gain their freshwater from a catchment area of 429 km^2, including the glacial river input from the glacier Jostedalsbre in the north. The riverine freshwater input leads to a stratified water column in the Inner Barsnesfjord from April to September, prevailing into the winter season during mild winters (Dale & Hovgaard 1993). Oxygen concentrations decrease in the fjord bottom water due to the sluggish water renewal. This favours the formation of periodic sediment anoxia in the Inner Barsnesfjord (Paetzel & Schrader 1991; Dale & Hovgaard 1993). Similarly, seasonal water stratification prevails in the Outer Barsnesfjord. Here, water circulation is stronger, leading to slightly oxic conditions in bottom waters and in the sediment; oxygen levels occasionally drop below the critical value of 2 ml l^{-1} (Paetzel & Schrader 1991; Dale & Hovgaard 1993).

Since 1983, the water of the Årøyelv has been supplied to the Inner Barsnesfjord mainly through a water tunnel, generating hydropower. The building of the hydropower plant changed the mode of freshwater supply into the Barsnesfjord. The annual freshwater flow rate is artificially kept at a similar level as before the regulation of the water course. Thus, only minor changes have been observed in the Inner Barsnesfjord sediment when comparing the deposits before and after the introduction of the hydropower plant in 1983 (Paetzel & Schrader 1991, 1992).

The Sogndalsfjord obtains its main freshwater input from the river Sogndalselv (Fig. 1b) with a catchment area of 178 km^2 including the lake Dalavatnet (395 m above sea level), located SE of Selseng (Fig. 1c). The Sogndalsfjord water column is also stratified. Oxygen concentrations stay above the critical value of 2 ml l^{-1} at the fjord bottom, although conditions have deteriorated throughout the last two decades (Dale & Hovgaard 1993). Tidal variation is on average 1.6 m per tidal cycle. A possible tidal effect on the sediment has not been investigated. The effect is assumed to be minor, as no major tidal flats are exposed during low water periods.

Together, the Inner and Outer Barsnesfjord reveal a water volume of 0.16 km^3 with 1.0 km^3 a^{-1} freshwater supply originating from the Årøyelv. The water volume of the Sogndalsfjord approximates 1.9 km^3 with 0.5 km^3 a^{-1} freshwater supply; 0.3 km^3 originate from the Sogndalselv (Dale & Hovgaard 1993).

Material and methods

During the autumn of 2007, sets of two parallel 0.5 m long sediment cores with intact sediment–water interfaces were retrieved from the deepest parts of the Inner Barsnesfjord (location 1 at 63 m water depth; number 1 in Fig. 1c, d), the Outer Barsnesfjord (location 2 at 80 m water depth; number 2 in Fig. 1c, d) and the Sogndalsfjord (location 3 at 260 m water depth; number 3 in Fig. 1c, d) using a Niemistö (1974) sediment corer.

Further, in summer 1988 sediment cores were retrieved and dated from nearby locations of the Inner Barsnesfjord, that is, BC-74-6 at 62 m water depth (Paetzel & Schrader 1992) and of the Outer Barsnesfjord, that is BC-74-2 at 75 m water depth (Paetzel & Schrader 1991), as summarized in Table 1. No additional sediment cores exist from location 3 of the Sogndalsfjord.

The 2007 sediment cores were generally described by visual inspection of structures, texture and particulate matter and by colour estimation using the colour code of Munsell (1994). Core quality was ascertained using sediment physical parameter determination, that is wet and dry bulk density, water content and porosity according to the method described in Paetzel & Schrader (1991). The same methods are used for the sediment core description of the 1988 investigation (Paetzel & Schrader 1991, 1992).

Sedimentation rates were determined for the 1988 cores BC-74-2 (Outer Barsnesfjord) and BC-74-6 (Inner Barsnesfjord) using ^{137}Cs from the 1963 maximum fallout of the world-wide nuclear weapon testing and from the fallout of the 1986 Chernobyl radioactive accident (Paetzel & Schrader

Table 1. *Sampling locations in longitude and latitude of the 2007 and 1988 core samples*

2007 core locations	Latitude N	Longitude E	Water depth (m)	1988 core locations	Latitude N	Longitude E	Water depth (m)
Location 1	61°15′18″	07°09′10″	63	BC-74-6	61°15′26″	07°09′50″	62
Location 2	61°14′31″	07°07′31″	80	BC-74-2	61°14′41″	07°07′45″	75
Location 3	61°12′32″	07°06′18″	260				

1991, 1992). In addition, sediment horizons of historically documented slide events were used for dating. The retrieval of the sediment surface remains uncertain in the 1988 cores due to the use of a box corer that did not preserve the sediment–water interface. Sedimentation rates of the uppermost sediments are therefore approximations.

The 2007 sediment cores (locations 1 and 2) were correlated with the 1988 cores using the visual facies distribution and the clay contents (Fig. 2). Sedimentation rates were transferred between the two sets of cores after adding the 2007 dating horizon of the sediment surface that was preserved using a Niemistö (1974) corer. Linear sedimentation rates were determined from the 2007 sediment surface down to the 1986 Chernobyl dating horizons. Sedimentation rates of location 3 (Sogndalsfjord) are estimated purely based on the visual correlation and the clay content between the cores of locations 2 and 3 (Fig. 2). Thus, location 3 sedimentation rates remain uncertain.

Sediment cores were opened into two halves. The sediment cores were subsampled continuously at frequencies based on the estimation of the averaged linear sedimentation rates. The subsampled sediments were homogenized across their range. The homogenized subsamples were then transferred onto smear slides.

Sediment particle composition was investigated using the smear slide technique described by Rothwell (1989). Naphrax (Brunel Microscopes Ltd; refractive index of 1.65–1.73) was used as a mounting medium. In smear slides, the percentage distribution of the respective particle fraction is estimated within three representative areas of each smear slide. Averaged percentages are determined across these three areas. The smear slide analysis was carried out for the following parameters.

(1) Mineral grain size: the grain sizes distinguished were clay (<2 μm), silt (2–63 μm) and sand (63 μm to 2 mm). The single grain size is presented as a percentage of the total mineral grains.
(2) Total organic matter: the surface area of the organic matter fraction was estimated in relation to the total surface area covered by

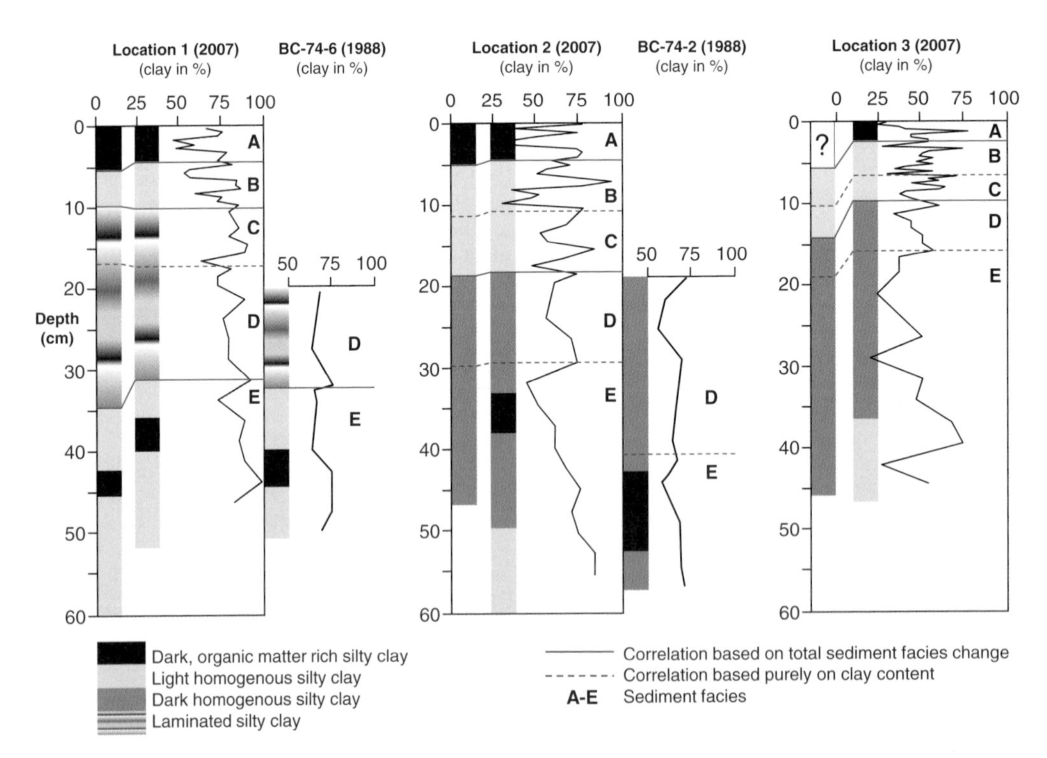

Fig. 2. Core-to-core correlation of the 2007 sediment cores of location 1–3 and the 1988 sediment cores of BC-74-6 and BC-74-2 (Paetzel & Schrader 1991, 1992). Lithology is plotted against the mineral clay fraction; letters A–E indicate successive sediment facies. Solid red lines mark facies boundaries following lithology units and dashed red lines mark facies boundaries following the mineral clay fraction. The question mark of location 3 refers to uncertainties about the recovery and correct description of the sediment surface.

the remaining particles. The total organic matter fraction is presented as a percentage of the surface area of the total particle fraction.

(3) Terrestrial organic matter: terrestrial organic matter represents all land-derived organic particles, that is pieces of higher plants (leaves, bark, grass, etc.) or pollen. Plant fragments occur in smear slides usually as dark brownish-black pieces with clear edges and shapes. They are easily distinguished from the primary produced marine organic matter that occurs in smear slides as light brownish, loosely packed aggregates without clear shapes. The terrestrial organic matter fraction is presented as a percentage of the total organic matter fraction.

Note that all smear slide analyses represent the surface area distribution of the respective particle fraction. The surface area distribution will deviate from the determination of weight percentages of the same fraction. The advantage of using smear slides (prior to weight determination) are: the method is non-destructive; the sediment material can be repeatedly analysed; analyses can be performed on a minimal amount of sediment material; and smear slides reveal the *in situ* distribution of single sediment particles and can be put in relation to the remaining sediment fraction.

The main limitation is the non-comparability of the smear-slide-estimated organic matter fraction with the instrumentally measured, weight-based organic matter determination; the relative changes should be similar, however. Further, mineral grains larger than sand cannot be accounted for and mineral grain sizes within the clay fraction cannot be distinguished.

The smear slides were also used for the diatom analysis. For the climate proxy study it was important to identify the parallel variation of different sediment particles within the same depth and thus time horizon. This parallel seasonal resolution of all sediment paramcters could only be guaranteed if all parameters were investigated within the same material. Thus, smear slide analysis was chosen for the diatom investigation prior to the analysis of regular diatom slides.

Smear slides have the disadvantage that single diatoms might be overlain by organic or mineral matter. On the other hand, diatoms are abundant in the fjord sediments. Counts of at least 150–200 specimens per slide assured a statistically robust number. Counting occurred continuously along at least three transects across each smear slide. Due to the use of smear slides, diatoms are presented as a percentage of total counts.

Some of the counted diatoms (e.g. *Skeletonema costatum*) are very thinly silicified and are thus easily dissolved. Dissolution could falsify the results of the diatom counting. This problem seemed to be negligible in the Barsnesfjord and Sognefjord sediments. Even the smallest specimens of *Skeletonema costatum* showed clearly visible internal structures; no dissolution patches were observed at the outer diatom skeletons. The preservation of the diatoms was in general excellent.

All diatoms were counted. Only three diatom taxa were identified on a species level: the dominating freshwater species was *Tabellaria flocculosa*, and the dominating marine species were *Skeletonema costatum* and *Chaetoceros* resting stages. The remaining species were grouped into: (a) circular diatoms mostly consisting of diverse marine *Thalassiosira* species; (b) benthic diatoms consisting of relatively large and heavily silicified skeletons, often belonging to the order pennales; and (c) other diatom species including diverse marine *Thalassionema* species, unidentified marine diatom species and freshwater species other than *Tabellaria flocculosa*.

Tabellaria flocculosa, *Skeletonema costatum* and the *Chaetoceros* resting stages were the most prominent species in the smear slides. Although it is uncommon (Stoermer & Smol 1999), these species were taken out of the total assemblage and presented on single taxa level. The reasoning is their significance for indicating the shift between favourably marine conditions (*Skeletonema costatum*) and favourably freshwater supply (*Tabellaria flocculosa*). The *Chaetoceros* resting stages indicate less favourable marine conditions following, for example, a major bloom (Hustedt 1930–1966). Their occurrence suggests deposition during the final stage of favourable marine conditions. In combination, these three species are expected to be useful indicators for either condition. They are therefore presented as a percentage ratio between freshwater and marine species.

Tabellaria flocculosa does not live in saline conditions as present in the Barsnesfjord or Sogndalsfjord (Hustedt 1930–1966). The occurrence of *Tabellaria* in fjord sediments thus indicates either phases of enhanced freshwater runoff from the surrounding lakes and rivers or phases of enhanced primary productivity in the main source area. These source areas are the river Årøyelv and the lake Hafslovatnet (for the Barsnesfjord) and the river Sogndalselv and the lake Dalavatnet (for the Sogndalsfjord). Its main occurrence is planktonic and benthic in littoral areas of lakes and in rivers originating from these lakes (Hustedt 1930–1966). *Tabellaria flocculosa* is acidophilus, occurring mainly in acid waters of pH <7 (Round *et al.* 1990). The Hafslovatnet reveals an acidity of *c.* pH $= 6.5$ during September. *Tabellaria flocculosa* is dominating the freshwater diatom fraction in the sediment of all sediment cores and is therefore used as a general freshwater supply indicator.

Skeletonema costatum and the *Chaetoceros* resting stages are used as a general marine condition indicator. Both species occur purely as marine phytoplankton (Hustedt 1930–1966), and suggest elevated primary productivity levels in Norwegian fjord environments (Jahnke *et al.* 1983). Both species are abundant in all sediment cores.

The E-KLIMA climate database of the Norwegian Meteorological Institute (DNMI 2009) has been used to retrieve regional precipitation, temperature, sun hour and cloud cover measurements between 1900 and 2007. Precipitation is given as a cumulative in total mm a^{-1}. Temperature is calculated as annual mean temperature (°C a^{-1}). Sun hours represent the cumulative annual hours of sunshine (hours a^{-1}). Cloud cover is a measured descriptive in numbers from 0 to 8; observation number 0 corresponds to no clouds at all and observation number 8 corresponds to full cloud cover. A cloud index is calculated from the arithmetic mean over three observations per day at 07 h, 13 h and 19 h and averaged on an annual basis (cloud index a^{-1}). Annual resolution of climate data is chosen to match the annual resolution of the sediment subsamples over the last 20 years.

The DNMI meteorological stations are listed in Table 2; their positions are illustrated in Figure 1. It is necessary to refer to different nearby meteorological stations as the meteorological data are inconsistent from station to station. For the Barsnesfjord, the closest meteorological stations are located in the Luster municipality (Fig. 1b). Station Hafslo (precipitation, Fig. 1b) is directly located within the catchment area of the Årøyelv. Station Bjørkehaug (sun hours, Fig. 1b) is located outside the catchment area; it is the only station close to the Barsnesfjord and Sogndalsfjord that reveals a measurement of sun hours.

For the Sogndalsfjord, the closest meteorological stations are located in the Sogndal municipality (Fig. 1b). Station Selseng (precipitation, Fig. 1b) is directly located within the catchment area of the Sogndalselv. Station Fjærland (Fjærland-Skarestad, cloud cover and temperature, Fig. 1b) is located outside the catchment area; it is the only station close to the Barsnesfjord and Sogndalsfjord that reveals observations of the cloud cover and temperature.

Due to the inconsistency between the local meteorological stations, the complete (apart from the cloud cover) meteorological dataset from Bergen (Fig. 1a) is used as a reference for Western Norway, ensuring the compatibility of the meteorological data between the different stations and in the region. For illustration of the compatibility, all meteorological data are correlated with each other in Table 3.

Another correlation is made between annual and winter precipitation and annual and summer temperature of the meteorological stations Hafslo and Selseng. Both are useful parameters for the description of glacier movement, and thus for the supply of glacially derived material into proglacial lake areas (e.g. Nesje *et al.* 2001; Nesje & Dahl 2003).

Results

Core description

Figure 2 reveals the different sediment facies (A–E) within the single cores of the 1988 (BC-74-6 and BC-74-2) and the 2007 (locations 1, 2 and 3)

Table 2. *The meteorological stations of the Norwegian Meteorological Institute (DNMI 2009), their references and recordings*

DNMI meteorological stations	DNMI site reference	Figure 1 map reference	Recorded parameters	Recorded years	Annual data point number
Bergen: Florida	50540	Bergen	Precipitation	1983–2008	$N = 26$
			Temperature	1949–2008	$N = 60$
Bergen: Fredriksberg	50560	Bergen	Precipitation	1904–1984	$N = 81$
			Temperature	1904–1984	$N = 81$
			Sun hours	1957–2005	$N = 49$
Sogndal: Fjærland-Skarestad	55840	Fjærland	Cloud cover	1954–2004	$N = 51$
			Temperature	1921–2008	$N = 85$*
Sogndal: Selseng	55730	Selseng	Precipitation	1900–2008	$N = 109$
Luster: Bjørkehaug	55430	Bjørkehaug	Sun hours	1964–2003	$N = 24$**
Luster: Hafslo	55550	Hafslo	Precipitation	1900–2008	$N = 109$

Note: N corresponds to the number of data points used for the correlation of Table 3; one asterisk* indicates no observations during the years from 1951 to 1952 and 2005; two asterisks** indicate no observations during the years from 1971 to 1972, 1982 to 1983 and 1991 to 2002; station Bergen-Fredriksberg was replaced by station Bergen-Florida in 1984.

Table 3. *Correlation factor* (r) *between climate parameters; location names correspond to DNMI meteorological stations of Figure 1 and Table 2; the Bergen data represent the correlation of the combined Bergen-Fredriksberg and Bergen-Florida record; number of data points is given in Table 2*

	Precipitation Bergen	Precipitation Hafslo	Precipitation Selseng	Sun hours Bergen	Sun hours Bjørkehaug	Cloud cover Fjærland	Temperature Bergen	Temperature Fjærland
Precipitation Bergen	1							
Precipitation Hafslo	0.8	1						
Precipitation Selseng	0.8	0.9	1					
Sun hours Bergen	−0.6	−0.6	−0.6	1				
Sun hours Bjørkehaug	−0.7	−0.5	−0.6	0.9	1			
Cloud cover Fjærland	0.5	0.4	0.6	−0.4	−0.5	1		
Temperature Bergen	0.4	0.3	0.4	0.1	0.2	0.3	1	
Temperature Fjærland	0.4	0.3	0.4	−0.1	0.0	0.3	0.9	1

sediment sampling. Sediment facies consist mainly of silty clay ranging from organic-matter-rich deposits and dark and light homogenous sediments to lamination. Note that facies C and D reveal laminated sediments in the Inner Barsnesfjord while these facies happen to be homogenous in the Outer Barsnesfjord and in the Sogndalsfjord (Fig. 2). This represents the different oxygen conditions at the bottom of the fjord basins.

Figure 2 also indicates the correlation between the cores based on visual description and the clay content. The correlation reveals a similar variation in the mineral clay particle content in the 2007 and the 1988 sediment cores. Sediments deposited prior to 1988 seem more compacted in the 2007 cores compared to the 1988 cores. The 1988 cores correlate with the 2007 cores below about 20 cm of location 1 (Inner Barsnesfjord), and below about 19 cm of location 2 (Outer Barsnesfjord). Note that the correlation horizon is different from the dating horizon. Compaction might be due to the accumulated sediment load over the last 20 years. The compressed sediment record might also be the result of using different sampling equipment.

Five main facies (from A to E) are distinguished. The top facies (facies A) correlates to the same facies boundary across the three cores taken in 2007. The boundary between facies B and C is only visible in the clay distribution of location 1. Boundaries of facies B and C are confirmed by the clay content and the sediment structure of locations 2 and 3. Sediment structures reveal laminated sediments at location 1 and homogenous sediments at location 2 and 3 across these two facies. The different behaviour of facies B and C can be explained by different oxygen levels during deposition in the three basins, suggesting periodically anoxic conditions in the Inner Barsnesfjord (location 1), while oxic conditions prevail in the Outer Barsnesfjord (location 2) and in the Sogndalsfjord (location 3).

The boundaries of the lowermost facies (D and E) are a bit uncertain due to the higher subsample frequency. Different structures also suggest variable oxygen conditions at equal time horizons between the fjord basins. The boundaries between D and E are visible in the clay fraction of the 2007 and the 1988 sediments. Both sets of sediment correlate well with each other, taking the compaction of the 2007 sediments into account.

This correlation is used to transfer sedimentation rates gained during the 1988 research (Paetzel & Schrader 1991, 1992) to the sediments of the 2007 investigation.

Sedimentation rates

Sedimentation rates are summarized in the age–depth relationship graph of Figure 3. Linear sedimentation rates are averaged across the combined Inner Barsnesfjord sediment cores retrieved at location 1 (Loc 1) in 1988 and 2007. The linear sedimentation rates range from 0.84 cm a^{-1} (Loc 1: 2007–1975, red line in Fig. 3a, b) and 0.68 cm a^{-1} (Loc 1: 1986–1975, green line in Fig. 3a, b) to 1.03 cm a^{-1} (Loc 1: 1975–1904, blue line in Fig. 3a, b), resulting in averaged linear sedimentation rates of 0.85 cm a^{-1} for the combined cores of this location (Fig. 3a, b).

The 1988 investigation revealed averaged linear sedimentation rates of 0.99 cm a^{-1} in the Outer Barsnesfjord sediment cores at location 2

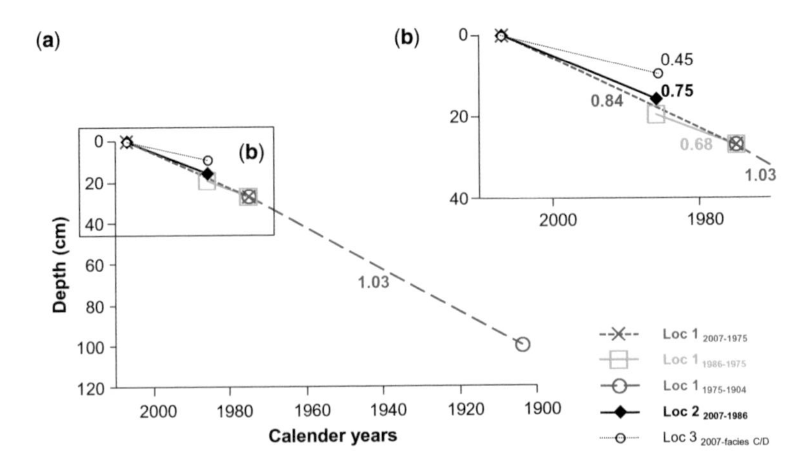

Fig. 3. (**a**) Age-depth relationship of the combined sediment cores taken at locations (Loc) 1, 2, and 3 during the 1988 and the 2007 investigation; numbers indicate sedimentation rates in cm a^{-1}. (**b**) An enlargement of the last 20 years of the age-depth relationship.

(Paetzel & Schrader 1991, 1992). These sedimentation rates were overestimated due to slide deposits that Paetzel and Schrader (1991, 1992) did not account for. Combining the results from the 1988 and 2007 investigations reveals an adjusted averaged linear sedimentation rate of 0.75 cm a^{-1} for the sediment cores retrieved at location 2 (Loc 2: dark black line in Fig. 3a, b).

Sedimentation rate estimation in the Sogndalsfjord core (location 3) is based on the facies correlation between the sediment cores of location 2 and 3 (Fig. 2). This correlation suggests averaged linear sedimentation rates of about 0.45 cm a^{-1} in the Sogndalsfjord (Loc 3: light black line in Fig. 3a, b). The sedimentation rate at location 3 is only a rough estimate as it is solely based on the visual correlation of similar sediment facies in neighbouring cores. On the other hand, this number is supported by the annual freshwater supply to the Sogndalsfjord. The freshwater supply to the Sogndalsfjord reveals about half of the amount of freshwater that is reaching the Barsnesfjord. This suggests that the Sogndalsfjord receives only half of the amount of nutrients and mineral matter from land compared to the Barsnesfjord. This restriction would result in sedimentation rates of more or less half of what is determined for the Barsnesfjord, suggesting sedimentation rates of 0.4–0.5 cm a^{-1} at location 3. Additional arguments for generally lower sedimentation rates in the Sogndalsfjord are the larger volume of the Sogndalsfjord, greater water depth and larger distance of the sampling location to the main freshwater source. Note that this is only a rough estimate and that the sedimentation of organic material based on marine nutrient supply is not accounted for.

Subsamples

The averaged linear sedimentation rates of location 1 (0.85 cm a^{-1}), location 2 (0.75 cm a^{-1}) and location 3 (0.45 cm a^{-1}) were used to decide on the thickness of the subsamples. At locations 1 and 2, subsample thickness is 0.5 cm across the upper 0–10 cm of the sediment core, resulting in a time resolution of minimum one year per subsample for the last 20–25 a. Beyond, subsample resolution increases to 1 cm thickness per subsample between 10 and 20 cm (averaging about 2 a per subsample) to 2.6 cm thickness per subsample at sediment depth greater than 20 cm (averaging about 5 a per subsample).

At location 3, a subsample thickness of 0.3 cm was chosen across the upper 0–9 cm of the sediment core, based on the averaged sedimentation rate of 0.45 cm a^{-1}. The subsamples also reveal a time resolution of minimum 1 a per subsample for the last 20–25 a. Corresponding to the subsampling of locations 1 and 2, a subsample thickness of 0.9 cm was chosen between 9 and 17.1 cm core depth and of 2.6 cm thickness at sediment depth greater than 17.1 cm.

Meteorological data

Table 3 shows a correlation of $r = 0.5$ to $r = -0.7$ between the different precipitation, cloud cover and sun hour records. A correlation between these records and temperature seems to be absent ($r = 0.0$ to $r = 0.4$). Although statistically not significant, a co-variation exists between the major temperature peaks, precipitation and cloud cover, as will be shown later.

Glacial variability influences the freshwater and particulate material supply into the fjords. Glacial variability depends on winter precipitation and summer temperature (e.g. Nesje *et al.* 2001; Nesje & Dahl 2003). These have to be taken into account to ensure the compatibility of the glacial record with the sediment record. Table 4 reveals a correlation of $r = 0.6$ between annual and winter precipitation and between annual and summer temperature at Hafslo and at Selseng. This indicates that the main input to the annual precipitation and annual temperature record occurs during winter (precipitation) and summer (temperature) respectively. It also ensures that the annual precipitation and temperature graphs can be used for further investigation as they basically reveal the same trends. This is of advantage in this study as the sediment data and all climate correlations (Figs 4–6) are based on annual averages.

The annual resolution of the subsamples allows the results from the smear slide analysis to be correlated with the record of annual precipitation in the respective areas (Fig. 1, meteorological stations Hafslo and Selseng for the Barsnesfjord and the Sogndalsfjord basin), gaining a high-resolution record of sediment change in relation to climate variation (here represented by precipitation) over the last 20 years (Figs 4–6).

The precipitation record of the Hafslo meteorological station can be divided into four major precipitation peaks over the last 20 years (Figs 4 & 5): peak 1 (from 2007 to 2006), peak 2 (from 2004 to 2003), peak 3 (from 2000 to 1997) and peak 4 consisting of three subpeaks (from 1995 to 1994; 1992; and from 1990 to 1988). The meteorological station at Selseng (Fig. 6) shows an additional subdivision of peak 1 (2007 and 2005). Peak 2 (from 2004 to 2003) occurs only as minor peak while peaks 3 and 4 coincide with the Hafslo record. Most precipitation maxima (here called precipitation peaks) and minima correlate with maxima and minima of the respective sediment parameters in all cores.

Table 4. *Correlation factor (r) between winter and annual precipitation and summer and annual temperature at the meteorological stations Hafslo, Selseng and Fjærland*

	Winter precipitation Hafslo	Annual precipitation Hafslo	Winter precipitation Selseng	Annual precipitation Selseng	Summer temperature Fjærland	Annual temperature Fjærland
Winter precipitation Hafslo	1					
Annual precipitation Hafslo	0.6	1				
Winter precipitation Selseng	0.9	0.5	1			
Annual precipitation Selseng	0.5	0.9	0.5	1		
Summer temperature Fjærland	−0.1	−0.1	0.0	−0.1	1	
Annual temperature Fjærland	0.3	0.3	0.4	0.4	0.6	1

Note: $r = 0.6$ and $r = 0.5$ for correlations between annual and winter precipitation, as well as for annual and summer temperature; number of data points is given in Table 2.

Inner Barsnesfjord

In the Inner Barsnesfjord (location 1, Fig. 4), peaks 1, 2 and 3 of increased precipitation coincide with concurrent peaks of increasing clay, total organic matter and the freshwater supply indicator. The freshwater supply indicator even resembles two of the peak 4 subpeaks. Clay content and total organic matter only shows a general increase across the peak 4 horizon. The percentage of

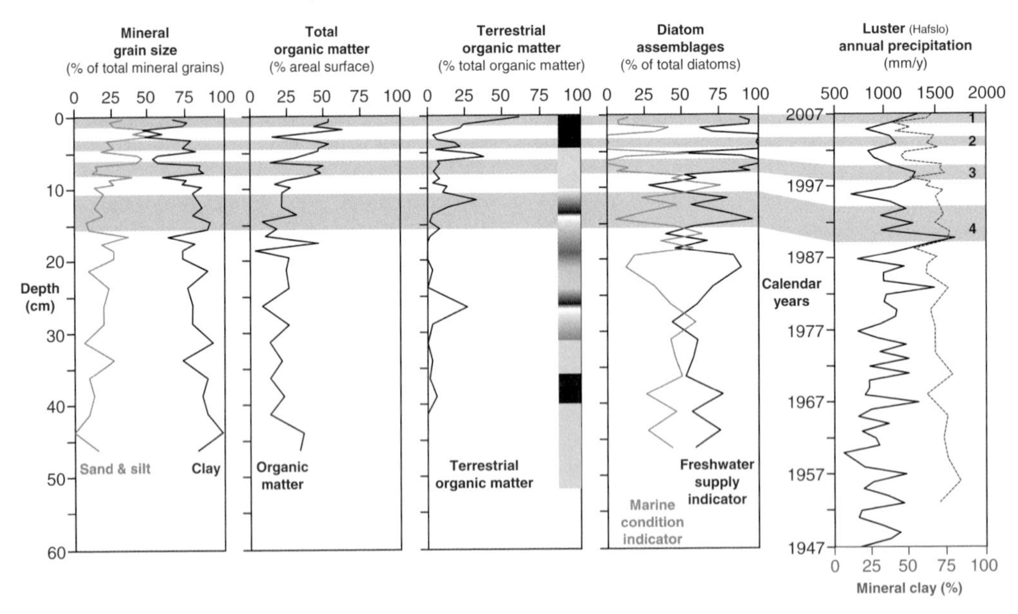

Fig. 4. Sediment composition for location 1 as analysed from smear slides. Core lithology is given in the middle graph; annual precipitation is from the meteorological station Hafslo of Figure 1. Grey shades indicate sediment correlation with peak precipitation horizons 1 to 4. Note that the sediment depth corresponds to calendar years of the precipitation graph; the depth scale has been converted into a timescale for the clay content and is added to the precipitation graph (red dashed line).

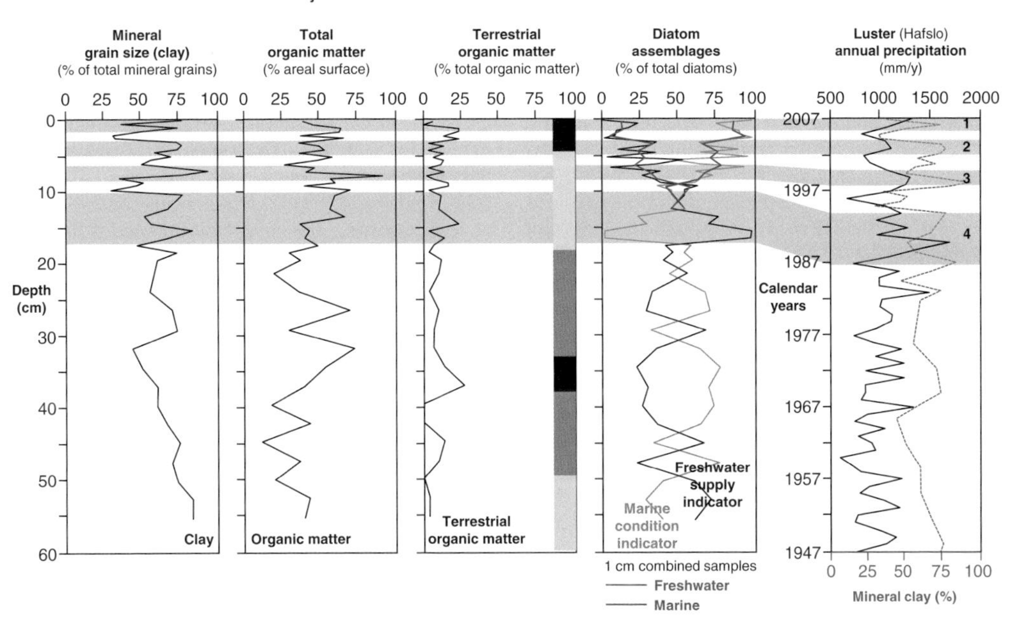

Fig. 5. As for Figure 4 for location 2. Solid red and blue lines represent composite graphs combining two sediment slices into one, hence gaining slices of annual resolution. Note that only the clay fraction is represented in the mineral grain-size graph to enhance visibility.

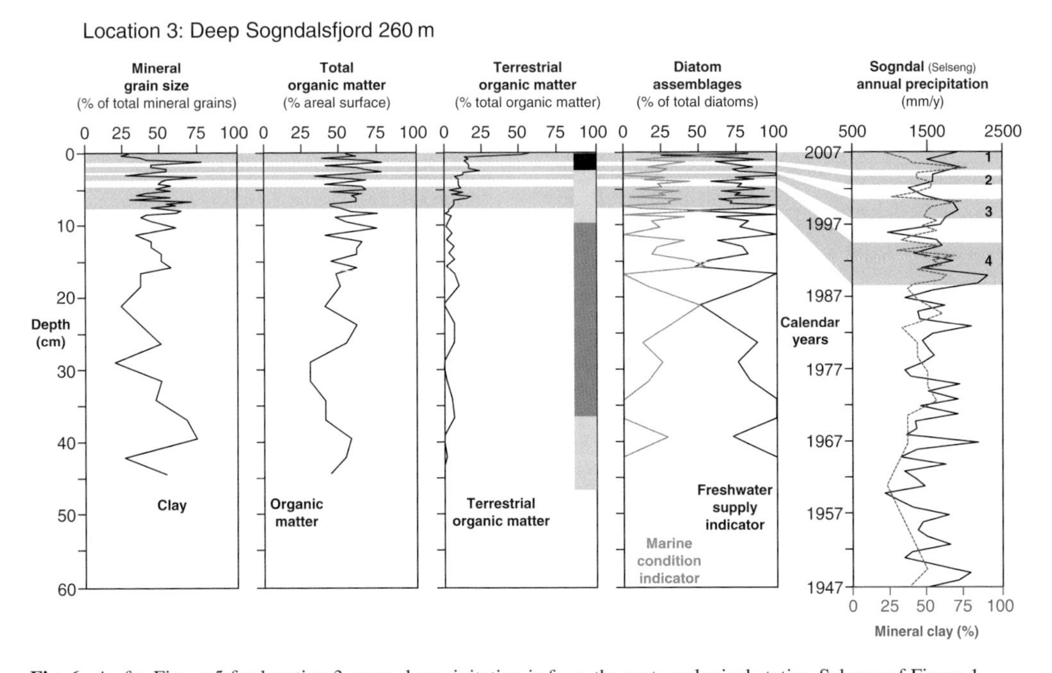

Fig. 6. As for Figure 5 for location 3; annual precipitation is from the meteorological station Selseng of Figure 1.

terrestrial organic matter indicates higher values at precipitation peaks 1, 2 and 4; it does not respond to the elevation of precipitation peak 3. The freshwater supply indicator clearly exceeds the marine diatom indicator at all sediment depths.

The clay content has been transferred from a depth graph into a time graph using the location 1 averaged sedimentation rate of 0.85 cm a^{-1}. The time graph is added to the precipitation graph as a dashed red line (Fig. 4). Although the peaks of precipitation and clay content clearly show a co-variation, they do not correlate in statistically significant numbers. This is due to the fact that sedimentation rates were impossible to determine on an exact annual resolution. The averaged linear sedimentation rates are estimated over dating horizons stretching over 5–21 a. On the other hand, there apparently exists a correlative relationship between the precipitation record and the clay content and the precipitation record and the freshwater supply indicator (Fig. 4), despite the missing statistically significant correlation.

Outer Barsnesfjord

In the Outer Barsnesfjord (location 2, Fig. 5), the same four-peak correlation with the precipitation record is visible for the mineral clay content. The assignment to precipitation peaks is more uncertain for the total organic matter, which also shows peaks during times of decreased precipitation; correlation is absent for the terrestrial organic matter content (apart from peak 3).

Elevated levels of all concentrations show additional subpeaks within the peak precipitation horizons. This is probably due to biasing when comparing 1-year precipitation records to almost half-year subsamples. Of special interest here is the sediment diatom record. The freshwater supply indicator follows the precipitation peaks 1, 2 and 4, including the major peak 4 elevations. However, the marine condition indicator also shows peak elevations within the time horizons of the precipitation peaks. To solve this problem, two 0.5-cm half-year measurements are combined to form a 1-cm single-year measurement (red and blue lines of the diatom graph in Fig. 5). The new combined record still shows an elevation of freshwater diatoms during precipitation peaks 1, 2 and 4 (including subpeaks). The marine diatom record of the Outer Barsnesfjord now shows the corresponding minima, as is the case in the Inner Barsnesfjord (Fig. 4). The only exemption is precipitation peak 3 where neither maxima nor minima occur in either diatom record.

A major difference compared to the Inner Barsnesfjord (Fig. 4) and the Sogndalsfjord (Fig. 6) is the overall increase of the marine diatom indicator and the corresponding overall decrease of the freshwater diatom indicator across the upper 10 cm of the sediment.

Also in the Outer Barsnesfjord, the clay content has been transferred from a depth graph into a time graph using the location 2 averaged sedimentation rate of 0.75 cm a^{-1}. The time graph is added to the precipitation graph as a dashed red line (Fig. 5). An apparent correlative relationship exists at least between the precipitation record and the clay content, despite the missing statistically significant correlation. Additional problems are the prevailing oxic (although occasionally low-oxic) conditions in the sediment giving rise to at least periodical bioturbation, disturbing the resolution and thus timing of the sediment signals.

Sogndalsfjord

In the Sogndalsfjord (location 3, Fig. 6), the subsample thickness (0.3 cm) has an approximate annual resolution across the upper 9 cm of the sediment. Sediment composition variability is similar to the Inner Barsnesfjord (Fig. 4). Mineral clay content, total organic matter and the freshwater supply indicator correlate well with the precipitation peaks 1, 2 and 3. In the Sogndalsfjord, we also observe: (a) the freshwater supply indicator resembles the peak 4 division into subpeaks; (b) clay content and total organic matter only show a general increase across the peak 4 horizon; and (c) the percentage of terrestrial organic matter indicates higher values at precipitation peaks 1, 2 and 4 and does not respond to the precipitation peak 3 elevation. There is a remarkable similarity between the sediment records of the terrestrial organic matter content of the Inner Barsnesfjord and the Sogndalsfjord, but not with the Outer Barsnesfjord.

As in locations 1 and 2, the clay content has been transferred from a depth graph into a time graph using the location 3 averaged sedimentation rate of 0.45 cm a^{-1}. The time graph is added to the precipitation graph as a dashed red line (Fig. 6). An apparent correlative relationship exists at least between the precipitation record and the clay content, despite the missing statistically significant correlation. Additional problems are the prevailing full oxic conditions in the sediment giving rise to bioturbation, disturbing the resolution and thus timing of the sediment signals.

Summarizing the smear slide records, there is a distinct dependency of the sediment signals on precipitation as indicated from the mineral clay content, the total organic matter percentage and the freshwater supply indicator. Enhanced precipitation yields the peaks of these parameters in the sediment; reduced precipitation reduces these signals.

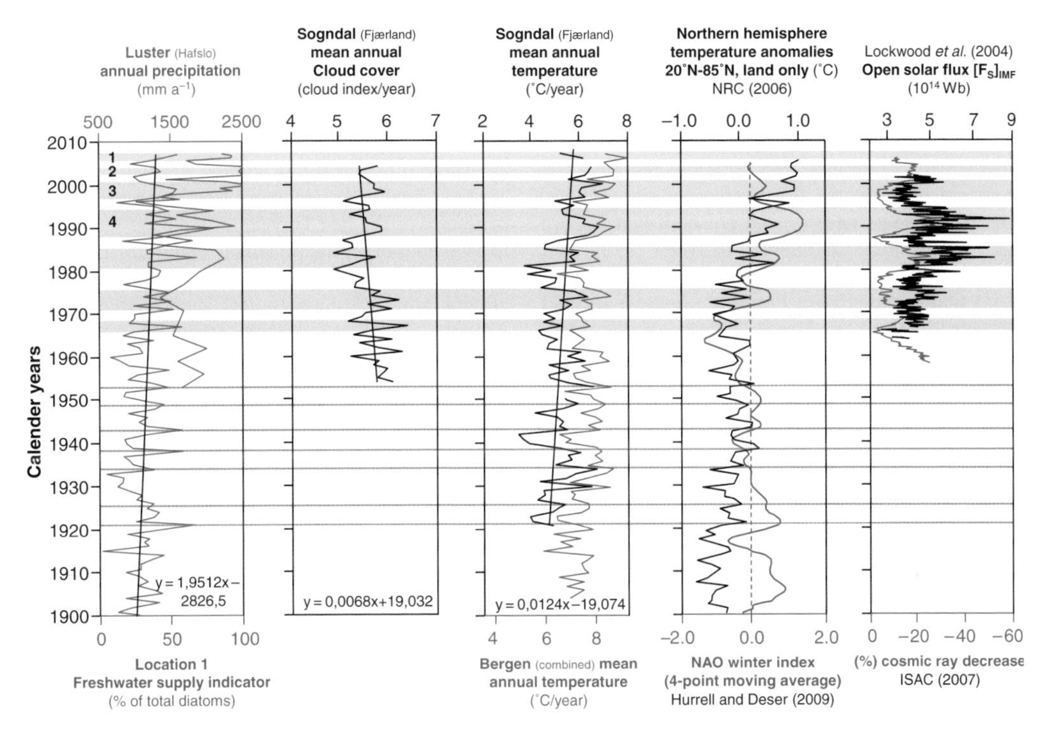

Fig. 7. The time relationship is shown between regional (Hafslo) meteorological data of precipitation (solid red line in the first graph) and the sediment record of the Barsnesfjord, here represented by the freshwater supply index of location 1 (solid blue line in the first graph). The precipitation and sediment relation is further transferred to regional cloud cover and temperature; it also co-varies with Northern hemisphere temperatures, NAO winter index and solar activity. Thin black lines in all graphs represent trend lines; the red dashed line in the fourth graph shows the zero-line of the NAO winter index. Grey shades indicate correlative horizons of concurrent parameter increase including precipitation peaks 1 to 4 of Figures 4–6; grey lines show peak correlation beyond the measured solar activity and cloud cover. Note that the annual precipitation and annual temperature correlate well with winter precipitation and summer temperature, respectively (Table 4) ($[F_s]$: solar force; IMF: interplanetary magnetic field; Wb: the Weber unit of magnetic flux).

Climate change indicators

Figure 7 shows that this relationship is also valid for other climate indicators, as stated earlier. Table 3 suggests a strong positive correlation ($r = 0.8$ and $r = 0.9$) between precipitation from coastal (Bergen) with the inner regions (Sogndal and Luster) of Western Norway. Correlation is intermediate and negative ($r = -0.6$ and -0.7) between precipitation and sun hours; correlation is weak and positive ($r = 0.4$ and 0.5) between precipitation, cloud cover and temperature; correlation is weak and negative ($r = -0.4$ and -0.5) between cloud cover and sun hours. No correlation ($r < 0.4$) is present between temperature and sun hours.

The annual precipitation record of Hafslo is plotted together with the freshwater supply index, annual temperature and the annual cloud cover record of the region in Figure 7. Although a correlation of $r < 6$ might generally be considered as less significant, the correlative relationship is obvious: all peak precipitation correlates with peak cloud cover and all time horizons of elevated precipitation correlate with time horizons of rising temperature over the last 100 years. The timing of the freshwater supply index of location 1 is used for representation of the fjord sediment record. The correlative relationship is predominant, with the limitations as discussed above. This correlation also includes peaks 1 to 4 of precipitation (Figs 4–6, indicated as shaded, numbered areas in Fig. 7). The co-variation of the graphs is also confirmed by the trend lines and their formulae, indicating a general increase of precipitation and temperature since the year 1900.

Discussion

It is suggested that the concurrent fjord sediment records of high clay percentage, elevated organic matter concentration and increased freshwater diatom content allows us to interpret the formation of these constituents during the concurrent climatic conditions of enhanced precipitation, increased cloud cover and rising temperature.

Furthermore, there is a distinct relation (Fig. 7) between the regional precipitation records and therefore between the fjord sediment signals over the last 60 years and: (a) the northern hemisphere land temperature anomalies between 20°N and 80°N latitude (NRC 2006); (b) the smoothed NAO winter index on a Portugal (Azores) Iceland gradient (Hurrell & Deser 2009); (c) the open solar magnetic flux, that is sunspot numbers (Lockwood *et al.* 2004); and (d) the decrease in cosmic rays (ISAC 2007).

The regional influence

The Barsnesfjord marks the southern end of the Årøyelv catchment area. In the north, the glacier Jostedalsbre strongly influences the amount of material supplied to the catchment. During the last decade, the glaciers of Norway have continuously retreated (Andreassen *et al.* 2005). Steiner *et al.* (2008) modelled prevailing glacial retreat in the Alps and in Norway for the next 25–50 years in case of ongoing (autumn) precipitation and (spring to summer) temperature increase, pointing out the importance of seasonal variations. Similar glacial retreat mechanisms are documented prior to and following the Little Ice Age (Nesje *et al.* 2001; Nesje & Dahl 2003; Steiner *et al.* 2008).

The glacial retreat leaves its imprint in the fjord sediments. Periods of increased winter precipitation and increased summer temperature (both represented within the annual precipitation and temperature pattern as discussed above and indicated in Table 4) lead to a higher glacial release of mineral clay particles due to the enhanced melting activity over the last 10 years. Over the same time period, the continuously decreasing volume of the glaciers explains the simultaneous overall decreasing trend of the relative amount of the mineral clay size in the Inner Barsnesfjord. Regardless, the clay volume seems to produce high enough turbidity in the water column to limit marine primary productivity in both Barsnesfjord basins.

The clear glacial signal of the Inner Barsnesfjord gets weaker in the Outer Barsnesfjord. The unexpectedly strong variation of sediment parameters in the Outer Barsnesfjord can be explained using the hydrographical conditions in the fjord basins (Dale & Hovgaard 1993). Freshwater outflow from the Barsnesfjord increases during times of enhanced precipitation. The stronger outflow from the Barsnesfjord is compensated with a stronger inflow of intermediate water from the Sogndalsfjord, thus enforcing estuarine circulation (Pritchard 1967). This compensation current might then transport additional marine diatoms from the Sogndalsfjord into at least the Outer Barsnesfjord, resulting in additional marine diatom, total organic matter and terrestrial organic matter peak signals.

Another explanation could be gradually improving light conditions and more effective nutrient utilization over the last 10 years, favouring increasing marine diatom production in the Outer Barsnesfjord. The turbidity effect of the clay might become diluted with distance to the source, that is the Årøyelv. This would then gradually ameliorate the conditions for marine primary productivity in the Outer Barsnesfjord. The possible productivity effect would then coincide with the last 10 years of increased precipitation and glacial retreat.

Glacial drainage is limited in the catchment area of the Sogndalselv. The main glacial influence on the Sogndalsfjord sediments is expected to originate from the outlet of the Barsnesfjord. On the other hand, the Barsnesfjord basins act as sediment traps due to their shallow sills. This suggests that the Sogndalsfjord sediment composition increasingly represents the direct runoff signal from precipitation. Mineral nutrient production from glacial activity is restricted, resulting in a decreased marine diatom productivity signal.

Thus, the Inner Barsnesfjord sediments are, in addition to precipitation runoff, strongly influenced by the glacial reaction on precipitation and temperature variations. The Outer Barsnesfjord reveals a combined record of glacial influence and fjord circulation patterns. The Sogndalsfjord mostly receives terrestrial input from precipitation runoff. All three environments show a similar sediment composition trend. It can be concluded from the sediment record that runoff from precipitation and the resulting (expected) variations in water turbidity and nutrient supply are one set of factors that determine the non-biotic and organic detritus fraction of the fjord sediments. The vicinity to glaciers enhances the runoff signal.

The Northern Hemisphere climate variability influence

Figure 7 suggests a strong relation between regional precipitation and temperature variation and the Northern Hemisphere temperature change (NRC 2006) and NAO variability (Hurrell & Deser 2009) over the last 100 years. The NAO variability represents one of the major climate forcing indices of the Northern Hemisphere, occurring on a decadal to century timescale and being most powerful during the winter months (for a review see Wanner *et al.* 2008). Nesje and Dahl (2003) and Steiner *et al.* (2008) suggest a direct link between a strong positive NAO index (wet and mild winters) and glacial advance. Andreassen *et al.* (2005) document a glacial advance for the period from 1990 to 1999. The advance corresponds to the precipitation peak 4 (Fig. 7) which correlates with a strong positive NAO index and the regional and Northern Hemisphere temperature increase. In a similar way, the negative NAO index of the last 10 years seems to coincide with the observed glacial retreat (Andreassen *et al.* 2005), suggesting a seasonal shift of the precipitation and temperature measurements across precipitation peaks 1 to 3 (Fig. 7).

Sediment variations are related to NAO forcing in Pliocene lacustrine deposits of the Villarroya Basin, Spain (Muñoz *et al.* 2002) and of Loch Ness, Scotland (Cooper *et al.* 2000). The NAO signal is also strong in the Nansen Fjord (Greenland) sediment foraminifer fraction (Overpeck *et al.* 1997). However, they could not find any land-related evidence for the NAO forcing over the last 400 years. Dean and Kemp (2004) found a corresponding Pacific Decadal Oscillation (PDA) record in the marine sediments of the Saanich Inlet, British Columbia. In the latter, the fluvial input seems to dominate the mineral clay fraction and allows climatic reconstructions over the last 2100 years.

The cosmic influence

In the 1950s, Berlage (1957) proposed a theory that the ENSO (El Niño Southern Oscillation) and the NAO variability are driven by the 11-year sunspot cycle. More recently, Bond *et al.* (2001) found that solar forcing exceeds the NAO forcing in the subpolar North Atlantic, while Jones and Mann (2004) claim that neither solar activity nor the NAO forcing could explain the late 20th century temperature increase. The 'Influence of Solar Activity Cycles on Earth's Climate' study (ISAC 2007) concludes that solar activity has a strong influence on the NAO forcing. Bard and Frank (2006) found that cosmic forcing is only valid on a century scale. Lockwood and Fröhlich (2007) doubt the application of the solar temperature forcing on a global scale, while this is heavily disputed by Svensmark & Friis-Christensen (2007) who argue for a direct solar–temperature relationship on a regional and global scale.

The Barsnesfjord and Sogndalsfjord sediment record reveals a clear relation between the timing of the clay content variation, freshwater diatoms and precipitation (Figs 4–7). Successive periods of high precipitation coincide with successive periods of increased cloud cover and higher regional and hemispheric temperature, including an overall increasingly positive NAO forcing (Fig. 7). The figure also indicates that high precipitation periods, and thus sediment parameters, have varied with maxima in open solar magnetic flux (Lockwood *et al.* 2004) and minima in cosmic ray variations (ISAC 2007) over the last 60 years. Thus, solar activity seems to affect increased regional precipitation which, in turn, influences the sediment record.

Figure 7 indicates a regional and northern hemispheric temperature increase during times of low cosmic ray measurements, which is consistent with the cosmic ray theory (Svensmark 1998; Marsh & Svensmark 2000*a*, *b*). The relation of cosmic rays to the NAO forcing (ISAC 2007) is also visible in Figure 7, and might link variations of the Barsnesfjord and Sogndalsfjord sediment record to variations in cosmic forcing.

Similar influences of solar forcing on climate change are suggested from a variety of glacial, moraine, tree ring and lacustrine sediment settings on a Holocene timescale across Scandinavia (Karlén & Kuylenstierna 1996) and from varved Lake Lehmilampi sediments, Finland deposited over the last 2000 a (Haltia-Hovi *et al.* 2007). Other indications are evident from the non-glacial Lake Holzmaar, Germany, relating varved sediments to the 11-year sunspot cycle on a Late Weichselian to Holocene time span (Zolitschka 1992; Vos *et al.* 1997). Patterson *et al.* (2004) related cosmic nuclides ^{14}C and ^{10}Be of marine-laminated clay-rich sediments of the Effingham Inlet, Vancouver Island (NE Pacific), to the sunspot activity of the middle Holocene Period. See also Wanner *et al.* (2008) for a recent review.

The sediment–climate model

Based on the above discussion, the combined and concurrent peak record of the mineral clay content, the total organic matter fraction and the freshwater supply indicator is suggested as a climate proxy in Western Norwegian fjord sediments. A simple box model is developed to illustrate the impact of solar and climate forcing on the formation processes of these deposits (Fig. 8). The model also indicates the possibility of interpreting climate variability from sediment change. As the climate record is of regional character, it can be predicted that such sediment–climate relations can be expected in areas where similar climate signals prevail. This basic model represents the first step towards a more sophisticated model that will include the fjord hydrographic variability and its effect on the biological communities in the fjord.

Another implication is the possibility of tracing the sediment–climate relation back in time, even beyond the boundaries of reliable dating on a 0–1000 a timescale. Correlation should be possible between the sediment record and reconstructed time series of climate change throughout the last millennium and the Holocene period. These implications

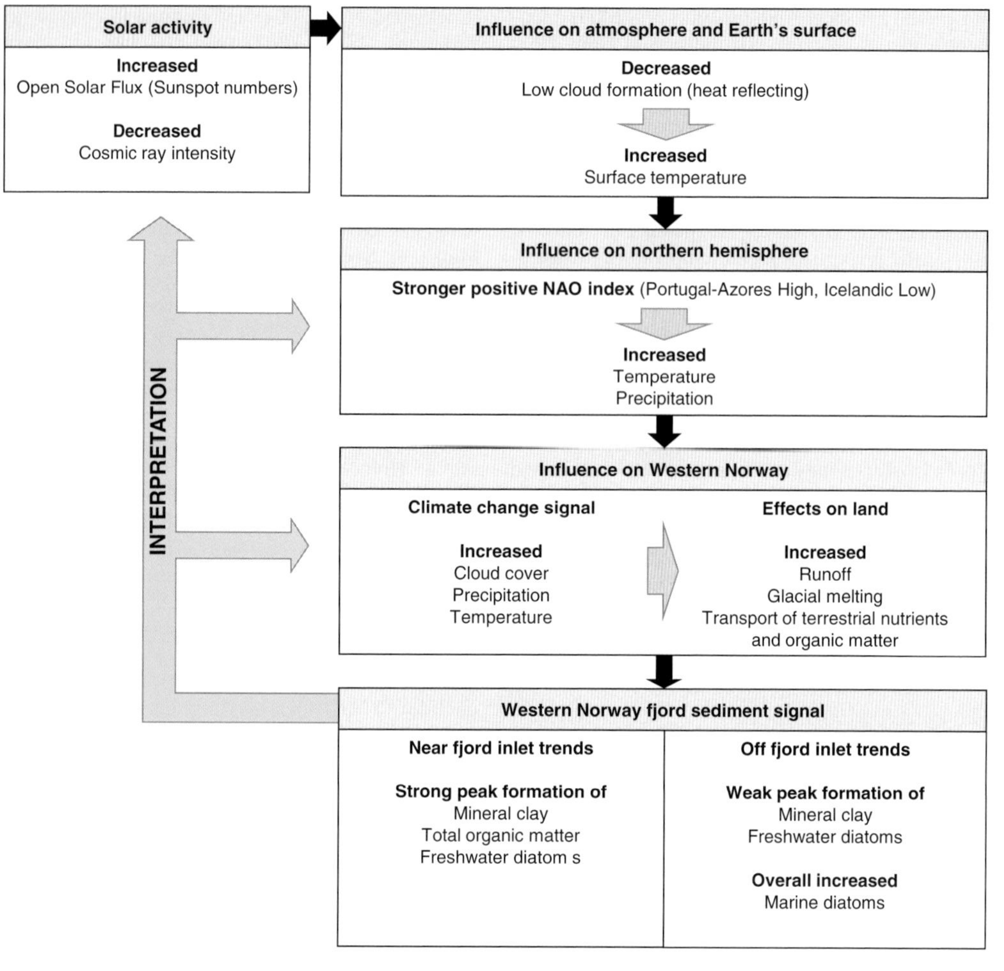

Fig. 8. The simple sediment–climate box model; the model illustrates the relation between solar forcing, northern hemisphere climate forcing (NAO), regional climate development and the fjord sediment signals of Western Norway. The model also illustrates the possibility of interpreting the sediment signal variation in terms of the primary forcing and should work across regions of similar climate input signals in space and time.

allow the sediment–climate relationship to be used to trace the impact of climate forcing on sediment formation processes, and thus cast new light on the driving mechanisms of major Holocene climate events such as the build up and retreat of the Little Ice Age. The model does not account for the additional climate influence from the CO_2 variability and/or sporadic (e.g. volcanic) events.

Conclusion

The major conclusions of this study are:

(1) Recent fjord sediments contain signals that can be related to regional climate change over the last 20 years.

(2) The fjord sediment record provides proxies that can reconstruct regional climate variations, including the NAO index and solar activity.

(3) The sediment–climate model implies the reconstruction of climatic and environmental events from fjord sediments on millennium to Holocene time scales.

The authors acknowledge with thanks the assistance given by the 2007 'From Mountain to Fjord' student group at the Sogn og Fjordane University College, Sogndal, Norway, who carried out the sediment sampling and most of the smear slide analysis. K. Ebert (Technical University Hamburg-Harburg, Germany) is especially thanked for her tremendous work in coordinating the working groups towards a common goal. P. Hovgaard (Skjersnes aquaculture station, Sogndal, Norway) is acknowledged for placing his research vessel *Knut* at the projects disposal and for assisting with sampling. The thorough reviews of A. Nesje and H. Cremer are highly appreciated.

References

ANDREASSEN, L. M., ELVERHØY, H., KJØLLMOEN, B., ENGESET, R. V. & HAAKENSEN, N. 2005. Glacier mass-balance and length variation in Norway. *Annals of Glaciology*, **42**, 317–325.

BARD, E. & FRANK, M. 2006. Climate change and solar variability: what's new under the sun? *Earth and Planetary Science Letters*, **248**, 1–14.

BERLAGE, H. P. 1957. *Fluctuations of the General Atmospheric Circulation of More Than One Year, Their Nature and Prognostic Value*. Royal Netherlands Meteorological Institute, Mededelingen Verhandelingen, **69**.

BOND, G., KROMER, B. *ET AL*. 2001. Persistent solar influence on North Atlantic climate during the Holocene Period. *Science*, **294**, 2130–2136.

COOPER, M. C., O'SULLIVAN, P. E. & SHINE, A. J. 2000. Climate and solar variability recorded in Holocene laminated sediments – a preliminary assessment. *Quaternary International*, **68–71**, 363–371.

DALE, T. & HOVGAARD, P. 1993. *En undersøkelse av resipientforholdene i Sogndalsfjorden, Barsnesfjorden og Kaupangerbukten i perioden 1991–1993*. [In Norwegian with English abstract and figure captions.] Sogn og Fjordane University College Scientific Report Series, 3/1993.

DEAN, J. M. & KEMP, A. E. S. 2004. A 2100 year BP record of the Pacific Decadal Oscillation, El Niño Southern Oscillation and Quasi-Biennial Oscillation in marine production and fluvial input from Saanich Inlet, British Columbia. *Palaeogeography, Plalaeoclimatology, Palaeoecology*, **213**, 207–229.

DNMI 2009. *The Official Website of the Norwegian Meteorological Institute (DNMI) for the Download of Official Climate Data from Norway*. Available online at: http://www.eklima.met.no

HALTIA-HOVI, E., SAARINEN, T. & KUKHONEN, M. 2007. A 2000-year record of solar forcing on varved lake sediment in eastern Finland. *Quaternary Science Reviews*, **26**, 678–689.

HUGHEN, K. A., BAILLIE, M. G. L. *ET AL*. 2004. Marine04 marine radiocarbon age calibration, 0–26 cal kyr bp. *Radiocarbon*, **46**, 1059–1086.

HURRELL, J. W. & DESER, C. 2009. North Atlantic climate variability: the role of the North Atlantic Oscillation. *Journal of Marine Systems*, doi: 10.1016/j.jmarsys.2008.11.026.

HUSTEDT, F. 1930–1966. Die Kieselalgen Deutschlands, Österreichs und der Schweiz unter der Berücksichtigung der übrigen Länder Europas, sowie der angrenzenden Meeresgebiete. *In*: RABENHORST, L. (ed.) *Kryptogamenflora von Deutschland, Österreich und der Schweiz*. Akademische Verlagsgesellschaft m.b.H., Leipzig.

IPCC 2007. Climate change 2007: the physical science basis. *In*: SOLOMON, S., QIN, D., MANNING, M., CHEN, Z., MARQUIS, M., AVERYT, K. B., TIGNOR, M. & MILLER, H. L. (eds) *Contribution of Working Group I to the Fourth Assessment Report of the Intergovernmental Panel on Climate Change*. Cambridge University Press, Cambridge.

ISAC 2007. *Influence of Solar Activity Cycles on Earth's Climate (ISAC)*. Danish National Space Center Scientific Report, Task 700 Summary Report.

JAHNKE, J., BROCKMANN, U. H., ALETSEE, L. & HAMMER, K. D. 1983. Phytoplankton activity in enclosed and free marine ecosystems in a southern Norwegian fjord during spring 1979. *Marine Ecology Progress Series*, **14**, 19–28.

JONES, P. D. & MANN, M. E. 2004. Climate over past millennia. *Reviews of Geophysics*, **42**, 1–42.

JONES, P. D., BRIFFAS, K. R. *ET AL*. 2009. High-resolution palaeoclimatology of the last millennium: a review of current status and future prospects. *Holocene*, **19**, 3–49.

KARLÉN, W. & KUYLENSTIERNA, J. 1996. On solar forcing of Holocene climate: evidence from Scandinavia. *The Holocene*, **6**, 359–365.

LOCKWOOD, M. & FRÖHLICH, C. 2007. Recent directed trends in solar climate forcings and the global mean surface air temperature. *Proceedings of the Royal Society*, **A463**, 2447–2460.

LOCKWOOD, M., FORSYTH, R. B., BALOGH, A. & MCCOMAS, D. J. 2004. Open solar flux estimates from near-Earth measurements of the interplanetary magnetic field: comparison of the first two perihelion passes of the Ulysses spacecraft. *Annales Geophysicae*, **22**, 1395–1405.

MARSH, N. & SVENSMARK, H. 2000*a*. Low cloud properties influenced by cosmic rays. *Physical Review Letters*, **85**, 5004–5007.

MARSH, N. & SVENSMARK, H. 2000*b*. Cosmic rays, clouds, and climate. *Space Science Reviews*, **94**, 215–230.

MUÑOZ, A., OJEDA, J. & SÁNCHEZ-VALVERDE, B. 2002. Sunspot-like and ENSO/NAO-like periodicities in lacustrinelaminated sediments of the Pliocene Villarroya Basin (La Rioja, Spain). *Journal of Palaeolimnology*, **27**, 453–463.

MUNSELL 1994. *Munsell Soil Color Charts*. Munsell Color 1994 revised edition, GretagMacbeth, NY, USA.

NESJE, A. & DAHL, S. O. 2003. The 'Little Ice Age' – only temperature? *The Holocene*, **13**, 139–145.

NESJE, A., MATTHEWS, J. A., DAHL, S. O., BERRISFORD, M. S. & ANDERSSON, C. 2001. Holocene glacier fluctuation and winter-precipitation changes in the Jostedalsbreen region, western Norway, base don glaciolacustrine sediment records. *The Holocene*, **11**, 267–280.

NIEMISTÖ, L. 1974. A gravity corer for studies of soft sediments. *Merentutkimuslait Julk/Havsforskningsinstituttets Skrifter*, **238**, 33–38.

NRC (NATIONAL RESEARCH COUNCIL) 2006. *Surface Temperature Reconstructions for the Last 2000 Years*. National Academic Press, Washington, DC.

OVERPECK, J., HUGHEN, K. *ET AL*. 1997. *Arctic environmental change of the last four centuries. Science*, **278**, 1251–1256.

PAETZEL, M. & SCHRADER, H. 1992. Recent environmental changes recorded in anoxic Barsnesfjord sediments: Western Norway. *Marine Geology*, **105**, 23–36.

Paetzel, M. & Schrader, H. 1991. Heavy metal (Zn, Cu, Pb) accumulation in the Barsnesfjord: Western Norway. *Norsk Geologisk Tidsskrift*, **71**, 65–73.

Patterson, R. T., Prokoph, A. & Chang, A. 2004. Late Holocene sedimentary response to solar and cosmic ray activity influenced climate variability in the NE Pacific. *Sedimentary Geology*, **172**, 67–84.

Pritchard, D. W. 1967. What is an estuary: physical viewpoint. *In*: Lauf, G. H. (ed.) *Estuaries*. American Association for the Applied Science, Washington, DC, **83**, 3–5.

Rothwell, R. G. 1989. *Minerals and Mineraloids in Marine Sediments: An Optical Identification Guide*. Elsevier Science Publishers Ltd, London.

Round, F. E., Crawford, R. M. & Mann, D. G. 1990. *The Diatoms: Biology and Morphology of the Genera*. Cambridge University Press, Cambridge.

Steiner, D., Pauling, A., Nussbaumer, S. U., Nesje, A., Luterbacher, J., Wanner, H. & Zumbühl, H. J. 2008. Sensitivity of European glaciers to precipitation and temperature – two case studies. *Climatic Change*, **90**, 413–441.

Stoermer, E. F. & Smol, J. P. 1999. *The Diatoms: Applications for the Environmental and Earth Sciences*. Cambridge University Press, Cambridge.

Svensmark, H. 1998. Influence of cosmic rays on Earth's climate. *Physical Review Letters*, **81**, 5027–5030.

Svensmark, H. & Friis-Christensen, E. 2007. *Reply to Lockwood and Fröhlich – The persistent role of the Sun in climate forcing*. Danish National Space Center, Scientific Report, **3/2007**.

Usoskin, I. G. & Kovaltsov, G. A. 2008. Cosmic rays and climate of the Earth: possible connection. *Comptes Rendus Geoscience*, **340**, 441–450.

Usoskin, I. G., Schüssler, M., Solanki, S. K. & Mursula, K. 2005. Solar activity, cosmic rays, and Earth's temperature: a millennium-scale comparison. *Journal of Geophysical Research*, **110**, A10102.

Vos, H., Sanchez, A., Zolitschka, B., Brauer, A. & Negendank, J. F. W. 1997. Solar activity variations recorded in varved sediments from the crater lake of Holzmaar – a maar lake in the Westeifel volcanic field, Germany. *Surveys in Geophysics*, **18**, 163–182.

Wanner, H., Beer, J. *et al.* 2008. Mid- to Late Holocene climate change: an overview. *Quaternary Science Reviews*, **27**, 1791–1828.

Zolitschka, B. 1992. Climatic change evidence and lacustrine varves from maar lakes, Germany. *Climate Dynamics*, **6**, 229–232.

Holocene climate variations at the entrance to a warm Arctic fjord: evidence from Kongsfjorden trough, Svalbard

KARI SKIRBEKK[1]*, DORTHE KLITGAARD KRISTENSEN[2], TINE L. RASMUSSEN[1], NALAN KOÇ[1,2] & MATTHIAS FORWICK[1]

[1]*Department of Geology, University of Tromsø, Norway*

[2]*Norwegian Polar Institute, Polar Environmental Centre, Tromsø, Norway*

**Corresponding author (e-mail: kari.skirbekk@uit.no)*

Abstract: The North Atlantic Current transports warm and salty water into the Nordic Seas and continues northwards into the Arctic Ocean as the West Spitsbergen Current. This current flows along the west coast of the Svalbard archipelago and into the fjords and troughs on the Svalbard shelf. We have investigated a core (NP05-11-21GC) from the Kongsfjorden trough, which spans the last 12 ka. The core site presently experiences seasonal inflow of Atlantic Water masses, and may therefore provide a record of past variations in Atlantic Water inflow to the Arctic. Lithological analysis and benthic foraminifera have been used to reconstruct the palaeoceanographic development in the area. The results show that cold and harsh conditions prevailed during the late part of the Younger Dryas and that the site was in proximity to glaciers. After 11.8 ka BP the first influence of Atlantic Water is seen in the fauna. This was followed by changes in glacial activity at 11.5 ka BP. During the period 11.5–10.6 ka BP the fauna indicate increased influence of Atlantic Water and the final deglaciation of the fjord after the Younger Dryas period. These initial ameliorated conditions were interrupted by a 250-year long cooling starting at 11.3 ka BP, corresponding to the Preboreal Oscillation. After 10.6 ka BP a marked change from an ice proximal to ice distal environment occurred accompanied by the strong influence of Atlantic Water masses. In the Mid-Holocene at 7 ka BP, the influence of Atlantic Water diminished but no sign of an immediate response of the glaciers to this are found in the core. The evidence suggests that intensification of glacial activity started as late as *c*. 3.5 ka BP. Comparing the core from Kongsfjorden trough with two shelf records from Svalbard generally shows similar faunal development although some differences exist. These are assumed to be related to the distance of each record to the position of the Arctic front.

A key component of the climate of the European Arctic is the inflow of warm and saline Atlantic Water masses to the Nordic and Arctic Seas. The strength of the northbound Atlantic Current has been known to fluctuate over time, changing the coastal climate conditions in northern Europe (e.g. Klitgaard-Kristensen *et al.* 2001; Husum & Hald 2004). However, natural variations in Atlantic Water inflow into the Arctic are not well known. Previous studies by Ślubowska *et al.* (2005) and Ślubowska-Woldengen *et al.* (2007) show high variability in the inflow of Atlantic Water to the shelf areas west and north of Svalbard during the past 17 ka.

The inflow of Atlantic Water along the west coast of Spitsbergen, Svalbard also affects many of the large fjord systems located there. One example is Kongsfjorden (Fig. 1), which is seasonally influenced by temperate and salty Atlantic Water masses derived from the North Atlantic and regarded as a warm Arctic fjord. The inflow of Atlantic Water into Kongsfjorden therefore makes it highly sensitive to any changes in this current. In recent years, the fjord has experienced a marked rise in both summer and winter temperatures. The rise is due to higher temperatures of the inflowing Atlantic Water (Cottier *et al.* 2007), which has resulted in minimum formation of seasonal sea ice both in 2006 and 2007 (Hop *et al.* 2006; Gerland & Renner 2007).

Previous investigations show that Kongsfjorden and the Kongsfjorden trough were occupied by an ice stream during the Last Glacial Maximum (LGM) (Landvik *et al.* 2005; Ottesen *et al.* 2007) which extended out to the shelf break (Vorren *et al.* 1998). The disintegration of the ice stream started at *c*. 15 ^{14}C ka BP; acoustic seismics reveal that the ice retreat of the glacier front occurred stepwise (Landvik *et al.* 2005; Ottesen *et al.* 2007). The glacier front reached the fjord mouth at 13 ^{14}C ka BP and a moraine was deposited (Lehman & Forman 1992). So far, no evidence of a Younger Dryas glacial re-advance in Kongsfjorden has been reported. The fjord was deglaciated at *c*. 9 ^{14}C ka BP (Lehman & Forman 1992).

In this study a sediment core from the Kongsfjorden trough, Svalbard covering the last 12 ka is

From: Howe, J. A., Austin, W. E. N., Forwick, M. & Paetzel, M. (eds) *Fjord Systems and Archives*. Geological Society, London, Special Publications, **344**, 289–304.
DOI: 10.1144/SP344.20 0305-8719/10/$15.00

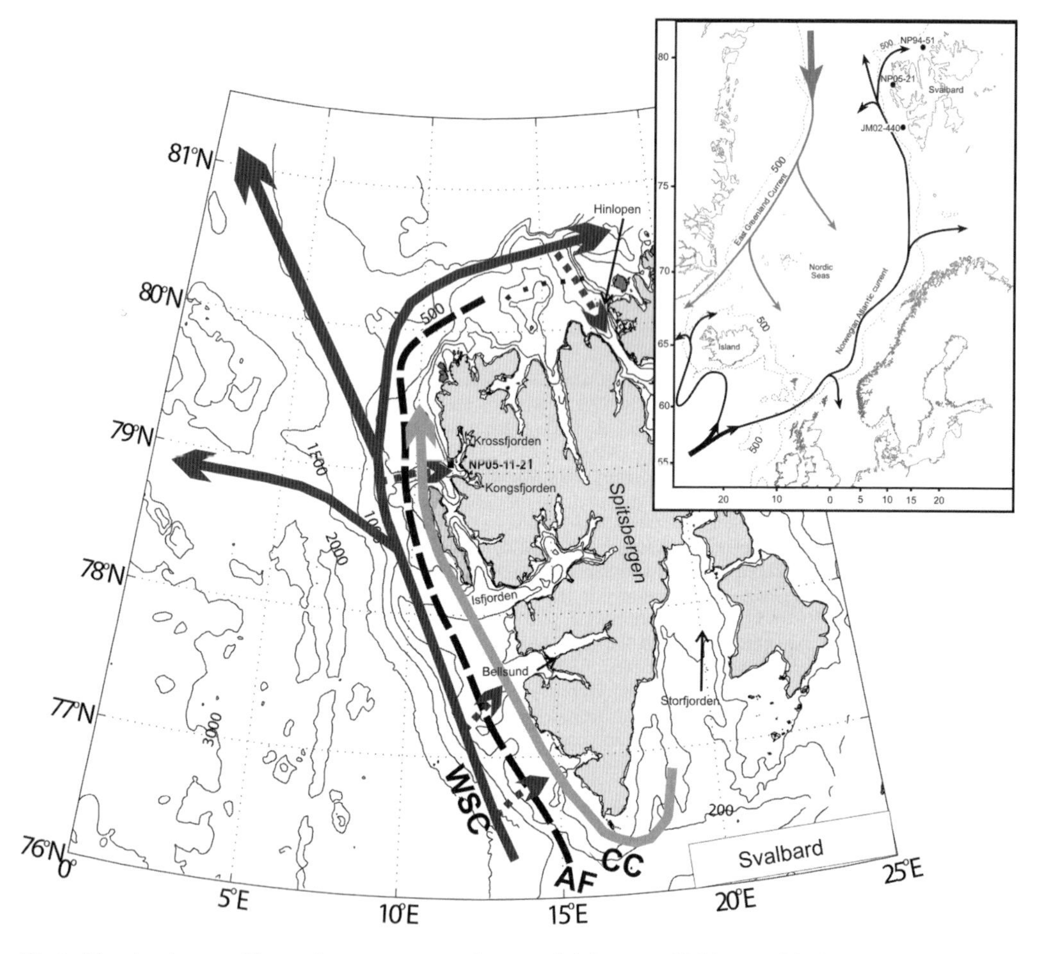

Fig. 1. Map showing prevailing surface currents around western Spitsbergen, which is part of the Svalbard archipelago. Location of the core NP05-11-21GC in Kongsfjorden trough is indicated by a black square. WSC, West Spitsbergen Current (dark); CC, Coastal Current (grey) and AF, Arctic Front (dashed line). The small inset map shows the main surface currents, Norwegian Atlantic Current and East Greenland Current, in the Nordic Seas. In addition to the Kongsfjorden core, two cores (black circles) in Hinlopen (NP94-51) and Bellsund (JM02-440) are indicated.

analysed for the content of benthic foraminifera and lithology. These proxies provide information on variations in Atlantic Water inflow into the fjord and glacial activity. We examine how the variations in Atlantic Water inflow into Kongsfjorden relate to changes on the shelf by comparing them to two records from the shelf (Ślubowska *et al.* 2005; Ślubowska-Woldengen *et al.* 2007). Finally, we evaluate the relationship (if any) between the variability of Atlantic Water inflow into Kongsfjorden and the glacial activity.

Physiographic–oceanographic setting

Kongsfjorden and Krossfjorden form one of the largest fjord systems on Svalbard (Fig. 1). On the shelf the two fjords merge and form the Kongsfjorden trough (Fig. 1). Nearly 60% of Svalbard is today glaciated (Kohler *et al.* 2007) and several tidewater glaciers drain into the Kongsfjorden–Krossfjorden fjord system. Kongsfjorden and the western margin of Svalbard are influenced by two major water currents, the West Spitsbergen Current (WSC) and the Coastal Current (CC). The WSC, a continuation of the North Atlantic Current, flows along the western margin of Svalbard bringing temperate and saline Atlantic Water to the Arctic (Fig. 1; Johannessen 1986). The CC drifts towards north along the coast of western Spitsbergen and brings Arctic Water from the eastern Svalbard shelf area (Fig. 1). The boundary between the CC and the WSC is marked by the Arctic Front at the shelf edge (Fig. 1; Svendsen *et al.* 2002). When the Arctic Front migrates towards the east and the south it means that Atlantic Water moves closer to the coast (Swift 1986).

Kongsfjorden is characterized by a seasonal circulation pattern (Svendsen *et al.* 2002). During the winter season the water masses are weakly stratified, covered by sea ice and Arctic Water masses prevail. The weaker stratification is mainly caused by sea-ice formation that leads to increased vertical convection and homogenization of the water column (Svendsen *et al.* 2002). Spring melting creates a freshwater layer that divides the water masses into two horizontal layers. This stratification is further strengthened by solar heating of the sea-surface layer (Svendsen *et al.* 2002). Intrusions of Atlantic Water onto the shelf and into the fjord normally take place in the deeper part of the water column, and often occur in spring and summer (Svendsen *et al.* 2002). However, in spring 2007, the inflow of Atlantic Water took place in the upper part of the water column (Tverberg & Nøst 2009). This was explained by differences in density across the front between the Atlantic Water and the Arctic Water at the shelf edge (Tverberg & Nøst 2009). This has previously not been considered as an important mechanism for the intrusion of Atlantic Water into the fjord (Svendsen *et al.* 2002). The amount of Atlantic Water in Kongsfjorden during summer varies annually (Hop *et al.* 2006). Only limited data are available on circulation and water-mass distribution in Krossfjorden, but it is assumed to be similar to Kongsfjorden (Svendsen *et al.* 2002).

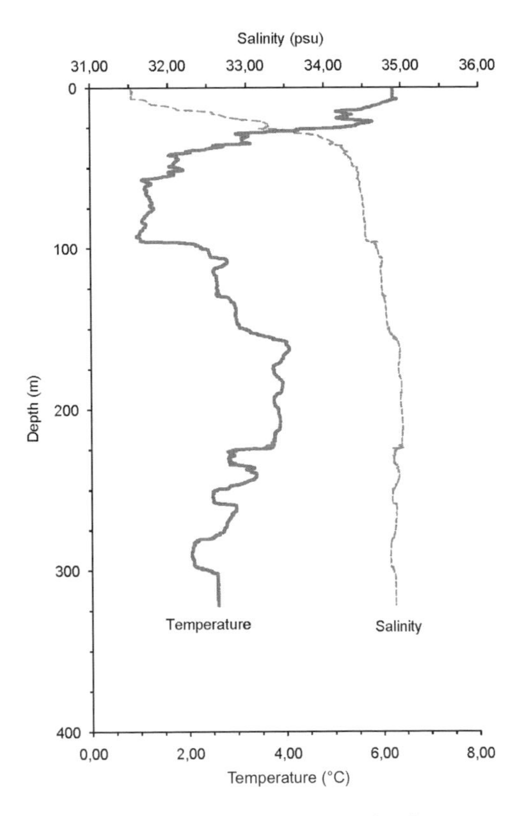

Fig. 2. CTD profile retrieved at the core location (NP05-11-21GC) in Kongsfjorden trough in August 2005. The profile clearly shows the vertical distribution of Atlantic Water masses at the site with the main core of Atlantic Water between 225 and 150 m depth and a zone of mixed Atlantic and cooler water masses 50 m above and below this main core.

Material and methods

The study is based on multiproxy analyses of the 515 cm long gravity core NP05-11-21GC. The core was retrieved from the Kongsfjorden trough at 327 m water depth in mid-August 2005. Concurrent with retrieving the core, a CTD profile was acquired at the same location and date using a Sea-Bird 911 (Fig. 2).

Prior to opening the core, the magnetic susceptibility was measured every centimetre with a GeoTek multisensor core logger (MSCL) using a loop sensor with a counting time of 10 s. The core was then split and half-core sections were X-rayed. A visual description was performed and colour information obtained from the Munsell soil colour chart. The core was then subsampled at 10 cm intervals in 1-cm thick slices. The visual description of the core revealed frequent changes in lithology at depths from 160–240 cm and 400–515 cm. The core was therefore sampled every 3–4 cm and every 2–3 cm, respectively. All samples (a total of 104) were freeze-dried. The samples were wet sieved over 63, 100 and 1000 μm sieves, and dried at low temperatures (<50 °C). Benthic foraminifera were counted and identified in the size fraction 100–1000 μm. In each sample, more than 300 specimens were identified to species level. The samples were then dry-sieved on a 500 μm mesh-size sieve and all minerogenic grains were counted.

Benthic foraminiferal concentrations and ice-rafted debris (IRD) are reported as flux values (no. of specimens $(cm^2\ ka)^{-1}$) using the bulk sediment density and sedimentation rate. Grains larger than 500 μm are regarded as IRD. For faunal diversity, the Walton (1964) diversity index was used. It is a count of the number of species that constitutes 95% of a fauna.

A total of 10 samples were AMS ^{14}C dated at the AMS ^{14}C Dating Centre, University of Aarhus, Denmark (Table 1). The dates were performed on monospecific samples of the benthic foraminifera

Table 1. *Radiocarbon dates and calibrated ages as used in age model for core NP05-11-21GC*

Lab reference	Core	Depth (cm)	Dated material	Raw AMS^{14}C	Calibrated age
AAR-10910	NP05-11/21GC	5–6	N. labradorica	443 ± 35	45
AAR-10480	NP05-11/21GC	20–21	N. labradorica	1008 ± 33	540
AAR-10911	NP05-11/21GC	110–111	N. labradorica	3668 ± 43	3500
AAR-10912	NP05-11/21GC	160–161	N. labradorica	4793 ± 44	4970
AAR-10913	NP05-11/21GC	240–241	N. labradorica	7545 ± 55	7940
AAR-10481	NP05-11/21GC	270–271	N. labradorica	8296 ± 45	8760
AAR-10914	NP05-11/21GC	320–321	Mixed benthic foraminifera	9345 ± 55	10 150
AAR-10915	NP05-11/21GC	410–411	N. labradorica	9920 ± 65	10 720
AAR-10916	NP05-11/21GC	446–557	Mixed benthic foraminifera	10 145 ± 65	11 130
AAR-10482	NP05-11/21GC	500–501	N. labradorica	10 535 ± 60	11 680

species *Nonionellina labradorica* and on samples of mixed benthic foraminifera species (Table 1). The ^{14}C ages were calibrated using the web-based program Calib 5.0.2 (Stuiver & Reimer 1993) and the calibration dataset Marine 04 (Hughen *et al.* 2004). A regional reservoir age ΔR of 105 ± 24 years was used in addition to a reservoir age of 360 ± 20 years (Bondevik *et al.* 2006; Mangerud *et al.* 2006). The reservoir effect may not have been constant through time (e.g. Bard *et al.* 1994; Bondevik *et al.* 2006). However, since no data are presently available from the Svalbard area, we used a constant reservoir correction throughout the studied time interval. The age model (Fig. 3; Table 1) is based on the peak of the probability curves which are all within the 1σ range. In the following, all ages are given as calibrated years BP (ka BP), unless stated otherwise.

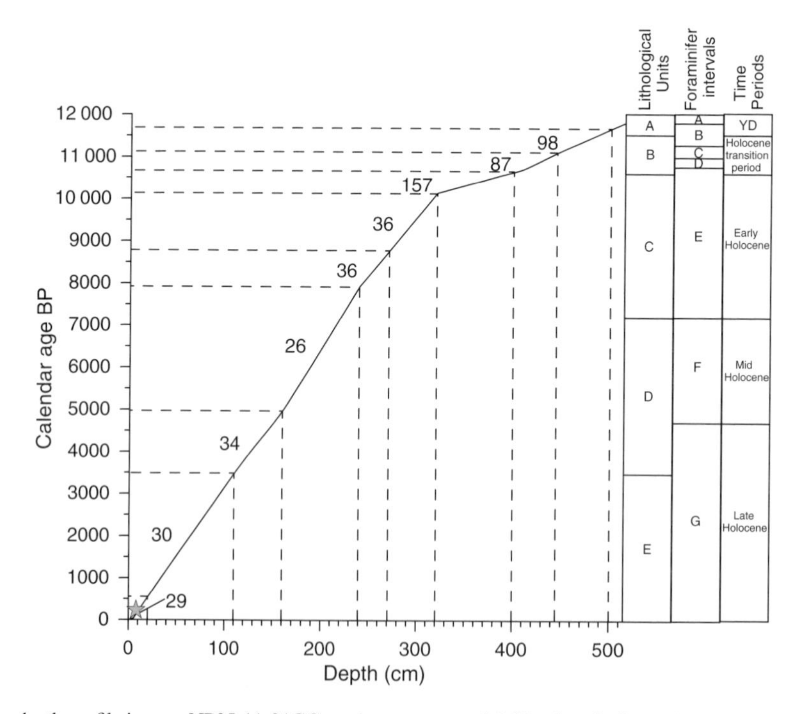

Fig. 3. Age–depth profile in core NP05-11-21GC used as an age model. Numbers indicate linear sedimentation rates (cm ka^{-1}) between AMS dates. More details on the AMS dates are listed in Table 1. The depth intervals of the lithological units and foraminifer intervals are shown along with the corresponding time periods. The star indicates the youngest age (not visible on the figure) and the corresponding sedimentation rate.

Results

Age model and sedimentation rates

The oldest date (10 535 $\pm$ 60 ^{14}C a) shows two distinct peaks in the probability distribution providing two calibrated ages of 11.68 ka BP and 11.46 ka BP, respectively (Table 1). The calibrated age of 11.68 ka BP has the highest probability peak and is therefore used in the age model. For the youngest date (443 $\pm$ 35 ^{14}C a) the use of a reservoir age of 360 $\pm$ 20 a and ΔR of 105 a yields a calibrated age younger than the first nuclear testing. The dated material has, however, not been affected by nuclear testing (pers. comm. J. Heinemeier 2007). Therefore, the calibrated age is set to be 1900 AD. The age model was constructed assuming linear sedimentation rates between the dated levels. The average sedimentation rate is 114 cm ka^{-1} before 10 ka BP and 32 cm ka^{-1} after 10 ka BP (Fig. 3). The age at the bottom of the core is calculated as *c.* 12 ka BP.

Lithostratigraphy

Core NP05-11-21GC is divided into the five units A–E based on major changes in sedimentologic properties (Figs 4 & 5).

Unit A (515–475 cm) was deposited during *c.* 12–11.5 ka BP (Fig. 3). It consists of dark mud with low IRD content (Fig. 4). The magnetic susceptibility is high (Fig. 5). The sedimentations rate is 98 cm ka^{-1} and the IRD flux has an average of 141 (cm ka^{-1})$^{-1}$, which is moderately high compared to the rest of the core (Fig. 5).

Unit B (475–395 cm) was deposited during 11.5–10.6 ka BP (Fig. 3). It comprises dark grey mud with colour changes to lighter grey from 460–445 cm (11.25–11.1 ka BP). At 430 cm (11 ka BP), the colour changes gradually from dark grey to greyish olive. Unit B shows high variability in IRD flux (Fig. 4). At 460–450 cm (11.25–11.10 ka BP), IRD is almost absent followed by an IRD flux of 677 (cm ka^{-1})$^{-1}$ at 440–405 cm. The top of the unit shows a marked drop in IRD flux. The magnetic susceptibility decreases upcore and the sedimentation rate increases to 157 cm ka^{-1} close to the upper boundary. The sediment is bioturbated and shell fragments were found in the uppermost part of the unit (Fig. 4).

Unit C (395–220 cm) was deposited during 10.6–7.2 ka BP (Fig. 3). It consists of mud of greyish olive colour that changes to dark greyish brown at 300 cm (9.5 ka BP). The unit has a

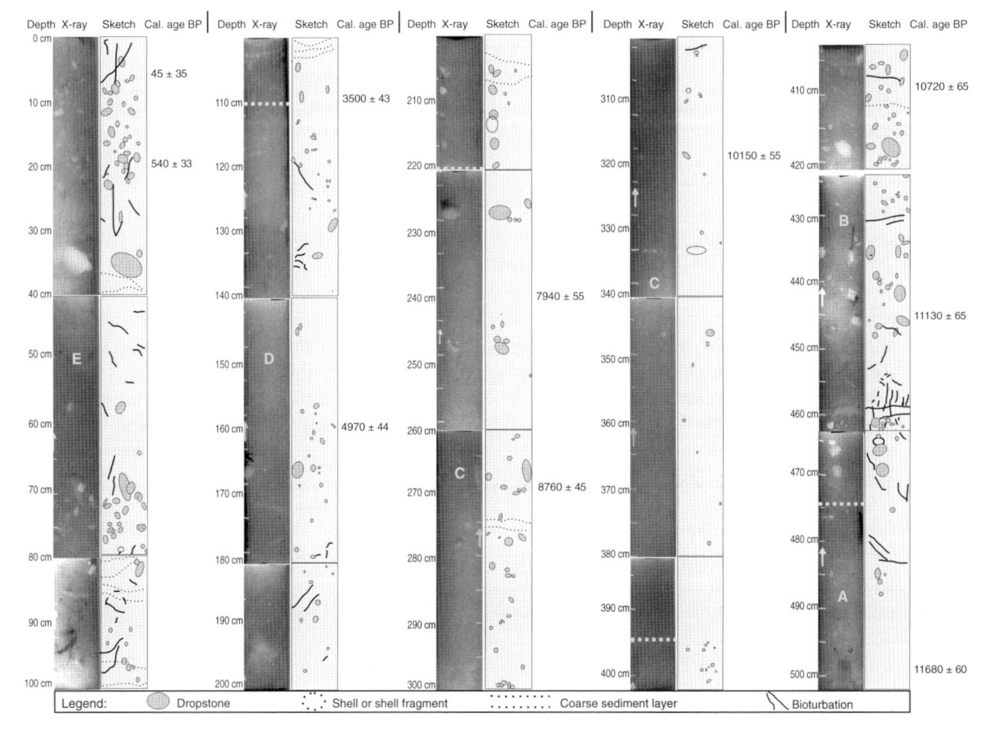

Fig. 4. X-ray photos of core NP05-11-21GC shown at the left hand-side and sedimentologic descriptions at the right hand-side plotted v. core depth (cm). The letters indicate lithological units and the thick dotted lines indicate the borders between them. Calibrated ages for the dated levels are marked.

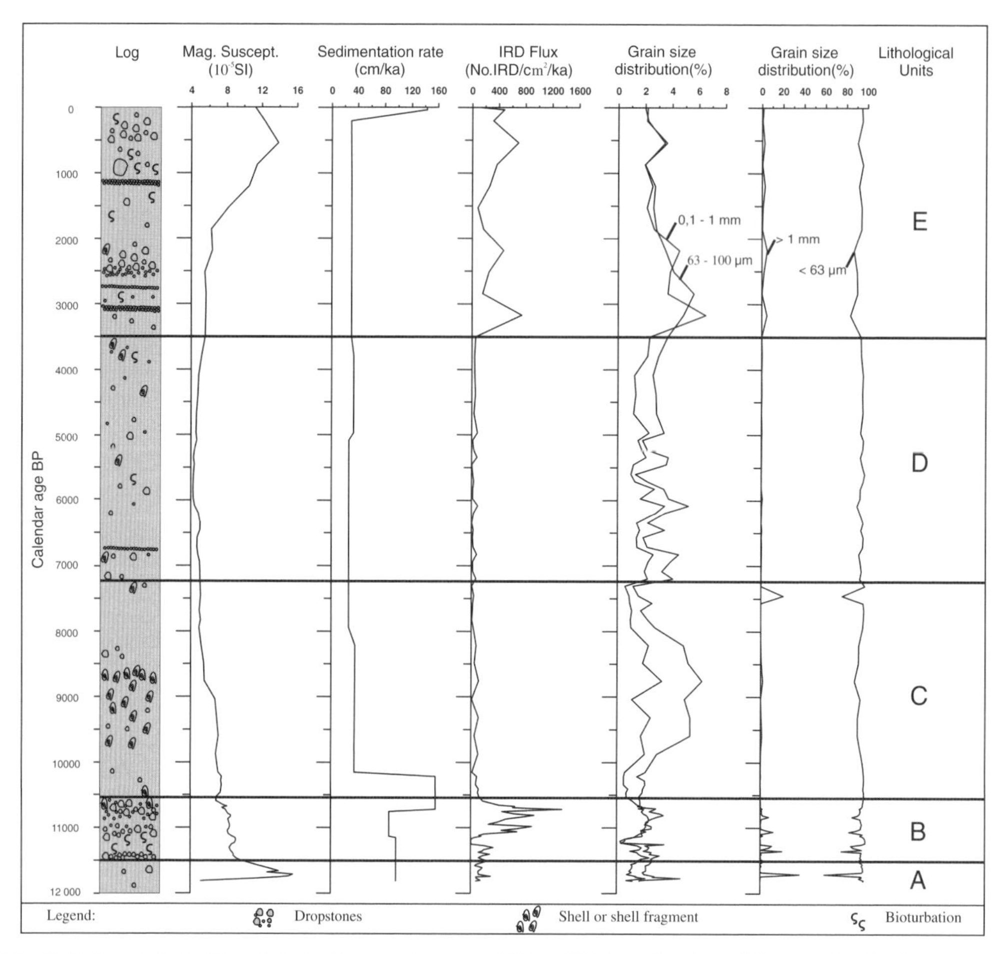

Fig. 5. Sedimentological log plotted with magnetic susceptibility, IRD flux of grains >0.5 mm and grain size distribution (dark line = 0.1–1 mm; light colour line = 0.063–0.1 mm) in core NP05-11-21GC v. age. Grain sizes are given as weight percentage. Lithologial units are indicated.

relatively high content of very fine to coarse sand (Fig. 5) while the IRD flux remains low. The magnetic susceptibility decreases upcore. Sedimentation rates drop from 157 cm ka^{-1} to 36 cm ka^{-1} at *c.* 330 cm depth (10.2 ka BP; Fig. 3). Comparatively large amounts of shells and shell fragments occur, in particular at 310–260 cm (9.9–8.5 ka BP; Fig. 4).

Unit D (220–110 cm) was deposited during 7.2–3.5 ka BP. It comprises dark brownish grey mud with low IRD flux (38 (cm ka^{-1})$^{-1}$) (Figs 3 & 5). The sedimentation rate remains constant and low (Fig. 5) and low magnetic susceptibility occurs at *c.* 190–170 cm (*c.* 6–5.3 ka BP) followed by a gradual increase. Shells and shell fragments occur sporadically and in lower numbers than in unit C. Traces of bioturbation occur occasionally.

Unit E (110–0 cm) was deposited during the last 3.5 ka BP. It consists of dark brownish grey mud with a general increase in IRD flux compared to unit D. The IRD flux and amounts of very fine to coarse sand vary throughout the unit (Fig. 5). The magnetic susceptibility starts to increase gradually at *c.* 1.8 ka BP and reach a maximum at *c.* 0.5 ka BP. This is followed by a decrease towards the top of the core. Few shells and shell fragments occur in the lower part of the unit and they are absent in the upper part. Traces of bioturbation occur occasionally (Fig. 5).

Fauna

Based on the most pronounced changes in percentage distribution of benthic foraminifera the core is divided into seven foraminifer intervals.

Foraminifer interval A (515–509 cm, 12–11.8 ka BP). The lowermost part of the core is dominated by *Elphidium excavatum* (*c.* 50%) and *Cassidulina reniforme* (*c.* 23%; Fig. 6a). Subdominant species are *Nonionellina labradorica* and *Islandiella helenae/norcrossi* each constituting *c.* 10% of the fauna (Fig. 6a). The flux of benthic foraminifera and faunal diversity are low (Fig. 6b).

Foraminifer interval B (509–463 cm, 11.8–11.3 ka BP). At 509 cm (11.8 ka BP), *N. labradorica* shows a sharp and rapid increase in percentage and dominates with *c.* 44% (Fig. 6a). Simultaneously, *E. excavatum* decreases rapidly and at 480–460 cm (11.5–11.3 ka BP) the percentage is below 2%. Associated species are *C. reniforme* and *I. helenae/norcrossi* with average percentages of 23% and 4%, respectively (Fig. 6a). At 11.5 ka BP (480 cm), *Buccella* spp. and *Cibicides lobatulus* begin to increase gradually (Fig. 6). The flux of benthic foraminifera and faunal diversity of benthic foraminifera remain low.

Foraminifer interval C (463–440 cm, 11.3–11 ka BP). The interval is characterized by an increase in percentage of *E. excavatum* (11%), *Stainforthia loeblichi* and the three Elphidium species; *E. albiumbilicatum, E. hallandence* and *E. bartletti* (Fig. 6), concurrent with a rapid decrease in *N. labradorica.* The three *Elphidum* species show an average distribution of 6%, 6% and 3%, respectively (Fig. 6b). The benthic foraminifera flux is low, while the faunal diversity is high (Fig. 6b).

Foraminifer interval D (440–420 cm, 11–10.8 ka BP). The interval is dominated by *C. reniforme* (up to 60% of the total fauna; Fig. 6a). In addition, high percentages of *Buccella* spp., a small increase in % *I. norcrossi/helenae* and a decrease in faunal diversity are seen compared to the previous interval (Fig. 6).

Foraminifer interval E (420–220 cm, 10.8–7.2 ka BP). This interval shows clear dominance of *C. reniforme* constituting an average of 39% as well as the almost absence of *E. excavatum* (Fig. 6a). *N. labradorica*, *C. lobatulus* and *A. gallowayi* vary in relative abundances throughout the interval (Fig. 6), whereas *Buccella* spp. and *I. norcrossi/helenae* are almost constant (Fig. 6). The benthic foraminifera flux is low in the beginning of the interval, but increases from 10 ka BP (310 cm).

Foraminifer interval F (220–150 cm, 7.2–4.7 ka BP). The interval is characterized by a sudden and rapid increase of *E. excavatum* starting at 7.2 ka BP, reaching a maximum of 40% by 6.9 ka BP (210 cm; Fig. 6a). *Buccella* spp. and *C. reniforme* show lower percentages compared to interval E. The benthic foraminifera flux and faunal diversity show generally high values compared to the entire core (Fig. 6).

Foraminifer interval G (150–0 cm, 4.7–0 ka). The interval is characterized by an additional increase in *E. excavatum*, which dominates the fauna with an average of 37%. This is followed by *N. labradorica*, constituting up to 20% (Fig. 6a). Throughout the interval *I. helenae/norcrossi* increases continuously, while *C. reniforme* decreases. The benthic foraminifera flux is generally low while the diversity is lower compared to interval F.

Correlations of lithologic units and foraminiferal intervals

In the following interpretation and discussion, the record from Kongsfjorden trough is divided into five time periods based on the age model and the units and intervals found in the lithological and foraminifera records. On this basis, the lithologic Unit A and foraminifera intervals A and partly B are combined to cover the time interval 12–11.5 ka BP. This corresponds to the late part of the Younger Dryas (Fig. 3). The remaining lithologic units and foraminifera intervals cover the Holocene. The Holocene is subdivided into four main time periods according to the foraminifera intervals; the Holocene transition period from 11.5–10.6 ka BP (lithological unit B and foraminifera intervals B, C, D and a small part of E); the Early Holocene 10.8–7.2 ka BP (lithological unit C and most of foraminifera unit E); the Mid-Holocene from 7.2–4.7 ka BP (partly lithological unit D and foraminifera unit F); and the Late Holocene from 4.7 ka BP to present (the remaining part of lithologial unit D, unit E and foraminifera inteval G; Fig. 3).

Environmental reconstructions from Kongsfjorden trough

Younger Dryas (c. 12–11.5 ka BP)

Only the youngest part of the Younger Dryas is present in the core. The combination of fine-grained sediments with low IRD content, low productivity and faunal diversity and dominance and co-occurrence of *E. excavatum* and *C. reniforme* indicate cold, unstable conditions and proximal glaciomarine environments (e.g. Jennings *et al.* 2004). The core site was probably under the influence of extensive sea-ice cover with occasional break-ups allowing ice rafting and the deposition of coarse grains. Grains larger than 500 μm are interpreted to be ice-rafted debris mainly deposited from icebergs. This is inferred from a study in proximity to our core site that shows IRD larger than 500 μm in the eastern Fram Strait primarily originates from icebergs and not sea ice in this area (Hebbeln 2000). Moreover, sea-ice sediments are often fine grained although in some cases they exhibit grain-size distributions similar to icebergs (Dowdeswell 1989; Pfirman *et al.* 1989; Gilbert 1990; Wollenburg 1991; Nürnberg *et al.* 1994; Dowdeswell *et al.* 1998). Different types of glacial activity, including active ice-streams (Dowdeswell *et al.* 1994), glacial retreats (Powell 1984; Dowdeswell *et al.* 1998; Vorren & Plassen 2002) or glacial advances (Dowdeswell *et al.* 1998; Vorren & Plassen 2002) generate icebergs. Icebergs can therefore reflect different types of glacial activity; the type of glacial activity should be evaluated in the context of additional available sedimentologic and biogenic properties.

During the final stages of the Younger Dryas period rapid changes in the fauna occur (Fig. 6) but these are not associated with changes in the lithology (Fig. 5). This suggests that water-mass alterations occurred earlier than changes on land or on glaciers. The decline of *E. excavatum* starting at 11.8 ka BP concurrent with an increase of *I. helenae/norcrossi* suggests a transition to a more glacier-distal environment (Osterman & Nelson 1989; Korsun & Hald 1998). After 11.8 ka BP the abundance of *N. labradorica* rapidly increases. This species has been associated with oceanic front areas where Atlantic Water presently mixes with Arctic Water in the Barents Sea (Steinsund

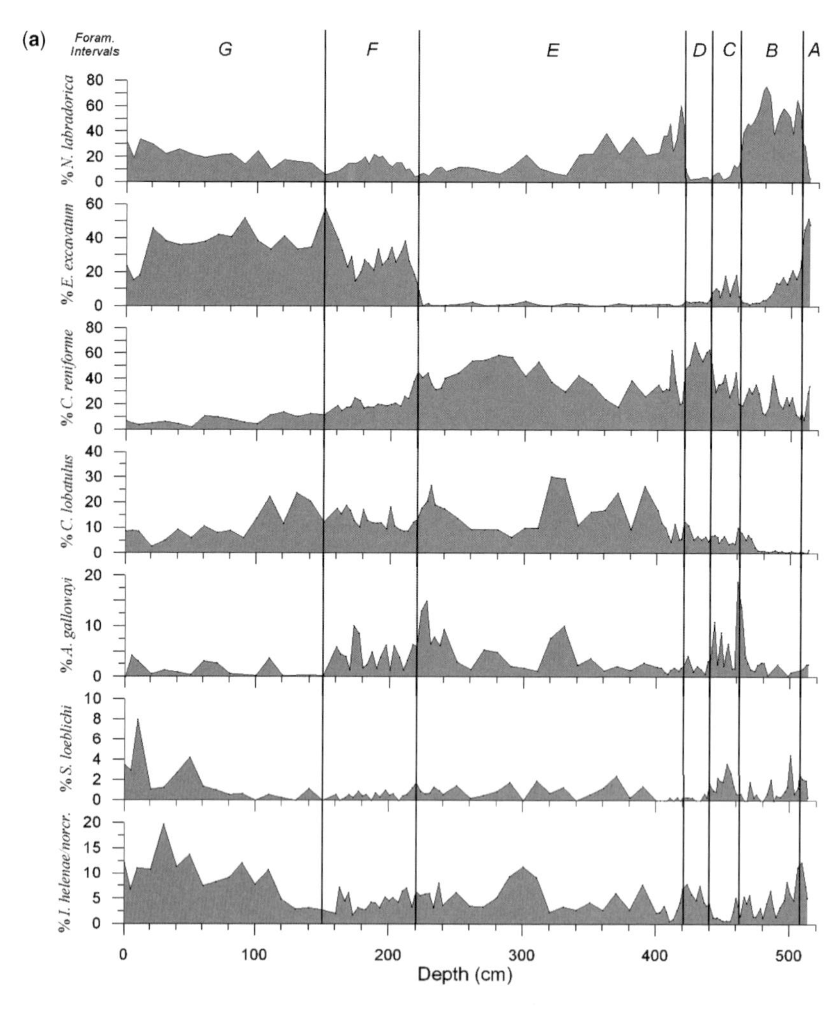

Fig. 6. (**a** & **b**) Relative abundances of benthic foraminifera species, flux of benthic foraminifera (no. individuals/cm^2/ka) and faunal diversity (number of species constituting 95% of the total fauna) in core NP05-11-21GC plotted v. depth. Foraminifer intervals are indicated.

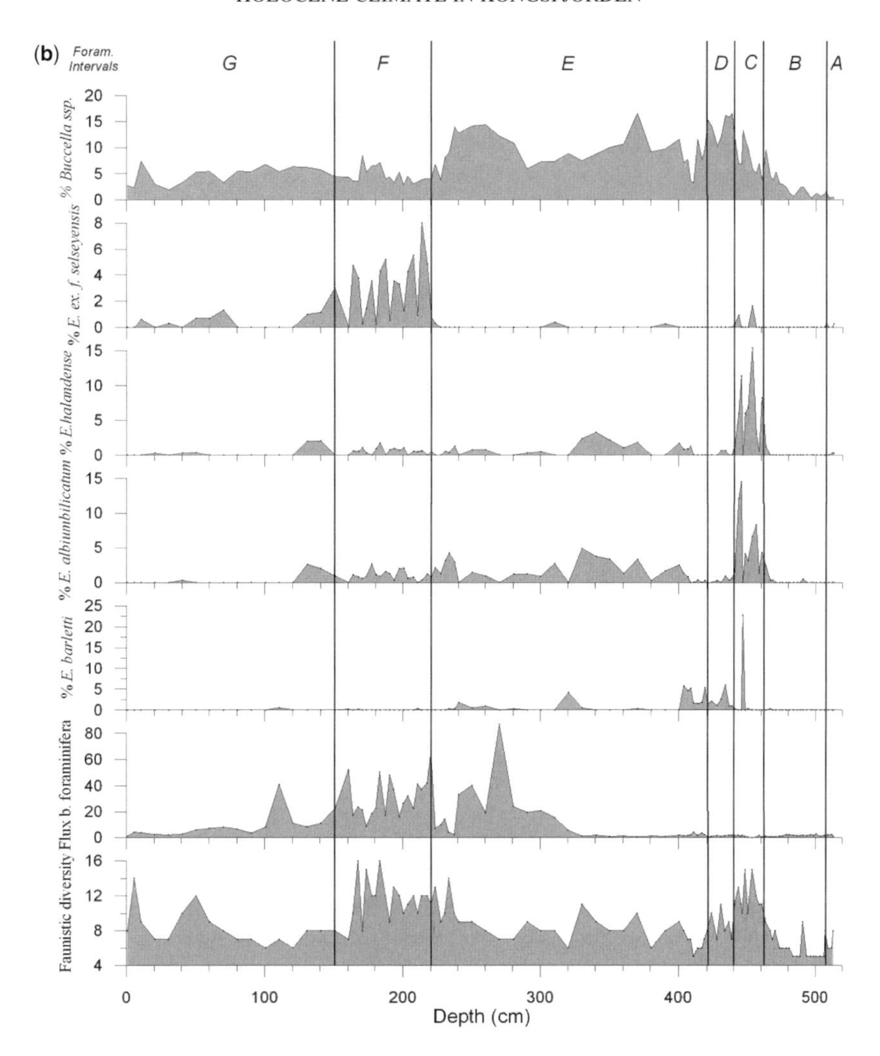

Fig. 6. *Continued.*

1994; Hald & Korsun 1997). The dominance of *N. labradorica* starting at 11.8 ka BP indicates increased influence of Atlantic-derived bottom waters. The concurrent decline of *E. excavatum* suggests diminished sea-ice cover, probably in combination with glacial retreat.

Holocene (11.5 ka BP to present)

The Holocene transition period (11.5–10.6 ka BP). The two peaks of *N. labradorica* at 11.8–11.3 ka BP and at 10.8 ka BP suggest a two-step warming during this period (Fig. 7). They probably reflect Atlantic-derived water and/or the proximity of oceanographic fronts (Steinsund 1994; Hald & Korsun 1997). The high IRD flux at 11.5 ka BP suggests rapid glacial retreat and deglaciation of the fjord. This is supported by the almost complete absence of *E. excavatum* from 11.8–11.3 ka BP, suggesting a glacier-distal environment. The initial warming was interrupted by a 250-year cooling starting at 11.3 ka BP. The benthic fauna indicates low salinity conditions, seen by the sudden appearance of the three *Elphidium* species *E. albiumbilicatum*, *E. subarcticum* and *E. bartletti*, which all show preference for environments characterized by lowered salinities (Lutze 1965; Feyling-Hanssen *et al.* 1971; Polyak *et al.* 2002). The concurrent re-introduction of *E. excavatum* and disappearance of *N. labradorica* suggest an abrupt shift to a colder climate and deteriorated environmental conditions dominated by Arctic Water of the coastal current. The decrease of IRD probably indicates permanent sea-ice cover that reduces the drift of icebergs and sea ice across the core site (Fig. 5). This is further supported by the small increase in

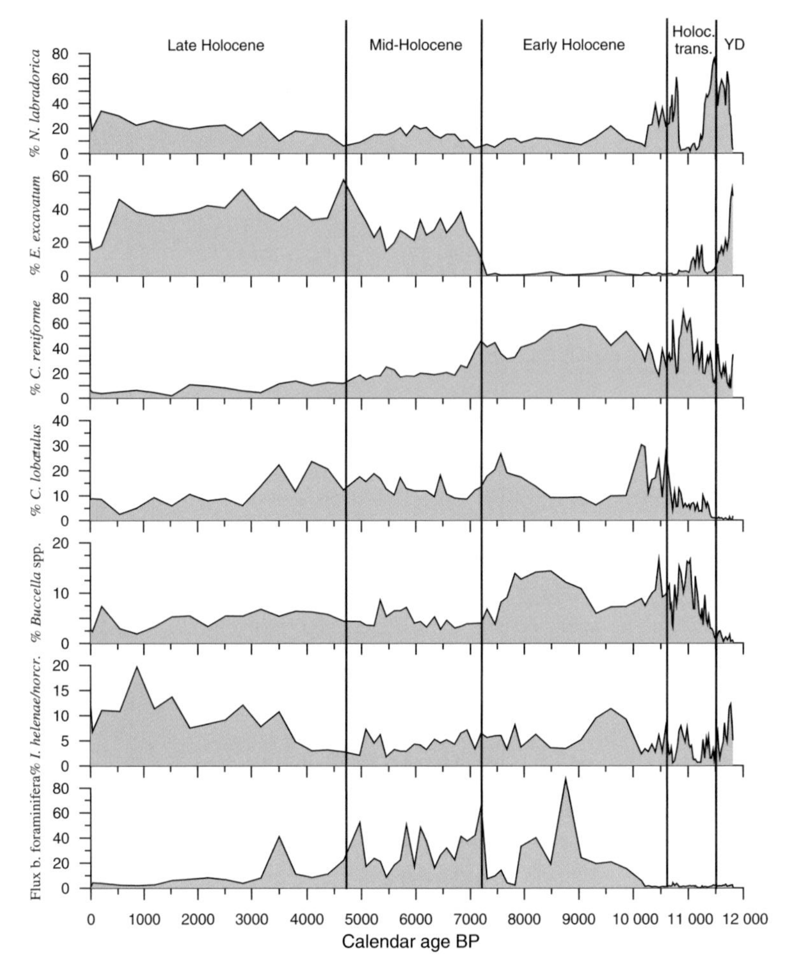

Fig. 7. Flux and relative abundances of dominating of benthic foraminifera species in core NP05-11-21GC plotted v. age. Main subdivisions of the Holocene and Younger Dryas are indicated.

percentage of *S. loeblichi*, a species related to sea-ice covered areas in the Barents Sea (Steinsund 1994). This inferred cooling from 11.3–11.05 ka BP can be correlated to the Preboreal Oscillation (PBO; Björck *et al.* 1997). The second warming at 10.8 ka BP has a similar fauna compared to the first warming, in addition to other species suggesting enhanced influence of Atlantic Water. This occurred concurrently with a high IRD flux, which represents the final stages of the disintegration of the Svalbard and Barents Sea ice sheet (e.g. Lehman & Forman 1992; Landvik *et al.* 2005).

The Early Holocene climate optimum (10.6–7.2 ka BP). The Early Holocene is characterized by the dominance of *C. reniforme* and the absence of *E. excavatum* and also high productivity in benthic foraminifera and diversity (Figs 6b & 7), high occurrences of molluscs and low IRD flux (Fig. 5). Altogether this indicates a strong influence of Atlantic Water masses at the core site (Ślubowska *et al.* 2005; Ślubowska-Woldengen *et al.* 2007) and a marked change from a glacier-proximal to glacier-distal environment. These changes indicate a significant climatic improvement that represents the Holocene climate optimum in this record.

Buccella spp. shows high values in the Holocene climate optimum (Fig. 7). This species has previously been shown to appear at the transition to the Holocene in shelf records from Svalbard, indicating that it may be associated with interglacial climate at high latitudes (Ślubowska *et al.* 2005; Ślubowska-Woldengen *et al.* 2007). It is also known to thrive in areas of high seasonal productivity connected to the algal blooms along the sea-ice margins in spring (Polyak & Solheim 1994; Steinsund 1994). Although there are no indications of a sea-ice edge in Kongsfjorden during the Early Holocene, we speculate that *Buccella* spp. may in general be an indicator of high productivity in these areas and, hence, indirectly an indicator of improved climate conditions.

Despite a considerable decrease in IRD flux, the sedimentation rate remains high until *c.* 10 ka BP (Figs 3 & 5). Similar findings in cores from the Norwegian margin have been explained by the availability of large amounts of fine-grained material in areas that have been subject to large glaciations (Vorren *et al.* 1983; Haflidason *et al.* 1998). The high sedimentation rate recorded in the basin, particularly from 11.5–10 ka BP, might be related to the deposition of winnowed material from the flanks of the trough (Vorren *et al.* 1983). The decline in sedimentation rate at *c.* 10 ka BP (Fig. 3) probably reflects reduced sediment delivery at this time and appears unrelated to diminishing current activity. This is inferred from high percentages of *C. lobatulus* and *A. gallowayi*, which indicate enhanced current activity in the bottom waters (Murray 1991). The elevated bottom-current speed is also supported by the increased amount of sandy material at this time (Fig. 5).

The Mid-Holocene (7.2–4.7 ka BP). The main change at the transition to the mid-Holocene is the rapid re-introduction of *E. excavatum* concurrent with a gradual decrease of *C. reniforme* (Fig. 7). The faunal shift suggests a change to a colder environment with lower salinity (Jennings *et al.* 2004; Hald *et al.* 2004), and can be related to a diminishing inflow of Atlantic Water compared to the early Holocene. However, this change is puzzling when seen in combination with the other proxies such as productivity and faunal diversity that are still high (Figs 6b & 7) and the reduced IRD supply (Fig. 5). Reductions in the inflow of Atlantic Water are expected to affect glacial activity, and thereby the IRD record, and also decrease faunal diversity and productivity; however, the IRD, productivity and faunal diversity records appear unaffected by this. This may indicate that the shift at 7.2 ka BP was related only to a minor reduction in the inflow of Atlantic Water. Alternatively, the inflow of Atlantic Water could have shifted to a circulation regime more similar to that of today with seasonal intrusions. In both cases, the reduction of inflow of Atlantic Water probably resulted in enhanced mixing of Atlantic and Arctic Waters, thereby lowering salinity slightly. The presence of the boreal subspecies *E. excavatum* forma *selseyensis* (Fig. 6b; Miller *et al.* 1982) in addition to the Arctic subspecies forma *clavatum* also suggest that the influence from glacial activity was low.

The Late Holocene (4.7–0 ka BP). In the Late Holocene the dominance of *E. excavatum* becomes even more pronounced (Fig. 7) and is accompanied by an increase in IRD flux (Fig. 5) and a significant drop in productivity (Fig. 7). This implies a colder environment with increased glacial influence (Jennings *et al.* 2004) and additional diminishing influence of Atlantic Water masses. This is also reflected in the decrease of foraminifera normally associated with enhanced current activity (Murray 1991). The increasing content of *I. norcrossi/helenae* (Fig. 7) may be indicative of more seasonal sea-ice cover (Steinsund 1994). Several distinct layers of coarse sediments are seen in this part of the core (Fig. 4). These may be comparable to similar sediment layers found in cores from the Greenland shelf that are related to the overturning of icebergs (Dowdeswell *et al.* 1994). A study of icebergs in Møllerfjorden, in the inner part of Krossfjorden, has shown that the IRD is often incorporated into the ice in thick layers or lenses (Dowdeswell *et al.* 1998); meltout of these could also be a plausible depositional mechanism for the coarse layers. The intervals of increased clast content may reflect random deposition from icebergs (Vorren *et al.* 1983), increased iceberg rafting during and after glacier surges (Dowdeswell *et al.* 1994) or enhanced iceberg rafting due to climatically induced glacier growth or retreat (Dowdeswell *et al.* 1998).

Palaeoceanographic implications

In order to examine how the findings in the core from Kongsfjorden trough relate to changes occurring on the shelf we compare our record to two Svalbard shelf records from Bellsund (Ślubowska-Woldengen *et al.* 2007) and Hinlopen Strait (Ślubowska *et al.* 2005; Fig. 1). The two core sites are located south (Bellsund) and north (Hinlopen) of Kongsfjord trough. Both sites are presently underlying the pathway of Atlantic Water flowing to the Arctic (Fig. 1). In the following, we particularly focus on how fluctuations in the position of the Arctic Front at the shelf edge affect the inflow of Atlantic Water to Kongsfjorden. Moreover, comparing these shelf records can also provide a better understanding of the regional climatic development along the Svalbard margin.

The three Svalbard shelf records generally show similar distribution patterns of the three dominant benthic foraminiferal species *C. reniforme*, *N. labradorica* and *E. excavatum* during the Holocene (Fig. 8). There are, however, some differences in timing of the onset of the Holocene warming; the amplitudes of faunal signals are also slightly different between the three sites.

The timing of the two-step warming in the Holocene Transition Period, indicated by the high

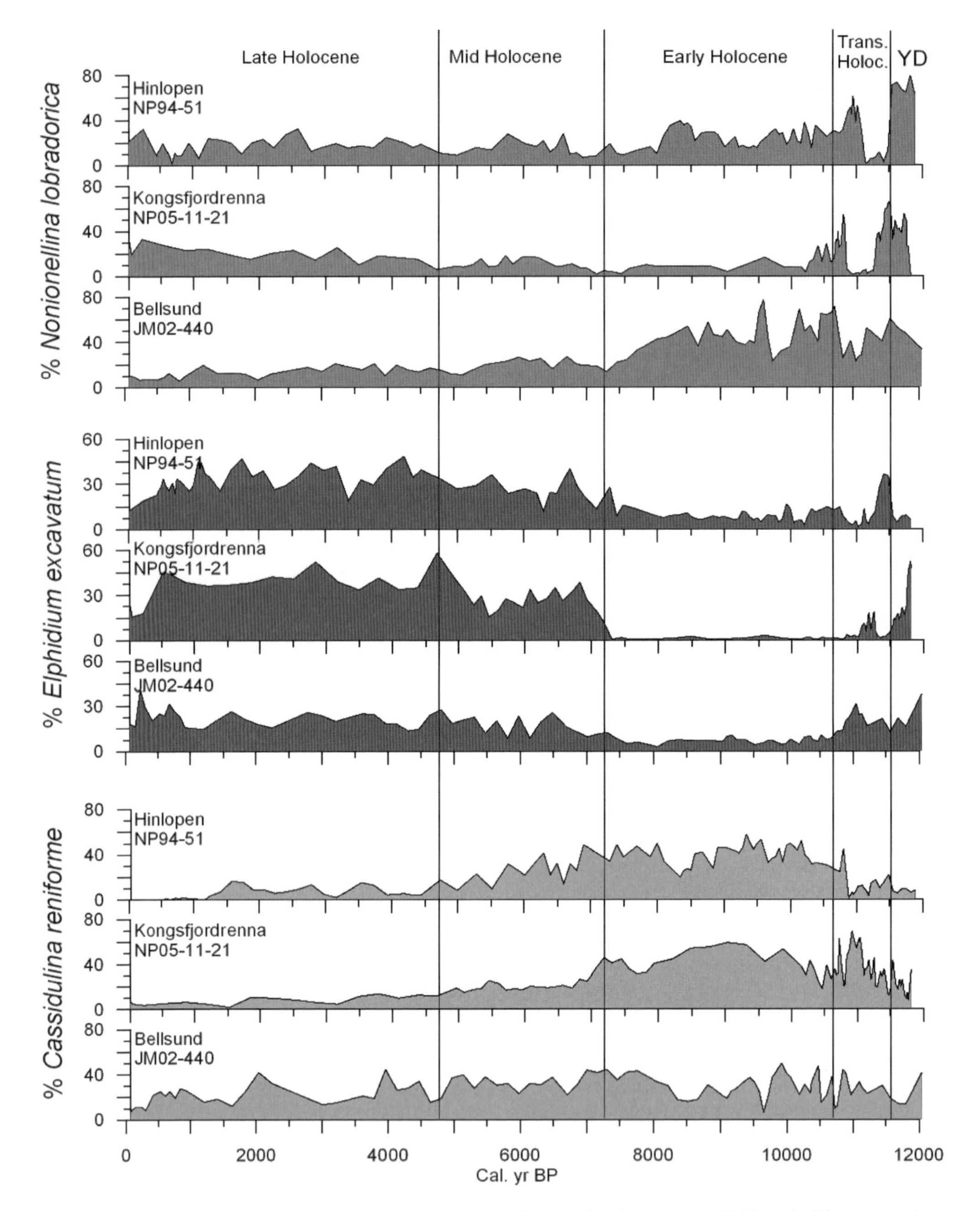

Fig. 8. Comparison of the three dominating species of benthic foraminifera in the core (NP05-11-21GC) presented in this study v. core NP94-51 (Hinlopen) and JM02-440 (Bellsund) (Ślubowska *et al.* 2005; Ślubowska-Woldengen *et al.* 2007), both from the Svalbard area (for core location see Fig. 1). Main subdivisions of the Holocene and Younger Dryas are indicated.

percentages of *N. labradorica*, shows a very similar development at the two northernmost sites (Hinlopen and Kongsfjorden). The southern site (Bellsund) shows a less pronounced shift in this species at the onset of the Holocene. The strong similarity between the two northern sites and Bellsund suggests that they reflect the same oceanographic changes related to the inflow of Atlantic Water to these sites at the onset of the Holocene. The slight delay in Kongsfjorden compared to Hinlopen can be attributed to uncertainties in the age models. The Hinlopen site is in close proximity to the shelf edge Arctic Front while the Kongsfjorden site is located further from the front. It is therefore likely that the Kongsfjorden site reflects the intrusion of Atlantic Water to the fjord and not the proximity of shelf edge Arctic Front.

The magnitude of faunal signals is also different between the three sites. In general, the two northernmost sites are more similar while the Bellsund site shows more subdued fluctuations in the fauna (Fig. 8). In Bellsund *N. labradorica* has a more even distribution pattern during the Holocene transition period. This indicates that this site was less affected by the cooling during the Preboreal oscillation. During the Holocene climatic optimum, the two northern sites are dominated by *C. reniforme* indicative of strong influence of Atlantic Water, and the Arctic Front is in distal positions to the two sites. During this time the shelf edge Arctic Front migrated south or east, which also implies a displacement of Atlantic Water masses closer to the coast. In the south at the Bellsund site *N. labradorica* dominates indicating proximity of the shelf edge Arctic Front during the Early Holocene.

In the Late Holocene, the relative abundance of *C. reniforme* decreases clearly at the two northernmost sites while this is not prominent at the Bellsund core site. Instead, this site shows a decrease in the relative abundance of *N. labradorica* suggesting less influence of the Arctic Front. It appears that the Arctic Front has been in proximity to the Bellsund site from the beginning of the Holocene, and then migrated westwards towards the end of the Early Holocene. At the Hinlopen site it migrated south after the Holocene transition period and north of the site in the Late Holocene. The inflow of Atlantic Water to Kongsfjorden is generally in phase with the changes seen on the Hinlopen site. This indicates that the shelf edge Arctic Front had a similar migration pattern on the shelf outside Kongsfjorden as in Hinlopen.

The glacial history in Kongsfjorden is poorly documented. According to this study the glacial front in the late part of the Younger Dryas was in proximity to our study site. This is in accordance with the ice extension during the Younger Dryas proposed by Mangerud & Landvik (2007). Our study also shows that ice rafting occurred (albeit with varying intensity) at the entrance of Kongsfjorden throughout the Holocene. This probably implies the presence of glaciers in the inner parts of Kongsfjorden during the Holocene, and agrees with previous studies of marine records from other fjords in Spitsbergen documenting a continuous presence of glaciers during the Holocene (Hald *et al.* 2004; Forwick & Vorren 2009). It should be noted, however, that some glaciers seem to have been smaller than present or even melted away during the Early Holocene, according to lacustrine records from Svalbard (Svendsen & Mangerud 1997; Snyder *et al.* 2000).

The advection of Atlantic Water to the Svalbard region provides moisture, which is known to be important for the growth of glaciers on land (Hebbeln *et al.* 1994). From the present study and other records, inflow of Atlantic Water to the Svalbard region during the Holocene is generally inferred to decrease at *c.* 7–7.5 ka BP with an additional reduction at *c.* 4–5 ka BP (Hald *et al.* 2004; Ślubowska *et al.* 2005; Ślubowska-Woldengen *et al.* 2007). However, in the core from Kongsfjorden trough, these reductions were not accompanied by increased glacial activity as inferred from the IRD record. In records from other west Spitsbergen fjords, the onset of intensified glacial activity occurs asynchronously between *c.* 9 and 4 ka BP (e.g. Svendsen & Mangerud 1997; Hald *et al.* 2004; Forwick & Vorren 2007; Baeten *et al.* 2010; Forwick *et al.* 2010). For example, in central Isfjorden (Fig. 1) ice rafting increased *c.* 9 ka BP (Forwick & Vorren 2009) while in Billefjorden, one of the tributary fjords of Isfjorden, an increase was not seen until 5.5 ka BP (Baeten *et al.* 2010), indicating large local variations. In the core from Kongsfjorden trough a marked increase in IRD occurs as late as 3.5 ka BP. The delayed onset of glacial activity might be the result of one or both of the following factors: (1) significantly delayed glacier growth in the Kongsfjorden area compared to glaciated areas further south; and (2) a combination of a relatively large distance to the glacier fronts and relatively warm water masses, resulting in the meltout of clasts before icebergs reached the core site.

The relationship between advection of Atlantic Water and glacial growth in the Svalbard region during the Holocene shows that time intervals with a strong inflow of Atlantic Water are generally associated with glacial retreat and warmer climate conditions (Birks 1991; Wohlfarth *et al.* 1995; Svendsen & Mangerud 1997; Hald *et al.* 2004; Ślubowska-Woldengen *et al.* 2007). Periods inferred to reflect glacier growth are generally connected to reduced inflow of Atlantic Water and cooler climate conditions on Svalbard (Birks 1991; Wohlfarth *et al.* 1995; Svendsen & Mangerud 1997; Hald *et al.* 2004; Ślubowska-Woldengen *et al.* 2007). This supports the theory, as previously concluded, that the climate on Svalbard is affected by the inflow of Atlantic Water (Svendsen & Mangerud 1997; Hald *et al.* 2004; Ślubowska-Woldengen *et al.* 2007). However, our findings also show that exceptions exist; this implies that

that the relationship is more complex and/or that other factors, such as atmospheric circulation and sea-ice distribution, also influence the relation. One factor that may be of importance to the coastal areas of Svalbard is the Arctic Front. In particular, the distance of the front with respect to the coast and its modifications on the inflow of Atlantic Water to the fjords, similar to the recent changes (Tverberg & Nøst 2009), affect the coastal climate conditions. A distal position of the front with respect to the coast will probably not affect the delivery of moisture to the glaciers to a large degree, provided the northern Norwegian Sea is ice free (Hebbeln *et al.* 1994).

Conclusions

The results of the faunal and sedimentological investigations in the sediment core covering the last *c.* 12 ka BP from Kongsfjorden trough show that the initial warming at 11.8 ka BP occurred simultaneously with the final phase of the deglaciation after the Younger Dryas. In the late Younger Dryas, the glacial front was located in proximity to the core site. During the Early Holocene, Atlantic Water masses prevailed in the fjord and glacial activity was diminished. In the Mid-Holocene transition period, the fjord was less influenced by Atlantic Water but productivity remained relatively high and no increase in glacial activity occurred. In the Late Holocene, Arctic Water from the Coastal Current dominated and glacial activity intensified. Comparing the record from Kongsfjorden trough to two shelf records north (Hinlopen, northen Svalbard) and south (Bellsund, southern Svalbard) shows that changes in the position of the Arctic Front at the latitude of Kongsfjorden are more in phase with the northern site (Hinlopen) than the southern site (Bellsund). The Arctic Front appears to remain in a more stable position at the Bellsund site than compared to stronger fluctuations at the two northern sites. The effect Atlantic Water has on glacial variations in Svalbard during the Holocene is more complex than simply stating that the strong inflow of Atlantic Water is related to periods with low glacial activity in this coastal setting. Other factors such as atmospheric circulation and sea-ice distribution should also be considered.

This is a contribution to the IPY-project 'SciencePub' (Arctic natural climate and environmental changes and human adaptation: from science to public awareness) funded by the Norwegian Research Council. Radiocarbon dates were performed at the Aarhus AMS 14C Dating Centre at the University of Aarhus, Denmark. We would like to thank the crew of RV *Lance* for help with sampling. The manuscript benefited from the constructive comments of anonymous reviewers.

References

Baeten, N. J., Forwick, M., Vogt, C. & Vorren, T. O. 2010. Late Weichselian and Holocene sedimentary environments and glacial activity in Billefjorden, Svalbard. *In*: Howe, J. A., Austin, W. E. N., Forwick, M. & Paetzel, M. (eds) *Fjord Systems and Archives*. Geological Society, London, Special Publications, **344**, 207–223.

Bard, E., Arnold, M. *et al.* 1994. The North Atlantic atmosphere-sea surface ^{14}C gradient during the Younger Dryas climatic event. *Earth and Planetary Science Letters*, **126**, 275–287.

Birks, H. H. 1991. Holocene vegetational history and climatic change in west Spitsbergen; plant macrofossils from Skardtjorna, an Arctic lake. *The Holocene*, **1**, 209–218.

Björck, S., Rundgren, M., Ingolfsson, O. & Funder, S. 1997. The Preboreal oscillation around the Nordic Seas: terrestrial and lacustrine responses. *Journal of Quaternary Science*, **12**, 455–465.

Bondevik, S., Mangerud, J., Birks, H. H., Gulliksen, S. & Reimer, P. 2006. Changes in North Atlantic radiocarbon reservoir ages during the Allerod and Younger Dryas. *Science*, **312**, 1514–1517.

Cottier, F., Nilsen, F., Inall, M. E., Gerland, S., Tverberg, V. & Svendsen, H. 2007. Wintertime warming of an Arctic shelf in response to large-scale atmospheric circulation. *Geophysical Research Letters*, **34**, L10607.

Dowdeswell, J. A. 1989. On the nature of Svalbard icebergs. *Journal of Glaciology*, **35**, 224–234.

Dowdeswell, J. A., Whittington, R. J. & Marienfeld, P. 1994. The origin of massive diamicton facies by iceberg rafting and scouring, Scoresby Sund, East Greenland. *Sedimentology*, **41**, 21–35.

Dowdeswell, J. A., Elverhøi, A. & Spielhagen, R. 1998. Glacimarine sedimentary processes and facies on the polar North Atlantic margins. *Quaternary Science Reviews*, **17**, 243–272.

Feyling-Hanssen, R. W., Joergensen, J. A., Knudsen, K. L. & Andersen, A. L. L. 1971. Late Quaternary foraminifera from Vendsyssel, Denmark and Sandnes, Norway. *Bulletin of the Geological Society of Denmark*, **21**, 67–71.

Forwick, M. & Vorren, T. O. 2007. Holocene mass-transport activity and climate in outer Isfjorden, Spitsbergen; marine and subsurface evidence. *The Holocene*, **17**, 707–716.

Forwick, M. & Vorren, T. O. 2009. Late Weichselian and Holocene sedimentary environments and ice rafting in Isfjorden, Spitsbergen. *Palaeogeography, Palaeoclimatology, Palaeoecology*, **280**, 258–274.

Forwick, M., Vorren, T. O., Hald, M., Korsun, S., Roh, Y., Vogt, C. & Yoo, K.-C. 2010. Spatial and temporal influence of glaciers and rivers on the sedimentary environment in Sassenfjorden and Tempelfjorden, Spitsbergen. *In*: Howe, J. A., Austin, W. E. N., Forwick, M. & Paetzel, M. (eds) *Fjord Systems and Archives*. Geological Society, London, Special Publications, **344**, 163–193.

Gerland, S. & Renner, A. H. H. 2007. Sea-ice mass-balance monitoring in an Arctic fjord. *Annals of Glaciology*, **46**, 435–442.

GILBERT, R. 1990. Rafting in glacimarine environments. *In*: DOWDESWELL, J. A. & SCOURSE, J. D. (eds) *Glacimarine Environments: Processes and Sediments*. Geological Society, London, Special Publications, **53**, 105–120.

HAFLIDASON, H., KING, E. L. & SEJRUP, H. P. 1998. Late Weichselian and Holocene sediment fluxes of the northern North Sea margin. *Marine Geology*, **152**, 189–215.

HALD, M. & KORSUN, S. 1997. Distribution of modern benthic foraminifera from fjords of Svalbard, European Arctic. *Journal of Foraminiferal Research*, **27**, 101–122.

HALD, M., EBBESEN, H. ET AL. 2004. Holocene paleoceanography and glacial history of the West Spitsbergen area, Euro-Arctic margin. *Quaternary Science Reviews*, **23**, 2075–2088.

HEBBELN, D. 2000. Flux of ice-rafted detritus from sea ice in the Fram Strait. *In*: GANSSEN, G. & WEFER, G. (eds) *Particle Flux and its Preservation in Deep-Sea Sediments*. Pergamon, Oxford, 1773– 1790.

HEBBELN, D., DOKKEN, T., ANDERSEN, E. S., HALD, M. & ELVERHØI, A. 1994. Moisture supply for northern ice-sheet growth during the last glacial maximum. *Nature*, **370**, 357–360.

HOP, H., FALK-PETERSEN, S., SVEIAN, H., KWASNIEWSKI, S., PAVLOV, V., PAVLOVA, O. & SØREIDE, J. E. 2006. Physical and biological characteristics of the pelagic system across Fram Strait to Kongsforden. *Progress in Oceanography*, **71**, 182–231.

HUGHEN, K. A., BAILLIE, M. G. L. ET AL. 2004. Marine04 marine radiocarbon age calibration, 0–26 cal kyr BP. *Radiocarbon*, **46**, 1059–1086.

HUSUM, K. & HALD, M. 2004. A continuous marine record 8000–1600 cal. Yr BP from the Malangenfjord, north Norway: foraminiferal and isotopic evidence. *The Holocene*, **14**, 877–887.

JENNINGS, A. E., WEINER, N. J., HELGADOTTIR, G. & ANDREWS, J. T. 2004. Modern foraminiferal faunas of the southwestern to northern Iceland Shelf; oceanographic and environmental controls. *Journal of Foraminiferal Research*, **34**, 180–207.

JOHANNESSEN, O. M. 1986. Brief overview of the physical oceanography. *In*: HURDLE, B. G. (ed.) *The Nordic Seas*. Springer, New York, 103–127.

KLITGAARD-KRISTENSEN, D., SEJRUP, H. P. & HAFLIDASON, H. 2001. The last 18 kyr fluctuations in Norwegian Sea surface conditions and implications for the magnitude of climatic change: evidence from the North Sea. *Paleoceanography*, **16**, 455–467.

KOHLER, J., JAMES, T. D., MURRAY, T., NUTH, C., BRANDT, O., BARRAND, N. E. & LUCKMAN, W. 2007. Acceleration in thinning rate on western Svalbard glaciers. *Geophysical Research Letters*, **34**, L18502, doi: 10.1029/2007GL030681.

KORSUN, S. & HALD, M. 1998. Modern benthic foraminifera off Novaya Zemlya tidewater glaciers. *Russian Arctic Arctic and Alpine Research*, **30**, 61–77.

LANDVIK, J. Y., INGÓLFSSON, Ó., MIENERT, J., LEHMAN, S. J., SOLHEIM, A., ELVERHOI, A. & OTTESEN, D. 2005. Rethinking late Weichselian ice-sheet dynamics in coastal NW Svalbard. *Boreas*, **34**, 7–24.

LEHMAN, S. J. & FORMAN, S. L. 1992. Late Weichselian glacier retreat in Kongsfjorden, West Spitsbergen, Svalbard. *Quaternary Research*, **37**, 139–154.

LUTZE, G. F. 1965. Foraminiferen-Fauna der Ostsee. *Meyniana*, **15**, 75–142.

MANGERUD, J. & LANDVIK, J. 2007. Younger Dryas cirque glaciers in western Spitsbergen: smaller than during the Little Ice Age. *Boreas*, **36**, 278–285.

MANGERUD, J., BONDEVIK, S., GULLIKSEN, S., HUFTHAMMER, A. K. & HØISÆTER, T. 2006. Marine ^{14}C reservoir ages for 19th century whales and molluscs from the North Atlantic. *Quaternary Science Reviews*, **25**, 3228–3245.

MILLER, A. A. L., SCOTT, D. B. & MEDIOLI, F. S. 1982. Elphidium excavatum (Terquem); ecophenotypic v. Subspecific variation. *Journal of Foraminiferal Research*, **12**, 116–144.

MURRAY, J. W. 1991. *Ecology and Palaoecology of Benthic Foraminifera*. Longman, London.

NÜRNBERG, D., WOLLENBURG, I. R. ET AL. 1994. Sediments in Arctic sea ice; implications for entrainment, transport and release. *Marine Geology*, **119**, 185–214.

OSTERMAN, L. E. & NELSON, A. R. 1989. Latest Quaternary and Holocene paleoceanography of the eastern Baffin Island continental shelf, Canada; benthic foraminiferal evidence. *Canadian Journal of Earth Sciences*, **26**, 2236–2248.

OTTESEN, D., DOWDESWELL, J. A., LANDVIK, J. Y. & MIENERT, J. 2007. Dynamics of the Late Weichselian ice sheet on Svalbard inferred from high-resolution sea-floor morphology. *Boreas*, **36**, 286–306.

PFIRMAN, S., GASCARD, J.-C., WOLLENBURG, I., MUDIE, P. & ABELMANN, A. 1989. Particle-laden eurasianarctic sea ice: observations from July and August 1987. *Polar Research*, **7**, 59–66.

POLYAK, L. & SOLHEIM, A. 1994. Late- and post-glacial environments in the northern Barents Sea west of Franz Josef Land. *Polar Research*, **13**, 197–207.

POLYAK, L., KORSUN, S. ET AL. 2002. Benthic foraminiferal assemblages from the southern Kara Sea, a river-influenced arctic marine environment. *Journal of Foraminiferal Research*, **32**, 252–273.

POWELL, R. D. 1984. Glacimarine processes and inductive lithofacies modelling of ice shelf and tidewater glacier sediments based on Quaternary examples. *Marine Geology*, **57**, 1–52.

ŚLUBOWSKA, M. A., KOÇ, N., RASMUSSEN, T. L. & KLITGAARD-KRISTENSEN, D. 2005. Changes in the flow of Atlantic water into the Arctic Ocean since the last deglaciation: evidence from the northern Svalbard continental margin, 80°N. *Paleoceanography*, **20**, PA4014, doi: 10.1029/2005PA001141.

ŚLUBOWSKA-WOLDENGEN, M., RASMUSSEN, T. L., KOÇ, N., KLITGAARD-KRISTENSEN, D., NILSEN, F. & SOLHEIM, A. 2007. Advection of Atlantic Water to the western and northern Svalbard shelf since 17,500 cal yr BP. *Quaternary Science Reviews*, **26**, 463–478.

SNYDER, J. A., WERNER, A. & MILLER, G. H. 2000. Holocene cirque glacier activity in western Spitsbergen, Svalbard; sediment records from proglacial Linnevatnet. *The Holocene*, **10**, 555–563.

STEINSUND, P. I. 1994. *Benthic foraminifera in surface sediments of the Barents and Kara Seas: modern and late Quaternary applications*. PhD thesis, University of Tromsø, Trosmø.

Stuiver, M. & Reimer, P. J. 1993. Extended ^{14}C database and revised CALIB radiocarbon calibration program. *Radiocarbon*, **35**, 215–230.

Svendsen, H., Beszczynska, M. A. *et al.* 2002. The physical environment of Kongsfjorden-Krossfjorden, an Arctic fjord system in Svalbard. *Polar Research*, **21**, 133–166.

Svendsen, J. I. & Mangerud, J. 1997. Holocene glacial and climatic variations on Spitsbergen, Svalbard. *The Holocene*, **7**, 45–57.

Swift, J. H. 1986. The Arctic Waters. *In*: Hurdle, B. G. (ed.) *The Nordic Seas*. Springer, New York, 129–153.

Tverberg, V. & Nøst, O. A. 2009. Eddy overturning across a shelf edge front: Kongsfjorden, west Spitsbergen. *Journal of Geophysical Research*, **114**, C04024, doi: 10.1029/2008JC005106.

Vorren, T. O. & Plassen, L. 2002. Deglaciation and palaeoclimate of the Andfjord-Vagsfjord area, North Norway. *Boreas*, **31**, 97–125.

Vorren, T. O., Hald, M., Edvardsen, M. & Lind-hansen, O.-W. 1983. Glacigenic sediments and sedimentary environments on continental shelves: general principles with a case study from the Norwegian shelf. *In*: Ehlers, J. (ed.) *Glacial Deposits in North-West Europe*. Balkema, Rotterdam, 61–73.

Vorren, T. O., Laberg, J. S. *et al.* 1998. The Norwegian-Greenland Sea continental margins; morphology and late Quaternary sedimentary processs and environment. *Quaternary Science Reviews*, **17**, 273–302.

Walton, W. R. 1964. Recent foraminiferal ecology and paleoecology; Part 1, Northeastern Gulf of Mexico Foraminifera; Part 2, Paleoecology of the subsurface Oligocene in coastal Texas. *In*: Imbrie, J. & Newell, N. D. (eds) *Approach to Paloecology*. Wiley & Sons, New York, 151–237.

Wohlfarth, B., Lemdahl, G., Olsson, S., Persson, T., Snowball, J. I. & Jones, V. 1995. Early Holocene environment on Bjørnøya (Svalbard) inferred from multidisciplinary lake sediment studies. *Polar Research*, **14**, 235–275.

Wollenburg, I. R. 1991. Sediment transport by Arctic sea ice – the recent load of lithogenic and biogenic material. *Berichte zur Polarforschung*, **127**, 159.

Scottish west coast fjords since the last glaciation: a review

KATE L. McINTYRE* & JOHN A. HOWE

Scottish Association for Marine Science, Scottish Marine Institute, Oban, Argyll, Scotland (UK), PA37 1QA

**Corresponding author (e-mail: Kate.McIntyre@sams.ac.uk)*

Abstract: Evidence from seismic profiles, sidescan and multibeam sonar surveys and sediment cores reveal important information about the depositional history of Scotland's fjords during the last deglacial transition and the subsequent Holocene. Devensian glaciation removed pre-existing sediments both inside and outside the fjord basins, and deposited a diamict over much of the subsequent erosional surface. Glaciomarine sequences deposited during the initial retreat of this ice margin are mainly preserved in the sea area outside the fjords. Younger glaciomarine units present within the fjord basins are often attributed to the later Younger Dryas re-advance (12.8–11.5 ka BP). Younger Dryas ice reworked Devensian sediments, depositing terminal moraines at the mouths of a number of fjords and glaciomarine units within the basins during its retreat. Ice margins oscillated throughout deglaciation, although the latter stages of post-Younger Dryas retreat may have occurred rapidly with topographic pinning an important factor in determining the style of retreat. Holocene records from the fjords are limited, but tentatively support the onshore evidence for a tundra-style landscape in the earliest Holocene, followed by a warming of the climate and, latterly (*c.* 2000 ka BP), increasing humidity. In the more recent past, human activities such as deforestation, fishing, aquaculture and industry have been recorded in the fjord sediments.

Fjords, generally known in Scotland as sea lochs, form an important transitional zone between marine and terrestrial environments. Their morphology, typically characterized by the presence of deep basins and shallow sills, leads to effective trapping of sediments originating from the land and from the sea surface. In the fjords of western Scotland, free from glacial ice for the past 11.5 ka, these sediments preserve valuable high-resolution records of deglacial and Holocene climate variability (Howe *et al.* 2002; Nørgaard-Pedersen *et al.* 2006; Mokeddem *et al.* 2007). Additionally, the mapping of submarine and sub-seabed depositional features in the fjords has provided key evidence for the reconstruction of glacial dynamics during the last glacial/deglacial transition, and for linking the onshore and offshore Scottish deglacial records (Boulton *et al.* 1981; Dix & Duck 2000; Howe *et al.* 2002; Stoker *et al.* 2006).

Scotland's west coast is characterized by the presence of more than 50 fjords (Syvitski & Shaw 1995), a legacy of its strongly glacially influenced past. Early research in these fjords focused mainly on hydrography (Milne 1972; Edwards & Edelsten 1977; Edwards & Sharples 1986) and biology (Pearson 1970; Gage 1972*a*, *b*; Gage & Geekie 1973; Ansell 1974; Davies 1975). The introduction of commercial fish-farming to western Scotland in the 1970s led to increased research effort on fjord circulation (Gillibrand *et al.* 1995). This was initially to assess the suitability of potential sites (e.g. Milne 1969) and subsequently to investigate the long-term ecological impacts of the industry (Brown *et al.* 1987; Gillibrand *et al.* 1995, 1996). Such research is concerned mainly with establishing the frequency and nature of deep-water renewal, which is also of interest to physical oceanographers seeking to understand vertical and horizontal water-mixing processes occurring in the fjord basins and across the sills (e.g. Booth 1987; Allen & Simpson 1998*a*, *b*; Inall *et al.* 2004).

The potential of Scottish fjords for palaeoclimate research was not realized until relatively recently, with a number of studies published over the past decade. Although onshore geomorphological features and uplifted sediments from around some of the fjords were incorporated into land-based studies of Scotland's late deglacial history – in particular, the limits and dynamics of glaciers associated with the Loch Lomond Re-advance of Younger Dryas age (e.g. Peacock 1971*a*, *b*; Gray 1975) – access to the seabed and its sediments was not possible until the introduction of marine geophysical surveying and coring techniques to the fjords. These tools were in general use by researchers on the continental shelf and margin since the 1970s (Andrews *et al.* 1996), but were not employed in the fjords until the seismic survey of Boulton *et al.* (1981) in and around Lochs Ailort and Nevis. Subsequently, acoustic surveying and/or sediment core analyses

From: Howe, J. A., Austin, W. E. N., Forwick, M. & Paetzel, M. (eds) *Fjord Systems and Archives.* Geological Society, London, Special Publications, **344**, 305–329.
DOI: 10.1144/SP344.21 0305-8719/10/$15.00

have been used in several Scottish fjords to obtain and interpret the deglacial and Holocene records of deposition they preserve. These studies, the results of which are summarized and discussed in this paper, have contributed substantially to our understanding of Scotland's glacial history and of its climate from late glacial to recent Holocene times.

Western Scotland: a regional overview

Scotland is part of the British Isles, situated on the passive oceanic margin of northwest Europe and the eastern North Atlantic. The places and features referred to in this paper are shown on the map in Figure 1.

Geology and physiography

Scotland is composed of a wide variety of bedrock types, reflecting its complex geological history. Along the fjordic west coast, the predominant rock types are the Moine and Dalradian metamorphic series of the Precambrian (quartzites, psammites, pelites, slates, phyllites and schists). These are intruded and overlain in places by plutonic/volcanic rocks mainly of Devonian age, for example the granitic intrusions of Etive-Cruachan, Rannoch Moor and Morvern, and the coeval Lorne Plateau Lavas which are extensive around Oban. Much later, Tertiary basaltic volcanism associated with the opening of the North Atlantic Rift *c.* 56 Ma ago produced many of the inner Hebridean islands, including Mull, Skye and Rum. Tertiary volcanism also led to the development of an extensive swarm of doleritic dykes intruding into the bedrock of many islands and the western mainland (Craig 1991).

To the NW, separated from the Moine series by the SSW–NNE trending Moine Thrust, the bedrock consists of Lewisian gneisses aged 1600–2900 Ma. On the mainland the Lewisian rocks are extensively overlain by Torridonian sandstones (800–1100 Ma). In the Outer Hebrides, however, they are generally exposed at the surface since the sandstones have been lost to erosion (Craig 1991). The wide variety of rock types and the presence of geological faults/thrusts within the bedrock have resulted in differential erosion of the landscape, as weathering

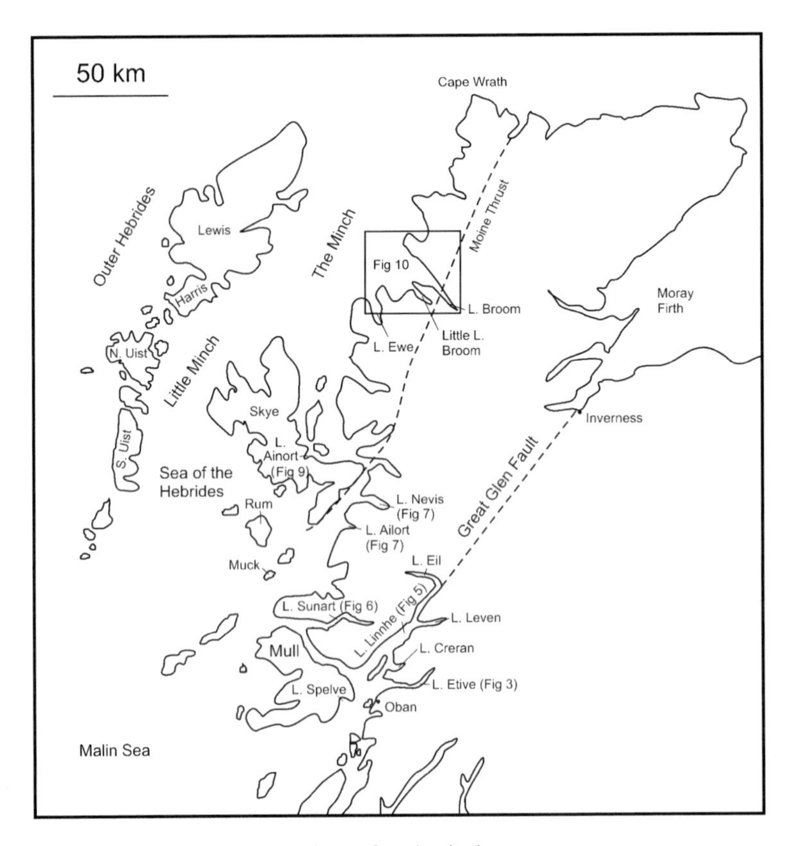

Fig. 1. Map of Western Scotland, showing locations referred to in the text.

processes have most effectively worn away the least resistant rocks. Scotland's west coast is characterized by its 'knock-and-lochan' topography of mountains, glens and fjords, the result of 500 ka of periodic glacial erosion on the landscape (Stoker *et al.* 2006).

Another important characteristic of the western Scottish coast are the numerous (>100) islands of widely varying sizes, which run broadly parallel to the mainland. These are divided into the Inner and Outer Hebrides, separated from one another by a body of water called the Minch and from the mainland by numerous sounds and channels. The Hebrides have a sheltering effect on the west coast mainland, protecting it from the worst of the North Atlantic swell and storms (Hall 1974).

Sea level

During the recent geological past Scotland's coastline has been much affected by changes in relative sea level, arising from isostatic adjustments to ice-sheet advance and retreat and the associated eustatic sea-level response. Isostatic uplift since the Last Glacial Maximum (*c.* 20 ka) has resulted in the presence of raised beaches along much of the coast, both on the mainland and around many of the Hebridean islands. Three raised shorelines were considered by Smith *et al.* (2000) to be sufficiently widely distributed to aid in the reconstruction of Holocene uplift patterns. These are the Storegga Tsunami Shoreline dated at 7.9 cal ka BP, the Main Postglacial Shoreline at 6.4–7.7 cal ka BP and the Blairdrummond Shoreline at 4.5–5.8 cal ka BP. A 'speculative' fourth described in a subsequent study, the Wigtown Shoreline, is dated at 1.52–3.7 cal ka BP (Smith *et al.* 2006). Differential uplift rates mean that over much of the Scottish coastline, the Blairdrummond Shoreline is at a higher level than the older Main Postglacial Shoreline (Smith *et al.* 2007). All of the postglacial shorelines are characterized by low gradients because of the relatively low rates of uplift during the Holocene (Sissons & Dawson 1981). Reconstructions carried out by Shennan *et al.* (1995, 1996, 1999, 2000) show that in northwest Scotland (Oban to Ullapool), relative sea level fell continuously throughout the Younger Dryas as isostatic uplift proceeded more rapidly than global sea-level rise. In the earliest Holocene, relative sea level remained stationary for a time and then rose to a mid-Holocene high at about 7 ka BP. This high was at its maximum in the Firth of Lorne, where it reached about 8 m O.D., and decreased northwards to about 2.5 m O.D. in Ullapool. After this time isostatic rebound took over again, and relative sea level has been falling steadily in the area ever since. Away from the main centres of uplift, however, it is recorded that the mid-to-late Holocene was a time of relative sea-level rise (Smith *et al.* 2007).

Sea-level fluctuations are of paramount influence in fjordic environments. Where a fjordic basin is separated from the sea by an entrance sill, the height of sea level relative to the sill determines whether the basin develops into a freshwater or a marine environment following withdrawal of the ice (Lloyd 2000; Howe *et al.* 2002; Nørgaard-Pedersen *et al.* 2006). In the marine lochs sea level strongly impacts upon fjordic circulation, because it determines the degree to which the sill restricts flow in and out of the loch and hence, along with rates of freshwater input, frequency of deep-water renewal events (Austin & Inall 2002; Nørgaard-Pedersen *et al.* 2006).

Shelf morphology and sediments

The western Scottish shelf varies in width from 40 km west of the Outer Hebrides to 190 km off Cape Wrath, with the shelf break lying at a depth of 140–200 m (Pantin 1991). The main water bodies are the Minch, which separates the Inner and Outer Hebrides, the Sea of the Hebrides south of Skye and the Minch and the Malin Sea between Northern Ireland and the Western Isles. Shelf morphology in this area is the most variable in the UK (Murray 2004) with numerous irregular banks, scarps and depressions resulting from differential glacial erosion of the sediments and underlying bedrock, enhanced locally by tidal scour (Pantin 1991). The deepest point on the shelf is the Muck Deep (320 m), a glacially scoured depression south of the Isle of Muck (Murray 2004). West of the Outer Hebrides, the shelf slopes westward towards the break and contains many small (*c.* 10 m) irregularities attributed to glacial erosion and sedimentation (Pantin 1991). The shelf break itself demarcates the maximum limit of the last (Devensian) glaciation (Bowen *et al.* 2002).

Sedimentation across the shelf consists mainly of terrigenous Quaternary deposits accumulated under glacial and stadial conditions (Davies *et al.* 1984). These deposits have an average thickness of 150 m, but there is considerable variation. Over the resistant topographic highs of Lewisian basement and Tertiary volcanics there can be little or no sediment cover, whereas in the glacially overdeepened basins thicknesses of up to 300 m have been preserved (Davies *et al.* 1984). The Quaternary sediments consist principally of muddy glacial and glaciomarine deposits, which presently act as a source for modern sedimentation (Pantin 1991). A thin diamict is also present in places, underlain by an extensive, shelf-wide erosion surface interpreted as Devensian in age, with underlying deposits dating back to previous glaciations (Davies *et al.* 1984).

Modern sediment is mainly derived from tide- and wave-mediated erosion of the underlying Quaternary beds. New supply is limited and localized, for example the basaltic sands around the Mull coast derived from weathering of the Tertiary bedrock (Pantin 1991). Transport occurs as a result of advective (e.g. tidal, wave, riverine) and diffusive (turbulent) processes. In general, sediment distribution reflects patterns in tidal current strength. Relatively fine sediments (sandy muds, muddy sands) predominate in the Minch and the Sea of the Hebrides, sands dominate further out in the Malin Sea and west of the Outer Hebrides and muds are generally restricted to deep/sheltered waters and seaward of the shelf edge. Exposed bedrock and gravel occur locally in areas of high tide/wave action, for example west of South Harris and the Uists (Pantin 1991).

Oceanography

Relatively warm and saline water of Gulf Stream origin is delivered to the shelf edge west of Scotland in the northeast-heading North Atlantic Drift, an offshoot of the main North Atlantic gyre (Hall 1974). The Atlantic water spreads eastward and northward across the shelf and meets the northward-flowing Scottish coastal current, which consists of relatively fresh Irish and Clyde Sea water from the North Channel (Ellett & Edwards 1983). A strong oceanic front between the two water masses has been reported to the west of Islay (Ellett 1979). As it travels north, the Scottish coastal current receives further freshwater inputs from land runoff via the fjords that indent the coastline (Ellett 1979). Scottish shelf waters therefore consist of a mixture of saline (>35.0, Ellett & Edwards 1983) Atlantic water, slightly dilute (34.0, Bowden 1980) Irish/Clyde Sea waters and coastal freshwater runoff. The relative proportions of these water masses and degree of mixing between them varies seasonally. In winter, the increased wind results in a well-mixed water column which allows Atlantic water to invade the outer shelf to the west of the islands (Ellett 1979). Mixing on the shelf is mainly by tidal/wave processes; tidal currents off the coast can be locally extremely strong (>500 cm s^{-1}) at spring highs where water flow is constricted by topography (Hall 1974). Homogeneous vertical temperature/salinity structures are common in shallow waters, particularly where tidal/wave-induced mixing is enhanced in the sounds; stratification develops in the summer over the deeper areas (Ellett & Edwards 1983). Seasonal thermocline fronts can inhibit mixing on the shelf (Austin & Kroon 1996). Mean sea-surface temperatures range from 7–8 °C (February) to 12–13 °C (August) (Hall 1974), with salinity values reflecting the mixture of water masses present: reduced compared to oceanic values, typically <35.0 (Hill *et al.* 1997).

The Scottish coastal current flows northwards with a velocity of 3–10 cm s^{-1}, and is unusual for the UK shelf which is generally characterized by weak and variable currents (Hill & Simpson 1988). Flow direction is significantly influenced by topography and by wind direction (Davies & Xing 2003), which is predominantly from the SW especially in winter (Hall 1974). Variability in the current speed has been explained in terms of fluctuations in an along-coast sea-level gradient, which in turn result from North Atlantic low-pressure systems changing track across the British Isles (Hill & Simpson 1988). The current flows northwards along the Hebridean shelf and into the Minch, where a part of it is recirculated south around Barra Head before continuing on a northwards path to the west of the Outer Hebrides (Ellett 1979).

Climate

The climate of western Scotland is a product of its proximity to the North Atlantic and its mountainous topography. Moisture and heat are delivered to the region from the North Atlantic Drift, resulting in a mild, maritime climate characterized by variable, often stormy weather and abundant rainfall (Hall 1974). Wind direction is mainly from the SW but varies seasonally, with northeasterlies/easterlies common in the spring. Air temperatures are very mild for the latitude, ranging from a mean of 5 °C in winter to 14 °C in summer (max 30 °C). Temperatures below 0 °C occur infrequently, and snowfall is generally slight. Mean annual precipitation is in the range 1000–2000 mm a^{-1}, with the mountains subject to particularly heavy rainfall hence leading to very high freshwater inputs to the fjord heads (Hall 1974). Highest rainfall values occur from September to January, with the driest months being April to June (Met Office 2009).

Northwest European climate is strongly dictated by variability in the North Atlantic Oscillation (NAO), a large-scale oscillation in atmospheric mass between Iceland and the Azores (Hurrell *et al.* 2003). By altering the paths taken by low-pressure systems over the North Atlantic and northwest Europe, the NAO influences westerlies and precipitation patterns on timescales ranging from <15 to 90 a (Appenzeller *et al.* 1998). The sea area west of Scotland is ideally positioned to monitor this variability; the northwest Scottish fjords have been identified as potential key study sites because of their sensitivity to freshwater runoff (Austin & Inall 2002; Gillibrand *et al.* 2005).

Over longer timescales, Scotland and the adjacent shelf area have been subject to repeated glaciations over the last 500 ka (Stoker *et al.* 2006). During the most recent glacial episode (the Devensian), ice covered much of the UK mainland and advanced as far as the modern shelf-break west of Scotland. Deglaciation occurred from the Last Glacial Maximum *c.* 20 ka ago until the onset of the Holocene *c.* 11.5 ka ago. It progressed in a non-linear fashion, with glacial margins oscillating intermittently in response to short-term climate fluctuations. The west coast fjords represent ice-sheet drainage conduits along which Quaternary glaciers made their way to the coast. During deglaciation, the basins filled with water and sediment from the retreating glaciers and were reinvaded by the sea, returning them to glaciomarine and eventually full marine conditions. Throughout the Holocene Scotland has remained completely free of glacial ice. The climate along its fjordic west coast is described by Syvitski & Shaw (1995) as 'temperate maritime'.

The Younger Dryas (Loch Lomond) ice cap

A major glacial re-advance occurred during the Younger Dryas stadial of 12.8–11.5 ka BP, widely known in Scotland as the Loch Lomond Advance or Re-advance after the freshwater loch that marks its southerly limit. During this period a substantial ice cap, up to 1000 m in altitude (Golledge 2007), was present over much of western Scotland. The limits and dynamics of this ice cap have been reconstructed using geomorphological mapping of relict glacial features such as moraines, till deposits, outwash fans/terraces, meltwater channels, ice-moulded/striated bedrock, glacial erratics, trimlines and periglacial features such as frost-shattering of unglaciated ground (e.g. Charlesworth 1956; McCann 1966; Sissons 1979; Thorp 1981, 1986; Ballantyne *et al.* 1998; Lukas & Benn 2006).

More recently, numerical models have been developed which seek to reproduce the palaeoclimatic conditions required to produce an ice cap which best fits the known mapped limits (Hubbard 1999; Golledge & Hubbard 2005; Golledge *et al.* 2008) (Fig. 2). The models suggest an initial build-up of discrete ice centres in the western Highlands which coalesced to form the Younger Dryas ice cap. A mean annual temperature of 10 °C below present and precipitation gradients decreasing from south to north and from west to east are required to meet best-fit limits. Low basal sliding is generally inferred with local exceptions, for example within the fjords. In addition, the model of Golledge *et al.* (2008) suggests that maximum limits in the north and west were reached earlier than those in the south and east. Mapped series of recessional moraines both on and offshore suggest that Devensian ice, including the Younger Dryas ice cap, retreated in an active, oscillatory fashion (e.g. Golledge & Hubbard 2005; Bradwell *et al.* 2008). Deglaciation towards the end of the stadial may have been more rapid and continuous, however (Dix & Duck 2000; Howe *et al.* 2002).

The extent to which deglaciation had progressed prior to the Younger Dryas is currently the subject of some controversy. The traditional view is that the British-Irish ice sheet of the last (Devensian) glaciation had largely disappeared by the time of the Loch Lomond Re-advance, leaving the landscape exposed to a period of 'new' glaciation (Ballantyne *et al.* 1998). This view is supported by Golledge's model (2008), which apparently predicts a short period of ice-free conditions prior to the onset of Younger Dryas cooling. However, this has recently been called into question by Bradwell *et al.* (2008) who used surface-exposure ^{10}Be dating of moraine ridges in northwest Scotland to demonstrate that substantial ice caps survived in this region throughout the Lateglacial Interstadial, a contention supported by the offshore evidence (Stoker *et al.* 2009). The authors suggest that this is also likely to have been the case elsewhere in Scotland, the inference being that the Scottish Younger Dryas ice cap may simply have represented a retreat stage of Devensian deglaciation rather than a 'new' episode of glaciation.

Modern Scottish fjord environments

In 1972, all the available hydrographic data from Scottish sea lochs was collated into a single report for the Department of Agriculture and Fisheries for Scotland (Milne 1972). The author divides the lochs into three categories: A-type, described as estuarine with deep water entrances and no sills; B-type, with a single entrance sill, and C-type with two or more sills. In the first category, the sea loch forms a simple marine inlet with no barrier to exchange between the loch and coastal deep waters, and hence cannot be considered a fjord. Milne (1972) describes the hydrography and circulation of B- and C-type fjords as being dependent on seabed topography, tidal range, freshwater runoff and wind speed/direction. This work was subsequently updated and built upon by Edwards and Sharples (1986), who produced a comprehensive catalogue of Scottish sea-loch morphology and hydrography for the Scottish Marine Biological Association (now the Scottish Association for Marine Science). Each of more than 100 lochs is described individually with respect to various parameters, including number of basins/sills, bathymetry, tidal range, catchment area and freshwater runoff. The authors note that sills and basins are important features of over half of the sea lochs described, and identify

Fig. 2. The modelled Younger Dryas ice cap over western Scotland, 12.8–11.5 ka BP. White line shows mapped limits. (Adapted from Golledge *et al.* 2008).

freshwater runoff and tidal mixing as the most important factors differentiating oceanography between lochs.

Several Scottish fjords have received more in-depth attention with respect to their circulation, in particular the physical processes associated with deep-water renewal which governs oxygen availability in the basins. In general, water circulation in the fjords is estuarine, with a seaward-flowing brackish layer typically 10 m thick at the surface and a denser, more saline landward flow at depth (e.g. Gage 1972*a*; Wood *et al.* 1973; Edwards & Edelsten 1977; Gillibrand *et al.* 1995, 1996; Allen & Simpson 1998*a*; Watts *et al.* 1998). Tidal currents can be locally extremely fast through narrows, for example 399 cm s^{-1} at the entrance to Loch Etive (Gage 1972*b*), and tidal fronts commonly form at the surface where inflowing coastal water meets the brackish outflow (Booth 1987). Water mixing is promoted over the sill regions (Allen & Simpson 1998*a*) and may be further enhanced by other factors such as internal tides (e.g. Allen & Simpson 1998*b*). Deep-water renewal is largely controlled by freshwater inputs, and tends to occur when surface salinity is increased during dry weather (Edwards & Edelsten 1977; Gillibrand *et al.* 1995; Allen & Simpson 1998*a*). Wind speed/direction has also been identified as a controlling factor in some cases, becoming more important when freshwater runoff is low (Allen & Simpson 1998*a*) and in the inner basins where coastal water density is less influential (Gillibrand *et al.* 1995). Renewal frequency varies widely between fjords; inner Loch Linnhe has a mean flushing time of 8 days (Hall *et al.* 1996) compared with 16 months for inner Loch Etive (Edwards & Edelsten 1977). Elsewhere, mean periods of 1 and 2 months are recorded for Lochs Sunart and Ailort, respectively (Gillibrand *et al.* 1995, 1996), and constant tidal flushing is reported for the shallow outer basins of Lochs Etive (Ridgeway & Price 1987) and Creran (Gage 1972*b*).

The nature and frequency of deep-water renewal governs oxygen availability in the deep basin water and sediments, with important consequences for benthic ecology and biogeochemical processes. Basin water in the Scottish fjords tends not to become completely anoxic, a fact attributed by Milne (1972) to the relatively high tidal range (2–4.5 m springs) along the Scottish west coast. However, oxygen does become depleted in the isolated deep waters between renewal events, particularly where organic inputs are high (e.g. around fish farm sites). In a study of Loch Ailort, Gillibrand *et al.* (1996) calculated that water deeper than 30 m would become anoxic within 9.5 weeks as a result of very high organic inputs from a fish farm; however, this could only rarely occur in practice as the fjord has a mean 8-week flushing time. In Loch Spelve, Brown *et al.* (1987) found that anoxic conditions did occur in sediments directly below salmon cages but were very localized with no discernable effects beyond 120 m. The deep inner basin of Loch Etive is particularly prone to oxygen depletion because of its long flushing time, combined with high terrestrial organic inputs (Gage 1972*b*; Ansell 1974; Howe *et al.* 2002; Loh *et al.* 2008*a*, *b*). The surficial sediments display reducing conditions just below the sediment–water interface, characterized by high manganese content (Overnell 2002), hydrogen sulphide generation (Gage 1972*b*) and restricted faunal activity compared to the well-oxygenated outer basin (Ridgeway & Price 1987). However, there is always enough oxygen to prevent anoxia at the interface; a faunal analysis by Gage (1972*b*) found no evidence of anaerobic conditions at any time.

Modern sedimentation patterns in the fjords are determined by supply and hydrodynamic regime. New input is mainly from rivers, with primary productivity contributing only a small amount of organic carbon compared with the generally high terrestrial inputs (Gage 1972*b*; Ansell 1974; Loh *et al.* 2008*a*, *b*; data from Loch Etive and Creran only). Resuspension and reworking of pre-existing sediments by bottom currents are important features, resulting in a net transfer of material into the deep basins (Ansell 1974; Green 1995). Particle size, as expected, is generally a function of depth and current speed; Gage (1972*b*) found an inverse relationship between particle size and depth in both Loch Creran and Loch Etive, but found sediments to be coarser at comparable depths in Creran because of the generally faster bottom currents in this fjord. Pearson (1970) found a similar relationship in Loch Linnhe between particle size and current speed/water depth, except for within the clay fraction which was unexpectedly high in a shallow area of high current velocity. The authors attribute this to deposition of fine particles in pockets between the boulders which predominated in the current-swept area. Bottom currents in deep basins tend to be slow except during deep-water renewal events (Pearson 1970; Edwards & Edelsten 1977; Gillibrand *et al.* 1996), allowing for the deposition and accumulation of fine-grained material. In shallower areas of faster currents, sediment cover is generally thinner and coarser (Pearson 1970; Gage 1972*b*; Green 1995; Howe *et al.* 2002).

Little information is available on modern sedimentation rates. Syvitski and Shaw (1995) describe accumulation rates for Scottish fjords as low (<1 mm a^{-1} averaged over the entire basin), due in part to the fact that glacial inputs have been absent since deglaciation was completed *c.* 11.5 ka BP. The available data suggest highly variable Holocene rates related to catchment area and riverine inputs, and to water depth and current regime, for example: <1.5 mm ka^{-1} on bathymetric highs in Little Loch Broom, where there are no major riverine sources (Wilson 2007); and 6000–7000 mm ka^{-1} in Loch Etive's deep inner basin (Shimmield 1993) which has the largest catchment area of any Scottish fjord (1400 km^2, Edwards & Edelsten 1977).

Late to postglacial sedimentary environments

The main research studies carried out in the fjords of western Scotland to date are summarized in Table 1. The results of each study are then described in more detail by area, progressing up the coast from south to north.

Loch Etive

Loch Etive (Fig. 3) has been particularly well studied with respect to its physical oceanography, benthic ecology and biogeochemistry (e.g. Gage

Table 1. *Summary of major research studies carried out in Scotland's west coast fjords*

Area	Method(s)	Focus of study	References
Loch Etive	Sidescan survey	Bathymetry and sedimentary processes since the Younger Dryas	Howe *et al.* (2001)
	Seismic survey and gravity core sedimentology	Post-Younger Dryas to present day depositional environments	Howe *et al.* (2002)
	Gravity/spear core analyses (foraminifera and sedimentology)	Effects of sea level on deep-water renewal	Nørgaard-Pedersen *et al.* (2006)
Loch Linnhe	Seismic survey	Deglacial sedimentation patterns since Last Glacial Maximum	Green (1995)
Loch Sunart	Seismic survey and long core analysis (pollen and grain size)	Local climate variability during deglaciation	Mokkedem *et al.* (2007)
Lochs Nevis and Ailort	Seismic survey	Younger Dryas glacial limits; sedimentary patterns inside and outside lochs since the Last Glacial Maximum	Boulton *et al.* (1981)
Loch Ainort	Seismic and sidescan surveys, gravity core analysis (grain size)	Glacial dynamics during post-Younger Dryas deglaciation	Dix & Duck (2000)
Summer Isles	Multibeam and seismic survey	Offshore record of geomorphological features associated with glacial activity	Stoker *et al.* (2006), Bradwell *et al.* (2008)

1972*a*, *b*; Wood *et al.* 1973; Ansell 1974; Edwards & Edelsten 1977; Ridgeway & Price 1987; Overnell *et al.* 1996; Austin & Inall 2002; Overnell 2002; Inall *et al.* 2004). This partly due to its close proximity to Dunstaffnage Marine Laboratory near Oban, but also because its long flushing time and consequent tendency to hypoxia make it unusual among Scottish fjord environments.

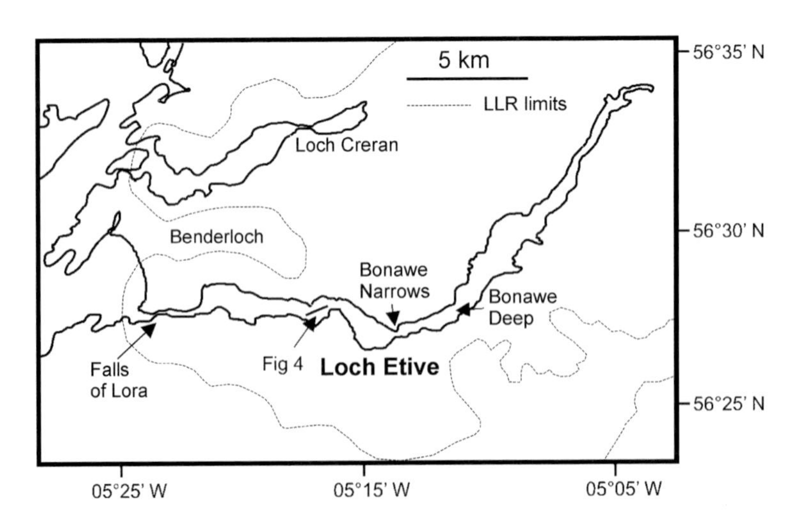

Fig. 3. Map showing location of Loch Etive and the adjacent Loch Creran. LLR, Loch Lomond Re-advance. These limits are taken from the online GIS database of Clark *et al.* 2004 (http://www.shef.ac.uk/geography/staff/clark_chris/britice.html), based on earlier publications. For wider setting, see Figure 1.

In recent years, Loch Etive has been subject to detailed sidescan sonar and seismic/gravity coring surveys (Howe *et al.* 2001, 2002). A high-resolution study has been carried out into the influence of catchment and sea level on its Holocene sedimentary record (Nørgaard-Pedersen *et al.* 2006). These studies have yielded much information on modern and ancient sedimentary processes in the fjord, on glacial dynamics during post-Younger Dryas deglaciation and on climate variability in the local area throughout the Holocene, which may be summarized as follows.

The sidescan survey of Howe *et al.* (2001) reveals several interesting geomorphological features on the floor of inner Loch Etive. Areas of high backscatter mainly located in shallow water are interpreted as outcrops of granite bedrock, with localized reflectors representing glacial erratics and/or rockfall debris. The latter are present throughout the fjord and in a concentrated area near the fjord head. Low backscatter in the glacially overdeepened basins and on the slopes is considered to represent an accumulation of fine-grained post-glacial sediments (presumed to be mainly riverine in origin) over the top of older glacial deposits, an interpretation subsequently borne out by the seismic and coring survey of Howe *et al.* (2002). Submarine outwash fans are identified at the mouths of rivers and streams, and there is evidence of downslope sediment creep in some areas. However, despite the steepness of the slopes revealed by the survey (5 to 15° at the SW end of the inner basin), there is no evidence of major slope failure in the fjord. The authors conclude from the relatively high sedimentation rates (7 mm a^{-1}) reported by Shimmield (1993) that a high-resolution deglacial and Holocene record might be preserved in the sediments of the inner fjord. This is confirmed by the subsequent studies of Howe *et al.* (2002) and Nørgaard-Pedersen *et al.* (2006), although the modern sedimentation rate of Shimmield (1993) proved to be significantly higher than during the earlier Holocene (*c.* 2 mm a^{-1}) as a result of anthropogenic land use and deforestation (Howe *et al.* 2002).

In the subsequent study of Howe *et al.* (2002), a seismic surveying and gravity coring programme was carried out in the inner basin and part of the outer basin of Loch Etive to investigate its deglacial and Holocene sedimentary infill. The seismic survey reveals a total sediment thickness of 30–60 m, presumed to be of post-Younger Dryas to Holocene age. Two seismic sequences are identified: the upper (B), a 5–10 m thick well-laminated unit interpreted as late Holocene river-derived sediment; and the lower (A), a well-laminated to transparent unit extending from directly beneath unit B to the seabed, interpreted as glaciolacustrine to glaciomarine inputs associated with post-Younger Dryas deglaciation. Chaotic facies within both units are interpreted as the result of gas produced during organic matter decomposition. The seismic units A and B are separated by a clear, apparently erosive reflector termed E1 by the authors (Fig. 4), presumed to represent an abrupt change in sedimentary regime from glacially influenced to open marine conditions dominated by fluvial inputs. The eight gravity cores analysed were found to consist mainly of organic-rich sandy muds and coarser riverine muddy sands, with a high water content at the surface. Shelly material was abundant throughout the cores, and terrestrial organic matter in the form of wood and plant fragments was present in many. All cores are <2.5 m long and hence represent the Holocene seismic unit B; however, the outermost core of the outer fjord (GC008) may have sampled the underlying deglacial unit A at its base, where grey compacted sands interpreted as glaciomarine in origin are present. Laminations in the lower parts of some of the cores are believed to be the result of bottom-current reworking, possibly as the result of deep-water renewal events which can periodically increase bottom-current velocities. This idea is expanded on in the subsequent work of Nørgaard-Pedersen *et al.* (2006), with the influence of sea level cited as a major control.

Benthic foraminifera were examined from the top and bottom of a single core (GC005, from the deep inner basin slope). The results revealed a well-preserved calcareous temperate assemblage at the base with very high abundance (>5000 specimens per g dry weight); this assemblage was dated to 9090 ± 90 ^{14}C a BP (reservoir-corrected age obtained from an *Arctica islandica* bivalve sample from the same horizon). The core top was characterized by a less diverse foraminiferal assemblage, with more agglutinated species dominating. This core and another from the same study (GC004) were subsequently subjected to more detailed analysis by Nørgaard-Pedersen *et al.* (2006) to construct a detailed high-resolution record of sea-level influence in the inner Etive basin during the Holocene. Their benthic foraminiferal analysis demonstrates that Loch Etive has been fully marine since the earliest Holocene, and hence does not support the lacustrine stage proposed by Walker *et al.* (1992) and referred to by Howe *et al.* (2002). Evidence from benthic foraminifera, grain-size analysis and sediment magnetic susceptibility are used in conjunction with 15 AMS ^{14}C dates from molluscs to show how changing sea levels over the Holocene have impacted on deep-water renewal in the basin. Renewal events appear to have been most frequent and vigorous towards the mid-Holocene sea-level high (8–9 m OD), a period characterized by incursions of coarse sandy material to the sediments

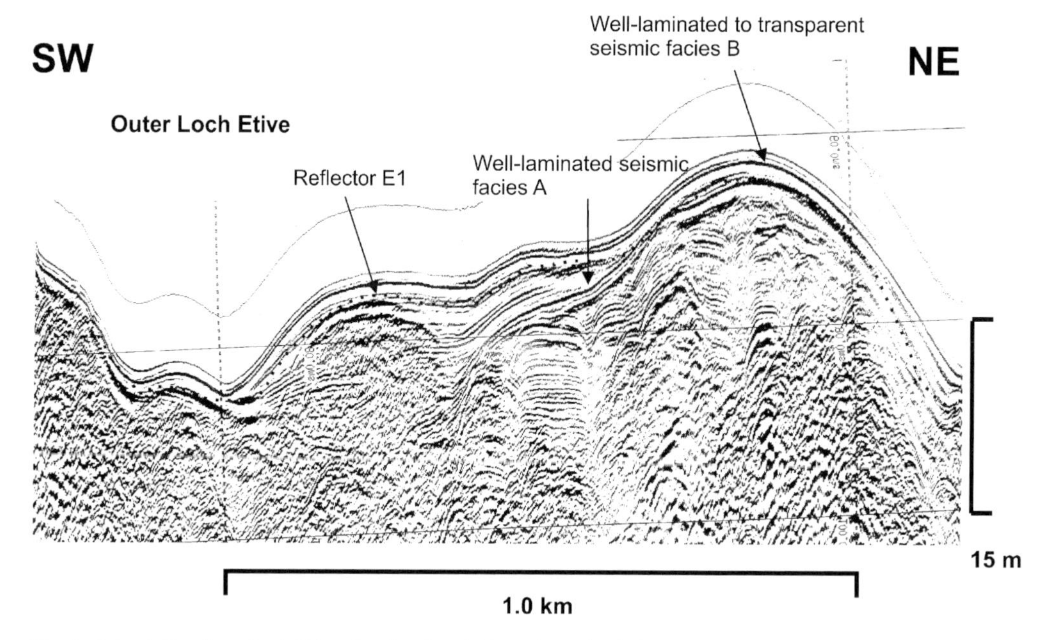

Fig. 4. Seismic line from Loch Etive showing the glaciomarine seismic facies A and full marine seismic facies B, separated by the reflector E1. Adapted from Howe *et al.* 2002.

and a peak abundance of epibenthic foraminiferal species such as *Cibicides lobatulus*. The effects are most marked in GC005, the core closest to the sill and hence most influenced by deep-water overflow. Following the mid-Holocene transgression there is a gradual increase in agglutinated taxa and organic carbon and an associated fining of the sediments; this is attributed to reduced oxygenation of the bottom waters as the sea level fell again, promoting water mixing at the sill and hence less frequent deep-water renewal. The late Holocene (1–0 ka BP) is marked by a reduction in grain size at both core sites, suggesting a further decrease in the vigour and frequency of deep-water circulation at this time. These findings are in broad agreement with the initial interpretations of Howe *et al.* (2002), and explain their observation that laminations of coarser material were restricted to the lower parts of the cores.

A pollen study of GC004 revealed that herbaceous species dominated in the lower part of the core (2.35–2.03 m). Tree pollen increased above this horizon and decreased again in the later Holocene, coinciding with an increase of heather pollen. This evidence supports earlier Holocene climate studies indicating that the early Holocene landscape in this area was characterized by pioneer vegetation and little tree cover, with deforestation and podzolization occurring around the fjord in the later Holocene.

Loch Linnhe

Loch Linnhe (Fig. 5) forms the SW part of the Great Glen fault, which runs SW–NE across Scotland to the Moray Firth on the east coast. It is positioned downstream of the former Younger Dryas ice cap centred on Rannoch Moor, and was a major outlet for seaward-flowing glacial ice at this time and during previous Quaternary glacial episodes. The inner fjord has been the subject of a number of studies with respect to its oceanography, physiography and ecology (e.g. Pearson 1970; Hall *et al.* 1996; Allen & Simpson 1998*a*, *b*; Watts *et al.* 1998); however, its palaeoenvironmental sedimentary record has only been investigated in a single, unpublished study by Green (1995) as part of a wider thesis on the Scottish Younger Dryas ice cap in this area.

Green's seismic survey of Loch Linnhe and the Firth of Lorne at its mouth reveals a highly irregular bedrock surface indicative of intense glacial erosion, covered by sediment thicknesses of up to 165 m. Sediment cover is generally thickest in the deep basins, accumulation being inhibited by strong tidal currents in the shallows/narrows. Five main basins are identified: the Inverscaddle basin in the inner fjord, the Kentallen, Shuna and Lismore basins in the outer fjord and the Don Basin in the Firth of Lorne. The seismic data are resolved into five distinct facies: A, comprising a

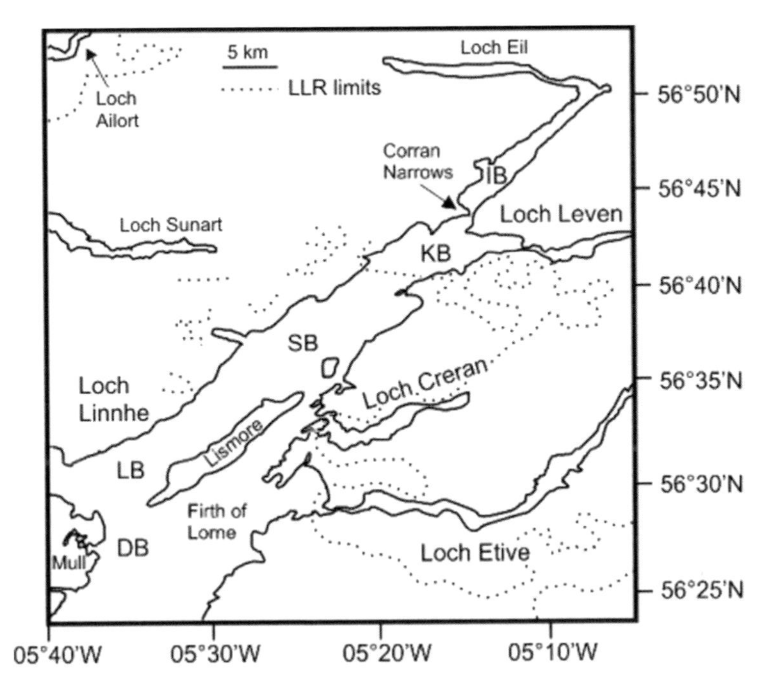

Fig. 5. Map of Loch Linnhe showing locations of basins identified by Green (1995). DB, Don Basin; KB, Kentallen Basin; IB, Inverscaddle Basin; LB, Lismore Basin; SB, Shuna Basin; LLR, Loch Lomond Re-advance (see Fig. 3). For wider setting, see Figure 1.

draped pattern with internal reflectors subparallel to the base, interpreted as sediments of riverine origin deposited from suspension; B, comprising subparallel, subhorizontal internal reflectors, interpreted as deposits from a higher energy environment than in facies A (e.g. slump-generated or current-controlled); C, wedge-shaped or hummocky masses with inclined surface and internal reflectors banked up against bedrock, interpreted as slumps; D, characterized by an irregular surface profile and lack of internal structure with incoherent reflectors, interpreted as diamict; and E, present at the base of many profiles, having a strong highly irregular surface reflector, interpreted as bedrock. The sedimentary infill of each basin is described in turn according to these interpretations.

The inner (Inverscaddle) basin contains Holocene sediments (Facies A and B) deposited from suspension and derived from rivers and incoming tidal currents. Deltas are present at river mouths, presumed to have accumulated throughout the Holocene. Diamict (Facies D) may be present in one location, although acoustic blanking makes this uncertain. The underlying units (Facies A and B) are interpreted as Devensian (post-Last Glacial Maximum, pre-Younger Dryas) glaciomarine sediments.

Seaward of the inner basin, and separated from it by the 180 m wide and 18 m deep Corran Narrows (Pearson 1970), lie the Kentallen and Shuna basins. The uppermost units, distributed throughout both basins except on topographic highs, are interpreted as marine Holocene sedimentation (Facies A and B) as in the Inverscaddle basin. They are underlain by a unit termed K2, belonging to Facies A in the Shuna basin and B in the Kentallen basin. This unit, acoustically transparent in places, is interpreted as representing proximal glaciomarine sediments accumulating rapidly from the Younger Dryas glacier, which must therefore have terminated just outside the Corran Narrows at the time of deposition. A possible submarine outwash fan is identified within K2 a little to the south where fans are also present onshore, lending support to this limit. Green (1995) reports ice-contact or proximal sediments (Facies D) just south of the Corran Narrows possibly forming submarine moraines, and cites Thorp (1986) for onshore evidence of a maximum limit for the Linnhe glacier in this basin. Underlying unit K2 in both the Kentallen and Shuna basins are further examples of Facies A and B, the reflectors sloping downwards to the SW implying a source from the Corran Narrows direction. These are interpreted as older glaciomarine sediments, deposited during Lateglacial deglaciation. There is evidence of slope failure within the unit in the form of slumps/debris flows (Facies C).

In the Lismore basin at the southern end of the fjord, acoustic blanking obscured much of the seismic record particularly in the northern part; interpretations here are therefore not well constrained. The uppermost unit (Facies A) is intepreted as Holocene marine and/or reworked distal glaciomarine sediments. Facies D (diamict) is widespread within the basin, possibly of Last Glacial Maximum (LGM) age. The underlying units belong to facies A, B and C, and are presumed to represent older glaciomarine deposits. Limited distribution of some units and an erosional upper surface on others suggests a major erosive event, possibly relating to pre-LGM ice advance.

Acoustic blanking also affected the seismic record in the Don basin just outside the fjord, and interpretations are hence provisional. The seismic units within the basin are interpreted as mainly glaciomarine in origin (Facies B), with the lower unit representing Devensian deposits and the upper unit of Lateglacial to Younger Dryas age. Units of slumped material (Facies C) also occur locally at all levels in sequences up to 12 m thick.

Loch Sunart

Situated within the Ardnamurchan peninsula, Loch Sunart (Fig. 6) has been subject to variable but generally high sedimentation rates of up to 0.5 cm a^{-1} within its main basin during the Holocene (Eiríksson *et al.* 2006). The fjord was included in a study by Eiríksson *et al.* (2006) which investigated late Holocene variability in the North Atlantic Current along a transect from the Iberian margin in the south to Iceland in the north. The 1100-year oxygen-isotope record from benthic foraminifera in Sunart revealed bottom-water temperatures of 10–13 °C during the first millennium AD, *c.* 2 °C warmer than at present, with a pronounced cooling recorded at AD 900. Carbon isotopes indicate that a change occurred in coastal and basin water exchange over the same period, perhaps as a result of enhanced westerlies, illustrating the potential of the fjords to record circulation changes driven by the North Atlantic Oscillation. The Loch Sunart isotope records suggest a multidecadal 1–2 °C variability in coastal temperatures over the past 2 ka, superimposed onto a longer-term (centennial) climate signal.

During a 2004 cruise on the French Polar Institute's research vessel *Marion du Fresne*, two piston cores were recovered from the main basin of Loch Sunart. The first of these (MD04-2832), obtained from a sediment drift, is the longest core ever to have been recovered from Scottish coastal waters (Howe & Austin 2005).A radiocarbon date of *c.* 8.0 cal ka BP was obtained from the core base, and stable isotope records from the benthic foraminifera *Ammonia beccarii* reveal correlation in some instances with the Holocene rapid climate change events described by Mayewski *et al.* (2004) and Austin *et al.* (2006).

The second piston core (MD04-2833), 12 m in length, was used in conjunction with a seismic survey of the fjord to reconstruct sedimentary environments over the last *c.* 20 ka (Mokeddem *et al.* 2007). In this study, the core was analysed with respect to its grain size and pollen record. The resulting lithofacies units are correlated with seismic sequences identified from an earlier (2002) survey in the fjord to produce a palaeoenvironmental record spanning the period from the LGM to the present (*c.* 20 ka), supported by three radiocarbon dates from mollusc shells (*Turitella* and *Pecten*). Five seismic facies are identified, summarized

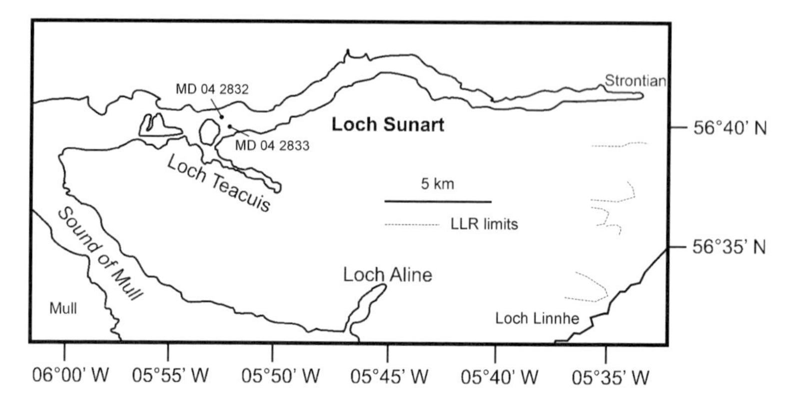

Fig. 6. Map of Loch Sunart showing locations of the piston cores MD04 2832 (Austin *et al.* 2006) and MD04 2833 (Mokeddem *et al.* 2007). LLR: Loch Lomond Re-advance (see Fig. 3). For wider setting, see Figure 1.

as: F1 and F2, disorganized acoustic facies interpreted as LGM basal moraine material and a coarse-grained Older Dryas deposit, respectively; F3, a semi-laminated facies in two phases, interpreted as representing proximal and distal glaciomarine deposits of Younger Dryas age; F4, a transparent facies interpreted as reflecting massive, rapid sediment discharge from the ice front during post-Younger Dryas deglaciation; and F5, characterized by thin laminated beds of silty clay, reflecting decreasing sediment supply during the recent Holocene. Pollen analysis reveals that a tundra landscape prevailed in the area during the Older Dryas (F2), with oak pollen appearing towards the end of this period as marshland began to regress. Within F4, an initial development of boreal forest and *Pinus*/*Corylus* expansion is recorded with a coeval decrease in *Betula* pollen, reflecting the onset of more temperate climate conditions. A reservoir-corrected radiocarbon age of 6910 $\pm$ 40 a BP is associated with this climate amelioration, correlating it with the relatively warm and humid Atlantic period. Three cooling events are identified within the facies, interpreted as corresponding to the 8200-year cold event (8.2–8.0 ka BP), the end of the Atlantic period (5.5–5.2 ka BP) and another cool period at 3.5–2.7 ka BP. The upper part of F4 and all of F5 are characterized by arboreal pollen, predominantly *Corylus*, *Quercus* and *Alnus* with meadows and fern moors increasing in the later Holocene, perhaps in response to increasing humidity.

Lochs Ailort and Nevis

These two fjords are situated on the mainland coast to the southeast of the Isle of Skye, opening to the sea in the Sounds of Arisaig and Sleat, respectively (Fig. 7). The fjords and the sea area outside were the subject of the earliest seismic survey of Scottish fjord systems (Boulton *et al.* 1981), which reveals a number of interesting sedimentary and geomorphological features relating to deglaciation of the area. Major submarine moraines are identified at the mouths of both fjords, presumed to mark the limits of Younger Dryas glaciers (the Mallaig Formation). A seismic unit interpreted as till (the Minch formation) is present throughout the survey area, directly overlying the bedrock. Outside the fjords, a draped unit characterized by well-defined, parallel internal reflectors overlies the till. This unit, termed the Muck Formation, is interpreted as representing glaciomarine sedimentation related to the retreat of Devensian ice from the shelf. It has a strongly eroded upper surface, on which Holocene-aged localized basin infills (the Arisaig Formation) have been deposited. There is no equivalent of the Muck Formation within the fjords, where apparent equivalents of the basin infills (the Nevis Formation) directly overlie the till. Possible morainic deposits were identified at sills and narrows inside both fjords (at Kylesknoydart in Nevis and Roshven in Ailort), corresponding to onshore proglacial outwash fans, thought to represent deglacial stillstands as a result of topographic

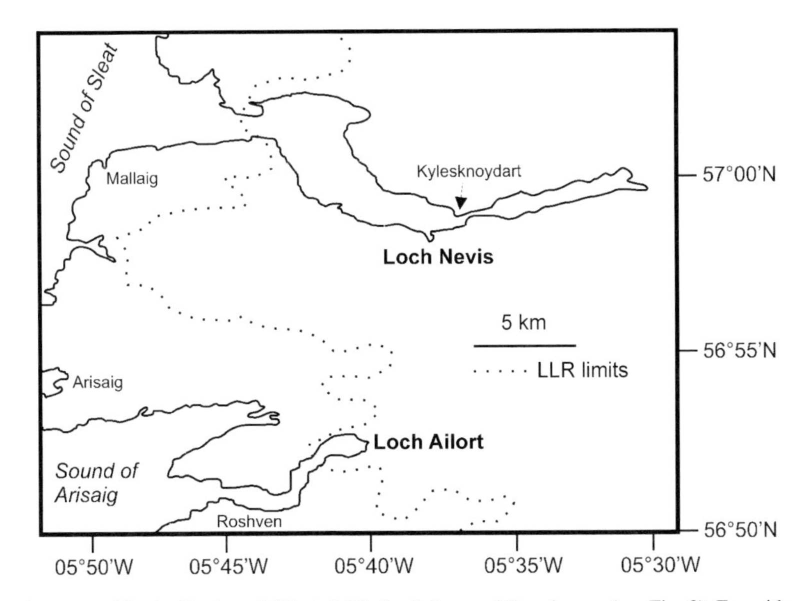

Fig. 7. Location map of Lochs Nevis and Ailort. LLR, Loch Lomond Re-advance (see Fig. 3). For wider setting, see Figure 1.

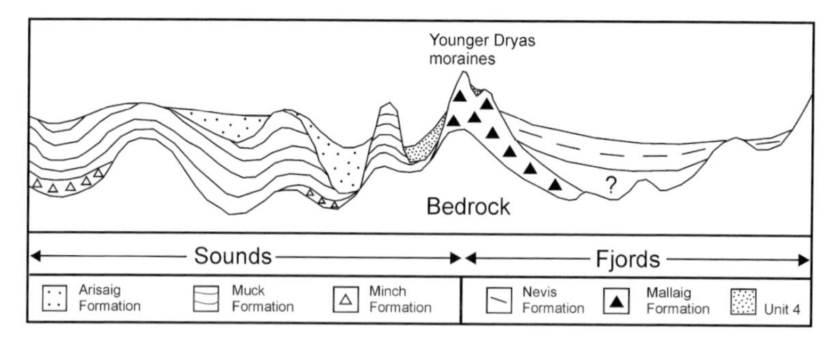

Fig. 8. Schematic diagram adapted from Boulton *et al.* (1981), summarizing the main findings of the authors' seismic survey in and around Lochs Ailort and Nevis. Unit 4 is of uncertain origin; the authors suggest either a proximal glaciomarine source or recent resedimentation.

pinning. The lack of Younger Dryas-aged deposits outside the terminal moraines at the fjord mouths is attributed to the relatively shallow water and fast tidal currents along the shelf in this area, which are thought to have resulted in the transport of glacial material across the shelf and its ultimate deposition in more distal basins. The main findings of this study are summarized in Figure 8.

Loch Ainort

Loch Ainort (Fig. 9), a SW–NE trending fjord on the east coast of the Isle of Skye, was subject to glaciation during the Younger Dryas stadial when a sizeable icefield was present in central Skye (Benn *et al.* 1992). The Lateglacial history of the region is well established from onshore geomorphology (e.g. Ballantyne 1988, 1989) and climate records (Benn *et al.* 1992). A high-resolution seismic survey of the fjord by Dix and Duck (2000) sought to obtain offshore evidence for the style and timing of the deglaciation, to be compared with these onshore records. The survey, supported by evidence from sidescan sonar and partially calibrated using examination and grain-size analysis of associated gravity cores, was used to investigate the sedimentary records of Loch Ainort and the sound just outside. Four seismic sequences were identified: LASS (Loch Ainort Seismic Sequence) 1 (14–40 m thick) characterized by high internal backscatter, numerous point-source hyperbolae and a strong surface reflector underlies the entire survey area

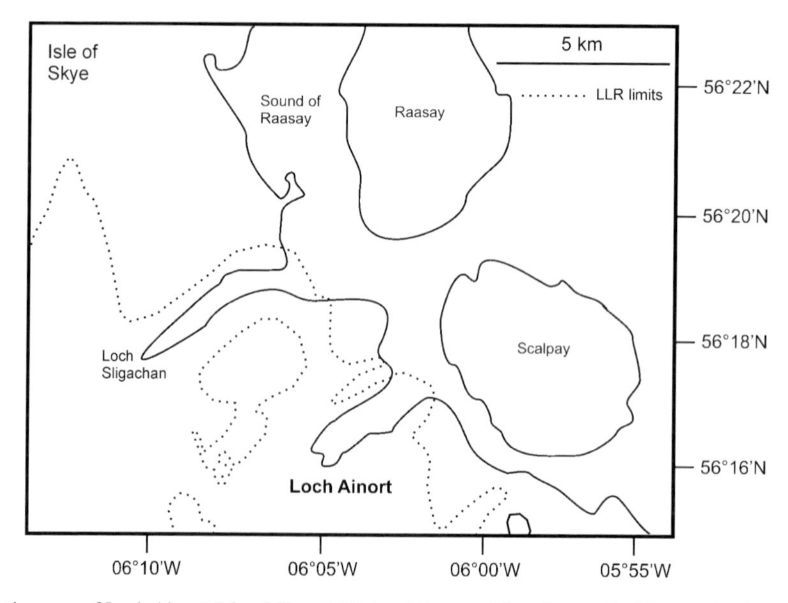

Fig. 9. Location map of Loch Ainort, Isle of Skye. LLR, Loch Lomond Re-advance. In this case, the limits have been taken from Benn *et al.* (1992). For wider setting, see Figure 1.

and is interpreted as a glaciogenic diamict (muddy gravel, verified in many of the core samples) of LGM age, subsequently reworked by Younger Dryas ice; LASS 2 (maximum 10 m) is acoustically similar but restricted in distribution to sites within the inner fjord, hence also interpreted as a glaciogenic diamict composed of muddy gravel; LASS 3 (maximum 12 m) unconformably overlies LASS 1/2 and is characterized by irregular but conformable, discontinuous internal reflectors which downlap onto the underlying units, interpreted as muddy/sandy gravel layers resulting from debris flows and/or proximal ice discharge; and LASS 4 (maximum 18 m) is the youngest sequence, comprising low to weak internal backscatter and weak, parallel, inclined reflectors with high lateral continuity and extent, interpreted as homogeneous fine-grained sediments (confirmed in the gravity cores) of Holocene age.

The seismic data revealed numerous curvilinear ridges within LASS 1, also visible on the sidescan sonograph, interpreted as moraines. The distribution of these moraines, concentrated around the mouth of the fjord, suggests that they were deposited during a period of rapid, punctuated retreat. In the inner fjord, morainal ridges are extensive and hummocky, interpreted as reflecting a period of more sustained retreat. The offshore evidence is thus largely in agreement with the onshore climate record of Benn *et al.* (1992), who proposed a two-step glaciation for this region. A first phase is characterized by oscillatory glacial retreat punctuated by stillstands and minor re-advances in response to decreasing precipitation, and a second phase of continuous retreat is a result of rapid climate amelioration. In Loch Ainort, the second stage is suggested by Dix & Duck (2000) to relate to basin morphology and differs from the onshore record in that minor stillstands/re-advances did occur (although infrequently). The survey also identified a series of moraines further out in the sound, presumed to mark the maximum limit of the Younger Dryas glacier although they lie some 800 m beyond the mapped onshore limits. This may demonstrate the unreliability of extrapolating onshore limits into offshore environments, although the authors acknowledge that the moraines could date to an earlier phase of glaciation.

Summer Isles

The northwest Highlands of Scotland are an outstanding example of a glacially eroded landscape, a fact that has been recognized in its designation as a UNESCO European Geopark (www.northwest-highlands-geopark.org.uk). A multibeam bathymetric survey of the adjacent offshore Summer Isles region (Fig. 10) was carried out by the British Geological Survey in 2005, revealing an equally striking and very well-preserved submarine landscape characterized by erosive and depositional features of glacial origin. The multibeam survey

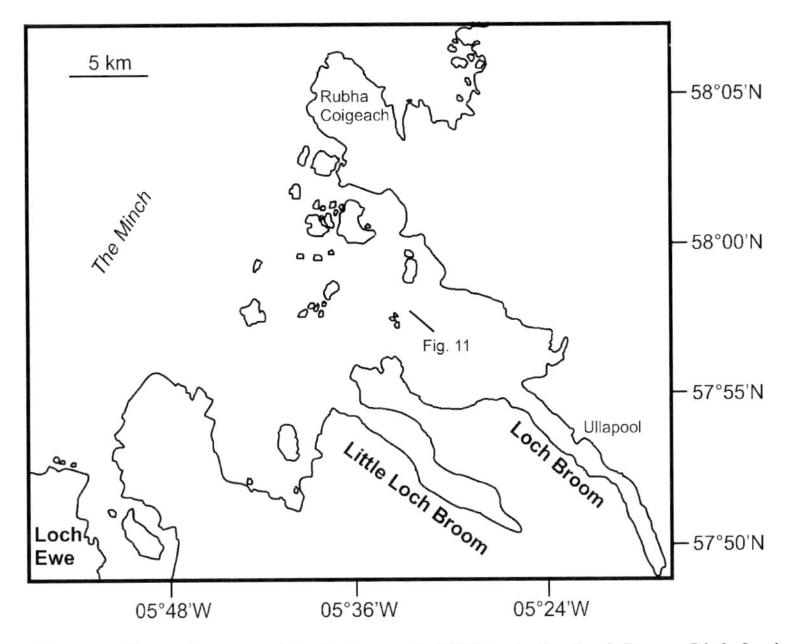

Fig. 10. Map of Summer Isles region mapped by Stoker *et al.* (2006) including Loch Broom, Little Loch Broom and Loch Ewe. The position of the seismic line in Figure 11 is indicated. For wider setting, see Figure 1.

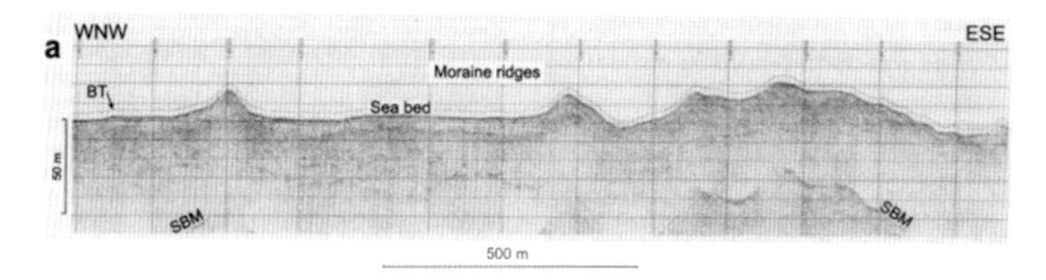

Fig. 11. BGS boomer profile 05/04–36 from Stoker *et al.* (2006), showing part of the recessional moraine series mapped by the authors. See Figure 10 for profile location. BT, bottom tracking indicator; SBM, seabed multiple. This Figure was originally published by Stoker *et al.* 2006 and is reproduced here with permission of the editorial board of the Scottish Journal of Geology.

covered a 200 km^2 area of seabed, including the west coast mainland fjords Loch Broom and Little Loch Broom. The data were supported by seismic profiles, also taken during the 2005 cruise. A subsequent cruise in 2006 obtained numerous gravity cores from the area which, along with the acoustic data, form the basis of an ongoing programme of research into the depositional history of the area.

The multibeam survey, described by Stoker *et al.* (2006), reveals a general pattern of deep (up to 180 m) troughs representing the offshore extensions of the fjord basins, separated by shallow (<40 m) marine banks. A number of superimposed geological and geomorphological features are identified and described in turn. Most striking is the suite of clearly defined recessional moraines, 10–20 m high and up to 3 km in length (Fig. 11), interpreted as reflecting punctuated ice-sheet retreat from the Minch where fjord glaciers coalesced during the LGM to form a northwest-flowing ice stream (Stoker & Bradwell 2005). A large moraine is also identified pinned between topographic highs at the mouth of Loch Ewe, an open (i.e. non-fjordic) sea loch (Davies 1975) at the SW end of the study area; smaller recessional moraines are also present behind it. These moraines are thought to relate to the Wester Ross Re-advance, cosmogenically dated using material from an adjacent onshore moraine to 15.5–17.9 ka BP (Everest *et al.* 2006). Elsewhere morainal ridges are identified in Loch Broom and Little Loch Broom and at the Rubha Coigeach headland to the north of the study area, where they form a north–south trending series of ridges superimposed by a younger, lower relief, SW–NE trending series. Other mapped features include erosional bedrock grooves, pockmarks from gas release and recent slide scars and slumps on the slopes of the fjord basins. Seismic profile interpretations describe a cover of diamict overlying the bedrock over much of the study area, both on the shallow banks and within the deep basins. Subsequent sedimentation is sparse on the banks, but forms thick deposits (50–70 m) of late to postglacial age (*c.* 17 ka to present) in the deep basins, overlying the diamict. Differential sedimentary infill between the basins is noted; outwith Loch Broom and Little Loch Broom a sheet-drape facies separates the diamict from younger, onlapping units; this facies is absent within the fjords. The prevalence of discrete point sources in the fjords, such as slump deposits, leads the authors to conclude that much of the sedimentation during the Lateglacial to postglacial period was locally controlled, with processes such as bottom-current erosion becoming more influential later on. A rapid rise in postglacial sea level is suggested as the reason for the very well-preserved condition of the submarine glacial landscape. The depositional environments of Little Loch Broom, as revealed by the survey in conjunction with sediment core analysis, are described in detail elsewhere in this volume (Stoker *et al.* 2010).

Discussion

The research studies described in this review demonstrate that Scotland's fjords contain a wealth of valuable information concerning late deglacial to Holocene environments and processes. Evidence from the fjord sediment records and adjacent sea area not only enhances and informs the terrestrial records but can be used to infer past glacial dynamics along the main drainage conduits (the fjord basins themselves), crucial in understanding the processes associated with the larger ice masses that fed these outlets. Additionally, maximal limits can be established for ice margins which terminated offshore. All of the available evidence is collated here to establish a broad-scale picture of the main events and processes associated with post-LGM deglaciation in western Scotland.

Late Devensian to Younger Dryas (c. 20.0–12.8 ka BP)

Glacial till deposited during the Devensian glaciation is widespread. It is reported as overlying bedrock in Loch Sunart, where it is described as basal moraine (Mokeddem *et al.* 2007), Lochs Nevis and Ailort (Boulton *et al.* 1981), Loch Ainort (Dix & Duck 2000) and over the entire Summer Isles region including Loch Broom and Little Loch Broom (Stoker *et al.* 2006). It is also recorded in the deep outer basins of Loch Linnhe and the Firth of Lorne, but is absent from the confines of the inner fjord (Green 1995) and also from the adjacent Loch Etive, where postglacial sedimentation directly overlies bedrock (Howe *et al.* 2002). This till is presumed to correlate to the thin diamict described by Davies *et al.* (1984), present across wide areas of the western Scottish shelf overlying an erosional surface associated with the advance of Devensian ice. Its absence in Lochs Linnhe and Etive may be the result of erosion by Younger Dryas ice. During the initial stages of the Younger Dryas stadial, ice build-up was concentrated over a site near Rannoch Moor, directly upstream of Lochs Linnhe and Etive (Hubbard 1999; Golledge *et al.* 2008). These fjords may therefore have been subjected to more intense erosion over a longer timescale than the more northerly sites, before post-Younger Dryas deglaciation began a new phase of sedimentation.

An accumulation of glaciomarine deposits from the retreating Devensian ice margin is reported in most of the study areas, overlying the till. These units, characterized on seismic profiles by a fairly uniform unit thickness with parallel internal reflectors, are mainly present outside the fjord basins. They are identified in the Muck Formation of Boulton *et al.* (1981) outside the terminal Younger Dryas moraines at the mouths of Lochs Ailort and Nevis; in the sheet-drape facies of Stoker *et al.* (2006) in the Summer Isles area (but not in Loch Broom or Little Loch Broom); in seismic subunit K4 in the presumed proglacial Kentallen and Shuna basins of Loch Linnhe (Green 1995); and possibly in the subunit LASF5 of Dix and Duck (2000), beyond the terminal Ailort moraine in the outer sound. These units may therefore be considered to correlate with one another and with the glaciomarine shelf sediments of Davies *et al.* (1984) which have a similar seismic character. Mokeddem *et al.* (2007) consider the seismic facies F3 in Loch Sunart to be correlated with the Muck Formation of Boulton *et al.* (1981). However, they subsequently interpret F3 to be a glaciomarine facies associated with Younger Dryas ice retreat; it is therefore younger than the units described above which are more likely to correlate to their F2 (14.02 ^{14}C ka BP). The absence of Devensian-aged (post-LGM) glaciomarine sediments within the fjord basins may in some instances be attributed to erosion by Younger Dryas ice (e.g. Green 1995). Stoker *et al.* (2006) suggest that low-energy environmental conditions during early deglaciation led to their deposition outside the fjords in the Summer Isles region. Here, deglaciation appears to have been associated with frequent stillstands and minor re-advances of the ice margin, resulting in deformation of the underlying till to form numerous recessional moraines. Bradwell *et al.* (2008) argue convincingly for the retention of glacial ice in this area throughout the Lateglacial Interstadial, and suggest that this may well have applied elsewhere in Scotland.

Younger Dryas (12.8–11.5 ka BP)

Younger Dryas limits are recorded in the presence of clear terminal moraines at the mouths of Lochs Ailort and Nevis and a presumed Younger Dryas terminal moraine in the strait beyond the mouth of Loch Ainort on Skye. In Loch Linnhe, the proposed limit lies within the Kentallen basin just south of the Corran Narrows, marked by terrestrial outwash deposits and the proglacial seismic subunit K2 of presumed Younger Dryas age seaward of this limit. However, a recent vibrocore obtained from the Shuna basin, *c.* 8 km further south, contains strongly compacted marine sediments at its base believed to belong to unit K2 and interpreted as having been glacially overridden (author's unpublished data). If this is the case, it implies a maximum limit for the Linnhe glacier at least 14 km south of the Corran Narrows, more in accordance with the model of Golledge *et al.* 2008 (Fig. 2). Compacted marine clays of Lateglacial Interstadial age are present in an onshore exposure at the entrance to the adjacent Loch Creran, believed to have been deposited in front of the advancing glacier and subsequently overridden by it (Peacock *et al.* 1989). The seismic subunit K2 in Loch Linnhe may have resulted from a similar process. Deformation of Younger Dryas glaciomarine sediments is also recorded in Loch Etive in the form of a small morainal bank just outside the sill at the Bonawe Narrows (Howe *et al.* 2002), presumably formed from a minor re-advance of the retreating glacier. Elsewhere, Younger Dryas ice advance effectively removed all pre-existing glaciomarine sediments, suggesting that overriding probably occurred during minor re-advances of relatively thin ice with low erosive power.

The Younger Dryas limit proposed by Green (1995) for Loch Linnhe appears to be supported by the presence of glacial outwash onshore around the Corran Narrows (Thorp 1986). The Ailort moraine of Dix and Duck (2000) suggests that terrestrial limits need not necessarily correspond to the fjord basin limit, however. More importantly, glacial outwash does not necessarily indicate a glacial maximum. Peacock (1971*b*) noted the presence of outwash associated with topographic constrictions at the mouths of Lochs Etive and Creran and at the Corran Narrows in Loch Linnhe, previously interpreted as demarcating Younger Dryas maximum limits (McCann 1966). Peacock (1971*b*) reinterpreted these deposits as retreat features, marking the positions of stillstands at topographic pinning points within the fjords. In his model, deglaciation results in the decoupling at topographic constrictions of an outer ice lobe, which becomes isolated from its upvalley source and thus stagnates *in situ*, and an inner lobe which remains active, grounded at the pinning point. He suggested that in fjords the presence of a topographic restriction, rather than climate, was the principal control on deglacial dynamics. Onshore outwash deposits also coincide with fjord constrictions at Roshven in Loch Ailort and Kylesknoydart in Loch Nevis, associated in both cases with submarine morainal structures identified by Boulton *et al.* (1981). The occurrence of these features well within the Younger Dryas limits, as defined by terminal moraines at both fjord mouths, strongly supports Peacock's contention that they are associated with the retreat of Younger Dryas glaciers formed during topographically forced stillstands. Their prevalence suggests that punctuated stillstands and minor re-advances of individual glaciers may have been a common feature of post-Younger Dryas deglaciation. Evidence from the Summer Isles survey of Stoker *et al.* (2006) supports the view that topography has a strong influence on basin deglaciation with moraines at the mouths of Loch Ewe, Loch Broom and Little Loch Broom all thought to represent glacial stillstands resulting from topographic pinning.

Glaciomarine units of Younger Dryas age are preserved within many fjord basins, for example Loch Etive (Howe *et al.* 2002), the Kentallen and Shuna basins of Loch Linnhe (Green 1995) and Loch Sunart (Mokeddem *et al.* 2007). Equivalent units are not present in Lochs Ailort and Nevis (Boulton *et al.* 1981), possibly because the Younger Dryas glaciers terminated at the coast rather than within the basins. Material from the glaciers was therefore carried away in the stronger coastal currents (Green 1995). The two-step deglaciation on the Isle of Skye, characterized by an initial, oscillatory phase and a second more continuous phase of glacial retreat (Dix & Duck 2000), does not appear to be replicated in any of the mainland records. However, the seismic evidence from Lochs Linnhe and Etive suggest that retreat during the later stages may have been very rapid (Green 1995; Howe *et al.* 2002), perhaps correlating with the second phase of deglaciation recorded on Skye. There is also strong evidence that retreat was punctuated by stillstands/minor re-advances in many of the fjords and in the outer sea area of the Summer Isles, suggesting that oscillating ice margins were the norm throughout post-Devensian deglaciation. Oscillations are also recorded onshore in the widespread presence of 'hummocky moraine', previously thought to have been deposited under stagnating ice but more recently reinterpreted as fragmentary recessional moraines marking stillstands/advances of the ice margin (Bennett & Boulton 1993; Benn & Lukas 2006).

Holocene (11.5 ka BP to present)

Increasing bottom-current activity and decreasing sediment supply during the Holocene resulted in reduced accumulation rates, especially at topographic highs, reported in the Summer Isles (Stoker *et al.* 2006; Wilson 2007), Loch Linnhe (Green 1995) and Loch Sunart (Mokeddem *et al.* 2007). Relatively thin Holocene sequences in Lochs Ailort and Nevis (Boulton *et al.* 1981) suggest a similar scenario in these fjords. In Loch Ainort the Holocene thickness is up to 18 m suggesting high accumulation rates. However, the authors attribute this to high glaciofluvial inputs in the early Holocene and suggest that contemporary rates are probably low in the Ainort basin. It is likely that total Holocene sedimentation elsewhere is also strongly enhanced by glaciogenic inputs in the early part of the record. In Loch Sunart, Mokeddem *et al.* (2007) obtained radiocarbon dates of 8710 ± 50 a BP and 6910 ± 40 a BP from a lithofacies interpreted as glaciomarine, implying glacially enhanced sedimentation rates until at least 7 ka BP. Brazier *et al.* (1988) suggest from onshore evidence in Glen Etive that maximum sediment supply in this area occurred between 10 and 4 ^{14}C ka BP, implying a slowing of accumulation rates after this period. The clearly defined seismic boundary between glaciomarine and marine sedimentation in Loch Etive (Howe *et al.* 2002) is not observed elsewhere and is therefore probably the result of basin topography and its effects on glacial retreat and postglacial fjord oceanography. Holocene sequences generally tend to be thickest in the deep basins (Green 1995; Wilson 2007). Inner Loch Etive is an exception, however, with Holocene sediment cover of just 3 m. This is the result of repeated winnowing by deep-water renewal events, enhanced during the high relative sea levels of the mid-Holocene (Nørgaard-Pedersen *et al.* 2006).

Deglaciation exposes an inherently unstable terrestrial landscape denuded of vegetation, and large volumes of uncompacted sediment which are readily eroded and transported into the fjord basins. The resulting high accumulation rates, combined with generally steep fjord slopes, has resulted in frequent sediment failure events such as slumps, slides and debris flows in most basins during the late to postglacial period (Green 1995; Dix & Duck 2000; Stoker *et al.* 2006, 2009; Wilson 2007). Loch Etive is again exceptional in this respect, with no evidence of major slope failure revealed by the side-scan and seismic surveys of Howe *et al.* (2001, 2002) despite the steepness of the slopes and high water content of the sediments. In Loch Ainort, Dix and Duck (2000) propose that mass-debris flow activity was common until the stabilization of the landscape by vegetation growth, thought to have occurred by *c.* 9.6 ka BP (Benn *et al.* 1992). Sea-level rise is suggested as the possible triggering mechanism for slope failure by Green (1995), sediments becoming destabilized by the increasing pore water pressure. Another possible cause is seismic activity which may occur as a result of isostatic readjustment following ice withdrawal, or independently of deglaciation (Green 1995; Stoker *et al.* 2006).

The palaeoecological response to Holocene climate amelioration is recorded in the sediments of Lochs Sunart and Etive (Nørgaard-Pedersen *et al.* 2006; Mokeddem *et al.* 2007). During the very early Holocene, the environments around these fjords were colonized by pioneer vegetation dominated by herbaceous species such as juniper and dwarf birch, and with limited tree cover. The more detailed pollen record from Loch Sunart documents the subsequent succession of colonizing species: an early expansion of pine and hazel was replaced in the mid-Holocene by mixed species woodland (hazel, oak and alder dominated) and in the later Holocene by the development of fern moorland and pasture. This pattern is in accordance with records from elsewhere in north and west Scotland (e.g. Birks 1972; Walker & Lowe 1977, 1979, 1981, 1985; Birks & Mathewes 1978; Boyd 1988; Jones *et al.* 1989; Bridge *et al.* 1990; Benn *et al.* 1992), indicative of initial postglacial climate warming and increasing humidity in the later Holocene.

Recent human impacts

During the most recent past, human activities have begun to impact upon the sedimentary records of Scotland's fjords. A marked increase in sediment accumulation in Loch Etive during the late Holocene (from 2 mm a^{-1} to 7 mm a^{-1}) has been attributed to massive deforestation which occurred around the loch to feed an industrial iron furnace in the 18th and 19th centuries (Howe *et al.* 2002; Nørgaard-Pedersen *et al.* 2006). More recently, concern has arisen over the impacts of fishing and fish-farming within the fjords. Trawling and dredging can cause long-lasting damage to sensitive habitats such as the serpulid reefs of Loch Creran (Moore *et al.* 2006) and the maerl beds in the Firth of Clyde (Hall-Spencer & Moore 2000), and it has been demonstrated that such activities alter benthic communities by decreasing diversity (Tuck *et al.* 1998). Likewise, aquaculture can impact negatively on the fjord environment. Fish farms are now present in most of Scotland's fjords, where they contribute significant quantities of organic carbon to the surrounding water and sediments (Gillibrand & Turrell 1997). They are also associated with water pollution through the discharge of chemicals used to treat parasites (Gillibrand & Turrell 1997) and with heavy metal contamination since the diet of farmed Atlantic salmon includes metals such as copper, zinc and cadmium (Dean *et al.* 2007). The impacts of fish farms tend to be very localized, however, and generally restricted to within a few hundred metres of the cages (e.g. Brown *et al.* 1987; Gillibrand *et al.* 1996; Dean *et al.* 2007).

Although the fjords of northwest Scotland are characterized by low population densities and little heavy industry, there is clear evidence of trace metal pollution (copper, zinc and lead) within their sediments. The main source of this pollution is identified as northward-flowing coastal water from the more industrial Clyde Sea/Irish Sea area (Krom *et al.* 2009). Heavy metals have deleterious effects on numerous marine species, and have been tentatively implicated in the inhibition of mussel growth in Loch Leven (Stirling & Okumuş 1995).

The main findings of the studies reviewed in this paper are summarized in Table 2.

Conclusions

The studies reviewed in this paper lead to the following conclusions.

(1) Devensian glaciation removed pre-existing sediments and left a diamict overlying bedrock over much of the study area, both inside and outside the fjordic basins. This till is presumed to correlate to the shelf-wide diamict reported to the west of Scotland by Davies *et al.* (1984).
(2) The retreat of the Devensian ice margin progressed in an oscillatory fashion and was punctuated by numerous stillstands/minor re-advances. Glaciomarine units associated with the initial (post-LGM) deglaciation are, in most cases, preserved outside the fjord basins only.
(3) Glacial re-advance associated with the Younger Dryas stadial removed pre-existing sediments from many basins. Elsewhere ice reworked Devensian sediments and deposited terminal moraines at the mouths of many fjords.
(4) The retreat of Younger Dryas ice resulted in the rapid accumulation of glaciomarine

Table 2. *Summary of major findings from studies reviewed in this paper*

Area	Pre-Younger Dryas	Younger Dryas	Holocene	References
Loch Etive	No record	Ice extended to sea; diamicton and deltaic sediments deposited at head of loch during deglaciation	Glaciomarine/ glaciolacustrine sedimentation abruptly replaced in the early Holocene by sedimentation of riverine inputs under marine conditions. Outwash fans developed at river mouths. Deep-water renewal frequency increased towards mid-Holocene and then decreased again, in response to sea level	Howe *et al.* (2001, 2002), Nørgaard-Pedersen *et al.* (2006)
Loch Linnhe	Glacial erosion of inner basins removed pre-existing sediments. Till deposited in deepest outer lochs. Glaciomarine deposits accumulated in Kentallen and Shuna basins and possibly deepest outer basins	Glacier reached maximum position south of Corran Narrows. Thick glaciomarine deposits accumulated in Kentallen and Shuna basins. Widespread debris flows and reworking by bottom currents	Sedimentation dominated by suspension fall-out of riverine inputs and bottom-current reworking. Accumulation rates locally high (up to 2.4 cm a^{-1}) in deep basins, but sedimentation low/ absent in shallower areas subject to strong tidal currents	Green (1995)
Loch Sunart	LGM basal moraine material deposited and overlain by coarse-grained sediments during the Older Dryas. Landscape dominated by tundra vegetation	Proximal and distal glaciomarine units deposited during retreat phase. Pollen absent	Three climatic cooling events identified at *c.* 8200, 5500 and 3500 a BP. Basin becoming sediment starved in later Holocene. Initial expansion of *Pinus/ Corylus* forest, subsequently replaced by mixed deciduous species (*Corylus*, *Quercus*, *Alnus*). Meadows and fern moors expanding in later Holocene	Mokeddem *et al.* (2007)
Lochs Nevis and Ailort	Devensian till deposited over bedrock; glaciomarine unit draped over the till (outside fjords only)	Terminal moraines deposited at fjord mouths; retreat punctuated by stillstands at topographic restrictions	Localized basin infill deposited over till in fjords, and over glaciomarine unit outside	Boulton *et al.* (1981)

(*Continued*)

Table 2. *Continued*

Area	Pre-Younger Dryas	Younger Dryas	Holocene	References
Loch Ainort	Devensian till deposited over survey area	Glacial advance reworked till. Terminal moraine formed 800 m beyond terrestrial limit. Retreat marked by rapid, oscillatory phase depositing de Geer moraines at fjord mouth, followed by more continuous deglaciation of inner basin. Localized debris flow deposition in inner basin	Deposition of fine-grained sediments, initially augmented by high glaciofluvial inputs	Dix & Duck (2000)
Summer Isles	Erosional grooves and till cover developed. Oscillatory retreat of ice sheet throughout Lateglacial Interstadial. Sheet-drape facies outside fjords deposited from melting ice. Glaciolacustrine/ glaciomarine units deposited containing slumps and debris flows	No record reported	Sediment inputs decreasing in the later Holocene. Regime becoming more erosional as a result of tidal bottom currents	Stoker *et al.* (2006), Bradwell *et al.* (2008)

units within many fjord basins. In some areas these were subsequently overridden by minor re-advances of the ice margin, producing compacted sediments and morainal structures.

(5) Topographic pinning played an important part in determining the style of glacial retreat within the fjord basins. Outwash deposits and moraines at narrows demonstrate that stillstands of the ice front frequently occurred in these areas.

(6) As with earlier Devensian deglaciation, Younger Dryas ice retreat was generally marked by oscillations of the glacial margins; however, within many inner basins there is evidence that the final phase of deglaciation may have been very rapid.

(7) Holocene climate records from the fjords are limited, but appear to support the onshore evidence that western Scotland was colonized by a tundra-style vegetation immediately after the withdrawal of Younger Dryas ice. This was followed by a warming of the climate, reflected in the succession of arboreal species assemblages and increased humidity in the later Holocene. The importance of post-glacial sea-level changes in regulating fjord circulation is also demonstrated.

(8) Human impacts recorded in the sediments of western Scottish fjords in the recent past include increased sediment runoff resulting from deforestation, mechanical damage from fishing gear, organic enrichment from fish farms and heavy metal pollution.

The authors would like to thank B. Austin and N. Nørgaard-Pedersen for constructive and thoughtful reviews which helped to improve the paper.

References

Allen, G. L. & Simpson, J. H. 1998*a*. Deep water inflows to upper Loch Linnhe. *Estuarine, Coastal and Shelf Science*, **47**, 487–498.

Allen, G. L. & Simpson, J. H. 1998*b*. Reflection of the internal tide in upper Loch Linnhe, a Scottish Fjord. *Estuarine, Coastal and Shelf Science*, **46**, 683–701.

Andrews, J. T., Austin, W. E. N., Bergsten, H. & Jennings, A. E. 1996. The Late Quaternary palaeoceanography of North Atlantic margins: an introduction. *In*: Andrews, J. T, Austin, W. E. N., Bergsten, H. & Jennings, A. E. (eds) *Late Quaternary Palaeoceanography of the North Atlantic Margins*. Geological Society, London, Special Publication, **111**, 1–6.

Ansell, A. D. 1974. Sedimentation of organic detritus in Lochs Etive and Creran, Argyll, Scotland. *Marine Biology*, **27**, 263–273.

Appenzeller, C., Stocker, T. F. & Anklin, M. 1998. North Atlantic oscillation dynamics recorded in Greenland ice cores. *Science*, **282**, 446–449.

Austin, W. E. N. & Kroon, D. 1996. Late glacial sedimentology, foraminifera and stable isotope stratigraphy of the Hebridean Continental Shelf, northwest Scotland. *In*: Andrews, J. T., Austin, W. E. N., Bergsten, H. & Jennings, A. E. (eds) *Late Quaternary Palaeoceanography of the North Atlantic Margins*. Geological Society, London, Special Publication, **111**, 187–213.

Austin, W. E. N. & Inall, M. E. 2002. Deep-water renewal in a Scottish fjord: temperature, salinity and oxygen isotopes. *Polar Research*, **21**, 251–258.

Austin, W. E. N., Howe, J. A. et al. 2006. *Scottish Sea Lochs (fjords): testing the evidence for Holocene rapid climate change events*. Geophysical Research Abstracts, European Geosciences Union, Vol. 8.

Ballantyne, C. K. 1988. Ice sheet moraines in southern Skye. *Scottish Journal of Geology*, **24**, 301–304.

Ballantyne, C. K. 1989. The Loch Lomond Readvance on the Isle of Skye, Scotland: glacier reconstruction and palaeoclimatic implications. *Journal of Quaternary Science*, **4**, 95–108.

Ballantyne, C. K., McCarroll, D., Nesje, A., Dahl, S. O., Stone, J. O. & Fifield, L. K. 1998. High-resolution reconstruction of the last ice sheet in NW Scotland. *Terra Nova*, **10**, 63–67.

Benn, D. I. & Lukas, S. 2006. Younger Dryas glacial landsystems in North West Scotland: an assessment of modern analogues and palaeoclimate implications. *Quaternary Science Reviews*, **25**, 2390–2408.

Benn, D. I., Lowe, J. J. & Walker, M. J. C. 1992. Glacier response to climatic change during the Loch Lomond Stadial and early Flandrian: geomorphological and palynological evidence from the Isle of Skye, Scotland. *Journal of Quaternary Science*, **7**, 125–144.

Bennett, M. R. & Boulton, G. S. 1993. Deglaciation of the Younger Dryas or Loch Lomond Stadial ice-field in the northern Highlands, Scotland. *Journal of Quaternary Science*, **8**, 133–145.

Birks, H. H. 1972. Studies in the vegetational history of Scotland. III. A radiocarbon-dated pollen diagram from Loch Maree, Ross and Cromarty. *New Phytology*, **71**, 731–754.

Birks, H. H. & Mathewes, R. W. 1978. Studies in the vegetational history of Scotland. V. Late Devensian and early Flandrian pollen and macrofossil stratigraphy at Abernethy Forest, Inverness-shire. *New Phytology*, **80**, 455–484.

Booth, D. A. 1987. Some consequences of a flood tide front in Loch Creran. *Estuarine, Coastal and Shelf Science*, **24**, 363–375.

Boulton, G. S., Chroston, P. N. & Jarvis, J. 1981. A marine seismic study of late Quaternary sedimentation and inferred glacier fluctuations along western Inverness-shire, Scotland. *Boreas*, **10**, 39–51.

Bowden, K. F. 1980. Physical and dynamical oceanography of the Irish Sea. *In*: Banner, F. T., Collins, M. B. & Massie, K. S. (eds) *The North-west Shelf Seas: The Seabed and the Sea in Motion II. Physical and Chemical Oceanography, and Physical Resources*. Elsevier Oceanography Series 24B, Elsevier Scientific Publishing Company, Amsterdam, 391–413.

Bowen, D. Q., Phillips, F. M., McCabe, A. M., Knutz, P. C. & Sykes, G. A. 2002. New data for the Last Glacial Maximum in Great Britain and Ireland. *Quaternary Science Reviews*, **21**, 89–101.

Boyd, W. E. 1988. Early Flandrian vegetational development on the coastal plain of north Ayrshire, Scotland: evidence from multiple pollen profiles. *Journal of Biogeography*, **15**, 325–337.

Bradwell, T., Fabel, D., Stoker, M. S., Mathers, H., McHargue, L. & Howe, J. A. 2008. Ice caps existed throughout the Lateglacial Interstadial in northern Scotland. *Journal of Quaternary Science*, **23**, 401–407.

Brazier, V., Whittington, G. & Ballantyne, C. K. 1988. Holocene debris cone evolution in Glen Etive, Western Grampian Highlands, Scotland. *Earth Surface Processes and Landforms*, **13**, 525–531.

Bridge, M. C., Haggart, B. A. & Lowe, J. J. 1990. The history and palaeoclimatic significance of subfossil remains of *Pinus sylvestris* in blanket peats from Scotland. *Journal of Ecology*, **78**, 77–79.

Brown, J. R., Gowen, R. J. & McLusky, D. S. 1987. The effect of salmon farming on the benthos of a Scottish sea loch. *Journal of Experimental Marine Biology and Ecology*, **109**, 39–51.

Charlesworth, J. K. 1956. The Late-glacial history of the highlands and the islands of Scotland. *Transactions of the Royal Society of Edinburgh*, **62**, 769–928.

Clark, C. D., Evans, D. J. A. et al. 2004. Map and GIS database of glacial landforms and features related to the last British Ice Sheet. *Boreas*, **33**, 359–375.

Craig, G. Y. 1991. *Geology of Scotland*. 3rd edn. The Geological Society, London.

Davies, A. M. & Xing, J. 2003. The influence of wind direction upon flow along the west coast of Britain and in the North Channel of the Irish Sea. *Journal of Physical Oceanography*, **33**, 69–82.

Davies, H. C., Dobson, R. M. & Whittington, R. J. 1984. A revised seismic stratigraphy for Quaternary deposits on the inner continental shelf west of Scotland between 55°30′N and 57°30′N. *Boreas*, **13**, 49–66.

Davies, J. M. 1975. Energy flow through the benthos in a Scottish sea loch. *Marine Biology*, **31**, 353–362.

DEAN, R. J., SHIMMIELD, T. M. & BLACK, K. D. 2007. Copper, zinc and cadmium in marine cage fish farm sediments: an extensive survey. *Environmental Pollution*, **145**, 84–95.

DIX, J. K. & DUCK, R. W. 2000. A high-resolution seismic stratigraphy from a Scottish sea loch and its implications for Loch Lomond Stadial deglaciation. *Journal of Quaternary Science*, **15**, 645–656.

EDWARDS, A. & EDELSTEN, D. J. 1977. Deep water renewal of Loch Etive: a three basin Scottish fjord. *Estuarine and Coastal Marine Science*, **5**, 575–595.

EDWARDS, A. & SHARPLES, F. 1986. *Scottish Sea Lochs: A Catalogue*. Scottish Marine Biological Association, Special Publication **134**.

EIRÍKSSON, J., BARTELS-JÓNSDÓTTIR, H. B. *ET AL*. 2006. Variability of the North Atlantic Current during the last 2000 years based on shelf bottom water and sea surface temperatures along an open ocean/shallow marine transect in western Europe. *The Holocene*, **16**, 1017–1029.

ELLETT, D. J. 1979. Some oceanographic features of Hebridean waters. *Proceedings of the Royal Society of Edinburgh*, **77B**, 61–74.

ELLETT, D. J. & EDWARDS, A. 1983. Oceanography and inshore hydrography of the Inner Hebrides. *Proceedings of the Royal Society of Edinburgh*, **83B**, 143–160.

EVEREST, J. D., BRADWELL, T., FOGWILL, C. J. & KUBIK, P. W. 2006. Cosmogenic ^{10}Be age constraints for the Wester Ross Readvance moraine: insights into British ice-sheet behaviour. *Geografiska Annaler*, **88A**, 9–17.

GAGE, J. 1972*a*. Community structure of the benthos in Scottish sea-lochs. I. Introduction and species diversity. *Marine Biology*, **14**, 281–297.

GAGE, J. 1972*b*. A preliminary survey of the benthic macrofauna and sediments in Lochs Etive and Creran, sea-lochs along the west coast of Scotland. *Journal of the Marine Biological Association of the UK*, **52**, 237–276.

GAGE, J. & GEEKIE, A. D. 1973. Community structure of the benthos in Scottish sea lochs. II. Spatial pattern. *Marine Biology*, **19**, 41–53.

GILLIBRAND, P. A. & TURRELL, W. R. 1997. The use of simple models in the regulation of the impact of fish farms on water quality in Scottish sea lochs. *Aquaculture*, **159**, 33–46.

GILLIBRAND, P. A., TURRELL, W. R. & ELLIOTT, A. J. 1995. Deep-water renewal in the upper basin of Loch Sunart, a Scottish fjord. *Journal of Physical Oceanography*, **26**, 1488–1503.

GILLIBRAND, P. A., TURRELL, W. R., MOORE, D. C. & ADAMS, R. D. 1996. Bottom water stagnation in a Scottish sea loch. *Estuarine, Coastal and Shelf Science*, **43**, 217–235.

GILLIBRAND, P. A., CAGE, A. G. & AUSTIN, W. E. N. 2005. A preliminary investigation of basin water response to climate forcing in a Scottish fjord: evaluating the influence of the NAO. *Continental Shelf Research*, **25**, 571–587.

GOLLEDGE, N. R. 2007. An ice cap landsystem for palaeoglaciological reconstructions: characterizing the Younger Dryas in Western Scotland. *Quaternary Science Reviews*, **26**, 213–229.

GOLLEDGE, N. R. & HUBBARD, A. 2005. Evaluating Younger Dryas glacier reconstructions in part of the western Scottish Highlands: a combined empirical and theoretical approach. *Boreas*, **34**, 274–286.

GOLLEDGE, N. R., HUBBARD, A. & SUGDEN, D. E. 2008. High-resolution numerical simulation of Younger Dryas glaciation in Scotland. *Quaternary Science Reviews*, **27**, 888–904.

GRAY, J. M. 1975. The Loch Lomond Readvance and contemporaneous sea-levels in Loch Etive and neighbouring areas of western Scotland. *Proceedings of the Geologists' Association* **86**, 227–238.

GREEN, D. R. 1995. *The glacial geomorphology of the Loch Lomond Advance in Lochaber*. PhD thesis, University of Edinburgh.

HALL, G. P. D. 1974. *West Coast of Scotland Pilot – West Coast of Scotland from Mull of Galloway to Cape Wrath including the Inner and Outer Hebrides and off-lying islands*. 11th edn. Hydrographer of the Navy, Ministry of Defence, England.

HALL, I. R., HYDES, D. J., STATHAM, P. J. & OVERNELL, J. 1996. Dissolved and particulate trace metals in a Scottish sea loch: an example of a pristine environment? *Marine Pollution Bulletin*, **32**, 846–854.

HALL-SPENCER, J. M. & MOORE, P. G. 2000. Scallop dredging has profound, long-term impacts on maerl habitats. *ICES Journal of Marine Science*, **57**, 1407–1415.

HILL, A. E. & SIMPSON, J. H. 1988. Low-frequency variability of the Scottish coastal current induced by alongshore pressure gradients. *Estuarine, Coastal and Shelf Science*, **27**, 163–180.

HILL, A. E., HORSBURGH, K. J., GARVINE, R. W., GILLIBRAND, P. A., SLESSER, G., TURRELL, W. R. & ADAMS, R. D. 1997. Observations of a density-driven recirculation of the Scottish coastal current in the Minch. *Estuarine, Coastal and Shelf Science*, **45**, 473–484.

HOWE, J. A. & AUSTIN, W. E. N. 2005. 10,000 Years of Environmental Change: Reading the Story Written in the Sediments of Loch Sunart. *Scottish Association for Marine Science Newsletter*, **31**, November 2005.

HOWE, J. A., OVERNELL, J., INALL, M. E. & WILBY, A. 2001. A side-scan sonar image of a glacially overdeepened sea loch, upper Loch Etive, Argyll. *Scottish Journal of Geology*, **37**, 3–10.

HOWE, J. A., SHIMMIELD, T., AUSTIN, W. E. N. & LONGVA, O. 2002. Post-glacial depositional environments in a mid-high latitude glacially-overdeepened sea loch, inner Loch Etive, western Scotland. *Marine Geology*, **185**, 417–433.

HUBBARD, A. 1999. High-resolution modeling of the advance of the Younger Dryas ice sheet and its climate in Scotland. *Quaternary Research*, **52**, 27–43.

HURRELL, J. W., KUSHNIR, Y., OTTERSEN, G. & VISBECK, M. 2003. An overview of the North Atlantic Oscillation. *In*: HURRELL, J. W., KUSHNIR, Y., OTTERSEN, G. & VISBECK, M. (eds) *The North Atlantic Oscillation: Climatic Significance and Environmental Impact*. Geophysical Monograph 134. American Geophysical Union, Washington, DC, 1–35.

INALL, M. E., COTTIER, F., GRIFFITHS, C. & RIPPETH, T. 2004. Sill dynamics and flow transformation in a jet fjord. *Ocean Dynamics*, **54**, 307–314.

JONES, V. J., STEVENSON, A. C. & BATTARBEE, R. W. 1989. Acidification of lakes in Galloway, south west Scotland: a diatom and pollen study of the post-glacial history of the Round Loch of Glen Head. *Journal of Ecology*, **77**, 1–23.

KROM, M. D., CARBO, P., CLERICI, S., CUNDY, A. B. & DAVIES, I. M. 2009. Sources and timing of trace metal contamination to sediments in remote sealochs, N. W. Scotland. *Estuarine, Coastal and Shelf Science*, **83**, 239–251.

LLOYD, J. 2000. Combined foraminiferal and thecamoebian environmental reconstruction from an isolation basin in NW Scotland: implications for sea-level studies. *The Journal of Foraminiferal Research*, **30**, 294–305.

LOH, P. S., MILLER, A. E. J., REEVES, A. D., HARVEY, S. M. & OVERNELL, J. 2008*a*. Assessing the biodegradability of terrestrially-derived organic matter in Scottish sea loch sediments. *Hydrology and Earth System Sciences*, **12**, 811–823.

LOH, P. S., REEVES, A. D., HARVEY, S. M., OVERNELL, J. & MILLER, A. E. J. 2008*b*. The fate of terrestrial organic matter in two Scottish sea lochs. *Estuarine, Coastal and Shelf Science*, **76**, 566–570.

LUKAS, S. & BENN, D. I. 2006. Retreat dynamics of Younger Dryas glaciers in the far NW Scottish Highlands reconstructed from moraine sequences. *Scottish Geographical Journal*, **122**, 308–325.

MAYEWSKI, P. A., ROHLING, E. E. ET AL. 2004. Holocene climate variability. *Quaternary Research* **62**, 243–255.

MCCANN, S. B. 1966. The limits of the Late-glacial Highland, or Loch Lomond, Readvance along the West Highland seaboard from Oban to Mallaig. *Scottish Journal of Geology*, **2**, 84–95.

MET OFFICE 2009 (www.metoffice.gov.uk/climate/uk/averages).

MILNE 1969. *Engineering aspects of marine fish culture*. PhD thesis, University of Strathclyde, 229 (in Milne 1972).

MILNE 1972. Hydrography of Scottish West Coast Sea Lochs. *Marine Research*, HMSO, **3**, 50.

MOKEDDEM, Z., BALTZER, A., CLET-PELLERIN, M., WALTER-SIMMONET, A. V., BATES, R., BALUT, Y. & BONNOT-COURTOIS, C. 2007. Climatic fluctuations recorded in the sediment fillings of Loch Sunart (northwestern Scotland) since 20,000 BP. *Comptes Rendus Géosciences*, **339**, 150–160.

MOORE, C. G., SAUNDERS, G. R., HARRIES, D. B., MAIR, J. M., BATES, C. R. & LYNDON, A. R. 2006. *The establishment of site condition monitoring of the subtidal reefs of Loch Creran Special Area of Conservation*. Scottish Natural Heritage, Commissioned Report No. **151** (ROAME No. F02AA409).

MURRAY, J. W. 2004. The Holocene palaeoceanographic history of Muck Deep, Hebridean shelf, Scotland: has there been a change of wave climate in the past 12 000 years? *Journal of Micropalaeontology*, **23**, 153–161.

NØRGAARD-PEDERSEN, N., AUSTIN, W. E. N., HOWE, J. A. & SHIMMIELD, T. 2006. The Holocene record of Loch Etive, western Scotland: influence of catchment and relative sea level changes. *Marine Geology* **228**, 55–71.

OVERNELL, J. 2002. Manganese and iron profiles during early diagenesis in Loch Etive, Scotland. Application of two diagenetic models. *Estuarine, Coastal and Shelf Science*, **54**, 33–44.

OVERNELL, J., HARVEY, S. M. & PARKES, R. J. 1996. A biogeochemical comparison of sea loch sediments. Manganese and iron contents, sulphate reduction and oxygen uptake rate. *Oceanologica Acta*, **19**, 41–55.

PANTIN, H. M. 1991. *The sea-bed sediments around the UK: their bathymetric and physical environment, grain-size, mineral composition and associated bedforms*. British Geological Survey, Research Report **SB/90/1**.

PEACOCK, J. D. 1971*a*. Marine shell radiocarbon dates and the chronology of deglaciation in Western Scotland. *Nature, Physical Science*, **230**, 43–45.

PEACOCK, J. D. 1971*b*. Terminal features of the Creran Glacier of Loch Lomond Readvance age in western Benderloch, Argyll, and their significance in the late-glacial history of the Loch Linnhe area. *Scottish Journal of Geology*, **7**, 349–356.

PEACOCK, J. D., HARKNESS, D. D., HOUSLEY, R. A., LITTLE, J. A. & PAUL, M. A. 1989. Radiocarbon ages for a glaciomarine bed associated with the maximum of the Loch Lomond Readvance in west Benderloch, Argyll. *Scottish Journal of Geology*, **25**, 69–79.

PEARSON, T. H. 1970. The benthic ecology of Loch Linnhe and Loch Eil, a sea-loch system on the west coast of Scotland. I. The physical environment and distribution of the macrobenthic fauna. *Journal of Experimental Marine Biology and Ecology*, **5**, 1–34.

RIDGEWAY, I. M. & PRICE, N. B. 1987. Geochemical associations and post-depositional mobility of heavy metals in coastal sediments: Loch Etive, Scotland. *Marine Chemistry*, **21**, 229–248.

SHENNAN, I., INNES, J. B., LONG, A. J. & ZONG, Y. 1995. Late Devensian and Holocene relative sea-level changes in northwestern Scotland: new data to test existing models. *Quaternary International*, **26**, 97–123.

SHENNAN, I., RUTHERFORD, M. M., INNES, J. B. & WALKER, K. J. 1996. Late glacial sea level and ocean margin environmental changes interpreted from biostratigraphic and lithostratigraphic studies of isolation basins in northwest Scotland. Geological Society, London, Special Publications, **111**, 229–244.

SHENNAN, I., TOOLEY, M. J., GREEN, F. M. L., INNES, J. B., KENNINGTON, K., LLOYD, J. M. & RUTHERFORD, M. M. 1999. Sea level, climate change and coastal evolution in Morar, northwest Scotland. *Geologie en Mijnbouw*, **77**, 247–262.

SHENNAN, I., LAMBECK, K. ET AL. 2000. Late Devensian and Holocene records of relative sea-level changes in northwest Scotland and their implications for glaci-hydro-isostatic modelling. *Quaternary Science Reviews*, **19**, 1103–1135.

SHIMMIELD, T. M. 1993. *A study of radionuclides, lead and lead isotope ratios in Scottish sea loch sediments*. PhD thesis, University of Edinburgh.

SISSONS, J. B. 1979. The limit of the Loch Lomond Advance in Glen Roy and vicinity. *Scottish Journal of Geology*, **15**, 31–42.

SISSONS, J. B. & DAWSON, A. G. 1981. Former Sea-levels and Ice Limits in Part of Wester Ross, Northwest

Scotland. *Proceedings of the Geologists' Association*, **92**, 115–124.

SMITH, D. E., CULLINGFORD, R. A. & FIRTH, C. R. 2000. Patterns of isostatic land uplift during the Holocene: evidence from mainland Scotland. *The Holocene*, **10**, 489–501.

SMITH, D. E., FRETWELL, P. T., CULLINGFORD, R. A. & FIRTH, C. R. 2006. Towards improved empirical isobase models of Holocene land uplift for mainland Scotland, UK. *Philosophical Transactions of the Royal Society*, **364**, 949–972.

SMITH, D. E., CULLINGFORD, R. A., MIGHALL, T. M., JORDAN, J. T. & FRETWELL, P. T. 2007. Holocene relative sea level changes in a glacio-isostatic area: new data from south-west Scotland, United Kingdom. *Marine Geology*, **242**, 5–26.

STIRLING, H. P. & OKUMUŞ, I. 1995. Growth and production of mussels (*Mytilus edulis* L.) Suspended at salmon cages and shellfish farms in two Scottish sea lochs. *Aquaculture*, **134**, 193–210.

STOKER, M. S. & BRADWELL, T. 2005. The Minch palaeo-ice stream, NW sector of the British-Irish Ice Sheet. *Journal of the Geological Society*, **162**, 425–428.

STOKER, M., BRADWELL, T., WILSON, C., HARPER, C., SMITH, D. & BRETT, C. 2006. Pristine fjord landsystem revealed on the sea bed in the Summer Isles region, NW Scotland. *Scottish Journal of Geology*, **42**, 89–99.

STOKER, M. S., BRADWELL, T., HOWE, J. A., WILKINSON, I. P. & MCINTYRE, K. L. 2009. Lateglacial ice-cap dynamics: evidence from the fjords of NW Scotland. *Quaternary Science Reviews* (in revision).

STOKER, M. S., WILSON, C., HOWE, J. A., BRADWELL, T. & LONG, D. 2010. Paraglacial slope instability in Scottish fjords: examples from Little Loch Broom, NW Scotland. *In*: HOWE, J. A., AUSTIN, W. E. N., FORWICK, M. & PAETZEL, M. (eds) *Fjord Systems and Archives*. Geological Society, London, Special Publications, **344**, 225–242.

SYVITSKI, J. P. M. & SHAW, J. 1995. Sedimentology and geomorphology of fjords. *In*: PERILLO, G. M. E. (ed.) *Geomorphology and Sedimentology of Estuaries*. Elsevier, Developments in Sedimentology, **53**.

THORP, P. W. 1981. A trimline method for defining the upper limit of Loch Lomond Advance glaciers: examples from the Loch Leven and Glen Coe areas. *Scottish Journal of Geology*, **17**, 49–64.

THORP, P. W. 1986. A mountain icefield of Loch Lomond Stadial age, western Grampians, Scotland. *Boreas*, **15**, 83–97.

TUCK, I. D., HALL, S. J., ROBERTSON, M. R., ARMSTRONG, E. & BASFORD, D. J. 1998. Effects of physical trawling disturbance in a previously unfished sheltered Scottish sea loch. *Marine Ecology, Progress Series* **162**, 227–242.

WALKER, M. J. C. & LOWE, J. J. 1977. Postglacial environmental history of Rannoch Moor, Scotland. I. Three pollen diagrams from the Kingshouse area. *Journal of Biogeography*, **4**, 333–351.

WALKER, M. J. C. & LOWE, J. J. 1979. Postglacial environmental history of Rannoch Moor, Scotland. II. Pollen diagrams and radiocarbon dates from the Rannoch Station and Corrour area. *Journal of Biogeography*, **6**, 349–362.

WALKER, M. J. C. & LOWE, J. J. 1981. Postglacial environmental history of Rannoch Moor, Scotland. III. Early- and mid-Flandrian pollen stratigraphic data from sites on western Rannoch Moor and near Fort William. *Journal of Biogeography*, **8**, 475–491.

WALKER, M. J. C. & LOWE, J. J. 1985. Flandrian environmental history of the Isle of Mull, Scotland. I. Pollen stratigraphic evidence and radiocarbon dates from Glen More, south-central Mull. *New Phytology*, **99**, 587–610.

WALKER, M. J. C., GRAY, J. M. & LOWE, J. J. (eds) 1992. *The South-West Scottish Highlands: Field Guide*. Quaternary Research Association, Cambridge.

WATTS, L. J., RIPPETH, T. P. & EDWARDS, A. 1998. The roles of hydrographic and biogeochemical processes in the distribution of dissolved inorganic nutrients in a Scottish sea-loch: consequences for the spring phytoplankton bloom. *Estuarine, Coastal and Shelf Science*, **46**, 39–50.

WILSON, C. R. 2007. *Deglacial depositional environments, Little Loch Broom, NW Scotland*. BSc (Hons) dissertation, UHI Millennium Institute.

WOOD, B. J. B., TETT, P. B. & EDWARDS, A. 1973. An introduction to the phytoplankton, primary production and relevant hydrography of Loch Etive. *Journal of Ecology*, **61**, 569–585.

Pollen analysis of Holocene sediments from Loch Etive, a Scottish fjord

PETER R. CUNDILL* & WILLIAM E. N. AUSTIN

School of Geography and Geosciences, University of St Andrews, St Andrews, Fife, KY16 9AL, Scotland, U.K.

**Corresponding author (e-mail: prc@st-andrews.ac.uk)*

Abstract: This paper outlines one of the first attempts to analyse the detailed pollen content of the marine sediments from a fjord (sea loch) located on the west coast of Scotland. The site chosen was in the Bonawe Deep, part of the upper basin of Loch Etive, and 2.3 m depth of sediments were sampled. These sediments were examined for their fossil pollen content and a pollen diagram was constructed from the results. The interpretation of the pollen record was supported by seven radiocarbon dates which demonstrated that the sediments accumulated over the last 8–9 ka of the Holocene. The pollen diagram was divided into three pollen assemblage zones. The lowest zone ET1 (*c.* 9–7.3 ka BP) was dominated by pollen of *Betula*, *Pinus* and *Corylus*, zone ET2 (*c.* 7.3–1.2 ka BP) contained a wide range of woodland pollen taxa (mainly *Betula*, *Pinus*, *Quercus*, *Alnus* and *Corylus*) and the uppermost zone ET3 (1.2–0 ka BP) was distinguished by fewer tree pollen taxa and increased values of open habitat taxa, particularly *Calluna*. The pollen diagram showed the development of vegetation during the Holocene which correlated well with existing results from studies of peat or freshwater lake sediments sampled in the west of Scotland. The only significant differences were the higher concentrations of *Pinus* pollen found in the Loch Etive site. These differences can be explained by the large catchment area of the loch which includes high ground, where it is proposed that extensive *Pinus* forest grew during much of the Holocene. The study demonstrates the practicality of using Scottish fjord sediments for Holocene pollen studies and potentially provides a link between terrestrial vegetation records and marine-based fossil records.

Fjordic environments are the result of overdeepening of mountain range river valleys by Pleistocene glaciation and the subsequent flooding of these valleys by the sea during interglacial stages (McKnight & Hess 2008). Fjords along the coast of Norway are the most well known and described, but they also occur in several other regions of the world such as western Canada, Alaska, Chile and South Island New Zealand. On a smaller scale, fjords are also found along the west coast of Scotland and are usually referred to as sea lochs although their mode of formation is the same as Norwegian fjords. They provide locations where both marine and freshwater sediments and fossils can mix and accumulate in the fjord basins. Loch Etive (Fig. 1) is one of the longest fjords in western Scotland, 30 km long and 1–2 km wide. It is a southwest–northeast trending fjord which has several sedimentation basins separated by shallow sills (Nørgaard-Pedersen *et al.* 2006). Loch Etive has one of the highest freshwater discharge to tidal flow ratios among Scottish fjords (Edwards & Sharples 1986), and a mean renewal period in the upper basin of 1.3 years (Edwards & Edelsten 1977).

Until recently there had been no attempt to investigate the sediments of Scottish fjords for their pollen content although elsewhere in the world marine sediments, generally in deep ocean locations, have successfully yielded pollen data (e.g. Rossignol-Strick 1995; Heusser *et al.* 2000; Scourse *et al.* 2005). A range of potential difficulties can be identified if such sediments are used to infer vegetation patterns on adjacent terrestrial environments. While Heusser *et al.* (2000) successfully used pollen from marine sediments to interpret large-scale vegetation patterns, De Busk (1997) argued that habitat level interpretations are not valid when using marine-based pollen data. However, examination of the surficial sediments of Loch Sunart (Cundill *et al.* 2006) demonstrated that, in the case of Scottish fjords, there was a correspondence between the pollen record and vegetation on the adjacent land surface. In addition, Mokeddem *et al.* (2007) showed in a limited way that a Holocene record could be obtained from Loch Sunart sediments. Berglund (1976) attempted to establish a link between pollen changes at freshwater and marine sediment sites in Sweden although he reported only partial success because of problems of analysing and interpreting the pollen record from marine sediments. Although the literature on the analysis of pollen in marine sediments appears to

From: Howe, J. A., Austin, W. E. N., Forwick, M. & Paetzel, M. (eds) *Fjord Systems and Archives.* Geological Society, London, Special Publications, **344**, 331–340.
DOI: 10.1144/SP344.22 0305-8719/10/$15.00

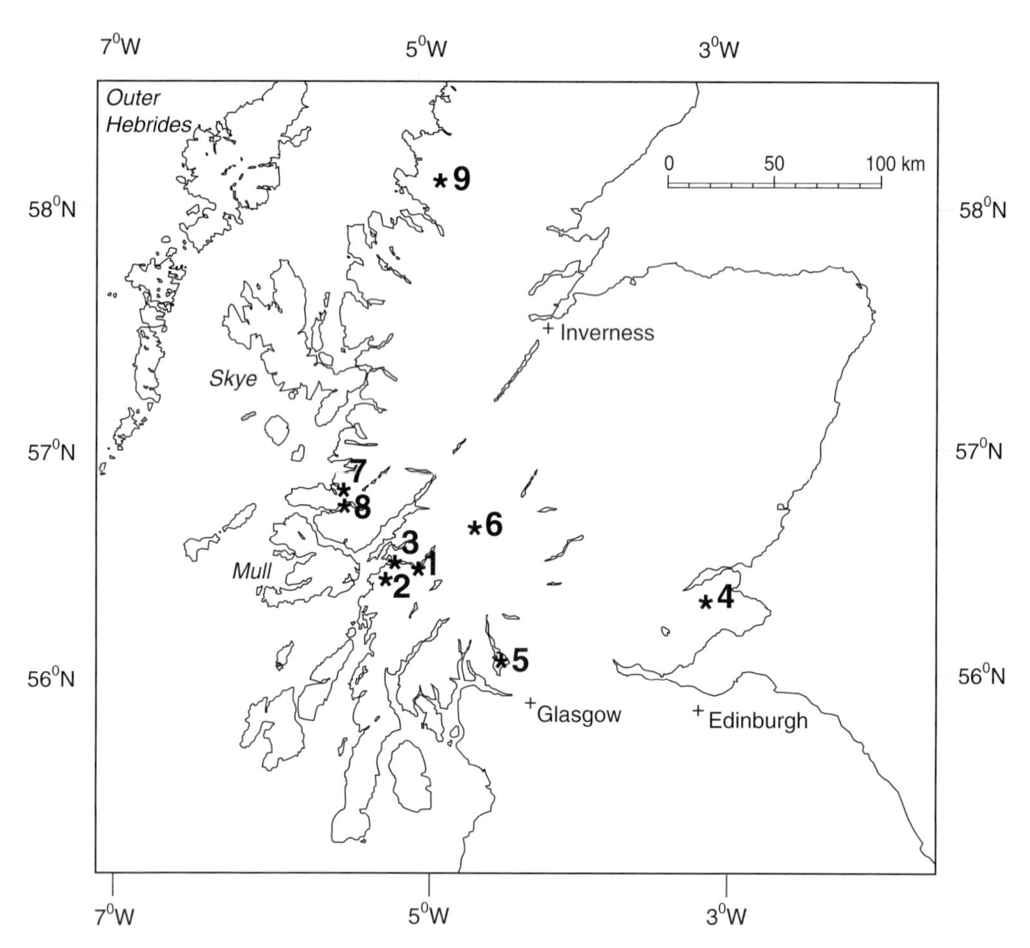

Fig. 1. Location of Loch Etive within Scotland together with other Scottish pollen sites either referred to in the text or in Table 3. 1, Loch Etive (this study); 2, Lochan a'Bhuil Bith (Donner 1957; Macklin *et al.* 2000); 3, Black Moss of Achnacree (Whittington 1983); 4, Black Loch (Whittington *et al.* 1993); 5, Loch Lomond (Dickson *et al.* 1978); 6, Rannoch Moor (Walker & Lowe 1977); 7, Kentra Moss (Ellis & Tallis 2000) and Claish Moss (Moore 1977); 8, Loch Sunart (Cundill *et al.* 2006) and 9, Loch Sionascaig (Pennington *et al.* 1972).

present results with varying degrees of success, it was considered worthwhile to attempt a more detailed pollen investigation of Holocene sediments from a fjord. Such sites provide a more confined and geographically limited sediment trap compared with an open ocean site, and they provide an opportunity to explore the comparability of the pollen record of these marine sediments with adjacent peat bog and freshwater lake sediments. By their physical shape and size, fjords may provide a valuable location for integrating terrestrial and marine pollen records (Nørgaard-Pedersen *et al.* 2006).

The core chosen for analysis was GC004 from the upper basin of Loch Etive (Nørgaard-Pedersen *et al.* 2006) located at 56°27.348′N, 05°11.239′W in the Bonawe Deep at a water depth of 148 m (Fig. 1). The 2.3 m length core was collected in 2000 using a standard British Geological Survey gravity corer (Howe *et al.* 2002; Nørgaard-Pedersen *et al.* 2006) and stored in a cold room at +4 °C before being analysed. Prior to pollen analysis, the core had already been examined for foraminifera, sediment and chemical elements and seven radiocarbon dates had been obtained; a detailed discussion of the stratigraphy and age control is provided by Nørgaard-Pedersen *et al.* (2006). The dates demonstrated that the core contained sediments with an almost complete record of Holocene accumulation. The aim of the study was therefore to explore the potential of obtaining a Holocene pollen record from Scottish fjord sediments and to attempt an evaluation of that record in the context of existing Scottish pollen records.

Methods

Preparation techniques for pollen analysis consisted of the extraction of samples at 5 cm intervals for most of the core; 10 cm intervals had to be employed in a few places because of the lack of material after the collection of samples for other analyses. A 0.5 cm^3 volume of material from each sample was measured by displacement, and a *Lycopodium* tablet (Stockmarr 1972) added to each sample. The chemical treatment of samples followed standard methods outlined in Faegri & Iversen (1989) and Moore *et al.* (1991) and included the use of hydrofluoric (HF) acid to remove much of the quartz present in marine samples. Each sample was then dehydrated before being mounted in silicone fluid. The introduced *Lycopodium* spores along with the fossil pollen and spores in each sample were identified and counted under a light microscope at $\times 600$ magnification. During counting, damaged pollen was differentiated on the basis of folded, split and corroded (Cushing 1964; Lowe 1982); these categories are grouped as 'damaged' for the purposes of this paper. The construction of the pollen diagram was based on percentages of total land pollen values (i.e. excluding spores and aquatic pollen) and was drawn and classified into three zones using the TILIA program (Grimm 1990).

Results

The summed results of the pollen counting presented in Table 1 show that the samples from the upper sediments were very polleniferous and could be counted relatively easily to a total of 400–500 dry land pollen; those from the lower sediments were much less polleniferous and lower pollen totals were accepted. Although there are recognized statistical difficulties with using low pollen totals, other studies of pollen from marine sediments (e.g. Rossignol-Strick 1995; Heusser *et al.* 2000) have accepted low totals for some samples because of very low pollen concentrations. In addition, the sediments from Loch Etive which contained the low total pollen counts demonstrate consistent pollen values between samples suggesting that the data is reliable. Counting of the introduced *Lycopodium* spores from a measured quantity of sediment facilitated the calculation of pollen concentrations because each tablet contained 11 267 $\pm$ 370 *Lycopodium* spores. The pollen concentration results (Table 1) indicate that pollen quantities in the core vary widely between samples with (in general) greater concentrations towards the upper sections of the core and lower concentrations near the base of the core. This is at least partly related to the characteristics of the sediments; coarser sediments are found at depth and finer, organic-rich, sediments are found near the top of the core (Nørgaard-Pedersen *et al.* 2006). The pollen throughout the core was predominantly straightforward to identify and therefore similar in condition to the modern pollen from Loch Sunart sediments (Cundill *et al.* 2006). Although between 53% and 77% of the pollen was damaged (Table 1), there was no substantially greater level of damage than that found in modern moss polsters from the Loch Sunart catchment (Cundill *et al.* 2006) or from freshwater lake sediments in general (Davis *et al.* 1984); very little pollen was unidentifiable. The radiocarbon dates used throughout this article are expressed as uncalibrated years BP. For ease of comparison with terrestrial ages the Loch Etive dates (Table 2), obtained from marine shells, have been corrected for a marine radiocarbon reservoir effect of approximately 400 a (e.g. Cage *et al.* 2006).

Three major pollen assemblage zones identified in the Loch Etive relative pollen diagram (Fig. 2) show significant changes to the vegetation growing in the Loch Etive catchment during the Holocene.

Zone ET1 2.30–1.75 m (c. 9–7.3 ka BP)

The pollen record is dominated by *Betula*, *Pinus* and *Corylus*. Low values of herbaceous taxa are present except for an increase in Poaceae at 1.78 m which may be the result of a low total pollen count.

Zone ET2 1.75–0.65 m (c. 7.3–1.2 ka BP)

This is a period when *Betula*, *Pinus*, *Quercus*, *Alnus* and *Corylus* pollen dominated the pollen record. The lowest numbers of herbaceous taxa within the core are seen in this zone, although values do increase from approximately the mid point of the zone. This is accompanied by a decline, in particular, of *Pinus* and *Quercus*.

Zone ET3 0.63–0.00 m (c. 1.2–0 ka BP)

Reduced tree pollen values (mainly *Pinus* and *Quercus*) and increased values of open habitat taxa, particularly *Calluna* and Poaceae along with the appearance of Cerealia, are the main distinguishing features of this zone.

Interpretation

Zone ET1 dates from soon after the final retreat of the ice from the Scottish Highlands (*c.* 10 ka BP; e.g. Gordon & Sutherland 1993; Edwards & Ralston 2003). Although the lowest two radiocarbon dates show a reversal, the dates themselves are consistent with the period after the rise in *Corylus*

Table 1. *Total land pollen counts together with the pollen concentration (based on the calculation of total pollen cm^{-3} of sediment using standard Lycopodium tablets each containing 11 267 $\pm$ 370 spores) at each sampled depth and the number of damaged pollen (folded, split and corroded – e.g. Lowe 1982) as a percentage of total land pollen*

Depth (cm)	Total land pollen	Pollen concentration	% Damaged pollen
8	424	106 302	66.6
13	517	117 450	59.5
18	436	104 605	54.9
23	520	168 039	59.2
28	490	118 243	58.8
33	505	185 360	62.7
38	468	103 612	56.3
43	527	106 089	69.7
48	180	48 640	65.5
53	505	63 574	59.9
58	301	61 106	76.9
68	269	37 284	71.4
73	269	42 671	75.7
78	170	18 006	62.0
83	281	32 979	70.1
88	185	26 723	68.1
93	248	28 368	59.7
98	94	8406	67.0
103	282	26 588	59.6
108	174	24 816	64.9
113	161	16 053	59.6
118	149	20 105	63.1
123	214	14 267	67.8
128	150	9038	62.7
138	126	13 024	56.3
143	204	9306	55.9
148	155	17 730	58.1
153	192	13 151	55.2
158	129	16 706	53.5
163	108	7397	61.1
168	135	10 243	57.8
173	136	5266	64.7
178	37	3309	70.2
183	176	10 986	69.9
188	51	4 061	64.7
198	56	13 716	67.9
210	108	7979	68.5
220	34	4975	55.9
230	44	7807	61.3

pollen which occurred throughout western Scotland by 8.5 ka BP (e.g. Boyd & Dickson 1986; Edwards & Ralston 2003). The persistent and relatively high values for *Salix* and *Juniperus* are also recorded by many researchers working on western Scotland cores, although the high values of *Pinus* are not recorded for any other site in the area and will be discussed further later.

The lower boundary of zone ET2 can be placed at the beginning of the Mid-Holocene period and is identified in this and all diagrams in Scotland by the distinct expansion of *Alnus* pollen. Various interpretations have been advanced for this feature including climate, human interference and soil changes (Edwards & Whittington 2003). Values of *Quercus* pollen also increase significantly at this time and ET2 overall indicates that a predominantly mixed deciduous forest formed in this area with persistently high values of *Pinus* pollen. The date for the start of the zone does not coincide with a radiocarbon date, but it does fall between 6870 $\pm$ 65 a BP and 8140 $\pm$ 70 a BP which fits with a date of 7.3 ka BP from Black Loch in Fife (Fig. 1). However, as with the Black Loch date, it is earlier than the 6.8 ka BP *Alnus* expansion in southern Scotland and the 6.2 ka BP age for the

Table 2. *Radiocarbon dates for Loch Etive. A marine radiocarbon age correction of 400 years is applied (Cage* et al. *2006; Nørgaard-Pedersen* et al. *2006)*

Depth (cm)	Laboratory number	Reservoir corrected ^{14}C age (^{14}C a BP)	Calibrated age (a BP) (1 standard deviation range)
1	AA-44941	>AD1950	>AD1950
59	AAR-7070	1020 ± 45	930 (965–925)
95	AAR-7069	2930 ± 60	3110 (2960–3210)
145	AA-44944	5110 ± 55	5895 (5925–5855)
163	AAR-7068	6870 ± 65	7680 (7790–7620)
202	AA-7067	8140 ± 70	9045 (9250–9015)
230	AA-44943	7430 ± 50	8315 (8350–8215)

same feature in western Scotland (Whittington *et al.* 1993). There is no clear decline in *Ulmus* pollen values although the radiocarbon date of 5110 ± 55 a BP coincides with a period of low pollen values for this taxon and is a date consistent with those obtained elsewhere in Scotland for this horizon (Edwards & Whittington 2003). The end of zone ET2 occurs where values of *Quercus* and *Pinus* decline and *Calluna* increase. This change just prior to 1020 ± 65 a BP is probably connected with major changes in the land use in the Loch Etive basin brought about by expansion in farming. There is evidence for farming activities in the Loch Etive area before this date with the discovery of prehistoric banks at Achnacree. While not appearing to be directly associated with agriculture, the banks contain pollen in the palaeosols buried beneath them which indicate that some (probably localized) pastoral farming occurred prior to 3309 ± 50 a BP (Whittington 1983).

Zone ET3 covers the last 1.2 ka of history in the area around Loch Etive and shows a pollen record which does not vary greatly. The pollen of *Betula, Alnus, Corylus, Calluna* and Poaceae dominate, indicating a mixed vegetation of scrub woodland with large areas of open ground. The reduced values of *Pinus* and *Quercus* pollen are most likely to be associated with woodland removal linked to an increase in farming activities. Soil deterioration and podsolization may also have influenced tree growth and this could have been accentuated by human activities (Davidson & Carter 2003). The zone is also distinguished by the presence of some Cerealia pollen and by substantial increases in the pollen of herbaceous species; these support the interpretation that farming had become a major activity within the landscape.

Discussion

The three pollen assemblage zones identified from the Loch Etive sequence show general changes in vegetation during the Holocene that can be identified in pollen diagrams throughout Scotland. Zone ET1 reflects the early stages of the Holocene with the dominance of *Betula*, *Pinus* and *Corylus* and the presence of *Juniperus* and *Salix*. The vegetation around Loch Etive then develops into the mixed deciduous woodland of zone ET2, typically found in the Mid-Holocene elsewhere. During this period, open taxa are reduced to the lowest levels found in the Holocene. Zone ET3 demonstrates the final stages of Holocene vegetation development with reduction in woodland and expansion of open habitat taxa reflecting the farming activities of an increased human population.

More detailed comparisons may be made between the Loch Etive results and those from other sites in western Scotland; a preliminary attempt to do so is illustrated in Table 3. However, there are difficulties when making more than generalized comparisons. Few studies have been conducted in the Loch Etive area and the majority of these studies have been carried out on peat bog sediments. From past studies it has been shown that peat bogs often reflect a pollen record dominated by local pollen sources (e.g. Caseldine 1981; Moore *et al.* 1991). Vegetation grows on the bog surface and can contribute substantial quantities of pollen to the pollen record. In contrast lake sediments, especially those derived from the centre of larger lakes, contain more of a regional picture of the vegetation. It should be noted that small lakes, as for peat bogs, may potentially contain over-representation of local pollen from the vegetation growing around the margins of the lake (e.g. Whittington *et al.* 1991; Seppa & Bennett 2003). However, all lakes regardless of their size have one significant difference compared to peat bogs: they receive pollen carried in by streams or overland flow. It has been demonstrated by Bonny & Allen (1984) that this results in a wider range of species represented in the lake pollen record. Plants which produce low quantities of pollen are less likely

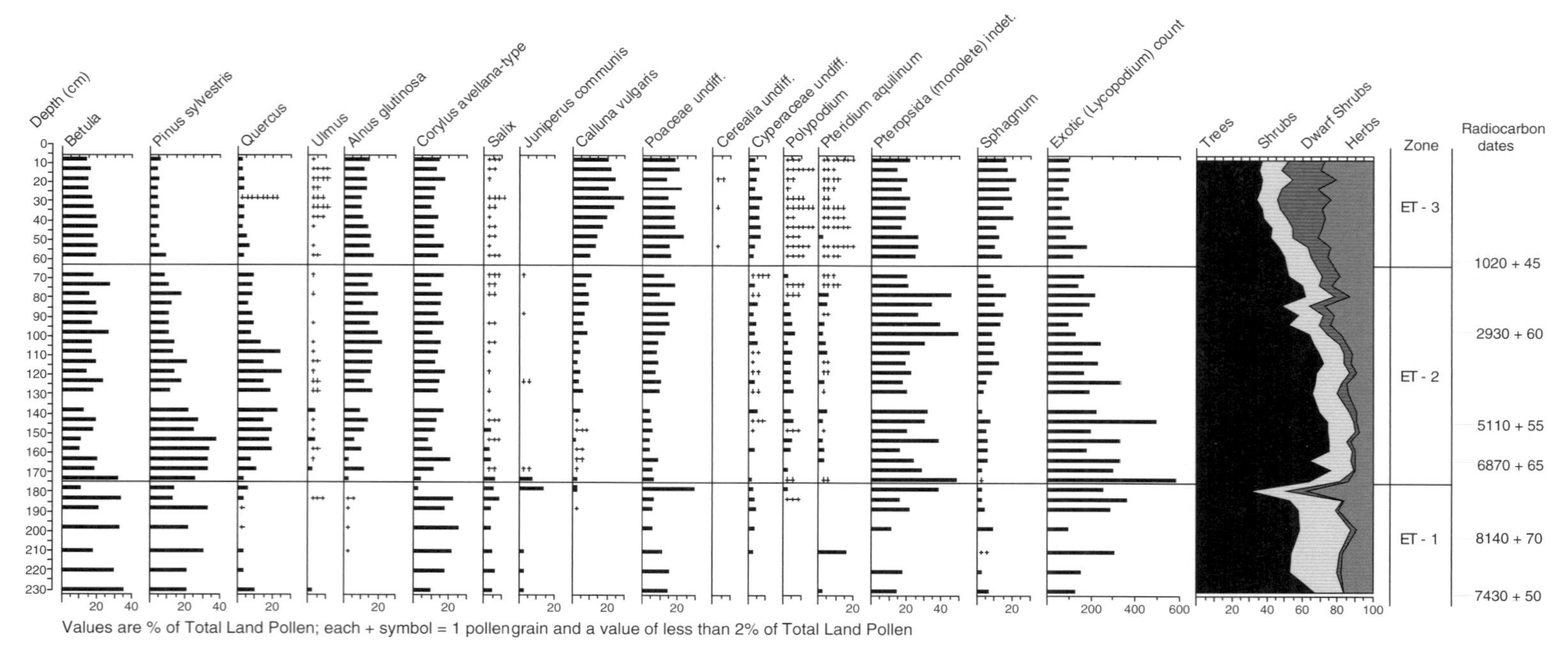

Fig. 2. Relative pollen diagram from Loch Etive. The pollen values are expressed as percentages of total land pollen. Only the major pollen taxa are shown in the diagram in order to reproduce the results in a simplified form. A diagram with all identified pollen taxa is available from the authors. Radiocarbon dates are corrected for a marine reservoir age of 400 years (Cage *et al.* 2006) and the stratigraphic order of these dates is discussed by Nørgaard-Pedersen *et al.* (2006).

Table 3. *Comparisons of pollen zones and radiocarbon dates between the Loch Etive results (this study) and other pollen sites in western Scotland. Pollen zone terminology is based on the original publications*

Loch Etive (this study)	Loch Sionascaig (Pennington *et al.* 1972)	Lochan a'Bhuil (Donner 1957)	Loch Lomond (Dickson *et al.* 1978)	Kentra Moss (Ellis & Tallis 2000)	Claish Moss (Moore 1977)	General vegetation characteristics in western Scotland
ET3		VIII	Highest Poaceae	AP-i AP-ii AP-iii	CM6	Increasingly open habitat as the present day is approached. Wide range of open habitat taxa, dominated by Poaceae, *Calluna* and Cyperacea. Rise in *Plantago lanceolata* at some sites. Lowest Holocene values for woodland taxa are reached during the last 1–2 ka.
1020	NWSVIiii & NWSVIii			1565		
			1700	2150		
2930		VIIb	*Plantago lanceolata* rise and gradual rise in open habitat taxa	AP-v 2245 3035		
ET2	4020		4205		CM5	
5110	NWSVIi NWSVi + ii	——			Ulmus decline	
6870	NWSIV	VIIa	6293		CM4	Maximum extent of woodland during the Holocene. Mixed deciduous taxa in west, central and southern Scotland, *Pinus* in the Highlands
——		——			Alnus rise	
ET1			High *Pinus*		CM3	Developing woodland with mainly *Betula* and *Corylus* but with significant values of *Pinus* at some sites. Open habitat taxa (dominated by Poaceae) also present especially early in the Holocene, alongside *Juniperus* and *Salix*.
7430	7880	VI	7823		CM2	
	NWSIII	V			CM1	
	9474	IV				

Note: All dates shown are conventional radiocarbon ages (a BP); Loch Etive ages are additionally corrected for a marine radiocarbon reservoir age of 400 years (Cage *et al.* 2006; Nørgaard-Pedersen *et al.* 2006).

to be represented in the pollen from a peat site because there is only one pathway (the atmosphere) through which they can reach the bog surface. In contrast, there are at least two pathways into a lake (stream and atmosphere). For example, Bonny (1976) cites results from Ennerdale in the English Lake District where trapping of *Calluna* pollen from the atmosphere gave values of 1% of total pollen whereas pollen trap samples in water yielded values of up to 12%.

Of the three freshwater lake sites relatively close to Loch Etive which have had their sediments analysed for pollen, only one – the site at Lochan a' Bhuilg Bith near Oban (Donner 1957; Macklin *et al.* 2000) – revealed a complete Holocene sequence which has the same general vegetation characteristics as Loch Etive (early Holocene development of a *Betula*, *Corylus* and *Pinus* woodland with *Salix* and a Mid-Holocene mixed deciduous woodland with low herbaceous pollen values). The significant difference in the Etive profile is the high Mid-Holocene values of *Pinus*, a feature which is explored later. The Lochan a' Bhuilg Bith core was radiocarbon dated by Macklin *et al.* (2000). Although it covers the majority of the Holocene it cannot be directly compared to Loch Etive because the full pollen record has not been published. The profile from the earlier paper by Donner (1957) has been tentatively correlated with Loch Etive results (Table 3).

The peat sites examined by Donner (1957) in the Oban area and Moore (1977) and Ellis & Tallis (2000) in the Ardnamurchan area are the only other sites with substantial Holocene deposits relatively close to Loch Etive. Although there are some differences in terms of the percentages of herbaceous and dwarf shrub pollen types, the general patterns of vegetation appear to be the same as Loch Etive. There are radiocarbon dates from Ellis & Tallis (2000) and although the earliest date from their site is only 3035 ± 45 a BP, the dates are consistent with both the pollen changes (reduced values of *Quercus* and *Alnus* and increased values of Poaceae and *Calluna*/Ericaeae) and radiocarbon dates from the upper section of the Loch Etive core.

Despite the fact that other pollen records from western Scotland are broadly the same as those from Loch Etive (Table 3), in one respect there is a significant difference. There are very much higher percentage values for *Pinus* throughout the Loch Etive core than at any other site in the region, except for a short period in the record from Site 3, Lochan Dubh (Donner 1957) when *Pinus* reaches 10–20%. These values are however based on an arboreal pollen sum, whereas Loch Etive values are based on the total land pollen sum. The high values of *Pinus* in the Loch Etive samples may be explained by the nature of the site. The substantial catchment that drains into the fjord includes a large area of highland and it is likely that pine grew in this upland area throughout the Holocene (Bennett 1984; Bridge *et al.* 1990; Tipping 1994). Walker & Lowe (1981) identify *Pinus* and *Betula* as dominating the Rannoch Moor plateau while *Betula*, *Corylus* and *Quercus* are the significant tree species on lower ground further to the west. In addition, the *Pinus* pollen is likely to have been transported into the fjord via streams as well as wind blown; it may have stayed on the surface of the fjord water for a long time before wetting and being incorporated into the fjord sediments. It is known that the bladders of *Pinus* pollen help it to float on water (Bonny & Allen 1984; Faegri & Iversen 1989) and it therefore may have been moved around the surface of the fjord by currents leading to a redistribution across the surface of the fjord. This was suggested in the interpretation of results from Loch Sunart (Cundill *et al.* 2006) where *Pinus* was also abundant in surficial sediments. The other (peat bog) pollen sites in the area are located close to the coast and in an area of predominantly westerly winds. This would have reduced the overall contribution of *Pinus* pollen being blown by the wind from the east. This fact, coupled with the local high pollen values derived from plants growing on or immediately around peat bogs, would have substantially reduced the relative amounts of *Pinus* pollen reaching most of the terrestrial pollen sites discussed above.

Conclusions

The Loch Etive pollen results potentially provide a valuable contribution to an understanding of the vegetation history of western Scotland. Despite the difficulties of counting pollen in the less polleniferous lower sediments from the site, the study has demonstrated a consistent and detailed Holocene record of the vegetation development around Loch Etive. Furthermore, this has been shown to be a regional record which appears to be as valuable as any that may be derived from freshwater loch sediments. Scottish fjords also contain marine fossils and sediments which, in some cases, have been investigated for marine events (see Nørgaard-Pedersen *et al.* 2006). This provides scope for further interpretation of palaeoenvironmental information that would not be possible with freshwater lake or peat bog sediments. The present study has highlighted the need for an integrated investigation of the pollen, marine fossils and sediments from Scottish fjord sediments to assess the possibility of linking the major marine and terrestrial events during the Holocene.

References

Bennett, K. D. 1984. The post-glacial history of *Pinus sylvestris* in the British Isles. *Quaternary Science Reviews*, **3**, 133–155.

Berglund, B. E. 1976. Pollen analysis of core B873 and an adjacent lacustrine section. *Boreas*, **5**, 221–225.

Bonny, A. P. 1976. Recruitment of pollen to the seston and sediment of some Lake District lakes. *Journal of Ecology*, **64**, 859–887.

Bonny, A. P. & Allen, P. V. 1984. Pollen recruitment to the sediments of an enclosed lake in Shropshire, England. *In*: Haworth, E. Y. & Lund, J. W. G. (eds) *Lake Sediments and Environmental History*. Leicester University Press, Leicester, 231–259.

Boyd, W. E. & Dickson, J. H. 1986. Patterns in the geographical distribution of the early Flandrian *Corylus* rise in south-west Scotland. *New Phytologist*, **102**, 615–623.

Bridge, K. D., Haggart, B. A. & Lowe, J. J. 1990. The history and palaeoclimatic significance of subfossil *Pinus sylvestris* in blanket peats from Scotland. *Journal of Ecology*, **78**, 77–99.

Cage, A. G., Heinemeir, J. & Austin, W. E. N. 2006. Marine radiocarbon reservoir ages in Scottish coastal and fjordic waters. *Radiocarbon*, **48**, 31–43.

Caseldine, C. J. 1981. Surface pollen studies across Bankhead Moss, Fife, Scotland. *Journal of Biogeography*, **8**, 7–25.

Cundill, P. R., Austin, W. E. N. & Davies, S. E. 2006. Modern pollen from the catchment and surficial sediments of a Scottish sea loch (fjord). *Grana*, **45**, 230–238.

Cushing, E. J. 1964. Redeposited pollen in late Wisconsin pollen spectra from East-Central Minnesota. *American Journal of Science*, **262**, 1075–1088.

Davidson, D. A. & Carter, S. P. 2003. Soils and their evolution. *In*: *Scotland After the Ice Age: Environment, Archaeology and History 8000BC – AD1000*. Edinburgh University Press, Edinburgh, 45–62.

Davis, M. B., Moeller, R. E. & Ford, J. 1984. Sediment focussing and pollen influx. *In*: Haworth, E. Y. & Lund, J. W. G. (eds) *Lake Sediments and Environmental History*. Leicester University Press, Leicester, 261–293.

De Busk, G. H. 1997. The distribution of pollen in the surface sediments of Lake Malawi, Africa, and the transport of pollen in large lakes. *Review of Palaeobotany and Palynology*, **97**, 123–153.

Dickson, J. H., Stewart, D. A., Baxter, M. S., Drndarsky, N. D., Thompson, R., Turner, G. & Rose, J. 1978. Palynology, palaeomagnetism and radiometric dating of Flandrian marine and freshwater sediments of Loch Lomond. *Nature*, **274**, 548–553.

Donner, J. J. 1957. The geology and vegetation of Late-glacial retreat stages in Scotland. *Transactions of the Royal Society of Edinburgh*, **63**, 221–264.

Edwards, A. & Edelsten, D. J. 1977. Deep water renewal of Loch Etive: a three basin Scottish fjord. *Estuarine and Coastal Marine Science*, **5**, 575–595.

Edwards, A. & Sharples, F. 1986. *Scottish Sea Lochs: A Catalogue*. Scottish Association of Marine Science, Oban, Scotland, Special Publications, **134**.

Edwards, K. E. & Ralston, I. B. M. 2003. *Scotland after the Ice Age: Environment, Archaeology and History 8000BC – AD1000*. Edinburgh University Press, Edinburgh.

Edwards, K. E. & Whittington, G. W. 2003. Vegetation change. *In*: *Scotland after the Ice Age: Environment, Archaeology and History 8000BC – AD1000*. Edinburgh University Press, Edinburgh, 63–82.

Ellis, C. J. & Tallis, J. H. 2000. Climatic control of blanket mire development at Kentra Moss, north-west Scotland. *Journal of Ecology*, **88**, 869–889.

Faegri, K. & Iversen, J. 1989. *Textbook of Pollen Analysis*. 4th edn. J. Wiley & Son, New York.

Gordon, J. E. & Sutherland, D. G. 1993. *Quaternary of Scotland*. Geological Conservation Review Series, Joint Nature Conservation Committee, Chapman and Hall, London.

Grimm, E. C. 1990. TILIA and TILIA GRAPH. PC spreadsheet and graphics software for pollen data. *INQUA, Working Group on Data Handling Methods, Newsletter* **4**, 5–7.

Heusser, L. E., Lyle, M. & Mix, A. 2000. Vegetation and climate of the northwest coast of North America during the last 500 K.Y.: high-level resolution pollen evidence from the northern California margin. *Proceedings of the Ocean Drilling Programme Science Research*, **167**, 217–226.

Howe, J. A., Shimmield, T., Austin, W. E. N. & Longva, O. 2002. Post-glacial depositional environments in a mid-latitude glacially-overdeepened sea loch, inner Loch Etive, western Scotland. *Marine Geology*, **185**, 417–433.

Lowe, J. J. 1982. Three Flandrian pollen profiles from the Teith Valley, Perthshire, Scotland. II Analysis of deteriorated pollen. *New Phytologist*, **90**, 371–385.

Macklin, M. G., Bonsall, C., Davies, F. M. & Robinson, M. R. 2000. Human-environment interactions during the Holocene: new data and interpretations from the Oban area, Argyll, Scotland. *The Holocene*, **10**, 109–121.

McKnight, T. L. & Hess, D. 2008. *Physical Geography; A Landscape Appreciation*. Pearson Education, London.

Mokeddem, Z., Baltzer, A., Clet-Pellerina, M., Walter-Simonnet, A. V., Bates, R., Balut, Y. & Bonnot-Courtoise, C. 2007. Climatic fluctuations recorded in the sediment fillings of Loch Sunart (northwestern Scotland) since 20,000 BP. *Comptes Rendus Geosciences*, **339**, 150–160.

Moore, P. D. 1977. Stratigraphy and pollen analysis of Claish Moss, north-west Scotland: significance for the origins of surface-pools and forest history. *Journal of Ecology*, **65**, 375–397.

Moore, P. D., Webb, J. A. & Collinson, M. E. 1991. *An Illustrated Guide to Pollen Analysis*. 2nd edn. Hodder & Stoughton, London.

Nørgaard-Pedersen, N., Austin, W. E. N., Howe, J. A. & Shimmield, T. 2006. A Holocene sea-loch record in Loch Etive, Western Scotland: hydrographic changes inferred from sediment properties and benthic foraminifera assemblages. *Marine Geology*, **228**, 55–71.

Pennington, W., Haworth, E. Y., Bonny, A. P. & Lishman, J. P. 1972. Lake sediments in northern

Scotland. *Philosophical Transactions of the Royal Society of London*, **B264**, 191–294.

ROSSIGNOL-STRICK, M. 1995. Sea-land correlation of pollen records in the eastern Mediterranean for the glacial-interglacial transition: Biostratigraphy v. Radiometric time-scale. *Quaternary Science Reviews*, **14**, 893–915.

SCOURSE, J., MARRET, F., VERSTEEGH, G. J. M., JANSEN, J. H. F., SCHEFU, B. E. & VAN DER PLICHT, J. 2005. High-resolution last deglaciation record from the Congo fan reveals significance of mangrove pollen and biomarkers as indicators of shelf transgression. *Quaternary Research*, **64**, 57–69.

SEPPA, H. & BENNETT, K. D. 2003. Quaternary pollen analysis: recent progress in palaeoecology and palaeoclimatology. *Progress in Physical Geography*, **27**, 548–579.

STOCKMARR, J. 1972. Tablets with spores used in absolute pollen analysis. *Pollen Spores*, **13**, 615–621.

TIPPING, R. 1994. The form and fate of Scotland's woodlands. *Proceedings of the Society of Antiquaries in Scotland*, **124**, 1–54.

WALKER, M. J. C. & LOWE, J. J. 1977. Postglacial environmental history of Rannoch Moor, Scotland, I. Three pollen diagrams from the Kingshouse area. *Journal of Biogeography*, **4**, 333–351.

WALKER, M. J. C. & LOWE, J. J. 1981. Postglacial environmental history of Rannoch Moor, Scotland, III. Early- and mid-Flandrian pollen stratigraphic data from sites on western Rannoch Moor and near Fort William. *Journal of Biogeography*, **8**, 475–491.

WHITTINGTON, G. W. 1983. A palynological investigation of a second millenium B.C. bank-system in the Black Moss of Achnacree. *Journal of Archaeological Science*, **10**, 283–291.

WHITTINGTON, G. W., EDWARDS, K. J. & CUNDILL, P. R. 1991. Late- and post-glacial vegetational change at Black Loch, Fife, eastern Scotland. *New Phytologist*, **118**, 147–166.

WHITTINGTON, G. W., EDWARDS, K. J. & CUNDILL, P. R. 1993. Black Loch. *In*: GORDON, J. E. & SUTHERLAND, D. G. (eds) *Quaternary of Scotland*. Geological Conservation Review Series, Chapman and Hall, London, **6**, 526–531.

A multiproxy palaeoenvironmental reconstruction of Loch Sunart (NW Scotland) since the Last Glacial Maximum

ZOHRA MOKEDDEM[1], AGNES BALTZER[1]*, EVELYNE GOUBERT[2] & MARTINE CLET-PELLERIN[1]

[1]*M2C, Université de Caen, UMR CNRS/INSU 6143, 2-4 rue des Tilleuls, 14000, Caen, France*

[2]*Université de Bretagne Sud, Centre de recherche Y. Coppens, 56017, Vannes, France*

**Corresponding author (e-mail: agnes.baltzer@unicaen.fr)*

Abstract: Loch Sunart is located on the NW coast of Scotland and contains a sedimentary sequence that records Lateglacial to Holocene climatic variations. A 12 m core MD04-2833 was acquired in the main basin of the loch sampling this sequence. We present the palaeoenvironmental data and palaeoclimatical scenario based on a multiproxy approach using pollen concentrations, sortable silt variation, lithic fraction and marine benthic foraminifera assemblages. These analyses allow the identification of major climate fluctuations such as cooling events. Global temperature decreases are discriminated from local water temperature decreases due to ice-melting processes by the presence of *Elphidium subarcticum* and the assemblage of *Cassidulina obtusa* and *Haynesina germanica*. Two meltwater pulses (MWP) are distinguished, which correspond to the MWP-Ia (15.5–13 cal ka BP) and MWP-Ib (12.2–10.1 cal ka BP). After the maximum water stratification occurred at 7.5 cal ka BP, full marine conditions were established around 6 cal ka BP, which correspond to the highest relative sea-level reached in the loch.

Fjords along the west coast of Scotland, locally known as sea lochs, represent areas of particular interest for the understanding of climatic and environmental variations on both long-term (thousands of years) and short-term (annual and decadal) scales (Cooper & O'Sullivan 1998; Howe *et al.* 2002; Hald *et al.* 2003; Nørgaard-Pedersen *et al.* 2006). Fjords are protected from swell and waves by sills, and therefore provide continuous and high-resolution late Pleistocene and Holocene sedimentary records since the last glacial maximum (Vorren *et al.* 1989; Syvitski & Shaw 1995; Aarseth 1997; Lyså *et al.* 2004). These sedimentary sequences reflect the interaction between terrestrial and marine processes, and therefore preserve past climatic and palaeoenvironmental changes (Syvitski *et al.* 1987; Desloges *et al.* 2002).

Despite its relatively isolated coastal location, the water body contained in Loch Sunart is strongly influenced by exchange with coastal water (Eiríksson *et al.* 2006). Furthermore, because of its location on the NW coast of Scotland, climatic signals within the loch are also closely related to large-scale events in the North Atlantic Ocean which has been shown as a major controller of global climate (Cooper & O'Sullivan 1998). The west coast of Scotland is therefore ideally located to monitor both changes in the influence of the adjacent North Atlantic Current and the impact of the North Atlantic Oscillation (NAO) (Gillibrand *et al.* 1995, 2005). Palaeoenvironmental studies in both Scotland and England which demonstrate this include Langdon *et al.* (2003), Dalton *et al.* (2005), Blundell & Barber (2005) and Macklin *et al.* (2005).

The aims of this paper are (1) to analyse the response of benthic foraminifera to marine environmental change and (2) to compare the benthic foraminifera responses to pollen response that has recorded terrestrial environment changes in order to determine the relative influence of climate forcing on both.

General settings

Loch Sunart is located on the southern side of the Ardnamurchan Peninsula on the west coast of Scotland (Fig. 1a). This east–west orientated loch is 33 km long and has an average width of 1.5 km. The bathymetry includes two major sills that divide the loch into three major basins. The main basin reaches a maximum depth of 124 m (Gillibrand *et al.* 2005) and is separated from the adjacent coastal waters by a 33 m deep sill (relative to mean sea level or MSL). The inner basin contains confined waters as it is separated from the main basin by a shallow sill (500 m long, 600 m wide and 8 m deep below MSL) at the Laudale Narrows. Deep water circulation is generally decoupled from the surface water circulation throughout the loch because of the sill at the entrance to the loch. Exchanges of

From: Howe, J. A., Austin, W. E. N., Forwick, M. & Paetzel, M. (eds) *Fjord Systems and Archives.* Geological Society, London, Special Publications, **344**, 341–353.
DOI: 10.1144/SP344.23 0305-8719/10/$15.00

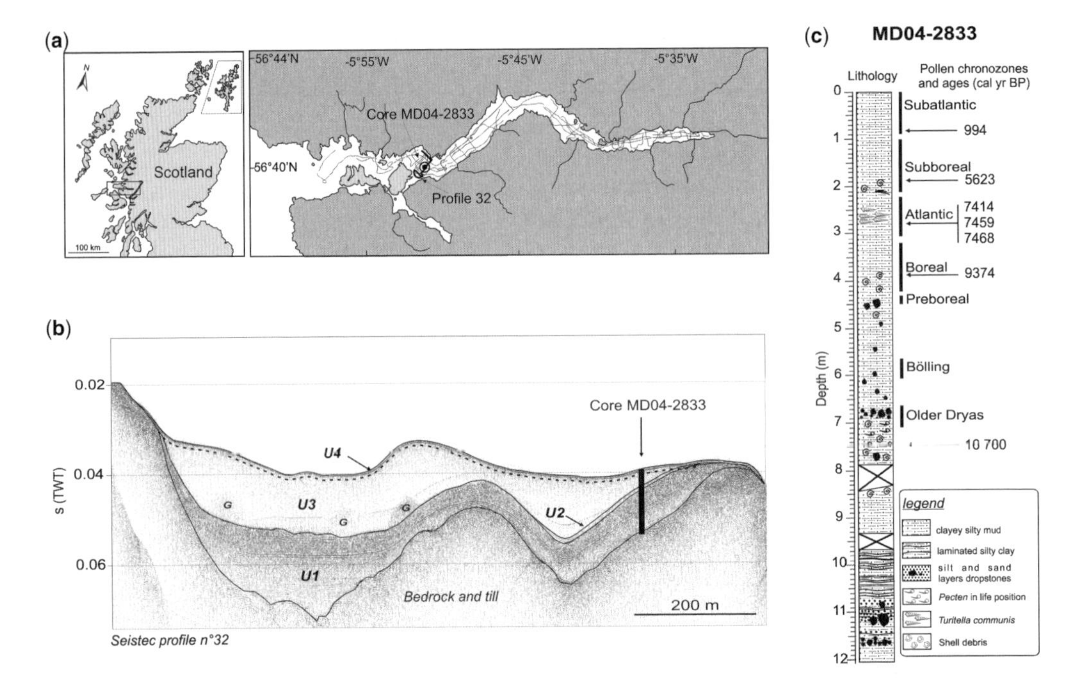

Fig. 1. (**a**) Location of the study area on the NW coast of Scotland, showing core site MD04-2833 in the main basin and the seismic survey profiles acquired by the Boomer Seistec in Loch Sunart. (**b**) Boomer Seistec profile No. 32 showing the four seismic facies (U1, U2, U3 and U4) and the gas blanking (G). Core MD04-2833 (12 m in length) is located at the end of this profile section and has been stopped on the till basement. (**c**) The lithology of the core is presented together with pollen chronozones based on the Mangerud *et al.* (1974) classification. Seven radiocarbon dates (given in calendar years) indicated by arrows along the core have been realized on shells and wood fragments (see also Table 1).

water within the loch occur mostly by diffusive processes and occasionally during advective renewal events (Gillibrand *et al.* 1995, 2005).

Seismic investigations in Loch Sunart conducted by Baltzer *et al.* (2010) reveal sediment sequences deposited since *c.* 22 ka BP where individual sequences represent the depositional history of glacial advance and retreat. The sedimentation history has been inferred from grain-size and pollen records obtained in core MD04-2833 (56°39.6′N; 5°51.6′W) sampled from a water depth of 38 m within the main loch by the giant Calypso core system on board the RV *Marion Dufresne.*

Results obtained from both the core and seismic analysis reveal abrupt climate shifts with vegetation evolution in the loch hinterland indicating five separate cooling intervals identified as rapid climate change events (O'Brien *et al.* 1995; Bond *et al.* 1997; Adams *et al.* 1999; Bond *et al.* 2001; Mayewski *et al.* 2004). In the core, Mokeddem *et al.* (2007) performed pollen analysis calibrated with ^{14}C dating which enabled identification of main chronozones and palaeoclimatic episodes such as: Older Dryas, Bölling, Preboreal, Boreal, Atlantic, Subboreal and Subatlantic (Fig. 1c). No pollen or spores occur below 710 cm; the first evidence of vegetation (tundra and shrubs) was therefore during the Older Dryas. Prior to this, a long section devoid of pollen was interpreted as representing glacial conditions at the Last Glacial Maximum (LGM) period (Mokeddem *et al.* 2007). The forests established after the warming during the early Preboreal, and continued to expand throughout the Holocene until the end of the Subboreal. The Subatlantic period shows the dominance of herbaceous vegetation replacing forest trees. Two sediment intervals completely lack pollen grains (intervals 660–610 cm and 420–560 cm), indicating a reduction in local vegetation and/or disappearance or destruction of pollen grains potentially due to the increase in hydrodynamic regime. In the seismic data, the acoustic substratum showed no internal structures and was interpreted as rock basement. The basal unit 1, characterized by short discontinuous reflections, corresponds to the basal glacial diamicton (till) related to Lateglacial deposits. This unit is described in Loch Ainort (Isle of Skye) by Dix & Duck (2000). The overlying unit 2 includes characteristics of fine acoustic continuous layers, present in the inner basin in particular. This unit is interpreted as glaciomarine sediments deposited during the Younger Dryas (Loch

Lomond Stadial; Binns *et al.* 1974; Boulton *et al.* 1981; Howe *et al.* 2002). Unit 3 corresponds to a thick transparent facies. This unit is related to a significant sediment supply and was deposited during the Early to Mid-Holocene (Binns *et al.* 1974; Boulton *et al.* 1981). Finally, the thin-laminations of unit 4 are correlated with the late Holocene with a decreasing sediment supply.

The seismic and radiocarbon calibrated pollen analyses indicate a complex series of palaeoclimatic changes in the loch since the Older Dryas (Baltzer *et al.* 2010). Five distinct cooling events are generally recognized at 9.8, 8.2, 5.8 and 1.2 cal ka BP and possibly the Little Ice Age, corresponding to the main rapid climate changes described in the literature (Bond *et al.* 1997; Mayewski *et al.* 2004). A previous study (Eiríksson *et al.* 2006) conducted in Loch Sunart (core GC023, 56°40.324′N; 5°50.328′W) shows a site of exceptionally high sediment accumulation (up to 0.5 cm a^{-1}) where proxy records reflect both variability in the physical properties (primarily temperature) of the North Atlantic and strength of westerlies in the region. Eiríksson *et al.* (2006) suggest that coastal temperatures over the last 2 ka were repeatedly influenced by a multidecadal high-frequency 1–2 °C temperature variability that overlies the longer term (centennial) climate signal, where a general cooling trend was seen from around AD 900.

Materials and methods

Radiocarbon dating

Seven samples of marine shells and wood fragments were selected from the core for radiocarbon dating. The dates are corrected for the modern marine reservoir effect of 405 ± 40 a for western Britain (Harkness 1983), calibrated using the CALIB Rev 5.0.1 program (Stuiver & Riemer 1993; Stuiver *et al.* 1998) and reported at two standard deviations (2σ) (Table 1). For the western coast of Scotland and the Orkney Islands, the difference $\Delta R = -33 \pm 93$ ^{14}C a between the 'regional' surface ocean ^{14}C age and the 'global' surface ocean ^{14}C age for the past 5.9 ka is calculated following the recommendation of Reimer *et al.* (2002). Ages are given in 'calibrated years Before Present' (cal a BP).

Grain-size analysis

Grain-size analysis was conducted using a Beckman-Coulter (LS230) laser particle sizer with sediment fractions ranging from 0.04 µm to 2 mm along the entire core. Three grain-size classes were distinguished according to the AFNOR (Association Française de Normalization) distribution (Chamley 2004): clays (<4 µm), silts (4–63 µm) and sands (63–2000 µm). The fractions were then combined to give the percentage weight per grain-size class for each sample using the Wentworth scale. This scale is defined by: $\Phi = -\log 2d$, where d is diameter of the grains in millimetres. Three important grain-size characteristics were chosen for this study: the sortable silt, the lithic fraction and the sorting (Fig. 2).

The sortable silt mean size ($\overline{SS}$) from 10 to 63 µm has been proposed as the most suitable parameter to reconstruct palaeocurrent speed (McCave *et al.* 1995*a*, *b*; Bianchi & McCave 1999; Hass 2002). This fraction is non-cohesive and easily sorted by deep currents during events of resuspension (McCave *et al.* 1995*a*); it is therefore more closely linked to hydrodynamic conditions than to sediment supply. High silt values therefore represent relatively high near-bottom current speeds and low values are related to near-bottom

Table 1. *Radiocarbon dates of mollusks shells and wood fragment in MD04-2833*

Laboratory code[a]	Depth (cm)	Material dated	Radiocarbon age (year BP)	Corrected[b] (calibrated) age (cal year BP) (Stuiver *et al.* 1998)	±2σ[c] (cal year BP)
Poz-23471	81	*Turritella communis*	1405 ± 30	994	777–1211
Poz-23648	198	Wood fragment	4885 ± 35	5623	5584–5663
Poz-1054	265	*Turritella communis*	6910 ± 40	7414	7325–7503
Poz-23579	270	Wood fragment	6520 ± 40	7459	7411–7507
Poz-23472	271	*Turritella communis*	6950 ± 40	7468	7389–7548
Poz-13368	385	*Pecten sp*	8710 ± 50	9374	9262–9447
UL-2853	776	*Pecten maximus*	14 020 ± 210	16 760	16 067–17 454

[a]Poz indicates Poznan Laboratory (Poland) and UL indicates radiocarbon dating laboratory of University of Laval (Canada).
[b]Marine reservoir correction: 405 ± 40 year (Harkness 1983).
[c]Standard deviation (Stuiver *et al.* 1998).

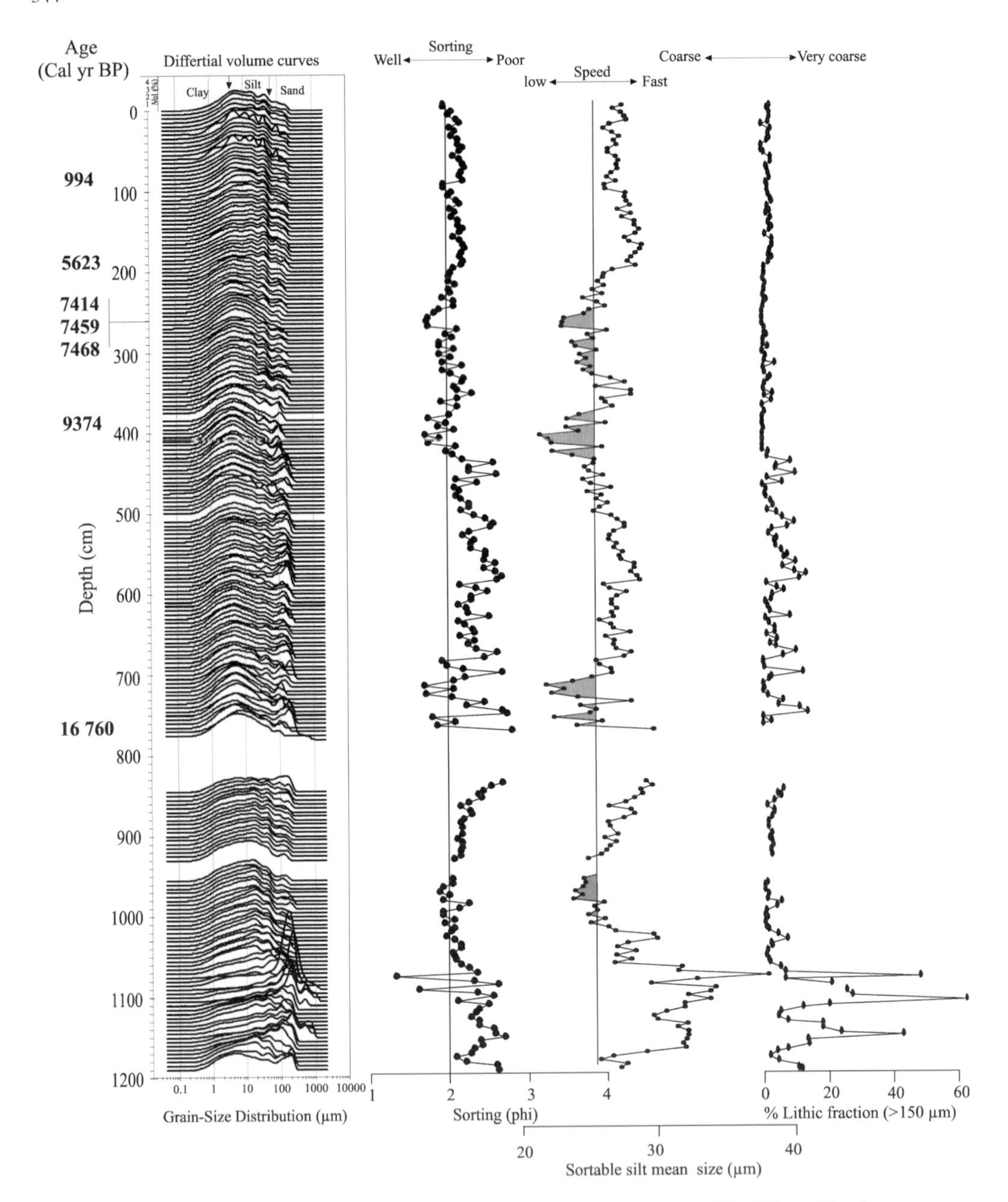

Fig. 2. Grain size distribution (Coulter Counter) data of core MD04-2833 sediments. The differential volume distribution curves have been stacked and positioned at their core depth and the volume (% scale) is indicated for the core top sample (0–1 cm) only. The sorting (phi) plotted curve is shown with well/poor sorting limit. This limit is used to determine the palaeocurrent speed limit within the sortable silt mean size (μm) curve. The lithic size fraction is plotted using the sandy fraction (>150 μm), and shows that the levels rich in coarse grains correspond to glacier melting periods.

flows with relatively low speed. McCave *et al.* (1995*a*) have further suggested that low-speed near-bottom currents characterize cold periods and high-speed currents warm periods.

The second important characteristic is the lithic fraction that corresponds to grains, which have a diameter greater than 150 μm. The percentage of this fraction was calculated for each sample in the core. This fraction represents the coarse terrestrial influx related to the glacial erosion during glacier retreat.

The third important characteristic, the sorting (*So*), is calculated using a method proposed by Chamley (2004). It is used to indicate the global sediment amount sorted during meltwater flows into the sea-loch basin.

Foraminifera analyses

The foraminifera study was based on selected species of 87 samples taken from the core. Approximately 20 g of sediment samples were washed through 125 μm and 63 μm sieves. Residues were oven dried (40 °C) and dry sample weights recorded. The 125 μm and 63 μm fractions were picked under a binocular at 60× magnification to achieve foraminifera counts in excess of 300 specimens where possible (Buzas 1990). Foraminifera determinations were based largely upon the taxonomic concepts (Murray 1971, 2000, 2003; Haynes 1973; Alve & Murray 2001). Palaeoenvironmental interpretations were based on the ecological affinities of the foraminifera (strategies of life including infaunal, epifaunal, epiphytic, oxic, dysoxic, anoxic, bathymetrical distribution, temperature and salinity) and on the ecomorphological characteristics associated with the morphogroups analysed following the work of Murray (1991) and Goubert *et al.* (2001). The ecomorphological groups were defined according to Severin (1983), Bernard (1986), Corliss & Chen (1988) and Duleba *et al.* (1999): epifaunal (biconvex, planoconvex trochospiral and rounded trochospiral forms); epifaunal or shallow infaunal (rounded trochospiral and rounded planispiral forms); and infaunal (elongate fattened, fattened tapered, tapering-flaring, tapered and cylindrical forms).

Results

Grain-size analysis and hydrodynamics

The grain-size analyses show a homogeneous distribution with a predominance of clay and silt throughout the core, except at its base where sand predominates (Fig. 2). Occasional gravel strata of different grain sizes are also present and are described by Mokeddem *et al.* (2007).

The size-frequency distribution reveals four different patterns (Fig. 2). The bottom section shows a multimodal distribution, with the main mode tending towards sand. The overlying segment (1020–740 cm) shows fine sand. In the next section (740–420 cm) the sand tends to be coarser. The uppermost segment shows a very homogeneous distribution where the main mode range lies between clay and fine silt.

The $\overline{SS}$ series (Fig. 2) show changes in current speed with a succession of high and low current speeds recorded within the core. Four intervals show clear decreases in current speed: 1020–960 cm, 770–710 cm, 430–380 cm and 330–240 cm. The lowest interval is difficult to interpret because it occurs immediately above an abrupt increase in grain size with a basal section rich in gravel, and is poorly sorted. The second interval (770–710 cm) covers the cooler period identified in the pollen assemblage as the Older Dryas chronozone. The third interval (430–370 cm) corresponds to the Boreal chronozone and the last interval (330–250 cm) is related to an abrupt event during the Atlantic period. The latter three low current speed intervals correspond to well-sorted sediments.

Benthic foraminifera assemblages

Benthic foraminifera data are plotted for species and assemblages with occurrences of >2% throughout the uppermost 1000 cm of the core (Fig. 3). As environmental indicators, we only considered and interpreted species and assemblages of foraminifera with a frequency >5%. Infrequently occurring taxa (<5% relative abundance) are omitted since they have an insignificant effect on the formation of the major groups (Kovach 1987, 1989).

The foraminifera diagram (Fig. 3) shows four main zones from the bottom of the core upwards.

The first zone (Z-1: 1000–710 cm) is relatively poor in species. *Elphidium clavatum* alone represents the major part of the calcareous species with a mean value of 48% and a maximum of 80% of the total species found in this zone. Porcelaneous species represented by *Quinqueloculina dunkerquiana* and *Quinqueloculina semunila* reach a minimum of 10% whereas agglutinated species are nearly absent.

In the second zone (Z-2: 710–430 cm), species richness is much higher and gradually increases upwards. *Elphidium clavatum* remains the predominant species with a major abundance peak at 600 cm. The second most abundant species, *Cassidulina obtusa*, varies between 20% and 40%. These variations are comparable to those shown by *Haynesina germanica*. The expansion levels of *Cassidulina obtusa* and *Haynesina germanica* alternate with the decrease in the frequencies of *Bulimina marginata, Ammonia tepida* and *Ammonia gp. beccarii.*

The third zone (Z-3: 430–190 cm) is marked by an expansion of *Bulimina marginata* accompanied by *Ammonia tepida, Ammonia gp. Beccarii* and *Hyalinea balthica*, whereas all species decrease and almost disappear.

The fourth zone (Z-4: 190–0 cm) is characterized by the development of agglutinated species

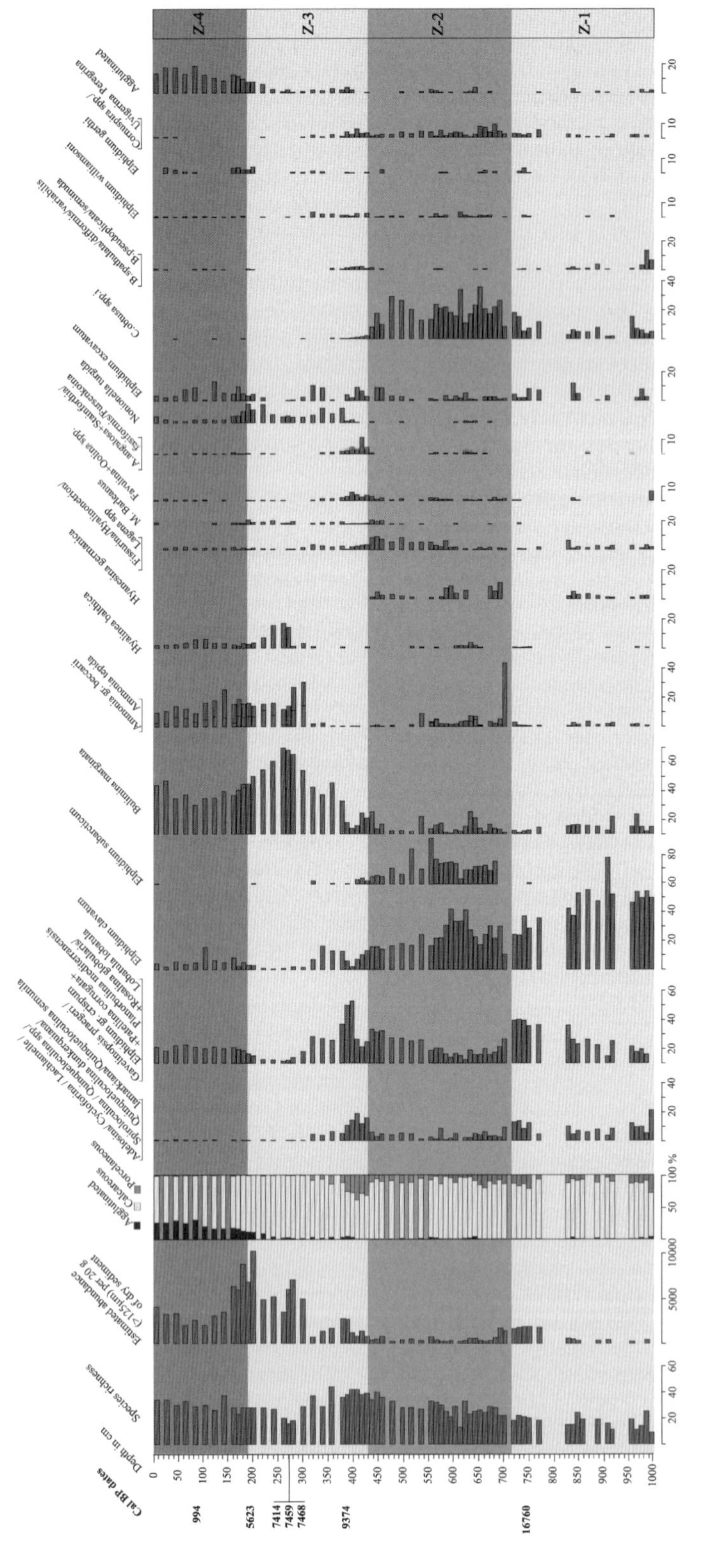

Fig. 3. Stratigraphic distribution of benthic foraminifera assemblages (percentages of total benthic faunas) in core MD04-2833. Four domains are delimited by global species changes.

(mainly *Textularia sagitulla* and *Textularia truncata*). Calcareous species with *Bulimina marginata*, *Ammonia tepida* and *Ammonia gp. beccarii* remain the most dominant taxa. The estimated fauna number per 20 g sediment exceeds 10 000 individuals at 200 cm depth in the core. Above this level, the number decreases and becomes stable from 140 cm to the top of the core. The dominant species are also stabilized at this level.

Palaeoenvironmental interpretation and discussion

The combination of the results obtained in this work with previous results allows us to better constrain the climate changes affecting both the continental and coastal domains on the west coast of Scotland from the LGM to the Late Holocene.

Results from the pollen concentration, three-benthic foraminifera assemblages and two of the grain-size characteristics are plotted on Figure 4.

The three-benthic foraminifera assemblages are established according to their ecological affinity. The first assemblage represents cooling indicators species: *Elphidium clavatum, Elphidium subarcticum* and *Hyalinea balthica* (e.g. Bock 1971; Lukina 2001; Murray 2006). The second assemblage represents confining indicators species: *Bulimina marginata, Ammonia tepida* and *Ammonia gp. Beccarii* (e.g. Fontanier *et al.* 2002; Scourse *et al.* 2002). The third assemblage represents coarse grain indicators species: *Haynesina germanica* and *Cassidiluna obtusa* (e.g. Gooday 1986; Alve & Murray 1994; Murray 2006). A comparison of the different methods presented here has allowed for differentiation of the core into four discrete intervals that are interpreted by four successive palaeoenvironments: (1) glacial (before *c.* 16 cal ka BP); (2) deglacial (Bølling/Allerød and Preboreal warming); (3) global warming (Boreal and Atlantic); and (4) transition to modern periods (Subboreal and Subatlantic) (Fig. 4).

The pollen data only provide details on the climate variations in the upper half of the core, whereas the foraminifera curves record climatic variations throughout the entire core providing information even during cold periods. An overall increase of the three cold species *Elphidium clavatum, Elphidium subarcticum* and *Hyalinea balthica* indicate sea-loch cooling *c.* 16–10 cal ka BP. The occurrence of numerous sharp diversions in the

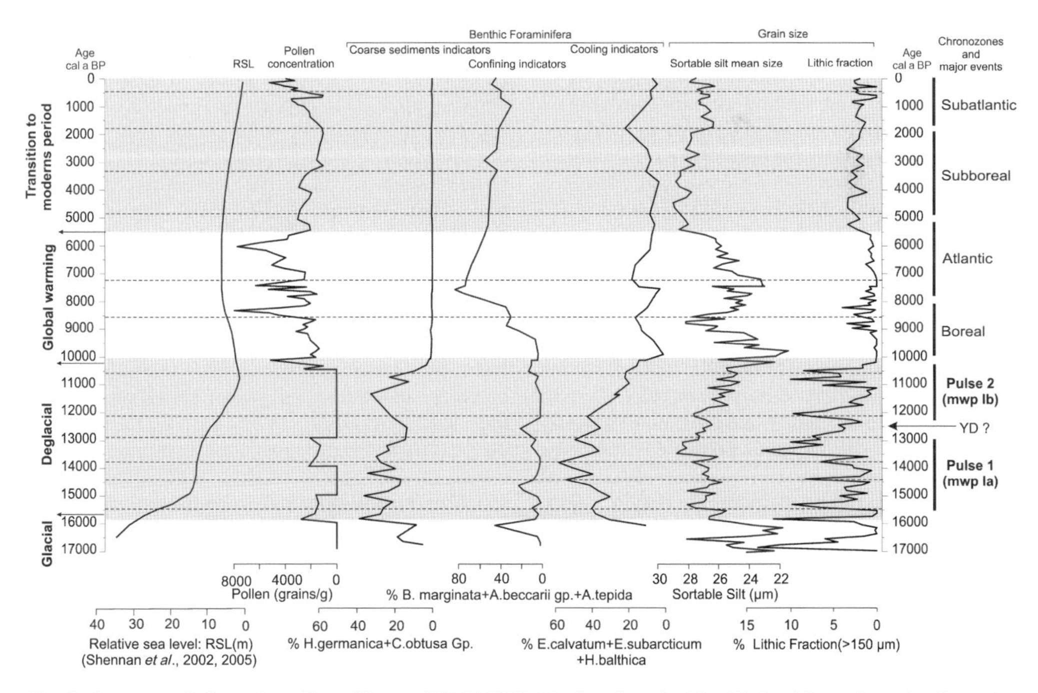

Fig. 4. Summary of all proxies collected in core MD04-2833. The four domains identified on Figure 3 are also shown in this figure. Benthic foraminifera are represented by three curves representing three assemblages: (1) indicating coarse sediments (*Cassidulina obtusa, Haynesina germanica*); (2) indicating cold temperature conditions (*Elphidium clavatum, Elphidium subarcticum* and *Hyalinea balthica*); (3) indicating confined conditions (*Bulimina marginata, Ammonia gp. beccarii* and *Ammonia tepida*). Dashed lines correspond to peaks of the curve illustrating benthic foraminifera cold species assemblage.

data (Fig. 4) probably implies abrupt and rapid decreases of water temperature.

Palaeoenvironment during c. 19–16 cal ka BP (before the Older Dryas)

The thick lithic bed observed at the base of the core (Fig. 4) is correlated with the basal diamicton deposited during the LGM (Baltzer *et al.* 2010). Neither pollen or faunal remains were found at this level, confirming that ice cover was probably the dominant landscape at this time.

The predominance of *Elphidium calvatum* above 1000 cm in the core (Fig. 4) reveals particularly low temperature conditions since this species can survive near or under ice (Febo 1999; Murray 2006). The occurrence of *Haynesina germanica* and *Cassidiluna obtusa* (Fig. 4) coincides with coarse sediments (sands and gravels) shown in the grain-size distribution (Fig. 2). This sequence is therefore interpreted as a time of ice-sheet melting, probably towards the end of this period (at *c.* 16 cal ka BP).

Palaeoenvironment during 16–10 cal ka BP (Older Dryas to Preboreal)

The pollen concentration curve shows the absence of pollen except at two intervals corresponding to the Older Dryas and Bølling chronozones (Figs 1 & 4). These periods are characterized by cold conditions as confirmed by the dominance of *Elphidium clavatum*, *Elphidium subarcticum* and *Hyalinea balthica*. However, these conditions were not sufficient to destroy tree species. Rather, the absence of pollen grains is explained by non-preservation where the pollen grains have been destroyed by strongly energetic currents and/or sediment remobilization rather than by low temperatures.

During this period (15.5–10.1 cal ka BP), the presence of *Elphidium subarcticum* (Fig. 3) indicates a reduction of salinity and creation of a brackish water layer as this species prefers environments such as estuaries with rich terrestrially derived organic matter input (Korsun & Polyak 1989; Febo 1999). This brackish water layer could be related to the important freshwater input caused by the increased runoff following the glacial retreat. The melting process may have triggered runoff processes enhanced by thin vegetation cover (tundra and shrubs), implying high terrestrial inputs to the sea-loch basins. The melting process is confirmed by the expansion of *Haynesina germanica* which lives in an area with high continental influences (Murray 2006). *Cassidulina obtusa* is also associated with coarse sediment input together with high organic input at a time of cold temperature conditions (Sejrup *et al.* 1980; Gooday 1986; Mackensen 1987; Gooday & Lambshead 1989; Kuijpers *et al.* 1998; Rasmussen *et al.* 1999; Murray 2006). This is in agreement with the destruction of pollen grains caused by a runoff event.

The assemblage of the two species *Haynesina germanica* and *Cassidulina obtusa* show many episodes of melting marked by numerous times of coarse sediment input (Fig. 4). These intervals can be grouped into periods. The first spans the time period 15.5–13 cal ka BP. It is comparable to the meltwater pulse Ia (MWP-Ia) (15–14 cal ka BP) identified in the North Atlantic (Keigwin & Lehman 1994; Maslin 1995) and also in Barbados (Atlantic Ocean, Caribbean Sea) by Fairbanks (1989). The pulse Ia clearly begins with the warming period of the Bølling determined by the pollen chronozone (Fig. 1c). A second period (12.2–10.1 cal ka BP, pulse 2), is observed in the sediment grain size data together with the foraminifera indicators. This period shows two major times of assemblage expansion for *Haynesina germanica* and *Cassidulina obtusa* at 11.3 and 10.7 cal ka BP (Fig. 4). We suggest a correlation of this second pulse to meltwater pulse Ib (MWP-Ib) described by Fairbanks (1989).

The transition between meltwater pulses Ia and Ib is marked by an abrupt disappearance of *Haynesina germanica* and *Cassidulina obtusa* that appears to correlate to a general reduction in terrestrially derived input. This hypothesis is reinforced by the decrease in *Elphidium subarcticum* (Fig. 3). Furthermore, the expansion of confined foraminifera species such as *Bulimina marginata*, *Ammonia tepida* and *Ammonia gp. Beccarii* (Fig. 4) during this period is an additional indication of oxygen depletion correlated with the stratification of the water column (Debenay *et al.* 1997; Evans *et al.* 2002; Fontanier *et al.* 2002; Scourse *et al.* 2002). This expansion also coincides with the transition from the melting process and re-establishment of colder conditions on glacier re-advance during the Younger Dryas (12.5 cal ka BP). After 10.1 cal ka BP, the deglaciation continues and the retreat of continental ice caps is illustrated by the synchronous disappearance of gravels and coarse-grained indicator species (*Haynesina germanica* and *Cassidulina obtusa*) (Fig. 4).

Palaeoenvironment during 10.1–0 cal ka BP (Boreal to Subatlantic)

From 10.1 cal ka BP to the present, the pollen data indicate a major transition to a continental environmental with general climatic warming during 10–5.5 cal ka BP (Boreal and Atlantic chronozones). This warming reaches a maximum at 7.5 cal ka BP (mid-Atlantic, Climatic Optimum), punctuated only by one abrupt cooling event interpreted as the 8.2 cal ka BP event (Baltzer *et al.* 2010).

The foraminifera data (Fig. 4) indicate that the confined species are dominant with maximum numbers noted at 7.5 cal ka BP, indicating a transition from normal to anoxic conditions suggesting water stratification. The water stratification caused the disappearance of many foraminifera species (Fig. 3) and was likely responsible for the decrease of bottom-current speeds indicated by the grain size $\overline{SS}$ data (Figs 2 & 4).

The maximum flooding sea level reached at 7.5 cal ka BP (Fig. 4) constituted the transition between the previous local influences (glacial/deglacial) and the modern oceanic global conditions. The expansion of agglutinated species, characteristic of all modern fjord systems, indicates the establishment of full marine conditions after 5.5 cal ka BP (Fig. 3). The interval from 5.5 cal ka BP to the present day does not demonstrate significant variations in foraminifera (Fig. 4), except at *c.* 1.7 cal ka BP for the assemblage of *Elphidium clavatum* and *Hyalinea balthica*. This abrupt expansion suggests a decrease in water temperature probably related to a rapid climatic change cooling event.

The interpretations are illustrated fully in Figure 5 with block diagrams, where each block shows the evolution of marine and continental domains though time.

The maximum extension of the British–Irish ice sheet is reached *c.* 20 cal ka BP and this glacial environment lasted until 17–16 cal ka BP (Fig. 5a). The main basin of Loch Sunart was probably covered by ice, and the relative sea level is estimated to be 20 m above present-day MSL for this period.

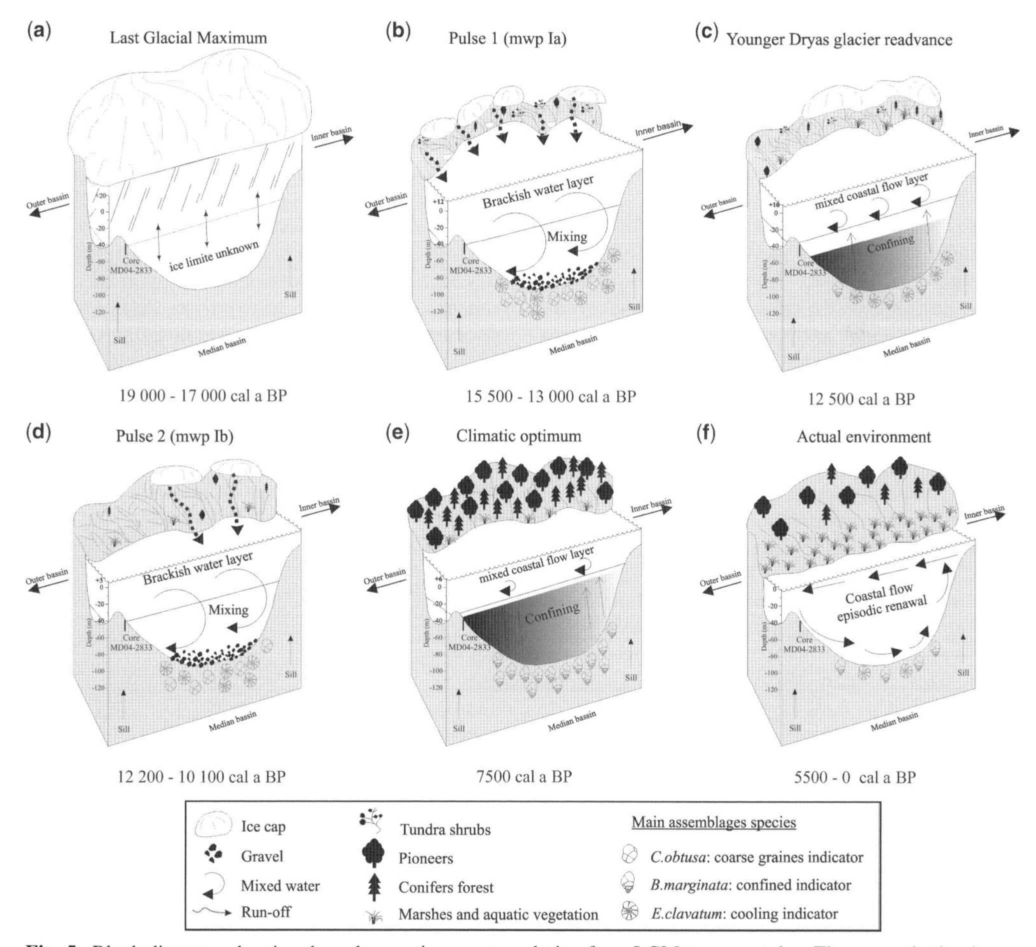

Fig. 5. Block diagrams showing the palaeoenvironment evolution from LGM to present day. The reconstitution is inferred using a combination of all proxies interpretation: ice cover, vegetation, fauna, hydrodynamism and sedimentation.

The final ice-sheet retreat occurred during 15.5–13 cal ka BP (Fig. 5b). This period corresponds to pulse 1 (MWP-Ia). The collapse of the glaciers caused a significant meltwater flow into the fjord basins. These freshwater flows created a thick brackish layer (below the outer sill) that interrupted the water renewal in the main basin. The relative sea-level decreased and reached 12 m above MSL.

At 12.5 cal ka BP (Fig. 5c), the deglaciation of the ice sheet is interrupted by the Younger Dryas glacier re-advance. The limit of the re-advance of the glacier reached the inner basin up to the Laudale Narrows. The interruption of the melting process and the presence of freshwater in the basin induced water stratification. The relative sea level reached 10 m above MSL due to the glacio-isostatic effect.

The retreat of the glaciers (Fig. 5d) at the end of the Younger Dryas was accompanied by significant meltwater flows. This episode corresponds to pulse 2 (MWP-Ib) which has environmental characteristics similar to those described for pulse 1 (MWP-Ia). The relative sea level (3 m above MSL) shows a sharp decline due to the acceleration of isostatic rebound after the gradual disappearance of glaciers.

At 10.1 cal ka BP (Fig. 5e), the permanent warming enhanced the development of a dense vegetation. The net expansion of the boreal forest and pioneer trees during the Atlantic chronozone is correlated with the Climatic Optimum. This period is synchronous to sea level rising (6 m above MSL) during 7.4–7.5 cal ka BP (Shennan *et al.* 2002, 2005). The increase of the water depth above the outer sill should have facilitated the deep-water renewal. However, the ice melting during the early Holocene generated large freshwater flows, leading to seawater dilution below the sill: the water stratification remains very important. The confined conditions probably persisted from 8 to 6 cal ka BP with a maximum at 7.4 cal ka BP.

At 6 cal ka BP (Fig. 5f), exchanges with seawater started again after a long confining period. The freshwater flows decreased significantly, enabling more frequent deep-water renewal. The basin system became definitely connected to the North Atlantic Ocean.

Conclusion

Data from Scottish sea lochs for pollen and foraminifera can provide valuable information for an assessment of the evolution of land and marine areas in response to climatic changes from the Older Dryas to present day. Pollen analysis highlights five cooling events during the Holocene (9.8, 8.2, 5.8 and 1.2 cal ka BP and the Little Ice Age). However, pollen records show discontinuous variations because of their scarcity during glacial period and non-preservation during the beginning of the deglaciation period. On the contrary, foraminifera provide continuous information on environmental and climatic variations from the Older Dryas to present day. Abrupt cooling of the water column suggesting cold conditions (at 8.5, 7.2, 4.8, 3.4 and 1.7 cal ka BP and 450 cal a BP) are expressed by the expansion of *Elphidium clavatum*, *Elphidium subarcticum* and *Hyalinea balthica* and likely correspond to cool periods recorded in North Atlantic records by Bond *et al.* (1997) and Mayewski *et al.* (2004). Moreover, it is possible to discriminate global temperature decreases from local temperature records resulting from glacial ice-melting processes. Ice melting is accompanied by a decrease in water salinity and an increase in the presence of *Elphidium subarcticum*. These periods are accompanied by an important supply of coarse sediment that containa an assemblage of *Cassidulina obtusa* and *Haynesina germanica*. Based on these criteria, two meltwater pulses have been classified: the first (15.5–13 cal ka BP) is comparable to the MWP-Ia and the second (12.2–10.1 cal ka BP) is comparable to the MWP-Ib, described by Fairbanks (1989).

After the maximum water stratification at 7.5 cal ka BP, full marine conditions are established around 6 cal ka BP which corresponds to the highest relative sea level reached in this loch during the Holocene. The fjord environment demonstrates the potential for similar places to provide data of significance in foraminifera analyses, pollen analyses and sedimentological character. These data can be combined to yield a complete record of the climatic changes within the marine and the terrestrial domains during both cool and warm periods.

The authors acknowledge significant contributions from A. V. Walter-Simonnet and J.-P. Simonnet in the grain-size analyses. We wish to thank C. Bonnot-Courtois for her very helpful comments on the manuscript. We are also grateful to M. Rabineau and P. L. Gibbard for their review and valuable observations. Finally, we thank the crew of the RV *Marion Dufresne* who were a great support during the core acquisition.

References

Aarseth, I. 1997. Western Norwegian fjord sediments: age, volume, stratigraphy, and role as temporary depository during glacial cycles. *Marine Geology*, **143**, 39–53.

Adams, J., Maslin, M. & Thomas, E. 1999. Sudden climate transitions during the Quaternary. *Progress in Physical Geography*, **23**, 1–36.

Alve, E. & Murray, J. W. 1994. Ecology and taphonomy of benthic foraminifera in a temperate mesotidal inlet. *Journal of Foraminiferal Research*, **24**, 18–27.

ALVE, E. & MURRAY, J. W. 2001. Temporal variability in vertical distributions of live (stained) intertidal foraminifera, southern England. *Journal of Foraminiferal Research*, **31**, 12–24.

BALTZER, A., BATES, R., MOKEDDEM, Z., CLET-PELLERIN, M., WALTER-SIMONNET, A.-V., BONNET-COURTOIS, C. & AUSTIN, W. E. N. 2010. Using seismic facies and pollen analyses to evaluate climatically driven change in a Scottish sea loch (fjord) over the last 20 ka. *In*: HOWE, J. A., AUSTIN, W. E. N., FORWICK, M. & PAETZEL, M. (eds) *Fjord Systems and Archives*. Geological Society, London, Special Publications, **344**, 355–370.

BERNARD, J. M. 1986. Characteristic assemblages and morphologies of benthic foraminifera from anox, organic-rich deposits: Jurassic through Holocene. *Journal of Foraminiferal Research*, **16**, 207–215.

BIANCHI, G. G. & MCCAVE, I. N. 1999. Holocene periodicity in North Atlantic climate and deep-ocean flow south of Iceland. *Nature*, **397**, 515–517.

BINNS, P. E., HARLAND, R. & HUGHES, M. J. 1974. Glacial and post glacial sedimentation in the sea of the Hebrides. *Nature*, **248**, 751–754.

BLUNDELL, A. & BARBER, K. 2005. A 2800-year palaeoclimatic record from Tore Hill Moss, Strathspey, Scotland: the need for a multi-proxy approach to peat-based climate reconstructions. *Quaternary Science Reviews*, **24**, 1261–1277.

BOCK, W. D. 1971. Paleoecology of section cored on the Nicaragua Rise, Caribbean Sea. *Micropaleontology*, **17**, 181–196.

BOND, G., SHOWERS, W. *ET AL.* 1997. A pervasive millennial-scale cycle in North Atlantic Holocene and glacial climates. *Science*, **278**, 1257–1266.

BOND, G., KROMER, B. *ET AL.* 2001. Persistent solar influence on North Atlantic climate during the Holocene. *Nature*, **294**, 2130–2136.

BOULTON, G. S., CHROSTON, P. N. & JARVIS, J. 1981. A marine seismic study of Late Quaternary sedimentation and inferred glacier fluctuations along western Inverness-shire, Scotland. *Boreas*, **10**, 39–51.

BUZAS, M. A. 1990. Another look at confidence limits for species proportions. *Journal of Paleontology*, **65**, 842–843.

CHAMLEY, H. 2004. *Bases de Sédimentologie*. Editions DUNOD, France, Sciences Supplement.

COOPER, M. C. & O'SULLIVAN, P. E. 1998. The laminated sediments of Loch Ness, Scotland: preliminary report on the construction of a chronology of sedimentation and its potential use in assessing Holocene climatic variability. *Palaeogeography, Palaeoclimatology, Palaeoecology*, **140**, 23–31.

CORLISS, B. H. & CHEN, C. 1988. Morphotype patterns of Norwegian Sea deep-sea benthic foraminifera and ecological implications. *Geology*, **16**, 716–719.

DALTON, C., BIRKS, H. J. B. *ET AL.* 2005. A multi-proxy study of lake-development in response to catchment changes during the Holocene at Lochnagar, north-east Scotland. *Palaeogeography, Palaeoclimatology, Palaeoecology*, **221**, 175–201.

DEBENAY, J.-P., BECK EICHLER, B., GUILLOU, J. J., EICHLER-COELHO, P., COELHO, C. & PORTO-FILHO, E. 1997. Comportements des peuplements de foraminifères et comparaison avec l'avifaune dans une lagune fortement stratifiée: la Lagoa da Conceiçao (SC, Brésil). *Revue Paléobilogie, Genève*, **16**, 55–75.

DESLOGES, J. R., GILBERT, R., NIELSEN, N., CHRISTIANSEN, C., RASCH, M. & ØHLENSCHLÆGER, R. 2002. Holocene glacimarine sedimentary environments in fiords of Disko Bugt, West Greenland. *Quaternary Sciences Reviews*, **21**, 947–963.

DIX, J. K. & DUCK, R. W. 2000. A high-resolution seismic stratigraphy from a Scottish sea loch and its implications for Loch Lomond stadial deglaciation. *Journal of Quaternary Science*, **15**, 645–656.

DULEBA, W., DEBENAY, J.-P., BECK EICHLER, B. & MICHAELOVITCH DE MAHIQUES, M. 1999. Holocene environmental and water circulation changes: foraminifer morphogroups evidence in Flamengo Bay (SP, Brazil). *Journal of Coastal Research*, **15**, 554–571.

EIRÍKSSON, J., BÁRA BARTELS-JÓNSDÓTTIR, H. *ET AL.* 2006. Variability of the North Atlantic Current during the last 2000 years based on shelf bottom water and sea surface temperatures along an open ocean shallow marine transect in western Europe. *The Holocene*, **16**, 1017–1029.

EVANS, J. R., AUSTIN, W. E. N., BREW, D. S., WILKINSON, I. P. & KENNEDY, H. A. 2002. Holocene shelf sea evolution offshore northeast England. *Marine Geology*, **191**, 147–164.

FAIRBANKS, R. G. 1989. A 17,000 year glacio-eustatic sea-level record: influence of glacial melting rates on the Younger Dryas event and deep ocean circulation. *Nature*, **342**, 637–642.

FEBO, L. A. 1999. *Taxonomy and modern distribution of elphidiid foraminifera in the Kara Sea, Arctic Russia*. Senior thesis, Department of Geological Science, The Ohio State University.

FONTANIER, C., JORISSEN, F. J. & LICARI, L. 2002. Live benthic foraminifera faunas from the Bay of Biscay: faunal density, composition, and microhabitats. *Deep-Sea Research I*, **49**, 751–785.

GILLIBRAND, P. A., TURRELL, W. R. & ELLIOTT, A. J. 1995. Deep-water renewal in the upper basin of Loch Sunart, a Scottish fjord. *Journal of Physical Oceanography*, **25**, 1488–1503.

GILLIBRAND, P. A., CAGE, A. G. & AUSTIN, W. E. N. 2005. A preliminary investigation of basin water response to climate forcing in a Scottish fjord: evaluating the influence of the NAO. *Continental Shelf Research*, **255**, 571–587.

GOODAY, A. J. 1986. Meiofaunal foraminiferans from the bathyal Porcupine Seabight (northeast Atlantic): size structure, standing stock, taxonomic composition, species diversity and vertical distribution in the sediment. *Deep-Sea Research*, **33**, 1345–1373.

GOODAY, A. J. & LAMBSHEAD, P. J. D. 1989. The influence of seasonally deposited phytodetritus on benthic foraminiferal populations in the bathyal northeast Atlantic. *Marine Ecology Progress Series*, **58**, 53–67.

GOUBERT, E., NERAUDEAU, D., ROUCHY, J. M. & LACOUR, D. 2001. Foraminiferal record of environmental changes: Messinian of the Los Yesos area (Sorbas Basin, SE Spain). *Palaeogeography, Palaeoclimatology, Palaeoecology*, **175**, 61–78.

HALD, M., HUSUM, K., VORREN, T. O., GROSFJELD, K., JENSEN, H. B. & SHARAPOVA, A. 2003. Holocene climate in the subarctic fjord Malangen, northern Norway: a multi-proxy study. *Boreas*, **32**, 543–59.

HARKNESS, D. D. 1983. The extent of the natural ^{14}C deficiency in the coastal environment of the United Kingdom. *Journal of the European Study Group on Physical, Chemical and Mathematical Techniques Applied to Archaeology*, **8**, 351–364.

HASS, H. C. 2002. A method to reduce the influence of icerafted debris on a grain size record from northern Fram Strait, Arctic Ocean. *Polar Research*, **21**, 299–306.

HAYNES, J. R. 1973. Cardigan Bay Recent foraminifera. *Bulletin of the British Museum (Natural History). Zoology Supplement*, **4**, 1–245.

HOWE, J. A., SHIMMIELD, T., AUSTIN, W. E. N. & LONGVA, O. 2002. Postglacial depositional environments in a mid-high latitude glacially-overdeepened sea-loch, inner Loch Etive, western Scotland. *Marine Geology*, **185**, 417–433.

KEIGWIN, L. D. & LEHMAN, S. J. 1994. Deep Circulation Change Linked to HEINRICH Event 1 and Younger Dryas in a Middepth North Atlantic Core. *Paleoceanography*, **9**, 185–194.

KORSUN, S. A. & POLYAK, L. V. 1989. Distribution of benthic morphogroups in the Barents Sea. *Oceanology*, **29**, 838–844.

KOVACH, W. 1987. Multivariate methods of analyzing paleoecological data. *Paleontological Society*, **3**, 72–104.

KOVACH, W. 1989. Comparisons of multivariate analytical techniques for use in Pre Quaternary plant paleoecology. *Review of Palaeobotany and Palynology*, **60**, 255–282.

KUIJPERS, A., TROELSTRA, S. R., WISSE, M., HEIER NIELSEN, S. & VAN WEERING, T. C. E. 1998. Norwegian Sea overflow variability and NE Atlantic surface hydrography during the past 150,000 years. *Marine Geology*, **12**, 75–99.

LANGDON, P. G., BREGER, K. E. & HUGHES, P. D. M. 2003. A 7500-years peat-based paleoclimatic reconstruction and evidence for an 1100-year cyclicity in bog surface wetness from Temple Hill Moss, Pentland Hills, southeast Scotland. *Quaternary Science Reviews*, **22**, 259–274.

LUKINA, T. G. 2001. Foraminifera of the Laptev Sea. *Protistology*, **2**, 105–122.

LYSÅ, A., SEJRUP, H. P. & AARSETH, I. 2004. The late glacial-Holocene seismic stratigraphy and sedimentary environment in Ranafjorden, northern Norway. *Marine Geology*, **211**, 45–78.

MACKENSEN, A. 1987. Bentische Foraminiferen auf dem Island- Schottland Rücken: Umwelt-Anzeiger an der Grenze zweier Ozeanischer Räume. *Paläont. Z.*, **61**, 149–179.

MACKLIN, M. G., JOHNSTONE, E. & LEWIN, J. 2005. Pervasive and long-term forcing of Holocene river instability and flooding in Great Britain by centennial-scale climate Change. *The Holocene*, **15**, 937–943.

MANGERUD, J., ANDERSEN, S. T., BERGLUND, B. E. & DONNER, J. J. 1974. Quaternary stratigraphy of Norden, a proposal for terminology and classification. *Boreas*, **3**, 109–128.

MASLIN, M. A. 1995. Changes in North Atlantic eep-water formation associated with the Heinrich Events. *Naturwissenschaften*, **82**, 330–333.

MAYEWSKI, P. A., ROHLING, E. ET AL. 2004. Holocene climate variability. *Quaternary Research*, **62**, 243–255.

MCCAVE, I. N., MANIGHETTI, B. & BEVERIDGE, N. A. S. 1995*a*. Changes in circulation of the North Atlantic during the last 25,000 years inferred from grain size measurements. *Nature*, **374**, 149–152.

MCCAVE, I. N., MANIGHETTI, B. & ROBINSON, S. G. 1995*b*. Sortable silt and fine sediment size/composition slicing: parameters for paleocurrent speed and paleoceanography. *Paleoceanography*, **10**, 593–610.

MOKEDDEM, Z., BALTZER, A., CLET-PELLERIN, M., WALTER-SIMONNET, A. V., BATES, R., BALUT, Y. & BONNOT-COURTOIS, C. 2007. Fluctuations climatiques enregistrées depuis 20 000 ans dans le remplissage sédimentaire du loch Sunart (Nord-Ouest de l'Ecosse). *Comptes Rendus Geosciences*, **339**, 150–160.

MURRAY, J. W. 1971. *An Atlas of British Recent Foraminiferids*. Heinemann, London.

MURRAY, J. W. 1991. *Ecology and Palaeoecology of Benthic Foraminifera*. Longmans, New York.

MURRAY, J. W. 2000. Revised taxonomy, Atlas of British Recent Foraminiferids. *Journal of Micropalaeontology*, **19**, 44.

MURRAY, J. W. 2003. An illustrated guide to the benthic foraminifera of the Hebridean Shelf, West of Scotland, with notes on their mode of life. *Palaeontologia Electronica*, **5**, 31.

MURRAY, J. W. 2006. *Ecology and Applications of Benthic Foraminifera*. Cambridge University Press, UK.

NØRGAARD-PEDERSEN, N., AUSTIN, W. E. N., HOWE, J. A. & SHIMMIELD, T. 2006. The Holocene record of Loch Etive, western Scotland: influence of catchment and relative sea level changes. *Marine Geology*, **228**, 55–71.

O'BRIEN, S. R., MAYEWSKI, P. A., MEEKER, L. D., MEESE, D. A., TWICKLER, M. S. & WHITLOW, S. I. 1995. Complexity of Holocene climate as reconstructed from a Greenland ice core. *Science*, **270**, 1962–1964.

RASMUSSEN, T. L., BALBON, E., THOMSEN, E., LABEYRIE, L. & VAN WEERING, T. C. E. 1999. Climate records and changes in deep outflow from the Norwegian Sea B150–55 ka. *Terra Nova*, **11**, 61–66.

REIMER, P. J., MCCORMACK, F. G., MOORE, J., MCCORMICK, F. & MURRAY, E. V. 2002. Marine radiocarbon reservoir corrections for the mid- to late Holocene in the eastern subpolar North Atlantic. *Holocene*, **12**, 129–135.

SCOURSE, J. D., AUSTIN, W. E. N., LONG, B. T., ASSINDER, D. J. & HUWS, D. 2002. Holocene evolution of seasonal stratifcation in the Celtic Sea: refined age model, mixing depths and foraminiferal stratigraphy. *Marine Geology*, **19**, 119–145.

SEJRUP, H. P., HOLTEDAHL, H., NORVIK, O. & MILJETEIG, I. 1980. Benthonic foraminifera as indicators of the paleoposition of the Subarctic Convenvergence in the Norwegian-Greeland Sea. *Boreas*, **9**, 203–207.

SEVERIN, K. P. 1983. Test morphology in benthic foraminifera as discriminator of biofacies. *Marine Micropaleontology*, **8**, 65–76.

SHENNAN, I., PELTIER, W. R., DRUMMOND, R. & HORTON, B. P. 2002. Global to local scale parameters determining relative sea-level changes and the post-glacial isostatic adjustment of Great Britain. *Quaternary Science Reviews*, **21**, 397–408.

SHENNAN, I., HAMILTON, S., HILLIER, C. & WOODROFFE, S. 2005. A 16 000-year record of near-field relative sea-level changes, northwest Scotland, United Kingdom. *Quaternary International*, **133–134**, 95–106.

STUIVER, M. & RIEMER, P. L. 1993. Extended ^{14}C data base and revised Calib 3.0 ^{14}C age calibration program. *Radiocarbon*, **35**, 215–230.

STUIVER, M., REIMER, P. J. ET AL. 1998. INTCAL98 radiocarbon age calibration, 24,000-0 cal BP. *Radiocarbon*, **40**, 1041–1084.

SYVITSKI, J. P. M. & SHAW, J. 1995. Sedimentology and geomorphology of fjords. *In*: PERILLO, G. M. E. (ed.) *Geomorphology and Sedimentology of Estuaries*. Developments in Sedimentology **53**, Elsevier, Amsterdam, 113–178.

SYVITSKI, J. P. M., BURREL, D. C. & SKEI, J. M. 1987. *Fjords Processes and Products*. Springer-Verlag, New York.

VORREN, T. O., LEBESBYE, E., ANDREASSEN, K. & LARSEN, K. B. 1989. Glacigenic sediments on a passive continental margin as exemplified by the Barents Sea. *Marine Geology*, **85**, 251–272.

Using seismic facies and pollen analyses to evaluate climatically driven change in a Scottish sea loch (fjord) over the last 20 ka

AGNÈS BALTZER[1]*, RICHARD BATES[2], ZOHRA MOKEDDEM[1], MARTINE CLET-PELLERIN[1], ANNE-VÉRONIQUE WALTER-SIMONNET[3], CHANTAL BONNOT-COURTOIS[4] & WILLIAM E. N. AUSTIN[2]

[1]*University of Caen, Centre de Géomorphologie du CNRS, Labo M2C, 24 rue des Tilleuls, 14 000 Caen, France*

[2]*School of Geography and Geosciences, University of St Andrews, St Andrews, Fife, Scotland KY16 9AL, UK*

[3]*CNRS-UMR 6565, Université de Franche-Comté, 25 000 Besançon, France*

[4]*EPHE, UMR, Labo de géomorphologie et du Littoral, 15 Bd de la Mer, 35 800 Dinard, France*

**Corresponding author (e-mail: agnes.agnes.baltzer@unicaen.fr)*

Abstract: Loch Sunart is a glacially over-deepened sea loch (fjord) on the west coast of Scotland, UK. The loch is divided into three sub-basins, separated by relatively shallow and narrow sills. A programme of data collection including high-resolution bathymetric sonar and sub-bottom seismic surveys were conducted in the loch as part of an investigation into the sedimentological and climatic change signatures preserved in western sea lochs since the Last Glacial Maximum. Very-high-resolution sub-bottom profiles were obtained using the SEISTEC boomer system. The seismic profiles revealed an igneous and metamorphic basement covered by a 10–70 m thick sediment sequence. Five different acoustic facies were recognized and interpreted in terms of glacial activity, ice retreat and subsequent Holocene sedimentation. These facies have been correlated to sediments sampled in a radiocarbon-dated 12 m long giant piston core (MD04-2833) acquired from the main basin of Loch Sunart. Pollen analyses conducted along the length of the core, together with ^{14}C dating, indicate a complex series of palaeoclimate changes in the loch. In particular, five distinct cooling events have been recognized *c.* 9.8, 8.2, 5.8, 1.2 cal ka BP and 771–1211 cal a BP (possibly the Little Ice Age), corresponding to phases of Holocene rapid climate change.

While the study of vegetation history using the technique of pollen analysis has been carried out on peat bog and lake sediments for at least 100 years and open ocean marine sediments for about 50 years, until recently very few investigations of the pollen content of fjord sediments have been reported. Nevertheless, in a recent communication, Cundill *et al.* (2006) demonstrated that Scottish sea-loch sediments hold the potential to provide a consistent and reliable record of regional vegetation history.

As the western Scottish sea lochs are located at an important junction between the North Atlantic Ocean and the European landmass these sites are particularly well situated for observations on climatic variations, in particular variations in sedimentation that have resulted from climatic changes since the Last Glacial Maximum (LGM) at *c.* 22 ka BP.

The history of glaciation and postglacial sedimentation in the Scottish lochs has been studied by a number of authors (Binns *et al.* 1974*a*, *b*; Boulton *et al.* 1981; Ballantyne & Gray 1984; Howe *et al.* 2001, 2002; Austin & Inall 2002; Nørgaard-Pedersen *et al.* 2006; Stoker *et al.* 2006). These authors have established the following general pattern of deglaciation: at the LGM, ice extended from the Scottish mainland to the outer shelf, west of the Outer Hebrides (Stoker 1995; Benn 1997; Smith *et al.* 2007). During this time, extensive ice eroded the land surface causing loch over-deepening and a 'cleaning' of the earlier sedimentary records. Subsequent to the LGM, sedimentation includes the accumulation of thick sequences of ice-contact and ice-marginal deposits in association with the retreating glaciers. During this phase of northern hemisphere deglaciation, the Loch Lomond Re-advance (Fig. 1a), described by numerous authors (e.g. Hubbard 1999), coincides with the cooling phase known as the Younger Dryas. This two-step sequence of glaciation, de-glaciation, local glacial regrowth and then final deglaciation in the early Holocene was tested in Loch Sunart using

From: Howe, J. A., Austin, W. E. N., Forwick, M. & Paetzel, M. (eds) *Fjord Systems and Archives*. Geological Society, London, Special Publications, **344**, 355–369.
DOI: 10.1144/SP344.24 0305-8719/10/$15.00

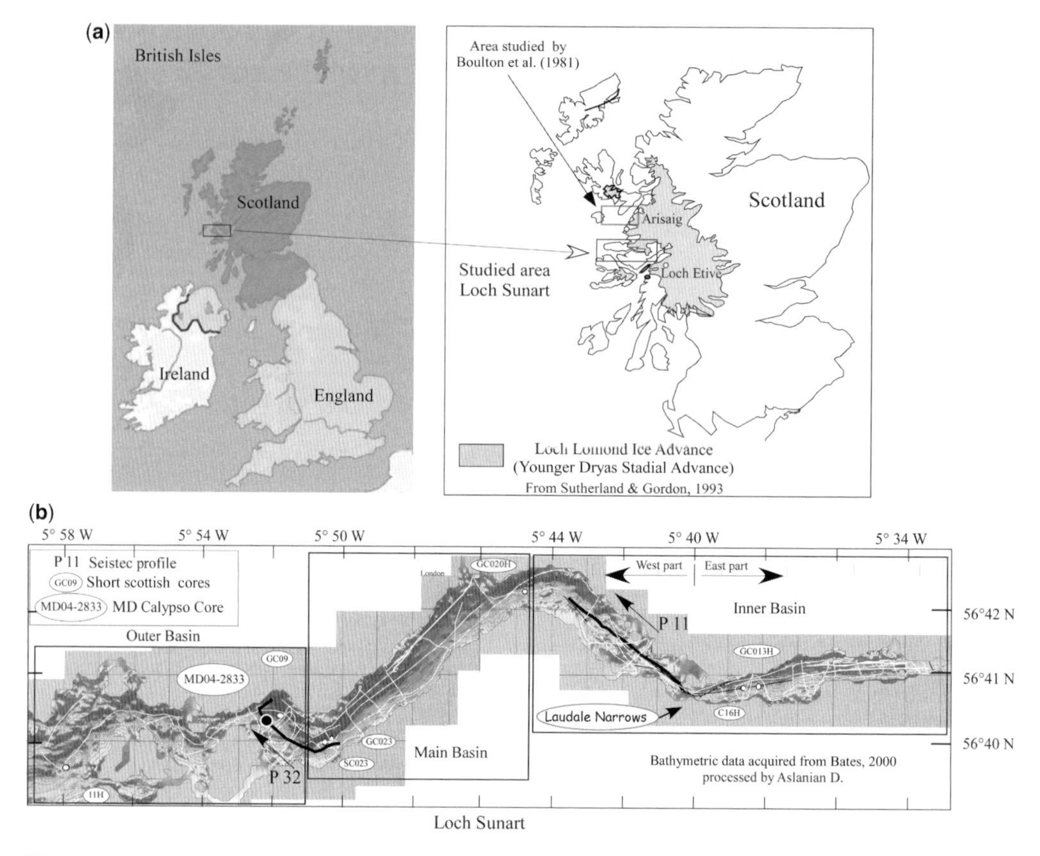

Fig. 1. Location of the field study area with the location of the area studied by Boulton *et al.* (1981). (**a**) NW coast of Scotland and Loch Sunart. Limits of the Loch Lomond Re-advance from Sutherland & Gordon (1983). (**b**) Bathymetry of Loch Sunart from Bates (2000) together with the SEISTEC seismic profiles and core MD04-2833 location.

high-resolution bathymetry together with very-high-resolution sub-bottom profiling. This paper presents these results which are correlated to sediment sequences sampled in a long (12 m) sediment core. The seismic acoustic facies are correlated with the sediment units and, for the first time, integrated with pollen analysis to reveal the complex history of regional vegetation responses and evidence for several significant rapid climate change events (Bond *et al.* 1997; Hall *et al.* 2004; Mayewski *et al.* 2004).

Loch Sunart (Fig. 1a) is situated on the southern shores of the Ardnamurchan Peninsula, Argyll (Scotland). At just over 31 km long, it is the second-longest sea loch in Scotland. The loch has an average width of 1.5 km and a maximum depth of 124 m (Gillibrand *et al.* 2005). The bathymetric profile of Loch Sunart is complex and the loch can be divided into an inner, a main and an outer basin which are separated by narrow and shallow sills (Fig. 1b).

The loch exhibits the characteristic morphology of fjords, with a steep-sided narrow cross-section and flat sea floor (Syvitski *et al.* 1987). The location of sills and the over-deepened trough profile results from increased glacial erosion at points where either the bedrock geology is more resistant to the erosive power of the glacier or at locations where multiple glaciers join and thus increase their cumulative erosive power. This typical morphology provides protection from swell and wave action and thus ideal conditions for the entrapment and preservation of long sediment records of deglaciation and late Holocene sediment accumulation (Howe *et al.* 2002; Cage *et al.* 2006). The length and the narrowness of this loch make it possible to meet a spectrum of hydrodynamic conditions including well-exposed conditions at the mouth to extremely calm conditions at the head of the loch. For example, surface currents reach up to 20 cm s^{-1} and can fall to below 5 cm s^{-1} at the head of the loch (Bates *et al.* 2004).

From the limited onshore mapping of the surface features around Loch Sunart, it has been proposed that the ice advance during the Younger Dryas stadial reached at least as far west as the Laudale Narrows (Figs 1b–2a) (Sutherland & Gordon 1983). This re-advance will likely have removed some if not all of the sediment sequences deposited within the inner loch basin; however, the pre-Younger Dryas sequence in the middle and outer loch is likely to be preserved.

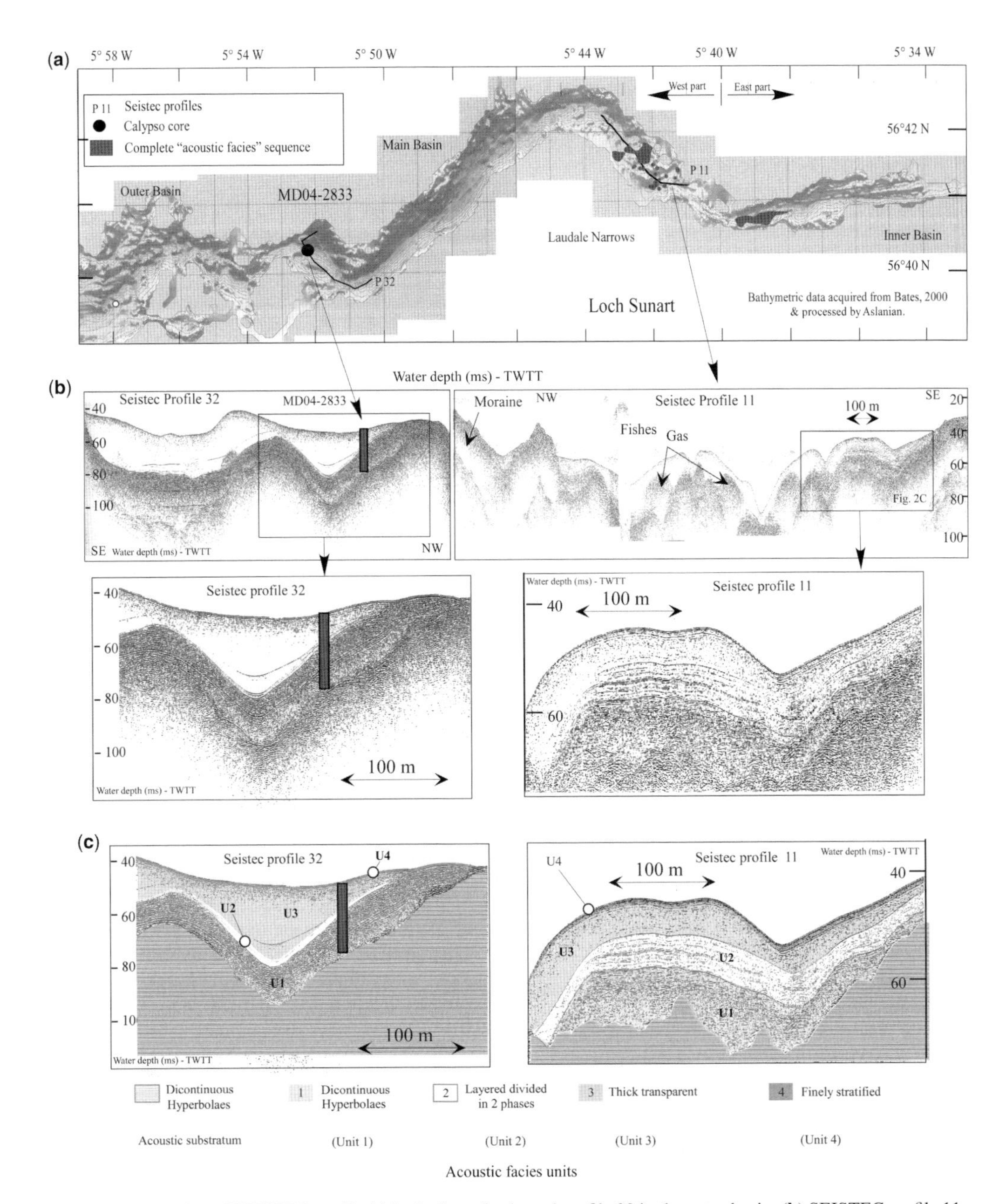

Fig. 2. (**a**) Location of SEISTEC profile 11 in the inner basin and profile 32 in the outer basin. (**b**) SEISTEC profile 11, characteristic of the inner basin, with gas facies and a complete sequence of acoustic Units. Seistec profile 32 at the long core MD04-2833 location. (**c**) SEISTEC profile 11 showing the five acoustic facies units (the acoustic substratum and the four sediment units).

Following the Younger Dryas Stadial, glacial retreat likely resulted in a sequence of ice-contact, ice-marginal and proglacial sediments similar to those associated with the decay of the LGM ice sheet. In many sea lochs, this includes numerous ice-contact ridges which are comparable with De Geer moraines (e.g. Stoker *et al.* 2006). In contrast, the Holocene post-glacial sediment record has probably been strongly controlled by bottom-current circulation within the sea lochs, as well as changes in the local catchment hydrology. Initial sedimentation rates were assumed to be high due to melting of the glaciers and erosion of the exposed, highly unstable land surfaces. This pattern follows that first described by Church & Ryder (1972).

Data acquisition and processing

In August 2002, geophysical data were acquired in Loch Sunart using a bathymetric sidescan system and a seismic sub-bottom system. In June 2004 a giant piston core, MD04-2833 (12 m), was acquired from the main basin. This core formed the basis of a novel approach to test for evidence of climate variations based on the extensive pollen record.

Bathymetric sidescan survey

The bathymetric sidescan is an extension of high-resolution digital sidescan that not only enables a picture of the seafloor to be produced across a swath sampled by the transducers along the boat track but also, through the use of multiple transducers, measures the bathymetry across the swath. The method has found widespread application for mapping large areas of seafloor as the width of survey is approximately 7 times the depth of water beneath the transducers (Bates & Byham 2001). The system used for this survey was a Submetrix System 2000 (SEA Ltd.) with 117 kHz bow-mounted transducers. These were deployed together with a TSS DMS05 (TSS Int. Ltd) motion reference unit linked to a differential global positioning system (DGPS) on the RV *Envoy*. During survey acquisition, navigation was provided by Hypack Max with electronic charting. This combination of instrumentation allows features on the seafloor to be mapped with near 100% coverage at a resolution of 50 cm or better. All data were processed using the Submetrix Grid2000 software with subsequent interpretations undertaken in ArcGIS (ESRI Inc.) and Fledermaus (IVS Inc.). The bathymetric sidescan investigation (Figs 1b & 2a) was conducted in the loch as a precursor for locating the core sites, grab sample locations and sub-bottom profiles lines.

Seismic acquisition

A Seistec Boomer System (IKB Ltd) was used to record sub-bottom seismic profile data in both deep (130 m of water depth) and shallow (5 m of water depth) water (Figs 1 & 2). The system has a frequency range from 1 to 10 kHz and a pulse duration ranging from 75 to 250 ms at a power of 150 J.

The system was towed at a speed of 3.5 knots and fired at a rate of 2 shots per second. The system has a depth resolution of 25 cm, with a penetration up to 100 m in soft sediments and 200 m in deep-water soft sediments (Simpkin & Davis 1983). Thirty-four profiles, recovering 60 km of line survey, were acquired in the loch (Fig. 1b) with an average penetration of 50 m (except in some areas where gas occurrence prevented any signal penetration). The data were recorded using an Elics-Delph acquisition system with positioning obtained by DGPS. The Elics-Delph system was used for subsequent processing and interpretation of the data.

MD04-2833 core acquisition and analysis

The long core MD04-2833 (Fig. 3) was acquired by the giant (Calypso) piston core system on board the RV *Marion Dufresne*. This coring system allows the acquisition of Kullenberg-type long piston cores up to a length of 57 m in soft marine sediments. This piston corer is similar to a gravity corer but, as its name indicates, a piston is used to reduce internal friction in the barrel thus enabling the corer to recover more complete and less disturbed sediment sequences than an open barrel gravity corer. The core was acquired in the middle loch to the NE of Carna Island (Figs 1b & 2a) at 56°39.93N, 5°51.58W in a water depth of 38 m.

Grain size and pollen analysis were conducted on core MD04-2833 at 5 cm intervals for a total of 228 samples. A Beckman-Coulter LS230 laser particle analyser was used for the grain size measurements on fraction ranges from 0.04 to 2 mm. Grains coarser than 2 mm, when encountered, were noted on the core log (Fig. 3b) but their occurrence was rare. For pollen analysis the samples were filtered between 10 and 125 μm after decalcification with hydrochloric acid (10%) and hydrofluoric acid (40%) to remove silicate minerals. Pollen samples were obtained from most depths with the exception of gravel sections at the base and in the middle part of the core.

Results

Seismic units

A typical example of Seistec sub-bottom profile acquired in the inner basin (profile 11, Fig. 2a)

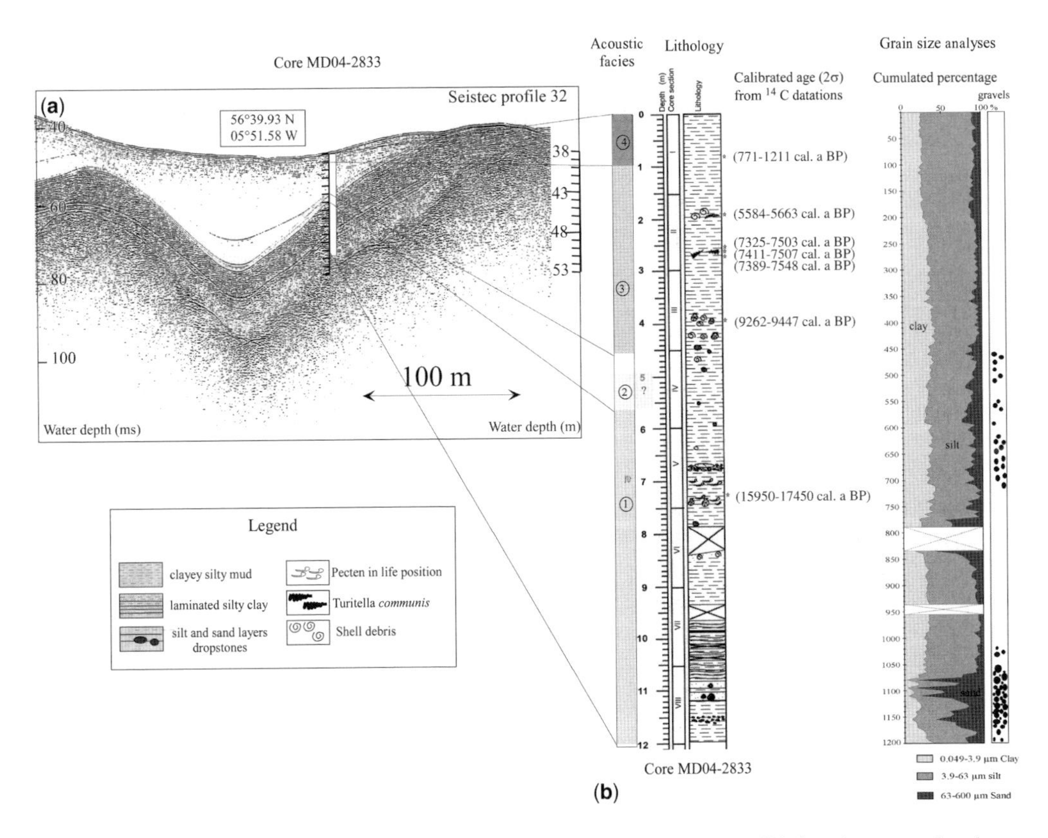

Fig. 3. (**a**) Seismic profile 32 with the location of the core MD04-2833. (**b**) The core lithology is compared to the acoustic facies identified on the seismic profile and the grain size cumulated percentages. Seven calibrated ^{14}C ages are presented along the core.

shows a representative section of all five seismic facies (Fig. 2b, c). The seismic units and a characteristic gas facies are described below in ascending stratigraphic order.

Acoustic substratum. The basal sequence is characterized by a dense reflectivity, homogeneity with no internal stratigraphic or structural reflections and a highly irregular surface (profile 11 on Fig. 2c) that displays a geometry consisting of mounds and depressions. In the inner basin, undulations or ridges of 3–5 m in height in the basal surface show more laterally continuous internal reflectors, sinuous crests and sides that are characterized by steep dips to the east and more gentle dips to the west (Fig. 2c). These features contrast with ridges of similar dimension but with no internal structure. It is postulated that those with internal structure consist of sediment while those without internal structure likely have a bedrock core. This constitutes the acoustic basement in the loch.

Unit 1. In the inner loch, a laterally discontinuous sequence of sediment overlies the basement and is characterized by short discontinuous reflections (Fig. 2b) with occasional hyperbolic reflectors, likely resulting from large boulders acting as point diffractors within the unit. In the middle loch this layer also occurs in isolated pockets where accumulations of sediment reach thicknesses of 2.5–8 m. In the deeper parts of the loch (greater than 60 ms two-way travel time or TWTT) the unit has a minimum thickness of 10 m on the bottom of the loch and at the mouth of the main and outer basins.

Unit 2. The sediment within this unit appears to drape over and around Unit 1 and is only identified in the inner basin between 20–40 ms TWTT in the east part of the basin and at depths greater than 40 ms TWTT in the deeper central part of the basin (Figs 2a, c). This unit is not clearly identified in the middle and outer basins, where strong bottom currents are likely to have limited deposition. The unit is characterized by fine acoustic layering that is 20 cm or less thick (Fig. 2c), is continuous and planar. The unit is divided into a lower and upper interval. The lower interval is most strongly laminated, whereas the upper interval is more transparent (Fig. 2c). No clear evidence of the unit is found in the main or outer basin as illustrated on Figure 2a. The top of this unit is marked in the inner basin by an unconformity that is especially clear in the shallow water sections (Fig. 2b).

Unit 3. The unit is ubiquitous throughout the loch and covers the previous units with a thickness varying from 5 to 10 m in the inner basin. In the main and outer basins, the unit has a maximum observable thickness of 75 m where it is not hidden by acoustic blanking derived from gas in the sediments. The basal part of this unit reveals a highly transparent acoustic signature. The external geometry of the unit is commonly undulated.

Unit 4. This unit is *c.* 1 m thick and is characterized by wavy to planar acoustic laminations that are 20 cm thick or less. It is present throughout the loch and directly overlies the acoustic substratum in the outer basin.

Shallow gas. The occurrence of shallow gas is evident throughout Loch Sunart and has resulted in acoustic blanking on some of the seismic profiles. This type of gas blanking is observed in the inner loch above the deeper parts of the lateral incised channels, corresponding to locations of high organic matter supply and the formation of biogenic methane. In Figure 2b the definition of Unit 3 is locally disturbed by the occurrence of gas patches, some of which display a domed character. This character is found where Unit 3 reaches a thickness of 20 m or more in the deeper parts of the outer loch. These gas blankets and gas curtains are similar to those described by Emery & Hoggan (1958), Davis (1992) and Baltzer *et al.* (2005).

Core MD04-2833 analysis

Sediment description: lithology and grain-size analysis. The analysis of the seismic profiles provided a context for selecting the optimal core site, where most of the units described above occur with a significant thickness. Figure 3a, b show the position of core MD04-2833 on seismic profile 32, which is situated in the main basin beyond the Younger Dryas limits of local glaciation yet still accessible to the RV *Marion Dufresne* (Fig. 2a). A mobile 4.5 kHz echosounder onboard the RV *Marion Dufresne* was placed close to the position of deployment of the Calypso coring system on the deck, in order to recover the core from the precise target location (Fig. 3a).

Core MD04-2833 recovered 12 m of sediments (Fig. 3a, c). The core stopped at a very stiff horizon; this is presumed to be the basal moraine observed on the seismic profiles (acoustic basement). As the water depth was very shallow (38 m) for the Calypso coring system, there is no sampling perturbation triggered by the elasticity of the cable. Two missing sections, between 790–840 cm and 930–970 cm, are likely due to the aggressive friction of the piston when the core abruptly stopped on the basal moraine.

Two significant intervals are recorded on a log of grain size (Fig. 3c): the lower core section (from bottom to 440 cm) is characterized by the presence of dropstones and the upper core section without dropstones.

The lithology comprises silty to clayey sediments of greyish colour (5Y 4/2; Munsell Soil Chart) throughout much of the sequence, with a horizon of laminated silty clay at *c.* 1000 cm core depth. Two significant layers of dropstones, buried in the silty clayey matrix are highlighted by the grain-size curves from 450 cm to 740 cm and from 1020 cm to the core bottom. Five levels with shells occur at 190, 270, 380, 670 and 840 cm (Fig. 3b). The following lithological descriptions are given with reference to their intersection with the seismic units.

Unit 1 (560–1200 cm). The core interval corresponding to Unit 1 exhibits a succession of different deposits with a decreasing grain size (Fig. 3c) from a clast-rich deposit at the base to sandy beds interlayered with silty beds. Up to 1060 cm, clasts ranging in size from 3 to 7 cm are intercalated with sandy lamination. The uppermost clast-rich bed at *c.* 1070 cm underlies a sequence of finely laminated beds (960 to 1070 cm), suggesting a marked change in the environment of deposition. The first shell debris occurs in a sandier unit with a sharp basal contact at 840 cm; numerous shells belonging to the genus *Pecten* are present from 700 to 750 cm. Above this level, the sediments become clayey-silty-mud-rich, containing sparse angular and black clasts of *c.* 2 cm of length.

Unit 2 (440–560 cm). Unit 2 is composed of a homogeneous sequence of greyish clayey-silty-mud with occasional shell fragments (530 and 570 cm) and some rounded pebbles and dropstones. The last dropstone observed, a fine-grained black lithology of 7 cm length, occurs at the top of this unit.

Unit 3 (100–440 cm). Unit 3 comprises a matrix of clayey-silty-mud with abundant shells and occasional wood fragments, but without dropstones. From 380 to 430 cm, a thick horizon of *Pecten* species shells and shell fragments occurs. A second shell-rich bed composed mainly of *Turitella communis* is found from 260 to 272 cm, including juveniles and mature individuals which are all extremely well preserved. The upper part of this unit is formed by a homogenous clayey-silty-mud with cracks, probably due to gas escape features, with wood fragments and sparse gastropods present between 160 and 200 cm.

Unit 4 (0–100 cm). Unit 4 consists of a clayey-silty-mud with occasional bivalve fragments. There is no visible bioturbation but some mud-filled cracks occur, probably related to gas escape features.

^{14}C ages

Five ^{14}C ages were acquired on different shell samples all assumed to be *in situ* (see Table 1) and two ^{14}C ages from wood fragments. A standard reservoir correction *R*(*t*) of 400 years was used for the shell marine samples according to the regional average value for western Britain (Harkness 1983; Cage *et al.* 2006) and calibrated calendar ages were obtained from the calibration tables in Stuiver & Reimer (1993) and Stuiver *et al.* (1998) by means of the CALIB 5.1.0b software. The correction of 400 years for the marine shells was confirmed by the correlation between ages from wood sample Poz23579 (at 270 cm) of 6520 ± 40 a BP and the *Turitella* sample (Poz 23472) of 6950 ± 40 a (at 271 cm).

Pollen analysis

Pollen analysis was conducted together with the grain-size analysis at 5 cm intervals along the core; however, pollen was only observed above 710 cm (Fig. 4). A general low pollen concentration together with dominance of *Pinus* pollen (transported by wind) and Filicales spores (transported by water) indicate a maritime character for the samples similar to patterns previously noted (Cundill *et al.* 2006; Sharapova *et al.* 2008). For each pollen analysis, an average of 600 grains were counted per sample with statistical analyses conducted using the GPALWIN software (Goeury 1988) and TILIA software (Grimm, Illinois State Museum, Springfield). Taxons classified in the different pollen spectrum of the diagram were grouped under larger sets according to their ecological affinities. Fourteen pollen classes were defined (Fig. 4b) and the different Pollen Assemblages Zones (PAZ) were numbered from F1 to F4 to indicate the evolution of the vegetation within the landscape represented by the top 710 cm of the core (Fig. 4a, b). The pollen analysis is once again described with reference to the seismic units.

Unit 1 (PAZ F1-1a to F1-2). From 710 cm to 660 cm, the palynological zonation (F1-1a and F1-1b) shows the development of a tundra landscape with marshes and *Sphagnum*. Some tall shrubs (*Juniperus*, *Populus*, *Betula*, *Pinus*) distinguish themselves from surrounding grasslands and herbs, suggesting improved climatic conditions in the context of deglaciation. A ^{14}C date of 14 420 ± 210 a BP (15 950–17 450 cal a BP) was obtained immediately beneath this horizon. From 660 cm to 610 cm the absence of pollen is probably due to cooler conditions, which enabled vegetation to develop. From 610 to 565 cm, the pollen record shows first a grassland vegetation then the appearance of trees such as *Quercus*, with an increase of *Alnus* combined with a decline of marsh species. Zone F1-2 is interpreted as a more temperate climate interval and is likely correlated to the Bölling period. This observation is in agreement with the occurrence of trees in the southern Hebrides as early as 12 ^{14}C ka BP (Bennett & Humphry 1995).

Unit 2 (no Pollen Assemblage Zone from 560 cm to 440 cm). This section (565–435 cm) corresponds to a quasi-sterile zone, linked to either the decrease in vegetation or dispersion (and non-preservation) of pollen grains due to an increase in hydrodynamical conditions. These highly energetic conditions are consistent with the Younger Dryas cooling event followed by warming and retreat of the ice fronts.

Unit 3 (PAZ F3-3 to F3-6d). Zone F3-3 (440–425 cm) shows the global development of pioneer trees such as *Corylus*, *Betula* with a general

Table 1. *^{14}C ages acquired from core MD04-2833*

Laboratory code*	Depth (cm)	Species	Age ^{14}C BP	Age (2σ)† age cal a BP
Poz 23471	81	*Turitella communis*	1405 ± 30	771–1211
Poz 23648	198	Wood fragment	4885 ± 35	5584–5663
Poz 10545	265	*Turitella communis*	6910 ± 40	7325–7503
Poz 23579	270	Wood fragment	6520 ± 40	7411–7507
Poz 23472	271	*Turitella communis*	6950 ± 40	7389–7548
Poz 13368	385	Pecten species	8710 ± 50	9262–9447
UL 2853	745	Pecten maximus	14 420 ± 210	15 950–17 450

*Poz (Poznan laboratory, Poland) and UL (Laval University, Canada).
†Standard deviation (Stuiver *et al.* 1998): Marine04 (shells) and Intercal04 (wood).

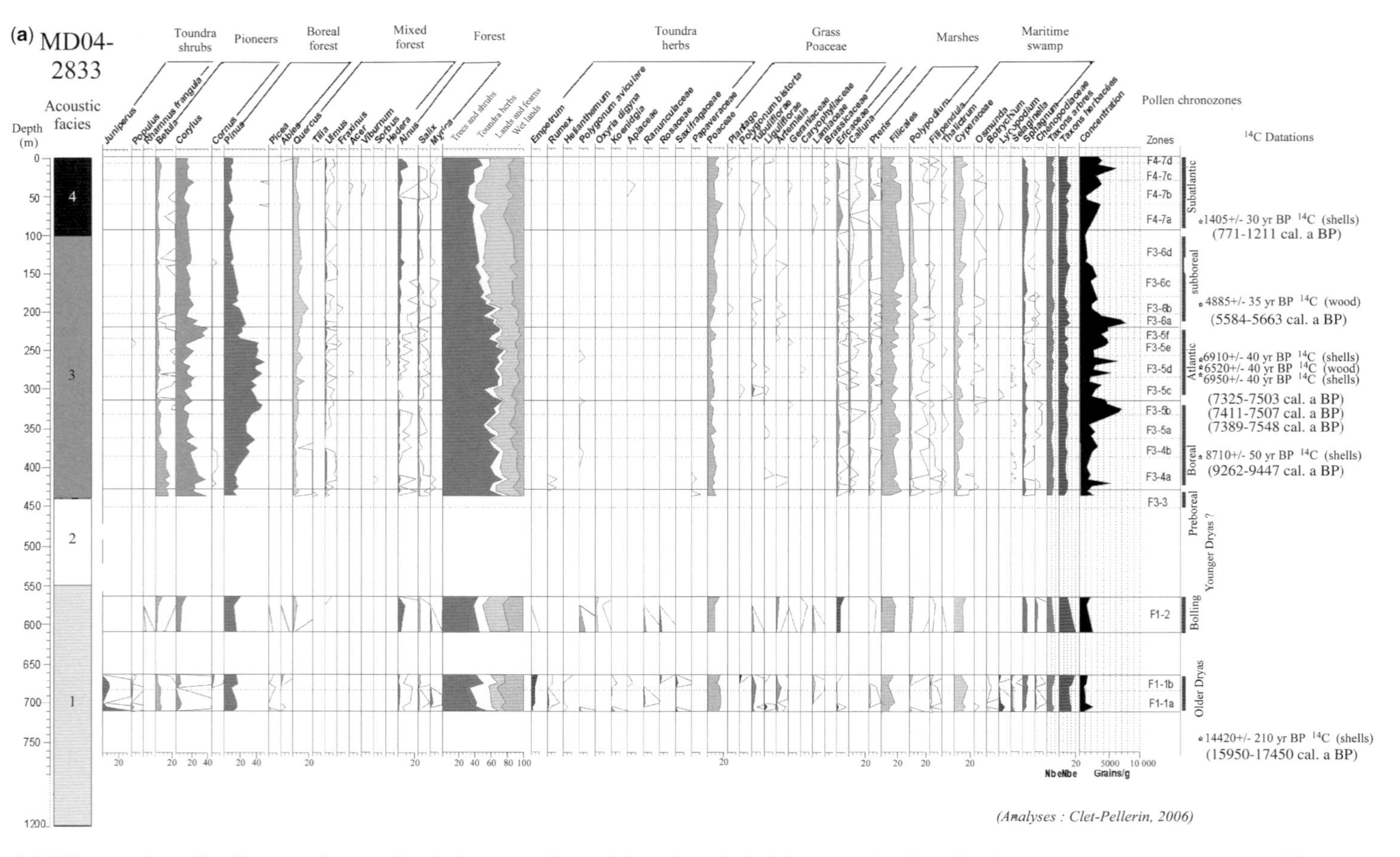

Fig. 4. (**a**) The complete pollen diagram v. the acoustic units is presented here, giving a chronological frame based on the pollen chronozones. For each chronozone, different Pollen Assemblage Zones (PAZ) have been defined from F1-1a to F4-7d.

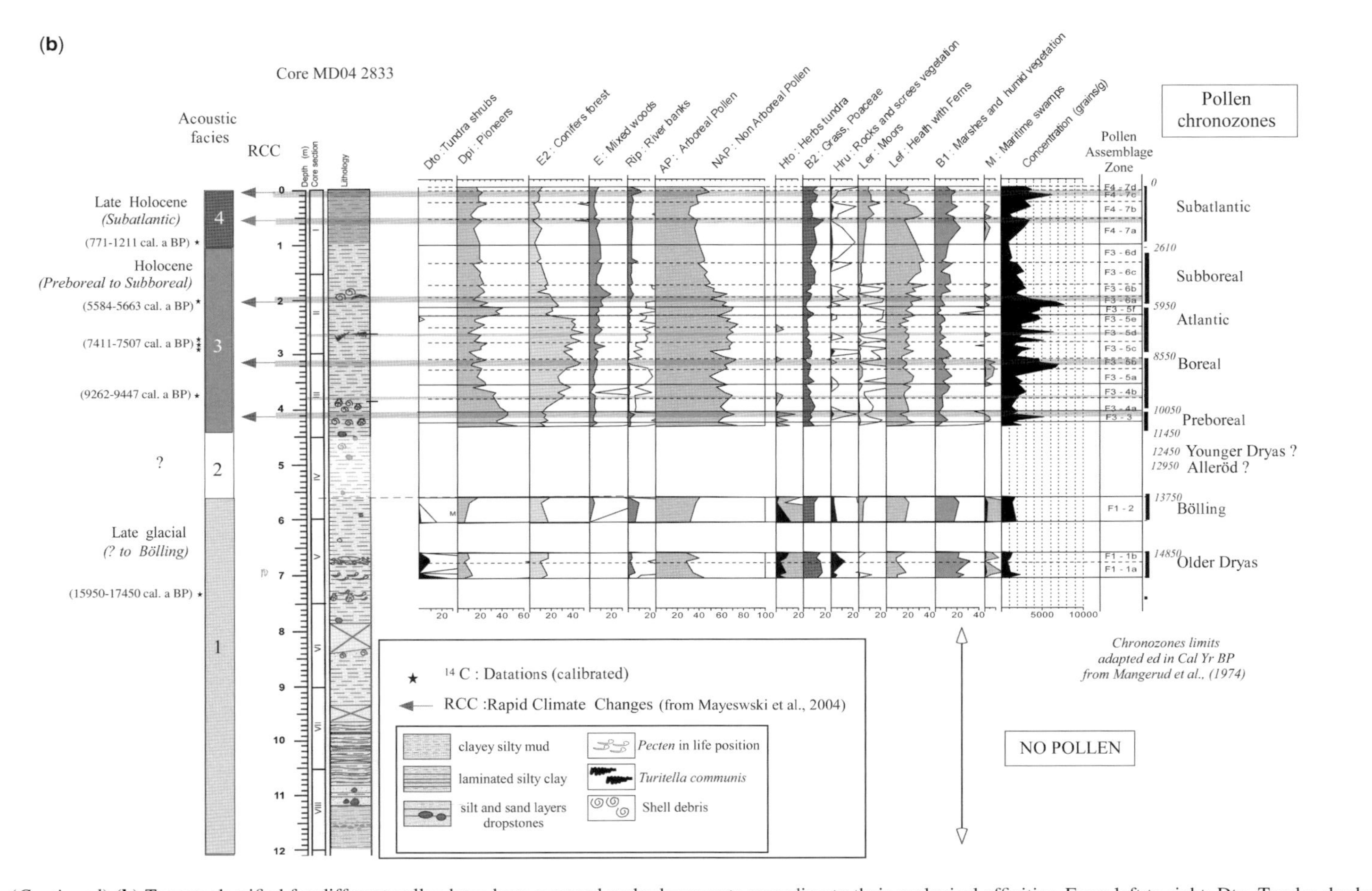

Fig. 4. (*Continued*) (**b**) Taxons classified for different pollen have been grouped under larger sets according to their ecological affinities. From left to right: Dto, Tundra shrubs (*Juniperus, Populus, Rhamnus frangula*); Dpi, pioneer trees and shrubs (*Betula*, *Corylus*, *Cornus*); E2, conifers forest (*Pinus silvestris*, *Picea*, *Abies*); E, leaves forest (*Quercus*, *Tilia*, *Ulmus*, *Fraxinus*, *Acer*, *Viburnum*, *Sorbus*, *Hedera*); Rip ripisylve (*Alnus*, *Salix*, *Myrica*); AP, trees and shrubs; NAP, herbs, filicales and bryophyte spores); Hto, herbs from tundra (*Empetrum*, *Rumex*, *Helianthemum*, *Polygonum aviculare*, *Oxyria digyna*, *Koenidgia islandica*); B2, terrestrial herbs; Hru, *Plantago lanceolata*, *Polygonum bistorta, Artemisia*; Ler, heathers and ferns (*Calluna*, *Polypodium*, *Pteris*); Lef, heath with ferns; B1, marshes (Filicales) and M, maritime swamps (*Filipendula*, *Thalictrum*, *Osmunda*, *Botrychium*, *Lycopodium selago*, *Selaginella, Sphagnum*). Seven ^{14}C dates are indicated by stars, and five rapid climatic change events are highlighted by grey lines with arrows.

transition to boreal forest species such as *Pinus silvestris*, indicating a progressive warming. Quercus pollen remains in low proportion. This PAZ corresponds to the Late Preboreal period and is followed by an increase of *Pinus* together with a decrease in *Corylus*. From 425 to 370 cm (F3-4a and F3-4b), the pollen record shows are interpreted as a mixed forest dominated by *Betula* and *Corylus*, accompanied by an increase of *Pinus*. On a regional basis a significant extension of the boreal forest at this period would correspond to the end of the Preboreal chronozone and is consistent with a ^{14}C date of 8710 a BP (9262–9447 cal ka BP) at 385 cm. From 370 to 310 cm, *Pinus* constitutes the main forest species (F3-5a to 5b) which is subsequently replaced by *Corylus* at the end of the Atlantic period (F3-5f). An increase in total pollen concentration through this interval is interpreted as an overall improvement in climate favouring the establishment of optimal climatic conditions for tree and shrub development (arboreal pollens). A ^{14}C date of 6520 a BP (7411–7507 cal a BP) for this zone allows a correlation with the Atlantic period, characterized by an average elevation of the temperature of 1 to 2 °C and an increase in humidity. From 220 to 130 cm, the record is dominated by *Corylus* with a decrease in *Pinus* (F3-6a to 6d) and a peak of *Quercus*. At 200 cm the index of arboreal pollen v. non-arboreal pollen (AP)/(NAP) is inverted with development of fern moors and vegetation such as *Sphagnum* and Cyperaceae which can be correlated to the Subboreal period (Adams 1997). The PAZ F3-6A marks the passage to a climatic environment similar to the actuel where marshes and humid vegetation become dominant.

Unit 4 (PAZ F4-7a to F4-7d). The upper part of the pollen record shows a widespread expansion of the fern moors and the development of marshes and humid vegetation, which correspond to the start of the Subatlantic period. This is in agreement with a ^{14}C date of 1405 $\pm$ 30 a BP (771–1211 cal a BP) made on a shell at the base of Unit 4. The transition from the PAZ F4-7a to PAZ F4-7b shows an abrupt decrease in trees and shrubs, tundra herbs and *Corylus* abundance, which could be correlated to the first stage of the Little Ice Age.

Discussion

This is the first pollen analysis conducted on a marine core spanning the entire history of deglaciation from Scottish coastal waters. The pollen data record (Fig. 4a, b) major environmental changes including two important periods.

(1) The transition between the Atlantic and the Subboreal chronozones at *c.* 5950 cal a BP (before 5663 cal a BP) appears clearly in the pollen record, suggesting a dramatic change in the regional vegetation history at this time.
(2) Five cooling events shown at 410, 310, 200, 60 and 10 cm are revealed by the conjunction of:
   - an abrupt and major decrease in the pollen concentrations which illustrates an important disappearance of vegetation related to climate deterioration;
   - a decrease of marshes and vegetation related to cooler conditions;
   - a peak of the non-arboreal pollen together with the decrease in proportion of the arboreal pollen revealing deterioration of climate; and
   - a decrease of mixed wood together with an increase in conifer forest type as a pioneer species.

Using a combination of the pollen and sedimentological analysis of core MD04-2833, a detailed interpretation can be made of the seismic records to include changes in the environment during sediment deposition. The seismic sections are discussed here with respect to the core data, their position in the loch and in comparison to previous work on the west coast of Scotland by Binns *et al.* (1974*a*, *b*), Boulton *et al.* (1981) and Howe *et al.* (2002). The correlation between the seismic units, core MD04-2833 and the pollen chronozones is illustrated on Figures 4b and 5.

Acoustic substratum

The acoustic basement is interpreted as either the rock basement where no internal structures are seen or, in some places, as stiff material most likely of glacial origin such as a till or moraine unit. Where internal structures are seen, the surface is marked by a smooth shape possibly associated with post-deposition erosion. Unfortunately, core MD04-2833 did not penetrate this unit. However, it is postulated that the till is a remnant from glaciations prior to the LGM. No clear association could be found for this unit with any of the sedimentary formations described by Boulton *et al.* (1981) apart from the rock basement.

Unit 1. Unit 1 corresponds to stiff material typical of a basal till sequence that contains numerous dropstones and gravel lenses in a clayey-silty-mud matrix. A continuous reflector within this unit corresponds to layering observed in the core at 10 m. This reflector may be related to local ice-sheet retreat and a melting event prior to regional deglaciation; however, the first evidence of effective and sustained deglaciation only appears after 15.95 cal ka BP (Fig. 4b) when the pollen record

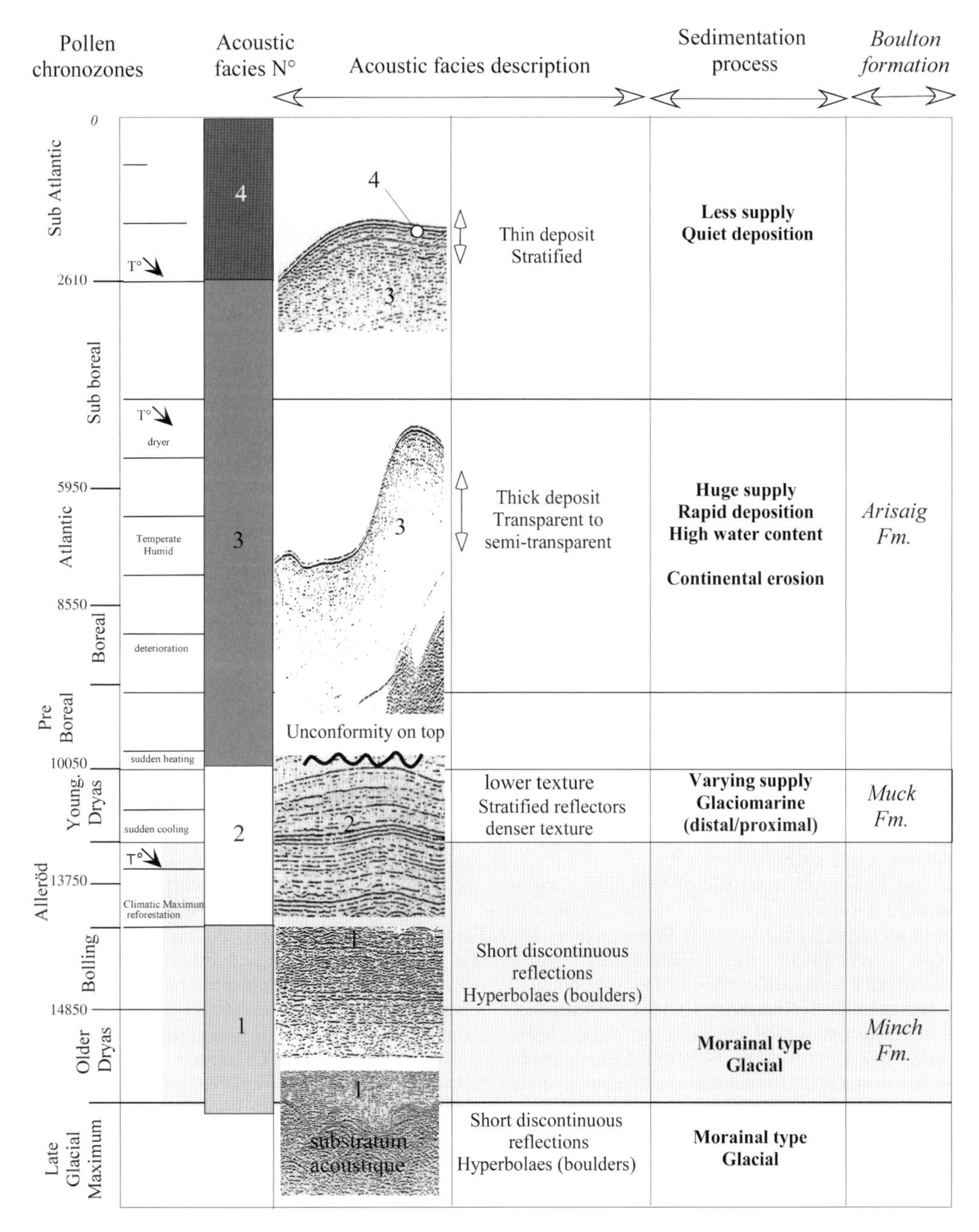

Fig. 5. Correlation between the four seismic acoustic units and the sediment of the core MD04-2833. The sedimentation process is deduced from the acoustic facies associated with the sediment analyses, and is compared to the formation described in the literature by Boulton *et al.* (1981). This figure allows a chronological frame for the acoustic facies to be defined: facies 4 corresponds to Subatlantic deposits, facies 3 corresponds to Preboreal and Subboreal deposits, facies 2 covers the period from Alleröd to the end of Younger Dryas and the facies 1 corresponds to older Dryas and Bolling deposits. Age limits of chronozones from Mangerud *et al.* (1974).

shows the development of a tundra landscape during an interval corresponding to the Older Dryas. This pattern of deglaciation and climatic amelioration was again halted by a return to a cold climate period without any pollen, corresponding with the end of the Older Dryas. The beginning of a more temperate period, most likely coinciding with the Bölling period (F1-2), occurs at the end of this unit.

Within a regional context this unit is comparable to the Minch formation identified by Boulton *et al.* (1981) (Fig. 5) and the Formation 1 of boulder clay identified by Binns *et al.* (1974*a*, *b*).

Unit 2. Unit 2 is well defined in the inner basin with fine acoustic layering suggesting a rhythmic and regular supply of sediment (Fig. 5), possibly of glaciomarine origin. It is not however observed in the main or outer basin (profiles 11 and 32; Figs. 2a–c). Glaciomarine sedimentation is typically prolific during immediate glacier retreat when large volumes of meltwater are available. The division between a lower darker and an upper lighter interval in this unit is interpreted as a change in sediment supply: the dark material is proximal to discharge areas (richer in coarse sand and pebbles), while the light material (richer in mud) is distal. It is therefore postulated that this unit corresponds to a period (somewhere between 13.75 and 10.05 cal ka BP) that incorporates Younger Dryas sedimentation with ice cap re-advance (proximal supply) and retreat (distal supply). This is supported by the ^{14}C age obtained above this interval at 9262–9447 cal a BP.

The second characteristic of Unit 2 is a major unconformity which cut across the top of it with up to 40 cm of vertical erosion (Figs 2c & 5). An erosive event of this magnitude requires either a substantial lowering of sea level, allowing sediment to be more easily transported under the influence of surface waves and currents, or a significant change in current patterns. Both of these conditions are likely met at this time interval as the relative sea level was low (*c.* 4 m below present level; Shennan *et al.* 2000, 2005) and the absence of pollen in the core indicates a potential for increased current activity and non-deposition of fine-grained material. Over a similar time interval, Traini (2006) observed bottom-current accelerations in cores acquired from Loch Creran and Etive. This unit is similar to the Muck formation (Fig. 5), described as a deposit subsequent to the advance and retreat of the last Scottish ice sheet by Boulton *et al.* (1981), and to the Formation 2 of Binns *et al.* (1974*a*, *b*). The distribution of this unit and interpretation in terms of a glacier advance only as far as the Laudale Narrows is in agreement with the terrestrial limits mapped for the Younger Dryas by Sutherland & Gordon (1983), but does not agree with the limit mapped by Sissons (1980*a*, *b*).

Unit 3. The sequence of sediments associated with Unit 3 is interpreted as representing a time of rapid and large volume sediment supply where significant reworking was likely taking place, even under weak current regimes. The base of this unit is marked by a highly transparent facies which suggest an unconsolidated state of the sediments, associated with rapid sedimentation. The unit contains features consistent with present day sedimentation, for example prograding deltas associated with modern stream inputs. Other features of the unit include ridges, undulations and drifts representing depositional or postdepositional deformation from bottom-current action (profile 11 on Figs 2b & 5). The occurrence of the dome features suggests that there was a significant reduction in bulk density within the sediments due to gas generation as a result of diagenesis (Schubel 1974).

Within this unit three distinct events are underlined by the pollen analysis: the first is situated *c.* 10.05 cal ka BP (Fig. 4b), the second *c.* 8.55 cal ka BP and the third *c.* 5.6 cal ka BP. The first date corresponds to the beginning of the Boreal period, characterized by a deterioration of climate. The second event could be correlated to the widespread climate cooling at *c.* 8.2 cal ka BP. This event has been related to changes in thermohaline circulation in the North Atlantic in response to enhanced meltwater production (Barber *et al.* 1999; Clark *et al.* 2001). In turn, a significant change in sea-loch circulation has been reported at this time by Nørgaard-Pedersen *et al.* (2006).

The interval *c.* 5600 cal a BP (5584–5663 cal ka BP) announces the end of the Atlantic period characterized by cooling conditions following the Climatic Optimum. This event appears as a major change in the pollen records (Fig. 4a, b), and could be detected on the grain size evolution at 200 cm. This unit is similar to the Arisaig Formation which was also identified by Boulton *et al.* (1981), and tends to correspond to Formations 3 and 4 of Binns *et al.* (1974*a*, *b*).

Unit 4. Unit 4 is characterized by thin laminated beds of silty-clay and is consistent with a relatively low sediment supply in a quiet environment (Figs 2c & 5). This deposit is similar to those observed in Loch Ness by Cooper & O'Sullivan (1998) and corresponds to recent (Subatlantic) sedimentation patterns with decreasing sediment supply (Fig. 5). This unit is interpreted as a period of lower sedimentation during the last phase of the Holocene. At the entrance of the loch, the high current action has inhibited thick Holocene sediment accumulation. The two last cooling events are noted in the pollen records after the 771 cal a BP date (Fig. 4b) and it is speculated from extrapolation of the sedimentation rate that they might correspond to the Little Ice Age *c.* 600 cal a BP.

The approach to palaeolandscape reconstruction described in this paper, based on a combination of seismic profiles, sediment and pollen analysis, allows four key elements of sea-loch history to be highlighted.

(1) Despite the low concentration of pollen grains (less than 5000 grains per gram of sediment), the sediments in Loch Sunart have recorded the landscape evolution from the Older Dryas until present. The five cooling events identified by the pollen records correspond to the time periods described by Mayewski *et al.* (2004) of significant rapid climate change: 9–8, 6–5 and 1.2–1 cal ka BP and 600–150 cal a BP. These events are characterized by polar cooling, tropical aridity and major atmospheric circulation changes which directly affect the vegetation and thus the pollen record.

(2) The correlation of the seismic units with the pollen records allows associated age assignments to the acoustic units: Unit 1 covers a period beginning before 16.76 cal ka BP and probably finishing around the end of the Bolling period; Unit 2 covers the period *c.* 13–11.5 cal ka BP; Unit 3 the period 11.5–2.6 cal ka BP and Unit 4 from 2.6 cal ka BP to present. The potential exists to extrapolate these results to other west coast sea lochs, where similar seismic facies have been recognized by various authors (Binn's *et al.* 1974*a*, *b*; Boulton *et al.* 1981; Howe *et al.* 2002).

(3) The preservation of a complete deglacial sedimentary sequence is best achieved in the main basin of Loch Sunart where the sequences can be matched to onshore ice advance limits for the Younger Dryas (Sutherland & Gordon 1993). Complete sequences are preserved in the numerous scour pits that are evident in the shallow water areas of the main basin, immediately to the west of the Laudale Narrows (Fig. 2a). The Laudale Narrows area marks the probable limit of ice re-advance during the Younger Dryas in agreement with the terrestrial limits mapped for the Younger Dryas by Sutherland & Gordon (1983).

(4) The date of 16.7 cal ka BP suggests that the material of Unit 1 is the basal marine sediment following regional deglaciation, which in turn suggests no significant age difference in the timing of offshore deglaciation of the continental shelf.

Conclusions

This study establishes for the first time a relative chronological framework for the regional deglaciation of Loch Sunart, based on pollen records and seismic facies interpretation. The pollen samples from this high-latitude loch recorded five cooling events, identified as rapid climate change events, which compare well with those previously described in the literature. For example, the age-calibrated seismic units can be compared to previous work conducted on the west coast of Scotland by Binns *et al.* (1974*a*, *b*), Boulton *et al.* (1981) and Howe *et al.* (2002). It is suggested that the characteristic seismic facies identified in Loch Sunart could be used as a mapping tool to determinate the spatial extension and the timescale evolution of the different palaeoenvironments since the LGM in other west coast lochs.

Critically, the date 17.45–15.95 cal ka BP, corresponding to an age of a *Pecten* shell (included in Unit 1) and providing the first evidence of marine inundation of Loch Sunart, suggests that the deglaciation process was (within the uncertainty of the dating method) simultaneous across the entire continental shelf of western Scotland. We conclude that this combined approach reveals the sensitivity of the pollen records within Scottish west coast sea lochs, and provides an approach which could be extended to other fjord environments.

The authors are especially grateful to the *Marion Dufresne* Captain J.-M. Lefèvre and the Chief Operator Y. Balut for the tenacity they showed in obtaining these incredible cores. The authors would like to thank the team including P. Bretel, J. Fournier, F. Lelong, E. Poizot, J. M. Rousset, J. & J.-C Rousset and A. Stépanian for their participation in data acquisition.

References

Adams, J. M. 1997. *Global Land Environments Since the Last Interglacial.* Oak Ridge National Laboratory, TN, USA. http://www.esd.ornl.gov/projects/qen/nerc.html

Austin, W. E. N. & Inall, M. E. 2002. Deep water renewal in a Scottish fjord: temperature, salinity and oxygen isotopes. *Polar Research*, **21**, 251–258.

Ballantyne, C. K. & Gray, J. M. 1984. The quaternary. *Geomorphology of Scotland: the Research contribution of J.B. Sissons. Quaternary Science Reviews*, **3**, 259–289.

Baltzer, A., Tessier, B., Nouzé, H., Bates, C. R., Moore, C. & Menier, D. 2005. Seistec Seismic profiles: a tool to differentiate gas signatures. *Marine Geophysical Researches* (MARI): in Subsurface imaging and sediment characterization in shallow water environment **26**, 235–245.

Barber, D. C., Dyke, A. *et al.* 1999. Forcing of the cold event of 8200 years ago by catastrophic drainage of Laurentide Lakes. *Nature*, **400**, 344–348.

BATES, C. R. & BYHAM, P. 2001. Swath-sounding techniques for near shore surveying. *Hydrographic Journal*, **100**, 13–18.

BATES, C. R., MOORE, C. G., HARRIES, D. B., AUSTIN, W. & LYNDON, A. R. 2004. *Broad scale mapping of sublittoral habitats in Loch Sunart, Scotland.* Scottish Natural Heritage, Commissioned Report No. 006 (ROAME No. F01AA401C).

BENN, D. I. 1997. Glacier fluctuations in western Scotland. *Quaternary International*, **38/39**, 137–147.

BENNETT, K. D. & HUMPHRY, R. W. 1995. Analysis of late-glacial and Holocene rates of vegetational change at two sites in the British Isles. *Review of Palaeobotanic. and Palynology*, **85**, 263–287.

BINNS, P. E., HARLAND, R. & HUGHES, M. J. 1974*a*. Glacial and post glacial sedimentation in the sea of the Hebrides. *Nature*, **248**, 751–754.

BINNS, P. E., MCQUILLIN, R. & KENOLTY, N. 1974*b*. *The geology of the sea of the Hebrides*. Institute of Geological Sciences 73/14.

BOND, G., SHOWERS, W. *ET AL.* 1997. A pervasive millennial-scale cycle in North Atlantic Holocene and glacial climates. *Science*, **278**, 1257–1266.

BOULTON, G. S., CHROSTON, P. N. & JARVIS, J. 1981. A marine seismic study of late Quaternary sedimentation and inferred glacier fluctuations along western Inverness–shire, Scotland. *Boreas*, **10**, 39–51.

CAGE, A. G., HEINEMEIER, J. & AUSTIN, W. E. N. 2006. Marine radiocarbon reservoir age calculations for Scottish coastal and fjordic waters. *Radiocarbon*, **48**, 31–43.

CHURCH, M. & RYDER, J. M. 1972. Paraglacial sedimentation: a consideration of fluvial processes conditioned by glaciation. *Bulletin Geological Society of America*, **83**, 3059–3072.

CLARK, P. U., MARSHALL, S. J., CLARKE, G. K. C., HOSTETLER, S. W., LICCIARDI, J. M. & TELLER, J. T. 2001. Freshwater forcing abrupt climate change during the last glaciation. *Science*, **293**, 283–287.

COOPER, M. C. & O'SULLIVAN, P. E. 1998. The laminated sediments of Loch Ness, Scotland: preliminary report on the construction of a chronology of sedimentation and its potential use in assessing Holocene climatic variability. *Palaeogeography, Palaeoclimatology, Palaeoecology*, **140**, 23–31.

CUNDILL, P. R., AUSTIN, W. E. N. & DAVIES, S. E. 2006. Modern pollen from the catchment and surficial sediments of a scottish sea loch (fjord). *Grana*, **45**, 230–238.

DAVIS, A. M. 1992. Shallow gas: an overview. *Continental Shelf Research*, **12**, 1077–1079.

EMERY, K. O. & HOGGAN, D. 1958. Gases in marine sediments. *AAPG Bulletin*, **42**, 2174–2188.

GILLIBRAND, P. A., CAGE, A. G. & AUSTIN, W. E. N. 2005. A preliminary investigation of the basin water response to climate change in a Scottish fjord: evaluating the influence of NAO. *Continental Shelf Research*, **25**, 571–587.

GOEURY, C. 1988. Acquisition, gestion et representation des données de l'analyse pollinique sur miecro-ordinateur. *Institut Français de Ponduchéry, ravaus de la section Scientifique et Technique*, **25**, 405–416.

GORDON, J. E. & SUTHERLAND, D. G. 1993. *Quaternary of Scotland.* Geological Conservation Review, 6.

HALL, I. R., BIANCHI, G. G. & EVANS, J. R. 2004. Centennial to millennial scale Holocene climate-deep water linkage in the North Atlantic. *Quaternary Science Reviews*, **23**, 1529–1536.

HARKNESS, D. D. 1983. The extent of the natural ^{14}C deficiency in the coastal environment of the United Kingdom. *Journal of European Study Group Phys., Chem. Math. Tech. Appl. Archaeol. PACT* 8 (IV.9), 351–364.

HOWE, J. A., OVERNELL, J., INALL, M. E. & WILBY, A. D. 2001. A side scan sonar image of a glacially-overdeepened sea loch, upper Loch Etive, Argyll. *Scottish Journal of Geology*, **37**, 3–10.

HOWE, J. A., SHIMMIELD, T., AUSTIN, W. E. N. & LONGVA, O. 2002. Post-glacial depositional environments in a mid latitude glacially-overdeepened sea loch, inner Loch Etive, western Scotland. *Marine Geology*, **185**, 417–433.

HUBBARD, A. 1999. High-resolution modeling of the advance of the Younger Dryas ice sheet and its climate in Scotland. *Quaternary Research*, **52**, 27–43.

MANGERUD, J., ANDERSEN, S. T., BERGLUND, B. E. & DONNER, J. J. 1974. Quaternary stratigraphy of Norden, a proposal for terminology and classification. *Boreas*, **3**, 109–128.

MAYEWSKI, P. A., ROHLING, E. *ET AL.* 2004. Holocene climate variability. *Quaternary Research*, **62**, 243–255.

NØRGAARD-PEDERSEN, N., AUSTIN, W. E. N., HOWE, J. A. & SHIMMIELD, T. 2006. The Holocene record of Loch Etive, western Scotland: influence of catchment and relative sea level changes. *Marine Geology*, **228**, 55–75.

SCHUBEL, J. R. 1974. Gas bubbles and the acoustically impenetrable, or turbid, character of some estuarine sediments. *In*: KAPLAN, I. R. (ed.) *Natural Gases in Marine Sediments*. Plenum Press, New York and London, 275–298.

SHARAPOVA, A., HALD, M. B., HUSUM, K. & JENSEN, J. 2008. Lateglacial and Holocene terrestrial and marine proxies reflecting climate changes in the Malangen fjord area, Norway, northeast North Atlantic. *Boreas*, **37**, 37445–457.

SHENNAN, I., LAMBECK, K. *ET AL.* 2000. Late Devensian and Holocene records of relative sea-level changes in northwest Scotland and their implications for glacio-hydro-isostatic modelling. *Quaternary Science Reviews*, **19**, 1103–1135.

SHENNAN, I., HAMILTON, S., HILLIER, C. & WOODROFFE, S. 2005. A 16 000-year record of near-field relative sea-level changes, northwest Scotland, United Kingdom. *Quaternary International*, **133–134**, 95–106.

SIMPKIN, P. G. & DAVIS, A. 1983. For seismic profiling in very shallow water, a novel receiver. *In Sea Technology*.

SISSONS, J. B. 1980*a*. The Loch Lomond Advance in the Lake District, northern England. Transactions of the Royal Society of Edinburgh. *Earth Sciences*, **71**, 13–27.

SISSONS, J. B. 1980*b*. Palaeoclimatic inference from Loch Lomond Advance glaciers. *In*: LOWE, J. J., GRAY, J. M.

& Robinson, J. E. (eds) *Studies in the Lateglacial of North-West Europe*. Pergamon Press, Oxford, 31–43.

Smith, D. E., Cullingford, R. A., Mighall, T. M., Jordan, J. T. & Fretwell, P. T. 2007. Holocene relative sea level changes in a glacio-isotatic area: new data from south west Scotland, United Kingdom. *Marine Geology*, **22**.

Stoker, M. S. 1995. The influence of glacigenic sedimentation on slope-apron development on the continental margin off Northwest Britain. *In*: Scrutton, R. A., Stoker, M. S., Shimmield, G. B. & Tudhope, A. W. (eds) *The Tectonics, Sedimentation and Paleoceanography of the North Atlantic Region 90*. Geological Society, London, 159–177.

Stoker, M., Bradwell, T., Wilson, C., Harper, C., Smith, D. & Brett, C. 2006. Pristine fjord landsystem revealed on the sea bed in the Summer Isles region, NW Scotland. *Scottish Journal of Geology*, **42**, 89–99.

Stuiver, M. & Reimer, P. J. 1993. Extended ^{14}C data base and revised CALIB3.0 ^{14}C age calibration program. *Radiocarbon*, **35**, 215–230.

Stuiver, M., Reimer, P. J. *et al.* 1998. INTCAL98 Radiocarbon Age Calibration, 24 000–0 cal BP. *Radiocarbon*, **40**, 1041–1083.

SUTHERLAND & GORDON 1983. In fluctuation in patterns of ice flow through time and the reasons for the divergent flow of Scottish ice across the islands. *Quaternary of Scotland Geological Conservation Review Series*, **6**, 1–9.

Syvitski, J. P. M., Burrell, D. C. & Skei, J. M. 1987. *Fjords: Processes and Products*. Springer-Verlag, New York.

Traini, C. 2006. *A comparison of post-glacial sedimentation in fjordic environment: Lochs Sunart, Etive and Creran Western Scotland*. MSc thesis, University of Caen.

Index

Note: Page numbers in *italic* denote figures. Page numbers in **bold** denote tables.